| 7A | 8A |
|---|---|
| | 1 **H** 1.00797 ± 0.00001 | 2 **He** 4.0026 ± 0.00005 |

3A 4A 5A 6A

| 5 **B** 10.811 ± 0.003 | 6 **C** 12.01115 ± 0.00005 | 7 **N** 14.0067 ± 0.00005 | 8 **O** 15.9994 + 0.0001 | 9 **F** 18.9984 ± 0.00005 | 10 **Ne** 20.183 ± 0.0005 |

1B 2B

| 13 **Al** 26.9815 ± 0.00005 | 14 **Si** 28.086 ± 0.001 | 15 **P** 30.9738 ± 0.00005 | 16 **S** 32.064 ± 0.003 | 17 **Cl** 35.453 ± 0.001 | 18 **Ar** 39.948 ± 0.0005 |

| 28 **Ni** 58.71 ± 0.005 | 29 **Cu** 63.54 ± 0.005 | 30 **Zn** 65.37 ± 0.005 | 31 **Ga** 69.72 ± 0.005 | 32 **Ge** 72.59 ± 0.005 | 33 **As** 74.9216 ± 0.00005 | 34 **Se** 78.96 ± 0.005 | 35 **Br** 79.909 ± 0.002 | 36 **Kr** 83.80 ± 0.005 |

| 46 **Pd** 106.4 ± 0.05 | 47 **Ag** 107.870 ± 0.003 | 48 **Cd** 112.40 ± 0.005 | 49 **In** 114.82 ± 0.005 | 50 **Sn** 118.69 ± 0.005 | 51 **Sb** 121.75 ± 0.005 | 52 **Te** 127.60 ± 0.005 | 53 **I** 126.9044 ± 0.00005 | 54 **Xe** 131.30 ± 0.005 |

| 78 **Pt** 195.09 ± 0.005 | 79 **Au** 196.967 ± 0.0005 | 80 **Hg** 200.59 ± 0.005 | 81 **Tl** 204.37 ± 0.005 | 82 **Pb** 207.19 + 0.005 | 83 **Bi** 208.980 ± 0.0005 | 84 **Po** (210) | 85 **At** (210) | 86 **Rn** (222) |

| 63 **Eu** 151.96 ± 0.005 | 64 **Gd** 157.25 ± 0.005 | 65 **Tb** 158.924 ± 0.0005 | 66 **Dy** 162.50 ± 0.005 | 67 **Ho** 164.930 ± 0.0005 | 68 **Er** 167.26 ± 0.005 | 69 **Tm** 168.934 ± 0.0005 | 70 **Yb** 173.04 + 0.005 | 71 **Lu** 174.97 ± 0.005 |

| 95 **Am** (243) | 96 **Cm** (247) | 97 **Bk** (249) | 98 **Cf** (249) | 99 **Es** (254) | 100 **Fm** (253) | 101 **Md** (256) | 102 **No** (253) | 103 **Lr** (257) |

**PETER A. ROCK, Ph.D.**

Associate Professor of Chemistry,
The University of California at Davis,
Davis, California

**GEORGE A. GERHOLD, Ph.D.**

Associate Professor of Chemistry,
Western Washington State College
Bellingham, Washington

SAUNDERS GOLDEN SERIES

# CHEMISTRY

## PRINCIPLES
## AND
## APPLICATIONS

**W. B. SAUNDERS COMPANY  1974**

PHILADELPHIA · LONDON · TORONTO

W. B. Saunders Company:   West Washington Square
                          Philadelphia, Pa. 19105

                          12 Dyott Street
                          London, WC1A  1DB

                          833 Oxford Street
                          Toronto, Ontario M8Z 5T9, Canada

Cover illustration courtesy of The Museum of Modern Art

BALLA, Giacomo
*Street Light* (Lampada—Studio di luce). (1909)
Oil on Canvas, 68-3/4 × 45-1/4
Collection, The Museum of Modern Art, New York
Hillman Periodicals Fund

CHEMISTRY — Principles and Applications                  ISBN 0-7216-7630-8

Last digit is the print number:     9    8    7    6    5    4    3    2    1

# PREFACE

This book is an introduction to the subject of chemistry. The level of presentation of the material was chosen to meet the requirements of students enrolled in a first-year, general chemistry course who are planning to major in science or engineering. A knowledge of calculus is not required to follow the presentation.

We have tried in our writing to avoid gross oversimplifications that lead to misunderstandings about chemical principles or behavior. We have tried to present the material in such a way that it will prepare students to apply their chemical knowledge on their own to the solution of real problems. We have striven in our presentation for scholarship without pedantry, simplicity without superficiality, and accuracy without affectation.

There is more material in this book than most instructors (and students) would want to cover in a one-year course. A broad coverage of topics has been included to provide the instructor with a good measure of flexibility in the emphasis of the course, and to eliminate the need for supplementary texts.

There are several novel features that, taken together, distinguish this book from other introductory chemistry texts. Among these features are:

(1) Extensive use is made of contour maps and correlation diagrams in the discussions of chemical bonding.

(2) The subject matter of chemical kinetics is developed to an extent that is commensurate with the importance of this area of chemistry.

(3) The subject matter of chemical thermodynamics is developed to a level that makes possible the analysis of chemical equilibria in a logical and straightforward manner without the need for introducing a myriad of *ad hoc, quasi*-thermodynamic concepts.

(4) An extensive section (about one-third of the book) treats the chemistry of the elements; it makes use of the material developed in the sections on chemical bonding, thermodynamics, and kinetics.

(5) The factual, experimental basis of chemistry is integrated with the development of the principles that enable us to understand the facts.

We have included numerous worked-out examples in the body of the text that serve to develop the reader's ability to apply the principles. A large number of problems are included at the end of each chapter to test and further develop the reader's mastery of the subject matter. The problems are arranged in order of increasing difficulty. We hope these problems will stimulate students to apply the principles on their own.

This book has benefited in many ways from the comments of numerous reviewers. Professors Eugene G. Rochow (Harvard University), William L. Masterton (University of Connecticut), Malcolm F. Nichol (UCLA), Donald Jicha (University of North Carolina), Dennis G. Peters (Indiana University), Ronald O. Ragsdale (University of Utah), and Jerry A. Bell (Simmons College) read the entire manuscript and provided many helpful comments. Professors Henry A. Bent (North Carolina State University) and William H. Fink, Robert K. Brinton, R. Bryan Miller, Dr. Martha Barrett, and Janice J. Kim (all of University of California, Davis) reviewed portions of the text and also made helpful comments.

Finally, we acknowledge the contributions of our many colleagues and mentors, for in many cases the former helped us to rediscover what the latter had earlier taught us.

PETER A. ROCK

GEORGE A. GERHOLD

# TABLE OF CONTENTS

# II. STATES AND PROPERTIES OF MATTER

# III.    THERMODYNAMICS

## IV. CHEMICAL KINETICS

## V. CHEMISTRY OF THE ELEMENTS

## APPENDICES

## INDEX

# CHEMISTRY

PRINCIPLES
AND
APPLICATIONS

"What is it that confers the noblest delight? What is that which swells a man's breast with pride above that which any other experience can bring to him? Discovery! To know that you are walking where none others have walked, that you are beholding what human eye has not seen before, that you are breathing a virgin atmosphere. To give birth to an idea — to discover a great thought — an intellectual nugget, right under the dust of a field that many a brain plow had gone over before. To find a new planet, to invent a new hinge, to find the way to make the lightnings carry your messages. To be the *first* — that is the idea. To do something, say something, see something, before *anybody* else — these are the things that confer a pleasure compared with which other pleasures are tame and commonplace, other ecstasies cheap and trivial. Morse, with his first message, brought by his servant, the lightning; Fulton, in that long-drawn century of suspense, when he placed his hand upon the throttle valve and lo, the steamboat moved; Jenner, when his patient with the cow's virus in his blood walked through the smallpox hospitals unscathed; Howe, when the idea shot through his brain that for a hundred and twenty generations the eye had been bored through the wrong end of the needle; the nameless lord of art who laid down his chisel in some old age that is forgotten now and gloated upon the finished Laocoön; Daguerre, when he commanded the sun, riding in the zenith, to print the landscape upon his insignificant silvered plate, and he obeyed; Columbus, in the *Pinta's* shrouds, when he swung his hat above a fabled sea and gazed abroad upon an unknown world! These are the men who have really *lived* — who have actually comprehended what pleasure is — who have crowded long lifetimes of ecstasy into a single moment."

*Mark Twain (1868)\**
*The Innocents Abroad*

## CHAPTER 1

# THE ATOM

"Chemistry is concerned with substances, with kinds of matter, the materials of which things are made. It is concerned with wood but not the log, with glass but not the bottle, with copper but not the coin, with clay but not the brick. Chemistry is concerned with identifying or characterizing properties of substances, . . . [and] with the processes in which substances are transformed into other substances."

*J. H. Hildebrand and R. E. Powell**

## 1.1  INTRODUCTION.

We are about to embark on the study of chemistry. Because at times a chemist studies the same matter as a physicist, and at other times he studies the same matter as a biologist, the distinction between chemistry and its sister sciences cannot be made on the basis of objects of study. If a distinction can be made, then it must be made on the basis of the methods of study. This text is a brief introduction to many of the methods of study that lie within the particular province of chemistry.

## 1.2  DEFINITIONS.

A central theme in science is simplification; break down a sample or a system into its simplest components, and discard those components that are not essential. The simplest components of matter—excluding those that survive only in isolation—are called *atoms*. Substances that contain only a single kind of atoms are called *elements*. Ninety-one elements have been found in nature. Physicists and chemists working together have been able to make and characterize an additional fifteen elements in the laboratory.

Atoms combine to form *molecules*. Several million different molecules are known. Fragments of molecules that are electrically charged are called *ions;* ions may be either atomic or molecular. A sample in which all of the molecules are identical is called a *compound.* Atoms can be combined to form molecules; molecules can be converted into other molecules, and molecules can be broken down into atoms. All these conversions are called *chemical reactions.*

*J. H. Hildebrand and R. E. Powell, *Principles of Chemistry*, 7th Edition (The Macmillan Co., New York, N. Y., 1964) pp. 2-3.

A *mixture* is a sample that can be separated by physical methods into two or more compounds or elements, or both. A *heterogeneous mixture* can be separated into components by mechanical means, for example, with tweezers, filters, or sieves. A *homogeneous mixture,* which is usually called a *solution,* is a more intimate mixture of compounds or elements; it cannot be separated into its components by mechanical means. Distillation, evaporation, and crystallization are a few examples of physical separation methods that can be used to separate the components of a homogeneous mixture.

## 1.3   METHODS FOR STUDYING INVISIBLE
## PARTICLES.

No one has ever seen an atom; no one will ever see an atom. In very rare circumstances it is possible to produce pictures of atoms, but these pictures tell us essentially nothing about atoms. Atoms are not the smallest or simplest form of matter; atoms are built from the elementary particles. Elementary particles are invisible as well.

Scientists interpret experiments on invisible particles in two ways. They develop mathematical equations that reproduce experimental results, and they play a serious game of model construction. The latter method, which is often referred to as "giving a physical picture," is widely used by chemists. There is one rule for this game that should be inviolable: the model must be capable of being used to predict the results of experiments. Deviation from agreement with the experimental results must cause the ultimate overhaul or replacement of the model. Any model in exact agreement with experiment (exact within experimental error) is allowable, but additional considerations determine its wide acceptance. An accepted model is simple and useful. By simple we do not mean mathematical simplicity; we mean simplicity in the set of axioms for the model. A model's usefulness is judged by the number and accuracy of the predictions that it facilitates.

The reader must gain familiarity with the current models if he wishes to avail himself of the knowledge of chemists. To understand a model we must understand the basic experiments upon which the model is based, because our model of the atom is influenced by the nature of experimentation itself.

Experiments on invisible particles require a particle detector. These experiments are all similar; the particle is subjected to a known force (for example, gravity), and the effect of the force on the motion of the particle is measured. The detector is needed to sense the motion of the particle.

Every particle is subject to an almost infinite number of forces because every particle interacts with every other particle in the universe. Fortunately, there are reasonable and simplifying assumptions that bring the number of forces to be considered within reason. The first assumption is that forces are *pairwise additive;* that is, the total force on particle A is the sum of those forces on A due to particle B, those forces on A due to particle C, and so on. Furthermore, the forces on particle A due to particle B can be represented by a sum of terms; for example,

Total Force = Gravitational Force + Electrical Force + Nuclear Force.

It must be remembered that, because forces are quantities with both magnitude and direction, opposing forces can add up to zero.*

The pairwise additivity of forces is useful when the total number of particles is small; with large systems the field approach is simpler. Newton's law of gravitation is an example. Newton's law states that the force of attraction between two bodies is proportional to the product of their masses divided by the square of their distance apart:

$$f \propto \frac{m_1 m_2}{d_{12}^2} \qquad (1.1)$$

where $\propto$ means "is proportional to," $m_1$ and $m_2$ are the masses, and $d_{12}$ is the separation. It is highly impractical to calculate the total gravitational force on, say, a satellite *via* repeated application of Equation (1.1). An attempt to do so might begin

$$F_{\text{total}} = f(\text{due to earth}) + f(\text{due to moon}) + f(\text{due to sun}) + \ldots$$

$$F_{\text{total}} \propto \left\{ \frac{m \cdot m_{\text{earth}}}{d^2_{\text{earth}}} + \frac{m \cdot m_{\text{moon}}}{d^2_{\text{moon}}} + \frac{m \cdot m_{\text{sun}}}{d^2_{\text{sun}}} + \ldots \right\}. \qquad (1.2)$$

Each term in this extended expression contains $m$, the mass of the satellite. We factor out this common term

$$F_{\text{total}} \propto mG, \text{ where } G = \left\{ \frac{m_{\text{earth}}}{d^2_{\text{earth}}} + \frac{m_{\text{moon}}}{d^2_{\text{moon}}} + \frac{m_{\text{sun}}}{d^2_{\text{sun}}} + \ldots \right\}. \qquad (1.3)$$

Here $G$ is the gravitational *field*. It is no easier to calculate the gravitational field $G$ from Equation (1.3) than to calculate the total force from Equation (1.2), but if we were to measure the total force on one satellite of known mass, we could then calculate the gravitational field

$$G \propto F_{\text{total}}/m \qquad (1.4)$$

Once the value of the gravitational field is known, we can use it to calculate forces on all other satellites at the same position.

We know from experiment that the magnitudes of all forces decrease rapidly with increasing separation of the interacting bodies. This means that interactions with distant bodies are vanishingly small and can be ignored. The precise definition of "distant" depends on the type of force: for nuclear forces, $10^{-11}$ cm is found to be distant; for gravitational forces, distant may mean millions of light years. Forces depend on the properties of the two particles: for gravitational forces, on the product of the two masses; for forces between stationary charges, on the product of the two charges; for nuclear forces, on as yet undiscovered properties. Forces with identical dependence on separation may differ greatly in magnitude; for example, the Coulombic force between two ions is much greater than the gravitational force. By the clever design of experiments and use of the field approach it is usually possible to write one- or two-term expressions for the forces acting on a particle during the experiments.

A "light year" is equal to the distance travelled by light in one year, namely, $9.46 \times 10^{12}$ km.

---

*A quantity which has both magnitude and direction is called a *vector* quantity. Force, velocity, momentum, and acceleration are common vector quantities.

## 1.4  PROPERTIES OF THE ELECTRON

Although visionaries speculated on the existence and nature of atoms, and experimental scientists like John Dalton attempted to deduce the nature of atoms, it was not until the end of the nineteenth century that anyone succeeded in measuring properties of individual atoms or of their component parts. These earliest results indicated that there were subatomic particles that were electrically charged. The characterization of these particles became the most exciting topic in physical science at the turn of the century.

The electron was the first subatomic particle to attract the attention of physicists. An advance in technology, the design of the mercury diffusion pump, led to new studies of a long known but little understood phenomenon, the discharge of electricity through gases at low pressure. The experiments showed that it was possible to obtain negatively charged particles from all the materials that were studied. The definitive experiments on these negative particles were those of J. J. Thomson and R. A. Millikan; their experiments were designed for measurement of the charge-to-mass ratio ($e/m$) and the charge ($e$) of these negative particles.

The apparatus that Thomson used to measure $e/m$ was similar to a black-and-white television picture tube (Figure 1.1). One end of an evacuated glass tube was coated with a phosphor which would emit tiny flashes of light when struck by the negatively charged particles. These negatively charged particles, which were named *electrons,* were emitted from a coated electrode at the opposite end of the tube.[*] Between the electron emitter and the electron detector there was a series of positively-charged metallic plates, each with a tiny hole at its center. Only those few electrons that happened to be travelling exactly down the axis of the tube could pass through these holes; the rest were trapped by the positively charged plates. These few remaining electrons came into the large end of the tube as a well-defined beam. When no additional forces were applied, the electrons in the beam continued in a straight line and struck the particle detector at $A$. A downward force on the beam was produced by a magnet. In a magnetic field that is uniform over a large area, electrons circle endlessly. The radius of this circle can be calculated by equating the magnetic force to the centrifugal force

---

[*]Zinc sulfide is a phosphor, and barium oxide emits electrons when heated in a vacuum.

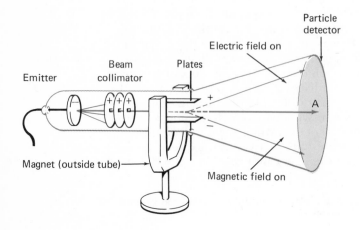

FIGURE 1.1  Thomson's $e/m$ apparatus. The beam of electrons strikes the particle detector at $A$ when there are no forces and when the electric and magnetic forces are balanced.

$$Bev = \frac{mv^2}{r} \tag{1.5}$$

where $B$ is the magnetic field strength, $e$ is the charge on an electron, $v$ is the velocity of the electrons in the beam, $m$ is the mass of the electron, and $r$ is the radius of the circle.

In Thomson's tube the magnetic field was not large enough to drive the electrons in a complete circle, but between the poles of the magnet the above equation was applicable, and in this region the electrons traveled along a segment of a complete circle. After leaving the magnetic field, the electrons flew straight down the tube and struck the screen somewhere below $A$ (Figure 1.1). This straight line tangent allowed calculation of the deflection within the field (and thus $r$) by simple geometry,* but there were still three unknown quantities—$e$, $m$, and $v$—in Equation (1.5).

The horizontal plates shown in the tube were charged to produce a Coulombic force on the electrons that opposed the magnetic force. With only the Coulombic force operative, the particles would have followed the upper path. With the magnetic field on, the charge on the plates was adjusted until the beam returned to its original position $A$. This exact balancing of two opposing forces and the resulting cancellation of their effects is an example of a *null procedure*. Null procedures are incorporated into the design of scientific experiments wherever possible because null procedures allow the use of extremely sensitive detectors, which are usually unstable and non-linear. In the present case it was easier to detect whether or not the spot on the screen had moved than to measure how far it had moved. At the null point (zero deflection) the electrostatic force equals the magnetic force

$$eE = Bev \tag{1.6}$$

where $E$ is the strength of the electric field. Measurement of the two fields was effectively a measurement of the velocity of the electrons in the beam, because (from Equation (1.6))

$$v = E/B \tag{1.7}$$

Equation (1.7) can be used to eliminate $v$ from Equation (1.5), leaving

$$Be = \frac{mE}{rB} \tag{1.8}$$

Equation (1.8) involves two unknown quantities, $e$ and $m$ (r was known from the magnetic deflection of the electrons), but these quantities are fundamental properties of electrons, whereas $v$ in Equation (1.5) is mainly a property of the particular tube. The charge-to-mass ratio of the electron is obtained from Equation (1.8) as

$$e/m = \frac{E}{B^2 r} \tag{1.9}$$

---

*Ideally, the distance from the magnetic field to the screen should be long, because then a small deflection in the field becomes a long distance on the screen. This distance acts as a lever to magnify the deflection (the signal). This is a simple example of an apparatus which amplifies a signal; such an apparatus is said to have *gain*.

Thomson obtained a value of about $10^8$ coulombs per gram (since corrected to $1.7588 \times 10^8$ C·g⁻¹). There was really no reason for Thomson to anticipate that all electrons should be identical, and this result does not prove that they are. However, it does indicate that all electrons have the same charge-to-mass ratio.

The determination of $e$, the charge of the electron, required an independent experiment; this was the oil-drop experiment (1909) of R. A. Millikan. This experiment also involved a null procedure: a downward gravitational force on a negatively charged particle was balanced by an upward Coulombic force. The gravitational force on an electron is vanishingly small, much smaller than the usual laboratory Coulombic force on electrons. Millikan circumvented this limitation by firing electrons into a relatively massive body, a tiny drop of oil (Figure 1.2). The electric field was adjusted until the magnitudes of the Coulombic and the gravitational forces were equal, or

$$eE = MG \tag{1.10}$$

where $M$ is the mass of the drop and $G$ is the gravitational field. The mass of a spherical drop of oil is equal to the volume of the sphere $(\frac{4}{3}\pi R^3)$ times the density of the oil. According to Archimedes' principle,* the measured mass is the true mass minus the buoyancy correction

$$M = \frac{4}{3}\pi R^3(\rho - \rho_0) \tag{1.11}$$

where $R$ is the radius of the drop, $\rho$ is the density of the oil, and $\rho_0$ is the density of air. Insertion of Equation (1.11) into Equation (1.10) gives

$$eE = \frac{4}{3}\pi R^3(\rho - \rho_0)G \tag{1.12}$$

All of the quantities in Equation (1.12) were known except for $e$ and $R$. The radius of a particular drop was measured by means of a second null procedure. When a body is dropped through a viscous fluid (in this case, air), it is not accelerated endlessly; it reaches a terminal velocity. At terminal velocity the force due to gravity is exactly balanced by a viscous

---

*A body immersed in a fluid is buoyed up by a force equal to the weight of the displaced fluid.

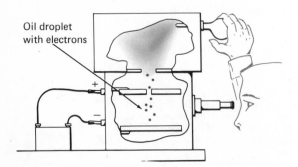

Oil droplet
with electrons

**FIGURE 1.2** Millikan's oil drop experiment. The charge on the plates is adjusted until a particular oil droplet is held motionless.

drag. The viscous drag is strongly dependent on the shape of the body; for example, a parachute is simply a device to alter the shape of a falling body, thereby increasing the viscous drag and decreasing the terminal velocity to a point where the impact on landing is seldom fatal. Millikan knew how to relate the viscous drag on a spherical oil drop to the radius of the drop (Stoke's equation), and he could equate the gravitational force to the viscous drag when the drop reached terminal velocity by

$$MG = 6\,\pi\eta RV_t \tag{1.13}$$

where $V_t$ is the terminal velocity, and $\eta$ is the viscosity of air. After a measurement of the electric field that held a selected drop motionless, the drop was released by discharging the electric field. The measured terminal velocity gave a value of $R$ from Equation (1.13), and $e$ was then obtained from Equation (1.12).

Unfortunately, there was no way to attach single electrons to individual drops. However, examination of a large number of drops showed that the charge on a drop was equal to

$$e = n(1.602 \times 10^{-19} \text{ coulomb}) = n(4.803 \times 10^{-10} \text{ esu})$$

where $n$ was an integer $(1, 2, 3, \ldots)$. The simplest interpretation of these results was that all electrons had exactly the smallest observed charge, $e = 1.6 \times 10^{-19}$ coulomb. The drops with twice this charge had picked up two electrons, drops with three times this charge had picked up three electrons, and so on. A multitude of later observations have confirmed this interpretation.

The mass of the electron was calculated from Millikan's value for the charge and Thomson's value for the charge-to-mass ratio. The result is

$$m = \frac{e}{(e/m)} = \frac{1.602 \times 10^{-19} \text{ C}}{1.759 \times 10^8 \text{ C} \cdot \text{g}^{-1}} = 9.11 \times 10^{-28} \text{ g}$$

This value of the electron mass is amazingly low; it is only one two-thousandth of the mass of the lightest known atom. This low value of $m$ led to the conclusion that there must be additional subatomic particles that carry most of the mass of the atom.

The next step in the characterization of electrons might seem to be a determination of the size of electrons. If any such attempts were made, we can only hope that they were made by staunch Calvinists, because, as we shall see in the next chapter, such attempts were predestined to failure.

## 1.5 THE NUCLEUS.

By 1900 it was clear that all atoms contained electrons, although the number of electrons within any particular atom was not known. The atomic masses of all the then discovered elements were known with moderate accuracy. Thomson had continued his research on electrons and he thought that there was some correlation between the atomic mass and the number of electrons in an atom. He suggested that the number of electrons was close to the atomic mass on the relative scale (on the relative scale the atomic mass of the lightest element, hydrogen, was taken as the unit mass). There had to be positive charge somewhere in a neutral atom to neutralize the negative charges carried by the electrons. Thomson assumed that the as yet unidentified positive charge and the unaccounted for mass were contained in the same material. He made the reasonable (and incorrect) assumption that atoms were spheres of uniform density and uniform positive charge which contained an appropriate number of imbedded electrons. Thomson's model was laid to rest by the Rutherford scattering experiment.

Alpha particles, which are emitted during certain radioactive disintegrations, were shown by Rutherford and his coworkers to have a charge opposite in sign and twice as great as that of the electron. Alpha particles have enough energy to pass through thin metal foils relatively unscathed, but Rutherford realized that the difference between relatively unscathed and absolutely unscathed could be used to probe the atom. The experiment was a simple one; those alpha particles from a radioactive source that happened to pass through a narrow channel in a thick lead shield formed a beam. The alpha particles in the beam passed through a gold foil and produced flashes of light when they struck a fluorescent screen (Figure 1.3). The effect of the electrons in the gold foil could be ignored, because any collision between electrons and the heavy alpha particles are comparable to collisions between tennis balls and high-speed freight trains. According to Thomson's model of the atom, as the alpha particle penetrated the foil, it should have been continuously and intimately surrounded by a pudding of positive charge. Each portion of this positive charge should have repelled the positively charged alpha particle as it passed by, and this repulsion should have produced a deflection. However, the uniform spread of positive charge would suggest that the alpha particle would be scattered one way as often as the other, and the net result would be little if any observable change in direction (Figure 1.4). Hans Geiger (of Geiger-counter fame) had been studying these small-angle scatterings at Rutherford's laboratory. In 1910 Geiger and Rutherford asked a student to look for large-angle scattering, apparently anticipating that he would find nothing. On the contrary, he found detectable scattering at all angles up to 180°. Rutherford has vividly recalled his reaction to this news:

"It was the most incredible event that has ever happened to me in my life. It was almost as incredible as if you fired a 15-inch shell at a piece of tissue paper, and it came back and hit you."

Almost as incredible, but not quite; for while no one would have ducked after firing the shell, Rutherford and Geiger had the perception to look for alpha particles at large angles.

Rutherford immediately realized that large-angle scattering could occur only if the alpha particle came *very* close to an object that had a massive positive charge. In Thomson's model of the atom there was no way to get very close to a massive positive charge because the positive

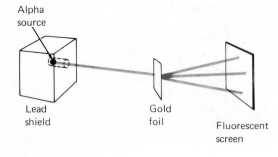

Alpha
source

Lead
shield

Gold
foil

Fluorescent
screen

**FIGURE 1.3** Rutherford's alpha scattering experiment.

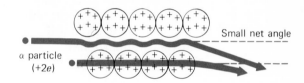

FIGURE 1.4 Two models of the gold atoms. According to the results of Rutherford, the model of Thomson was incorrect.

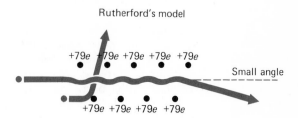

charge of the atom was too diffuse; the alpha particle could come *very* close to that fraction of the charge which was on one side of Thomson's atom, but it could not come very close to all of the positive charge (Figure 1.4). Rutherford's analysis of the trajectory of an alpha particle passing a single, heavy (and therefore fixed), positively charged particle showed that, in order for the scattering angle $\theta$ to be large, the impact parameter $b$ (see Figure 1.5) had to be very small and the charge $Q$ had to be very large. Atoms were known to have diameters of about $10^{-8}$ cm, yet the scattering angles observed with a gold target could only be reproduced by setting $b \leq 10^{-11}$ cm. The inescapable conclusion was that all of the positive charge and most of the mass of the atom are concentrated in a very small part of the whole atom. We call this part the *nucleus*.

Rutherford used the results of these scattering experiments to determine the radius of the nucleus. The trajectories are a consequence of the Coulomb's-Law repulsions between the nuclei and the alpha particles.

FIGURE 1.5 Deflection of an alpha particle by a massive nucleus.

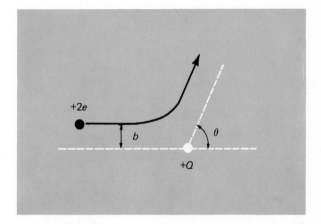

During an alpha particle's approach to a nucleus, its initial kinetic energy is converted into the potential energy of a system of two positive charges. In the case of a head-on collision, the alpha particle is stopped dead at some distance $r_{min}$; at this distance the energy conversion is complete, and the original kinetic energy ($mv^2/2$) can be equated to the resulting potential energy at the minimum separation (Coulomb's Law)

$$\frac{mv^2}{2} = \frac{(2e)Q}{r_{min}} \tag{1.14}$$

where $2e$ is the charge on the alpha particle and $Q$ is the nuclear charge.

The velocity of alpha particles is fixed by the selection of the radioactive source. Rutherford's analysis was sufficient for low-velocity alphas, but serious deficiencies appeared for high-velocity alphas. Higher-velocity alphas penetrated closer to the nucleus. The failure of the theory for close penetration indicated that the repulsion between the positive charges was no longer the sole source of the scattering forces, and that higher-velocity alpha particles actually were striking the edge of the nucleus itself (Figure 1.6). A variety of alpha sources were used to probe for this edge. For example, alpha particles with velocity $1.50 \times 10^9$ cm/sec and mass $6.68 \times 10^{-24}$ g were scattered perpendicularly to the foil surface. Such alphas got to within

$$r_{min} = \frac{2eQ}{mv^2/2} = \frac{2 \times 79(4.80 \times 10^{-10}\ \text{esu})^2}{(6.68 \times 10^{-24}\ \text{g})(1.50 \times 10^9\ \text{cm/sec})^2/2} = 4.8 \times 10^{-12}\ \text{cm} \tag{1.15}$$

of the center of a gold nucleus in a head-on collision.* Normal scattering implied that the alphas never reached the edges of the nuclei. Alphas with somewhat higher velocities were not scattered normally. Therefore, it was clear that the edge of the nucleus is about $10^{-12}$ cm from the center; that is, the radius of the gold nucleus is about $10^{-12}$ cm.

The preceding discussion assumed a knowledge of the total positive charge on the nucleus. From the neutrality of the atom we know that we can write this positive charge as $+Ze$, where $e$ is the charge of a single electron, and $Z$ is the number of electrons in the atom (which must be an integer). $Z$ is, of course, a property of the nucleus as well; it is called the *atomic number*. Careful analysis of a Rutherford scattering pattern yields

---

*The charge on the alpha particle is $2 \times 4.80 \times 10^{-10}$ esu, and the charge on the gold nucleus is $79 \times 4.80 \times 10^{-10}$ esu.

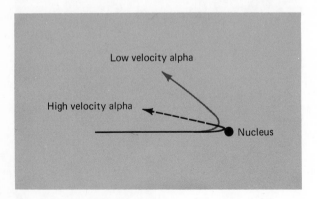

**FIGURE 1.6** Normal and abnormal scattering. The high velocity alpha particle actually contacts the nucleus; Coulomb's law does not account for its trajectory.

a value of Z. In principle, from a single, elegant experiment Rutherford was able to determine the size and the atomic number of the nucleus and, in turn, the number of electrons and the approximate structure of the atom. In practice these determinations of atomic numbers were too crude to be of much use, but accurate values soon became available from Moseley's studies on X-ray emission.

Rutherford's work with alpha particles also provided measurement of another fundamental constant, *Avogadro's number*. Each atom and each molecule has a definite mass. Moreover, an atom of helium weighs about four times as much as an atom of hydrogen. However, in most chemical applications it is not possible to use or weigh single atoms. Avogadro's number has been chosen to be the number of atoms (or molecules) in a sample that has a mass in grams equal to the relative atomic (or molecular) weight, for example, the number of atoms in 1.008 g of H or in 4.002 g of He. The amount of matter in Avogadro's number of particles is referred to as a *mole* (1.008 g H = 1 mole H; 4.002 g He = 1 mole He; the abbreviation for mole is mol).

Rutherford found that radium (Ra) produced alpha particles by the reaction

$$Ra \rightarrow Rn + \alpha$$

(where Rn is the element radon) and that the alpha particles (helium nuclei) quickly picked up two electrons to become neutral helium atoms

$$\alpha + 2e^- \rightarrow He$$

This process was observed to produce $5.40 \times 10^{-11}$ grams of He per second from a certain radium sample. The Geiger counter indicated that the same sample of radium was producing $0.815 \times 10^{13}$ alpha particles per second. Therefore, there must be

$$\frac{0.815 \times 10^{13}}{5.40 \times 10^{-11}} = 1.51 \times 10^{23} \text{ particles/g of He}$$

By definition, 4.002 g of He contains Avogadro's number of atoms, which is

$$\left(1.51 \times 10^{23} \frac{\text{particles}}{\text{g}}\right) (4.002 \text{ g}) = 6.04 \times 10^{23} \text{ particles}$$

This was not the earliest determination of Avogadro's number, but it was a very good one. The presently accepted value is

$$\boxed{N_0 = 6.022 \times 10^{23} \text{ mol}^{-1}}$$

Integral atomic numbers (i.e., the number of positive charges in a nucleus) led naturally to the idea of subnuclear elementary particles, some number of which are combined to form a nucleus. Such particles would have to be massive enough to provide the atomic mass as well. The early chemical determinations of accurate atomic weights were crucial to the evolution of modern nuclear theory. The original stimulus for this work was provided by Prout, a physician, who in 1816 suggested that the predominance of atomic weights that were almost exactly integral multiples of the smallest atomic weight (hydrogen) was no coincidence. In other words, there must be subnuclear particles, each of which contributes a unit of mass. However, there were a number of non-integral multiples which could not be explained this way (e.g., Cl(35.45) and Cu(63.54)). Furthermore, there is not a simple proportionality between the atomic number and the atomic weight. The atomic weight increases more rapidly than the atomic number.

A partial explanation of the non-integral atomic weights was provided by the discovery of *isotopes*. All the atoms in a chemically pure sample of a particular element have the same atomic number; all the atoms in an isotopically pure sample of a particular element have the same atomic number *and* atomic weight. These isotopic atomic weights are all slightly less than integer multiples of the atomic weight of the lightest isotope of hydrogen, but these integers are equal to or greater than the atomic numbers. The atomic weights of the elements are determined by the isotopic masses and the isotopic composition.

We now know that the nucleus of any atom is an aggregation of two kinds of massive elementary particles, consisting of $Z$ protons, each of which contributes a unit of positive charge, and $N$ neutrons (neutral particles), each of which contributes only mass. A particular isotope of an element is designated by the symbol for the element with the integer $(N + Z)$ as a superscript; the atomic number is often included as a subscript, such as hydrogen-one, $^1_1H$, carbon-twelve, $^{12}_6C$, and uranium-two-thirty-eight, $^{238}_{92}U$.

The neutron is very slightly heavier than the proton, so the sum of masses of the $N$ neutrons and $Z$ protons is a bit greater than $(N + Z)$ times the mass of one proton. The atomic weight is less than this; the difference between this mass sum and the actual nuclear mass is given by Einstein's mass-energy relation as

$$\Delta E = (\Delta m)c^2 \qquad (1.16)$$

For $^2_1H$ the mass sum is 1.00866 + 1.00780 = 2.01646. The actual mass is 2.01407, so $\Delta m =$ 0.00239, and $\Delta E =$ 2.15 $\times 10^{18}$ ergs.

where $\Delta m$ is the mass difference in grams, $c$ is the velocity of light (3.00 $\times$ $10^{10}$ cm/sec), and $\Delta E$ is the energy (in ergs) released upon binding together the neutrons and protons into a nucleus. These binding energies are about six orders of magnitude greater than those encountered in non-nuclear chemical reactions. For this reason the study of changes in the structure or composition of nuclei is a specialized topic; it will be treated in Chapter 23. Our present knowledge of the forces involved in nuclear binding is incomplete, but it is known that the neutrons play a vital role in the process. An isotope with too many or too few neutrons is unstable and will decompose spontaneously. At least one neutron is required to balance the Coulombic repulsion of each additional proton; extra neutrons are needed to hold the heavier nuclei together (Figure 1.7).

Separation of different isotopes of the same element via differences in chemical reactivity is difficult; it seldom occurs in the preparation of samples for atomic-weight determinations. Chemical atomic weights are weighted averages of the atomic weights of the naturally-occurring mixtures of isotopes. The refinement of Thomson's apparatus into the modern mass spectrometer provided the tool for the determination of isotopic atomic weights. A gaseous sample of the element in question is bombarded with an electron beam. This bombardment produces a large number of unipositive ions of the element; the charge-to-mass ratio measurement for these unipositive ions is effectively a mass measurement. Of course, these masses are for atoms minus one electron; the mass of the neutral atom is essentially the measured mass plus the mass of the missing elec-

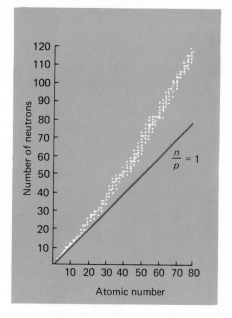

**FIGURE 1.7**  Composition of the stable nuclei. The heavy nuclei have more neutrons than protons (the solid line indicates equal numbers of neutrons ($n$) and protons ($p$)).

tron. A typical result for xenon is shown in Figure 1.8; naturally-occurring xenon is a mixture of nine stable isotopes.

The accuracy with which the atomic weights of light isotopes can be measured is remarkable. If we assign the neutral carbon isotope with six protons and six neutrons ($^{12}_{6}C$) a mass of *exactly* 12 atomic mass units,*

---

*This is a definition of a new unit of mass, the atomic mass unit amu or u:

$$1u = \frac{1g \cdot mol^{-1}}{6.022 \times 10^{23} \ mol^{-1}} = 1.661 \times 10^{-24} \ g = 1.661 \times 10^{-27} \ kg$$

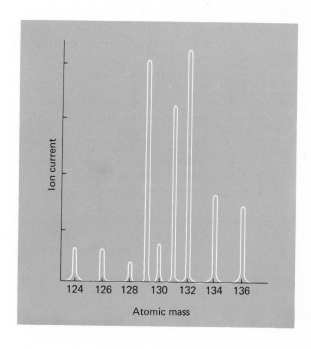

**FIGURE 1.8**  Mass spectrum of xenon. The height of the peaks is proportional to the abundance of the particular isotopes; the first two peaks have been magnified by a factor of 40.

then the oxygen isotope with eight protons and eight neutrons ($^{16}_{8}O$) has an atomic mass of $15.99491502 \pm 0.00000028$. The choice of $^{12}_{6}C$ as the standard for the relative scale of atomic weights is a recent one (1961). There were two older scales that differ in the fourth significant figure from the present scale; older compilations by physicists took $^{16}_{8}O$ to have an atomic weight of exactly 16, whereas older compilations by chemists were based on the value of exactly 16 for the *naturally-occurring mixture* of oxygen isotopes.

Chemical atomic weights may be obtained from isotopic weights by computing the weighted average. For example, a typical (normal-isotopic-abundance) sample of oxygen contains 99.759% $^{16}_{8}O$, 0.037% $^{17}_{8}O$, and 0.204% $^{18}_{8}O$. The appropriate atomic weights of these three isotopes on the modern scale are 15.99491, 16.99914, and 17.99916, respectively. The chemical atomic weight of oxygen is then

$$15.99491(0.99759) + 16.99914(0.00037) + 17.99916(0.00204) = 15.9994.$$

It is obvious that chemical atomic weights can change if the isotopic compositions of elements vary. These variations are usually insignificant, but there are a few significant departures from the average. For example, Israel has become a major source of the heavy isotopes of oxygen because of the abnormally high amounts of $^{18}_{8}O$ and $^{17}_{8}O$ in the water of the Dead Sea (this excess gives them a head start on the separation of isotopes). Some commercially available lithium compounds in the U.S.A. are unusually rich in $^{7}_{3}Li$ because some of the $^{6}_{3}Li$ has been extracted for use in nuclear reactions. Old lead-storage batteries that have been charged and discharged many times have a higher than normal percentage of deuterium in the battery electrolyte (aqueous sulfuric acid), because of the preferential reduction of $^{1}_{1}H^{+}(aq)$ (over $^{2}_{1}H^{+}(aq)$) to $H_2(g)$.

## PROBLEMS*

1. List your reasons for registering for this course. If this list includes "to fulfill a requirement for a degree in . . .," try to list reasons why this course is a requirement.

2. How many protons and how many neutrons are there in the following nuclei: $^{2}_{1}H$, $^{70}_{31}Ga$, $^{102}_{45}Rh$, $^{124}_{54}Xe$?

3. The chemical atomic weight of chlorine is 35.453 amu. The mass spectrum of chlorine shows only two isotopes, $^{35}_{17}Cl$ with mass 34.969 and $^{37}_{17}Cl$ with mass 36.966. What is the relative abundance of each isotope in a typical sample?

4. If the diameter of a nucleus is of the order of $10^{-13}$ cm and the diameter of an atom is of the order of $10^{-8}$ cm, what fraction of the "volume of the atom" is occupied by the nucleus?

5. The velocity of light is $3.0 \times 10^{10}$ cm/sec. Compute the distance (in centimeters, meters, and miles) traveled by light in one year.

---

*Problems are arranged roughly in order of increasing difficulty.

6. Given the following masses,

$^1_0n$: 1.00866     $^2_1D$: 2.01407

$^1_1H$: 1.00780     $^4_2He$: 4.00256

e⁻: 0.00055

compute the energy evolved per nucleus and per mole in each of the following processes:

$$^1_0n = {}^1_1H + {}^0_{-1}e$$

$$^0_{-1}e + 2\,{}^1_1H = {}^2_1D$$

$$2\,{}^2_1D = {}^4_2He$$

$$^1_1H + {}^1_0n = {}^2_1D$$

Express your results in joules ($1\,J = 1\,kg. \cdot m^2 \cdot sec^{-2}$).

7. How many grams are there in one atomic mass unit?

8. What is the mass of an electron in amu?

9. The isotopes $^6_3Li$ (6.015 amu) and $^7_3Li$ (7.016 amu) have the normal relative abundances 7.40% and 92.60%. What is the chemical atomic weight of lithium? What is the atomic weight of a sample of lithium from which 1% of the original $^6_3Li$ has been extracted?

10. Compute the energy (in joules) of an electron subjected to a potential of 1 volt.

11. Avogadro's number is $6.0225 \times 10^{23}$ on the $^{12}_6C$ scale. What was it on the chemists' old scale?

12. What is the ratio of the atomic weights on the modern scale and (a) the chemist's old scale; (b) the physicist's old scale?

13. What would Avogadro's number be if we took pounds as our fundamental unit of weight instead of grams?

14. The human body is largely water and organic molecules. Let us make the assumption that the "average" atom in the body has atomic weight 10.0. Use this to estimate the total number of electrons in a 125 pound body.

15. The energy required to remove an electron from a hydrogen atom is 13.3 eV. Compute the difference in mass between a hydrogen atom and a separated proton plus an electron. ($1\,eV = 1.6022 \times 10^{-12}$ erg.)

16. How close can an alpha particle with velocity of $10^9$ cm/sec approach the nucleus of an oxygen atom?

17. In a typical Thomson experiment the magnetic field is large enough

to displace the electron beam about 20° downward within the field. Suppose we wished to check the value for $e/m$ to five significant figures and we had a measuring device which could sense a beam movement on the screen of 0.01 mm. How long must we make the distance from the field to the screen to give the desired accuracy?

18. The Millikan oil-drop experiment relies on a balance between gravitational force and electrostatic force. There is a problem with units; gravitational force is usually calculated in the cgs (cm·g·sec) system, so that charge must be measured in esu units and the electric field in statvolts/cm (1 statvolt = 300 volts). Assume you had an apparatus with plates 1 cm apart and a standard 1.5 volt battery with a voltage adjusting device. To what accuracy must you regulate this voltage to balance the gravitational force on a free electron to within 1%? Now attach this electron to an oil droplet. How heavy should this drop be to allow use of the whole 1.5 V?

19. List all of the physical properties that you can think of that would possibly be of use in the characterization of a substance. Use your general list and the *Handbook of Chemistry and Physics* to prepare specific lists for the following pure substances:

(a) table salt                    (b) sucrose (table sugar)
(c) water                         (d) grain alcohol
(e) ascorbic acid (Vitamin C)

## References

1.1. Shamos, M. H., *Great Experiments in Physics* (Holt, New York, 1959).
1.2. Birks, J. B., *Rutherford at Manchester* (Benjamin, New York, 1963).
1.3. Weeks, M. E., *Discovery of the Elements* (6th Ed., Journal of Chemical Education, Easton, Pa., 1956).
1.4. Nier, A. O. C., *Mass Spectroscopy—An Old Field in a New World,* American Scientist, *54*, 359–384 (1966).
1.5. Sanders, J. H., *The Fundamental Atomic Constants* (Oxford University Press, Fair Lawn, N. J., 1961).
1.6. Hildebrand, J. E., and Powell, R. E., *Avogadro's Number,* Chapter 21 in *Principles of Chemistry* (7th Ed., Macmillan, New York, 1964).
1.7. Feynman, R. P., R. B. Leighton, and M. Sands, *The Feynman Lectures on Physics* (Addison-Wesley, Reading, Mass., 1965).
     Note: In spite of their difficulty, these volumes contain such unusual insight into basic physics that every physical scientist should eventually read them.

## CHAPTER 2

# ATOMIC SPECTRA AND ATOMIC STRUCTURE

*"The first attempt which was made in this direction was the following: I chose substances of the smallest atomic weight and arranged them in order of the value of their atomic weights. From this, it appeared that there seemed to be a periodicity in the properties of the individual substances, and indeed that, if one considered the valences, one element followed the other in the arithmetical sequence of their atomic weights."*

*D. Mendeleev\**

Rutherford showed that the atom is a massive, positively-charged nucleus surrounded, at distances that are large compared to the size of the nucleus, by negatively-charged electrons. Without a force to counteract the attraction between opposite charges, atoms would collapse. The similarity in the force laws for electrical and gravitational attraction (both are inversely proportional to the square of the separation of the interacting bodies) led scientists to consider a solar-system model of the atom in which the counter force was the centrifugal force on orbiting electrons. There were fatal flaws in this model. Force produces acceleration in classical mechanics, and in contrast to neutral planets, every classical charged particle radiates energy when accelerated. As the electrons in the solar-system model radiated away their energy, they would spiral toward the nucleus, and atoms would collapse. All atomic models based on classical mechanics (including the solar-system model) lead to unstable atoms. The unraveling of the mysteries of atomic structure required a new, more general mechanics, *quantum mechanics*, which was capable of dealing with objects that exhibit both particle-like and wave-like properties.

## 2.1 WAVES AND PARTICLES.

Careful observation of their surroundings convinced early scientists that there were two classes of motion, particle motion and wave motion.

---

\*D. Mendeleev, J. Russ. Chem. Soc., *1*, 1869. (Quote taken from G. B. Kauffman, J. Chem. Educ., *46*, 135 (1969).) The arrangement of the elements in the modern periodic table is by *atomic number*, rather than by *atomic weight*. Mendeleev's great contribution was his arrangement of the elements in terms of the *periodicities* in their chemical properties.

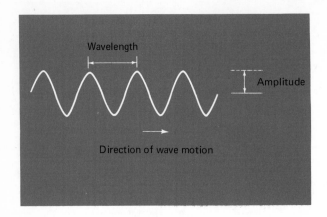

**FIGURE 2.1**　A simple wave. The amplitude is measured from the average to the peak.

The only wave which is seen easily, a water wave, exhibits most of the characteristics of light, sound, radio, and all other waves. The only waves which need concern us are periodic waves, waves that repeat themselves every so many seconds.

A segment of the cross section of the simplest periodic wave in water might look like the wave in Figure 2.1. This segment of the wave *moves as a whole with time;* if we sit at one point and count, we can determine the frequency in peaks, or troughs, or cycles per second.* The frequency $\nu$ (in cycles/sec), times the wavelength $\lambda$ (the distance between peaks in cm/cycle) is equal to the velocity $v$ (in cm/sec) of the wave, or

$$\nu\lambda = v \tag{2.1}$$

The impossibility of direct visual observation of many types of waves (e.g., radio waves) makes it imperative that we find distinguishing characteristics of wave motion. A basic characteristic of wave behavior is *interference*, which is the combination of two or more waves into a single wave. What happens if two waves with the same wavelength and amplitude (wave height) come together at some point in a medium? The amplitudes of waves are additive (think of the water waves), so the result depends only on the relative positions of the peaks in the two waves, called the *phase*. Three of the many possible phases and the resultant waves, obtained by adding the two separate wave amplitudes at every point, are shown in Figure 2.2. The in-phase waves exhibit *constructive interference* (increased total amplitude), whereas the out-of-phase waves exhibit *destructive interference* (decreased total amplitude).

In classical mechanics, interference is an indication of wave motion. You may have observed interference of either water or sound waves. Two stones thrown into a pond will produce an interference pattern wherever their separate waves cross; similarly, "out-of-phase" speakers in a stereo set produce destructive interference patterns of sound waves at some locations.

*Diffraction* is a more convenient indicator of wave motion. If a wave of wavelength $\lambda$ strikes a flat mirror, then the wave is reflected at a definite angle. The heavy cross lines in Figure 2.3 indicate the positions of successive peaks in the wave. Strange things happen with tiny mirrors. When the width of the mirror is similar to the wavelength of the wave, the wave

---

*The unit *cycles per second* has been given the name *hertz*.

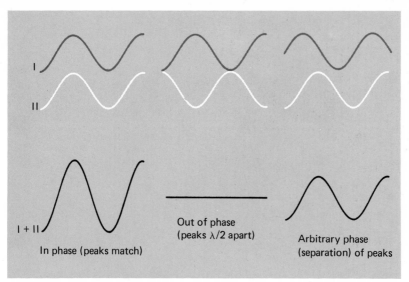

**FIGURE 2.2** Interfering waves. The amplitude is doubled when the waves are in phase and is destroyed when the waves are out of phase.

is not reflected; it is scattered (Figure 2.4). When the wave is scattered it has amplitude at all positions above the mirror, and it appears to originate at the mirror.* A second scattering mirror within a few wavelengths of the first will produce an identical pattern of waves which will interfere with the waves scattered by the first mirror (Figure 2.5). Wherever the arcs from the scattered waves cross on the diagram the peaks match perfectly, and constructive interference occurs. Constructive interference occurs only at definite angles to the plane of the mirrors. It is worthwhile to derive

---

*The difference can be seen in the reflection of a water wave from a breakwater and the scattering of a water wave from a piling.

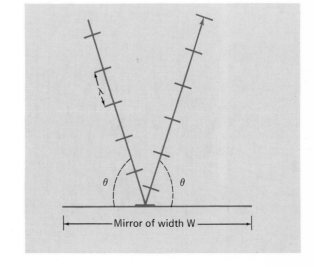

**FIGURE 2.3** Reflection at a mirror.

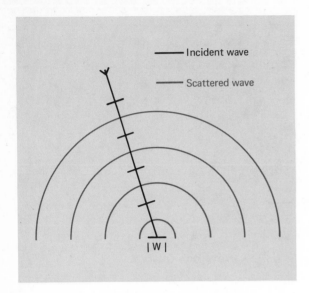

**FIGURE 2.4**   Scattering from a small mirror.

an expression for these angles. Let the distance between the two scattering mirrors be $d$ (see Figure 2.6). Two peaks that were together in the incoming beam (say at $A$) are no longer together after scattering because the wave scattered on the left must travel a distance $d \sin \theta$ greater than the wave scattered on the right. If $d \sin \theta$ is exactly one, or two, or any integral number of wavelengths, the peak from the left will match a peak from the right which was one wavelength, or two wavelengths, or any integral number of wavelengths behind it in the incident beam. Thus, the condition for constructive interference is

$$d \sin \theta = n\lambda \qquad (2.2)$$

where $n$ *is any positive integer.* At any other value of $\theta$ the peaks will not match precisely, and the intensity is reduced, as is seen in Figure 2.5. This

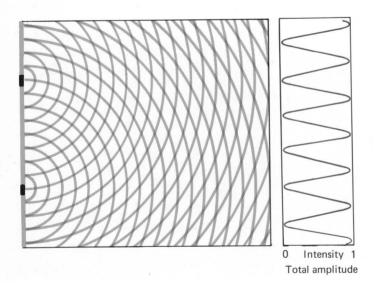

0     Intensity   1
Total amplitude

**FIGURE 2.5**   Diffraction. The intensity is high along directions where the two waves are in phase.

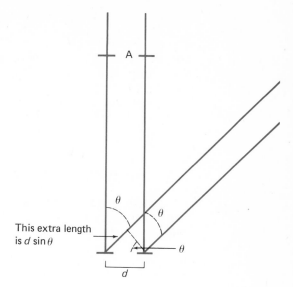

**FIGURE 2.6**  Geometric condition for diffraction. The extra length must be an integral number of wavelengths.

This extra length is $d \sin \theta$

alternating pattern of intensities is called a *diffraction pattern.*[*] It is also unmistakable evidence of wave motion.

The inclusion of a series of scattering centers, each separated from its neighbors by exactly the distance $d$, will sharpen the intensity pattern. The condition for constructive interference is unchanged, but completely destructive interference (zero intensity) occurs at all intermediate angles. Under such conditions the effect can be used to determine $d$ from a known $\lambda$ and measured $\theta$, or to measure or resolve different $\lambda$ values from known $d$ and $\theta$ values. The former is the method used to measure interatomic distances in crystals, and the latter is the method used to measure wavelengths in diffraction-grating spectrometers.

The energy carried by a wave is independent of its velocity; it is directly proportional to the square of the wave's amplitude. The spatial extension of a wave implies that this energy cannot be delivered instantaneously, but only continuously.

The foregoing brief discussion of waves has stressed those characteristics that serve to distinguish waves from particles. A particle can be given any desired velocity (within very wide limits). Particles exhibit no destructive interference. The combination of two particles gives new particles, or an aggregate of the original particles, but it never gives zero particles. Particles exhibit no diffraction effects; in a diffraction experiment, either they strike one of the scattering centers and rebound at a definite angle, or they miss. Finally, particles do have definite positions, and a rigid particle can transmit all of its kinetic energy to the target at the instant of collision.

The mechanics that were known at the beginning of this century provided a perfect rationale for all particles and waves that were subject to visual observation. How then might one handle particles that were too small to be seen (e.g., nuclei and electrons) or waves that were invisible

---

[*]The diffraction pattern of light waves from a phonograph record is readily observable because the color (i.e., wavelength) of the scattered light changes with the angle of observation.

(e.g., light*)? The obvious answer is, and was, by extrapolation. Apply the mechanics of large particles to very small particles, and apply the mechanics of visible waves to invisible waves. This approach was initially successful. Light was shown to exhibit diffraction effects, and this result was regarded as evidence against Newton's particle theory of light. However, at the turn of the century, evidence began to accumulate that these apparently reasonable extrapolations were in fact erroneous. The failure of the solar-system model for the atom was evidence of trouble with particles; the photoelectric effect was evidence of trouble with waves.

## 2.2   THE PHOTOELECTRIC EFFECT.

Under certain conditions, light, which shows every characteristic of wave motion, can cause the ejection of electrons from the surfaces of metal plates. If light is a wave, then the kinetic energy of the ejected electrons should be proportional to the light intensity (amplitude squared is intensity). Actually, the observed electron kinetic energy is independent of the intensity of the light. If the frequency is below a certain threshold value, no electrons are ejected regardless of the light intensity. When the frequency is above this threshold, the kinetic energy of the electrons is a linear function of the frequency,† and the light intensity regulates the number of electrons which are ejected (Figure 2.7).

By 1905, Albert Einstein realized that the incident light in the photoelectric effect was behaving like a stream of particles, rather than like a wave. He proposed that light is indeed a stream of a vast number of individual particles (around $10^{18}$/sec in a flashlight beam), which are called *photons*. Moreover, he suggested that the energy of the photons is proportional to the frequency of the light. The photons, as particles, transmit their energy instantaneously upon collision with an electron. If the photon provides enough energy, the electron overcomes whatever forces bind

---

*Although our eyes respond to light in the visible region of the spectrum, we cannot see the wave motion of light. We see light in the same sense that a blind man might sense water waves by standing in the surf.

†According to the classical theory of waves, the energy of a wave is proportional to the square of the frequency.

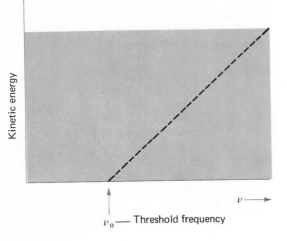

**FIGURE 2.7**  Results of the photoelectric experiment.

it to the metal and escapes, with any excess energy as residual kinetic energy. The principle of conservation of energy applied to this process gives

$$E = h\nu = h\nu_0 + \frac{1}{2} mv^2 \tag{2.3}$$

where $h$ is a proportionality constant between energy and frequency, $h\nu$ is the energy of the incoming photon, $h\nu_0$ is the minimum energy for an electron to escape from the metal, $m$ is the electron mass, and $v$ is the electron velocity. Rearrangement of Equation (2.3) gives

$$\frac{1}{2} mv^2 = h\nu - h\nu_0 \tag{2.4}$$

This is the equation for the straight line in Figure 2.7.

It is difficult for anyone who does not have an extensive training in classical physics to imagine the impact of Einstein's suggestion. Although it accounted for the photoelectric effect in a simple and satisfying manner, it was based on what, to a classical physicist, seemed to be a mass of contradictions. Maxwell's equations, which treat light as a combination of electric and magnetic fields propagating via wave motion, were and still are regarded as the greatest triumph of 19th Century physics. Light was unquestionably a wave. But Einstein, by treating light as a stream of particles, had succeeded where the wave theory had failed. Worse yet, Einstein's particles had frequencies. In retrospect we can see that the dilemma was caused by the extrapolation from the visible to the invisible. Macroscopic laws of motion are not applicable to sub-microscopic objects; for these latter objects distinctions between particles and waves are no longer valid. The laws of motion of sub-microscopic objects, such as electrons or photons, are different from those of macroscopic particles or waves.

A proportionality between frequency and energy for radiant-energy (resonators) had been suggested first in 1900 by Max Planck. Planck was concerned with the frequency distribution of light emitted by a hot, glowing body. Classical theory predicted a ridiculous result: the energy emitted at any frequency should be proportional to the frequency squared. This is ridiculous because there is no upper limit on frequency, so the body must continually emit an infinite amount of energy. The predicted and experimental results at several temperatures are shown in Figure 2.8. Planck first found an empirical formula which reproduced the experimental curves, but which contained an adjustable constant $h$ (called, naturally, Planck's constant). He was subsequently able to derive his formula with the aid of the — at that time wild — assumption that the radiant body (treated as a sub-microscopic set of resonators) emitted energy as light only in tiny packets, and that the energy of each packet was $h\nu$. These packets of energy were called *quanta*. It was easy to dismiss Planck's hypothesis as a scientific curiosity until Einstein's explanation of an unrelated phenomenon agreed with Planck's hypothesis in both approach and detail. Specifically, each photon carries a quantum of energy $h\nu$, and the value for the proportionality constant $h$ is exactly the same in the Planck equation ($E = h\nu$) and the equation used by Einstein to analyze the photoelectric effect. The value of $h$ is

$$\boxed{h = 6.626 \times 10^{-34} \text{ J} \cdot \text{sec}}$$

$$\boxed{E = h\nu}$$

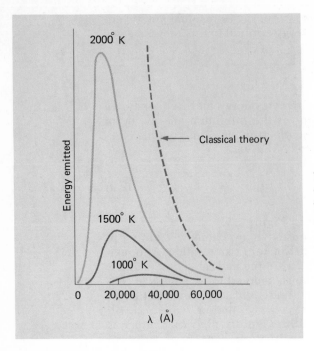

**FIGURE 2.8**  Wavelengths of energy emitted by a glowing object. The amount and distribution of energy emitted is dependent on the temperature.

Further experiments demonstrated that under certain conditions light had to be treated as a stream of *quantized photons* (i.e., as an integral number of photons). These experiments led directly to two breakthroughs in the development of quantum mechanics. The first of these came in the area of atomic spectroscopy.

## 2.3  ATOMIC SPECTROSCOPY AND THE BOHR ATOM.

According to Figure 2.8, the intensity of light emitted by a glowing body varies with frequency. Notice that the intensity is a smooth function of frequency; there are no wild fluctuations in any narrow frequency range. This distribution of intensity is typical of many light sources. The intensity shift of such a source with increasing temperature is quite obvious; compare the color of the coil on an electric range ($\sim$ 700°C), a light bulb (4000°C), and a photoflood lamp (6000°C).* However, there are many sources of light whose intensity patterns are radically different from that shown in Figure 2.8. These sources emit light only at selected frequencies. Examples of these are "neon lamps" [which in fact contain neon (red), or krypton (green), or xenon (blue)], mercury-vapor lamps (blue-white), sodium-vapor lamps (yellow), and hydrogen lamps (faint purple). These lamps contain an easily vaporizable element, and the light is produced by an electric discharge through the vapor. Under suitable conditions a lamp of this type can be made from any element. A section of the intensity distributions of the light emission due to hydrogen is shown in Figure

---

*Different colors of light are simply different responses of the human eye to different frequencies of light; that is, different colors mean different frequencies. The human eye responds only to wavelengths in the 4000-7000Å range. The response is not uniform.

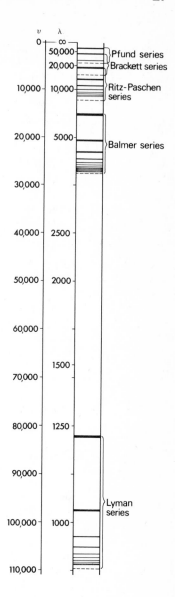

**FIGURE 2.9** Schematic representation of the H-atom spectrum. The intensity is indicated roughly by the thickness of the lines. The dotted lines correspond to the series limits, at which a continuous spectrum sometimes joins the series. [Adapted from G. Herzberg, *Atomic Spectra and Atomic Structure* (2nd edition, Dover Publications, 1944).] The various series can all be fit by the equation

$$\nu = R\left(\frac{1}{n_2^2} - \frac{1}{n_1^2}\right).$$

A specific series is obtained when the value of $n_2$ is fixed: $n_2 = 2$ (Balmer, visible); $n_2 = 1$ (Lyman, far ultraviolet); $n_2 = 3$, $n_2 = 4$, $n_2 = 5$ (Paschen, Brackett, and Pfund, all in the infrared). For $n_2$ fixed and $n_1 > n_2$, as $n_1$ increases $\nu$ approaches the limiting value $\nu_\infty = R/n_2^2$.

2.9. The study of these intensity patterns is called *atomic spectroscopy;* the pattern itself (that is, a plot of intensity versus wavelength, frequency, or energy) is called a *spectrum.*

The atom with the simplest spectrum is the simplest atom, hydrogen. As early as 1885 it was known that the positions of some of the lines in this spectrum could be calculated from the empirical (Balmer Series) formula

$$\nu = 3.29 \times 10^{15}\left(\frac{1}{2^2} - \frac{1}{n^2}\right) \text{ cycles/sec} \qquad (2.5)$$

where $n$ is an integer larger than two. Subsequent work on the hydrogen spectrum showed that other series of emission lines existed in spectral regions that are invisible to the human eye, and that these series also could be represented by formulae of comparable simplicity (Figure 2.9).

By 1913, Niels Bohr had realized that the proportionality of energy to frequency was the key to the interpretation of the hydrogen atom spectrum. The Bohr model of the atom is not drastically different from the solar-system model. However, it involves the *assumptions* that electrons exist in certain stationary (time-independent) orbits or states, that each stationary state has a particular energy associated with it, and that electrons cannot radiate energy and spiral toward the nucleus. The stationary states are called *energy levels*. Light is emitted in order to conserve energy when an electron falls from a stationary state of high energy to one of lower energy. This conservation equation is usually called the Bohr frequency condition:

$$E_2 - E_1 = h\nu \tag{2.6}$$

where $E_2$ is the energy of the higher state, $E_1$ is that of the lower state, and $h\nu$ is the energy of the emitted photon.

Bohr chose the simplest possible model for hydrogen, an electron in a circular orbit around a proton. The particular circular orbits which were allowed were determined by the *postulate* that only those orbits with angular momentum ($m\nu r$) equal to an *integral multiple* of $h/2\pi$ were permissible:

$$m\nu r = nh/2\pi \tag{2.7}$$

where $n$ is an integer, and $r$ is the radius of electron's orbit. In other words, the angular momentum was quantized ($n = 1, 2, 3, \ldots$).

The initial successes of Bohr's theory were impressive. According to the theory, the energy of an election in an orbit characterized by the integer $n$ is given by (see problem 25)

$$E_n = -\frac{2\pi^2 m e^4 Z^2}{n^2 h^2} \tag{2.8}$$

where $Z$ is the atomic number, $e$ and $m$ are the charge and mass of the electron, and $h$ is Planck's constant. The Bohr frequency condition (Equation (2.6)) can be rewritten for the particular case where $n$ in the lower level is equal to 2, as

$$E_n - E_2 = h\nu = \frac{2\pi^2 m e^4 Z^2}{h^2} \left(\frac{1}{2^2} - \frac{1}{n^2}\right) \tag{2.9}$$

This has the same form as the empirical equation for the hydrogen spectrum (Equation (2.5)); moreover, the empirical coefficient is given exactly as

$$3.29 \times 10^{15} = \frac{2\pi^2 m e^4 Z^2}{h^3}$$

The success of the Bohr theory was short lived. It accurately predicted the spectra for the H atom and the *one-electron* ions He$^+$ and Li$^{+2}$, but all attempts to extend the theory to atoms with two or more electrons failed.

## 2.4  WAVE-PARTICLE DUALITY.

In 1924, Louis de Broglie took the next step forward.

"I was convinced that the wave-particle duality discovered by Einstein in his theory of light quanta was absolutely general, . . . and it seemed certain to me, therefore, that the propagation of a wave is associated with the motion of a particle of any sort."

Before this idea could be tested by experiment, an estimate of the expected wavelength of a particle was needed. From Equations (2.3) and (2.1) for photons

$$E = h\nu \qquad (2.3)$$

$$\nu\lambda = c \qquad (2.1)$$

The symbol $c$ represents the velocity of light in vacuum.

and from Einstein's relation between energy and mass

$$E = mc^2 \qquad (2.10)$$

we have

$$mc^2 = h\nu = hc/\lambda \qquad (2.11)$$

Therefore,

$$\lambda = h/mc = h/p$$

where $p$ is the usual symbol for momentum (momentum = mass × velocity). De Broglie believed that this equation was a general one which applied to all matter, including electrons. *All moving objects, whatever their mass, have a wavelength associated with them of*

$$\boxed{\lambda = h/mv} \qquad (2.12)$$

where $m$ is the mass and $v$ is the velocity. The diffraction of electron waves was observed soon after De Broglie put forth his hypothesis, and today, microscopes (electron microscopes) using electrons rather than photons are found in laboratories throughout the world.

The confirmation of de Broglie's hypothesis removed the question of whether electrons and photons are particles or waves. They, and every other object, are neither, and both. That is, every object has both wave properties, such as wavelength, and particle properties, such as mass. In certain experiments the characteristics we traditionally associate with wave motion will predominate, while in other experiments the characteristics we associate with particle motion will predominate. Table 2.1 shows that the wavelengths for massive macroscopic objects are extremely small. When we remember that our tests for wave motion (for example, diffraction experiments) require scattering centers that are small compared to the wavelength, it is clear why the experimental conditions for diffraction of baseballs are unachievable. Although baseballs are both particles and waves, they will always act as particles under any experimental conditions. It is unfortunate for our intuitive understanding of quantum mechanics that we can see only those objects which have short wavelengths.

TABLE 2.1

WAVELENGTHS OF VARIOUS OBJECTS

| PARTICLE | MASS (G) | VELOCITY (CM/SEC) | WAVELENGTH (Å) |
|---|---|---|---|
| Electron (1 volt) | $9.1 \times 10^{-28}$ | $5.9 \times 10^{7}$ | 12 |
| Electron (100 volt) | $9.1 \times 10^{-28}$ | $5.9 \times 10^{8}$ | 1.2 |
| Electron (10,000 volt) | $9.1 \times 10^{-28}$ | $5.9 \times 10^{9}$ | 0.12 |
| Proton (100 volt) | $1.67 \times 10^{-24}$ | $1.38 \times 10^{7}$ | 0.029 |
| $\alpha$ Particle (100 volt) | $6.6 \times 10^{-24}$ | $6.9 \times 10^{6}$ | 0.015 |
| $\alpha$ Particle from Radium | $6.6 \times 10^{-24}$ | $1.51 \times 10^{9}$ | $6.6 \times 10^{-5}$ |
| Rifle Bullet (.22) | 1.9 | $3.2 \times 10^{4}$ | $1.1 \times 10^{-23}$ |
| Golf Ball | 45 | $3 \times 10^{3}$ | $4.9 \times 10^{-24}$ |
| Baseball | 140 | $2.5 \times 10^{3}$ | $1.9 \times 10^{-24}$ |

$1\text{Å} = 10^{-10}\text{m} = 0.1$ nanometer $= 0.1$ nm

From J. D. Stranathan, *The Particles of Modern Physics* (Philadelphia: Blakiston, 1942).

## The Uncertainty Principle.

One of the classical distinctions between particles and waves is that particles occupy a definite position, whereas waves are spread through space. Which description is correct for subatomic particles? Consider the following experiment: We wish to measure the position of an electron (at least the $x$ coordinate) by observation through a hypothetical, super-power microscope. There is a limit on the accuracy with which the position of an object can be determined by its interaction with a wave. If an object is smaller than one wavelength of the light used to probe for its position, scattering will occur regardless of *how* much smaller it is. In other words, once the object is smaller than one wavelength of the light used, there is no observable change in the scattering pattern if the object is moved a distance less than one wavelength.* Therefore, if we wish to see the position of the electron accurately, we must use very short-wavelength light. But each photon has momentum $p = h/\lambda$ which it imparts to the electron upon collision. In attempting to measure the $x$ coordinate to an accuracy of $\Delta x \simeq \lambda$, we have given the electron an additional momentum in the $x$ direction that lies somewhere between zero and $h/\lambda$. Note that, although the total momentum given the electron is $h/\lambda$, there is also an uncertainty $\Delta p_x = h/\lambda$ in the $x$ component because we know essentially nothing about the direction of the scattered photon. The product of the uncertainty in the momentum of the object and the uncertainty in the position of the object is

$$\Delta p_x \cdot \Delta x \simeq h \qquad (2.13)$$

This expression is called the *Heisenberg Uncertainty Relation*. It is a limitation on the accuracy of certain experiments, which arises because of the unavoidable interaction between the measuring device (the photon) and the subject (the electron). It does not imply that the position of an elec-

---

*It is less obvious, but nonetheless true, that the edge of a larger object also causes scattering of light. The effect can be observed at the edge of a breakwater.

tron cannot be measured with any desired accuracy; it simply says that in so doing the experimenter introduces such a large uncertainty in the momentum of the electron that the measurement tells where the electron *was*, but not where it *is*.

Classical mechanics (Newton's laws) is based on the assumption that a simultaneous determination of position and momentum is possible. The momentum is needed for the calculation of the trajectory (the position at all future times) of the object. The uncertainty relation says that the initial determination is of limited accuracy. Is the limitation serious? Let us assume that we are satisfied with knowing the position of an electron in a 1 Å diameter atom to about 50% or 0.5 Å accuracy. This would require a photon which would produce a minimum change in the momentum of[*]

$$\Delta p_x = \frac{h}{\Delta x} = \frac{6.6 \times 10^{-27} \text{ erg} \cdot \text{sec}}{5 \times 10^{-9} \text{ cm}} = 1.3 \times 10^{-18} \text{ g} \cdot \text{cm/sec}.$$

Because the electronic mass is $9.1 \times 10^{-28}$ g, the change in the velocity is

$$\Delta v = \Delta p/m = 1.3 \times 10^{-18}/9.1 \times 10^{-28} = 1.4 \times 10^9 \text{ cm/sec}$$

This is an incredible velocity; such an electron has sufficient kinetic energy to blast apart any atom. Thus, we see that the limitation is extremely serious, for the first crude measurement to determine a trajectory has destroyed the atom. It is clear that Newton's equations of motion are useless for electrons in atoms.

## 2.5 THE SCHRÖDINGER EQUATION.

Frequently it is useful to study light via a wave equation. Schrödinger thought that a corresponding wave equation for an electron might be useful in explaining some hitherto unexplained properties of the electron, and in particular, its behavior in the neighborhood of a positively-charged nucleus. The Schrödinger equation for the simplest possible system, which is a one-dimensional system with no time dependence, is

$$-\frac{h^2}{8\pi^2 m} \frac{d^2\psi}{dx^2} + V \psi = E\psi \tag{2.14}$$

This is an equation for a particle with mass $(m)$ and energy $(E)$, which is under the influence of a potential $(V)$. The expression $\frac{d^2}{dx^2}$ means the second derivative, or the rate of change of the rate of change, of $\psi$, the amplitude of the electron wave, with respect to $x$.[†] The solutions to this equation ($\psi$—called *psi*), of which there are very many, are functions— called *wavefunctions*—which give the amplitude of the electron wave as a function of position. The absence of time as a variable means that the solutions correspond to stationary states.

The amplitude of an electron wave is useless unless it can be related

---

[*]$10^7$ erg = 1 joule, therefore,
$\qquad h = (6.626 \times 10^{-34} \text{ J} \cdot \text{sec}) (10^7 \text{ erg} \cdot \text{J}^{-1}) = 6.626 \times 10^{-27} \text{ erg} \cdot \text{sec}$
[†]You are probably familiar with acceleration, which is the rate of change of velocity with respect to time. Because velocity is the rate of change of position with respect to time, acceleration is the second derivative of position with respect to time.

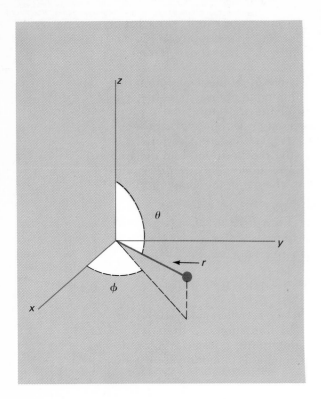

**FIGURE 2.10**   Spherical coordinates.

to a physically measurable quantity. The standard interpretation is that the square of the wavefunction, $\psi^2$, indicates how the *probability* of finding the particle varies with position, that is, $\psi^2$ is a probability density. The *probability* of finding the particle within a small volume $(dv)$ is $\psi^2 dv$. This probability interpretation is within the spirit of the uncertainty principle; as long as there is a finite probability of the electron being elsewhere, the required uncertainty in its position exists.

The interpretation of $\psi^2$ as a probability density imposes two limitations on the possible wavefunctions. The first is that $\psi$ must be uniquely defined for all space. By this we mean the following: The angle $\theta$ (Fig. 2.10) can be used as a position variable; it represents rotation in some plane about a nucleus. A uniquely defined wavefunction must have a unique value at every position; it must have the same value for, say, $\theta = 37°$ and $\theta = 37° + 360°$, because these are both exactly the same position. The second limitation is that the total probability of finding the electron somewhere in space must be unity. This is, in practice, a requirement that the probability, and thus the wavefunction, be zero at $x = \infty$. These limitations are examples of boundary conditions. *The imposition of boundary conditions on the solutions to wave equations reduces the number of possible frequencies to a few allowed values.* This is true whether the solutions apply to guitar strings (where the boundary conditions are imposed by the frets, and the frequencies allowed are the pitch and its overtones) or to electrons. In the latter case, restricted frequencies mean restricted energies.

In principle (but not in practice), all physical properties of any system can be calculated from its wavefunction by a surprisingly straightforward procedure.* Our interest here is only in electron energies and positions in atoms and molecules.

Whenever a physical quantity is restricted to a set of discrete values, we say that the quantity is *quantized*. For example, the angular momentum and the energy of the electron in the Bohr theory are quantized, and the total energy in a light beam is quantized. Because boundary conditions always introduce quantization, the mechanics based on the Schrödinger equation is called *quantum mechanics*.

## 2.6 THE HYDROGEN ATOM.

The Schrödinger equation cannot be solved exactly for any atom but hydrogen. There is an analogy with the Bohr theory. In that case, no one knew what to do for atoms more complex than hydrogen. We now know what to do for complex atoms—solve the appropriate Schrödinger equation—but no one knows how to do it.

However, a knowledge of the correct equation to be solved allows us to seek approximate solutions (approximate wavefunctions) for more complex atoms. Fortunately, the wavefunctions for the hydrogen atom are the basis for good *approximate* solutions. We will examine the hydrogen wavefunctions in detail. The process of calculating these wavefunctions is complex; we simply present results.

The initial step in the solution of any Schrödinger equation is the selection of a suitable coordinate system; that is, one selects the coordinate system in which the potential energy has the simplest algebraic form. The potential energy for the hydrogen atom is $-e^2/r$, where r is the electron-proton separation. Spherical coordinates are most suitable. The spherical coordinates of any point are $r$, $\theta$, and $\phi$, as shown in Figure 2.10.

The imposition of boundary conditions on the wavefunctions leads to restrictions on the possible energies of the electron. These restrictions give rise to variables in the wavefunctions; the values that can be assumed by these variables are restricted to a set of integers. These variables and their integral values are called *quantum numbers*. The most important quantum number, called the *principal quantum number* and denoted by the symbol $n$, is a consequence of the requirement that the wavefunction approach zero at large distances from the nucleus. This quantum number restricts the possible electron energies to the values

$$E_n = -\frac{2\pi^2 m e^4 Z^2}{n^2 h^2} \tag{2.15}$$

which is identical to the Bohr result (Equation (2.8)). The energy in Equation (2.15) is defined relative to a zero of energy where the electron and the nucleus are completely separated. Consequently, the energies of states with the electron bound to the nucleus are negative. The principal quantum number $n$ can be any positive non-zero integer. The restriction to integers means the *energy is quantized*.

The expression for $E_n$ in Equation (2.15) accounts for the regular series

---

*The interested reader should consult any standard quantum mechanics text. *Quantum Mechanics in Chemistry* by M. Hanna (Benjamin, 1966) contains several simple examples.

observed in the hydrogen spectrum. The energy difference between two
energy levels is

$$E_n - E_{n'} = -\frac{2\pi^2 m e^4 Z^2}{h^2}\left[\frac{1}{n^2} - \frac{1}{n'^2}\right] = h\nu \qquad (2.16)$$

In Figure 2.11 the distinction between the energy level pattern for an atom
or molecule and the associated spectrum is illustrated. Notice that every
line in the spectrum is associated with the energy *difference* between two
energy levels. The energy of a free electron is not quantized[*] and can have
any positive value. Therefore, at frequencies that correspond to energies
greater than the largest energy difference between the bound stationary
states of an electron in an atom, we find a continuum in the spectrum. The
Bohr frequency condition still applies, but the upper energy in the difference is no longer quantized. The energy at which the spectrum changes
from discrete lines to a continuum is called the *ionization limit.* Light
of frequency greater than the ionization limit can knock an electron free
from the atom; that is, it can *ionize* the atom. Every electron bound to any
atom or molecule has an ionization limit. The energy required to remove
an electron is called the *ionization energy.*

The results embodied in Figure 2.11 and Equation (2.16) are identical
to those obtained from the Bohr theory, but this identity is misleading; the
quantum-mechanical solution is much more complex. There are two more
quantum numbers that appear in the wavefunctions as a result of the requirement that the wavefunctions be uniquely defined (two, because there
are two distinct angles which can be incremented to return to the original
position). One of these quantum numbers is the *azimuthal quantum number, $\ell$.* Although $\ell$ does not appear in the energy expression for the hydrogen atom, it does appear in the expression for the angular momentum of the
electron, which, as a consequence, also is quantized:

$$\text{angular momentum} = \sqrt{\ell(\ell + 1)}h/2\pi.$$

As the angular momentum $(mvr)$ and the energy $\left(\frac{1}{2}mv^2 - \frac{e^2}{r}\right)$ are not independent, it is reasonable that the possible values of $\ell$ are limited by the
total energy, that is, by *n.* *The quantum number $\ell$ can assume any positive
integer value, zero included, up to and including $(n - 1)$.* The third quantum number is the *magnetic quantum number, m.* It specifies the component of the total angular momentum of the electron along a particular axis.
*The magnetic quantum number can assume any integer value between
$m = +\ell$ and $m = -\ell$, inclusive.*

There is a fourth quantum number for the electron, the *electron spin
quantum number, s.* The electron spin can assume only the values $+\frac{1}{2}$ and
$-\frac{1}{2}$. Electron spin enters naturally only when relativity is introduced
into the quantum theory. It is not fortuitous that a three-dimensional problem leads to three quantum numbers, and that introduction of a fourth
dimension (time) via relativity leads to a fourth quantum number. One-

---

[*]Because the electron is free, it can be infinitely far away from the nucleus, so the boundary condition which produces the energy quantization is inoperative.

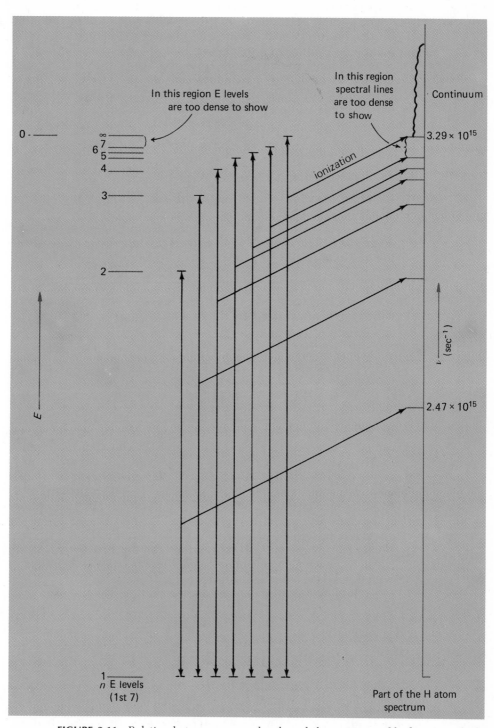

**FIGURE 2.11**  Relation between energy levels and the spectrum of hydrogen. At high energies the levels lie too close to show in the energy level diagram or to resolve in the spectrum.

and two-dimensional problems contain one and two quantum numbers, respectively.

The various restrictions on the possible values of the quantum numbers for the electrons in an atom exclude most combinations of the quantum numbers. As an example, take the lowest energy state — *the ground state* — of the hydrogen atom. From Equation (2.15), this is the state with minimum $n$, or $n = 1$. The maximum value of $\ell$ is $\ell = n - 1 = 1 - 1 = 0$, and the minimum value is also $\ell = 0$. Thus $\ell$ can have only one value, $\ell = 0$. If the only possible value of $\ell$ is $\ell = 0$, then the only possible value of $m$ is $m = 0$, but $s$ always can be either $+\frac{1}{2}$ or $-\frac{1}{2}$. There are then only two possible wavefunctions for the ground state of hydrogen; they correspond to the two quantum number sets $1, 0, 0, \frac{1}{2}$ and $1, 0, 0, -\frac{1}{2}$.[*] These two wavefunctions are for states with exactly the same energy; therefore, the ground state is said to be *doubly degenerate.*

Any state with energy above the ground-state energy is called an *excited state.* The first excited state of hydrogen is one with $n = 2$. Here the maximum value of $\ell$ is $\ell = n - 1 = 1$, and $\ell$ can also be 0. When $\ell = 1$, $m$ can be $-1$, or 0, or $+1$. Again, $s = \pm\frac{1}{2}$. We therefore have eight different wavefunctions and states with the same energy (eightfold degeneracy) which are characterized by the quantum-number sets

$$2, 1, -1, \tfrac{1}{2} \qquad 2, 1, 1, \tfrac{1}{2} \qquad 2, 1, 0, \tfrac{1}{2} \qquad 2, 0, 0, \tfrac{1}{2}$$

$$2, 1, -1, -\tfrac{1}{2} \qquad 2, 1, 1, -\tfrac{1}{2} \qquad 2, 1, 0, -\tfrac{1}{2} \qquad 2, 0, 0, -\tfrac{1}{2}$$

By similar application of the rules, the chart of permitted combinations can be generated (Table 2.2). The "alternate notation" column of Table 2.2 involves the use of the letters $s$, $p$, $d$, and $f$ to denote values of $\ell$. This practice dates from earlier spectral designations of associated lines in atomic spectra: *s*harp ($\ell = 0$), *p*rincipal ($\ell = 1$), *d*iffuse ($\ell = 2$), and *f*ine ($\ell = 3$). The chart could be extended indefinitely, but we shall see that this is unnecessary.

Thus far we have indexed and discussed energies for the wavefunctions. The mathematical expressions for the first few wavefunctions are given in Table 2.3. No provision for electron spin is included, an omission which will be continued throughout this text.

The lowest energy wavefunction (1, 0, 0) has the form

$$\psi\,(1s) = N e^{-Zr/a_0}$$

where

$$N = \left(\frac{1}{\pi}\right)^{\frac{1}{2}} \left(\frac{Z}{a_0}\right)^{3/2}$$

The quantity $N$ is called the *normalization factor;* the value of $N$ is determined by the requirement that the *total* probability of finding the electron over all space (adding up $\psi^2$ everywhere) must be exactly one. Normaliza-

---

[*]By convention, the order is always $n$, $\ell$, $m$, $s$.

**TABLE 2.2**

PERMITTED COMBINATIONS OF QUANTUM NUMBERS

| $n$ | $\ell$ | ALTERNATE NOTATION | $m$ | $s$ | NUMBER OF POSSIBILITIES WITH THESE VALUES OF $n$ AND $\ell$ |
|---|---|---|---|---|---|
| 1 | 0 | $1s$ | 0 | $\pm\frac{1}{2}$ | 2 |
| 2 | 0 | $2s$ | 0 | $\pm\frac{1}{2}$ | 2 ⎫ |
| 2 | 1 | $2p$ | $-1, 0, 1$ | $\pm\frac{1}{2}$ | 6 ⎬ 8 |
| 3 | 0 | $3s$ | 0 | $\pm\frac{1}{2}$ | 2 ⎫ |
| 3 | 1 | $3p$ | $-1, 0, 1$ | $\pm\frac{1}{2}$ | 6 ⎬ 18 |
| 3 | 2 | $3d$ | $-2, -1, 0, 1, 2$ | $\pm\frac{1}{2}$ | 10 ⎭ |
| 4 | 0 | $4s$ | 0 | $\pm\frac{1}{2}$ | 2 ⎫ |
| 4 | 1 | $4p$ | $-1, 0, 1$ | $\pm\frac{1}{2}$ | 6 ⎬ 32 |
| 4 | 2 | $4d$ | $-2, -1, 0, 1, 2$ | $\pm\frac{1}{2}$ | 10 |
| 4 | 3 | $4f$ | $-3, -2, -1, 0, 1, 2, 3$ | $\pm\frac{1}{2}$ | 14 ⎭ |

**TABLE 2.3**

SOME WAVEFUNCTIONS FOR ONE-ELECTRON ATOMS

$$\psi\,(1s)\; =\left(\frac{1}{\pi}\right)^{1/2}\left(\frac{Z}{a_0}\right)^{3/2}\,e^{-Zr/a_0}$$

$$\psi\,(2s)\; =\left(\frac{1}{32\pi}\right)^{1/2}\left(\frac{Z}{a_0}\right)^{3/2}\left(\frac{Zr}{a_0}-2\right)e^{-Zr/a_0}$$

$$\psi\,(2p_z)=\left(\frac{1}{32\pi}\right)^{1/2}\left(\frac{Z}{a_0}\right)^{3/2}\left(\frac{Zr}{a_0}\right)e^{-Zr/2a_0}\cos\theta$$

$$\psi\,(2p_x)=\left(\frac{1}{32\pi}\right)^{1/2}\left(\frac{Z}{a_0}\right)^{3/2}\left(\frac{Zr}{a_0}\right)e^{-Zr/2a_0}\sin\theta\sin\phi$$

$$\psi\,(2p_y)=\left(\frac{1}{32\pi}\right)^{1/2}\left(\frac{Z}{a_0}\right)^{3/2}\left(\frac{Zr}{a_0}\right)e^{-Zr/2a_0}\sin\theta\cos\phi$$

$$a_0 = 0.53\ \text{Å} = \frac{h^2}{4\pi^2 m e^2}$$

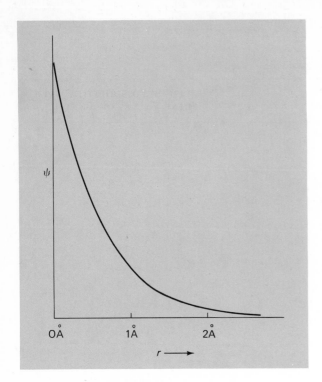

**FIGURE 2.12**   Plot of $\psi$ versus $r$ for the $1s$ orbital of hydrogen.

tion factors appear in all complete wavefunctions. Note that here $e$ is the base of the natural logarithms, rather than the electron charge.

Visual representations of the wavefunctions are more useful than mathematical formulas for a qualitative discussion. Representations of wavefunctions are called *orbitals*, and they can be drawn in a variety of ways. The simplest representation is a plot of $\psi$ versus $r$; the example shown in Figure 2.12 is the $1s$ orbital. This is not an informative representation. A more useful one is a plot of the probability of finding the electron at a distance $r$ from the nucleus versus $r$. This is not simply a plot of $\psi^2$ versus $r$, for the following reason. All points that lie at a distance $r$ from a central point lie on the surface of a sphere of radius $r$. The surface area of a sphere of radius $r$ is $4\pi r^2$. The probability of finding an electron anywhere on the surface of a sphere of radius $r$ is* $4\pi r^2 \psi^2$. The function $4\pi r^2 \psi^2$ is called a *radial distribution function*. The radial distribution function for a $1s$ orbital of the hydrogen atom is plotted as a function of $r$ in Figure 2.13. Notice that, although $\psi^2$ itself is large at the nucleus ($r = 0$), the probability of finding an electron there is very small. Although there

---

*We need to calculate the probability that the electron is at $r = r_0$; $r_0$ can never be specified exactly (nor in this case do we wish to do so). What is really implied is the probability that the electron is between $r_0 - \delta$ and $r_0 + \delta$, where $\delta$ is much smaller than $r_0$. If the electron is between these two radii, it is in a volume $\frac{4}{3}\pi(r_0 + \delta)^3 - \frac{4}{3}\pi(r_0 - \delta)^3$. Since $\delta$ is very small, we write $(r_0 + \delta)^3 \simeq r_0^3 + 3r_0^2 \delta$. Then the volume is

$$(4/3) \pi [r_0^3 + 3\,r_0^2\,\delta - r_0^3 + 3\,r_0^2\,\delta] = 4\pi r_0^2\,(2\,\delta)$$

The factor $2\,\delta$ is arbitrary and can be the same at every value of $r$, so it is not included.

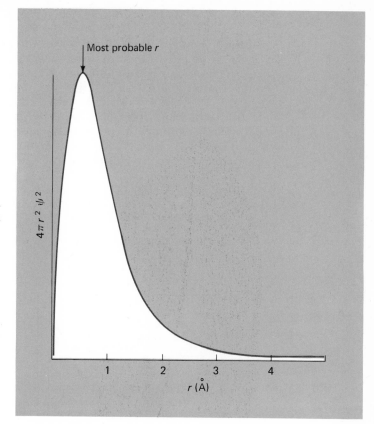

**FIGURE 2.13** Probability of finding a 1s electron in a hydrogen atom at a distance r from the nucleus.

is not a discrete radius for the H atom in the classical sense, there is a distance where the probability of finding the electron is a maximum.

Wavefunctions are three-dimensional; three-dimensional orbitals are also useful representations. There are two types of three-dimensional representations in use. In one (Figure 2.14), the wavefunction (or its square) is represented by a varying density of printed points. Stereo pairs of these representations give a three-dimensional picture of the orbital

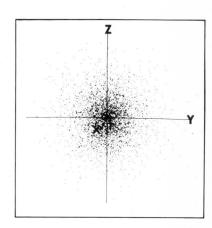

**FIGURE 2.14** Representation of the 1s orbital of hydrogen. The density of dots in a region indicates the electron density in that region. The axes extend 2.1 Å from the nucleus. (From D. T. Cromer, *J. Chem. Educ.* **45**, 629 (1968).)

(Figure 2.15). If stereo viewers are not available,* either of the pairs can be viewed (without serious error) as a projection of the electron-density distribution onto a plane which passes through the nucleus. A *contour map* is a more practical three-dimensional representation. Instead of altitude, each contour line is a boundary corresponding to some value of $\psi$ or $\psi^2$. Notice that $\psi^2$ is not a probability function, but a probability-*density* function; the quantity $\psi^2$ is usually called the *electron-density function*. The electron will most likely be found where the electron density is high *over a large volume*. We shall rely extensively on contour maps in this text. These will be of two types: elaborate maps reproduced from the research literature, and crude maps. The crude maps will have contour

---

*Here are two methods for viewing stereo pictures.
1. View the left picture with the left eye and the right with the right. Focus on a distant point until the central image merges. A card between the images and a viewing distance of 8 in. is suitable.
2. Buy or construct a viewer from a pair of simple lenses. Inexpensive viewers are available from Stereo Magniscope, Inc., 40-31 81st Street, Elmhurst, New York 11373.

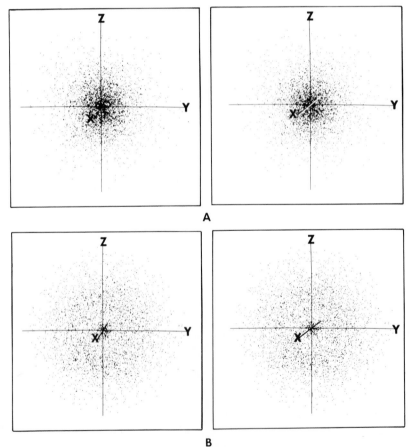

**FIGURE 2.15**   Stereo pictures of orbitals: (a) $1s$; (b) $2s$; (c) $2p_z$; (d) $3s$; (e) $3p_z$; (f) $3d_{xy}$; (g) $3d_{x^2-y^2}$; (h) $3d_{z^2}$. In part (a) the axes extend 2.1 Å out from the nucleus; in parts (b) and (c) they extend 5.3 Å out from the nucleus. The spherical node in (b) is visible only in the stereo view; the $xy$ plane in (c) is a node. In parts (d), (e), and (h) the axes extend 13 Å out from the nucleus, while in parts (f) and (g) they extend 10.6 Å. Part (c) is a slab 4 Å thick. (From D. T. Cromer, *J. Chem. Educ.* **45**, 629 (1968).)

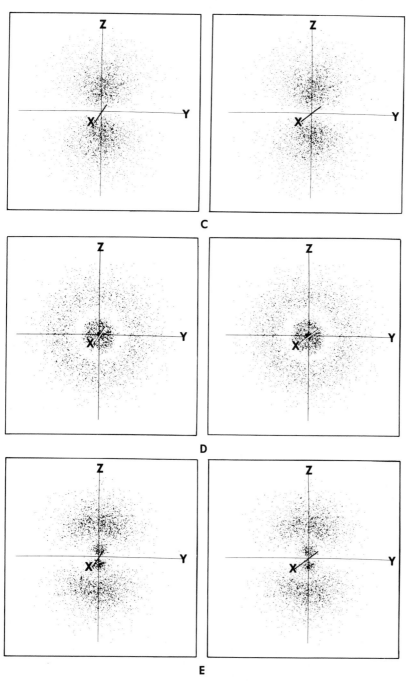

C

D

E

FIGURE 2.15 Continued.

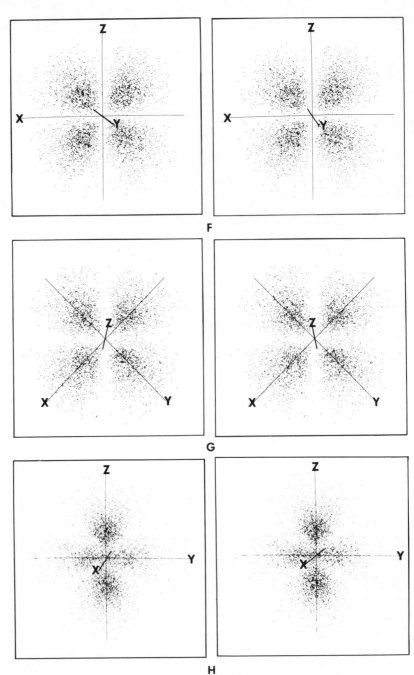

**FIGURE 2.15** Continued.

**FIGURE 2.16** A contour map of the electron density in a plane through the center of the hydrogen 1s orbital. The contour line marked H surrounds about 40% of the total electron density, and the contour line marked L surrounds about 80% of the total electron density.

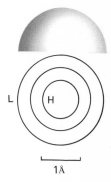

lines marked H and L which enclose—very approximately—40% (H, for region of high density) and 80% (L) of the total electron density.

In order to provide both a feeling for the shape of orbitals and an accurate electron-density map, we shall depict orbitals in the following way: A three-dimensional drawing of the orbital will be sliced open to reveal a contour map of the electron density; the slice will be along a plane which is especially informative (for atoms this plane will always pass through the nucleus). Our first example is the 1s wavefunction (Figure 2.16). This orbital could be likened to an onion; when we slice it open we see layers of constant electron density.

Examination of the one-electron wavefunctions with $n = 2$ shows that there are certain finite values of the coordinates for which the wavefunction is zero. For the $\psi(2s)$ wavefunction this occurs when $(Zr/a_0) - 2 = 0$; for $\psi(2p_z)$ it occurs when $\cos \theta = 0$. For $\psi(2s)$, the points where the wavefunction is zero form a sphere; the $\psi(2s)$ wavefunction has a *spherical node*. For $\psi(2p_z)$, the $xy$-plane corresponds to all points where $\theta = 90°$, and consequently ($\cos 90° = 0$) this wavefunction has a *nodal plane* (an angular node). The $\psi(2p_x)$ and $\psi(2p_y)$ wavefunctions also have nodal planes. We thus reach the important conclusion that all wavefunctions with $n = 2$ have one node; not surprisingly, all atomic wavefunctions have $n - 1$ nodes of all types.[*]

In Figure 2.17 we illustrate the meaning of the contour maps in more detail. The electron density in a plane through each of the three orbitals shown is represented by peaks in a deformed plane. By slicing the peaks off these figures, we generate the contour maps.

Orbital representations for the $n = 2$ and $n = 3$ orbitals are given in

---

[*] These are $n - \ell - 1$ spherical nodes; the other $\ell$ nodes are planar or conical nodes.

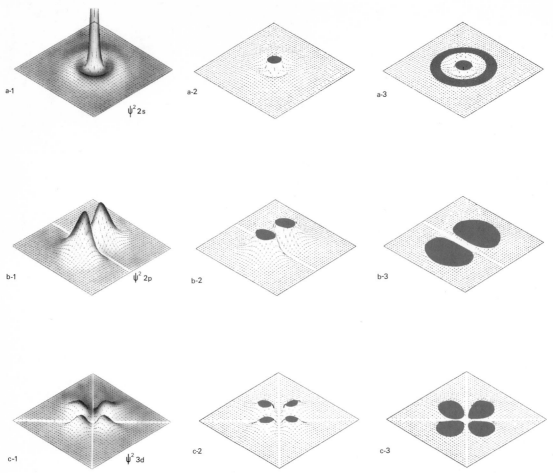

**FIGURE 2.17** Electron density in a plane, and contour representations. In part (1) the value of $\psi^2$ in the plane is indicated by the height *above* the plane for three selected orbitals. (Note that in a-1 the top of the peak has been left off. The peak actually extends about $2\frac{1}{2}$ times as high as the chopped peak.) In part (2) we sliced off the peaks, thereby showing the contour lines which surround regions of high electron density. In part (3) we slice much deeper, leaving a contour line which fences out regions of low electron density. Compare with Figures 2–18 and 2–19(c). Part (1) from W. T. Bordass and J. W. Linnett, J. Chem. Educ. *47,* 672 (1970) by permission.

Figures 2.18 and 2.19, respectively. The signs designate the mathematical sign of the wavefunction in the indicated regions. *The signs do not indicate charges.* Nodes are shown as dashed lines on the contour maps. The $3p_x$, $3p_y$, $3d_{yz}$, and $3d_{xy}$ orbitals differ from those included only in the labeling of axes. Note that all $\ell = 2$ ($d$) orbitals have two angular nodes; the $n = 3$, $\ell = 1$ orbitals have one angular node and one spherical node, and the $n = 3$, $\ell = 0$ orbital has two spherical nodes.

The wavefunctions for hydrogen are separable into radial ($r$) and angular ($\theta$ and $\phi$) parts. An orbital representation for a $p$ orbital that includes only the angular portion of the wavefunction is shown in Figure 2.20. There is an unfortunate tendency to confuse a plot of the angular portion of the wavefunction with a contour map of the *complete* wavefunction. The differences between the two are obvious (compare Figures 2.20 and

(*Text continued on page 51.*)

More elaborate contour maps of the atomic orbitals are given in reference 2.8.

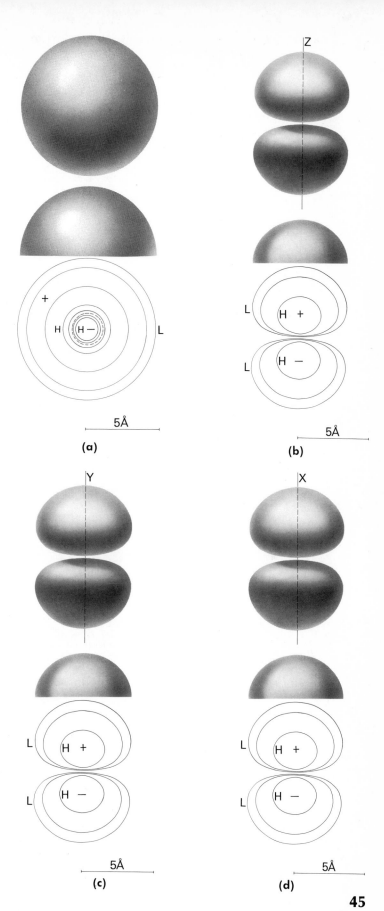

**FIGURE 2.18** Contour maps of hydrogen orbitals: (a) $2s$; (b) $2p_z$; (c) $2p_y$; (d) $2p_x$. The dashed contour lines designate nodes.

5Å

**(a)**

5Å

**(b)**

5Å

**(c)**

5Å

**(d)**

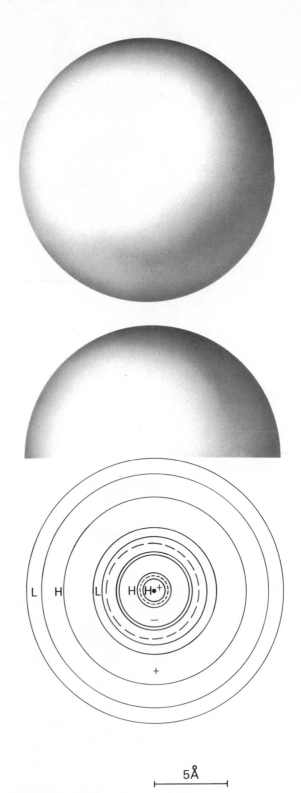

**FIGURE 2.19**   Contour maps of hydrogen orbitals: **A** 3*s*:

*Ill. continued on opposite page.*

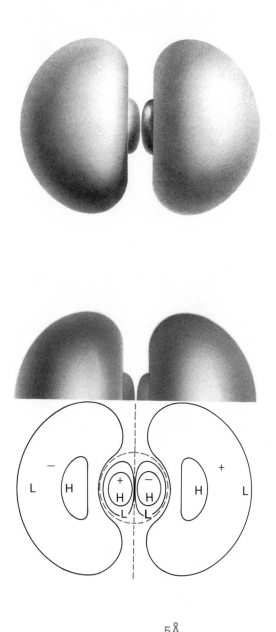

**FIGURE 2.19B**  Contour map of hydrogen $3p$ orbital.

*Ill. continued on following page.*

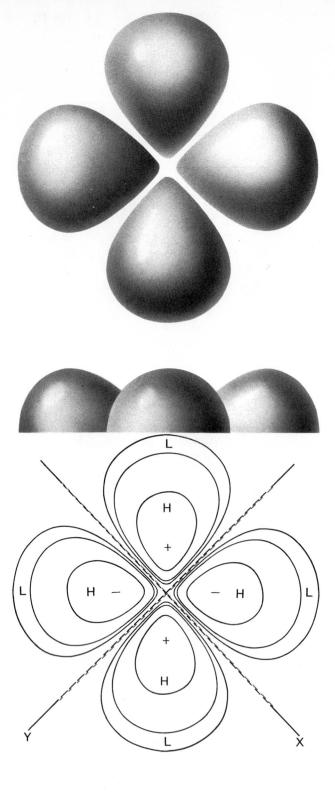

**FIGURE 2.19C**   Contour map of hydrogen $3d_{xy}$ orbital.

5Å

*Ill. continued on opposite page.*

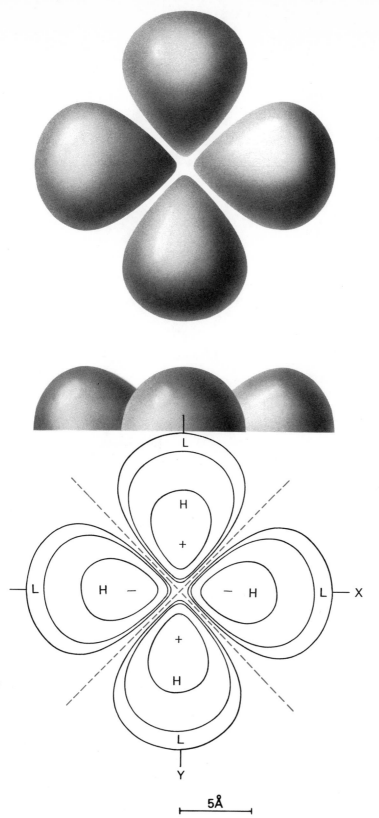

**FIGURE 2.19D** Contour map of hydrogen $3d_{x^2-y^2}$ orbital.

*Ill. continued on following page.*

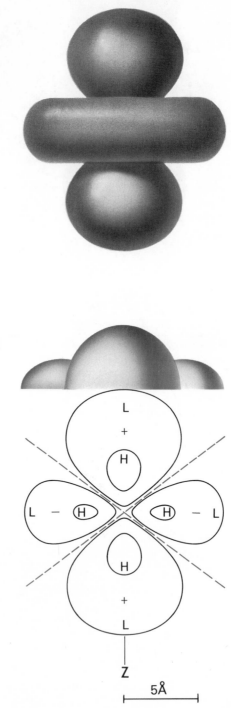

**FIGURE 2.19E**   Contour map of hydrogen $3d_{z^2}$ orbital.

(a)  (b)

**FIGURE 2.20** Common (incomplete) representations of $p$ orbitals: (a) Angular part of a $p$ orbital; (b) the square of the angular part of a $p$ orbital.

2.18b). The square of the angular part of the wavefunction is also used as an orbital representation. This is even further removed from an accurate electron-density representation. Both drawings may be misleading and henceforth will be avoided.

## 2.7 HYBRIDIZATION.

Certain orbitals are labeled with Cartesian-coordinate subscripts, for example, $z$ in $2p_z$. The number of different subscripts for orbitals of a given $\ell$ value is the same as the number of permitted values of $m$ for each value of $\ell$. However, these subscripts do not correspond to the different values of $m$. Although the $2p_z$ orbital does have $m = 0$, the $2p_x$ orbital does not have $m$ equal to either $+1$ or $-1$. The latter two orbitals have identical shapes and orientations (Figure 2.21).

Differential equations like the Schrödinger equation usually have many solutions. In particular, any linear combination of solutions is also

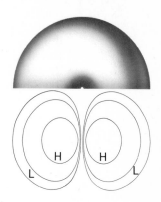

**FIGURE 2.21** Shape of orbitals with $m = 1$ or $m = -1$.

a solution. In quantum mechanics these new linear combinations are called *hybrid orbitals*. The $2p_x$ and $2p_y$ orbitals are *hybrid orbitals*, because[*]

$$\psi(2p_x) = \frac{\sqrt{2}}{2}\Big[\psi(2,\,1,\,1) + \psi(2,\,1,\,-1)\Big]$$

$$\psi(2p_y) = \frac{\sqrt{2}}{2}\cdot\sqrt{-1}\,\Big[\psi(2,\,1,\,-1) - \psi(2,\,1,\,1)\Big]$$

(2.17)

Hybridization plays a central role in qualitative theories of molecular structure. One essential feature of hybridization is that the *number of resulting orbitals* (or coordinates) *must be exactly the same as the number of original orbitals (orbital conservation).* It is possible to form sets of two, and only two, new $p$ orbitals from the $(2, 1, 1)$ and $(2, 1, -1)$ wavefunctions; the orbitals $2p_x$ and $2p_y$ are one possible set.

Agreed that there is the possibility of forming hybrids, why are hybrid orbitals needed? We use hybrid orbitals because the solution to some problems is (or at least appears to be) simpler with the hybrid orbitals than with the original orbitals. If we wish to calculate the behavior of a hydrogen atom in a magnetic field, the orbitals that contain only a single value of the magnetic quantum number are the best. If we are interested in chemical bonding, the hybrid orbitals $2p_x$ and $2p_y$ may be more useful, for they, like chemical bonds, are highly directional.

## 2.8  MULTI-ELECTRON ATOMS AND THE PERIODIC TABLE.

The massive accumulation of data in chemistry would be absolutely indigestible if there were no patterns in the properties of the elements. This was as obvious to early chemists as it is today, and the search for helpful patterns began early in the previous century and reached its culmination in the Periodic Table of the Elements.

Dmitri Mendeleev was the first to suggest a version of this table that had predictive value. He arranged the elements then known (about 70 of them) in horizontal rows according to increasing atomic weights; elements with similar chemical properties were placed in vertical columns (e.g., bromine under chlorine, potassium under sodium). The insistence on chemical similarity among vertical neighbors implied the existence of undiscovered elements, and the predicted chemical properties of these undiscovered elements were of great assistance in their discovery.[†] Moreover, the first periodic chart stimulated an extensive re-examination of those elements whose atomic weights did not fit into the emerging pattern. Eventually it was discovered that the atomic number, and consequently the number of electrons, was the key parameter in the ordering. There have been countless proposals as to the best layout for the periodic chart, and it is still possible to find disagreement today. We shall use the con-

---

[*]The convention is $\psi(n,\,\ell,\,m)$.

[†]This process continues today. The chemical properties of the laboratory-produced trans-uranium elements, which can be predicted from the periodic table, have made possible the characterization of some elements, the total existent amount of which is far too small to be visible. See Chapter 23.

ventional arrangement, because it can be rationalized in terms of the wave-functions of hydrogen.

Before generating a periodic chart from our knowledge of the hydrogen wavefunctions, we must introduce two additional rules for multi-electron atoms:

1. *Pauli Exclusion Principle: No two electrons in a particular atom can have identical sets of four quantum numbers.* Either the spin quantum numbers must be different, or at least one of the three quantum numbers $n$, $\ell$, and $m$ must be different in the two sets. As far as we now know, the exclusion principle is as fundamental a property of the electron as the electronic charge; it cannot be derived from a more basic theory.

2. *Hund's Rule:* The electrons in the ground state of an atom will have the same spin quantum number as long as this does not require extra energy.

Hund's Rule can be rationalized. If two electrons have the same spin, by the Pauli Principle they must be in different orbitals. Electrons in different orbitals are on the average a bit farther apart than electrons in the same orbital. Increased separation means reduced electron-electron repulsion energy. The saving in energy is small, and Hund's Rule only applies to degenerate (or near-degenerate) energy levels.*

The Schrödinger equation cannot be solved exactly for any system with two or more electrons. Therefore, we must make do with approximate solutions. Although approximate solutions of any desired accuracy can be generated for any atom or molecule, the resulting wavefunctions are complicated. For qualitative, pictorial wavefunctions we settle for less accuracy. We use the wavefunctions from the simplest related system—a hydrogenic atom. By *hydrogenic* we mean a one-electron atom with a nuclear charge $Z$ (as always, in atomic units, so that the true $Z$ is the atomic number).

It is not necessary for our purposes to discuss the wavefunction for the atom; it is enough to specify an orbital for each electron in the atom. This list of orbitals for all of the electrons in an atom is called the *electron configuration*. Qualitative treatments of molecular structure and chemical bonding are largely assignments of electron configurations. Usually we will be concerned only with lowest energy configuration, the ground-state configuration. This can be deduced simply. The available orbitals are arranged in order of increasing energy; each electron in turn is assigned to that unfilled† orbital that has the lowest energy. The electron assignments must be consistent with the Pauli Principle and Hund's Rule.

The simplest multi-electron atom is helium (He), which has $Z = 2$. In the ground state of helium both electrons are in the hydrogenic orbital that has quantum numbers $n = 1$, $\ell = 0$, $m = 0$, or more simply 1, 0, 0. The two electrons obey the Pauli Principle by aligning their spins in opposite directions, that is, $s = 1/2$ and $-1/2$ respectively. Frequently the ground-state electron configuration of He is written in the abbreviated form $1s^2$; the superscript is the *occupation number* for the orbital (i.e., the number of electrons in the orbital), the 1 designates $n = 1$, and the $s$ designates $\ell = 0$.

Although the shapes of the simple hydrogenic orbitals are basically correct, the energies calculated from Equation (2.15) are grossly in error because no provision has been made for the mutual repulsion between the

---

*If two electrons were placed in different orbitals with different spins, the rationalization would still apply. Hund's Rule says that this does not happen for the ground state.

†Strictly speaking, orbitals do not exist without electrons in them.

two electrons in the 1s orbital. Equation (2.15) can be preserved if Z is given an arbitrary value, often called the effective nuclear charge ($Z_{eff}$). For He, $Z_{eff} = 1.4$, a value which can be justified by saying that about 60% of the time one electron electrostatically screens the other electron from the nuclear charge. There are a large number of excited-state electron configurations of the helium atom; we will consider only two of the lowest energy ones, namely, those in which one electron has been raised from an orbital with $n = 1$ to an orbital with $n = 2$. In hydrogen, all $n = 2$ orbitals have the same energy, but that is not true for multi-electron atoms. How do we know this? In the spectrum of atomic hydrogen there is a line that is caused by the transition of an electron from an orbital with $n = 2$ to an orbital with $n = 1$ (Figure 2.11). In the corresponding section of the helium spectrum there are two lines, one intense and one very weak (Figure 2.22). We find two lines because the electronic configurations $1s^1 2s^1$ and $1s^1 2p^1$ do not have the same energy.

A general equation for the average radius $\bar{r}$ of any hydrogenic orbital is

$$\bar{r} = \frac{n^2}{Z} \left\{ 1 + \frac{1}{2} \left[ 1 - \frac{\ell(\ell + 1)}{n^2} \right] \right\} a_0 \qquad (2.19)$$

For example, an electron in a $2p$ ($\ell = 1$) orbital is on the average closer to

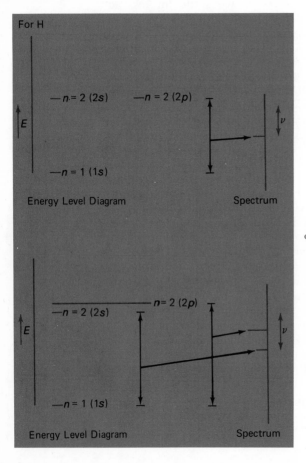

**FIGURE 2.22**  Spectral evidence for order of orbital energies.

the nucleus than one in a $2s$ ($\ell = 0$) orbital. For a $2s$ electron in He ($n = 2$, $\ell = 0$).

$$\bar{r} = \frac{2^2}{2} \left[ 1 + \frac{1}{2} \left( 1 - 0 \right) \right] 0.53 \text{ Å} = 1.59 \text{ Å}$$

and for a $2p$ electron in He ($n = 2$, $\ell = 1$)

$$\bar{r} = \frac{2^2}{2} \left[ 1 + \frac{1}{2} \left( 1 - \frac{1 \cdot 2}{2^2} \right) \right] 0.53 \text{ Å} = 1.33 \text{ Å}$$

Although the $2s$ and $2p$ electron energies are identical in the absence of the $1s$ electron (Equation (2.15)), this is not true in the presence of a $1s$ electron, because an electron in a $2p$ orbital is on the average closer to the nucleus and therefore closer to the $1s$ electron than an electron in a $2s$ orbital. Consequently, the interelectronic repulsion energy for the configuration $1s^1 2p^1$ is larger than for the configuration $1s^1 2s^1$, and the latter configuration has the lower energy. By convention, all the repulsion energy between any two electrons in an atom is assigned arbitrarily to the second (higher energy) electron of the pair. When discussing the excited states of He we say that the $2p$ orbitals lie higher on the energy scale than the $2s$ orbital. A consideration of Equation (2.19) leads to the general conclusion that *in multi-electron atoms, given a value of n, the higher $\ell$ value orbitals have higher energies.*

The next element is lithium (Li, Z = 3). The $1s$ orbital of a lithium atom will accept two electrons with spins opposed, but the Pauli Principle states that the third electron must be assigned to another orbital. The next lowest energy orbital is the $2s$, and the ground-state lithium atom configuration is $1s^2 2s^1$. Proceeding to more complex atoms, beryllium (Be, Z = 4) has the ground state configuration $1s^2 2s^2$. Boron (B, Z = 5) is the first element with $p$ electrons. The choice between $x$, $y$, or $z$ is meaningless unless there is some natural or external axis.* It is sufficient to say that a boron atom has the ground-state electron configuration $1s^2 2s^2 2p^1$. Hund's Rule applies to carbon (C, Z = 6) and nitrogen (N, Z = 7) because the three $2p$ orbitals are degenerate. The configurations are written: carbon$-1s^2 2s^2 2p^2$ and nitrogen$-1s^2 2s^2 2p^3$. In the nitrogen configuration the symbol $2p^3$ means $2p_x^1$, $2p_y^1$, $2p_z^1$ with three identical values of the electron-spin, $s$. The configurations of the next three elements oxygen (O, Z = 8), fluorine (F, Z = 9) and neon (Ne, Z = 10) are generated by adding electrons with opposed spins to the three $2p$ orbitals. Thus, oxygen has the configuration $1s^2 2s^2 2p^4$, fluorine has the configuration $1s^2 2s^2 2p^5$, and neon has the configuration $1s^2 2s^2 2p^6$. In neon, electrons have been placed in all available orbitals with principal quantum numbers one and two. Notice that the maximum permitted occupation numbers are the same as the numbers in the last column of Table 2.2.

One final complication arises as the number of electrons increases. The energy gap between orbitals with the same value of $n$ but different values of $\ell$ increases as the number of electrons increases. Moreover, the energy gap between orbitals with successive values of $n$ decreases as $n$ increases (Figure 2.11). The net result is that orbitals with high $n$ and low

---

*There is never a natural axis in an isolated atom. Presumably, an isolated atom is always spherical. This can be reconciled with non-spherical orbitals by assuming that the axes of the orbitals are shifting very rapidly until tied down by an external field, or by assuming that the electron is rapidly jumping between the three degenerate orbitals.

**TABLE 2.4**

ENERGY ORDER OF THE ORBITALS

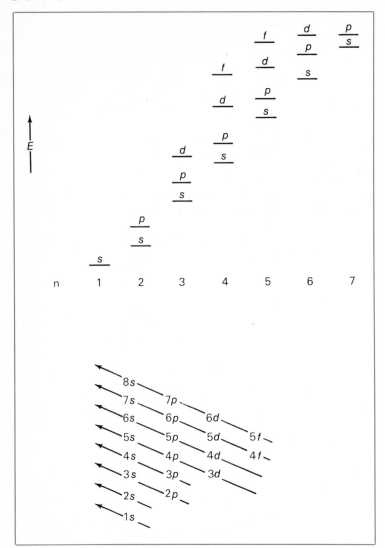

$\ell$ can actually lie below orbitals with lower $n$ and higher $\ell$. For example, the 4$s$ orbital is lower in energy than the 3$d$ orbital for an atom with nineteen electrons.

Table 2.4 is useful for working out configurations of complex atoms. The position of an orbital on the diagram is correct for the atomic number at which that orbital is first occupied. For example, the 4$s$ orbital is shown to lie below the 3$d$ orbital because that is the position at $Z = 19$ when it is first filled. At lower and higher atomic numbers the positions of these orbitals are reversed. The ground-state configurations of all the elements (as monatomic gases) can be obtained with the aid of this diagram. The correct configurations are shown in Table 2.5. Notice that there are occasional question marks. As the levels shift their relative positions with increasing Z, there are some near coincidences in orbital energies that make clear-cut predictions impossible, and detailed spectroscopic studies are required to determine the actual configurations.

It is surprising that the electron configurations for the heavy elements bear any resemblance to predictions based on solutions for the hydrogen atom. This is especially puzzling when we consider that the energy of the second electron in the He atom is less than one-half that calculated for a hydrogenic atom with $Z = 2$. Apparently electron-electron repulsion is of considerable importance, even in this case.

Why, then, are the hydrogenic orbitals of any validity for complex, heavy atoms? The answer is rather simple. The chemical and physical properties of an atom are dominated by those few electrons with the largest principal quantum number. For example, the properties of a sodium atom (configuration $1s^2 2s^2 2p^6 3s^1$) are dominated by the single $3s$ electron. What kind of an orbital does this electron occupy? To a very good approximation, this electron is held to the remainder of the atom not by a nuclear charge of $Z = +11$, but by an *effective* nuclear charge which is less than 11 because of the intervening $1s$, $2s$, and $2p$ electrons. A charged sphere appears from the outside to have all its charge at its center.* The $1s$ and $2s$ orbitals are charged spheres centered on the nucleus, but the $2p$ orbitals are functions of the angles $\theta$ and $\phi$. The electron-density distribution arising from three filled $2p$ orbitals is easy to calculate; it is

$$\text{Density} = \psi^2_{2px} + \psi^2_{2py} + \psi^2_{2pz}$$

Our only concern here is the angular dependence; from Table 2.3 we obtain $\psi^2_{2px} \propto (\sin^2\theta)(\cos^2\phi)$, $\psi^2_{2py} \propto (\sin^2\theta)(\sin^2\phi)$, and $\psi^2_{2pz} \propto \cos^2\theta$. Therefore

$$\text{Density} \propto \{(\sin^2\theta)(\cos^2\phi) + (\sin^2\theta)(\sin^2\phi) + \cos^2\theta\}$$
$$\text{Density} \propto \{\sin^2\theta\,(\cos^2\phi + \sin^2\phi) + \cos^2\theta\}$$
$$\text{Density} \propto \{\sin^2\theta + \cos^2\theta\} = \text{constant at fixed radius}$$

where we have used the trigonometric identity $\sin^2 x + \cos^2 x = 1$. The electron-density distribution from three filled $p$ orbitals is thus independent of angle; it is a sphere of charge. To the extent that a $3s$ orbital is outside $1s$, $2s$, and $2p$ orbitals, the effective charge is the nuclear charge minus the electronic charge of the inner shells, which appears to be centered at the nucleus. This is $11 - 10 = 1$. Thus, to a reasonable approximation, the problem is a repeat of the hydrogen atom problem—an electron in the neighborhood of a positive charge—and the $3s$ orbital of sodium is rather similar (in shape, but not in size) to the $3s$ orbital of hydrogen.

Every filled subshell forms a sphere of charge.† Thus, as Z increases, the outer orbitals continue to resemble the orbitals for hydrogen in shape and number, although the latter are smaller. Because chemical properties are largely determined by the outer (valence) electrons, and because the outer electrons all occupy hydrogen-like orbitals, the heavy elements have chemical properties similar to those of lighter elements with the same number of electrons in the valence orbitals.

---

*In the same way, the mass of an object appears to be concentrated at its center of gravity.
†A subshell includes all the possible orbitals for given values of $n$ and $\ell$.

**TABLE 2.5**

ELECTRONIC CONFIGURATIONS

| Z | | n=1 s | 2 s | p | 3 s | p | d | 4 s | p | d | f | 5 s | p | d | f |
|---|---|---|---|---|---|---|---|---|---|---|---|---|---|---|---|
| 1 | H | 1 | | | | | | | | | | | | | |
| 2 | He | 2 | | | | | | | | | | | | | |
| 3 | Li | 2 | 1 | | | | | | | | | | | | |
| 4 | Be | 2 | 2 | | | | | | | | | | | | |
| 5 | B | 2 | 2 | 1 | | | | | | | | | | | |
| 6 | C | 2 | 2 | 2 | | | | | | | | | | | |
| 7 | N | 2 | 2 | 3 | | | | | | | | | | | |
| 8 | O | 2 | 2 | 4 | | | | | | | | | | | |
| 9 | F | 2 | 2 | 5 | | | | | | | | | | | |
| 10 | Ne | 2 | 2 | 6 | | | | | | | | | | | |
| 11 | Na | 2 | 2 | 6 | 1 | | | | | | | | | | |
| 12 | Mg | 2 | 2 | 6 | 2 | | | | | | | | | | |
| 13 | Al | 2 | 2 | 6 | 2 | 1 | | | | | | | | | |
| 14 | Si | 2 | 2 | 6 | 2 | 2 | | | | | | | | | |
| 15 | P | 2 | 2 | 6 | 2 | 3 | | | | | | | | | |
| 16 | S | 2 | 2 | 6 | 2 | 4 | | | | | | | | | |
| 17 | Cl | 2 | 2 | 6 | 2 | 5 | | | | | | | | | |
| 18 | Ar | 2 | 2 | 6 | 2 | 6 | | | | | | | | | |
| 19 | K | 2 | 2 | 6 | 2 | 6 | | 1 | | | | | | | |
| 20 | Ca | 2 | 2 | 6 | 2 | 6 | | 2 | | | | | | | |
| 21 | Sc | 2 | 2 | 6 | 2 | 6 | 1 | 2 | | | | | | | |
| 22 | Ti | 2 | 2 | 6 | 2 | 6 | 2 | 2 | | | | | | | |
| 23 | V | 2 | 2 | 6 | 2 | 6 | 3 | 2 | | | | | | | |
| 24 | Cr | 2 | 2 | 6 | 2 | 6 | 5 | 1 | | | | | | | |
| 25 | Mn | 2 | 2 | 6 | 2 | 6 | 5 | 2 | | | | | | | |
| 26 | Fe | 2 | 2 | 6 | 2 | 6 | 6 | 2 | | | | | | | |
| 27 | Co | 2 | 2 | 6 | 2 | 6 | 7 | 2 | | | | | | | |
| 28 | Ni | 2 | 2 | 6 | 2 | 6 | 8 | 2 | | | | | | | |
| 29 | Cu | 2 | 2 | 6 | 2 | 6 | 10 | 1 | | | | | | | |
| 30 | Zn | 2 | 2 | 6 | 2 | 6 | 10 | 2 | | | | | | | |
| 31 | Ga | 2 | 2 | 6 | 2 | 6 | 10 | 2 | 1 | | | | | | |
| 32 | Ge | 2 | 2 | 6 | 2 | 6 | 10 | 2 | 2 | | | | | | |
| 33 | As | 2 | 2 | 6 | 2 | 6 | 10 | 2 | 3 | | | | | | |
| 34 | Se | 2 | 2 | 6 | 2 | 6 | 10 | 2 | 4 | | | | | | |
| 35 | Br | 2 | 2 | 6 | 2 | 6 | 10 | 2 | 5 | | | | | | |
| 36 | Kr | 2 | 2 | 6 | 2 | 6 | 10 | 2 | 6 | | | | | | |
| 37 | Rb | 2 | 2 | 6 | 2 | 6 | 10 | 2 | 6 | | | 1 | | | |
| 38 | Sr | 2 | 2 | 6 | 2 | 6 | 10 | 2 | 6 | | | 2 | | | |
| 39 | Y | 2 | 2 | 6 | 2 | 6 | 10 | 2 | 6 | 1 | | 2 | | | |
| 40 | Zr | 2 | 2 | 6 | 2 | 6 | 10 | 2 | 6 | 2 | | 2 | | | |
| 41 | Nb | 2 | 2 | 6 | 2 | 6 | 10 | 2 | 6 | 4 | | 1 | | | |
| 42 | Mo | 2 | 2 | 6 | 2 | 6 | 10 | 2 | 6 | 5 | | 1 | | | |
| 43 | Tc | 2 | 2 | 6 | 2 | 6 | 10 | 2 | 6 | 6(5?) | | 1(2?) | | | |
| 44 | Ru | 2 | 2 | 6 | 2 | 6 | 10 | 2 | 6 | 7 | | 1 | | | |
| 45 | Rh | 2 | 2 | 6 | 2 | 6 | 10 | 2 | 6 | 8 | | 1 | | | |
| 46 | Pd | 2 | 2 | 6 | 2 | 6 | 10 | 2 | 6 | 10 | | | | | |
| 47 | Ag | 2 | 2 | 6 | 2 | 6 | 10 | 2 | 6 | 10 | | 1 | | | |
| 48 | Cd | 2 | 2 | 6 | 2 | 6 | 10 | 2 | 6 | 10 | | 2 | | | |
| 49 | In | 2 | 2 | 6 | 2 | 6 | 10 | 2 | 6 | 10 | | 2 | 1 | | |
| 50 | Sn | 2 | 2 | 6 | 2 | 6 | 10 | 2 | 6 | 10 | | 2 | 2 | | |

**TABLE 2.5**

ELECTRONIC CONFIGURATIONS (Continued)

| Z | | $n=$ 4 $s$ | $p$ | $d$ | $f$ | 5 $s$ | $p$ | $d$ | $f$ | 6 $s$ | $p$ | $d$ | $f$ | 7 $s$ |
|---|---|---|---|---|---|---|---|---|---|---|---|---|---|---|
| 51 | Sb | 2 | 6 | 10 | | 2 | 3 | | | | | | | |
| 52 | Te | 2 | 6 | 10 | | 2 | 4 | | | | | | | |
| 53 | I | 2 | 6 | 10 | | 2 | 5 | | | | | | | |
| 54 | Xe | 2 | 6 | 10 | | 2 | 6 | | | | | | | |
| 55 | Cs | 2 | 6 | 10 | | 2 | 6 | | | 1 | | | | |
| 56 | Ba | 2 | 6 | 10 | | 2 | 6 | | | 2 | | | | |
| 57 | La | 2 | 6 | 10 | | 2 | 6 | 1 | | 2 | | | | |
| 58 | Ce | 2 | 6 | 10 | 2(3?) | 2 | 6 | | | 2(1?) | | | | |
| 59 | Pr | 2 | 6 | 10 | 3(4?) | 2 | 6 | | | 2(1?) | | | | |
| 60 | Nd | 2 | 6 | 10 | 4 | 2 | 6 | | | 2 | | | | |
| 61 | Pm | 2 | 6 | 10 | 5(6?) | 2 | 6 | | | 2(1?) | | | | |
| 62 | Sm | 2 | 6 | 10 | 6 | 2 | 6 | | | 2 | | | | |
| 63 | Eu | 2 | 6 | 10 | 7 | 2 | 6 | | | 2 | | | | |
| 64 | Gd | 2 | 6 | 10 | 7 | 2 | 6 | 1 | | 2 | | | | |
| 65 | Tb | 2 | 6 | 10 | 9(10?) | 2 | 6 | | | 2(1?) | | | | |
| 66 | Dy | 2 | 6 | 10 | 10(11?) | 2 | 6 | | | 2(1?) | | | | |
| 67 | Ho | 2 | 6 | 10 | 11(12?) | 2 | 6 | | | 2(1?) | | | | |
| 68 | Er | 2 | 6 | 10 | 12(13?) | 2 | 6 | | | 2(1?) | | | | |
| 69 | Tm | 2 | 6 | 10 | 13 | 2 | 6 | | | 2 | | | | |
| 70 | Yb | 2 | 6 | 10 | 14 | 2 | 6 | | | 2 | | | | |
| 71 | Lu | 2 | 6 | 10 | 14 | 2 | 6 | 1 | | 2 | | | | |
| 72 | Hf | 2 | 6 | 10 | 14 | 2 | 6 | 2 | | 2 | | | | |
| 73 | Ta | 2 | 6 | 10 | 14 | 2 | 6 | 3 | | 2 | | | | |
| 74 | W | 2 | 6 | 10 | 14 | 2 | 6 | 4 | | 2 | | | | |
| 75 | Re | 2 | 6 | 10 | 14 | 2 | 6 | 5 | | 2 | | | | |
| 76 | Os | 2 | 6 | 10 | 14 | 2 | 6 | 6 | | 2 | | | | |
| 77 | Ir | 2 | 6 | 10 | 14 | 2 | 6 | 7 | | 2 | | | | |
| 78 | Pt | 2 | 6 | 10 | 14 | 2 | 6 | 9 | | 1 | | | | |
| 79 | Au | 2 | 6 | 10 | 14 | 2 | 6 | 10 | | 1 | | | | |
| 80 | Hg | 2 | 6 | 10 | 14 | 2 | 6 | 10 | | 2 | | | | |
| 81 | Tl | 2 | 6 | 10 | 14 | 2 | 6 | 10 | | 2 | 1 | | | |
| 82 | Pb | 2 | 6 | 10 | 14 | 2 | 6 | 10 | | 2 | 2 | | | |
| 83 | Bi | 2 | 6 | 10 | 14 | 2 | 6 | 10 | | 2 | 3 | | | |
| 84 | Po | 2 | 6 | 10 | 14 | 2 | 6 | 10 | | 2 | 4 | | | |
| 85 | At | 2 | 6 | 10 | 14 | 2 | 6 | 10 | | 2 | 5 | | | |
| 86 | Rn | 2 | 6 | 10 | 14 | 2 | 6 | 10 | | 2 | 6 | | | |
| 87 | Fr | 2 | 6 | 10 | 14 | 2 | 6 | 10 | | 2 | 6 | | | 1 |
| 88 | Ra | 2 | 6 | 10 | 14 | 2 | 6 | 10 | | 2 | 6 | | | 2 |
| 89 | Ac | 2 | 6 | 10 | 14 | 2 | 6 | 10 | (1?) | 2 | 6 | 1 | | 2(1?) |
| 90 | Th | 2 | 6 | 10 | 14 | 2 | 6 | 10 | | 2 | 6 | 2 | | 2 |
| 91 | Pa | 2 | 6 | 10 | 14 | 2 | 6 | 10 | 2(3?) | 2 | 6 | 1 | | 2(1?) |
| 92 | U | 2 | 6 | 10 | 14 | 2 | 6 | 10 | 3 | 2 | 6 | 1 | | 2 |
| 93 | Np | 2 | 6 | 10 | 14 | 2 | 6 | 10 | 4(5?) | 2 | 6 | 1 | | 2(1?) |
| 94 | Pu | 2 | 6 | 10 | 14 | 2 | 6 | 10 | 6(7?) | 2 | 6 | 1 | | 2(1?) |
| 95 | Am | 2 | 6 | 10 | 14 | 2 | 6 | 10 | 7(8?) | 2 | 6 | 1 | | 2(1?) |
| 96 | Cm | 2 | 6 | 10 | 14 | 2 | 6 | 10 | 7(9?) | 2 | 6 | 1 | | 2(1?) |
| 97 | Bk | 2 | 6 | 10 | 14 | 2 | 6 | 10 | 8(10?) | 2 | 6 | 1 | | 2(1?) |
| 98 | Cf | 2 | 6 | 10 | 14 | 2 | 6 | 10 | 10(11?) | 2 | 6 | | | 2(1?) |
| 99 | Es | 2 | 6 | 10 | 14 | 2 | 6 | 10 | 11(12?) | 2 | 6 | | | 2(1?) |
| 100 | Fm | 2 | 6 | 10 | 14 | 2 | 6 | 10 | 12(13?) | 2 | 6 | | | 2(1?) |
| 101 | Md | 2 | 6 | 10 | 14 | 2 | 6 | 10 | 13(14?) | 2 | 6 | | | 2(1?) |
| 102 | No | 2 | 6 | 10 | 14 | 2 | 6 | 10 | 14 | 2 | 6 | (1?) | | 2(1?) |
| 103 | Lw | 2 | 6 | 10 | 14 | 2 | 6 | 10 | 14 | 2 | 6 | 1(2?) | | 2(1?) |

We can now appreciate the basis for the periodic table (Figure 2.23), where the elements are arranged by atomic number in horizontal rows (called *periods*). Succeeding periods are constituted and aligned so that vertical columns (called *groups*) contain elements that have similar chemical properties, which in most cases means atoms whose incomplete shells have identical electron configurations. For example, the first group contains the elements hydrogen (H), lithium (Li), sodium (Na), potassium (K), rubidium (Rb), cesium (Cs), and francium (Fr), all of which have the *valence-shell* configurations $ns^1$. Similarly, the halogens fluorine (F), chlorine (Cl), bromine (Br), iodine (I), and astatine (At) are in a single group because they all have the valence-shell configuration $ns^2np^5$. When there are no antecedents for a configuration, a new group is introduced; for example, boron heads a new group with configuration $2s^22p^1$ because there are no $p$ orbitals in the first shell. Scandium (Sc), which has the valence-shell (notice that this is not just the outermost shell) configuration $3s^23p^64s^23d^1$, is also the first of a kind, as are all the elements with $Z = 21$ to 30. Each of these elements heads a new group, as do the elements 58 through 71. The latter groups are separated from the body of the chart solely for convenience in printing. There is one exception to the system. Helium is placed above neon, rather than above beryllium; the complete-shell configuration is more fundamental than the two-electron configuration.

The arrangement of the elements described in the preceding paragraph places the elements with similar properties in the same group. For some groups these similarities are striking. Thus we speak of the chemistry of the *alkali metals* (1A group, the elements in the column beginning with, but not including, hydrogen), the *alkaline earths* (2A group, headed by beryllium), the *halogens* (7A group, headed by fluorine), and the *noble gases* (8A or O group, headed by helium). Toward the center of the periodic table, the group (vertical) similarities are not as strong as for the 1A, 2A, 7A, and 8A groups. The elements in the groups headed by elements 21 through 30 show some horizontal similarities. This is because all of these elements have outer-shell configurations $ns^2$. These elements will be discussed as a single group, the *transition* or *d-block elements*. The short groups headed by elements 58 through 71 contain elements whose properties show strong horizontal relationships, so strong that chemical separation of two horizontal neighbors is difficult. They all have identical configurations in the two outer shells, $ns^2$, $(n-1)s^2$, $(n-1)p^6$, $(n-1)d^1$; the differences in configuration lie in the number of $f$ electrons in the $(n-2)$ shell. The elements in the period, starting with cerium, are usually called the *lanthanides;* those in the period, starting with thorium, are called *actinides* (together called the $f$-block elements).

## 2.9  ATOMIC PROPERTIES.

Except for mass, which increases fairly regularly with atomic number, the important atomic properties show the periodicity implied by the periodic table. Most chemical and physical properties of the elements are not the properties of the individual atoms but of aggregates of atoms, either molecules (e.g., $H_2$, $N_2$, $I_2$) or crystalline solids (e.g., the metals such as iron and the semi-metals such as germanium). Here we are concerned with properties of individual atoms.

There is no unambiguous method for measuring atomic sizes because the electron density around a nucleus simply falls off gradually with in-

| 1A | 2A | 3B | 4B | 5B | 6B | 7B | 8 | 8 | 8 | 1B | 2B | 3A | 4A | 5A | 6A | 7A | 0 |
|---|---|---|---|---|---|---|---|---|---|---|---|---|---|---|---|---|---|
| 1 **H** 1.00797 ±0.00001 | | | | | | | | | | | | | | | | | 2 **He** 4.0026 ±0.00005 |
| 3 **Li** 6.939 ±0.0005 | 4 **Be** 9.0122 ±0.00005 | | | | | | | | | | | 5 **B** 10.811 ±0.003 | 6 **C** 12.01115 ±0.00005 | 7 **N** 14.0067 ±0.00005 | 8 **O** 15.9994 ±0.0001 | 9 **F** 18.9984 ±0.00005 | 10 **Ne** 20.183 ±0.0005 |
| 11 **Na** 22.9898 ±0.00005 | 12 **Mg** 24.312 ±0.0005 | | | | | | | | | | | 13 **Al** 26.9815 ±0.00005 | 14 **Si** 28.086 ±0.001 | 15 **P** 30.9738 ±0.00005 | 16 **S** 32.064 ±0.003 | 17 **Cl** 35.453 ±0.001 | 18 **Ar** 39.948 ±0.0005 |
| 19 **K** 39.102 ±0.0005 | 20 **Ca** 40.08 ±0.005 | 21 **Sc** 44.956 ±0.0005 | 22 **Ti** 47.90 ±0.005 | 23 **V** 50.942 ±0.0005 | 24 **Cr** 51.996 ±0.001 | 25 **Mn** 54.9380 ±0.00005 | 26 **Fe** 55.847 ±0.003 | 27 **Co** 58.9332 ±0.00005 | 28 **Ni** 58.71 ±0.005 | 29 **Cu** 63.54 ±0.005 | 30 **Zn** 65.37 ±0.005 | 31 **Ga** 69.72 ±0.005 | 32 **Ge** 72.59 ±0.005 | 33 **As** 74.9216 ±0.00005 | 34 **Se** 78.96 ±0.005 | 35 **Br** 79.909 ±0.002 | 36 **Kr** 83.80 ±0.005 |
| 37 **Rb** 85.47 ±0.005 | 38 **Sr** 87.62 ±0.005 | 39 **Y** 88.905 ±0.0005 | 40 **Zr** 91.22 ±0.005 | 41 **Nb** 92.906 ±0.0005 | 42 **Mo** 95.94 ±0.005 | 43 **Tc** (99) | 44 **Ru** 101.07 ±0.005 | 45 **Rh** 102.905 ±0.0005 | 46 **Pd** 106.4 ±0.05 | 47 **Ag** 107.870 ±0.003 | 48 **Cd** 112.40 ±0.005 | 49 **In** 114.82 ±0.005 | 50 **Sn** 118.69 ±0.005 | 51 **Sb** 121.75 ±0.005 | 52 **Te** 127.60 ±0.005 | 53 **I** 126.9044 ±0.00005 | 54 **Xe** 131.30 ±0.005 |
| 55 **Cs** 132.905 ±0.0005 | 56 **Ba** 137.34 ±0.005 | 57 ***La** 138.91 ±0.005 | 72 **Hf** 178.49 ±0.005 | 73 **Ta** 180.948 ±0.0005 | 74 **W** 183.85 ±0.005 | 75 **Re** 186.2 ±0.05 | 76 **Os** 190.2 ±0.05 | 77 **Ir** 192.2 ±0.05 | 78 **Pt** 195.09 ±0.005 | 79 **Au** 196.967 ±0.0005 | 80 **Hg** 200.59 ±0.005 | 81 **Tl** 204.37 ±0.005 | 82 **Pb** 207.19 ±0.005 | 83 **Bi** 208.980 ±0.0005 | 84 **Po** (210) | 85 **At** (210) | 86 **Rn** (222) |
| 87 **Fr** (223) | 88 **Ra** (226) | 89 †**Ac** (227) | 104 **Ku** (257) | 105 **Ha** (260) | | | | | | | | | | | | | |

*Lanthanum Series

| | | | | | | | | | | | | | |
|---|---|---|---|---|---|---|---|---|---|---|---|---|---|
| 58 **Ce** 140.12 ±0.005 | 59 **Pr** 140.907 ±0.0005 | 60 **Nd** 144.24 ±0.005 | 61 **Pm** (147) | 62 **Sm** 150.35 ±0.005 | 63 **Eu** 151.96 ±0.005 | 64 **Gd** 157.25 ±0.005 | 65 **Tb** 158.924 ±0.0005 | 66 **Dy** 162.50 ±0.005 | 67 **Ho** 164.930 ±0.0005 | 68 **Er** 167.26 ±0.005 | 69 **Tm** 168.934 ±0.0005 | 70 **Yb** 173.04 ±0.005 | 71 **Lu** 174.97 ±0.005 |

†Actinium Series

| | | | | | | | | | | | | | |
|---|---|---|---|---|---|---|---|---|---|---|---|---|---|
| 90 **Th** 232.038 ±0.0005 | 91 **Pa** (231) | 92 **U** 238.03 ±0.005 | 93 **Np** (237) | 94 **Pu** (242) | 95 **Am** (243) | 96 **Cm** (247) | 97 **Bk** (247) | 98 **Cf** (249) | 99 **Es** (254) | 100 **Fm** (253) | 101 **Md** (256) | 102 **No** (253) | 103 **Lw** (257) |

Atomic Weights are based on $C^{12}$—12.0000 and Conform to the 1961 Values

**FIGURE 2.23** Periodic table.

creasing distance from the nucleus. An arbitrary scale of atomic sizes can be established by first assuming that each atom is a sphere. By methods which will be discussed later, the distance between nuclei in a molecule can be measured. The internuclear distance in the $Cl_2$ molecule is 1.98 Å, so each Cl atom is assigned a radius* of $1.98/2 = 0.99$ Å. The carbon-carbon single-bond internuclear distance for a number of compounds averages 1.54 Å, so carbon is assigned a radius of $1.54/2 = 0.77$ Å. The carbon-chlorine distance in $CCl_4$ is 1.76 Å, which is equal to $0.99 + 0.77$.

Values of single-bond covalent radii (Table 2.6) are useful in estimating internuclear distances in molecules; they are defined for just this purpose. The single-bond covalent radii are much smaller than the valence-shell orbitals; a sphere of radius 0.30 Å about a proton includes less than 1/3 of the total electron density arising from a $1s$ electron.

The periodic trends in the single-bond covalent radii are obvious. The radii decrease toward the right, and generally increase toward the bottom of the periodic table. In Equation (2.19) we saw that the average radius of any orbital was a function of $n^2/Z$. Within a period (a horizontal row) the radii should decrease, because all the valence-shell electrons have the same value of $n$, whereas $Z$ increases; that is, the radius is proportional to $1/Z_{eff}$. The trend as we go down the table can be rationalized as a delicate balance between increasing $n^2$ and increasing $Z_{eff}$. The rationalization is more complicated than the fact, which is that the heavier atoms within a group are usually larger than the lighter ones.

The strengths of the forces binding electrons into the atom are determined by measuring ionization energies. The *ionization energies* for a particular atom are the minimum energies required to remove the first, the second, and succeeding electrons from a ground-state atom in a gaseous sample. For example, the removal of one electron from a sodium atom to produce a free electron and a sodium ion† requires at least 5.14 eV (1 eV = 1 electron volt = $1.602 \times 10^{-19}$ J):

$$Na(1s^2 2s^2 2p^6 3s^1) + 5.14 \text{ eV} \rightarrow Na^+(1s^2 2s^2 2p^6) + e^-$$

---

*More precisely, a *single-bond covalent radius;* values are usually good to within ±0.03 Å. The value for H is 0.30 Å for all bonds other than HH, which has a bond length of 0.74 Å, corresponding to a radius of 0.37 Å.

†The orbital *energies* in an ion are not the same as in the neutral atom.

**TABLE 2.6**

SINGLE-BOND COVALENT RADII (Å)

| C | 0.77 | N | 0.70 | O | 0.66 | F | 0.64 |
|---|------|----|------|----|------|----|------|
| Si | 1.17 | P | 1.10 | S | 1.04 | Cl | 0.99 |
| Ge | 1.22 | As | 1.21 | Se | 1.17 | Br | 1.14 |
| Sn | 1.40 | Sb | 1.41 | Te | 1.37 | I | 1.33 |
| | | | (H = 0.30) | | | | |

1 Å = 0.1 nm

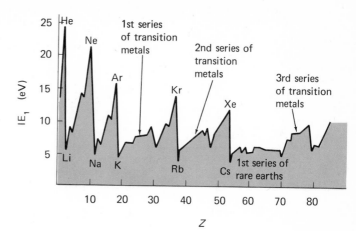

**FIGURE 2.29** First ionization energies of the elements.

Any energy greater than 5.14 eV also can ionize sodium; the electron and the ion carry the excess energy as kinetic energy.

There is also a *second ionization energy* for sodium. It is 47.29 eV:

$$Na^+(1s^22s^22p^6) + 47.29\ eV \rightarrow Na^{+2}(1s^22s^22p^5) + e^-$$

It is possible to conceive of, although increasingly difficult to measure, eleven ionization energies for sodium. However, only the ionization energies of the valence electrons are of major importance in chemistry.

The periodicities summarized in the periodic chart are also evident in a plot of the first ionization energy versus atomic number (Figure 2.24). The first ionization energies for the noble gases appear as maxima; the alkali metals are at minima. The *s* and *p* electrons are differentiated clearly between Li and Ne and between Na and Ar, and breaks also are visible in these regions.* Many of the details in Figure 2.24 can be rationalized on the basis of effective charges and electron-electron repulsions. As an example, the ionization energy of potassium is less than that of sodium because the effective charges are about the same ($Z_{eff} \simeq +1$), and the potassium electron is from a higher energy orbital ($n = 4$ rather than $n = 3$).

We can write the electron configurations for ions using the same procedure as for atoms, for example, $Na^+(1s^22s^22p^6)$. However, there will be some difficulties with some transition metals because of the proximity of the energies of the 3*d* and 4*s* orbitals; the removal of a single electron reduces the repulsion energy enough to restore the "normal" order 3*d*, 4*s*. For example, the electronic configuration of $Fe^{2+}$ ($Z = 26$; $26 - 2 = 24$) is $1s^22s^22p^63s^23p^63d^6$.

The chemical and physical properties of an element are determined largely by the number of valence electrons. These properties are consequences of the ability of atoms to hold and attract electrons. The ionization energies in Table 2.7 indicate why the inner-shell electrons are not directly involved in the chemical and physical properties. For example, the ionization energies of boron are: 8.3 eV to remove the first electron (a 2*p*), 25.2 eV to remove the second (a 2*s*), 37.9 eV to remove the third (the second 2*s*), and 259 eV to remove the fourth (a 1*s*). There is a drastic increase in the energy necessary to remove an inner shell electron. Similar jumps

---

*Can you explain this break using Hund's Rule?

TABLE 2.7

IONIZATION ENERGIES (in eV)

| ATOMIC NUMBER | ELEMENT | $IE_1$ | $IE_2$ | $IE_3$ | $IE_4$ |
|:---:|:---:|:---:|:---:|:---:|:---:|
| 1 | H | 13.60 | | | |
| 2 | He | 24.59 | 54.40 | | |
| 3 | Li | 5.39 | 75.65 | 122.4 | |
| 4 | Be | 9.32 | 18.21 | 153.9 | 217.8 |
| 5 | B | 8.30 | 25.16 | 37.93 | 259.4 |
| 6 | C | 11.26 | 24.39 | 47.89 | 64.50 |
| 7 | N | 14.54 | 29.62 | 47.46 | 77.47 |
| 8 | O | 13.62 | 35.16 | 54.96 | 77.43 |
| 9 | F | 17.43 | 34.99 | 62.68 | 87.28 |
| 10 | Ne | 21.57 | 41.09 | 65.07 | 97.21 |
| 11 | Na | 5.14 | 47.33 | 71.66 | 98.90 |
| 12 | Mg | 7.65 | 15.03 | 80.16 | 109.4 |
| 13 | Al | 5.99 | 18.83 | 28.45 | 120.0 |
| 14 | Si | 8.15 | 16.34 | 33.48 | 45.16 |
| 15 | P | 11.02 | 19.66 | 30.17 | 51.36 |
| 16 | S | 10.36 | 23.42 | 35.01 | 47.33 |
| 17 | Cl | 13.01 | 23.81 | 39.92 | 53.36 |
| 18 | Ar | 15.76 | 27.63 | 40.92 | 59.82 |
| 19 | K | 4.34 | 31.82 | 47.72 | 60.95 |
| 20 | Ca | 6.11 | 11.88 | 51.23 | 67.24 |
| 21 | Sc | 6.56 | 12.90 | 24.76 | 73.74 |
| 22 | Ti | 6.85 | 13.63 | 28.15 | 43.26 |
| 23 | V | 6.72 | 14.23 | 29.71 | 47.7 |
| 24 | Cr | 6.77 | 16.50 | 30.96 | 49.5 |
| 25 | Mn | 7.44 | 15.65 | 33.70 | — |
| 26 | Fe | 7.89 | 16.19 | 30.66 | — |
| 27 | Co | 7.85 | 17.06 | 33.51 | — |
| 28 | Ni | 7.63 | 18.16 | 35.18 | — |
| 29 | Cu | 7.72 | 20.30 | 36.85 | — |
| 30 | Zn | 9.40 | 17.97 | 39.72 | — |
| 31 | Ga | 5.99 | 20.52 | 30.71 | 64.2 |
| 32 | Ge | 7.89 | 15.94 | 34.23 | 45.5 |
| 33 | As | 9.80 | 20.2 | 28.3 | 50.3 |
| 34 | Se | 9.76 | 21.5 | 32.0 | 42.9 |
| 35 | Br | 11.8 | 21.6 | 35.9 | — |
| 36 | Kr | 14.0 | 24.6 | 36.9 | — |
| 37 | Rb | 4.18 | 27.5 | 39.9 | — |
| 38 | Sr | 5.69 | 11.0 | — | — |
| 52 | Te | 9.02 | 18.6 | 31.2 | 38.2 |
| 53 | I | 10.46 | 19.1 | — | — |
| 54 | Xe | 12.1 | 21.2 | 32.1 | — |
| 55 | Cs | 3.89 | 25.1 | — | — |
| 56 | Ba | 5.21 | 10.0 | — | — |

1 eV = $1.602 \times 10^{-19}$ J

**TABLE 2.8**

ELECTRON AFFINITIES (in eV)

| ATOMIC NUMBER | ELEMENT | ELECTRON AFFINITY |
|:---:|:---:|:---:|
| 1 | H | .75 |
| 3 | Li | (.54) |
| 4 | Be | (−.6) |
| 5 | B | (.2) |
| 6 | C | 1.25 |
| 7 | N | (−.1) |
| 8 | O | 1.47 |
| 9 | F | 3.45 |
| 11 | Na | (.74) |
| 12 | Mg | (−.3) |
| 13 | Al | (.6) |
| 14 | Si | (1.63) |
| 15 | P | (.7) |
| 16 | S | 2.07 |
| 17 | Cl | 3.61 |
| 19 | K | (.7) |
| 30 | Zn | (−.9) |
| 31 | Ga | (.18) |
| 32 | Ge | (1.2) |
| 33 | As | (.6) |
| 34 | Se | (1.7) |
| 35 | Br | 3.36 |
| 48 | Cd | (−.6) |
| 49 | In | (.2) |
| 52 | Te | (2.2) |
| 53 | I | 3.06 |

$1 \text{ eV} = 1.602 \times 10^{-19} \text{ J}$

occur throughout the list whenever an inner shell is opened. Experimental conditions in chemistry are usually mild, and energies of the magnitude required to launch a successful attack on an inner shell are not generally available. The noble gases, which have complete outer shells, have large first ionization energies and are relatively invulnerable to chemical attack.

In general, a neutral atom will pick up extra electrons if they are available; the *electron affinity* is the energy released when this happens:

$$\text{atom} + \text{electron} \rightarrow \text{negative ion} + \text{energy (E.A.)}$$

A positive value of E.A. indicates that energy is released upon formation of the negative ion. The more positive the value of E.A., the greater the affinity of the neutral atom for an additional electron. Only a few electron affinities have been measured accurately. It is often easier to calculate them via quantum mechanics than to measure them. A partial list is given in Table 2.8; the calculated values are in parentheses.

Periodicity is also found for electron affinities. The halogens have the highest electron affinities. The lowest electron affinities appear not in the elements with the lowest ionization energies, but in the elements

that attain a filled or half-filled subshell* by acquiring an additional electron. This anomalous stability of complete or half-complete subshells
also is evident in the ionization energies. Look, for instance, at the breaks
in the curve (Figure 2.24) between $Be(2s^2)$ and $B(2s^22p)$, and between
$N(2s^22p^3)$ and $O(2s^22p^4)$.

---

*The $2s$ orbital is a subshell of the $n = 2$ shell; the three $2p$ orbitals are also a subshell
of this shell.

## PROBLEMS

1. What is the velocity of an electron with wavelength 2 Å?

2. Give the ground-state configurations of the following atoms (pref-
erably without consulting Table 2.5): $B(Z = 5)$, $N(Z = 7)$, $P(Z = 15)$,
$Kr(Z = 36)$, $Ru(Z = 44)$, $Ca(Z = 55)$.

3. Give the ground-state configurations of the ions $K^+$, $Cl^-$, $O^{-2}$,
$Ba^{+2}$, $Fe^{+2}$, $Fe^{+3}$, $Ag^+$. Which atoms have the same configurations as these
ions?

4. The removal of an electron from the surface of solid potassium
requires 2.26 eV of energy ($1 \text{ eV} = 1.60 \times 10^{-12}$ erg). What is the threshold
frequency for potassium in a photoelectric effect experiment? Will light
with a wavelength of 3800 Å produce photoelectrons? What kinetic energy
will the electrons have? How fast will they be traveling?

5. The intensity of a light wave is proportional to the square of the
amplitude. What determines the intensity of a beam of photons?

6. Given that the earth's mass is $5.98 \times 10^{27}$ g and the average linear
velocity is $2.98 \times 10^6$ cm/sec, what is the minimum uncertainty in the
earth's position?

7. Calculate the frequency of the photon emitted when the electron
in a hydrogen atom falls from an $n = 4$ to an $n = 2$ orbital. Wavelengths
between 4000 Å and 7000 Å are visible. Would this photon be visible?

8. Calculate the wavelength for the $n=4$ to $n=2$ electronic transition
in the He atom ($Z_{eff} = 1.4$). The hydrogen spectrum has one line in the
spectrum for this transition. How many lines are there in the He spectrum
for this same jump?

9. Use Equation (2.15) to calculate the ionization energy for a ground-
state hydrogen atom. Repeat the calculation for a $2s$ electron.

10. Why might one expect the electron affinities of the noble gases to
be zero?

11. Calculate the effective charge for the valence electron in Li from
the first ionization energy.

12. Complete Table 2.2 for the $n = 5$ orbitals.

13. How would the configurations of the elements in the second row differ if $2s$ and $2p$ orbitals were always degenerate? Where would the breaks come in the plot of ionization energy versus atomic number?

14. Make a plot of the second ionization energy versus atomic number. Indicate the places where ionization requires breakage of complete shells, complete subshells, and half-complete subshells.

15. Element A has a greater effective charge for its valence electrons than does element B. Will element A have a larger or smaller (a) electron affinity, (b) first ionization potential, than element B?

16. Calculate the eleventh ionization energy of sodium.

17. List the number and type (spherical or angular) of nodes in: (a) a $4p$ orbital, (b) a $5s$ orbital, (c) a $6f$ orbital.

18. Show that the electron density for a filled pair of $2p_x$ and $2p_y$ orbitals is independent of $\phi$. This electron density has the symmetry of a cylinder about the Z axis.

19. Which of the following could be described as quantized: (a) electron charge, (b) electron mass, (c) nuclear charge, (d) nuclear mass?

20. Two waves with the same wavelength interfere. Can the resulting wave ever have a different wavelength? Answer this by adding two waves with various phases and amplitudes.

21. Construct a figure like Figure 2.5 using the same $d$ but longer wavelength. Use the two figures to verify that

$$d \sin \theta = n\lambda$$

That is, measure $d$, $\theta$, and $\lambda$ and show that the intensity peaks come at integral values of $n$.

22. Make a plot of atomic radius versus atomic number on the same scale as Figure 2.24. Discuss the similarities of the two figures.

23. Use the plots from Figure 2.24 and Problem 22 to estimate the atomic radius and first two ionization potentials of element 87 (Fr, francium). Francium is so rare that samples large enough for these determinations have never been prepared.

24. List the elements in Table 2.5 whose electron configurations differ from those that you would predict using the rationale developed in this chapter.

25. The Bohr expression for the energy of the hydrogen atom can be derived from Equation (2.7) and elementary physics. Carry out this derivation by the following steps:
    a. Equate the Coulombic force holding the electron in a circular orbit to the centrifugal force, i.e., $Ze^2/r^2 = mv^2/r$.

b. Write an expression for the total energy of the electron as a sum of its potential energy and its kinetic energy, $E = \frac{1}{2} mv^2 - \frac{Ze^2}{r}$.

c. Show that $E = -Ze^2/2r$.

d. Use the Bohr assumption (Equation (2.7)) and the results above to show that

$$r = n^2 h^2 / 4\pi^2 m e^2 Z$$

e. Use the above results to obtain Equation (2.15).

26. Find a transition in the He spectrum which would produce visible photons.

27. Why is the electron affinity of carbon larger than that of nitrogen?

28. The flame test is a sensitive qualitative test for the presence of many alkali metals. A sample of the material is introduced into a flame, and a characteristic color is produced for each element. A lithium flame is bright red ($\lambda = 6708$ Å); a sodium flame is yellow ($\lambda = 5897$ Å), and a potassium flame is faint purple ($\lambda = 7680$ Å and some small amount of 4040 Å). All of these transitions are similar and do not involve the inner-shell electrons. What shells do the electrons jump between in each of these transitions (excepting the 4040 Å transition)? Is it necessary to invoke changes in effective charge to explain these colors?

29. Assume that it was never possible for an orbital with high $n$ and low $\ell$ to have lower energy than any orbital with lower $n$; that is, $3d$ would always be lower than $4s$. Draw a periodic chart for this situation.

## References

2.1. "Light," special issue of Scientific American, September, 1968.

2.2. Hochstrasser, R. M., *Behavior of Electrons in Atoms* (Benjamin, New York, 1964).

2.3. Cromer, D. T., *Stereo Plots of Hydrogen-like Electron Densities*, Journal of Chemical Education, *45*, 626 (1968).

2.4. Ogryzlo, E. A., and Porter, G. B., *Contour Surfaces for Atomic and Molecular Orbitals*, Journal of Chemical Education, *40*, 256 (1963).

2.5. Gamow, G., *Mr. Tompkins Explores the Atom* (Macmillan, New York, 1945).

2.6. Feynman, R. P., Leighton, R. B., and Sands, M., *The Feynman Lectures on Physics* (Addison-Wesley, Reading, Mass., 1965).

2.7. Hinshelwood, C. N., *The Structure of Physical Chemistry* (Oxford University Press, 1951).

2.8. Gerhold, G. A., *Percentage Contour Maps of Electron Densities in Atoms*, American Journal of Physics, *40*, 988 (1972).

## CHAPTER 3

# MOLECULES

"Two atoms may conform to the rule of eight, or the octet rule, not only by the transfer of electrons from one atom to another, but also by sharing one or more pairs of electrons. These electrons which are held in common by two atoms may be considered to belong to the outer shells of both atoms.

"The simplest explanation of the predominant occurrence of an even number of electrons in the valence shells of molecules is that the electrons are definitely paired with one another.

"Two electrons thus coupled together, when lying between two atomic centers, and held jointly in the shells of the two atoms, I have considered to be the chemical bond.

"The pair of electrons which constitutes the bond may lie between two atomic centers in such a position that there is no electric polarization, or it may be shifted toward one or the other atom in order to give to that atom a negative, and consequently to the other atom a positive charge. But we can no longer speak of any atom as having an integral number of units of charge, except in the case where one atom takes exclusive possession of the bonding pair, and forms an ion."

*Gilbert Newton Lewis\**

## 3.1 ATOMIC INTERACTIONS.

Atoms interact. The remainder of this text is a brief introduction to the myriad consequences of this fact. We, and our environment, are consequences of atomic interactions.

Atomic interactions are arbitrarily classified as strong or weak. Our interest is mainly in strong, attractive interactions that produce identifiable aggregates of atoms called molecules. By strong interactions we mean those in which attractive forces between the atoms are of sufficient magnitude to withstand thermal disruptions at temperatures commonly accessible in the laboratory. A small molecule that requires the input of at least 0.2 eV of energy per atom to break it apart will usually survive in isolation.

As atoms approach each other, interactions begin. If these interactions are attractive, the potential energy of the molecule is lower than the potential energy of the separated atoms. If this energy-lowering is greater than 0.2 eV per atom, the molecule will not fly apart under normal conditions. This does not necessarily mean that the molecule itself is stable, for there may be other lower-energy molecules containing the same atoms in a different arrangement, or two or more molecules that are stable when kept apart may undergo changes when they are brought together. Stability is a

---

°G. N. Lewis, *Valence* (Chemical Catalog Co., Inc., New York, 1923) pp. 79–83.

relative term. For example, A may be stable with respect to B, but unstable with respect to C. One final complication remains. Although aggregate C has lower potential energy than A, the process of rearrangement from A to C may require more than 0.2 eV of energy.[*] That is, both A and C may represent energy holes, and C may be a deeper hole than A, but the walls of both holes may be more than 0.2 eV high. In this case, both groups of atoms will survive all but the most violent disruptions.

Unfortunately, an exact quantum-mechanical treatment of even the simplest molecule is a complex task, far beyond the interest or training of the average professional chemist. He, and we, must content ourselves with crude, approximate treatments which lead to useful qualitative predictions about the composition and structure of molecules. Ideally, we specify the composition of a molecule by giving the absolute number of atoms of each element in the molecule; for example, we describe carbon dioxide as $CO_2$, where each molecule contains one carbon atom and two oxygen atoms. However, we must sometimes be satisfied to give relative numbers of atoms because of ambiguity in the meaning of single molecules for many condensed-phase species. As for structure, the first thing we must specify is how the atoms are joined to one another. The structure of carbon dioxide is O—C—O (linear, equal C—O distances) rather than O—O—C,

or O⟍C⟋O, O⟍C⟋O, or O—C-O. In this chapter we shall concentrate on physical methods for determining structure and other relevant properties of molecules.

## 3.2 MOLECULAR PARAMETERS.

### a. Electronic Energy Levels.

Just as in the case of atoms, electron configurations of molecules can be specified. These configurations are of central importance in theories of molecular structure. There is a large number of possible electron configurations even for simple molecules, and there is a definite energy corresponding to each configuration. Therefore, electronic spectra, in which the energy differences between electronic configurations are determined by the measurement of the frequencies at which the molecule absorbs radiation, are complex. Fortunately, certain simplifications are possible.

The valence-shell electrons in a molecule are excited (raised to higher energy) by light in the wavelength range from 1200 to 7000 Å.[†] These electrons are of primary interest to us because they are the electrons that hold the molecule together. The wavelength at which absorption occurs can help us deduce the energy levels occupied by electrons in molecules. This is quite analogous to the analysis of electronic spectra in atoms (see Figure 2.22). Most of the details of the electron distribution in a molecule

---

[*]Because $1 \text{ eV} = 1.602 \times 10^{-19} \text{ J}$, 0.2 eV per molecule corresponds to $(0.2 \text{ eV/molecule})(1.6 \times 10^{-19} \text{ J/eV})(6.0 \times 10^{23} \text{ molecule/mole}) = 2 \times 10^4 \text{ J/mole}$, or 20 kJ/mole, or $(20 \text{ kJ/mole})\left(\dfrac{1 \text{ calorie}}{4.184 \text{ J}}\right) \approx 5 \text{ kcal/mole}$.

[†]Visible light (4000–7000 Å) is at the high wavelength end of this range. Molecules with loosely restricted electrons can absorb selected frequencies of visible light and thus are colored.

can be tested only by an extensive study of the electronic absorption spectra of these molecules.

Electrons can be removed from molecules, thereby producing ions. Ionization energies of molecules are difficult to measure because the vigorous conditions used to ionize atoms often lead to fragmentation of molecules. However, a recent adaptation of the photoelectric-effect experiment has provided a method for measuring molecular ionization energies. A high-energy photon strikes a molecule and knocks an electron loose; the difference between the energy of the incoming photon and the kinetic energy of the ejected electron is the ionization energy of this particular electron:

$$\text{I.E.} = h\nu - \frac{1}{2}\,mv^2. \tag{3.1}$$

## b.  Vibrational Energy Levels.

Two component parts of any aggregate of matter will remain together only if there is some force holding them together. That is, there must be some restoring force which opposes small disturbances. The simplest model for an aggregate with a restoring force is two masses and a connecting spring; the restoring force is proportional to the distance the spring is stretched or compressed. An example of such a system is an automobile; the axles are connected to the rest of the car by springs. When the wheels are displaced from their normal positions relative to the car body, say by a pothole, the whole system is set in motion and tends to oscillate or vibrate (shock absorbers are used to damp these oscillations). The massive nuclei in a molecule, which are bound together by strong forces, oscillate at a fixed frequency when disturbed. Molecules do not have shock absorbers, and they continue to vibrate until further disturbed.

The combination of two massive bodies and an ideal (Hooke's Law) spring is called a *harmonic oscillator.* In a molecule the massive bodies are nuclei, and the spring is provided by the electrons. The Schrödinger equation can be solved exactly for the harmonic oscillator. The solution is a quantized series of allowable vibrational motions with a corresponding series of equally spaced energy levels. Figure 3.1 is a conventional presen-

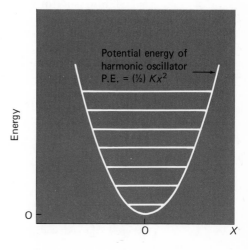

**FIGURE 3.1**   Harmonic oscillator. The energy levels of the harmonic oscillator are drawn within the potential energy well; $E_n = \left(n + \dfrac{1}{2}\right)h\nu$, where $n = 0, 1, 2, \ldots$.

tation of these energy levels. We first plot the potential energy of the system versus the distance of separation of the nuclei, with the zero of the distance scale set at the equilibrium (average) internuclear separation. The allowed energy levels are then inscribed within the potential-energy well. As stated earlier, this potential-energy well must be at least 0.2 eV deep in a stable molecule. The separation between any two adjacent vibrational energy levels is proportional to the strength of the bond holding the nuclei together. The frequency of absorption indicates the spacing between two vibrational energy levels ($\Delta E = h\nu$). The ability of the bond to resist stretching is governed by the *force constant* ($k$) of the bond.[*] These absorptions lie in the infrared region of the spectrum ($> 7000$ Å).

Certain molecular vibrational frequencies are characteristic of certain subgroups and are relatively insensitive to the composition of the rest of the molecule. For example, an oxygen atom which is connected only to one carbon atom always produces absorption with a wavelength near 60,000 Å, whereas a hydrogen atom connected to a carbon atom produces absorption at about 35,000 Å. Thus, infrared spectra provide information ("fingerprints") about the presence or absence of certain subgroups within the molecule. These fingerprints are often of great utility in unraveling the structure of the molecule.

The natural vibrations (normal vibrational modes) of two molecules with identical nuclei at their ends are diagrammed in Figure 3.2.

It is significant that the lowest energy for the quantum-mechanical harmonic oscillator does not lie at the minimum potential energy. Instead it lies uphill by an amount equal to $\frac{1}{2} h\nu_0$, where $\nu_0$ is the oscillator frequency. This means that the harmonic oscillator (and the molecule) never comes to rest, but instead always vibrates. The energy $\frac{1}{2} h\nu_0$ is called the *zero-point* energy (Figure 3.1). It must be stressed that there is absolutely no way to extract this energy from the intact molecule; its removal would lead to a violation of the Heisenberg uncertainty principle because both the momentum ($p = 0$) and the position (the equilibrium separation, $x = 0$) of each nucleus would be known exactly if the atoms did not vibrate.

## c.  Bond Dissociation Energies.

There is a limit beyond which a real spring cannot be stretched without permanent deformation or rupture. It is possible to rupture a chemical bond as well. The minimum energy required to rupture a bond is called the *bond dissociation energy* or (less precisely) the *bond energy*. For example, the bond dissociation energy, $D_0$, for the hydrogen molecule is the energy required to dissociate a mole of $H_2$ molecules into atoms:

$$H_2 \rightarrow H + H; \qquad D_0 = 104 \text{ kcal/mole}$$

---

[*]For a diatomic molecule, the stretching frequency is related to the force constant by the expression

$$\nu = \frac{1}{2\pi} \left( \frac{k}{\mu} \right)^{\frac{1}{2}}$$

where $\mu$, the reduced mass, is given by $\mu = m_1 m_2 / (m_1 + m_2)$.

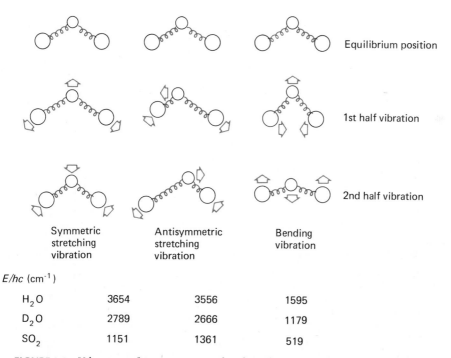

Linear triatomic molecule

| | | | |
|---|---|---|---|
| | | | Equilibrium position |
| | | | 1st half vibration |
| | | | 2nd half vibration |
| Antisymmetric stretching vibration | Symmetric stretching vibration | Bending vibration (also another one perpendicular to page) | |

$E/hc$ (cm$^{-1}$)

| | | | |
|---|---|---|---|
| $CO_2$ | 2349 | 1337 | 667 |
| $CS_2$ | 1523 | 657 | 397 |

Non-linear triatomic molecule

| | | | |
|---|---|---|---|
| | | | Equilibrium position |
| | | | 1st half vibration |
| | | | 2nd half vibration |
| Symmetric stretching vibration | Antisymmetric stretching vibration | Bending vibration | |

$E/hc$ (cm$^{-1}$)

| | | | |
|---|---|---|---|
| $H_2O$ | 3654 | 3556 | 1595 |
| $D_2O$ | 2789 | 2666 | 1179 |
| $SO_2$ | 1151 | 1361 | 519 |

**FIGURE 3.2** Vibrations of two triatomic molecules. The motions during one half vibration are shown; in the other half, all motions are reversed.

The determination of bond energies lies within the realm of thermo-dynamics, where the usual energy units are kcal/mole. Because 1 eV/molecule is equal to 23.06 kcal/mole, we have

$$D_0 \text{ (for } H_2) = 104 \text{ kcal/mole} \left( \frac{1 \text{ eV/molecule}}{23.06 \text{ kcal/mole}} \right) = 4.51 \text{ eV/molecule}$$

$$D_0 \text{ (for } H_2) = (104 \text{ kcal/mole})(4.184 \text{ J/cal}) = 435 \text{ kJ} \cdot \text{mol}^{-1}$$

### d.  Bond Lengths and Bond Angles.

The average distance between two connected nuclei in a molecule is called the *bond length*; this is a necessary parameter in a detailed specification of the structure of any molecule. A description of the structure of a molecule containing two or more bonds necessarily requires the value of the angles between the bonds. These angles are called *bond angles*. The incessant vibrations of the nuclei mean that the bond angles and bond lengths are changing continually, but they oscillate about average values. It is these average values that are measured and specified. For example, water ($H_2O$) has the (average) structure

### e.  Dipole Moments.

In a neutral molecule the total number of electrons is equal to the total positive charge on the nuclei. However, if the centers of positive and negative charge in the molecule do not coincide, then the molecule has a *dipole moment*. A dipole moment is a vector quantity because it has both direction (by convention from minus to plus) and magnitude. The magnitude is the product of the net charge and the separation of the centers of charge; for a sodium ion ($Na^+$) and a chloride ion ($Cl^-$) 2.5 Å apart, the dipole moment is

$$\mu = 4.8 \times 10^{-10} \text{ esu} \times 2.5 \times 10^{-8} \text{ cm} = 12.0 \times 10^{-18} \text{ esu} \cdot \text{cm} = 12 \text{ debye (D)}$$

where $10^{-18}$ esu $\cdot$ cm is defined as one debye (1 debye = 1 D $\simeq 3.34 \times 10^{-30}$ Amp $\cdot$ m $\cdot$ sec).

The presence of a non-zero dipole moment in a molecule often enables us to distinguish between various possible structures for the molecule. For example, water ($H_2O$) has a dipole moment (1.86 D), whereas carbon dioxide ($CO_2$) does not. The fact that water has a dipole moment, together with the additional fact (from vibrational spectroscopy) that the two O—H bonds in water are equivalent, requires that the water molecule be bent rather than linear:

The absence of a net dipole moment for $CO_2$ requires that the molecule be linear and symmetric:

$$\overset{\delta^-}{O}\underset{\rightarrow}{\phantom{-}}\overset{\delta^+}{C}\underset{\leftarrow}{\phantom{-}}\overset{\delta^-}{O}$$

$$\text{net } \mu = 0$$

because any other arrangement of the C and two O atoms will give rise to a net dipole moment.

## 3.3 CHEMICAL BONDS.

The interactions (forces) that hold atoms together in molecules are referred to as *chemical bonds*. We shall discuss two *idealized* types of chemical bonds, namely, *ionic bonds* and *covalent bonds*. These bond types will be discussed separately, but it should be realized that examples of pure ionic or pure covalent bonds are very rare. Most molecules contain bonds that are intermediate between the two extreme types; intermediate bonds can be thought of as composites of both types.

### a. Ionic Bonds.

The simplest bonds that hold atoms together are ionic bonds. As the name implies, the forces which produce these bonds are the Coulomb's-law attractions between opposite charges. The net energy change for the process

$$A + B \rightarrow A^+B^-$$

can be broken down into a sequence of imaginary steps:
  a. An electron is pulled off one atom and taken to infinity; the energy absorbed is the ionization energy ($A \rightarrow A^+ + e^-$).
  b. An electron is brought from infinity and added to another atom; the energy released is the electron affinity ($B + e^- \rightarrow B^-$).
  c. The ions are brought together; the energy released is the coulombic interaction energy ($A^+ + B^- \rightarrow A^+B^-$).
This sequence bears no relation to what actually happens, but each step produces an energy change which can be calculated. Because the starting point (two atoms) and the end point (two ions at their equilibrium separation) are the same in the true process and in the arbitrary sequence, the energy change calculated for the latter will apply to the former (the *net* energy change is independent of the path).

The most favorable case for the formation of an ionic bond is the one that involves the removal of an electron from the atom with the lowest ionization energy (cesium, Cs; I.E. = 3.89 eV) and the addition of the electron to the atom with the highest electron affinity (chlorine, Cl; E.A. = 3.61 eV). In so doing, we invest 3.89 eV of energy (step a) and receive back only 3.61 eV (step b) for a net loss of 0.28 eV. The two separate ions are then brought to their equilibrium separation of 3.50 Å* (step c), which, according to Coulomb's law, liberates (lowers) energy, calculated as

$$E = 14.4 q_1 q_2 / r \qquad \text{(for } E \text{ in eV, } r \text{ in Å)} \qquad (3.2)$$

---

*Calculated from the table of ionic radii (Appendix 7).

$$E = 14.4(1)(-1)/3.50 = -4.11 \text{ eV}$$

Thus, the arrangement

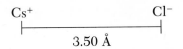

| Cs⁺ | Cl⁻ |

has a potential energy 4.11 eV lower than that of the separated ions, and total potential energy of the ionic arrangement is 3.83 eV *lower* $(0.28 - 4.11 = -3.83 \text{ eV})$ than that for

Cs                     Cl

3.50 Å

Because the molecule is of lower energy than the atoms, the molecule is energetically stable with respect to the atoms. The energy situation is even more favorable (except at very high temperatures) for the formation of *crystalline* CsCl than for gaseous cesium chloride, because each ion in the solid is surrounded by eight ions of opposite charge. In other words, we do not have $Cs^+Cl^-$ molecules in the solid; the solid involves a three-dimensional array of interacting ions. The above calculation is over-simplified, because there is nothing included which explains why the ions are not closer together. Closer approach is prevented by the mutual repulsion between interpenetrating orbitals of the two ions.

Next we consider a triatomic molecule made from a dipositive ion and two negative ions. Again we take the most favorable case: chlorine for the negative ion, and the atom with the lowest second ionization energy (barium, Ba) for the dipositive ion. Formation of the dipositive ion requires energy

$$Ba \rightarrow Ba^+ + e^-; \text{ I.E.}_1 = 5.21 \text{ eV}$$

$$Ba^+ \rightarrow Ba^{+2} + e^-; \text{ I.E.}_2 = \underline{9.95 \text{ eV}}$$
$$15.16 \text{ eV}$$

and the formation of the negative ions returns energy

$$2Cl + 2e^- \rightarrow 2Cl^-; 2 \text{ E.A.} = 2(-3.61 \text{ eV}) = -7.22 \text{ eV}$$

for a net increase of 7.94 eV. We then bring the ions together in the arrangement

Cl⁻                Ba⁺²                Cl⁻
├────3.15 Å────┼────3.15 Å────┤

The two $Ba^{+2} - Cl^-$ interactions return

$$E = 2(14.4)(2)(-1)/3.15 = -18.29 \text{ eV}$$

of energy, but the $Cl^- - Cl^-$ repulsion costs

$$E = 14.4(-1)(-1)/6.30 = 2.29 \text{ eV}$$

Thus the ionic arrangement is $-8.06 \text{ eV} = (-18.29 + 2.29 + 7.94)$ lower in potential energy than the atoms. Notice that the Coulomb's-law energies do not depend explicitly on angle. The molecule is linear because a linear molecule has the two negative chlorine ions at the maximum separation. Although the $Ba^{+2} - Cl^-$ subgroup of this molecule definitely has a dipole moment, the linear molecule as a whole does not, because the two $Cl^- - Ba^{+2}$ dipole moments exactly oppose, and therefore cancel, each other.

Finally, consider the diatomic molecule carbon monoxide (CO). Because carbon has the lower ionization energy, we invest

$$C \rightarrow C^+ + e^-; \text{ I.E.} = 11.26 \text{ eV}$$

and receive

$$O + e^- \rightarrow O^-; \text{ E.A.} = -1.47 \text{ eV}$$

for a net increase of 9.79 eV. The radii of these ions must be estimated; the radius of the $O^-$ ion is about 1.03 Å, and a reasonable value for the $C^+$ radius is about 0.62 Å, which gives an equilibrium $C^+$—$O^-$ separation of 1.65 Å. This returns only

$$E = (14.4)(1)(-1)/1.65 = -8.73 \text{ eV}$$

which is not enough to overcome the deficit of 9.79 eV.* Yet the molecule exists. Thus, we are forced to the conclusion that ionic attractions are not adequate for an explanation of the chemical bond in carbon monoxide. There are a vast number of molecules in which the bonds are not explicable in terms of ionic attractions. Some other types of interatomic forces must be involved in bonding atoms together.

## b. Covalent Bonds.

We know from experiment that the total energy of the hydrogen molecule ion, $H_2^+$, is considerably lower than the total energy of a hydrogen atom and a proton; attractions between oppositely charged ions cannot account for this fact. It is tempting to say that an electron which is close to two nuclei (e.g., between the two nuclei in $H_2^+$) will have lower energy than an electron close to one nucleus, and that this lowering of energy accounts for the bond. A consideration of the $H_2$ molecule shows that this picture is inadequate. Two electrons between two nuclei would be very close to each other. The resulting electron-electron repulsion, which is not present in the atoms, would make a large, unfavorable contribution to the total energy of the $H_2$ molecule. However, the energy released upon formation of the $H_2$ molecule from two H atoms is almost double that released upon formation of $H_2^+$. Where does the extra energy lowering (i.e., stability) come from? To answer this question we must discuss the orbitals in molecules. For the moment we simply state that the compositions of many, many molecules indicate that a pair of electrons that is

---

*This conclusion could be reversed by bringing the nuclei closer together; the observed bond distance is 1.2 Å. This distance implies a severe interpenetration of the electron clouds of the two ions, which destroys the applicability of Coulomb's law. Moreover, the dipole moment of CO is very small (~0.1 D), which is inconsistent with ions 1.2 Å apart.

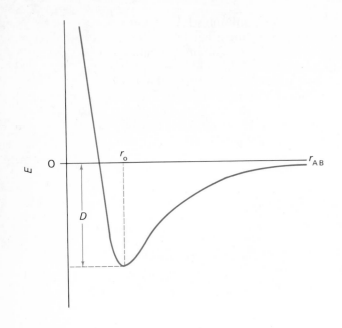

*shared* by two nuclei can bond those two nuclei together. We call that kind of a bond a *covalent bond.*

The release of energy during formation of a chemical bond can be illustrated by a plot of the total electronic energy versus the bond distance. A typical plot for a covalent bond is shown in Figure 3.3. It is extremely difficult to calculate energy curves for real molecules from the Schrödinger equation, so it is useful to have an approximate equation that generates a bond-energy curve from appropriate bond parameters. Several equations have been found. One of them is the Morse potential,

$$E = D \left[ 1 - e^{-\beta(r_{AB} - r_0)} \right]^2 \tag{3.3}$$

The significance of $D$ and $r_0$ is indicated in Figure 3.3. The distance $r_0$ is called the *equilibrium bond distance,* and $D$ is related to the experimental bond dissociation energy

$$D = D_0 + \frac{1}{2} h\nu_0 \tag{3.4}$$

where $D_0$ is the experimental bond-dissociation energy, and $\frac{1}{2} h\nu_0$ is the zero-point energy of an oscillator with frequency $\nu_0$. The constant $\beta$ in Equation (3.3) is equal to $(k/2D)^{1/2}$ where $k$ is the force constant of the bond. Thus, values of $r_0$, $k$, and $D$ enable us to compute an approximate potential energy curve for the stretching of a bond.

## 3.4 GENERATION OF EMPIRICAL FORMULAS FOR STABLE MOLECULES.

How many covalent bonds can an atom form? How many electrons can one atom share with its neighbors? The Pauli principle is a fundamental property of electrons and applies in molecules as well as in atoms. Therefore, a maximum of two electrons can be in a region of space, and these two electrons must have their spins opposed.[*] The shared pair of electrons is so common that the term is almost synonymous with covalent bond. The Pauli principle limits the number of electron pairs which can form about any one atom within a given shell. The maximum number of covalent bonds is the number that will provide sufficient electron pairs for the element to "complete its valence shell." For the elements in the second period, a complete valence shell is the configuration $2s^2 2p^6$. Elements of the second period usually form covalent bonds so as to surround the nuclei with valence shells containing eight electrons (the *octet rule*). The octet rule applies best to elements of the second period.

## a. Lewis Diagrams.

There is a useful notation for the description of the bonding situations that are most frequently encountered, the Lewis diagram.[†] *A Lewis diagram designates only the electrons in the valence shell of each atom, the valence electrons.* The atoms of the second period have the Lewis diagrams:

$$\text{Li·} \quad \text{Be:} \quad \text{·B:} \quad \text{·C:} \quad \text{·N:} \quad \text{:O:} \quad \text{:F:} \quad \text{:Ne:}$$

The valence shell ($n = 2$) electrons are represented by dots. The pairing of electrons is indicated (Be is diamagnetic), but the placement of the dots has no further significance.

Lewis diagrams for molecules are constructed in the following manner:
(1) Compute the total number of valence electrons in the molecule by adding up the number of valence electrons on each atom in the molecule. For example, in $CH_4$ we have a total of $4 + 4 \times (1) = 8$ valence electrons, whereas for $NH_3$ we have $5 + 3 \times (1) = 8$.
(2) Place the atoms that are bonded together in the molecule next to one another. For $CH_4$ and $NH_3$ the arrangements are

$$
\begin{array}{ccc}
 & \text{H} & \\
\text{H} & \text{C} & \text{H} \\
 & \text{H} &
\end{array}
\qquad\qquad
\begin{array}{ccc}
\text{H} & \text{N} & \text{H} \\
 & \text{H} &
\end{array}
$$

(3) Place the electrons *in pairs* either between bonded atoms (bond pairs) or in non-bonding positions (lone pairs) in such a way that the valence shell of each atom in the molecule is completed:

$$
\begin{array}{c}
\text{H} \\
\text{H:}\overset{..}{\text{C}}\text{:H} \\
\overset{..}{\text{H}}
\end{array}
\qquad\qquad
\begin{array}{c}
\text{H:}\overset{..}{\text{N}}\text{:H} \\
\overset{..}{\text{H}}
\end{array}
$$

---

[*]In the next chapter we shall give a precise definition of "a region of space" in terms of orbitals.

[†]G. N. Lewis originated the idea of the electron-pair bond. Lewis diagrams are most useful in simple descriptions of the bonding in compounds and ions involving the light elements.

Note that the valence shell of H ($n = 1$) requires two electrons, whereas the valence shells for C and N ($n = 2$) require eight electrons.

In the hydrogen molecule ($H_2$), which has a total of two electrons, the bond is represented by a pair of dots (electrons) between the nuclei:

$$H:H$$

Notice that, by use of both members of the shared pair, both hydrogen atoms have complete valence shells. (Can hydrogen be bonded to two atoms in a Lewis structure?)

Further examples of Lewis diagrams for molecules are as follows:[*]

1. Fluorine.

$$:\ddot{F}\cdot \qquad\qquad :\ddot{F}:\ddot{F}: \qquad\qquad F_2$$

atom                molecule              fluorine

The Lewis diagrams for $Cl_2$, $Br_2$, and $I_2$ are analogous. Why?

2. Compound between Hydrogen and Fluorine.

$$H\cdot \qquad\qquad \cdot\ddot{F}: \qquad\qquad H:\ddot{F}: \qquad\qquad HF$$

atoms                        molecule      hydrogen fluoride

3. Compound between Hydrogen and Oxgen.

$$H\cdot \qquad\qquad \cdot\ddot{O}\cdot \qquad\qquad H:\ddot{O}:H \qquad\qquad H_2O$$

atoms                        molecule              water

The geometry of the Lewis diagram has no significance; water is not a linear molecule. An equally good Lewis diagram for water is

$$H:\ddot{O}:$$
$$\ddot{H}$$

Water is not the only possible compound between hydrogen and oxygen. From Lewis diagrams alone we would predict an infinite series of compounds, of which the next two are

$$H:\ddot{O}:\ddot{O}:H \qquad\qquad\qquad H:\ddot{O}:\ddot{O}:\ddot{O}:H$$

hydrogen peroxide ($H_2O_2$)              $H_2O_3$

Members of this series with more than two oxygen atoms have never been observed. It is possible to draw satisfactory Lewis diagrams for very

---

[*]In simple molecules and ions the unique atom is often the central one. For example, in methane ($CH_4$) the carbon atom is the central atom, and in perchlorate ion ($ClO_4^-$) the chlorine atom is the central atom.

unstable molecules, and some additional knowledge (experimental results) is needed to eliminate these "rotten apples."

4. Nitrogen Trifluoride.

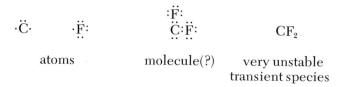

$$\quad\text{atoms} \qquad\qquad \text{molecule} \qquad \text{nitrogen trifluoride}$$

Why is the Lewis diagram for phosphorous trifluoride similar to this one? Give the Lewis diagram for $NF_2^-$.

5. Compound between Carbon and Fluoride.

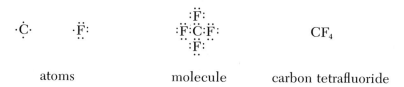

$$\quad\text{atoms} \qquad \text{molecule(?)} \qquad \text{very unstable}$$
$$\text{transient species}$$

Although this structure has all the electrons paired, the valence shell of carbon contains only six electrons. In order to account for the chemistry of carbon, we must excite the carbon atom to the configuration $1s^2 2s^1 2p^3$. This excited configuration is represented crudely by the Lewis diagram ·Ċ· Using this, we have

$$\qquad\qquad\qquad\qquad\qquad \begin{array}{c} :\ddot{F}: \\ :\ddot{F}:\overset{\displaystyle .}{\underset{\displaystyle .}{C}}:\ddot{F}: \\ :\ddot{F}: \end{array} \qquad\qquad CF_4$$

$$\quad\text{atoms} \qquad\qquad \text{molecule} \qquad\qquad \text{carbon tetrafluoride}$$

The carbon tetrafluoride molecule is a well-characterized species; evidently this molecule has lower energy than the arrangement $CF_2 + 2F$ because the energy returned by the formation of two additional bonds easily compensates for the energy necessary to excite the carbon atom.

## b. Multiple Bonds.

In all the preceding examples there was at least one atom which needed only a single electron to complete its valence shell. We now treat more complex cases. Consider the $N_2$ molecule,

$$\qquad\qquad ·\ddot{N}· \qquad\qquad\qquad ·\ddot{N}:\ddot{N}·$$

$$\text{atom} \qquad \text{possible structure of } N_2$$

This diagram is unacceptable because it has several unpaired electrons and incomplete valence shells. The solution is to place more than one pair of electrons between the nuclei. The octet rule can be satisfied in $N_2$ by placing three pairs of electrons between the nuclei. The three covalent bonds constitute a *triple bond*:

$$:N:::N:$$

molecule

Note that each N has attained the octet by sharing three pairs.

The carbon monoxide molecule (CO) is *isoelectronic* with $N_2$; it has exactly the same number of electrons. The Lewis diagrams for the two molecules have the same electron arrangement:

$$\cdot \ddot{C} \cdot \qquad \cdot \ddot{O} \cdot \qquad\qquad :C:::O:$$

atoms                              molecule

## c. Compounds between Carbon and Hydrogen.

Carbon has the ability to bond to itself in chains of unlimited length. As a result, there are a very large number of compounds between carbon and hydrogen. Here we restrict ourselves to the hydrogen-carbon compounds with one or two carbon atoms. In some of these compounds, multiple bonds are required to complete the carbon octets.

$$H \cdot \qquad \ddot{C} \cdot \qquad\qquad H:\overset{\displaystyle \ddot{}}{\underset{\displaystyle \ddot{}}{C}}:H \qquad\qquad CH_4$$

atoms                         methane

$$H:\overset{\displaystyle H}{\underset{\displaystyle H}{\ddot{C}}}:\overset{\displaystyle H}{\underset{\displaystyle H}{\ddot{C}}}:H \qquad\qquad C_2H_6$$

ethane

$$\overset{\displaystyle H}{\underset{\displaystyle H}{\ddot{C}}}::\overset{\displaystyle H}{\underset{\displaystyle H}{\ddot{C}}} \qquad\qquad C_2H_4$$

ethylene

$$H:C:::C:H \qquad\qquad C_2H_2$$

acetylene

## d. Ionic Compounds.

Lewis diagrams can also be written for ions. For example,

$$\dot{Cs} \qquad \cdot \ddot{\underset{\displaystyle \cdot\cdot}{Cl}}: \qquad\qquad Cs^+ \left[ :\ddot{\underset{\displaystyle \cdot\cdot}{Cl}}: \right]^- \qquad\qquad Cs^+Cl^-$$

atoms                         ions                    cesium chloride

The brackets are included to stress the ionic nature of the bond; the electrons placed between the nuclei are on the chlorine atom and are not shared. In ionic compounds the nuclei are usually too far apart for effective sharing of electrons.

## e.  Compounds of Nitrogen and Hydrogen.

$\cdot \ddot{N} \cdot$          $\cdot H$                    $H \colon \ddot{N} \colon H$                    $NH_3$

atoms                                                                      ammonia

$H \; H$
$H \colon \ddot{N} \colon \ddot{N} \colon H$                    $N_2H_4$
hydrazine

In addition to these molecules, it is possible to form the ammonium ion by addition of a proton to the ammonia molecule; this ion satisfies the octet rule and is a familiar chemical entity. The ammonium halides, such as ammonium fluoride, are interesting in that they contain both covalent and ionic bonds.

$\left[ \begin{array}{c} H \\ H \colon \ddot{N} \colon H \\ H \end{array} \right]^{+}$                    $\colon \ddot{F} \colon ^{-}$                    $NH_4F$

ammonium ion          fluoride ion   ammonium fluoride

The amide ion, $NH_2^-$, can be obtained from $NH_3$ by removal of a proton.

The Lewis diagram notation can be streamlined by replacing each shared electron pair by a dash, and omitting the *non-bonding electrons* (the valence-shell electrons which are not shared by two atoms). In this streamlined notation we have

$$H—H \qquad H—\underset{|}{\overset{H}{N}}—H \qquad N \equiv N \qquad \underset{\underset{H}{|}}{\overset{\overset{H}{|}}{H—N—H}} \; +$$

hydrogen          ammonia          nitrogen     ammonium ion

$$\underset{O}{\overset{O}{O—S—O}} \; ^{-2} \qquad H—\underset{\underset{H}{|}}{\overset{\overset{H}{|}}{C}}—H \qquad \underset{H \; H}{\overset{H \; H}{C = C}} \qquad H—C \equiv C—H$$

sulfate ion          methane          ethylene          acetylene

## f.  Inadequacies of the Lewis-Diagram Method.

Even though the existence of many simple molecules can be predicted correctly with the Lewis diagrams, there are many exceptions (e.g., NO, $HF_2^-$, $B_2H_6$). Furthermore, there are cases involving second-period elements where the method leads to qualitatively incorrect descriptions of molecules. The oxygen molecule is an interesting case:

$\cdot \ddot{O} \cdot$          $\ddot{O} \colon \colon \ddot{O}$          $O = O$          $O_2$

atom          molecule                              oxygen

The oxygen molecule does indeed have the composition $O_2$, but it is known to have unpaired electrons (paramagnetic), whereas the Lewis diagram has all the electrons paired.* There are molecules which have an odd number of electrons in their valence shells and also are paramagnetic. The Lewis method does not predict the existence of these molecules. Examples are nitrogen monoxide, NO, and chlorine dioxide, $ClO_2$.

A different type of inadequacy is illustrated by the molecule sulfur dioxide, $SO_2$.

$$:\!\overset{\displaystyle\cdot}{S}\!: \qquad :\!\overset{\displaystyle\cdot}{\underset{\displaystyle\cdot}{O}}\!: \qquad\qquad :\!\overset{\displaystyle\cdot\cdot}{O}\!=\!\overset{\displaystyle\cdot\cdot}{S}\!-\!\overset{\displaystyle\cdot\cdot}{\underset{\displaystyle\cdot\cdot}{O}}\!:$$

atoms                molecule

This diagram is not in agreement with nature, for sulfur dioxide does not have two distinct kinds of sulfur-oxygen bonds (one single, one double); rather, it has two identical bonds. A somewhat more realistic representation of the $SO_2$ molecule is intermediate between the two Lewis diagrams

$$O\!=\!S\!-\!O \leftrightarrow O\!-\!S\!=\!O$$

The intermediate structure has each oxygen atom bonded to sulfur by a bond and a half. By an unbelievably bad choice of terms this situation is called *resonance,* and the two Lewis diagrams are called *resonance structures.* It must be stressed repeatedly that there is absolutely no evidence that the molecule actually resonates back and forth between the two structures. Frequently, several resonance structures can be drawn for a molecule.† The classic example of resonance occurs in the benzene molecule ($C_6H_6$). A wealth of structural information indicates that the structure is intermediate between

and

---

*Note that a Lewis diagram for a paramagnetic oxygen of the type $:\!\overset{\displaystyle\cdot}{O}\!=\!\overset{\displaystyle\cdot\cdot}{O}\!:$ requires five orbitals in the valence shell of oxygen, and there are only four.

†All nuclei must be in exactly the same position in all resonance structures of one molecule.

Other resonance structures, for example,

are of minor importance.

Most molecules adopt the most symmetrical configuration. Therefore, resonance should be expected whenever two or more equivalent structures can be drawn whose intermediate will be more symmetric than any individual structure.

## g. Linnett Diagrams.

Some of the difficulties inherent in the Lewis diagram approach can be traced to the insistence on electron pairs. J. W. Linnett has suggested that the octet of electrons should be regarded as two *quartets* (rather than four pairs); all electrons in a quartet have the same spin. These quartets are regarded as having the four electrons at the corners of a tetrahedron. Bonds are formed by making electrons members of quartets around more than one nucleus. The tetrahedra can share a *corner* (one electron in common), an *edge* (two electrons in common) or a *face* (three electrons in common). According to this theory, the fluorine atom has a quartet of electrons with $\alpha$ spin and a triad with $\beta$ (opposed) spin.

$\alpha$ spin          $\beta$ spin triad          atom

The fluorine molecule can be represented as two quartets of $\alpha$-spin sharing a corner and two quartets of $\beta$ spin also sharing a corner

$\alpha$ spin          $\beta$ spin          molecule

Note the similarity in this case to the Lewis diagram.

The advantage of the Linnett approach is that the orbitals in the two spin sets do not have to match exactly. For example, oxygen can be represented as

$\alpha$ spin          $\beta$ spin          molecule
(corner share)     (face share)     (2 unpaired electrons)

In this case the Linnett diagram is not the same as the Lewis diagram, because there are two more electrons with the $\alpha$ spin than with the $\beta$ spin. However, both approaches predict an $O_2$ molecule held together by four electrons. The observed paramagnetism of oxygen implies that the Linnett representation is closer to reality.

The Linnett method also reduces (but does not always eliminate) the need for resonance structures. The sulfur dioxide molecule is a case in point.

$$\underset{\alpha \text{ spin}}{\overset{x\quad x\quad x}{\underset{x}{\times}\,O\,\underset{x}{\times}\,S\,\underset{x}{\overset{x}{\times}}\,O\,\times}} \qquad\qquad \underset{\beta \text{ spin}}{\overset{o\quad o\quad o}{\circ\,O\,\underset{o}{\overset{}{\circ}}\,S\,\circ\,\underset{o}{O}\,\circ}} \qquad\qquad \underset{\text{molecule}}{\overset{xo\quad xo\quad xo}{\times\,O\,\underset{x}{\overset{o}{\times}}\,S\,\underset{x}{\times}\,O\,\underset{o}{\overset{x}{\circ}}}}$$

Here abandonment of the electron-pair bond has allowed representation of the sulfur-oxygen bonds as three-electron bonds, both with two electrons of one spin and one of the other. This is quite analogous to the bond and a half represented by two resonance structures. The Linnett diagram for $SO_2$ predicts that $SO_2$ has no unpaired electrons, because the number of electrons with $\alpha$ and $\beta$ spin are equal.

The introduction of resonance and the Linnett diagrams have patched some of the weak points of the simple Lewis method. There still remains one fatal flaw. The approach is inalterably qualitative; there is no way to extract quantitative information from these diagrams. Only the crudest estimates of bond energies and lengths are possible (large-small, short-long), and few estimates of bond angles can be made.* More quantitative results require the use of quantum mechanics.

## PROBLEMS

1. Convert 0.2 eV to kcal/mole and kjoule/mole.

2. Give the Lewis diagrams for the following molecules. Give resonance structures where they are appropriate.

| | | | | |
|---|---|---|---|---|
| HBr | ClF | $N_2F_2$ | $CO_2$ | $H_2SO_4$ |
| $NF_3$ | $N_2O$ | $P_4$ | $CCl_4$ | $N_2O_5$ |
| $O_3$ | $NO_2$ | $O_2NCl$ | $HNO_3$ | $CS_2$ |

3. Give the Linnett structures for all the molecules in problem 2. For which molecules do these differ from the Lewis diagrams?

4. Give the Lewis diagrams for the following ions. Give resonance structures where appropriate.

| | | | | |
|---|---|---|---|---|
| $NO_3^-$ | $SO_3^{-2}$ | $ClO_4^-$ | $IO_3^-$ | $CN^-$ |
| $NO_2^-$ | $CO_3^{-2}$ | $N_3^-$ | $ClO^-$ | $HCO_3^-$ |

---

*Bond angles can be estimated via the Linnett procedure by a consideration of the orientation of the interconnected tetrahedra.

5. Which atoms of the fourth period are paramagnetic (see Chapter 2)?

6. The light source used in the photo-ionization apparatus is the He lamp with a strong emission line at $\lambda = 586$ Å. What is the largest ionization energy that can be measured with this apparatus?

7. Show by calculation that the ionic compound $Cs^{+2}(Cl^-)_2$ is unstable. Assume a Cs-Cl distance of 3.3 Å.

8. If a CsCl molecule were pulled apart, at what separation would the ions revert to the atoms?

9. Give Lewis diagrams for the following molecules, which do not obey the octet rule.

$$BF_3, PCl_5, SF_6, BeH_2$$

10. In the text we showed that the $Ba^{+2}(Cl^-)_2$ "molecule" was stable relative to the atoms. Is it stable compared to $Ba^+$-$Cl^-$ and Cl atom? (Take the $Ba^+$-$Cl^-$ distance as 3.50 Å.)

11. Give Lewis diagrams for the two molecules with the empirical formula $C_2H_6O$.

12. Estimate bond energies for the following molecules on the assumption that they are ionic.

$$CaO, HCl, LiF, BaBr_2$$

Ionic radii: $Ca^{2+}$ (0.99 Å); $O^{2-}$(1.40); $Cl^-$(1.81); $Li^+$ (0.60); $F^-$(1.36); $Ba^{2+}$ (1.35); $Br^-$ (1.95).

13. Explain why the octet rule can be relaxed for phosphorus and sulfur, but not for nitrogen and oxygen.

14. Why is it difficult to remove electrons from complete shells? Calculate energies for the formation of $He^+Cl^-$ and $Cs^+He^-$. Use the following radii:

$$Cs^+(1.69 \text{ Å}); Cl^-(1.81); He^+(0.30); He^-(1.50).$$

15. Studies of the HCl molecule have provided the following data: $D_0 = 103$ kcal/mole, $r_0 = 1.27$ Å, $k = 4.82 \times 10^5$ dynes/cm, and $\nu_0 = 2886$ cm$^{-1}$. Make a plot of the potential energy versus internuclear distance for this molecule. Draw the potential energy for the harmonic oscillator on the same graph.

16. Calculate the ion separation necessary to give a net gain in energy of 0.2 eV for the process

$$F + F \rightarrow F^+ + F^-$$

Is this a reasonable bond distance for $F_2$?

17. Show that for small displacements from $r_0$ the Morse potential energy expression reduces to the harmonic oscillator energy expression. (Hint: for small $x$, $e^{-x} \simeq 1 - x + \dfrac{x^2}{2} - \cdots$).

## References

3.1. Pauling, L., *The Nature of the Chemical Bond* (3rd Ed., Cornell University Press, Ithaca, N.Y., 1960).

3.2. Linnett, J. W., *The Electronic Structure of Molecules* (John Wiley and Sons, Inc., New York, 1964).

3.3. Lewis, G. N., *Valence* (Chemical Catalog Co., New York, 1923). An interesting pre-quantum-mechanical description of the chemical bond.

3.4. Gray, H. B., *Chemical Bonds* (W. A. Benjamin, Inc., Menlo Park, Calif., 1973).

## CHAPTER 4

# MODERN TREATMENTS OF CHEMICAL BONDING

"... an 'orbital' is simply an abbreviation for one-electron orbital wavefunction. ... a set of orbitals represents a housing arrangement for electrons. A very strict rule (Pauli's exclusion principle) applies to every orbital, whether atomic or molecular, namely, that not more than two electrons can occupy it. A molecular orbital is defined in exactly the same way as an atomic orbital, except that its one-electron Schrödinger equation is based on the attractions of two or more nuclei plus the averaged repulsions of the other electrons. ... the form of the orbital tells us, among other things, what fraction of time the electron in it can spend in different regions of space around the nucleus, or nuclei. A definite energy is associated with each orbital, either atomic or molecular. ... it is the energy required to take the electron entirely out of the orbital, out into free space."

*R. S. Mulliken**

Our inability to solve the Schrödinger equation for most chemical problems of interest forces us to seek useful approximate solutions. The common approximation methods used to analyze the bonding in molecules are the *valence-bond* method and the *molecular-orbital* method. The *valence-bond* method is an attempt to translate Lewis diagrams into a usable mathematical form. The distinguishing feature of this approach is the assignment of the bonding electrons to orbitals that are localized around two bonded nuclei. J. Slater and L. Pauling made major contributions to the development of this method. We will not discuss the valence-bond method; instead we will use what is called a *localized-molecular-orbital* method. The molecular-orbital method is an attempt to find wavefunctions for the molecule as a whole. Molecular orbitals are not necessarily localized around any two nuclei. F. Hund and R. Mulliken made major contributions to the development of this method.

## 4.1 THE MOLECULAR-ORBITAL METHOD AND THE COVALENT BOND.

The electron configuration of an atom is generated by successively assigning electrons to the atomic orbitals, which are arranged in ascending order of energy. The electron configuration of a molecule is generated in the same way; the possible assignments are limited in both cases by the

---

*R. S. Mulliken, Science, *157*, 13 (1967).

Pauli principle and by Hund's rule. However, the orbitals are quite different; atomic orbitals are centered on a single nucleus, whereas molecular orbitals may extend over all the nuclei in the molecule. For this reason, solution of molecular problems is much more difficult than solution of atomic problems.

### a. Formation of Molecular Orbitals from Atomic Orbitals.

The simplest molecule is hydrogen, $H_2$. We use $H_2$ to illustrate the most common procedure for the generation of approximate molecular orbitals. We assume that the molecular orbitals for $H_2$ can be written as a linear combination (sum or difference) of atomic orbitals centered on the two nuclei. For example, one molecular orbital is the sum of two $1s$ atomic orbitals, one on each nucleus:

$$\psi = 1s_A + 1s_B \tag{4.1}$$

Experience shows that qualitative wavefunctions for the ground states of molecules require the inclusion of only the valence-shell atomic orbitals in the linear combinations.

Since the $H_2$ molecule is known to have a bond distance of 0.74 Å, when computing the molecular-orbital wavefunction (Equation (4.1)) we separate the hydrogen nuclei by 0.74 Å. In Figure 4.1 we have plotted the two atomic $1s$ orbitals and their sum at this nuclear separation versus distance along the internuclear axis. Every wavefunction must be normalized; that is, the value of $\psi^2$ totaled over all volume must be unity. Introduction of the correct normalization factor, $N_+$, gives the dotted curve for the molecular orbital

$$\psi_{1\sigma} = N_+(1s_A + 1s_B) \tag{4.2}$$

This orbital is called a molecular orbital because it extends over the whole molecule.

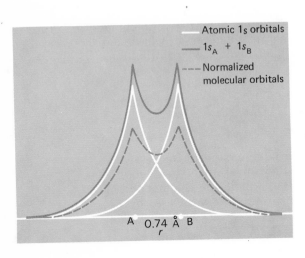

Atomic $1s$ orbitals
$1s_A + 1s_B$
Normalized molecular orbitals

A   0.74 Å   B
$r$

**FIGURE 4.1** Molecular orbital for $H_2$, a plot of the wavefunction along the internuclear axis.

## b. Bonding Orbitals.

If the energy of the molecular orbital in Equation (4.2) is less than the energy of the atomic $1s$ orbitals, electrons in the orbital will hold the molecule together. An orbital that holds a molecule together is called a *bonding orbital*.

Does Equation (4.2) represent a bonding orbital? Before we can decide, we must take two facts into account. First, an electron whose wavefunction changes rapidly as a function of position has a higher kinetic energy than an electron whose wavefunction changes slowly as a function of position. Second, when the forces in a system are inversely proportional to the square of the distance (e.g., Coulomb's law and the law of gravitation), the potential energy ($V$) and the kinetic energy ($T$) are not independent at equilibrium. The relation is

$$V = -2T \qquad (4.3a)$$

so that their sum is

$$E_{\text{sum}} = V + T = \frac{V}{2} \qquad (4.3b)$$

With these two facts we can qualitatively describe the molecular orbital that is responsible for the chemical bond.

In Figure 4.1 we see that the molecular-orbital wavefunction changes less rapidly between the nuclei than do the atomic-orbital wavefunctions. Therefore, formation of the molecular orbital has reduced the kinetic energy of the electrons between the nuclei. The reduction is analogous to the reduction in kinetic energy caused by firing retrorockets on a satellite that is in a stable orbit around the earth; the electron in the stable $1s$ atomic orbital is like the satellite; both the satellite and the electron are held by inverse square forces, and Equation (4.3) applies to both systems. Firing retrorockets causes the satellite to drop toward the earth until it achieves a new stable orbit. In the new orbit its potential energy is more negative (i.e., lower), and its final kinetic energy is higher (Equation (4.3a)). Because the potential energy has double the magnitude of the kinetic energy, the sum of the two is decreased (Equation (4.3b)). In a molecule the reduction of kinetic energy between the nuclei causes the orbitals to contract around the two nuclei, thereby lowering the potential energy.* Again the final kinetic energy is increased, but the sum of potential and kinetic energy is decreased, and a bond is formed.†

The changes noted above can be seen in the contour maps of Figure 4.2. In part (a), the two hydrogen atoms are at a separation of about 4 Å; part (b) is a hydrogen molecule. Notice that the contour lines are few and far between along the internuclear axis in the molecule; the kinetic energy is very low in this region. Notice that most of the contour lines are closer together in the molecule than in the atoms; this shows the contraction that

---

*Mathematically, the effective charge in the $1s$ orbitals is made greater than one. This is not shown in Figure 4.1.

†This analogy and extensive mathematical justification was given by K. Rudenberg and M. Feinberg, J. Chem. Phys., *54*, 1495 (1971).

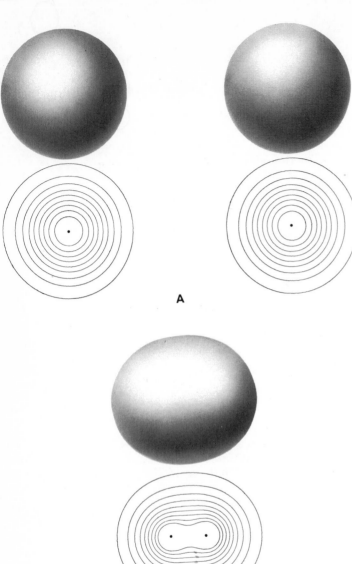

**A**

**B**

**FIGURE 4.2** Molecular orbitals for two H atoms, and for the $H_2$ molecule. In the molecule the atoms are 0.74 Å apart; the separate atoms are 4.0 Å apart.

lowers the potential energy and the sum of energies (while raising the total kinetic energy).

The molecular orbital in Figure 4.2b is unchanged by rotation about the internuclear axis; in other words, it has cylindrical symmetry about the bond axis. Molecular orbitals with cylindrical symmetry about the bond axis are called *sigma* ($\sigma$) *orbitals*. The molecular orbital formed from the $1s$ atomic orbitals is the first (lowest energy) sigma orbital; the electron configuration of $H_2$ is $1\sigma^2$.

The electron in the hydrogen molecule-ion, $H_2^+$, will occupy a $1\sigma$ orbital. However, the $1\sigma$ orbital is not identical to the $1\sigma$ orbital in $H_2$, because the bond lengths of the two molecules are not the same. The arguments given above apply, and the $1\sigma$ orbital is a bonding orbital. The electron configuration of $H_2^+$ is $1\sigma^1$.

## c. Delocalization Energy.

The terms *delocalization* and *delocalization energy* are encountered frequently in the discussion of molecular orbitals for polyatomic molecules. These terms are used to express the idea that, all other factors being equal, the more space (nuclear centers) an electron has available to it, the lower its total energy will be. Delocalization is achieved by constructing molecular orbitals that extend over several (or all) of the nuclei in the molecule.

As in $H_2$, the spreading of the orbital over several nuclei leads to a contraction of the orbital toward the nuclei and an associated lowering of potential and total energy.

## d. Antibonding Orbitals.

The expression for the molecular orbitals (Equation (4.2)) is reminiscent of the expression for a hybrid atomic orbital (Equation (2.17)); the difference is that *a molecular orbital is a combination of atomic orbitals on different nuclei*. As in the atomic case, the number of orbitals is fixed. Two atomic orbitals must lead to two molecular orbitals (*orbital conservation*). The most suitable choice for the other linear combination of $1s$ orbitals is

$$\psi_{1\sigma^\circ} = N_-(1s_A - 1s_B) \tag{4.4}$$

Figure 4.3 shows the atomic orbitals, their difference, and the normalized molecular orbital. With this orbital we have the reverse of the situation just discussed. There is an increase in the variation of the wavefunction between the nuclei, and consequently an increase in the kinetic energy in this region. The increase is analogous to accelerating a satellite, which raises it to a higher-energy, more distant orbit. The molecular orbital has a higher total energy than the original atomic orbitals. Therefore, it is called an *antibonding orbital*. This particular antibonding orbital is denoted $1\sigma^*$ — read one-*sigma*-star — because it is the first (lowest energy) antibonding orbital with cylindrical symmetry along the molecular axis.

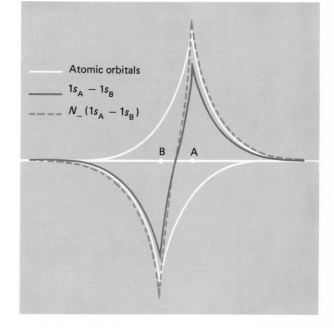

**FIGURE 4.3** Antibonding molecular orbital for $H_2$, a plot of the wavefunction along the internuclear axis.

## e. Correlation Diagrams.

Assignment of electrons to orbitals requires prior ordering of the orbitals according to energy. The ordering is presented conveniently on an energy-level diagram. For simple molecules a special energy-level diagram is useful. This diagram, called a *correlation diagram*, contains the molecular orbitals, the atomic orbitals of the component atoms, and the correlations between them. As usual in energy-level diagrams, the vertical axis indicates energy.

In Figure 4.4 we have a simple correlation diagram for a homonuclear diatomic molecule. The energy levels of the orbitals for the two separated atoms are located above the labels A and B (for simplicity we include only the 1s orbital). We imagine a sequence of steps to take us from the atoms to the molecule. These steps simplify energy calculations but bear no resemblance to what actually happens. The sequence is:

1. Form an appropriate set of orbitals on the atoms. When this step is necessary, the process used is called *hybridization*, and that portion of the correlation diagram is labeled H.
2. Bring the atomic orbitals to their equilibrium separation in the molecule. The electron-electron repulsion between overlapping atomic orbitals raises the energy of the orbitals. The overlap energy is introduced in the region S.*
3. Form the molecular orbitals from the overlapping atomic orbitals.

For the correlation diagram in Figure 4.4, step (1) is unnecessary; steps (2) and (3) are labeled. The dashed lines indicate which atomic orbitals are used to construct the molecular orbitals. *The energy gained in forming the $1\sigma$ orbital is about equal to the energy lost in forming the $1\sigma^*$ orbital when these energies are compared to those of the atomic orbitals at the equilibrium separation* (the S regions). When compared to the energies of the atomic orbitals at large separation (A and B regions), the decrease in energy for the $1\sigma$ orbital is significantly less than the increase in energy for the $1\sigma^*$ orbital. This is a general and important result; *when compared to the separated atoms, the bonding orbitals are invariably less bonding than the antibonding orbitals are antibonding.* It is the latter comparison that is the more important.

The central portion (AB) of the correlation diagram can be used to obtain the ground-state electron configurations of the simplest molecules

---

*The use of S for overlap is related vaguely to standard quantum mechanical notation, but it is best regarded here as an arbitrary label.

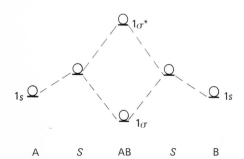

**FIGURE 4.4**  Partial correlation diagram for two identical atoms. The atoms are labeled A and B, and the molecule is labeled AB. See the text for an explanation of the overlap region, labeled S.

TABLE 4.1

| MOLECULE | CON-FIGURATION | NET BONDING ELEC-TRONS | BOND LENGTH | BOND ENERGY | MAGNETIC PROPERTIES |
|---|---|---|---|---|---|
| $H_2^+$ | $1\sigma^1$ | 1 | 1.06 Å | 61.1 kcal $\cdot$ mol$^{-1}$ | paramagnetic |
| $H_2$ | $1\sigma^2$ | 2 | 0.74 Å | 103.2 kcal $\cdot$ mol$^{-1}$ | diamagnetic |
| $He_2^+$ | $1\sigma^2 1\sigma^{\circ 1}$ | 1 | 1.08 Å | 60 kcal $\cdot$ mol$^{-1}$ | paramagnetic |
| $He_2$ | $1\sigma^2 1\sigma^{\circ 2}$ | 0 | | -------unstable------- | |

1 Å = 0.1 nm; 1 kcal = 4.184 kJ.

in exactly the same way that atomic energy level diagrams are used (Table 4.1). Notice that, as the excess of bonding electrons over antibonding electrons increases, the bond is shortened and strengthened. The $He_2$ molecule is unstable because the two antibonding electrons are more antibonding than the two bonding electrons are bonding.

## 4.2 HOMONUCLEAR DIATOMIC MOLECULES.

### a. Molecular Orbitals for Second-Period Diatomic Molecules.

Any molecule containing five or more electrons must utilize molecular orbitals constructed from atomic orbitals with $n = 2$. The appropriate linear combinations of $2s$ orbitals are

$$\psi_{2\sigma} = N_+(2s_A + 2s_B)$$
$$\psi_{2\sigma^\circ} = N_-(2s_A - 2s_B) \tag{4.5}$$

The combination of the various $2p$ orbitals requires closer attention because one pair of orbitals, by convention the $2p_z$ orbitals, point along the internuclear axis, whereas the other two pairs of orbitals ($2p_x$ and $2p_y$) are perpendicular to this axis. Figure 4.5 shows how a bonding orbital is formed from the $2p_z$ orbitals by the linear combination

$$\psi_{3\sigma} = N_-(2p_{zA} - 2p_{zB}) \tag{4.6}$$

This molecular orbital has cylindrical symmetry about the bond axis (Figure 4.5); it is the third *sigma* bonding orbital. The corresponding antibonding orbital is

$$\psi_{3\sigma^\circ} = N_+(2p_{zA} + 2p_{zB}) \tag{4.7}$$

Another bonding molecular orbital is formed by the "plus" combination of the $2p_x$ orbitals, which are perpendicular to the bond axis (Figure 4.6). This molecular orbital has a nodal plane when viewed down the bond axis. When viewed axially, this orbital appears to be identical to an atomic $p$ orbital; it is the first *pi* orbital, $1\pi$. The "minus" combination of

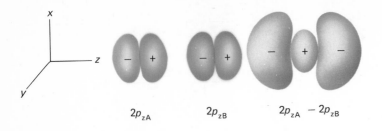

**FIGURE 4.5** Combination of $p$ orbitals into a sigma bonding orbital.

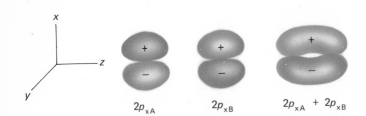

**FIGURE 4.6** Combination of $p$ orbitals into a pi bonding orbital.

these same atomic orbitals yields the $1\pi^*$ orbital. The molecular orbitals formed from the $2p_y$ atomic orbitals are identical in form and energy to these, so *the $1\pi$ and $1\pi^*$ orbitals are doubly degenerate.*

The energy order of the new molecular orbitals must be known before they can be used to predict configurations and properties of diatomic species. Because $2s$ orbitals always lie below the $2p$ orbitals, the $2\sigma$ and $2\sigma^*$ molecular orbitals would be expected to lie below those derived from the $2p$ atomic orbitals. This is consistent with experiment. The alignment of the $2p_z$ orbitals should be more favorable for bond formation, so the remaining orbitals should be in the order $3\sigma$, $1\pi$, $1\pi^*$, $3\sigma^*$. This is not invariably correct; for some diatomic molecules the order is $1\pi$, $3\sigma$, $1\pi^*$, $3\sigma^*$. A rationalization of the latter order of the $3\sigma$ and $1\pi$ orbitals requires hybridization of the atomic orbitals.

## b. s-p Hybridization.

Molecular orbitals are not always formed from a single atomic orbital on each atom. In fact, we have introduced one molecular orbital that "contains" two atomic orbitals from each atom, in the sense that the $2p_x$ orbitals are themselves hybrid orbitals formed from two hydrogenic orbitals†

$$\psi_{1\pi} = N_+(2p_{xA} + 2p_{xB}) = \frac{N_+}{2}(2p_{+1,A} + 2p_{-1,A} + 2p_{+1,B} + 2p_{-1,B}). \quad (4.8)$$

There are two ways of including several orbitals per atom. If we were performing quantitative calculations on a computer, we would use wavefunctions composed of sums of atomic orbitals. For a qualitative discussion it is simpler to group all the orbitals from one atom that are used in a particular molecular orbital into a single *hybrid orbital*. We then discuss the molecular orbital as a linear combination of the two hybrid orbitals. For example, the $1\pi$ molecular orbital (Equation (4.8)) is written as a combination of the hybrid $2p_x$ orbitals.

---

†The subscripts are values of the magnetic quantum number $m$.

The question then becomes, what hybrid orbitals are especially suitable for formation of bonding molecular orbitals? The $2s$ orbitals are poorly adapted to formation of chemical bonds because they are non-directional. Addition of a $p$ orbital to an $s$ orbital will increase the directionality of the orbital relative to an $s$ orbital. A set of $sp$-hybrid orbitals is shown in Figure 4.7. Notice that, as the fraction of $p$ in the hybrid orbital is increased, the orbital becomes more directional and more suitable for formation of sigma bonding orbitals. However, the energy of these hybrid orbitals is higher than the energy of a simple $2s$ orbital, because the $2p$ orbital that is added is of higher energy. The lowest energy molecular orbital will be

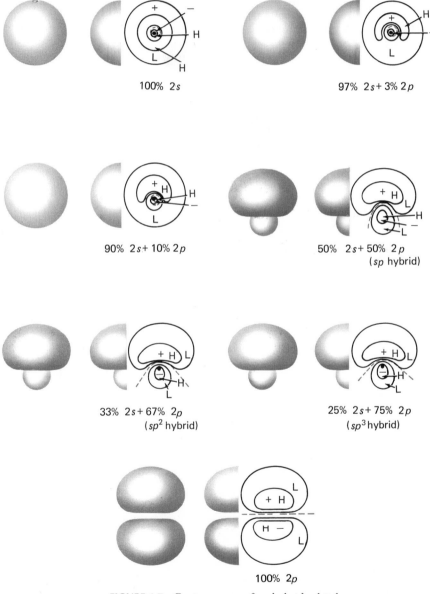

FIGURE 4.7   Contour maps of $sp$ hybrid orbitals.

formed from hybrid orbitals that strike an appropriate balance between the return in bonding energy from addition of $2p$ and the original cost in energy necessary to add this $2p$. The cost will depend on the energy separation between $2s$ and $2p$; this separation increases with increasing nuclear charge. Presumably, then, hybridization will be most important for lithium and least important for fluorine.

If we use two orbitals to form hybrid orbitals, we must form two hybrid orbitals. Each hybrid orbital shown has a partner. For example, the hybrid $90\% \ 2s + 10\% \ 2p$ has a partner hybrid $10\% \ 2s + 90\% \ 2p$.

*The true electron density in the ground state of a molecule is the electron density that gives the lowest energy.* The electrons know nothing of our mathematical manipulations. They automatically adopt the lowest-energy configuration; electrons do not "hybridize." We attempt to describe this lowest-energy configuration in terms of the simple atomic orbitals at our disposal. In the frequent cases where a description in terms of a single hydrogenic orbital on each nucleus is not close enough to reality to be useful, we still attempt to use only one orbital per nucleus, but to compensate for our rigidity we abandon the simple hydrogenic orbitals in favor of the more flexible hybrid orbitals. *Hybridization is a mathematical construction* rather than a real process.

What changes does hybridization cause in the correlation diagram? The only hybridization that is useful in a discussion of diatomic molecules is a mixture of the $2p_z$ with the $2s$, which makes the hybrid orbitals point along the bond axis. This hybridization is indicated in the regions labeled H in the correlation diagrams (Figure 4.8). Hybridization produces three

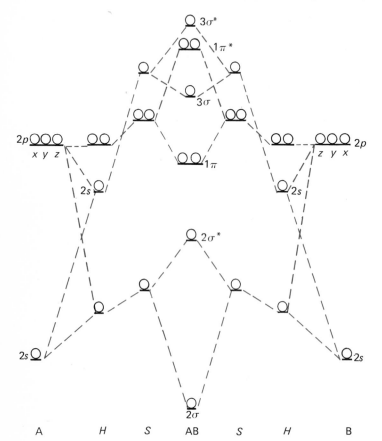

**FIGURE 4.8** Correlation diagram for second period diatomic molecules with hybridization.

changes in the correlation diagram. The energy of the $2s$ orbital is raised by adding $2p_z$, and the energy of the orbital which was originally $2p_z$ is lowered accordingly. The overlap causes a sharp increase in the energy of the higher hybrid orbital. Finally, the $2\sigma$ and $2\sigma^*$ orbitals are much farther apart in energy when formed from hybrid orbitals (stronger bonds), but the $3\sigma$ and $3\sigma^*$ orbitals are much closer in energy. The net result of these effects is that the energy order is $1\pi, 3\sigma, 1\pi^*, 3\sigma^*$ when $s$-$p$ hybridization is invoked, whereas the order is $3\sigma, 1\pi, 1\pi^*, 3\sigma^*$ without $s$-$p$ hybridization.

## c. Bonding in Diatomic Molecules of Second-Period Elements.

**Lithium.**    The lithium atom has three electrons. Therefore, the diatomic molecule $Li_2$ is predicted to have the six-electron configuration $1\sigma^2, 1\sigma^{*2}, 2\sigma^2$ with a net of two bonding electrons. As predicted, the molecule is stable; it has a bond length of 2.67 Å and a bond energy of about 25 kcal/mole. Both of these values are consistent with the large average radii of $n = 2$ atomic orbitals in Li. Contour maps of the orbitals and the total electron density for $Li_2$ are shown in Figure 4.9. There is little difference between the $1\sigma$, the $1\sigma^*$, and the two $1s$ atomic orbitals. This demonstrates the validity of the usual assumption that *only the electrons in the valence shell need be included in qualitative treatments of bonding.*

**Beryllium.**    The beryllium atom has four electrons, so the diatomic molecule $Be_2$ should have the configuration $1\sigma^2, 1\sigma^{*2}, 2\sigma^2, 2\sigma^{*2}$, and no net bonding electrons. As predicted, $Be_2$ is not a stable molecule.

**Boron.**    The boron diatomic molecule, $B_2$, has the predicted configuration $1\sigma^2, 1\sigma^{*2}, 2\sigma^2, 2\sigma^{*2}, 1\pi^2$. Thus, we expect (and find) a stable molecule. The bond length is 1.59 Å, and the bond energy is 69 kcal/mole. Remember that there are two $1\pi$ orbitals, and that by Hund's rule the two electrons are assigned to different orbitals with their spins parallel.* Therefore, we predict (and find experimentally) two unpaired electrons. The contour maps for $B_2$ valence orbitals are shown in Figure 4.10. Note especially the overall contraction of the orbitals compared to $Li_2$. This is due to the increased nuclear charge, which also accounts for the increase in bond energy.

**Carbon.**    The $C_2$ molecule has the twelve-electron configuration $1\sigma^2, 1\sigma^{*2}, 2\sigma^2, 2\sigma^{*2}, 1\pi^4$. The molecule has four net bonding electrons (a double bond in the Lewis picture) and no unpaired electrons. The contour maps are shown in Figure 4.11.

**Nitrogen.**    Here the diatomic molecule has configuration $1\sigma^2, 1\sigma^{*2}, 2\sigma^2, 2\sigma^{*2}, 1\pi^4, 3\sigma^2$ with six net bonding electrons, or a triple bond. The large excess of bonding electrons gives a short (1.10 Å), strong (225 kcal/mole) bond. The contour maps are shown in Figure 4.12. Diatomic nitrogen, $N_2$, is the last molecule in which the experimental results require us to invoke $s$-$p$ hybridization. The $2s$-$2p$ energy separation increases with increasing nuclear charge until the initial energy investment for effective hybridization is too high, and there is little advantage in raising the energy

---

*Would we predict $B_2$ to be paramagnetic in the absence of $s$-$p$ hybridization?

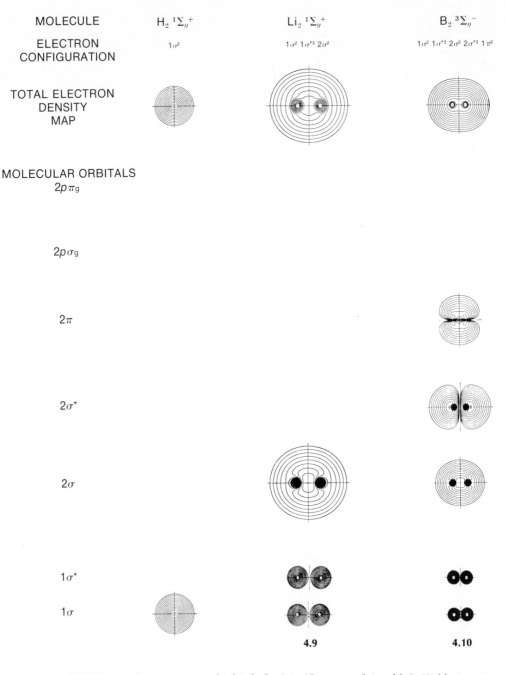

**FIGURE 4.9**  Contour maps of orbitals for $Li_2$. (Courtesy of Arnold C. Wahl, Argonne National Laboratory.)

**FIGURE 4.10**  Contour maps of valence orbitals for $B_2$. (Courtesy of Arnold C. Wahl, Argonne National Laboratory.)

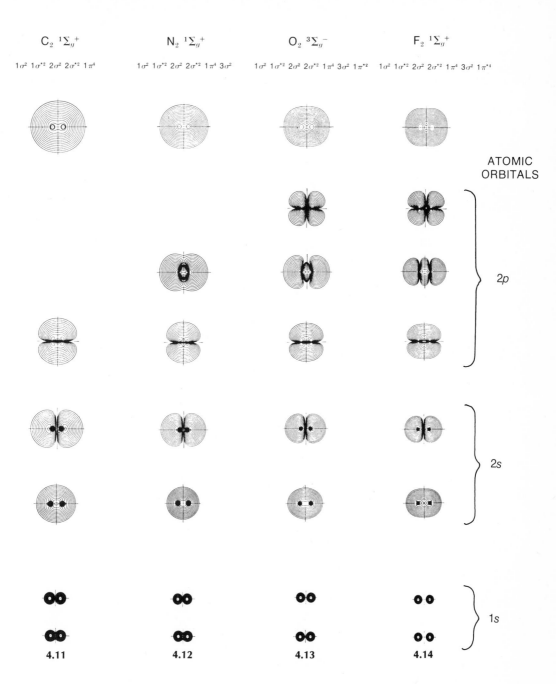

$C_2$ $^1\Sigma_g^+$

$1\sigma^2 1\sigma^{*2} 2\sigma^2 2\sigma^{*2} 1\pi^4$

$N_2$ $^1\Sigma_g^+$

$1\sigma^2 1\sigma^{*2} 2\sigma^2 2\sigma^{*2} 1\pi^4 3\sigma^2$

$O_2$ $^3\Sigma_g^-$

$1\sigma^2 1\sigma^{*2} 2\sigma^2 2\sigma^{*2} 1\pi^4 3\sigma^2 1\pi^{*2}$

$F_2$ $^1\Sigma_g^+$

$1\sigma^2 1\sigma^{*2} 2\sigma^2 2\sigma^{*2} 1\pi^4 3\sigma^2 1\pi^{*4}$

ATOMIC
ORBITALS

$2p$

$2s$

$1s$

4.11　　　4.12　　　4.13　　　4.14

**FIGURE 4.11** Contour maps of valence orbitals for $C_2$. (Courtesy of Arnold C. Wahl, Argonne National Laboratory.)

**FIGURE 4.12** Contour maps of valence orbitals for $N_2$. (Courtesy of Arnold C. Wahl, Argonne National Laboratory.)

**FIGURE 4.13** Contour maps of valence orbitals for $O_2$. (Courtesy of Arnold C. Wahl, Argonne National Laboratory.)

**FIGURE 4.14** Contour maps of orbitals for $F_2$. (Courtesy of Arnold C. Wahl, Argonne National Laboratory.)

of the $3\sigma$ to lower the energy of the $2\sigma$ *if both are going to contain electrons*. By experiment, we know that beyond $N_2$ the $3\sigma$ orbital lies below the $1\pi$ orbital.

**Oxygen.** The inability to explain the paramagnetism of the $O_2$ molecule is a major defect in the Lewis theory. The molecular orbital configuration of $O_2$ is $1\sigma^2, 1\sigma^{*2}, 2\sigma^2, 2\sigma^{*2}, 3\sigma^2, 1\pi^4, 1\pi^{*2}$. Thus, $O_2$ has four net bonding electrons, or a double bond, as in the simple theory. However, by Hund's rule the last two electrons are unpaired in the $1\pi^*$ orbitals, thus explaining the observed paramagnetism.

Photoelectron spectra are obtained from an experiment similar to the photoelectric effect experiment, but using a gas sample instead of a metal one.

The $O_2$ contour maps are shown in Figure 4.13. The photoelectron spectrum of $O_2$ is shown in Figure 4.15. This spectrum shows four groups of sharp lines. Each group of lines can be interpreted as arising from the ionization of an electron from a different molecular orbital in the $O_2$ molecule. This is a beautiful confirmation of the molecular-orbital approach. The sharp lines in the spectrum arise from transitions to different vibrational levels of $O_2^+$ ions.

**Fluorine.** The diatomic molecule $F_2$ has the configuration $1\sigma^2, 1\sigma^{*2}, 2\sigma^2, 2\sigma^{*2}, 3\sigma^2, 1\pi^4, 1\pi^{*4}$ with two net bonding electrons, a single bond. Examination of the contour maps (Figure 4.14) shows that all the previously mentioned trends are maintained. Note especially that the $1\sigma$ and $1\sigma^*$ orbitals are almost indistinguishable. Thus, we often write the configuration $1s^2, 1s^2, 2\sigma^2, 2\sigma^{*2}, 3\sigma^2, 1\pi^4, 1\pi^{*4}$. Similar notation is used for the other diatomic molecules.

**Neon.** The neon diatomic molecule would have configuration $1s^2, 1s^2, 2\sigma^2, 2\sigma^{*2}, 3\sigma^2, 1\pi^4, 1\pi^{*4}, 3\sigma^{*2}$ and hence no net bonding electrons. The molecule is unknown.

The properties of the diatomic molecules formed by second-period elements are summarized in Table 4.2. You should note the correlations

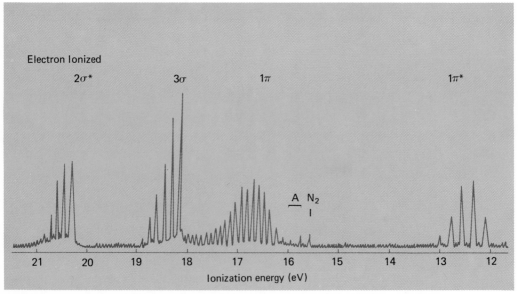

**FIGURE 4.15** Photoelectron spectrum of $O_2$. The peaks marked A and $N_2$ are due to impurities.

**TABLE 4.2**

DIATOMIC MOLECULES OF SECOND-PERIOD ELEMENTS

| SPECIES | ELECTRON CONFIG- URATION | NET BONDING ELECTRONS | BOND LENGTH (Å) | BOND ENERGY (kcal · mol$^{-1}$) | MAGNETIC PROPERTY |
|---------|-----------------|-----------------|-----------------|-----------------|-----------------|
| Li$_2$ | $2\sigma^2$ | 2 | 2.67 | 25 | diamagnetic |
| Be$_2$ | $2\sigma^2, 2\sigma^{\circ2}$ | 0 | — | — | — |
| B$_2$ | $2\sigma^2, 2\sigma^{\circ2}, 1\pi^2$ | 2 | 1.59 | 69 | paramagnetic |
| C$_2$ | $2\sigma^2, 2\sigma^{\circ2}, 1\pi^4$ | 4 | 1.31 | 150 | diamagnetic |
| N$_2$ | $2\sigma^2, 2\sigma^{\circ2}, 1\pi^4, 3\sigma^2$ | 6 | 1.10 | 225 | diamagnetic |
| O$_2$ | $2\sigma^2, 2\sigma^{\circ2}, 1\pi^4,$ $3\sigma^2, 1\pi^{\circ2}$ | 4 | 1.21 | 118 | paramagnetic |
| F$_2$ | $2\sigma^2, 2\sigma^{\circ2}, 3\sigma^2,$ $1\pi^4, 1\pi^{\circ4}$ | 2 | 1.42 | 36 | diamagnetic |
| Ne$_2$ | $2\sigma^2, 2\sigma^{\circ2}, 3\sigma^2,$ $1\pi^4, 1\pi^{\circ4}, 3\sigma^{\circ2}$ | 0 | — | — | — |

1 Å = 0.1 nm; 1 kcal = 4.184 kJ.

between numbers of net bonding electrons and molecular properties. Finally, you should look again at the contour maps for the total electron density of these molecules. It is convenient to represent molecules by balls (representing nuclei and inner-shell electrons) connected by springs (representing bonds). The limitations of these representations should be very clear from the contour maps. Molecules are not balls connected by springs; they are better thought of as *blobs of electron density* (with imbedded nuclei) whose dimensions are large compared with the internuclear separation.

## d. Diatomic Molecules with Electrons in Orbitals with n = 3.

The configurations for heavier homonuclear diatomic molecules are analogous to those of the second-period diatomic molecules. The *d* orbitals contribute relatively little to the bonding in diatomic molecules. The orbital level diagram for third-period diatomic molecules is the same as that shown in Figure 4.8 except for numbering. The bond parameters of some additional homonuclear diatomic molecules are given in Table 4.3.

## 4.3 HETERONUCLEAR DIATOMIC MOLECULES.

The simplest molecules that contain different types of atoms are the heteronuclear diatomic molecules. In previous examples, the symmetry of the molecule restricted the form of the molecular orbital because the electron density had to be as symmetric as the arrangement of the nuclei. For example, the molecular orbital

$$2s_A + 2p_{zB}$$

is not suitable for a homonuclear diatomic molecule because the two ends of the molecule would be different. This restriction is no longer operative in a heteronuclear molecule, because the two nuclei are different.

TABLE 4.3

SOME HOMONUCLEAR DIATOMIC MOLECULES AND IONS

| MOLECULE | BOND LENGTH (Å) | DISSOCIATION ENERGY (kcal · mol$^{-1}$) | MOLECULE | BOND LENGTH (Å) | DISSOCIATION ENERGY (kcal · mol$^{-1}$) |
|---|---|---|---|---|---|
| $Ag_2$ | — | 39 | $Na_2$ | 3.08 | 17.3 |
| $Br_2$ | 2.28 | 46 | $O_2^+$ | 1.12 | — |
| $Cl_2$ | 1.99 | 58 | $O_2^-$ | 1.26 | — |
| $Cl_2^+$ | 1.89 | — | $O_2^{2-}$ | 1.49 | — |
| $Cd_2$ | — | 2.0 | $P_2$ | 1.89 | 117 |
| $Cu_2$ | — | 47 | $S_2$ | 1.89 | 101 |
| $Ge_2$ | — | 65 | $Sb_2$ | — | 70 |
| $Hg_2$ | 3.3 | 1.4 | $Se_2$ | 2.15 | 65 |
| $I_2$ | 2.66 | 36 | $Si_2$ | 2.25 | 81 |
| $K_2$ | 3.92 | 12.2 | $Te_2$ | 2.59 | 53 |
| $N_2^+$ | 1.12 | — | | | |

1 Å = 0.1 nm; 1 kcal = 4.184 kJ.

## a. Electronegativity.

Different nuclei attract electrons differently. In a molecule the situation is complex because a nucleus is not attracting a free electron; it is attempting to pull an electron away from a second nucleus while attempting to hold its own electrons against the pull of the second nucleus.

*Electronegativity* is the attraction for an electron exerted by a nucleus *within a molecule.* There are several electronegativity scales. The Mulliken scale defines the electronegativity of an element as the average of the electron affinity and the first ionization energy (in electron volts). Because of a lack of accurately measured electron affinities, the Pauling electronegativity scale, which is based on bond energies, is usually employed. On the Pauling scale the electronegativity difference, $X_A - X_B$, for two elements A and B is *defined* as

$$|X_A - X_B| = 0.208 \, [D_{AB} - (D_{AA}D_{BB})^{1/2}] \qquad (4.9)$$

where $D_{AB}$, $D_{AA}$, and $D_{BB}$ are the dissociation energies of the respective diatomic molecules in kcal/mole.* Unequal competition for electron density produces a molecule with a positive end and a negative end (a dipole). In general, $D_{AB}$ will be larger than the mean of the dissociation energies for the two homonuclear molecules because of the attractive energy between the two oppositely charged ends.

It has been possible to generate a scale of electronegativity for almost all of the elements. These are shown in Table 4.4, which is arranged so as to stress the periodic relationships in electronegativities. Remember that, from the Pauling definition, only *differences* in electronegativities are meaningful. It is significant that the electronegativity difference between fluorine and hydrogen is 1.9, but it is irrelevant that the electronegativity of fluorine is twice that of hydrogen.

---

*1 eV = 23.06 kcal/mole; $(1/23.06)^{1/2} = 0.208$.

**TABLE 4.4**

PAULING ELECTRONEGATIVITY SCALE

| 1 H 2.1 | | | | | | | | | | | | | | | | | 2 He - |
|---|---|---|---|---|---|---|---|---|---|---|---|---|---|---|---|---|---|
| 3 Li 1.0 | 4 Be 1.5 | | | | | | | | | | | 5 B 1.9 | 6 C 2.5 | 7 N 3.0 | 8 O 3.5 | 9 F 4.0 | 10 N - |
| 11 Na 0.9 | 12 Mg 1.2 | | | | | | | | | | | 13 Al 1.5 | 14 Si 1.8 | 15 P 2.1 | 16 S 2.5 | 17 Cl 3.0 | 18 Ar - |
| 19 K 0.8 | 20 Ca 1.0 | 21 Sc 1.3 | 22 Ti 1.5 | 23 V 1.6 | 24 Cr 1.6 | 25 Mn 1.5 | 26 Fe 1.8 | 27 Co 1.8 | 28 Ni 1.8 | 29 Cu 1.9 | 30 Zn 1.5 | 31 Ga 1.6 | 32 Ge 1.8 | 33 As 2.0 | 34 Se 2.4 | 35 Br 2.8 | 36 Kr - |
| 37 Rb 0.8 | 38 Sr 1.0 | 39 Y 1.2 | 40 Zr 1.4 | 41 Nb 1.6 | 42 Mo 1.8 | 43 Tc 1.9 | 44 Ru 2.2 | 45 Rh 2.2 | 46 Pd 2.2 | 47 Ag 1.7 | 48 Cd 1.4 | 49 In 1.7 | 50 Sn 1.8 | 51 Sb 1.9 | 52 Te 2.1 | 53 I 2.5 | 54 Xe - |
| 55 Cs 0.7 | 56 Ba 0.9 | 57-71 - 1.1-1.2 | 72 Hf 1.3 | 73 Ta 1.5 | 74 W 1.7 | 75 Re 1.9 | 76 Os 2.2 | 77 Ir 2.2 | 78 Pt 2.2 | 79 Au 2.4 | 80 Hg 1.9 | 81 Tl 1.8 | 82 Pb 1.8 | 83 Bi 1.8 | 84 Po 2.0 | 85 At 2.2 | 86 Rn - |
| 87 Fr 0.7 | 88 Ra 0.9 | 89 Ac 1.1 | 90 Th 1.3 | 91 Pa 1.5 | 92 U 1.7 | 93-103 Np-Lw 1.3 | | | | | | | | | | | |

## b. Orbital Diagrams for Heteronuclear Diatomic Molecules.

Crude molecular orbitals for heteronuclear diatomic molecules can be formed from a linear combination of those atomic orbitals that lie closest in energy. The relative energies of the valence orbitals are indicated roughly by the electronegativities; the valence orbitals of the more electronegative atom will lie lower in energy. In the carbon monoxide molecule, oxygen is the more electronegative atom. The correlation diagram for CO is shown in Figure 4.16.

In Figure 4.16 we see that CO has the configuration $1s_C^2$, $1s_O^2$, $2\sigma^2$, $2\sigma^{*2}$, $1\pi^4$, $3\sigma^2$. This is *formally* the same as the $N_2$ molecule, but the orbitals themselves are different. The electron density of each bonding orbital is shifted toward the more electronegative oxygen. Compare the contour maps in Figure 4.17 with those for $N_2$ in Figure 4.12. The orbital diagram in Figure 4.16 can be used for all diatomic molecules formed from two atoms of the same period. The bond parameters for many such molecules can be found in Table 4.5.

A different orbital diagram is needed for diatomic molecules formed from atoms of different periods. As illustrations we consider two molecules, hydrogen fluoride (HF) and lithium hydride (LiH). The hydrogen 1s

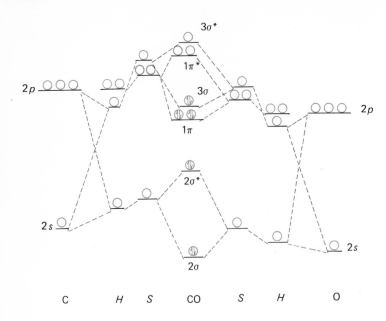

orbital is of higher energy than either the $2s$ or $2p$ fluorine orbitals. Molecular orbitals are formed from the $1s$ orbital on hydrogen and the closest (in energy) orbital on fluorine, the $2p$ (Figure 4.18). Of the three $2p$ orbitals, only one will combine effectively with a $1s_H$ orbital to form bonding and antibonding molecular orbitals.

Bonding orbitals have electron densities that change less rapidly between the nuclei than the electron densities of the parent atomic orbitals. Electron densities are proportional to the square of the wavefunctions:

$$(1s_H \pm 2p_F)^2 = (1s_H)^2 \pm 2(1s_H \cdot 2p_F) + (2p_F)^2$$

The first and last terms on the right hand side are the electron densities of a $1s$ orbital on a hydrogen atom and a $2p$ orbital on a fluorine atom, respectively. The electron density in the molecular orbital will deviate from that of two independent atomic orbitals only when the cross term $(1s_H \cdot 2p_F)$ is

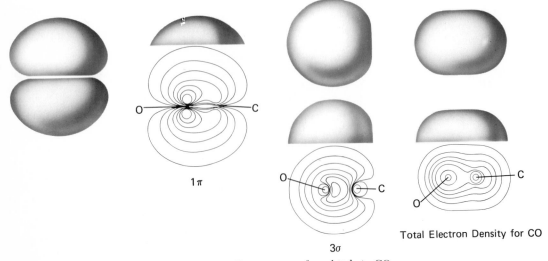

Total Electron Density for CO

**FIGURE 4.17** Contour maps for orbitals in CO.

**TABLE 4.5**

HETERONUCLEAR DIATOMIC MOLECULES AND IONS

| Molecule | Bond Length (Å) | Dissociation Energy (kcal · mol$^{-1}$) | Molecule | Bond Length (Å) | Dissociation Energy (kcal · mol$^{-1}$) |
|---|---|---|---|---|---|
| AlO | 1.62 | 85 | CO | 1.13 | 256 |
| BBr | 1.89 | 95 | CO$^+$ | 1.12 | 147 |
| BCl | 1.72 | 97 | CS | 1.53 | 180 |
| BeF | 1.36 | 122 | HCl | 1.23 | 102 |
| BeH | 1.34 | 51 | HCl$^+$ | 1.33 | 103 |
| BeH$^+$ | 1.31 | 74 | HF | 0.92 | 134 |
| BeO | 1.33 | 85 | HS | 1.35 | 80 |
| BF | 1.26 | 99 | KH | 2.24 | 43 |
| BN | 1.28 | 115 | LiH | 1.60 | 58 |
| BO | 1.20 | 210 | NH | 1.04 | 88 |
| CH | 1.12 | 80 | NO | 1.15 | 162 |
| CH$^+$ | 1.13 | 83 | NO$^+$ | 1.06 | 244 |
| ClF | 1.63 | 60 | OH | 0.97 | 100 |
| CN | 1.17 | 188 | OH$^+$ | 1.03 | 101 |

1 Å = 0.1 nm; 1 kcal = 4.184 kJ.

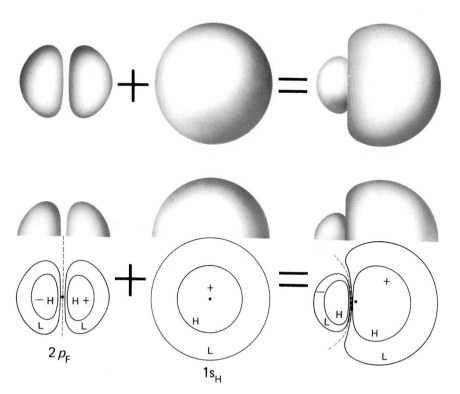

**FIGURE 4.18**  Combination of a $1s$ and a $2p$ orbital for bonding in HF.

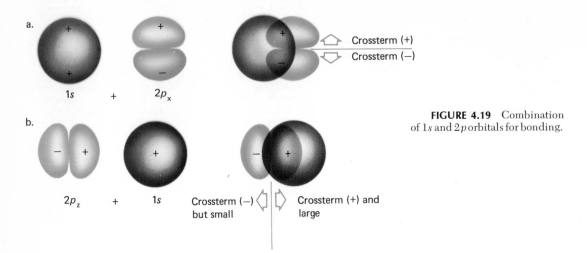

a.

$1s$   $+$   $2p_x$

Crossterm $(+)$
Crossterm $(-)$

b.

$2p_z$   $+$   $1s$   Crossterm $(-)$
but small

Crossterm $(+)$ and
large

**FIGURE 4.19**  Combination
of $1s$ and $2p$ orbitals for bonding.

non-zero. Moreover, it is not sufficient that this term be non-zero at some
point; rather, the net value from all space must be non-zero. There are cases
where the cross term is positive at some points, and negative at others. If
the positive and negative contributions exactly cancel, the resulting orbital
is neither bonding or antibonding. Molecular orbitals of this type are called
*non-bonding* orbitals, and electrons in *non-bonding orbitals* are called
*non-bonding electrons*. Consider, for example, the combination of a $2p$
orbital with a $1s$ orbital. If the $2p$ orbital axis is perpendicular to the bond
axis (a $2p_x$ or $2p_y$ orbital), then the cross term is positive above the axis;
below the axis the cross term has exactly the same magnitude, but it is
negative. The net result is zero, and the molecular orbitals constructed
from these two atomic orbitals are non-bonding orbitals. If the $2p$ orbital
and bond axes coincide, we again have a region where the cross term is
positive and a region where it is negative. However, the net result is not
zero, because the $1s$ wavefunction is large in the positive region but small
in the negative region. Therefore, in HF a bonding molecular orbital can
only be formed from the $2p_{z\mathrm{F}}$ orbital (Figure 4.19).

The correlation diagram for HF is simple (Figure 4.20). Hybridization
is not important for the fluorine atom. The configuration is $1s_\mathrm{F}^2$, $2s_\mathrm{F}^2$, $\sigma^2$,
$2p_{x\mathrm{F}}^2$, $2p_{y\mathrm{F}}^2$. The molecule has a single bond and three pairs of non-bonding

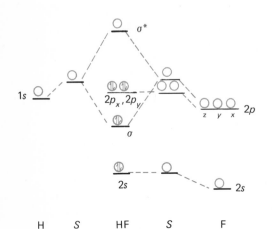

$\sigma^*$

$1s$

$2p_x, 2p_y$

$\sigma$

$2p$
$z \quad y \quad x$

$2s$

$2s$

H     $S$     HF     $S$     F

**FIGURE 4.20**  Correlation diagram for HF; the $1s$
orbital on fluorine is not shown. The molecular orbital
configuration is $1s_\mathrm{F}^2\ 2s_\mathrm{F}^2\ \sigma^2\ 2p_{x\mathrm{F}}^2\ 2p_{y\mathrm{F}}^2$.

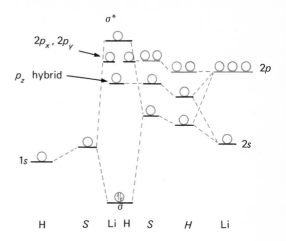

**FIGURE 4.21** Correlation diagram for LiH. The molecular orbital configuration for LiH is $1s^2_{Li}\,\sigma^2$. (The $1s$ orbital on Li is not shown.)

electrons (lone pairs) on the fluorine atom. Inner-shell electrons are not included in the list of non-bonding electrons. The correspondence of the configuration with the Lewis diagram for HF is obvious and exact.

It would be more realistic to write the bonding molecular orbital as $1s_H + \lambda 2p_F$, where $\lambda$ is a parameter that in HF is greater than unity. This parameter has the effect of producing a slight shift of the electron density toward the more electronegative fluorine atom, consistent with the known dipole moment of the molecule. Such a bond is said to possess *partial ionic character,* because the bonding electrons are not shared equally.

Another example of a heteronuclear diatomic molecule is lithium hydride (LiH). Here the atomic $1s$ orbital of the hydrogen atom lies below the $2s$ lithium orbital, but far above the $1s$ orbital in Li. However, in lithium the $2s$ and $2p$ orbitals are so close in energy that hybridization is important. Therefore, the correlation diagram is constructed as shown in Figure 4.21. The configuration of LiH is $1s^2_{Li},\,\sigma^2$.

The configurations for other heteronuclear diatomic molecules can be derived in a similar fashion. Valence-shell atomic orbitals of similar energy are combined to form molecular orbitals, if the symmetries are such that bonding and antibonding orbitals result. Introduction of hybrid orbitals avoids questions as to whether $s$ orbitals or $p$ orbitals should be used when the valence-shell orbital of one atom has energy between the $s$ and $p$ energies of its partner.

## 4.4 MOLECULAR ORBITALS IN POLYATOMIC MOLECULES.

The distinguishing feature of the molecular-orbital approach is the formation of orbitals that extend over the whole molecule.

Beryllium hydride ($BeH_2$) is a linear triatomic molecule. From the electronegativities we know that the $1s$ orbital of hydrogen lies lower in energy than the valence-shell orbitals of beryllium. Bonding and antibonding molecular orbitals can be formed from the two $1s$ hydrogen atomic orbitals together with the $2s$ and $2p_z$ beryllium atomic orbitals. The $2p_x$ and $2p_y$ beryllium atomic orbitals have the wrong symmetry to combine with $1s$ orbitals in a linear molecule (see Figure 4.18). The four atomic orbitals are combined to form the four molecular orbitals shown in Figure 4.22. The sigma bond formed from the $2p_z$ orbital is unusually strong because of the linear molecular geometry. These considerations lead to the correlation diagram given in Figure 4.23, and the configuration $1s^2_{Be}$,

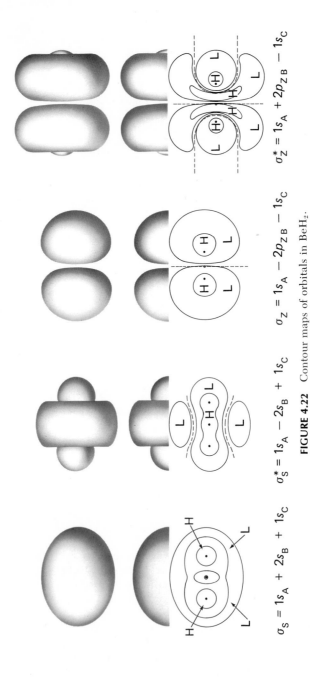

**FIGURE 4.22**  Contour maps of orbitals in $BeH_2$.

$\sigma_S = 1s_A + 2s_B + 1s_C$

$\sigma_S^* = 1s_A - 2s_B + 1s_C$

$\sigma_Z = 1s_A - 2p_{ZB} - 1s_C$

$\sigma_Z^* = 1s_A + 2p_{ZB} - 1s_C$

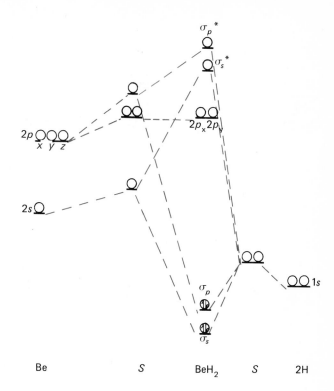

**FIGURE 4.23** Correlation diagram for BeH$_2$. The 1s orbital on Be is not shown. The molecular-orbital electron configuration is $1s^2_{Be}\,\sigma_s^2\,\sigma_p^2$.

$2\sigma_s^2$, $2\sigma_p^2$. There are four bonding electrons spread throughout the molecule; each hydrogen atom is bound to the central atom by a single bond. Notice that in the Lewis formalism this molecule would have to be ionic Be$^{+2}$(H$^-$)$_2$ or violate the octet rule. The ionic diagram is in sharp contrast to the electron densities of the molecular orbitals.

Water, HOH, is a nonlinear triatomic molecule. The atomic valence orbitals for water are the same as for BeH$_2$, but the difference in geometry requires serious alteration of the molecular orbitals. The sketches in Figure 4.24 show that the 2s and two of the three 2p oxygen orbitals could be used to form bonding molecular orbitals, but further consideration shows that the best choices are the $2p_x$ and an equal hybrid of the 2s and $2p_z$ oxygen orbitals (called an sp orbital). This choice produces the molecular orbitals in Figure 4.25 and the correlation diagram in Figure 4.26. The electron configuration of water is $1s_O^2$, $\sigma_s^2$, $\sigma_x^2$. $2sp^2_{zO}$, $2p^2_{yO}$. Water has four bonding electrons or two single bonds, and two lone pairs of electrons. This corresponds to the Lewis diagram.

The generation of molecular orbitals for larger molecules is a problem of ever increasing complexity. Not only does it become more difficult to visualize the molecular orbitals, but the increasing number of available orbitals makes *qualitative* predictions of their energy order impossible. For these more complex molecules we must adopt a different approach.

## 4.5 BOND ORBITALS.

The valence-bond approach is an attempt to recast the Lewis concept of the electron-pair bond in the language of quantum mechanics. In this

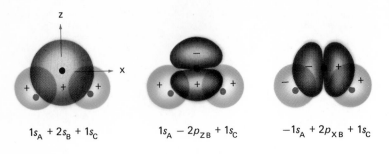

$1s_A + 2s_B + 1s_C$          $1s_A - 2p_{ZB} + 1s_C$          $-1s_A + 2p_{XB} + 1s_C$

(a)

**FIGURE 4.24** Possible molecular orbitals in $H_2O$ (*a*) using *p* orbitals, (*b*) using *sp* hybrid orbitals.

Bonding effective with this hybrid

No effective bonding with this hybrid

(b)

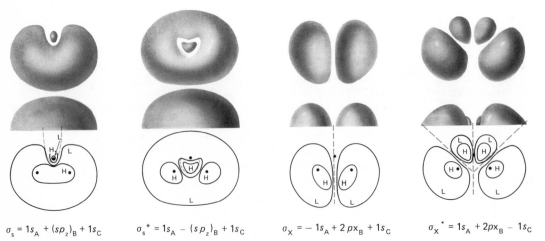

$\sigma_s = 1s_A + (sp_z)_B + 1s_C$     $\sigma_s{}^* = 1s_A - (sp_z)_B + 1s_C$     $\sigma_X = -1s_A + 2px_B + 1s_C$     $\sigma_X{}^* = 1s_A + 2px_B - 1s_C$

**FIGURE 4.25** Contour maps for orbitals in $H_2O$. Two orbitals include *sp* hybrid orbitals on oxygen. The hydrogen atoms are designated A and C, and the oxygen atom is designated B.

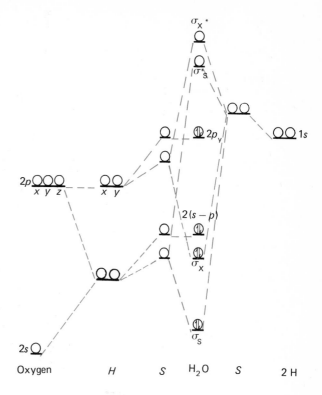

**FIGURE 4.26** Correlation diagram for $H_2O$. In the hybridization region $H$ on the left we have formed two $sp$ hybrids on oxygen using the $2s_0$ and $2p_{z0}$ orbitals. One of these hybrids is used to form $\sigma_s$ (with the $1s$ orbitals on hydrogen) and the other becomes an $sp$ non-bonding orbital, designated $2(s-p)$ in the diagram.

approach the orbitals are not delocalized throughout the molecule; they are restricted to a region about two nuclei which consequently are bonded together. We shall not use the valence-bond method; instead we shall discuss a method that is similar to the valence-bond method but which is more amenable to qualitative discussion, namely, the *localized molecular-orbital method*.

There is no difference between the conventional and the localized molecular orbital methods for diatomic molecules. In polyatomic molecules the restriction to two nuclei imposes additional demands on the directional properties of the component atomic orbitals. For example, in water the $sp_z$ hybrid orbital of oxygen, which was suitable in the conventional molecular-orbital description because it formed a bonding orbital which included both hydrogen atoms, is not a wise choice for a localized orbital which forms a bond between the oxygen atom and a single hydrogen atom.

Atomic orbitals with suitable directional properties for localized bond orbitals usually can be constructed by hybridization. Again we emphasize that hybrid orbitals are "after the fact." The geometry of the hybrid orbitals does not determine the molecular geometry; rather, hybrid orbitals are constructed so as to approximate roughly the molecular geometry, which is fixed by the requirement of minimum energy.

Let us first consider those hybrid orbitals that can be made from a single $s$ orbital and one, two, or three $p$ orbitals. The $s$ atomic orbital can be combined with a single $p$ orbital to form two $sp$-hybrid orbitals, with two $p$ orbitals to form three $sp^2$ hybrid orbitals, or with three $p$ orbitals to

TABLE 4.6

PROPERTIES OF HYBRID ORBITALS

| HYBRID | NUMBER OF ORBITALS | BOND ANGLE | MOLECULAR GEOMETRY |
|---|---|---|---|
| $sp$ | 2 | 180° | linear |
| $sp^2$ | 3 | 120° | planar |
| $sp^3$ | 4 | 109°28′ | tetrahedral |
| $dsp^2$ | 4 | 90° | planar |
| $d^2sp^3$ | 6 | 90° | octahedral |

Consult reference 2.8 for maps that show the subtle differences more clearly.

form four $sp^3$ hybrid orbitals.[*] The differences in electron densities between these three types are rather subtle, as can be seen in Figure 4.7. The important difference is in the geometry. The major axes of two $sp$ hybrids are 180° apart on a line perpendicular to the other two unused $p$ orbitals. The three $sp^2$ hybrids lie in a plane that is perpendicular to the remaining unused $p$ orbital; their axes are 120° apart. The four $sp^3$ hybrid orbitals point toward the corners of a regular tetrahedron; the angle between the axes of any two $sp^3$ hybrid orbitals is 109°28′. The three types of orbitals are suitable for use in linear, planar, and tetrahedral molecules, respectively (Table 4.6). The angular relations between the various orbitals are diagrammed in Figure 4.27 by showing the axes of the various orbitals.[†]

The $sp$, $sp^2$, and $sp^3$ cases in no sense exhaust the catalog of possible molecular geometries. However, the majority of molecules which do not involve $d$ orbitals come rather close to one of these three. Bond angles are fixed by electron-electron repulsions; the minimum energy will occur in that geometry in which the distance between the different orbitals is the greatest. At this point you should use a sphere and several toothpicks to show that the $sp$ arrangement produces the maximum possible separation of two orbitals, the $sp^2$ arrangement produces the maximum possible separation of three orbitals, and the $sp^3$ arrangement produces the maximum possible separation of four orbitals. Maximum separation is the reason for the ubiquity of these three molecular geometries.

Additional hybrid orbitals can be formed when there are $d$ orbitals in the valence shell. Six $d^2sp^3$ hybrid orbitals point toward the corners of an octahedron; this geometry gives maximum separation of the six orbitals. There are a few elements that form square-planar compounds. The bonding in square-planar compounds can be described using four $dsp^2$ orbitals which point towards the corners of a square. Obviously, $dsp^2$ orbitals do not give maximum separation of the orbitals; square-planar compounds will be discussed in Chapter 19.

---

[*]The convention of indicating the number of orbitals of a type (e.g., $p$) by a superscript is unfortunate, in that a superscript might be confused with an occupation number. In practice there is seldom ambiguity.

[†]The authors believe that accurate electron density maps for multiple orbitals on a single center are not useful, and that the conventional diagrams using long-lobed orbitals can lead to serious misconceptions about electron density distributions in molecules.

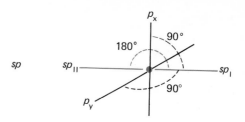

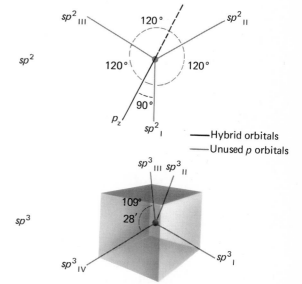

FIGURE 4.27   Geometries of hybrid orbitals.

Only the axes of the orbitals are shown for clarity

## a. Bond-Orbital Representations of Simple Molecules.

There are two atoms attached to the central atom in beryllium hydride. The maximum separation of the electrons in the two bond orbitals will be obtained by using $sp$ hybrid orbitals at 180°. The absence of a dipole moment indicates that the $BeH_2$ molecule is linear (i.e., bond angle 180°). If we adopt the notation $H_A - Be - H_B$, then the two localized molecular orbitals (unnormalized) are

$$\psi_I = 1s_A + sp_{IBe}$$

$$\psi_{II} = sp_{IIBe} + 1s_B$$

An orbital similar to one of these orbitals is shown in Figure 4.28. The sum of electron densities from these two orbitals is very similar to the sum of $\sigma_s$ and $\sigma_p$ in the molecular orbital approach.

Methane ($CH_4$) has four bonds around a central atom, so $sp^3$ hybrid orbitals can be used to describe the bonding. The bonding orbitals are of the general form

$$\psi = 1s_H + sp_C^3$$

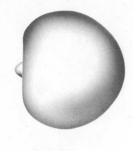

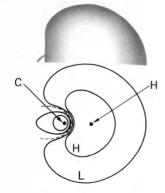

**FIGURE 4.28** Contour map of a bonding orbital between a $1s$ hydrogen orbital and an $sp$-hybrid carbon orbital.

and they are shown in Figure 4.28. All bond angles in $CH_4$ are exactly 109°28', the tetrahedral angle. The tetrahedron is formed by the planes connecting any three hydrogen atoms; the carbon atom sits at the center of the tetrahedron (Figure 4.29).

The next second-period hydrogen compound is ammonia, $NH_3$. Ammonia is known to be non-planar from a dipole-moment measurement. This result can be rationalized as follows. Comparison of the electron densities of a hybrid atomic orbital (Figure 4.7) and a bond orbital formed from a hybrid atomic orbital (Figure 4.28) shows that the two are almost identical close to the central nucleus. Therefore, the principle of maximum separation of orbitals should apply to all valence-shell electrons, whether

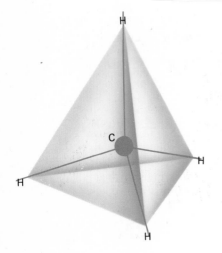

**FIGURE 4.29** Tetrahedral carbon. The carbon nucleus is at the center of the tetrahedron.

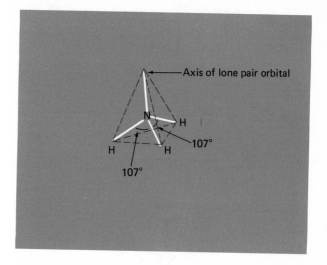

FIGURE 4.30 Bond angles in $NH_3$.

or not they participate in bonding. By this argument the ammonia molecule, where the central atom is surrounded by three bonding orbitals and a non-bonding orbital, should be treated as though it were surrounded by four roughly equivalent orbitals. Therefore, the geometry and bonding should be similar to that in methane, with $sp^3$ hybrid orbitals and $109°28'$ bond angles. This is almost correct; the molecule is pyramidal with $107°$ bond angles (Figure 4.30).

It is possible to rationalize the $2\frac{1}{2}°$ difference between the observed bond angle and the tetrahedral angle. The kinetic-energy reduction along the bond which allows the contraction of the orbital (Section 4.1b) does not apply to the non-bonding orbital. Therefore, *a non-bonded orbital is bigger than a bonding orbital,* and the bonding orbitals are squeezed together a bit (in $NH_3$ by $2\frac{1}{2}°$).

Unfortunately, it is not always possible to describe accurately the bonding in even simple molecules with a single type of hybrid orbital. An example of the problem is the series of molecules: water ($H_2O$); hydrogen sulfide ($H_2S$); and hydrogen selenide ($H_2Se$). All of these molecules have a central atom whose valence-shell configuration is $ns^2np^4$. We expect two bond orbitals and two lone pair orbitals in the molecules which, by analogy with the ammonia molecule, should be distributed at approximately the tetrahedral angle about the nucleus. Experimentally we find that this approach is not unreasonable for water, where the bond angle is $105°$. However, the logic should also apply to $H_2S$ and $H_2Se$, but these bond angles are $92°$ and $91°$, respectively. A $90°$ bond angle results from bond formation using two simple $p$ orbitals instead of hybrid orbitals. Which is the correct description for this series of molecules? *Neither one by itself* is correct; either we use both descriptions or we elaborate on the rationalization given earlier.

## b. Bond-Orbital Representations of Complex Molecules.

The localized molecular-orbital approach can be summarized. For molecules that involve only single bonds between atoms, the molecular geometry about an atom can be predicted by counting the number of orbitals that will be required to accommodate the bonding and non-

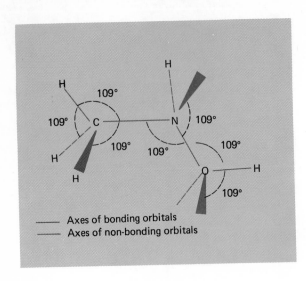

**FIGURE 4.31**   Molecular geometry predicted for $CH_3NHOH$.

_____ Axes of bonding orbitals
_____ Axes of non-bonding orbitals

bonding electrons. Those hybrid orbitals that maximize the separations of this number of orbitals are the appropriate ones.

As an example, consider the molecule

$$H-\overset{\overset{\displaystyle H}{|}}{\underset{\underset{\displaystyle H}{|}}{C}}-\overset{\overset{\displaystyle H}{|}}{\underset{\underset{\displaystyle ..}{N}}{}}-\overset{..}{\underset{..}{O}}-H$$

Carbon is surrounded by four bonding orbitals, nitrogen by three bonding and one non-bonding orbitals, and oxygen by two bonding and two non-bonding orbitals. Therefore, each of these atoms is surrounded by four occupied orbitals, which implies $sp^3$ hybridization and the molecular geometry shown in Figure 4.31. The angles around nitrogen and oxygen could be adjusted toward the values in ammonia and water, for the same reasons used in the simple molecules.

As a second example, consider the molecule $SF_6$. The sulfur atom is surrounded by six F atoms at the apices of an octahedron. This geometry requires six equivalent bonding orbitals. The presence of six equivalent orbitals implies $d^2sp^3$ hybridization on sulfur. Each of the $d^2sp^3$ hybrids is combined with the appropriate $p$ orbital on one of the F atoms to yield a bonding orbital, and these six bonding orbitals yield the octahedral geometry shown in Figure 4.32.

The reader might wonder what has happened to antibonding orbitals in the localized molecular orbital method. Notice that in this method the only orbitals that are introduced are those which are occupied. If both a bonding and an antibonding localized molecular orbital are occupied, the resulting electron density is the electron density of the two atomic orbitals, because electron densities are additive:[*]

_____

[*]It can be shown that, to a good approximation, $N_+^2 + N_-^2 = 1$.

**FIGURE 4.32**   Structure of $SF_6$.

$$\psi^2(\text{bonding}) = (2s_A)^2 + 2(2s_A \cdot 2s_B) + (2s_B)^2$$

$$\psi^2(\text{antibonding}) = (2s_A)^2 - 2(2s_A \cdot 2s_B) + (2s_B)^2 \qquad (4.11)$$

$$\psi^2(\text{bonding}) + \psi^2(\text{antibonding}) = 2(2s_A)^2 + 2(2s_B)^2$$

Instead of antibonding orbitals, in the bond-orbital approach we have non-bonding orbitals and fewer bonding orbitals.

## 4.6 ELECTRON DEFICIENT COMPOUNDS.

Application of the method of localized bond orbitals to the compound of boron and hydrogen would lead to the prediction of a compound with composition $BH_3$. The bonding in the molecule should be based on $sp^2$ hybrid orbitals centered on the boron atom; the molecule should be planar. In fact, there is no evidence for the existence of a stable $BH_3$ molecule. The $BH_4^-$ ion does exist, and it has the expected tetrahedral geometry.

The simplest boron hydride that can be prepared is diborane, $B_2H_6$. Structural studies show that diborane has the geometry shown in Figure 4.33. Two boron atoms and six hydrogen atoms provide a total of twelve valence electrons. The boron-hydrogen bonds at the ends of the molecule are normal, with two electrons in bonding orbitals that can be thought of as being formed from an $sp^3$ hybrid orbital on the boron atom and a $1s$ orbital on the hydrogen atom. The four terminal B–H bonds use eight of the 12 valence electrons. The boron-hydrogen bonds in the center of the molecule are quite different. The bonding orbitals extend over three atoms, two boron atoms and a hydrogen atom: $(sp^3)_{B1} + (1s)_H + (sp^3)_{B2}$ (Figure 4.34). These are still localized bond orbitals, not molecular orbitals. They extend over three nuclei, rather than over the normal number of two, but they do not extend over the whole molecule. The central B—H—B bonds are called *three-center bonds*. Each of the two B—H—B bonds uses two valence electrons.

The formation of three-center bonds has completed the outer shell of each nucleus in the diborane molecule. Some of the electrons must be shared between two boron atoms and a hydrogen atom because there is an electron deficiency in the molecule. We might expect that electron deficient compounds would be rather unstable, and they usually are. Many electron deficient compounds react violently with compounds that contain lone-pair electrons.

Many electron deficient compounds are known. A few of the more complex boron hydrides are shown in Figure 17.4. In each of these mole-

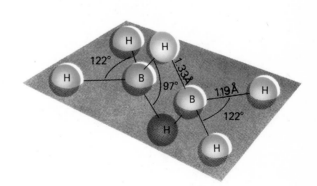

**FIGURE 4.33**   The structure of diborane.

**FIGURE 4.34**   A localized, three-center bonding orbital in diborane.

cules the complex geometry is a result of the necessity for the nuclei to make do with an insufficient number of electrons. Some of these molecules are based on three-center boron orbitals (Figure 4.35); some contain even more complex multi-center bonds, such as the five-center bond of the apical boron atom in $B_5H_9$. A detailed discussion of the geometries and bonding in electron deficient molecules is a task of some complexity. However, our simple theory does at least allow for the existence of these compounds, and the earlier theories do not.

## 4.7 $\pi$ BONDING IN LOCALIZED MOLECULAR-ORBITAL THEORY.

The concepts of *sigma* and *pi* bonds can be extended to polyatomic molecules by referring all symmetries to the bond axis rather than a molecular axis. All of the bond orbitals discussed to this point are $\sigma$ orbitals, and the bonds are $\sigma$ bonds. We find that $\pi$ bonds and $\pi$ orbitals occur in multiply-bonded molecules. The simplest illustration of $\pi$ bonding is found in the series ethane ($C_2H_6$), ethylene ($C_2H_4$), and acetylene ($C_2H_2$). These molecules have the Lewis diagrams

$$
\begin{array}{ccc}
\begin{array}{c}
\text{H} \quad \text{H} \\
| \quad\; | \\
\text{H---C---C---H} \\
| \quad\; | \\
\text{H} \quad \text{H}
\end{array}
&
\begin{array}{c}
\text{H} \qquad\quad \text{H} \\
\diagdown \qquad \diagup \\
\text{C}=\text{C} \\
\diagup \qquad \diagdown \\
\text{H} \qquad\quad \text{H}
\end{array}
&
\text{H---C}\equiv\text{C---H}
\end{array}
$$

    ethane       ethylene       acetylene

Single bonds between nuclei are almost always $\sigma$ bonds. Ethane, which has four bonding orbitals around each carbon, can be described using $sp^3$ hybrid orbitals on carbon. This leads to the correct structure (Figure 4.36). All of the bond angles are close to 109°28′. The C—H bond length is the same as in methane (1.09 Å); the same orbitals are used to form the CH bonds in the two molecules. The C—C bond length is 1.54 Å.

**FIGURE 4.35**   A three-center bonding orbital involving three boron atoms as found in various boron hydrides.

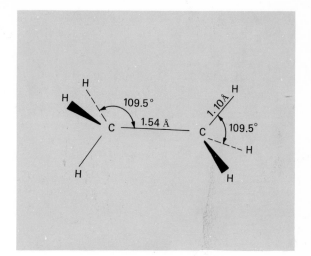

**FIGURE 4.36** Predicted structure of ethane.

In molecules with multiple bonds, we first assign electrons to the $\sigma$ bonds; that is, we construct the *sigma* framework. A multiple bond always involves a *sigma* bond plus one or two *pi* bonds. In ethylene, each carbon is bonded to three other atoms by at least a single bond. Three *sigma* bonds can be formed from the three $sp^2$ hybrid orbitals. This gives a *sigma* framework with the geometry shown in Figure 4–37. The five *sigma*-bonding orbitals (4 C—H and 1 C—C) can accept a total of ten electrons. However, two carbon atoms and four hydrogen atoms have a total of twelve valence electrons. The only unused valence-shell orbitals are the $2p$ orbitals perpendicular to the planes defined by the three $sp^2$ hybrids at each carbon. These two atomic orbitals can be combined to form a *pi* bonding orbital similar to that shown in Figure 4.11. This can happen only if the two planes defined by the $sp^2$ hybrid orbitals at the two carbon atoms are themselves co-planar. Formation of the *pi* bond eliminates free rotation around the C—C bond and forces the molecule into a planar configuration. Ethylene is planar with the structure shown in Figure 4.38. This structure is in good agreement with our expectations.

In the acetylene molecule, each carbon atom is bonded to two atoms,

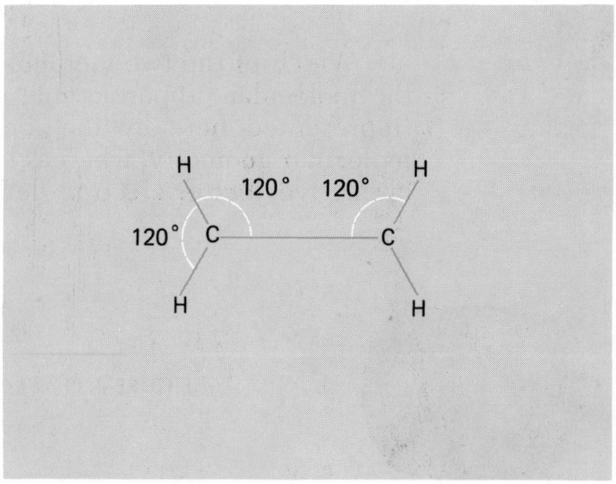

**FIGURE 4.37** Predicted structure of ethylene.

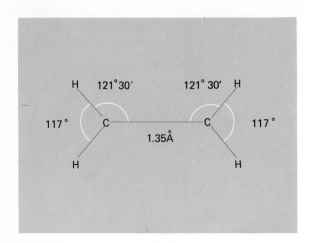

FIGURE 4.38   Actual structure of ethylene.

so the *sigma*-bond framework requires *sp*-hybrid orbitals on the two carbon atoms. There are two C—H *sigma* orbitals and one C—C *sigma* orbital. These three orbitals can accept six electrons, which leaves four electrons to be accommodated in bond orbitals formed from the unused *p* orbitals. The two pairs of *p* orbitals which are perpendicular to the C—C bond can be used to form two *pi* bonding orbitals for the four electrons. Acetylene is a linear molecule with the two carbon atoms bonded by one *sigma* and two *pi* bonds. The electron density from the two *pi* orbitals is roughly a barrel of charge with a hole down the axis (Figure 4.39).

Table 4.7 contains some representative bond energies and lengths. As noted earlier, the bond lengths usually decrease with increasing nuclear charge, but they increase as the principal quantum number of the valence shell increases. The bond lengths also decrease as the number of bonding electrons between the two nuclei increases (as in the series C—C, C=C, and C≡C). The bond energies roughly follow the bond lengths, being greater for shorter bonds, but some exceptions occur. These exceptions can usually be rationalized on the basis of electronegativity differences within the subgroup.

## 4.8   COMPARISON OF THE LOCALIZED AND CONVENTIONAL (EXTENDED) MOLECULAR-ORBITAL METHODS.

Which of the two methods is better? It depends on the molecule and the molecular properties that are in question. The valence-bond method (represented here by the localized molecular-orbital method) predicts molecular geometry, whereas the conventional molecular-orbital approach often gives better electron densities. As an example of the utility of both

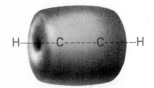

FIGURE 4.39   *Pi* orbitals in acetylene.

TABLE 4.7

AVERAGE BOND ENERGIES

| MOLECULE | BOND | LENGTH (Å) | ENERGY (kcal · mol⁻¹) |
|:---:|:---:|:---:|:---:|
| $CH_4$ | C—H | 1.09 | 99.3 |
| $NH_3$ | N—H | 1.01 | 93.4 |
| $H_2O$ | O—H | 0.96 | 110.6 |
| HF | F—H | 0.92 | 135.0 |
| $SiH_4$ | Si—H | | 81 |
| $PH_3$ | P—H | | 78 |
| $H_2S$ | S—H | 1.33 | 87 |
| HCl | Cl—H | 1.27 | 103 |
| $C_2H_6$ | C—C | 1.54 | 83 |
| $C_2H_4$ | C=C | 1.35 | 125 |
| $C_2H_2$ | C≡C | 1.21 | 230 |

1 Å = 0.1 nm; 1 kcal = 4.184 kJ.

approaches, we consider another molecule, benzene ($C_6H_6$). Benzene is a ring compound which can be represented by the pair of Lewis diagrams

A discussion of the bonding in this molecule in terms of the general molecular-orbital method is complex. There are six $1s$ hydrogen orbitals, six $2s$ carbon orbitals, and eighteen $2p$ carbon orbitals in the valence shells. The simplest molecular-orbital set contains thirty molecular orbitals, and sophisticated calculations are needed even to attempt to order these orbitals correctly. Moreover, the resulting molecular orbitals are not useful in predicting the molecular geometry; in fact, this latter problem is so great that the correct geometry must be introduced before even the most complex calculations now possible can be initiated.

The valence-bond treatment of this molecule is also complex. The *sigma* framework is no problem because it is identical in the two structures. Each carbon atom is bonded to three other atoms; the hybridization should be $sp^2$. Therefore, all bond angles are 120°, and the molecule is planar. This is all true. The difficulties arise with the *pi* bonds. Bond orbitals are closely related to Lewis diagrams; if several Lewis diagrams are needed to describe the molecule (resonance), then no single valence-bond structure will be adequate either. Thus we have a dilemma: the molecular-orbital method, which is suited to handle the *pi* orbitals that extend over the whole molecule, becomes too complex to be of qualitative use for the *sigma* electrons, whereas the valence-bond method, which gives immediately useful results for the *sigma* framework, is ill-adapted for treatment of the *pi* electrons.

This dilemma can be resolved in the following way: the *sigma* electrons are handled *via* bond orbitals, thus fixing the molecular geometry correctly, and then the remaining six $2p$ orbitals (one on each of the six carbon atoms) are combined into *pi* molecular orbitals, *which give reasonably accurate electron densities* in the vulnerable outer regions of the molecule. Thus, we use $sp^2$ hybrid orbitals on carbon atoms to construct the *sigma* framework, and we combine the $2p_z$ orbitals into three bonding and three antibonding molecular orbitals (Figure 4.40) that extend over the carbon framework of the molecule. The *pi* electron configuration of benzene is $\pi_1^2$, $\pi_2^2$, $\pi_3^2$. It is possible to relate most of the properties of benzene, and a host of related compounds, to the *pi* electron densities of the delocalized molecular orbitals.

The valence-bond and molecular-orbital approaches lead toward the same result (the true electron distribution), when extended to their utmost accuracy, although the extensions are formally different. In molecular-orbital calculations, more and more atomic orbitals are included in the linear combinations,* whereas in valence-bond calculations more and more Lewis diagrams (resonance structures) are introduced. An interesting example of the convergence of the two methods is a recent analysis of extensive molecular-orbital calculations on methane ($CH_4$) and methanol ($CH_3OH$). These calculations showed that the electron densities between carbon and hydrogen were almost identical in the two molecules, thus verifying the idea of bond orbitals that are unchanged from molecule to molecule. Further, the bond orbital was found to be similar to a localized molecular orbital formed from an $sp^3$ hybrid on carbon and a $1s$ on hydrogen.

---

*Configuration interaction, that is, orbitals that describe ionic-type excited states, must also be included.

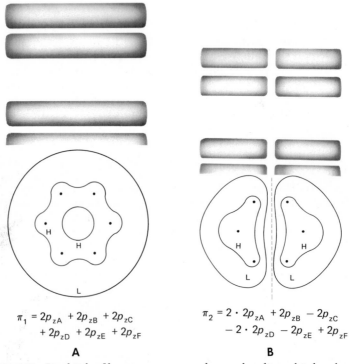

$$\pi_1 = 2p_{zA} + 2p_{zB} + 2p_{zC} + 2p_{zD} + 2p_{zE} + 2p_{zF}$$

**A**

$$\pi_2 = 2 \cdot 2p_{zA} + 2p_{zB} - 2p_{zC} - 2 \cdot 2p_{zD} - 2p_{zE} + 2p_{zF}$$

**B**

**FIGURE 4.40**   *Pi* orbitals of benzene; $\pi_1$, $\pi_2$, and $\pi_3$ are bonding orbitals, whereas $\pi_4$, $\pi_5$, and $\pi_6$ are antibonding orbitals.

*Ill. continued on opposite page.*

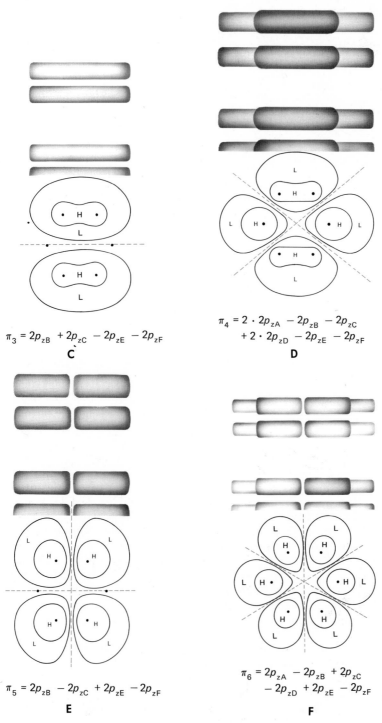

$$\pi_3 = 2p_{zB} + 2p_{zC} - 2p_{zE} - 2p_{zF}$$

C

$$\pi_4 = 2 \cdot 2p_{zA} - 2p_{zB} - 2p_{zC}$$
$$+ 2 \cdot 2p_{zD} - 2p_{zE} - 2p_{zF}$$

D

$$\pi_5 = 2p_{zB} - 2p_{zC} + 2p_{zE} - 2p_{zF}$$

E

$$\pi_6 = 2p_{zA} - 2p_{zB} + 2p_{zC}$$
$$- 2p_{zD} + 2p_{zE} - 2p_{zF}$$

F

**FIGURE 4.40** *Continued.*

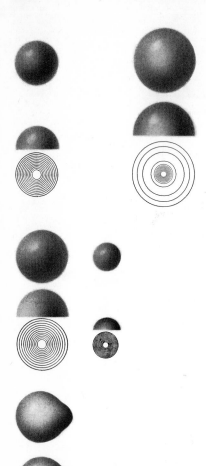

**FIGURE 4.41**    Contour maps for LiF. Li is on the right and F is on the left. The contour map for an ionic molecule is quite similar to those for covalent molecules. (Courtesy of Arnold C. Wahl, Argonne National Laboratory.)

This chapter has been devoted entirely to the subject of covalent bonding. We now present one last contour map (Figure 4.41), a map for an ionic molecule, LiF. Figure 4.41 consists of three parts; the separated atoms, the separated ions, and the final molecule. Notice that the final molecule does not consist of two separated ions. There is severe interpenetration of the electron clouds, as indeed there must be to account for the repulsive force which holds the two "ions" apart. It should be clear from this map that only slight changes in electron density are required to effect a smooth transition from ionic to covalent bonding.

## 4.9   VALENCE-SHELL-ELECTRON-PAIR-REPULSION (VSEPR) THEORY FOR MOLECULAR GEOMETRY.*

R. J. Gillespie has combined the localized electron-pair concept of G. N. Lewis with some simple assumptions regarding the relative effective

---

*This discussion closely follows that given by R. J. Gillespie in J. Chem. Educ., **47**, 18–23 (1970).

| NUMBER OF ELECTRON PAIRS | PREDICTED ARRANGEMENT OF PAIRS |
|:---:|:---|
| 2 | linear |
| 3 | equilateral triangle |
| 4 | tetrahedron |
| 5 | trigonal bipyramid |
| 6 | octahedron |
| 7 | distorted octahedron |
| 8 | square antiprism |

sizes of lone pairs and bonding pairs of electrons in the valence shell of an atom into a simple theory that is remarkably successful in predicting the molecular geometry of a wide variety of compounds. A localized electron pair is regarded as an electron pair housed in a localized molecular orbital that extends over only one (lone pair) atomic center or two (bond pair) atomic centers. (In some electron-deficient systems the electron pair may occupy a three- or four-center localized molecular orbital.)

The Valence-Shell-Electron-Pair-Repulsion Theory says that the arrangement of the bonds around an atom is determined by the number of electron pairs (lone and bond) surrounding the atom and by the relative sizes and shapes of the localized molecular orbitals in which the electron pairs are housed. The preferred (lowest-energy) arrangement of a given number of valence-shell electron pairs is assumed to be that which maximizes the distance between the pairs. In other words, the electron pairs in localized molecular orbitals occupy reasonably well defined regions of space, and the electron-electron repulsions between the pairs are minimized by the adoption of a molecular geometry that keeps the electron pairs as far apart as possible. The preferred arrangements for two to eight electron pairs in the valence shell of an atom are shown above. The preferred electron-pair arrangements allow us to predict the geometry of any molecule of the type $AX_n$ (where A is the central atom and each of the X's, the *ligands*, are bonded to A by a single pair of electrons), as is shown in Figure 4.42. For example, with *four* electron pairs we have the following possibilities (where E is a lone pair) shown below.

The simple theory outlined above can be extended to give a qualitative understanding of some of the finer details of molecular geometries by explicitly taking into account the effects of the size and shape of electron pairs in different types of localized molecular orbitals. The extension of the theory involves the following assumptions:

(1) A non-bonding pair (lone pair) of electrons is larger and takes up more room on the surface of an atom than a bonding pair of electrons.

| GENERAL CASE | # LONE PAIRS | # BOND PAIRS | MOLECULAR GEOMETRY | EXAMPLE |
|:---:|:---:|:---:|:---|:---:|
| $AX_4$ | 0 | 4 | tetrahedral | $CH_4$ |
| $AX_3E$ | 1 | 3 | pyramidal | $NH_3$ |
| $AX_2E_2$ | 2 | 2 | angular | $H_2O$ |

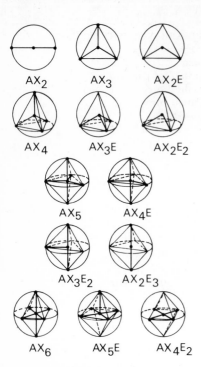

AX₂        AX₃        AX₂E

AX₄        AX₃E        AX₂E₂

AX₅        AX₄E

AX₃E₂        AX₂E₃

AX₆        AX₅E        AX₄E₂

**FIGURE 4.42**  Shapes of singly-bonded molecules containing up to six electron pairs in the valency shell. A, central atom; X, ligand; E, unshared electron-pair. (From Gillespie, R. J.: *J. Chem. Educ.,* 47, 18–23, 1970).

A ligand is an ion or group of atoms bound to the atom in question.

(2) The space that a bonding electron pair takes up on the surface of an atom decreases with increasing electronegativity of the ligand.

(3) The two electron pairs of a double bond (or the three electron pairs of a triple bond) take up more space on the surface of an atom than the one electron pair of a single bond.

A simple illustration of the effect of lone pairs on bond angles is provided by the series $CH_4$, $NH_3$, $OH_2$ shown below. In this series the successive replacement of bonding pairs by the larger lone pairs squeezes the remaining bonding pairs closer together and decreases the H-A-H angle.

Another example of the effect of lone pairs on bond angles is found in $BrF_5$. This molecule involves 12 electrons $(7+5)$, and therefore six electron pairs, in the valence shell of Br. The predicted geometry of the electron pairs in an $AX_5E$-type molecule is octahedral, and therefore we predict $BrF_5$ to have a square-pyramidal structure

| MOLECULE | NUMBER OF LONE PAIRS | H-A-H ANGLE |
|----------|----------------------|-------------|
| $CH_4$ | 0 | 109.5° |
| $NH_3$ | 1 | 107.3° |
| $OH_2$ | 2 | 104.5° |

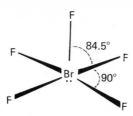

with the lone pair in the sixth octahedral position. The larger size of the lone pair as compared to the Br-F bond pairs is invoked to explain the 84.5° F-Br-F bond angles.

The effect of a change in electronegativity on the size of bond pairs is used to explain the different bond angles in $NH_3$ and $NF_3$:

The much greater electronegativity of F as compared to H makes the N—F bond orbital smaller than the N—H bond orbital. The large lone pair of electrons on N can thus push the smaller N—F bond orbitals closer together than it can the N—H bond orbitals.

The effective size of a double-bond orbital is about the same as that of a lone-pair orbital. The larger size of a multiple-bond orbital as compared to a single-bond orbital is seen: (a) in *planar* molecules of the type $X_2C{=}O$ and $X_2C{=}CH_2$; (b) in *pyramidal* molecules of the type $X_2S{=}O$; and (c) in *tetrahedral* molecules of the type $X_3P{=}O$. The X—C—X angle in $X_2C{=}O$ and $X_2C{=}CH_2$ is always less than 120°; the X—S—O angle in $X_2SO$ compounds is always greater than the X—S—X angle; and the X—P—X angle in $X_3PO$ compounds is less than 109.5°. These effects can be seen in Table 4.8. Note also (Table 4.8) that in most cases the angles between the X ligands decrease with increasing electronegativity of X.

A particularly interesting series of molecules for which the VSEPR Theory is successful in rationalizing the observed geometries is $PF_5$, $SF_4$, and $ClF_3$. These molecules all involve five electron pairs around the central atom. The preferred geometry for five electron pairs around an atom in a molecule without double bonds is trigonal bipyramidal. The structures of these three molecules are shown in the following diagrams.

| MOLECULE | TYPE | STRUCTURE |
|---|---|---|
| $PF_5$ | $AX_5$ | 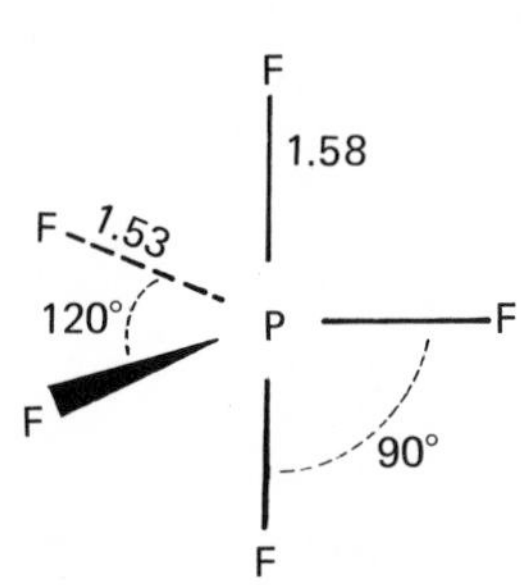 |

| MOLE | TYPE | STRUCTURE |
|------|------|-----------|
| | | |

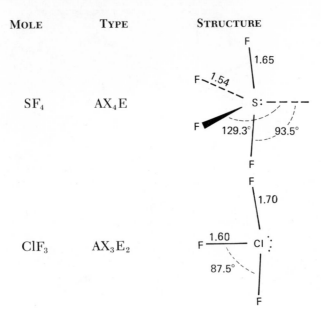

| SF₄ | AX₄E | |
| ClF₃ | AX₃E₂ | |

SF$_4$    AX$_4$E

ClF$_3$    AX$_3$E$_2$

**TABLE 4.8**

| 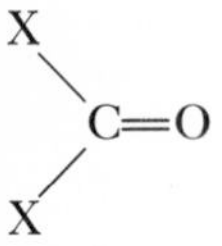 | ∠ X—C—X | ∠ X—C=O |
|---|---|---|
| H$_2$CO | 115.8° | 122.1° |
| Cl$_2$CO | 111.3° | 124.3° |
| F$_2$CO | 108.0° | 126.0° |

| 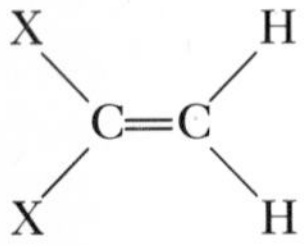 | ∠ X—C—X | ∠ X—C—C |
|---|---|---|
| H$_2$C=CH$_2$ | 116.8° | 122° |
| Cl$_2$C=CH$_2$ | 114.0° | 123° |
| F$_2$C=CH$_2$ | 110.0° | 125° |

| 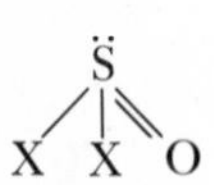 | ∠ X—S—X | ∠ X—S—O |
|---|---|---|
| (CH$_3$)$_2$SO | 100° | 107° |
| Br$_2$SO | 96° | 108° |
| F$_2$SO | 92.8° | 106.8° |

| 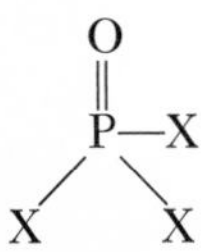 | ∠ X—P—X |
|---|---|
| Br$_3$PO | 106° |
| Cl$_3$PO | 103.6° |
| F$_3$PO | 102.5° |

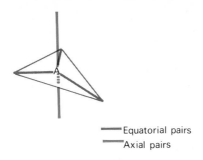

————Equatorial pairs
————Axial pairs

The *axial* electron pairs in the trigonal bipyramidal arrangement are not equivalent to the *equatorial* electron pairs, because the axial electron pairs have *three* nearest neighbors at 90°, whereas the equatorial pairs have only *two* nearest neighbors at 90°. Consequently, the lowest energy configuration has the axial pairs at a greater distance from the central atom. In all such trigonal bipyramidal molecules, the axial A—X bond lengths are greater than the equatorial A—X bond lengths. Furthermore, the lone pairs, which are larger than the bond pairs, always occupy equatorial positions, because this arrangement minimizes the lone pair-bond pair repulsions (there is more room for the lone pairs in equatorial positions). By analogous reasoning, the most electronegative ligands (those with the smallest bond pairs) always go in the equatorial positions (e.g., $Cl_2PF_3$ has F atoms in the equatorial positions). The greater space requirements for a lone pair as compared with a bond pair can be invoked to rationalize the small distortions of the F—A—F bond angles in $SF_4$ and $BrF_3$ relative to those in $PF_5$.

The VSEPR Theory is simple and useful. The structures of many molecules and ions can be predicted from this theory. The extension of the theory to molecules with multi-center bonds such as $B_2H_6$ is simple (multi-center bonds require less room than an ordinary single bond because of the greater electron delocalization.) There are, however, a few cases where the theory fails. Two examples are: (a) although $BeCl_2$ is, as predicted, linear, the molecule $BaF_2$, which is also predicted to be linear, is bent; (b) although $XeF_6$ (7 pairs) has the predicted distorted octahedral geometry, the ion $TeCl_6^{-2}$ (7 pairs) has a regular octahedral geometry.

---

**PROBLEMS**

1. Give the electron configurations for all the diatomic hydrides of the second row atoms, i.e., LiH, BeH, ... HF.

2. Give the electron configurations, the net number of bonding electrons, and the magnetic properties for the following molecules.

| | | | |
|---|---|---|---|
| BF | CN | NO | BeO |
| BN | CO | NF | LiF |
| BO | CF | BeF | LiO |

Add bond lengths and energies to your table when possible.

3. Construct the molecular-orbital correlation diagram applicable to the molecule NO. Indicate the separated atomic orbitals, the overlap region, and the resulting molecular orbitals. Label all the orbitals in your diagram. Place the appropriate number of electrons in the appropriate orbitals on your diagram.

4. Use your results in problem 3 to answer the following questions:
   (a) Is NO paramagnetic?
   (b) How many net bonding electrons does NO have?
   (c) How many $\sigma$ bonds does NO have?
   (d) How many $\pi$ bonds does NO have?
   (e) What is the *complete* (i.e., including all the electrons) molecular orbital configuration of NO?

5. The effective nuclear charge for He is 1.4, so the He atom is smaller than the H atom. Yet the bond in $He_2^+$ is longer and weaker than $H_2^+$, although they are both one-electron bonds. How can this happen?

6. Arrange the following diatomic molecules in order of increasing bond length: $O_2^{+2}$, $O_2^{+1}$, $O_2$, $O_2^-$, $O_2^{-2}$.

7. It is usually true that bonding orbitals have high electron density between the nuclei and that antibonding orbitals have a node that crosses the internuclear axis. Find an exception to this rule in the text.

8. Predict the geometry of the following molecules and ions. In all cases describe the hybridization on the central atom.

| | | | |
|---|---|---|---|
| $BCl_3$ | $BF_4^-$ | $ClO_2$ | $HCOO^-$ |
| $N_3^-$ | $CO_3^{-2}$ | $SO_4^{-2}$ | $NO_2Cl$ |
| $SiF_4$ | $CF_4$ | $NH_4^+$ | $CN^-$ |
| $NO_2$ | $CS_2$ | $OF_2$ | $NO_2^-$ |
| $SO_3$ | $H_2Te$ | $SO_2$ | $PCl_3$ |

9. Which of the molecules and ions of problem 8 have dipole moments?

10. From your knowledge of the bonding in ammonia and the bond angles in the series $H_2O$, $H_2S$, $H_2Se$, predict the bond angles in phosphine $(PH_3)$ and arsine $(AsH_3)$.

11. Predict structures of the following molecules. List the number of *sigma* and *pi* bonds for each species.

$$CO_2 \qquad\qquad H_2N\!-\!\overset{\displaystyle \|}{\underset{\displaystyle O}{C}}\!-\!NH_2 \qquad\qquad H_3C\!-\!\overset{\displaystyle H}{\overset{\displaystyle |}{C}}\!=\!CH_2$$

$H_2CO$

$HNO_3 \qquad\qquad\qquad H_2C\!=\!C\!=\!CH_2 \qquad\qquad H_2NNH_2$

$H_2SO_4$

$$H_3C-\underset{\displaystyle\overset{\displaystyle H}{|}}{\underset{\displaystyle |}{\underset{\displaystyle H}{C}}}-OH \qquad H_3C-O-CH_3 \qquad H_2N-OH$$

$$\underset{\displaystyle HC-OH}{\overset{\displaystyle O}{\overset{\displaystyle \|}{}}}$$

12. Predict the structures of the following two molecules.

    (a) $C_6H_5OH$         (b) $C_{10}H_8$

     phenol              naphthalene

    (Hints: Both compounds are based on the benzene ring structure, (a) on a single ring and (b) on two rings with a side in common.)

    13. Use the localized molecular orbital (MO) approach to describe the *sigma* framework in diazomethane ($H_2CNN$, planar) and use MO theory to describe the $\pi$-bonding. Make rough sketches of the various $\pi$-MO's used and construct a molecular-orbital energy-level diagram.

    14. Give the electron configurations of the homonuclear, third-period, diatomic molecules. Compare the bond lengths of the resulting molecules (Table 4.3) with the number of net bonding electrons. Is there a correlation? Is there a correlation between the atomic radii and bond distance for those molecules with the same number of net bonding electrons? Why is this so?

    15. Give the electron configurations for the $O_2^+$ ions produced in each of the four regions of the photoelectron spectrum (Figure 4.15). The spacing between the vibrational levels is related to the force constant of the bond; the stronger the bond, the larger the spacing. Discuss the spacings observed in terms of your configurations for the ion.

    16. What would we call the compound of lithium and hydrogen if the energy of the $1s$ orbital in hydrogen were greater than the $2s$ and $2p$ in lithium? Would this compound be stable?

    17. Water has a dipole moment of 1.84 D. What is the dipole moment of each O—H bond when the bond angle is 104.5°?

    18. The $BF_3$ molecule is planar, but the $NF_3$ molecule is not. How would you explain this difference?

    19. Arrange the species NO, $NO^+$, $NO^-$, and $NO^{+2}$ in order of *increasing* bond length. Explain very briefly, using MO electronic configurations, how you arrived at your results.

    20. Construct the MO correlation diagram applicable to $OH^-$. Indicate the separated AO's, the overlap region, and the resulting MO's. Label all the orbitals in your diagram. (Note $2p_O < 1s_H$ in energy).

    21. Electronegativity is a measure of an atom's ability to attract extra

electrons. Yet fluorine, with the highest electronegativity, has a lower electron affinity than chlorine. How is this possible?

22. Table 4.5 contains several pairs of diatomic molecules of the type BeH, $BeH^+$, OH, $OH^+$, etc. In some cases the ion has a longer, weaker bond and in some cases a stronger, shorter bond. Can you explain these differences?

23. Sketch the bonding orbital for the molecule-ion $H_3^+$ (linear *and* ring). What kinds of bonds do we expect in the ring $H_3^+$? Are there any precedents that you know of for such a bond?

24. Using your MO diagram from problem 20, offer a short explanation for the experimental observation that the bond dissociation energies of the species $OH^-$, OH, and $OH^+$ are all within a few kilocalories of one another.

25. Use the VSEPR Theory to predict the structures of the following molecules:

| | | |
|---|---|---|
| a) $PCl_5$ | g) $(CH_3)_2C{=}O$ | m) $XeF_2$ |
| b) $PCl_3$ | h) $(CH_3)_2S{=}O$ | n) $XeF_4$ |
| c) $F_3P(CH_3)_2$ | i) $(NH_2)_2C{=}O$ | o) $XeF_6$ |
| d) $F_4PCH_3$ | j) $F_2C{=}CCl_2$ | p) $XeO_3$ |
| e) $FPCl_4$ | k) $F_2C{=}CFCl$ | q) $BF_3$ |
| f) $F_2PCl_3$ | l) $IF_7^-$ | r) $BrCl_3$ |
| | | s) $Re_2Cl_8^{-2}$ |

**References**

4.1. Pauling, L., *The Nature of the Chemical Bond* (3rd Edition, Cornell University Press, Ithaca, N.Y., 1960).

4.2. Wahl, A. C., and Blukis, U., *Educational Film Loops on Atomic and Molecular Structure*, Journal of Chemical Education, *45*, 787 (1968).

4.3. Wahl, A. C., *Molecular Orbital Densities: Pictorial Studies*, Science, *151*, 961 (1966).

4.4. Wahl, A. C., *Chemistry by Computer*, Scientific American, *222–4*, 54 (1970).

4.5. Coulson, C. A., *Valence* (Oxford, New York, 1961).

4.6. Benson, S. W., *Bond Energies*, Journal of Chemical Education, *42*, 502 (1965).

4.7. R. J. Gillespie, *The Electron-Pair Repulsion Model for Molecular Geometries*, Journal of Chemical Education, *47*, 18 (1970).

## CHAPTER 5

# GASES

The fraction of an equilibrium collection of gas molecules with speeds between $c$ and $c + \Delta c$ ($\Delta c \ll c$) is given by

$$4\pi \left( \frac{m}{2\pi kT} \right)^{3/2} c^2 e^{-mc^2/2kT} \Delta c$$

where $m$ is the molecular mass, $T$ is the absolute temperature, and $k = 1.38 \times 10^{-23}$ J·K$^{-1}$.

from the works of
*J. C. Maxwell and L. Boltzmann*

## 5.1 INTRODUCTION.

Scientists seldom have the opportunity to experiment on individual molecules. Practical considerations force us to use samples that contain enormous numbers of molecules. Most properties of bulk samples are not solely the properties of the individual molecules; the bulk properties also are consequences of interactions between molecules.

In Chapter 3 we discussed the reaction

$$Cs(g) + Cl(g) = CsCl(g)$$

and we saw that the energy change for the reaction could be calculated from the ionization energy of the cesium atom, the electron affinity of the chlorine atom, and the bond distance of the diatomic molecule. However, the above reaction does not describe a practical laboratory procedure, because around 25°C elemental chlorine is a diatomic gas, $Cl_2$, and cesium and cesium chloride are crystalline materials. Hence, at 25°C we are forced to study the reaction

$$2Cs(s) + Cl_2(g) = 2CsCl(s)$$

We see that an interpretation of the laboratory results requires a knowledge of the collective properties of large samples.

## 5.2 GASES: DEFINITIONS AND MEASURING DEVICES.

Any sample of matter that fills the whole volume of its container *regardless of the container size or shape* is a gas. It is always possible to restrict the volume of a gas by reducing the volume of the container, although the reduction will require the action of a force. The relation between the volume restriction and the force needed to cause it is the starting point for our study of gases.

**135**

It is convenient to divide the properties of bulk samples of matter into two classes, those which depend upon the size of the sample (extensive properties) and those which do not depend upon the size of the sample (intensive properties). The force needed to restrict the volume of any gas is dependent upon the size of the sample. Force can be replaced by an intensive parameter by dividing the force by the area over which the force is applied. This new parameter is called the *pressure*; it has the units of force per unit area:

$$P(\text{pressure}) = f(\text{force})/A(\text{area}) \qquad (5.1)$$

The *barometer* is a simple device for the measurement of pressure. The underlying principle is the equality of the pressures at all points on any horizontal plane within a liquid. For simplicity we shall discuss a barometer made from a tube with a cross-section of 1 cm². If such a tube (which is sealed at one end) is completely filled with a liquid and then inserted into a container of the liquid, we arrive at the situation shown in Figure 5.1. For convenience we take the horizontal plane defined by the surface of the liquid in the container as our reference height. The pressure within the inverted tube is the pressure on the top of the column of liquid plus the weight of the liquid in the column (remember that the tube has unit cross section). The pressure on the surface of the liquid in the container is the pressure of the atmosphere. Because all points in a horizontal

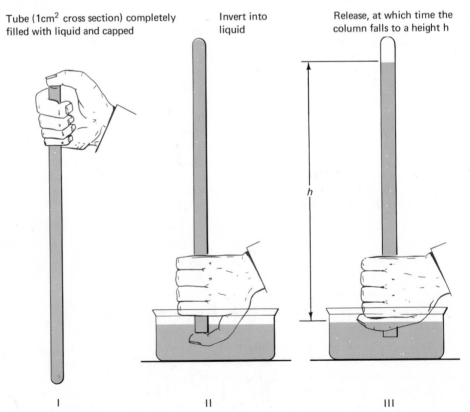

Tube (1cm² cross section) completely filled with liquid and capped

Invert into liquid

Release, at which time the column falls to a height h

*h*

I

II

III

**FIGURE 5.1**  Barometer. The atmosphere supports a column of height *h*.

plane, both those under the inverted tube and those on the surface, are subject to the same pressure, we can write

$$P(\text{atm}) = \text{weight of column}/1 \text{ cm}^2$$

The method of filling the tube has insured that the only pressure on the top of the column is that due to the vapor of the liquid, a pressure which can be made negligible by a judicious choice of the liquid.

The weight of the liquid column is proportional to the height $h$. The height should be large enough that it can be accurately measured, but small enough to avoid the necessity for holes in the ceiling. Mercury is the only liquid which is dense enough to avoid the latter problem for measurements of atmospheric pressure at room temperature. Various organic liquids, such as $n$-butyl phthalate, are occasionally used for measurements of low pressures.

Atmospheric pressure is useless as a standard unit of pressure, because it changes with weather, humidity, temperature, and altitude. At sea level on an "average" day the atmosphere will support a column of mercury about 760 mm high. One *standard atmosphere* is taken as that pressure which will support the weight of a column of mercury exactly 760 mm high at 0°C:[*]

$$1 \text{ atm} \equiv 760 \text{ mm of Hg} = 29.92 \text{ inches of Hg}$$

One must specify a temperature for the mercury column, because the density of mercury changes with temperature; the standard temperature is taken as 0°C, where the density of mercury is $13.59 \text{ g} \cdot \text{cm}^{-3}$. The unit mm of mercury (at 0°C) is now called the Torr in honor of Torricelli, the inventor of the barometer:

$$1 \text{ mm of Hg} \equiv 1 \text{ Torr}$$

The units of pressure must be defined ultimately in terms of the fundamental units of mass, length, and time:

$$P = f/A \tag{5.2}$$
$$= \text{mass} \times \text{gravitational acceleration}/\text{area}$$
$$= \text{density} \times \text{height} \times \text{area} \times \text{gravitational acceleration}/\text{area}$$

$$1 \text{ Torr} = \frac{13.59 \text{ g} \cdot \text{cm}^{-3} \times 0.1000 \text{ cm} \times 1.000 \text{ cm}^2 \times 980.7 \text{ cm} \cdot \text{sec}^{-2}}{1.000 \text{ cm}^2}$$

$$1 \text{ Torr} = 1.333 \times 10^3 \text{ g} \cdot \text{cm}^{-1} \cdot \text{sec}^{-2} = 1.333 \times 10^3 \text{ dyne} \cdot \text{cm}^{-2} \tag{5.3}$$

---

[*]A definition is indicated by the sign of equivalence $\equiv$. Significant figures are not involved in definitions; it is assumed that the equality is exact. The SI unit of force is the newton

$$1 \text{ newton} = 1 \text{ N} = 1 \text{ kg} \cdot \text{m} \cdot \text{s}^{-2} = 10^5 \text{ dyne} = 10^5 \text{ dyn}$$

The SI unit of pressure is the pascal

$$1 \text{ pascal} = 1 \text{ Pa} = 1 \text{ N} \cdot \text{m}^{-2} = 10 \text{ dyne} \cdot \text{cm}^{-2}$$

$$1 \text{ Torr} = (1.333 \times 10^3 \text{ dyne} \cdot \text{cm}^{-2}) \left(\frac{1 \text{ newton}}{10^5 \text{ dyne}}\right) \left(\frac{100 \text{ cm}}{\text{m}}\right)^2$$

$$1 \text{ Torr} = 1.333 \times 10^2 \text{ N} \cdot \text{m}^{-2}$$

The definition of the atmosphere as a unit of pressure is then

$$1 \text{ atm} \equiv 760.0 \text{ Torr} \times 1.333 \times 10^3 \text{ dyne} \cdot \text{cm}^{-2} \cdot \text{Torr}^{-1}$$

$$1 \text{ atm} \equiv 1.013 \times 10^6 \text{ dyne} \cdot \text{cm}^{-2} = 1.013 \times 10^5 \text{ N} \cdot \text{m}^{-2}$$

The pressure unit *bar* is sometimes used. It is defined as

$$1 \text{ bar} \equiv 10^6 \text{ dyne} \cdot \text{cm}^{-2} = 10^5 \text{ N} \cdot \text{m}^{-2}$$

so that

$$1 \text{ atm} = 1.013 \text{ bar}$$

Laboratory pressures can range between the extremes of a few megabars and "vacuums" of $10^{-16}$ Torr around 25°C. It has been estimated that the pressure inside a vessel free of helium which is immersed in liquid helium (at 4.2°K) is about $10^{-30}$ Torr.

The *manometer* is a device for measuring pressures. It is a U-tube partially filled with liquid. The sample to be measured is trapped on one side, and a known pressure is applied to the other arm of the U-tube. The known pressure is usually either zero (as in the barometer) or atmospheric pressure. Figure 5.2 shows the same pressure measured with the two types of manometer. In both cases we equate pressures in the plane defined by the lower surface. In I the pressure on the plane is 1 atm, so

$$P_{\text{sample}} + h_{\text{I}} = 760 \text{ Torr}$$

whereas in II

$$P_{\text{sample}} = h_{\text{II}}$$

Here, $h$ is measured in mm and $P_{\text{sample}}$ is in Torr.

I. Exposed to air    II. Vacuum manometer

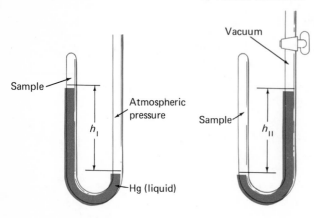

FIGURE 5.2   Two types of manometer.

## 5.3  BOYLE'S LAW.

The first systematic study of the pressure needed to restrict the volume of a gas was carried out by Robert Boyle in 1662. Boyle trapped a sample of gas in a manometer and varied the pressure by adding mercury to the open arm. He found that the volume of the gas was inversely proportional to the pressure on the gas. Under a fixed set of experimental conditions (where only the pressure and the volume are allowed to vary) his results could be represented satisfactorily by the equation

$$V = k/P \qquad\qquad (5.4)$$

where $k$ was some constant that depended on the amount of gas taken and on the temperature. For example, $k = 24.4$ $\ell \cdot$ atm for a 29.0 g sample of air at 25°C. Some results for this hypothetical sample are listed in Table 5.1 and plotted in Figure 5.3.

Rearrangement of Equation (5.4) gives

$$P = k/V \qquad\qquad (5.5)$$

A plot of $P$ versus $1/V$ will be a straight line passing through the origin if the data correspond to Equation (5.5). The human eye and brain are quite adept at recognizing straight lines. For this reason we always try to re-arrange any equation to be tested so that the plot is a straight line. If the points above and below the line are both distributed randomly along the line, and if their deviations from the line are not beyond experimental error, then we conclude that the equation fits the data. It is obvious that Equation (5.5) is correct from Figure 5.4; Equation (5.5) (and (5.4)) is a valid summary of the data in Table 5.1. Equations (5.4) and (5.5) are ex-pressions of Boyle's law.

If the deviations at one end of the line had been all of one type (high or low), the equation would probably have been judged inapplicable. If Boyle had extended his measurements to higher pressures, he would have found that high pressure points would have fallen below the straight line. In this way he would have discovered the limitation of his law; it does not apply accurately at high pressures. The evidence of the failure of the equa-tion is the *systematic* deviation of the data from the straight line at high pressures.

TABLE 5.1

P-V DATA FOR 29 g OF AIR AT 25°C

| P (atm) | V (liters) | 1/V |
|---------|-----------|--------|
| 0.47 | 51.0 | 0.0196 |
| 0.97 | 25.0 | 0.0400 |
| 1.22 | 20.0 | 0.0500 |
| 2.04 | 12.0 | 0.0834 |
| 2.44 | 10.0 | 0.1000 |
| 3.05 | 8.0 | 0.1250 |

1 atm = $1.013 \times 10^5$ N $\cdot$ m$^{-2}$; 1 $\ell$ = $10^{-3}$ m$^3$ = 1 dm$^3$.

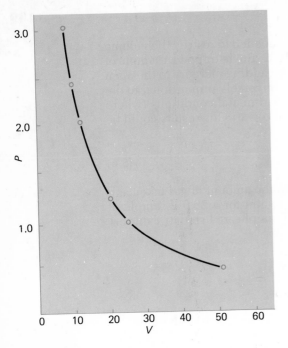

**FIGURE 5.3**   *P*-*V* plot for 29 g of air.

The limitation on Boyle's law does not mean that it must be discarded. Although it is only valid over a limited pressure range, it is a most useful range of pressure, and Boyle's law is less complex than any equation which covers a wider range of pressure. We use the simplest law that is valid within the required accuracy for the conditions under which we are working, but we must not casually extend the law into regions where its validity is untested.

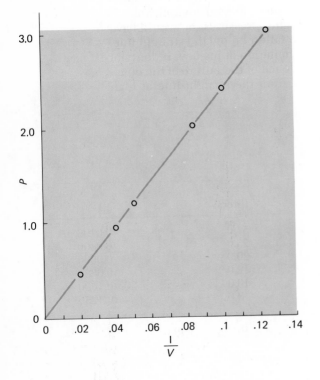

**FIGURE 5.4**   *P* versus 1/*V* plot for 29 g of air.

## 5.4  CHARLES' LAW AND TEMPERATURE SCALES.

Boyle noticed that his results were not reproducible unless he held the temperature constant, but he was unable to investigate this effect because he had no good device for measurement of temperature. It was shown later that, if the pressure on a gas sample was maintained constant, then the volume of the sample increased as the temperature increased. Experiments by Jacques Charles and others led to the following simple equation relating the Celsius temperature and the volume:

$$V = V_0(1 + \alpha t) \qquad (5.6)$$

In Equation (5.6), $t$ is the temperature, $\alpha$ is an experimental proportionality constant, and $V_0$ is the volume of the sample when $t = 0°C$.

Although the concept of temperature as an indicator of hot and cold is a very familiar one, the definition and measurement of temperature require care. Consequently, we here assume that reliable thermometers exist and postpone a discussion of the temperature concept until Chapter 9. We assume that you are familiar with the conventional Celsius temperature scale, on which the melting point of ice is 0°C and the boiling point of water is 100°C. The size of the Celsius degree is 1/100 of the difference between the normal boiling and melting points of water.

**Absolute Temperature.** The constant $\alpha$ in Equation (5.6) has the dimension of reciprocal temperature, so we can write Equation (5.6) as

$$V = V_0 \left(1 + \frac{t}{t_0}\right) = \frac{V_0}{t_0}(t_0 + t) \qquad (5.7)$$

where $t_0 = 1/\alpha$. Careful experiments yield the value

$$t_0 = 273.15 \pm .01$$

Substitution of this value for $t_0$ into Equation (5.7) gives

$$V = \frac{V_0}{273.15}(273.15 + t) \qquad (5.8)$$

Equation (5.8) suggests that the temperature-volume relation would be simplified if we shifted our temperature scale by 273.15 degrees. The new temperature scale is called the *absolute temperature scale* or the Kelvin scale; and the new temperatures are specified as K. As we shall see in Chapter 9, the Kelvin scale is the fundamental scale. The relation between the Celsius and Kelvin scales is

$$T(K) = 273.15° + t(°C) \qquad (5.9)$$

Equation (5.8) can now be rewritten as

$$\frac{V}{T} = \frac{V_0}{273.15} \qquad (5.10)$$

Remember that $V_0$ is the volume of the gas sample at 0.000°C, which is 273.15 K. Therefore, Equation (5.10) is

$$\frac{V}{T} = \frac{V_0}{T_0}$$

which is one way of writing

$$\frac{V}{T} = \text{constant} \tag{5.11}$$

This is Charles' Law in its simplest form; the volume of a sample of gas at fixed pressure is directly proportional to the absolute temperature. Equation (5.11) implies that *the absolute temperature can never be negative*, for negative volume is meaningless. The lowest possible temperature on the absolute scale is 0 K, which corresponds to −273.15°C.

We can construct a thermometer based on Charles' Law by trapping a sample of air with a drop of mercury in a capillary tube which is sealed at the bottom and open to the atmosphere at the top. The height of the mercury in the column is then directly proportional to the absolute temperature of the air trapped in the column:

$$\begin{pmatrix} \text{volume of air} \\ \text{trapped in the} \\ \text{capillary} \end{pmatrix} = \begin{pmatrix} \text{height of the} \\ \text{column of} \\ \text{trapped air} \end{pmatrix} \times \begin{pmatrix} \text{cross-sectional} \\ \text{area of the} \\ \text{capillary} \end{pmatrix}$$

or

$$V = hA \tag{5.12}$$

where $A$ is a known constant for a given capillary. Combination of Equations (5.12) and (5.10) yields

$$hA = \left(\frac{V_0}{273.15}\right) T \tag{5.13}$$

A measurement of $h$ at 273.15 K (0°C) gives $V_0$, and thus

$$T = \left(\frac{273.15\,A}{V_0}\right) h \tag{5.14}$$

## 5.5   THE IDEAL-GAS EQUATION.

Experimental measurements of the $PV$ product for different amounts of the same gas held at the same temperature show that $PV$ is directly proportional to the number of moles, $n$, of gas; that is,

$$PV = (\text{constant})n \qquad \text{(at fixed } T) \tag{5.15}$$

With the use of a Charles' Law-type thermometer to measure temperature, it has also been shown experimentally that the constant in Equation (5.15) is proportional to the absolute temperature, $T$. In other words

$$PV = nRT$$  (5.16)

where $R$ is a constant (called *the gas constant*), which is independent of $P$, $V$, or $T$. Equation (5.16) is called the *ideal-gas equation*. Many gases obey the ideal-gas equation to within a few per cent at all pressures up to one atmosphere; the deviations increase with increasing pressure.

## 5.6   USE OF THE IDEAL-GAS EQUATION.

The ideal-gas equation is basic to the study of gases. Several examples of its use follow.

(a) Given that 1 mole (4.0 g) of He at 25°C occupies 25 $\ell$ at $P = 0.98$ atm, what is the value of $R$? Rearrangement of Equation (5.16) gives

$$R = \frac{PV}{nT} = \frac{(0.98 \text{ atm})(25 \text{ } \ell)}{(1.0 \text{ mole})(298 \text{ K})} = 0.082 \text{ } \frac{\ell \cdot \text{atm}}{\text{mole} \cdot \text{K}}$$  (5.17)

More accurate experiments give

$$R = 0.08205 \text{ } \frac{\ell \cdot \text{atm}}{\text{mole} \cdot \text{K}}$$

(b) What volume does 1.000 mole of an ideal gas occupy at 0.00°C and $P = 1.000$ atm?

$$V = \frac{nRT}{P} = \frac{(1.000 \text{ mole}) \left(0.08205 \frac{\ell \cdot \text{atm}}{\text{mole} \cdot \text{K}}\right) (273.2 \text{ K})}{1.000 \text{ atm}} = 22.41 \text{ } \ell$$

These conditions are referred to as the standard temperature and pressure, *STP*.

(c) One mole of hydrogen (2.0 g) occupies 22.4 $\ell$ at 1.00 atm pressure and 0°C. What volume does it occupy at 2.00 atm and 100°C? We write the ideal gas equation as

$$\frac{PV}{T} = nR$$

Because $R$ is a constant, and $n$ is a constant for this problem, we can write

$$\frac{P_1 V_1}{T_1} = nR = \frac{P_2 V_2}{T_2}$$  (5.18)

where the subscripts refer to the initial and final conditions. Simple algebra gives

$$V_2 = V_1 (P_1/P_2) (T_2/T_1) = (22.4 \text{ } \ell) \left(\frac{1.00 \text{ atm}}{2.00 \text{ atm}}\right) \left(\frac{373 \text{ K}}{273 \text{ K}}\right) = 15.3 \text{ } \ell$$

There is another way to work this problem. Once we know the gas constant $R$, we are ready to calculate any new volume

$$V = \frac{nRT}{P} = \frac{(1.00 \text{ mole}) \left( 0.0821 \dfrac{\ell \text{ atm}}{\text{mole K}} \right) (373 \text{ K})}{(2.00 \text{ atm})} = 15.3 \ell$$

Notice that in the first method the fact that $n = 1.00$ mole was not used, and in the second the original $P$, $T$, and $V$ were not used.

(d) The gas constant $R$ is so useful that it is worth computing its value in several sets of units. Because pressure has the dimensions of force per unit area, the product $PV$ has the dimensions of force × distance, or energy. The gas constant, therefore, has the dimensions of energy/mole · deg. If we use the common pressure unit Torr, then

$$R = \frac{PV}{nT} = \frac{(760.0 \text{ Torr})(22.41 \ell)}{(1.000 \text{ mole})(273.15 \text{ K})} = 62.36 \frac{\text{Torr} \cdot \ell}{\text{mole} \cdot \text{K}} \qquad (5.19)$$

In a similar manner, we can evaluate $R$ in any desired units (Table 5.2).

(e) A volume of 22.41 liters of any (ideal) gas at STP contains one mole, and Avogadro's number of molecules. Each molecule in the gas occupies an effective volume of

$$\frac{22.4 \ \ell/\text{mole at STP}}{6.02 \times 10^{23} (\text{molecule/mole})} = 3.72 \times 10^{-23} \ \ell/\text{molecule at STP}$$

which is $37.2 \times 10^3 \text{ Å}^3/\text{molecule}$. A typical molecule is approximately a sphere of about 3 Å diameter, which has a volume of about 38 Å³. It follows that a gas is mostly empty space; the molecules at STP occupy only about 0.1% of the volume. This conclusion will be important in discussing the properties of gases.

(f) The ideal-gas equation also can be used in a determination of molecular weight. A 2.20 g sample of gas occupies $1.55 \ell$ at 600.0 Torr pressure and room temperature (25°C). What is the molecular weight?

$$n = \frac{PV}{RT} = \frac{(600.0 \text{ Torr})(1.55 \ell)}{\left( 62.4 \dfrac{\text{Torr} \cdot \ell}{\text{mole} \cdot \text{K}} \right) (298 \text{ K})} = 0.0500 \text{ mole}$$

TABLE 5.2

R, THE GAS CONSTANT

| R | UNITS |
|---|---|
| 0.08205 | $\ell \cdot \text{atm/mole} \cdot \text{K}$ |
| $8.314 \times 10^7$ | $\text{erg/mole} \cdot \text{K}$ |
| 8.314 | $\text{joule/mole} \cdot \text{K}$ |
| 62.36 | $\ell \cdot \text{Torr/mole} \cdot \text{K}$ |
| 1.987 | $\text{cal/mole} \cdot \text{K}$ |

Then because

$$0.0500 \text{ mole} = 2.20 \text{ g},$$

$$1 \text{ mole} = 2.20 \text{ g}/0.0500 = 44.0 \text{ g}$$

(g) In the same way we can calculate the number of moles of gas in 1 liter of air at STP, if we treat air as an ideal gas. Compute the number of moles of gas that occupy 1.00 $\ell$ at STP:

$$n = \frac{PV}{RT} = \frac{(1.00 \text{ atm})(1.00 \ \ell)}{\left(0.0821 \ \dfrac{\ell \cdot \text{atm}}{\text{mole} \cdot \text{K}}\right)(273 \text{ k})} = 0.045 \text{ mole}$$

There is an equivalent solution. Since 1 mole of gas occupies 22.4 liters at STP, one liter is

$$\frac{1.00 \ \ell}{22.4 \ \ell/\text{mole}} = 0.045 \text{ mole}$$

## 5.7 DALTON'S LAW.

We have just calculated the number of moles, and hence the number of molecules, in a liter of air at STP. Yet we know that dry air is a mixture of gases, predominantly nitrogen (79%), oxygen (20%), and argon (1%) with minor amounts of other gases (Table 5.3). Air also contains variable percentages of water, carbon dioxide, and numerous pollutants, the amounts of which are strongly dependent on location. It is surely correct to write the total number of molecules, and thus the number of moles, as a sum of the numbers of component molecules:

$$n_{\text{air}} = n_{O_2} + n_{N_2} + \text{minor amounts of others.}$$

The assumption that air is an ideal gas then allows us to write (ignoring the minor components of air for simplicity)

TABLE 5.3

COMPOSITION OF DRY AIR

| SUBSTANCE | MOLE FRACTION PERCENTAGE |
|---|---|
| Nitrogen ($N_2$) | 78.09 |
| Oxygen ($O_2$) | 20.95 |
| Argon (Ar) | 0.93 |
| Carbon Dioxide ($CO_2$) | 0.03 |
| Neon (Ne) | $1.8 \times 10^{-3}$ |
| Helium (He) | $5.24 \times 10^{-4}$ |
| Krypton (Kr) | $1.0 \times 10^{-4}$ |
| Hydrogen ($H_2$) | $5.0 \times 10^{-5}$ |
| Xenon (Xe) | $8.0 \times 10^{-6}$ |
| Ozone ($O_3$) | $1.0 \times 10^{-6}$ |
| Radon (Rn) | $6.0 \times 10^{-18}$ |

$$n = n_{O_2} + n_{N_2} = \frac{PV}{RT}$$

Because $R$ is a fundamental constant, because the temperature is uniform throughout the sample, and because the volume is the same for both gases (a liter of air does not have the oxygen, say, in the top layer or along the left side), we postulate that

$$n = n_{O_2} + n_{N_2} = \frac{V}{RT}(P_{O_2} + P_{N_2}) \tag{5.20}$$

The pressures $P_{O_2}$ and $P_{N_2}$ are called the *partial pressures* of oxygen and nitrogen. They are the pressures we would measure for the individual gases if all other components were absent:

$$n_{O_2} = \frac{V}{RT}P_{O_2} \qquad\qquad n_{N_2} = \frac{V}{RT}P_{N_2}$$

This example shows that the application of the ideal-gas equation to a mixture of gases includes the assumption that the total pressure is the sum of the partial pressures. In our example,

$$P_{\text{tot}} = P_{O_2} + P_{N_2} \tag{5.21}$$

This is an example of *Dalton's law.* It is, of course, subject to experimental test. It is found to be valid over the same range of pressure in which the component gases obey the ideal-gas law.

When does Dalton's law fail? One system in which it fails is the mixture of the gases ammonia, $NH_3$, and hydrogen chloride, $HCl$. In a mixture of these two gases the total pressure is far below the sum of the partial pressures of the two components. In fact, at pressures close to atmospheric pressure a mixture of the two gases $NH_3$ and $HCl$ gives a solid compound with the composition $NH_4Cl$ (ammonium chloride). The compound formation is evidence that the molecules of the two gases interact strongly. In practice, we observe large deviations from Dalton's law in all cases where other evidence of intermolecular interactions is available. Therefore, we suspect that Dalton's law (and in turn the ideal-gas equation) is valid only for non-interacting molecules. To prove this, we turn to the kinetic theory.

## 5.8  KINETIC THEORY OF GASES.

The wide validity of Dalton's law and the ideal-gas law implies that interactions between molecules of gases generally can be ignored at low pressures. The absence of large interactions at low pressures means that the properties of bulk samples of gas are rather simply related to those of the molecules; the only complications appear in the statistical methods required to handle the immense numbers of independent molecules. The study of gases at the molecular level is called the kinetic theory of gases, for the molecules in the gas are in motion.

## a. Boyle's Law.

Our first task in the kinetic theory is a derivation of Boyle's law. We begin with a single molecule inside a rectilinear box with sides $a$, $b$, and $c$ and volume

$$V = abc$$

The pressure within the box is produced by collisions of the molecule with the walls. The molecule is assumed to be a point-mass ($m$) with velocity $u$. The second assumption is that the molecule rebounds elastically every time it collides with a wall. That is, its kinetic energy is unchanged by the collision, although its direction is changed. Let us concentrate on the motion back and forth between faces $A$ and $A'$ (Figure 5.5); then we need only know the $x$ component of the molecule's velocity,[*] $u_x$. The molecule will strike face $A$ every $\Delta t$ seconds

$$\Delta t = 2a/u_x \qquad \text{(sec/collision)} \qquad (5.22)$$

for it must cover a distance parallel to the $x$ axis of $2a$ between successive collisions with one face. A perfect rebound requires conservation of momentum:

$$p(\text{momentum before collision}) = p(\text{momentum after collision})$$

$$mu_x \quad + \quad 0 \quad = \quad m(-u_x) \quad + \quad p \qquad (5.23)$$

molecule    wall    molecule    wall

The collision simply reverses the direction of $u_x$, and the conservation equation requires transfer of momentum $2mu_x$ to the wall for this reversal. If we call $\Delta p$ the change in momentum of the wall, then

$$\Delta p = 2mu_x \qquad (\text{g} \cdot \text{cm/sec} \cdot \text{collision}) \qquad (5.24)$$

---

[*]From elementary physics we know that the components of velocity are independent. Motion in the $y$ and $z$ direction and collisions with the other faces are irrelevant to our immediate discussion.

**FIGURE 5.5** Co-ordinates for derivation of Boyle's law.

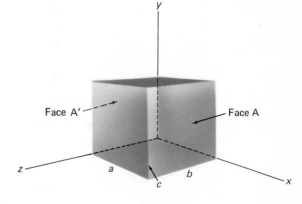

In one second the molecule transfers to the wall a total momentum of

$$\frac{\Delta p}{\Delta t} = \frac{2mu_x}{2a/u_x} = \frac{mu_x^2}{a} \qquad \left(\frac{\text{g} \cdot \text{cm}}{\text{sec}^2}\right) \qquad (5.25)$$

where we have used Equations (5.24) and (5.22). Momentum transfer per unit time has the units of force; in fact, Newton's first law defines force in just this way:

$$f = \frac{\Delta p}{\Delta t} = \frac{mu_x^2}{a} \qquad \left(\frac{\text{g} \cdot \text{cm}}{\text{sec}^2}\right) \qquad (5.26)$$

Pressure is defined as force per unit area, and the area of face $A$ is $bc$, so

$$P = f/bc = \frac{mu_x^2}{abc} = \frac{mu_x^2}{V} \qquad (5.27)$$

or

$$PV = mu_x^2 \qquad (5.28)$$

Collisions with the wall change the sign, but not the magnitude, of $u_x$, so $mu_x^2$ is a constant, and Equation (5.28) is Boyle's law for a single molecule.

The transfers of momentum from a large number of independent molecules are simply additive. That is,[*]

$$P = \sum_{i=1}^{N} P_i = \sum_{i=1}^{N} \frac{m_i u_{xi}^2}{V} \qquad (5.29)$$

Equation (5.29) is Dalton's law; its derivation required the assumption that each molecule acted independently of the others so that the transfers of momentum could be added.

Identical molecules all have the same mass, in which case Equation (5.29) can be written as

$$P = \frac{m}{V} \sum_{i=1}^{N} u_{xi}^2 \qquad (5.30)$$

We cannot assume that all molecules in the sample have the same velocity components, and in fact they most definitely do not. However, we can introduce an average value for $u_x^2$, which is defined in the usual way for averages:

$$\overline{u_x^2} = \frac{\sum\limits_{i=1}^{N} u_{xi}^2}{N} \qquad (5.31)$$

---

[*]The symbol $\sum\limits_{i=1}^{N}$ means the sum over all the values of any variable which is written with a subscript $i$. With three molecules,

$$\sum_{i=1}^{3} P_i = P_1 + P_2 + P_3$$

and

$$\sum_{i=1}^{3} \frac{m_i u_{xi}^2}{V} = \frac{1}{V} \sum_{i=1}^{3} m_i u_{xi}^2 = \frac{1}{V}(m_1 u_{x1}^2 + m_2 u_{x2}^2 + m_3 u_{x3}^2)$$

Equation (5.31) is the $x$ component of the *mean-square velocity*. It is not the same as the $x$ component of the mean velocity squared (written $\overline{u_x}^2$). The latter quantity is zero, because there are as many molecules going backward as forward.

It is more convenient to write our equation in terms of molecular speeds instead of velocity components. The magnitude of the velocity is the speed, $c$; the speed is related to the velocity components by the equation

$$c^2 = u_x^2 + u_y^2 + u_z^2 \qquad (5.32)$$

Speed can never be negative because it has no direction associated with it; that is, by convention

$$c = +\sqrt{c^2}$$

Equation (5.32) is valid for mean-square values

$$\overline{c^2} = \overline{u_x^2} + \overline{u_y^2} + \overline{u_z^2} \qquad (5.33)$$

The labeling of axes is arbitrary and, therefore, the mean-square velocity components of the gas molecules must be the same in all directions; hence

$$\overline{u_x^2} = \overline{u_y^2} = \overline{u_z^2} = \frac{\overline{c^2}}{3} \qquad (5.34)$$

Combination of Equations (5.30), (5.31), and (5.34) yields

$$PV = m \sum_{i=1}^{N} u_{xi}^2 = Nm\,\overline{u_x^2} = \frac{Nm}{3}\,\overline{c^2} \qquad (5.35)$$

This equation is Boyle's law for a collection of $N$ molecules of mass $m$.

Collisions between molecules become inevitable as the number of molecules increases. We now show that billiard-ball (perfectly rebounding) collisions have no effect on the equations just derived. Consider two very special cases, both of which involve only two molecules with identical velocities bouncing back and forth between $A$ and $A'$. In case I, the molecules start side by side, and in case II they fly in line. As time passes, the sequence diagrammed in Figure 5.6 will occur. In the time interval covered, the molecules have completed a path and returned to their original positions. In case II the molecules collided twice, but in both cases there were two collisions with each wall. The collisions between molecules have no effect on the collisions with walls, and thus no effect on the pressure. The result is a general one; with many molecules in the container, molecules in the middle seldom make it to the wall; but molecules near the wall seldom get away from the wall, and hence they strike it with great frequency.

## b.  Kinetic Energy and Temperature.

The translational kinetic energy of a single molecule is

$$\epsilon_i = (1/2)\,m_i c_i^2$$

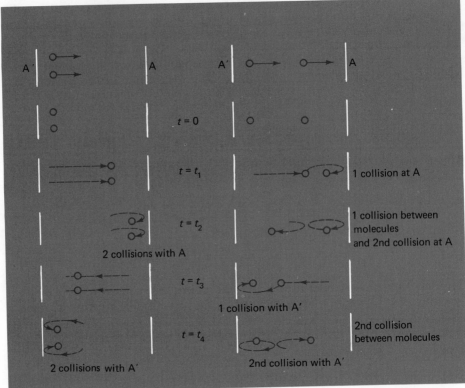

**FIGURE 5.6**   Illustration of the fact that billiard-ball collisions do not affect the pressure.

In the absence of interactions between particles[*] the total energy for a large number of identical particles is the sum of individual particle energies:

$$E = \sum_{i=1}^{N} \epsilon_i = (1/2)\, m \sum_{i=1}^{N} c_i^2 = (1/2)\, Nm \, \overline{c^2} \qquad (5.36)$$

Combination of Equations (5.36) and (5.35) yields

$$Nm \, \overline{c^2} = 2E = 3PV$$

or

$$PV = (2/3)\, E \qquad (5.37)$$

The ideal-gas equation gives $PV = nRT$, and thus

$$PV = (2/3)\, E = nRT \qquad (5.38)$$

The total translational kinetic energy per mole of gas $(E/n)$ is

$$\boxed{\dfrac{E}{n} = (3/2)\, RT} \qquad (5.39)$$

---

[*]As used in the kinetic theory, "interactions" does not include billiard-ball collisions.

The translational kinetic energy of a given amount of gas depends only on the temperature of the gas. Equation (5.39) gives us a molecular interpretation of the temperature; in an ideal gas the temperature is a measure of the average translational kinetic energy of the molecules.

In Chapter 3 we stated that an aggregate has to have a minimum binding energy of about 0.2 eV in order to survive the thermal disruptions which it encounters. One mole of an ideal gas has a translational kinetic energy of $3RT/2$, and one molecule has an energy of

$$\bar{\epsilon} = \frac{3}{2} \frac{R}{N_0} T$$

where $N_0$ is Avogadro's number. At 25°C, $\bar{\epsilon}$ is equal to

$$\bar{\epsilon} = \frac{3}{2} \frac{(8.31 \times 10^7 \text{ erg/mole} \cdot \text{K})(298 \text{ K})}{(6.02 \times 10^{23} \text{ molecule/mole})} = 6.17 \times 10^{-14} \text{ erg/molecule}$$

In electron volts, this is

$$\bar{\epsilon} = (6.17 \times 10^{-14} \text{ erg}) (6.24 \times 10^{11} \text{ eV/erg}) = 0.038 \text{ eV/molecule}$$

It is clear that a molecule with a binding energy of 0.2 eV will easily survive a normal collision at room temperature in which, at worst, all the translational kinetic energy of two molecules can be converted into potential energy by deforming the molecules.

### c. Diffusion.

If we take pure samples of two different gases at the same pressures and connect their containers, the two gases will become intimately intermixed (Figure 5.7). Spontaneous mixing occurs by a process called dif-

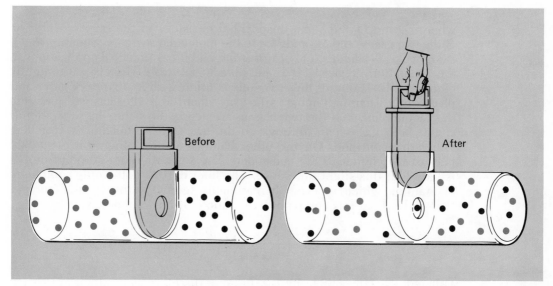

**FIGURE 5.7** Diffusion.

fusion. The rate at which diffusion occurs should be related to the speeds at which two kinds of molecules move. The (square) root (of the) mean square speed, $c_{rms} = (\overline{c^2})^{1/2}$, can be obtained by substitution of Equation (5.36) into Equation (5.39) (for 1 mole, $n = 1$ and $N = N_0$)

$$\frac{1}{2} N_0 m \overline{c^2} = \frac{3}{2} RT \qquad (5.40)$$

where $N_0 m$ is the molecular weight in grams (call it $M$). Then

$$c_{rms}^2 = \overline{c^2} = \frac{3RT}{M} \qquad (5.41)$$

and the root-mean-square speed is

$$\boxed{c_{rms} = \left(\frac{3RT}{M}\right)^{1/2}} \qquad (5.42)$$

Equation (5.42) can be used to estimate the magnitudes of molecular speeds in gases. For $N_2$ at room temperature,

$$c_{rms} = \left[\frac{3 (8.3 \times 10^7 \text{ erg/mole} \cdot \text{K}) (298 \text{ K})}{(28 \text{ g/mole})}\right]^{1/2} = 5.1 \times 10^4 \text{ cm/sec}$$

$$= 1{,}140 \text{ miles/hr}$$

For comparison, a golf ball can be hit with a velocity of $3.0 \times 10^3$ cm/sec, and a rifle bullet reaches a velocity of $3.2 \times 10^4$ cm/sec.

The rate of diffusion (escape of the gas through a small hole) of a gas is directly proportional to the root-mean-square speed of the molecules. From Equation (5.42), the ratio of diffusion rates of two different gases $D_1/D_2$ is therefore given by

$$\frac{D_1}{D_2} = \frac{c_{rms,1}}{c_{rms,2}} = \left(\frac{M_2}{M_1}\right)^{1/2} \qquad (5.43)$$

where $D_1$ and $D_2$ are some unspecified measures of the rates of diffusion of the two gases, and $M_2$ and $M_1$ are the molecular weights. Equation (5.43) is known as *Graham's law of diffusion*, and it has practical applications in the separation of gases. It is possible to fabricate vessels with slightly porous walls. If one of these vessels is filled with a mixture of gases and placed in a vacuum, Graham's law says that the lighter gas will pass into the vacuum faster. If the two masses are nearly the same, the separation is slight; both gases pass through so the gas mixture outside is enriched, rather than purified. On the other hand, it is an economical procedure because no chemicals are consumed. The separation of uranium isotopes is based on this slight difference. The fluoride $UF_6$ is the only compound of uranium which is a gas at reasonable temperatures; for the isotopes $^{238}_{92}U$ and $^{235}_{92}U$ in this compound, the ratio of diffusion rates is

$$\left(\frac{238 + 114}{235 + 114}\right)^{1/2} = 1.004$$

Although the $UF_6$ which diffuses through the barrier initially is only 0.4% richer in $^{235}U$ than the starting mixture, a sequence of vessels and recycling can be used to obtain the pure isotope.

## 5.9 DISTRIBUTION FUNCTIONS.

In the derivation of Boyle's law both the velocity and the momentum of the molecule entered into the calculation of the pressure, and their combination led to the appearance of $\overline{c^2}$ in the equation. Many other properties of gases depend only on the first power of $c$, and thus $\bar{c}$ appears in equations for some properties of gases. There are other important parameters which can have a wide range of values in a gaseous sample, such as translational kinetic energy. For many purposes it is not sufficient to know an average value for these parameters; in a complete theory we must know how the molecules are distributed among the possible values. The algebraic function which contains this information is called the *distribution function*. It is generally written in the form:

$$\left\{ \begin{array}{l} \text{the fraction of the total molecules} \\ \text{with a value of } y \text{ between} \\ y_0 \text{ and } y_0 + \Delta y \end{array} \right\} = \left\{ \begin{array}{l} \text{a function of } y \text{ (and other} \\ \text{variables) times the allowable} \\ \text{uncertainty in } y, \text{ which is } \Delta y \end{array} \right\}$$

We have already encountered one distribution function, the electron distribution function. In that case we wrote

$$\left\{ \begin{array}{l} \text{probability that the} \\ \text{electron is between} \\ r_0 \text{ and } r_0 + \Delta r \end{array} \right\} = 4\pi r^2 \, \psi^2(r, \theta, \phi) \, \Delta r$$

We are concerned now with other distribution functions, those for molecular velocities and speeds.

The probability that a molecule has a speed between $c$ and $c + \Delta c$ (where $\Delta c \ll c$), or equivalently, the *fraction* of molecules with speeds between $c$ and $c + \Delta c$, is given by

$$\left\{ \begin{array}{l} \text{fraction of molecules} \\ \text{with speeds between} \\ c \text{ and } c + \Delta c \end{array} \right\} = p(c) \, \Delta c$$

where $p(c)$ is the distribution function for molecular speeds. The derivation of an expression for $p(c)$ is complex, and we shall present here only an heuristic development of the expression for $p(c)$.

The speed is related to the velocity components by the equation

$$c^2 = u_x^2 + u_y^2 + u_z^2$$

This is the equation for a sphere of radius $c$, if we use a coordinate system with $u_x$, $u_y$, and $u_z$ as axes. The volume of a very thin spherical shell bounded by the surfaces at $c$ and $c + \Delta c$ is equal to the surface area of the sphere, $4\pi c^2$, times the thickness, $\Delta c$, of the shell; that is, $4\pi c^2 \Delta c$. The probability that a molecule of mass $m$ at temperature $T$ has a speed $c$ is proportional to the Boltzmann factor[*] $e^{-mc^2/2kT}$, where $k$ is Boltzmann's constant. The Boltzmann constant is the gas constant per molecule,

$$k = R/N_0 = \frac{8.314 \text{ joule} \cdot \text{mole}^{-1} \cdot \text{K}^{-1}}{6.022 \times 10^{23} \text{ molecules} \cdot \text{mole}^{-1}}$$

---

[*]Note that since $\epsilon = \frac{1}{2} mc^2$, the Boltzmann factor is of the form $e^{-\epsilon/kT}$.

$$k = 1.381 \times 10^{-23} \text{ J} \cdot \text{K}^{-1} \tag{5.44}$$

The fraction of the molecules that lies within the shell of volume $4\pi c^2 \Delta c$ is proportional to the product of the volume of the shell and the probability that a molecule has a speed $c$:

$$p(c)\Delta c = A(4\pi c^2 \Delta c)e^{-mc^2/2kT} \tag{5.45}$$

The constant $A$ is a normalization constant; the value of $A$ is fixed by the requirement that the sum of the fractions of molecules in all of the intervals must be unity. The expression for $A$ is

$$A = (m/2\pi kT)^{3/2} \tag{5.46}$$

and thus

$$p(c)\Delta c = 4\pi \left(\frac{m}{2\pi kT}\right)^{3/2} c^2 e^{-mc^2/2kT} \Delta c \tag{5.47}$$

This is the Maxwell-Boltzmann distribution law for molecular speeds. Equation (5.47) is plotted in Figure 5.8 for several temperatures. The units (Figure 5.8) are chosen so that one figure will serve all gases.

It is interesting that the distribution function broadens as the temperature rises (Figure 5.8); the fraction of very fast (often called "hot") molecules increases much more quickly than the mean speed of all the molecules. This has important implications in chemical reactions, for a high-speed molecule has extra kinetic energy which can be used to deform itself and another molecule with which it collides. The speed distribution function can easily be converted to another useful distribution function, that for translational kinetic energy. Because

$$E = \frac{1}{2}mc^2$$

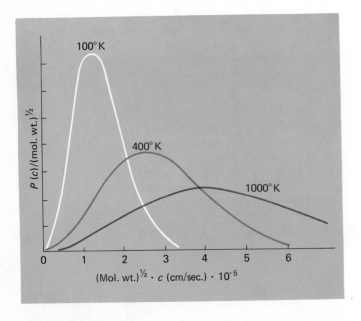

**FIGURE 5.8** Distribution functions for molecular speeds. The units are chosen so that one plot serves for all gases.

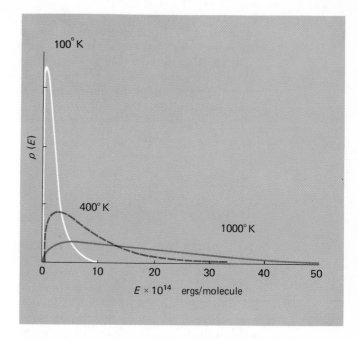

**FIGURE 5.9** Kinetic energy distribution function.

and*

$$\Delta E = mc\Delta c$$

Equation (5.47) becomes

$$p(E)\Delta E = \left(\frac{4}{\pi}\right)^{1/2} \frac{1}{(kT)^{3/2}} E^{1/2} e^{-E/kT} \Delta E \qquad (5.48)$$

The translational kinetic energy distribution function is plotted for several temperatures in Figure 5.9.

The distribution functions describe the fact that a gas is an assembly of molecules with a wide *range* of velocities and energies.

## 5.10 HEAT CAPACITY AND THE EQUIPARTITION OF ENERGY.

The heat capacity of a gas is a quantity that is of great utility in thermodynamic calculations. It is crudely defined as the quantity of heat in calories that must be added to a sample to cause a temperature rise of 1 K. We convert this to an intensive quantity by taking a standard sample size of one mole; thus, we have the *molar heat capacity* with units of calories per mole per K.†

---

*An increase in speed of $\Delta c$ produces an increase in energy $\Delta E$

$$\Delta E = \frac{1}{2} m(c + \Delta c)^2 - \frac{1}{2} mc^2 = \frac{m}{2} (c^2 + 2c\Delta c + \Delta c^2 - c^2) = \frac{m}{2} (2c\Delta c + \Delta c^2)$$

which is, for small $\Delta c$ ($\Delta c \ll c$),

$$\Delta E = mc\Delta c$$

†In physics, the standard sample size is 1 g, and the heat capacity is called the *specific heat*. The SI units for heat capacity are joule per mole per degree Kelvin or $J \cdot mol^{-1} \cdot K^{-1}$.

The heat capacity of a substance depends on the conditions under which the heat is transferred to the sample. Consequently, it is necessary to define two heat capacities, one for processes carried out with constant volume ($C_V$), and one for processes carried out with constant pressure ($C_P$). When the volume is held constant, all of the energy added is invested in increased motion of the molecules in the gas. If the increased motion is all translational, then the addition of an increment of energy $\Delta E$ under these conditions produces a temperature rise $\Delta T$; from Equation (5.39),

$$\frac{\Delta E}{n} = (3/2)\,R\Delta T \qquad\qquad (5.49)$$

Then, for 1 mole ($n = 1$),

$$C_V = \Delta E/\Delta T = (3/2)R = (3/2)\,(1.99\ \text{cal/mole} \cdot \text{K}) = 2.99\ \text{cal/mole} \cdot \text{K} \quad (5.50)$$

All monatomic gases do have this heat capacity at ordinary temperatures.

Polyatomic gas molecules have heat capacities that are larger than 3.0 cal/mole · K and are temperature dependent. More energy must be added to produce a given temperature rise in a polyatomic gas than in a monatomic gas. The kinetic theory as we have developed it does not allow for rotation and vibration of the molecules, because these motions do not contribute to the pressure or to the average translational kinetic energy (and in turn the temperature of the gas). The classical assumption is that the energy of a molecule is equally partitioned among all of the possible motions (degrees of freedom) of the molecule; this is called the equipartition of energy. This assumption leads to the value

$$\frac{C_V}{n} = \frac{3}{2}R \;\; + \;\; \frac{3}{2}R \;\; + \;\; (3N\text{-}6)R \qquad (5.51)$$

$$\text{(translation)} \qquad \text{(rotation)} \qquad \text{(vibration)}$$

for a nonlinear $N$-atom molecule.* Each of the three translational degrees of freedom (in the $x$, $y$, and $z$ directions) contributes $RT/2$ to the translational energy and $R/2$ to the translational heat capacity. Each of the three rotational degrees of freedom (one around each axis) contributes $RT/2$ to the rotational energy and $R/2$ to the rotational heat capacity. An $N$-atom molecule has a total of $3N$ degrees of freedom, and thus there are $3N$-3(trans)-3(rot) = $3N$-6 vibrational degrees of freedom. Each vibrational degree of freedom contributes $RT$ to the energy and $R$ to the heat capacity.

In practice we find that $C_V/n$ is usually less than the value predicted by Equation (5.51), because the vibrational motions are not fully excited at ordinary temperature. The calculated value of $C_V$ given by Equation (5.51) is approached at high temperatures.

---

*For a linear $N$-atom molecule,

$$\frac{C_V}{n} = \frac{3}{2}R + R + (3N\text{-}5)R.$$

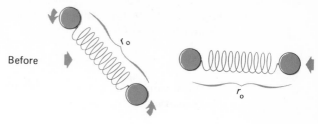

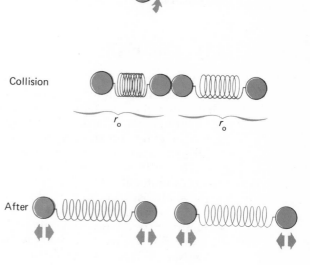

**FIGURE 5.10** Example of the conversion of translation and rotation to vibration by a collision.

Equipartition requires a free exchange of energy among the different types of molecular motions; for example, rotational energy must easily convert to vibrational energy upon collision (Figure 5.10). An average molecule has translational kinetic energy of $3kT/2$. This energy is quantized (the boundary conditions are provided by the walls of the container), but the gaps between translational energy levels are so small that they can be ignored, and the translational energy can be treated as a continuous function. The rotational and vibrational motions are quantized as well, but here the gaps are much larger. Can these gaps be ignored? It depends on the temperature. For example, the geometry of a particular collision might allow the conversion of 1/3 of the kinetic energy of translation into rotation. Is it possible to increase the rotational energy by exactly $(1/3)(3kT/2)$? If the spacing between the rotational energy levels is much smaller than $kT$, there is very likely to be some transition between rotational levels which matches $kT/2$ with near perfection, and the conversion is likely. But, if the rotational level spacing is comparable to or larger than $kT$, it is very unlikely that any rotational transition will exactly match $kT/2$. In the latter case it is unlikely that the kinetic energy will convert into rotational energy, and consequently the rotational motion of such a molecule will not contribute to the heat capacity.

At room temperature the rotational levels for all but the lightest molecule ($H_2$) are closely spaced (compared to $kT$), and their rotations contribute the classical value to the heat capacity. It is usually necessary to exceed 1000 K before $kT$ becomes much larger than the vibrational level spacings. Therefore, molecular gases with a nonlinear structure have heat capacities of about[*]

$$C_V = (3/2)R + (3/2)R = 3R$$

per mole under common conditions. The vibrational motion accounts for the observed temperature dependence of $C_V$.

## 5.11 COLLISIONS.

A gas consists of widely separated molecules that are moving with high velocities. Molecules must collide occasionally. Collisions are of vital importance in chemistry, because two molecules must come together in order to react. An exact derivation of the expression for the frequency of intermolecular collisions is quite involved, but an approximate expression can be obtained by a simple argument.

The collision-frequency problem is simplified if one of two different types of molecules in a gas mixture is moving much faster than the other. This is almost true for a mixture of light and heavy molecules, such as $H_2 + I_2$ (see Equation (5.42)). As a $H_2$ molecule moves through a volume containing (relatively motionless) $I_2$ molecules, it sweeps out a cylinder (Figure 5.11). The cylinder has base area of $\pi\sigma^2$ (cm²) and height of $c$ (cm/sec).[†] The light molecule sweeps out a volume of $\pi\sigma^2 \bar{c}$ (cm³/sec). The base diameter of the cylinder ($=2\sigma$) is chosen so that, if a heavy molecule happens to be within this volume, it gets hit. The number of collisions per second is then the volume swept out per second times the density of the heavy molecules expressed in molecules per unit volume:

$$Z = \pi\sigma^2 \bar{c} n_B \qquad \text{collisions/sec} \tag{5.52}$$

---

[*]A linear molecule can rotate only about two axes (why?), so its expected heat capacity is $(5/2)R$.

[†]The *average* or *mean speed*, $\bar{c}$, and the root-mean-square speed, $c_{rms}$, are not quite the same:

$$\bar{c} = (8RT/\pi M)^{1/2}; \quad c_{rms} = (3RT/M)^{1/2}; \quad \bar{c} = 0.915\, c_{rms}.$$

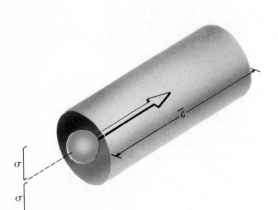

**FIGURE 5.11**   Cylinder swept out by a moving gas molecule.

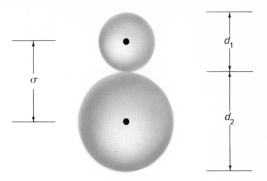

**FIGURE 5.12** Calculation of the diameter of the cylinder swept out.

Of course, the crucial step is the selection of $\sigma$. The marginal case is the glancing blow (Figure 5.12). If the distance between the centers of the molecules (which are assumed to be hard spheres) is less than $(d_1 + d_2)/2$, then the molecules will collide; if it is greater, then there will be no collision. Therefore, we set

$$\sigma = (d_1 + d_2)/2 \qquad (5.53)$$

If we had $n_A$ light molecules per unit volume, and if each light molecule had the mean speed $\bar{c}$, then there would be

$$Z_{AB} = n_A Z = \pi \sigma^2 \bar{c} n_A n_B = \pi \sigma^2 n_A n_B (8RT/\pi M_A)^{1/2} \text{ collisions/cm}^3\text{sec} \quad (5.54)$$

A correct treatment takes into account: (a) that the molecules do not all have the same speed (velocity distribution); and (b) that it is the *relative* velocity of the colliding molecules that is important. The result of this much more complex analysis is

$$Z_{AB} = \pi \sigma^2 n_A n_B \left( \frac{8RT}{\pi \mu} \right)^{1/2}. \qquad (5.55)$$

The only difference between Equations (5.54) and (5.55) is that (5.54) contains $M_A$, whereas (5.55) contains $\mu$. The quantity $\mu$ is the *reduced mass* of the A-B pair,

$$\frac{1}{\mu} = \frac{1}{M_A} + \frac{1}{M_B} \qquad (5.56)$$

Note that $\mu \simeq M_A$, if $M_B \gg M_A$. For collisions between like molecules (A = B) the right hand side of Equation (5.55) must be divided by 2 to avoid counting each collision twice; i.e., $Z_{AA} = \dfrac{Z_{AB}}{2}$.

The indefinite boundaries of real molecules make the calculation of hard-sphere diameters all but impossible. Values can be deduced from measurement of properties of gases that involve collisions. Examples of such properties are diffusion speeds, flow rates, and heat conductivities of gases. A representative set of values is given in Table 5.4. There is a range for each molecule because the different methods give different results.

As an example of the application of Equation (5.55), we shall calculate

**TABLE 5.4**

COLLISION DIAMETERS FOR COMMON GASES*

| | |
|---|---|
| He | 2.1–2.2 Å |
| Ne | 2.4–2.6 Å |
| Ar | 3.5–3.7 Å |
| Kr | 4.0–4.2 Å |
| Xe | 4.6–4.9 Å |
| $H_2$ | 2.5–3.1 Å |
| $N_2$ | 3.5–4.3 Å |
| $O_2$ | 3.4–4.0 Å |
| $CO_2$ | 4.2–5.8 Å |
| $H_2O$ | 4.7–6.0 Å |
| $CH_4$ | 3.8–4.8 Å |
| $C_2H_6$ | 5.3–7.0 Å |
| HCl | 4.3–4.5 Å |

1 Å = 0.1 nm.

*Adapted from W. Kauzmann, *Kinetic Theory of Gases* (W. A. Benjamin, New York, 1966).

the number of $N_2$–$O_2$ collisions in a cubic centimeter of air at 0°C and 1 atm pressure. For simplicity we assume that air is 20% oxygen and 80% nitrogen. At STP, 1 cm³ of ideal gas contains

$$\frac{6.02 \times 10^{23} \text{ molecules/mole}}{22{,}400 \text{ cm}^3/\text{mole}} = 2.69 \times 10^{19} \text{ molecules/cm}^3$$

For collisions between nitrogen and oxygen,

$$\sigma = (1/2)(3.6 + 3.8) = 3.7 \text{ Å}$$

Because only 1/5 of the $2.69 \times 10^{19}$ molecules/cm³ are $O_2$ molecules, and 4/5 are $N_2$ molecules, we have

$$Z_{N_2,O_2} = 3.14 \left( \frac{2.69 \times 10^{19}}{5} \frac{O_2 \text{ molecules}}{\text{cm}^3} \right) \left( \frac{4 \times 2.69 \times 10^{19}}{5} \frac{N_2 \text{ molecules}}{\text{cm}^3} \right)$$

$$\times (3.7 \times 10^{-8} \text{ cm})^2 \times \left( \frac{8 \times 8.3 \times 10^7 \text{ ergs/mole} \cdot \text{K} \times 273 \text{ K}}{3.14 \times (32 \times 28/60) \text{ g/mole}} \right)^{1/2}$$

$$Z_{N_2,O_2} = 3.1 \times 10^{28} \text{ collisions/cm}^3 \text{ sec}$$

Collisions are very frequent occurrences in a gas.

## 5.12   REAL GASES.

All real gases are imperfect gases; they do not obey the ideal-gas equation or Dalton's law exactly. An understanding of the causes of the deviations from ideal-gas behavior make it possible to modify the ideal-gas equation to achieve a more widely useful equation of state.

**van der Waals Equation.**    The fact that gases condense to liquids or solids shows that attractions between molecules exist. In Figure 5.13 we

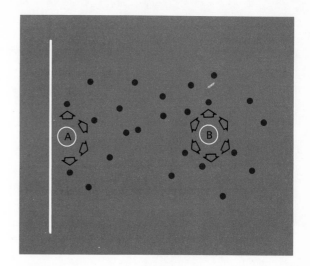

**FIGURE 5.13** Attractive forces between molecules. Molecule A feels a net force toward the right; molecule B feels no net force.

see two molecules in a dense gas, one on the surface and one within the gas. Both molecules are attracted by their neighbors; the molecule within the gas is pulled in all directions equally (on the average), but the molecule on the surface is pulled away from the wall. Thus, the surface molecules are always slowing as they hit the wall; their momentum transfer is less than it would be in the absence of attractive forces.

Because attractive forces lower the pressure (the momentum transfer), the observed pressure should be corrected upwards to give an ideal pressure for use in the zero attraction, ideal-gas equation. From experiments at high pressure on real gases we find that a reasonable expression for the correction is

$$P(\text{ideal}) = P(\text{observed}) + a\left(\frac{n}{V}\right)^2 \tag{5.57}$$

where $a$ is an experimentally determined proportionality constant.

The derivation of Boyle's law from the kinetic theory used the assumption that each molecule could fly around the complete volume of the container. This is an assumption that the molecules are points, for if they have finite volume, then that volume is not free for another molecule to pass through. We must correct for this unavailable volume; we write the volume available to the molecules as

$$V_{\text{container}} - n\,b \tag{5.58}$$

where $n$ is the number of moles of gas, and $b$ is a parameter which is related to the volume actually occupied by the molecules in a mole of gas.

The ideal-gas equation then becomes

$$\left(P + a\left(\frac{n}{V}\right)^2\right)(V - n\,b) = nRT \tag{5.59}$$

This equation is called the van der Waals equation. It can be used to describe the $P$, $V$, $T$ behavior of a gas over a wider range of these variables than can the ideal-gas equation. Values of the van der Waals constants $a$ and $b$ for a few gases are given in Table 5.5.

**TABLE 5.5**

VAN DER WAALS CONSTANTS FOR COMMON GASES*

| GAS | $a(\ell^2 \cdot atm/mole^2)$ | $b(\ell/mole)$ |
|---|---|---|
| He | 0.0341 | 0.0237 |
| $H_2$ | 0.244 | 0.0266 |
| $N_2$ | 1.39 | 0.0391 |
| CO | 1.49 | 0.0399 |
| $O_2$ | 1.36 | 0.0318 |
| $C_2H_4$ | 4.47 | 0.0571 |
| $CO_2$ | 3.59 | 0.0427 |
| $NH_3$ | 4.17 | 0.0371 |
| $H_2O$ | 5.46 | 0.0305 |

*From W. J. Moore, *Physical Chemistry* (Prentice-Hall, Inc., Englewood Cliffs, New Jersey, 1962).

## 5.13  CONCLUSION.

Our initial goal was to obtain a familiarity with the basic properties of bulk samples of gases. Most of the essential properties of gases are summarized by the equation relating the pressure, volume, and temperature of the gas, that is, the equation of state for the gas. The study of the properties of gases is in large part a search for equations of state and for distribution functions. The search is moderately successful because of the weak interactions between molecules. There is no perfect equation of state, and the price for greater accuracy is always greater complexity. For this reason we always use the simplest equation of state which is sufficiently accurate for the problem at hand.

## PROBLEMS

1. A sample of gas occupies $1.39 \ \ell$ at STP. What volume does it occupy at $P = 740$ Torr and $t = 25°C$?

2. What volume does 2.00 g of $O_2$ occupy at 0.75 atm and 170°C? (a) Assume ideal-gas behavior; (b) assume van der Waals-gas behavior.

3. Classify the following as intensive or extensive properties: pressure, temperature, volume, number of moles, density, root-mean-square speed.

4. What is the total pressure if 0.75 g of $H_2$ is mixed with 3.00 g of $NH_3$ in a 10.0 $\ell$ flask at 25°C?

5. A mixture of $N_2$ and $H_2$ has a density of 0.600 g/$\ell$ at 22°C and 730 Torr. How many moles of $N_2$ and $H_2$ are there in the mixture?

6. One liter of air at 1.0 atm and 20°C weighs 1.2 g. Calculate an average molecular weight for air. Is this consistent with the composition of air given in Table 5–3?

7. Calculate the standard atmospheric pressure in pounds per square inch.

8. Calculate the average translational energy and root-mean-square speeds for each of the following molecules: $H_2$, $CO$, $I_2$, $UF_6$.

9. Which gas should have the greater diffusion rate, $H_2$ or He? He gas passes through thin glass walls rather easily, but $H_2$ does not. Can you think of a reason for this?

10. Recent measurements have shown that the atmosphere of Venus is mostly carbon dioxide. At the surface the temperature is about 800°C, and the pressure is about 75 atm. In the unlikely chance that a resident of Venus defined an STP and took those values, what value would he find for the volume of a mole of ideal gas?

11. A tire is inflated to a volume of 12.0 $\ell$ at 25°C and a *gauge* pressure of 32 psi. In use, the tire temperature reaches 50°C. What is the absolute pressure in the tire at 50°C?

12. A 1 $\ell$ sample of gas that contains only one compound of nitrogen and oxygen weighs 4.82 g at STP. What is its molecular weight? What is its formula?

13. Sonic booms can cause structural damage. Calculate the total force in pounds due to a sonic boom with a pressure pulse of $10^{-3}$ atm on a wall 12 ft × 36 ft.

14. Calculate the averages $\bar{u}_x$, $\overline{u_x^2}$, and $(\overline{u_x^2})^{1/2}$ for the following set of molecules:

$$u_{x1} = 1, \ u_{x2} = 2, \ u_{x3} = 1, \ u_{x4} = 0, \ u_{x5} = 3$$

15. Estimate the heat capacities of the following gases at room temperature and at 1000 K: Ar, $CO_2$, $N_2$, Ne, $NH_3$, $H_2O$.

16. Calculate the number of collisions per $cm^3$ between oxygen and nitrogen molecules in a container which contains air at a good laboratory vacuum ($P = 10^{-6}$ Torr).

17. Equation (5.54) is an approximate expression for the number of collisions between light molecules ($H_2$) and heavy ones ($I_2$). Compare the number calculated via the approximation with the correct one (Equation (5.55)).

18. A skilled skin diver can dive without tanks to depths of about 50 ft. Assume that the lungs of such a diver hold 6 $\ell$ of air. What volume does this air occupy when the diver is 50 ft under water?

19. (a) Archimedes' principle states that the buoyancy of an object immersed in a fluid is equal to the mass of fluid displaced. What volume of He would be needed to lift a man on a day when the temperature was 17°C? Assume that the mass of the man, basket, and balloon is 300 kg, and the pressure is 1 atm. (b) Repeat the calculations of part (a) to find the

amount of He needed at 60,000 ft. Assume the pressure is 0.065 atm and the temperature is 225 K.

20. A large cylinder of gas contains about 2000 ft³ of gas measured at STP. Calculate, both from the ideal-gas law and from the van der Waals equation, the pressure inside a full cylinder of $N_2$ gas at 25°C. Assume that the volume inside the cylinder is 4 ft³.

21. About 10% of the molecules in a sample of $N_2$ have a speed of $800 \pm 50$ meters/sec at 1250 K. What percentage of these molecules lies within this speed range if the temperature is lowered to 750 K?

22. The speed of sound in a gas is given by

$$u_{\text{sound}} = \left\{ \frac{C_V + R}{C_V} \left( \frac{RT}{M} \right) \right\}^{1/2}$$

Calculate the speed of sound in air. Explain how your result can be used to compute how far away from you a lightning bolt strikes in an electrical storm.

## References

5.1. Hildebrand, J. H., *An Introduction to Molecular Kinetic Theory* (Reinhold, New York, 1963).
5.2. Kauzmann, W., *Kinetic Theory of Gases* (W. A. Benjamin, Inc., New York, 1966).

## CHAPTER 6

# CRYSTALS

"The application of the scientific method does not consist solely of the routine use of logical rules and procedures. Often a generalization that encompasses many facts has escaped notice until a scientist with unusual insight has discovered it. Intuition and imagination play an important part in the scientific method."

*Linus Pauling**

## 6.1 INTRODUCTION.

From gases, in which the molecular distribution is completely chaotic, we turn to crystals, which are characterized by near perfect order. We are interested in solids for many reasons. Molecular structures can be determined from the diffraction pattern of x-rays scattered by a crystal, and this is the most direct method for the determination of the structures of molecules. Also, in some solids we can influence the behavior of electrons within the solid by changing the electronic energy levels. The amazing variety of solid-state electronic devices is based on this capability.

A solid is defined as any sample of matter whose volume and shape are independent of the volume and shape of its container. There are two types of solids, amorphous solids and crystals. Amorphous solids can be visualized as extremely dense, frozen gases; there is no long-range order. Glass is an example of an amorphous solid.

In a crystal each molecule or ion has a definite position and orientation relative to its neighbors.† The whole crystal can be reproduced by replicating a simple three-dimensional unit, the *unit cell*, in all directions throughout the sample (Figure 6.1). Unit cells usually contain only a small number (1 to 12) of atoms, ions, or molecules.‡ The regular array of molecules, atoms, or ions is called the crystal lattice.

---

*Linus Pauling, *General Chemistry*, 3rd Ed. (W. H. Freeman and Co., San Francisco, 1970) pp. 14–15.

†We shall ignore for the present those crystals that involve disorder in the orientation of molecules or ions that are located on fixed lattice positions. Also, there are some crystals in which species rotate about a fixed lattice position (e.g., $H_2(s)$).

‡The number of molecules per unit cell, $z$, is given by

$$z = N_0 \rho V / M$$

where $N_0$ is Avogadro's Number, $\rho$ is the density, $V$ is the volume of the unit cell, and $M$ is the molecular weight.

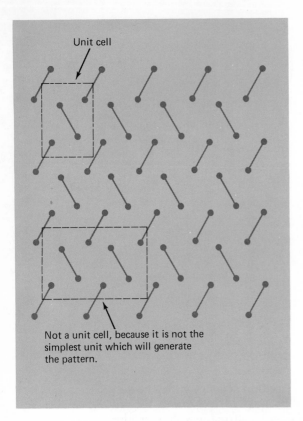

Unit cell

Not a unit cell, because it is not the simplest unit which will generate the pattern.

**FIGURE 6.1**   Two-dimensional crystal of diatomic molecules.

The repetitive pattern of unit cells accounts for many of the properties of crystals. Consider a simple, regular array of the most symmetrical units, spherical ions. Figure 6.2 shows that a crystal based on this simple unit cell may have different properties along different directions. The regular pattern of the lattice produces an *anisotropic* material; with less regularity, as in a glass, the distributions average out the same in all directions, so that amorphous solids are *isotropic*.

The characteristic shape of a crystal is governed by the structure of the unit cell, but the relation is a subtle one; we cannot predict the shape of a crystal solely from the structure of the unit cell.

The structure of sodium chloride is shown in Figure 6.3. The unit cell of sodium chloride is a cube. A NaCl crystal grown from a tiny seed crystal suspended in a solution is a perfect cube (I), reflecting, it appears, the unit cell structure. But a crystal grown on the base of the container from the same solution is a rectangular parallelopiped (II). This crystal can grow up and out, but not down; it is, therefore, one half of a cube. The shape of a crystal depends upon the conditions under which it is grown. Sodium chloride crystals grown from a solution containing a trace of urea are octahedra (III). An octahedron is related to a cube; it can be visualized as the result of chipping away the corners of a cube along certain planes (Figure 6.4). In fact, crystals do tend to fracture preferentially along certain planes, which is an example of anisotropy. The skilled gem cutter relies on this anisotropy.

The adoption of two or more different crystal types by a single compound is called *polymorphism*. Polymorphism is common in crystalline substances.

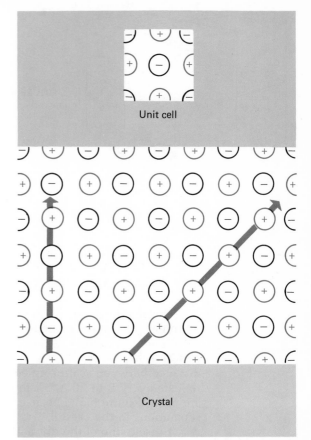

**FIGURE 6.2** Anisotropy as a consequence of order. The properties of the crystal are different along the two directions indicated.

Unit cell

Crystal

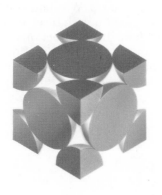

Unit cell

**FIGURE 6.3** NaCl crystal. One unit cell can result in several different crystal shapes.

Crystal shapes

Cube (I)

Rectangle (II)

Octahedron (III)

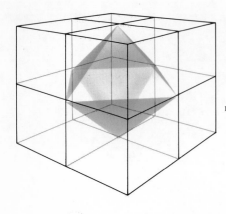

**FIGURE 6.4**  Relation of the octahedron to the cube: 45° cuts remove the corners of a cube to give an octahedron.

## 6.2   X-RAY DIFFRACTION.

The regularity of the atomic or molecular spacings in a crystal is the reason why crystals can be used as diffraction gratings for light of the proper wavelength. A regular sequence of scattering centers produces a diffraction pattern when exposed to waves whose wavelength is comparable to the distance between scattering centers. The pattern of atoms in most crystals repeats every 2 to 50 Å.* X-rays with wavelengths in the range 0.5 to 4 Å are used (Figure 6.5), because this range is of the order of the internuclear separations in the crystal.

A discussion of the diffraction of X-rays by crystals requires an elaboration of our discussion of diffraction in Chapter 2. We shall restrict our discussion to one-dimensional examples and begin with three equally spaced scattering centers. The problem is to find those directions along which the three scattered waves are exactly in phase. The directions are found by drawing lines which are simultaneously tangent to three circles (Figure 6.6). Each circle represents a peak in one wave scattered by one of the

---

*Very large molecules, such as enzymes, are often much larger than this, and so are their unit cells.

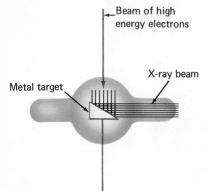

Beam of high energy electrons

X-ray beam

Metal target

**FIGURE 6.5**   X-ray tube. If the target is Al, a few of the high energy electrons will ionize Al in the following way:

$$Al(1s^2,2s^2,2p^1) + e^- \rightarrow Al^+(1s^1,2s^2,2p^1) + 2e^-$$

X-rays of a definite wavelength will be emitted by the process:

$$Al^+(1s^1,2s^2,2p^1) \rightarrow Al^+(1s^2,2s^2) + h\nu,$$

where $h\nu$ represents the x-ray photon.

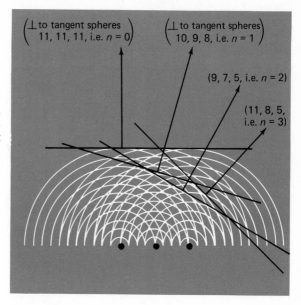

$\left(\begin{array}{c}\perp \text{ to tangent spheres} \\ 11, 11, 11, \text{ i.e. } n = 0\end{array}\right)$ $\left(\begin{array}{c}\perp \text{ to tangent spheres} \\ 10, 9, 8, \text{ i.e. } n = 1\end{array}\right)$

$(9, 7, 5, \text{ i.e. } n = 2)$

$(11, 8, 5, \text{ i.e. } n = 3)$

**FIGURE 6.6** Three-center construction for diffraction conditions. The numbers in parentheses indicate numbers of peaks in the wave since striking the scattering centers.

three centers. The perpendiculars to these tangent lines point along the direction in which there will be constructive interference of the X-rays.*

A detector, placed so as to intercept the scattered waves, will respond when positioned along directions indicated by the arrows in Figure 6.6. The directions (angles) are used to calculate the spacing between the scattering centers. The diffraction condition (Bragg equation) is (see Equation (2.2))

$$n \lambda = A d \sin \theta \qquad (6.1)$$

where $n$ is an integer, $\lambda$ is a known wavelength, and $A$ is a numerical factor determined by the geometry of the apparatus used.†

Let us now add a second trio of scattering centers spaced equidistantly between the first set. The diffraction condition here is more stringent; the lines must be simultaneously tangent to six circles, and this requirement is met by only half of the lines in Figure 6.6. This result could be calculated from Equation (6.1), for the addition of the second trio has halved $d$ and the direction which corresponded to $n = 2$ in Figure 6.6 corresponds to $n = 1$ in Figure 6.8. The important conclusion is that those directions which corresponded to odd values of $n$ in Figure 6.6 are directions for destructive interference in Figure 6.8. Line $A$ in Figure 6.9 is the tangent line for one of these "missing" directions. Along this direction all the waves from the original trio are in phase with each other, as are all the waves from the second trio. But the tangent line for one trio falls exactly halfway between the two circles that indicate maxima for the second trio. Thus, the maxima in the diffracted waves from the first set match the minima in

---

*The distance scale in Figure 6.6 is fixed by the separation of the centers. If they are, say, 10 Å apart, then the whole figure includes a distance of about 100 Å. On this scale the detector is huge and very far away. By the time the scattered waves reach the detector the circles are huge, and the pattern ⌢⌢ is ——. That is, the peaks truly match.

†The diffraction discussion in Chapter 2 was for the geometry where $A = 1$.

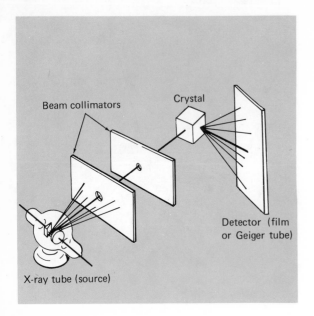

**FIGURE 6.7**  X-ray diffraction apparatus.

the waves from the second, and this gives completely destructive interference. The addition of the second trio of scattering centers doubles the intensity along certain directions (twice as many waves interfering constructively) and removes all the intensity along others (exact cancellation of the two sets of scattered waves).

The chemically interesting question is, "What if the two trios of scattering centers do not have equal efficiencies for scattering x-rays?" Along the directions where the two sets of waves constructively interfere, there is still increased, but not double, intensity. Along those directions typified by Figure 6.9 there is reduced, *but not zero*, intensity. The detector senses a series of signals with alternating intensities, high, low, high, low, and so on. The intensities and spacings of this pattern can be analyzed to find the spacings and the relative scattering efficiencies of the scattering centers.

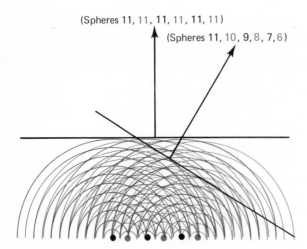

**FIGURE 6.8**  Six-center diffraction construction.

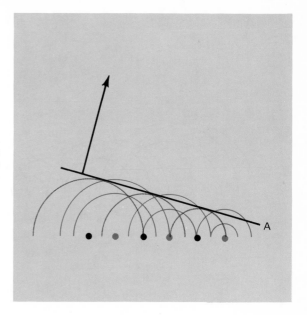

**FIGURE 6.9** Detail of Figure 6.8. The $n = 1$ direction of Figure 6.6; line A falls exactly between the tangents to spheres (4,3,2) and (3,2,1) from the colored set of centers.

Finally, what if the spacings between centers alternate as they would in, say, a crystal of diatomic molecules? Again we look at only a portion of the complete geometric construction (Figure 6.10). It is unlikely that the waves from the two sets of scattering centers will ever match perfectly.[*] The directions for diffraction intensity can be deduced by consideration of only one set of centers. Along some of these directions (case A in Figure 6.10) the peaks will almost match and the diffracted intensity will be large. Along others (case B in Figure 6.10) the two sets of peaks will be nearly out of phase and the intensity will be low. Again the angles yield $d$, and the intensity pattern gives $r$ and the relative scattering efficiencies of the two types of centers.

The determination of the relative scattering efficiencies is extremely useful, for the scattering efficiencies of atoms and ions are proportional to the electron densities surrounding the nuclei. Each atom and ion has a unique scattering efficiency. Hence, the diffraction pattern allows determination of the interatomic spacings and the identification of the atom or ion. However, because the scattering efficiency is determined by the total electron density, it is quite difficult to detect hydrogen atoms, which have no dense, inner shells. Very precise data and special statistical methods are needed to extract this information from the pattern.

For clarity in the geometric constructions, we have restricted ourselves to (hypothetical) one-dimensional crystals. Atoms are really quite inefficient scatterers for x-rays; most of the x-rays pass completely through the crystal. All planes, whether deep in the crystal or on the surface, contribute equally to the diffraction pattern. The equations for constructive interference in three dimensions are complicated, and the analysis of the intensities of the diffracted beams is difficult. However, the principles of computation are the same as those illustrated in the one-dimensional examples, and a computer is not bothered by the tedium of the calculations.

---

[*]This will occur only in the rare case when $d$ is an exact integral multiple of $r$.

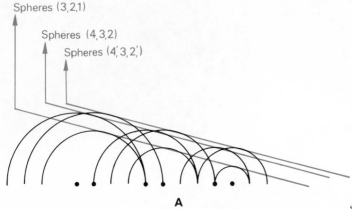

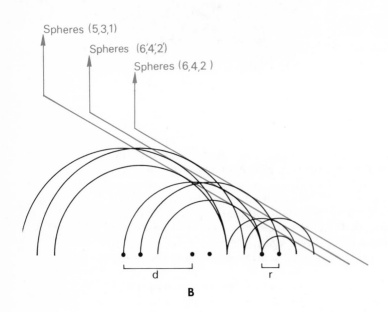

**FIGURE 6.10**  Diffraction construction for a diatomic molecule. In case A the intensity is high because tangents to spheres (4,3,2) and spheres (4',3',2') almost coincide. In case B the intensity is low because the tangent to spheres (6',4',2') is almost halfway between the tangents to spheres (6,4,2) and spheres (5,3,1).

The determination of a crystal structure usually ends with computer-generated electron-density maps of the unit cell. A result of a crystal structure determination is shown in Figure 6.11. This compound is somewhat unusual in that it can crystallize in two different unit cells, only one of which is shown. The electron density contour map enables us to determine which scattering centers are carbon and which are oxygen. Oxygen has more electrons than carbon and therefore scatters more x-rays. The nuclei lie at the centers of the concentric circles; the bond angles and lengths can be read directly off the map. This electron density map bears little resemblance to those we saw in Chapter 4; here the electron density is localized about the nuclei. The earlier maps showed only the valence electron density, but x-ray scattering efficiencies depend on the total electron density, and the associated electron density maps are dominated by inner-shell electrons. We see no evidence of the hydrogen atoms in this map; their positions are assigned with our knowledge of valence and of bond angles. It is quite easy to decide which oxygen atoms are bonded to hydrogen, because the other oxygen atoms are double-bonded to carbon,

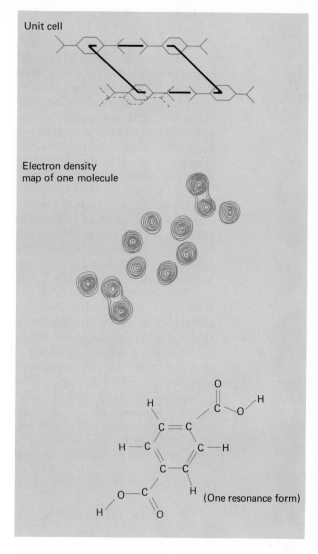

Unit cell

Electron density
map of one molecule

(One resonance form)

**FIGURE 6.11**  Crystal structure of terephth-
alic acid ($C_8H_6O_4$). The dashed lines in the unit
cell represent a molecule in the next layer.
(Reproduced by permission from *Acta Cryst.*
*22*, 387 (1967)).

and the double bond is appreciably shorter than the single bond. It is this
reduced separation which produces the C—O overlap of several contours
in the map, rather than the pi-bonding electrons. Of course, it would be
desirable to extract maps of valence-electron density from the x-ray measurements, perhaps by subtraction of the contributions of the inner-shell
electrons. Efforts along these lines are in progress, but the details of the
method remain to be worked out.

## 6.3  FACTORS GOVERNING CRYSTAL STRUCTURE.

Crystal structures are governed by the size of the molecules, atoms,
or ions, and by the nature of the forces which hold these units together into
a crystal. Crystals are conveniently classified according to the forces which
hold them together. The classifications are based on extremes, but all
gradations of types exist.

## a. Ionic Crystals.

Ionic crystals are held together by the electrostatic attractions between ions of opposite charge. Ionic attractions are non-directional, and they are stronger when the ions of opposite charge are closer to each other. Therefore, the volumes of the ions are the main determinant in the crystal structure, for they limit the distance of closest approach. Monatomic ions are spherical, and many other ions are roughly spherical. How tightly can spheres be packed? Spheres with equal radii whose centers all lie in a single plane can be packed regularly in a cubic or in a hexagonal array (Figure 6.12). It is apparent from Figure 6.12 that the hexagonal array is the more compact.

If a second layer of spheres in a cubic array is placed directly over the first layer, and similarly for additional layers, then the resulting structure is referred to as *simple cubic*. On the other hand, if the second layer of spheres is placed directly over the holes in the first layer, and the third layer directly over the spheres in the first layer, and the fourth layer directly over the second layer, and so on, then the resulting structure is referred to as *body-centered cubic* (BCC).

The most compact arrangement for the second hexagonal layer of spheres is achieved by placing each sphere in a well in the lower layer. Only half of the wells can be filled by the spheres in the second layer (Figure 6.13). It makes no difference which half is filled. If the third hexagonal layer of spheres is placed directly over the spheres in the first layer, and the fourth layer of spheres is placed directly over the spheres in the second layer, and so forth, then the resulting structure is referred to as *hexagonal close-packed* (HCP). On the other hand, if the third layer of spheres is placed directly above the holes in the first layer not filled by the second layer, and this three-layer pattern repeats throughout the crystal, then the resulting structure is referred to either as *face-centered cubic* (FCC), or less frequently, as *cubic-close-packed* (CCP). The fraction of the

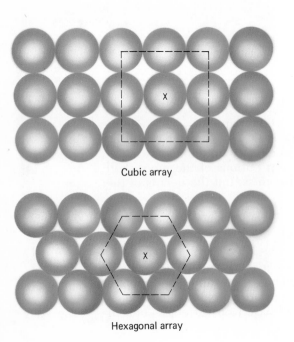

Cubic array

Hexagonal array

**FIGURE 6.12**  Packing of spheres.

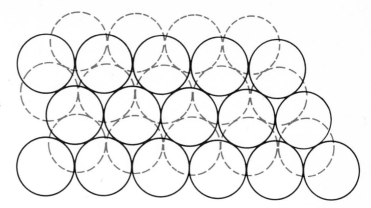

**FIGURE 6.13** Two layers of a closest-packed array of spheres.

total volume filled by the spheres is the same for HCP and FCC lattices (74% of the total).

The different ions in an ionic compound are almost never the same size. It is often possible to predict the crystal structure of an ionic compound by starting with a *closest-packed* array of the larger ion (almost always the negative ion) and then by fitting the smaller ions into the empty spaces.

The spaces between any two layers of *closest-packed* spheres are of two types. Tetrahedral holes are surrounded by four touching spheres, and octahedral holes are surrounded by six touching spheres. Careful counting (Figure 6.15) shows that each sphere forms part of the boundary of six octahedral holes.[*] Because six spheres are required to bound each

---

[*]The holes are between the layers of spheres, so that only half of those surrounding any one atom appear in a two-layer drawing.

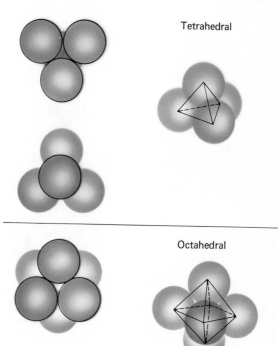

Tetrahedral

Octahedral

**FIGURE 6.14** Holes in closest-packed arrays.

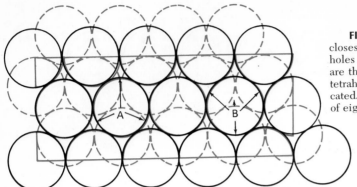

**FIGURE 6.15** Incidence of holes in closest-packed arrays. The three octahedral holes below sphere A are indicated. There are three more above for a total of six. Four tetrahedral holes below sphere B are indicated. There are four more above for a total of eight.

octahedral hole, the number of octahedral holes is $N$, the number of large ions in the crystal. Each atom also forms part of the boundary of eight tetrahedral holes (Figure 6.15). Because four spheres are needed to form a tetrahedral hole, there are $2N$ such holes in the array. A few crystals are based on the simple cubic array of Figure 6.16. The spheres in the second layer are placed directly above those in the first. There are $N$ cubic holes in this structure, for each sphere is surrounded by eight holes, and each hole is surrounded by eight spheres.

Each type of hole has a definite size. It is useful to compute the radius of the largest sphere that will fit inside the hole. Refer to the octahedral hole shown in Figure 6.17. By the Pythagorean theorem,

$$2(r_1 + r_2)^2 = (2r_2)^2$$
$$r_1 + r_2 = \sqrt{2}\, r_2 \tag{6.2}$$
$$r_1 = (\sqrt{2} - 1)r_2 = 0.414\, r_2$$

Therefore, any small sphere with radius ($r_1$) less than $0.414\, r_2$ ($r_2$ is the radius of a large sphere) will fit inside an octahedral hole. The corresponding numbers for tetrahedral and cubic holes are $0.225\, r_2$ and $0.73\, r_2$, respectively. These results are summarized in Table 6.1.

Of course, ions are not really hard spheres, as was assumed in the foregoing discussion. Therefore, we must examine a few crystals to see how to translate the packing of spheres into structures for ionic crystals. We find that small positive ions almost never occupy a hole with a radius equal to or greater than the radius of the ions. Instead, the cations squeeze into a smaller hole. This is not too surprising, for this arrangement moves the large negative ions slightly apart. A positive ion in a hole larger than itself would tend to pull the negative ions into even closer contact with

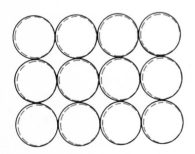

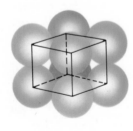

**FIGURE 6.16** Cubic packing and the geometry around a hole.

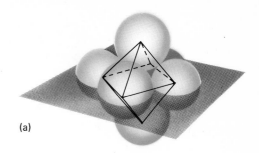

(a)

**FIGURE 6.17**   Size of an octahedral hole. Part (b) shows the geometry in the plane shown in part (a).

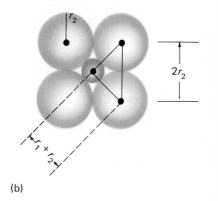

(b)

each other, surely an energetically unfavorable structure. As examples, consider the two compounds NaCl and CsCl. The ionic radii (Å) are $Na^+$ (0.95), $Cl^-$ (1.81), and $Cs^+$ (1.60); therefore,

$$r_{Na^+} = 0.54\ r_{Cl^-}$$

$$r_{Cs^+} = 0.93\ r_{Cl^-}$$

The sodium ion is too small to fill the cubic holes in a cubic array of chloride ions ($r_1 < 0.73\ r_2$) completely, so it squeezes into an expanded octahedral hole ($r_1 > 0.414\ r_2$). The cesium ion is larger than a cubic hole in a cubic lattice, so it squeezes into the cubic holes (Figure 6.18). These rules are summarized in Table 6.2. From the examples, we see that the stoichiometries of ionic compounds can be accommodated by partial filling of the available sites.

Compounds are found in which both the tetrahedral and octahedral holes are occupied. These materials are called *spinels*, after the mineral

**TABLE 6.1**

| STRUCTURE | HOLE TYPE | NUMBER AVAILABLE | SIZE LIMIT FOR $r_1$ |
|---|---|---|---|
| Closest-packed | Tetrahedral | $2N$ | $0.225\ r_2$ |
| Closest-packed | Octahedral | $N$ | $0.414\ r_2$ |
| Cubic | Cubic | $N$ | $0.73\ r_2$ |

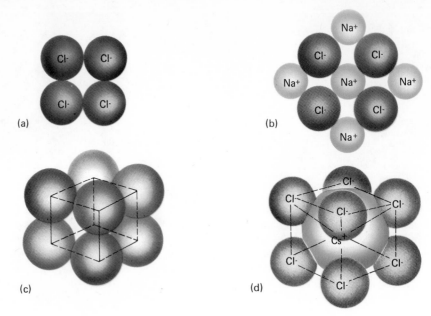

**FIGURE 6.18** Crystal structures of NaCl and CsCl. (a) An octahedral site in the $Cl^-$ lattice. (b) $Na^+$ ions crowded into an expanded octahedral site. (c) Cubic site in a $Cl^-$ lattice. (d) $Cs^+$ ion crowded into an expanded cubic site.

*spinel*, $MgAl_2O_4$. The crystal consists of a closest-packed array of $O^{-2}$ ions, in which $Mg^{+2}$ ions occupy one-eighth of the tetrahedral holes, and $Al^{+3}$ ions occupy one-half of the octahedral holes. Reverse spinels, in which the positions of the dipositive and tripositive ions are reversed, are found as well. One interesting example is the magnetic iron ore, *magnetite*.[*] Magnetite has the formula $Fe_3O_4$, which by our usual rules of valence would be written $Fe^{+2}(Fe^{+3})_2(O^{-2})_4$. Again the $O^{-2}$ ions form a closest-

---

[*]Early navigators used the response of magnetite lodestones to the earth's magnetic field to guide their ships.

**TABLE 6.2**

| SITE | RADIUS-RATIO RANGE | FRACTION OF SITES FILLED | FORMULA | TYPICAL EXAMPLES |
|---|---|---|---|---|
| Cubic | 0.73–1.00 | 100% | $A^+B^-, A^{+2}B^{-2}$ | $CsCl, CsBr, TlCl$ |
| Octahedral | 0.41–0.73 | 100% | $A^+B^-$ | $LiBr, NaCl, NH_4Cl$ |
| | | | | $AgF, AgBr,$ |
| | | | $A^{+2}B^{-2}$ | $BaS, MgO$ |
| Tetrahedral | 0.22–0.41 | 50% | $A^+B^-, A^{+2}B^{-2}$ | $CuI, AgI, ZnS$ |
| Tetrahedral | 0.22–0.41 | 100% | $A^{+2}B_2^-$ | $SrF_2$ |
| | | | $A^{+4}B_2^{-2}$ | $UO_2$ |
| Tetrahedral | 0.22–0.41 | 100% | $A_2^+B^{-2}$ | $Li_2O, Na_2S$ |

packed array. The $Fe^{+2}$ ions occupy octahedral holes, but the $Fe^{+3}$ ions are evenly divided between tetrahedral and octahedral holes. Magnetite is called a *reverse spinel*, because the tetrahedral holes contain the higher charged ions.

The structures of ionic crystals show that there are no simple molecules in these compounds. Each ion in the crystal is surrounded by four, or six, or eight equivalent ions of the opposite charge, no one of which can be singled out as a unique partner of the original ion. The absence of individual molecules within ionic crystals does account for many of their properties. For example, it is difficult to break individual molecules away from an ionic crystal, so that ionic crystals have high melting points and low vapor pressures.

## b. Molecular Crystals.

Crystals composed of neutral molecules are called *molecular crystals*. The forces that hold together the molecules in molecular crystals arise from the attractions between molecules with permanent or induced dipole moments. Two helium atoms do not interact at large separation, because from the outside both their electrons appear to be centered at the nucleus. When the two He atoms are very close, they constitute the diatomic molecule $He_2$, which is unstable. At some intermediate separation there is an attractive interaction. If one of the He atoms is polarized momentarily, the resulting instantaneous dipole moment produces an electric field which induces a dipole moment in the second atom. Once formed, these two dipoles tend to maintain each other and to align so as to reduce the energy of the system (Figure 6.19). So long as they maintain this mutual alignment, the two atoms attract each other.

A molecular crystal is a vast array of molecules, either with their permanent dipole moments aligned or with their instantaneous dipole moments in phase. These forces are much weaker than those between ions or those involved in the formation of chemical bonds. This does not mean that it is necessarily easier to pull apart a molecular crystal than an ionic crystal, for the weak attractions between pairs of atoms can total to very strong attractions between large molecules. It does mean that, at any point in the molecule, the permanent or induced dipole forces are less strong than the chemical bonds, so that molecules can be separated as units without destroying chemical bonds.

Just as in gases, the agitation of the molecules in a solid increases with increasing temperature. Therefore, the melting point is an indicator of the strength of the forces between molecules in the crystal. The forces holding the $H_2$ crystal together are so weak that the melting point is only 14 K. For larger molecules the forces become appreciable because the loosely held valence electrons are quite polarizable. Melting points of molecular crystals range from 14 K ($H_2$ at 1 atm) to about 600 K. There are many molecules with permanent dipole moments, and they form molecular crystals whose crystal structures maximize the attraction of the permanent dipoles.

The weak forces between molecules in a molecular crystal do not distort the individual molecules significantly, so that the electron density map of a molecular crystal yields a structure for the molecule which is usually accurate for an individual molecule in the gas phase as well. It is possible to carry out a complete structure determination on a moderately

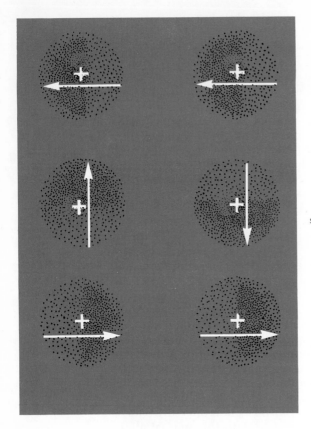

**FIGURE 6.19** Two close He atoms with instantaneous dipoles in phase.

complex molecule (∼30 atoms, not counting hydrogen) using a sample too small for any one of the chemical or spectral methods of structure determination. For example, Figure 6.20 is an electron density map of a glyceryl tricarboxylate, and Figure 6.21 is an electron density map for 5-ethyl-6-methyluracil.

## c. Metals.

All elements with valence shells partially but less than half filled are metals (boron is an exception). Because of the order of orbital energies, well over two-thirds of the elements are metals. Metallic crystals are of great practical importance because many of them are strong without being brittle, and most metals are good conductors of heat and electricity.

In metallic crystals there are always more valence-shell orbitals available for bonding than there are electrons to fill them. As usual, when there are many equivalent orbitals for an electron, the lowest energy is achieved by assigning the electron to a bonding orbital that is a linear combination of all the available orbitals. In a molecule such as benzene, the bonding orbital extends over the whole molecule; in a metallic crystal the bonding orbital extends over the whole crystal. For this reason, a metal can be treated as an array of closest-packed spheres (the nuclei and the inner-shell electrons) immersed in a sea of delocalized valence electrons (Figure 6.22). Because electrons which are not bound to a particular

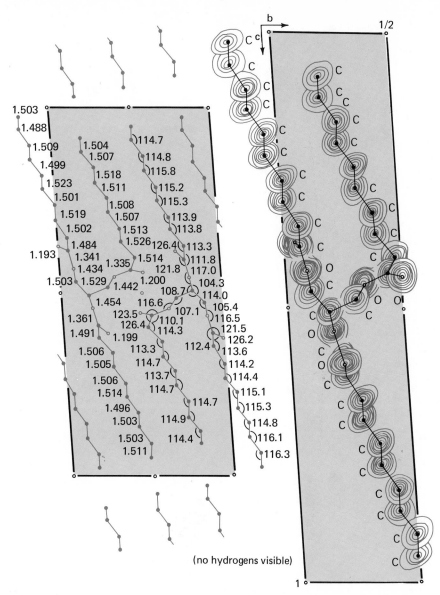

**FIGURE 6.20** Structure of a lipid ($\beta$-tricaprin). The bond lengths and bond angles can be read from the electron density map on the right. (Reproduced by permission *Acta Cryst. 21*, 770 (1966)).

nucleus are easily displaced by an electric field, the electrical conductivity of metals is high. Substances that have color absorb some fraction of visible light; many substances with very delocalized electrons are colored. Metals contain a vast number of delocalized electrons, and they absorb and reflect all wavelengths in the visible region. This produces the metallic color exhibited by almost all metals.*

---

*The exceptions are gold and copper, which are more or less yellow. These two metals do not absorb all the blue light; the blue is absorbed while the yellow is reflected, hence the color.

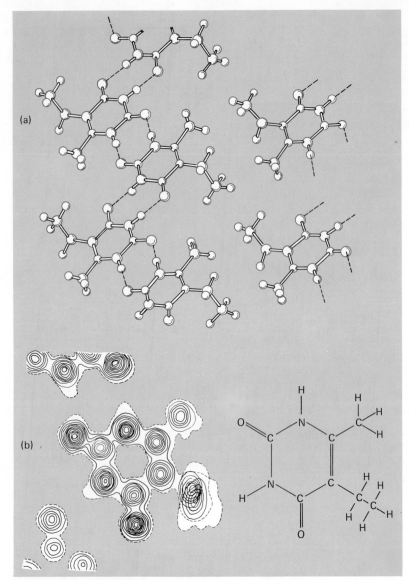

**FIGURE 6.21** Structure of 5-ethyl-6-methyluracil ($C_7N_2O_2H_{10}$). (a) Crystal structure; (b) electron density of a molecule projected on the plane of the molecule. (Reproduced by permission from *Acta Cryst.* **20**, 703 (1966).)

## d. Bonded Crystals.

Elements with exactly half-filled valence shells can complete their octets by forming bonds with their neighbors in the solid phase. In this way, an interlocking array which extends throughout the lattice can be built up.

Carbon is a good example of this phenomenon. Carbon crystallizes in three forms. The diamond form is built upon the tetrahedral carbon atom; each carbon atom in the crystal is chemically bonded to four neighbors (Figure 6.23). The bonds are typical carbon-carbon bonds formed from

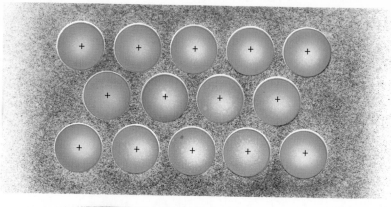

Electron density of
valence electrons

Nucleus and inner
shell electrons

**FIGURE 6.22** Metallic crystal.

$sp^3$ hybrid orbitals; they are very strong and very directional. The C—C bond distance is 1.54 Å, which is the same as that in ethane. The whole crystal is a single molecule. Because the removal of material from the crystal requires rupture of many strong bonds, diamond is one of the hardest substances known. It is widely used as a cutting edge, in phonograph needles, and for other applications requiring high resistance to wear.

**FIGURE 6.23** Unit cell of diamond. Note the interlocking tetrahedra around atoms A, B, C, and D.

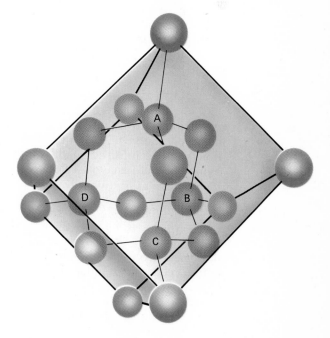

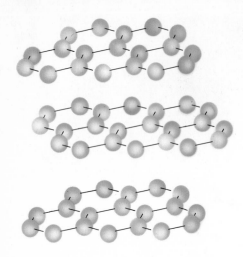

**FIGURE 6.24**    Graphite.

Similar crystals are formed by silicon, germanium, and tin, and by the compounds silicon carbide (SiC) and boron nitride (BN).

Graphite is a second crystal form of carbon. In graphite, each carbon atom is bonded to three others by three *sigma* bonds formed from $sp^2$ orbitals. This produces a planar array. The C—C bond distance is 1.42 Å, which is about the same as in benzene (1.39 Å). As in benzene, the remaining $p$ orbital perpendicular to the plane is used in $pi$ bonding, but here the $pi$ orbitals extend over vast sheets in the crystal. The sheets are held together by induced dipole attractions like those that bind together molecular crystals (Figure 6.24). Within a given plane, graphite is a chemically bonded crystal (all the atoms in one plane are bonded together into a "super" molecule) and as such has a high melting point. It also shows metallic bonding (the $pi$ electrons above and below the planes), and it is a fair electrical conductor if the electric field is applied parallel to the planes. Graphite is so easily deformed in directions parallel to the planes that it is frequently used as a lubricant. Finally, it is a molecular crystal in a direction perpendicular to the planes (evidenced by the poor electrical conductivity in this direction).

*Graphite is black rather than metallic in color, because of surface irregularities.*

### e. Hydrogen-Bonded Crystals.

The electron density map of the uracil derivative in Figure 6.21 does not show the positions of the hydrogen atoms very clearly, but their positions can be inferred readily from our knowledge of bonding or from a more elaborate analysis of the x-ray diffraction pattern. The packing of the molecules in the unit cell shows that the hydrogen atoms bonded to nitrogen are very close to, and aligned with, the oxygen atoms in a neighboring molecule. The molecules are closest at these same positions, so close that there appears to be a weak chemical bond between the hydrogen atoms of one molecule and the oxygen atoms of a neighbor. This is our first example of a special type of chemical bond, *the hydrogen bond.* A vast amount of evidence has accumulated which indicates that a hydrogen atom bonded to a very electro-negative atom has such a low electron density that it can be attracted to a lone pair of electrons on another electronegative atom. Hydrogen bonds are much weaker than most other chemical bonds; their

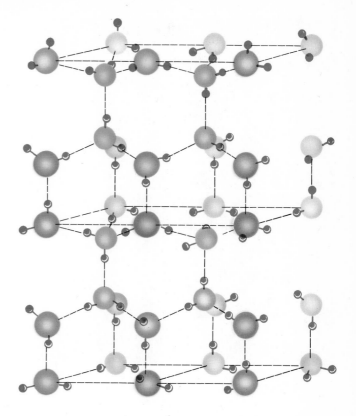

**FIGURE 6.25**  Structure of an ice crystal.

strengths run between 0.20 and 0.50 eV (4 to 10 kcal) per bond. However, this is usually enough to control the packing of molecules in the crystal. Ice is an important example; the oxygen atoms of the water molecules align in a tetrahedral geometry which is perfect for hydrogen-bond formation (Figure 6.25). The resulting structure is a very open one. Many closer packings of water are possible, but none of them allow such efficient hydrogen bonding.

The distance between atoms in a molecular crystal can be used to define van der Waals radii for the atoms. Van der Waals radii give the separations which the induced dipole attractions can maintain. Separations less than the sum of these radii are indicators of stronger attractive forces, e.g., a chemical bond.

Hydrogen bonds, by virtue of their intermediate strength, occupy a unique position in chemistry. They are not individually strong enough to hold groups of atoms together to form new molecules, but in concert they fix the spatial configurations of molecules. This is especially true for most large molecules found in living cells. The best known examples are the nucleic acids, which are intimately involved in the processes of cell reproduction and heredity. Deoxyribonucleic acid (DNA) is the molecule that contains the genetic information in the cell. Each individual has cells with a unique DNA molecule, but all DNA molecules have certain features in common. They are built on two molecular backbones which at regular intervals have attached one of four simple molecules, *guanine, adenine, thymine,* or *cytosine*. These four molecules can pair up in only two ways, because of precise geometric requirements on the formation of hydrogen bonds (Figure 6.26). Notice that the guanine-thymine (GT) and adenine-cytosine (AC) pairs can form hydrogen bonds that lock them together, whereas the other combinations (GC and AT) neither fit together, nor have the same dimensions as the GT and AC pairs. The hydrogen bonds between the pairs hold the two backbones of the molecule together in a double helix in the manner shown in Figure 6.27. The important feature of this structure is that, because of the unique geometry of the hydrogen

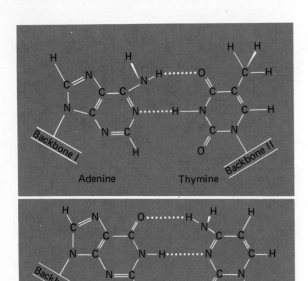

Adenine          Thymine

Guanine          Cytosine

Note: All four molecules are planar.

**FIGURE 6.26** Basic units of DNA. Notice the specific pairing due to hydrogen bonds.

bonds, a single backbone can serve as a template for the reconstruction of its partner. Thus, two single backbones from one molecule can produce two identical new molecules; these molecules can reproduce. This is but one example of many in which hydrogen bonding is crucial in biological processes.

## 6.4   CRYSTAL DEFECTS.

The perfection implied by the description of a crystal as a simple unit cell replicated endlessly in three dimensions is an exaggeration. All crystals contain imperfections or defects. The defects are of two types, chemical impurities and defects in the formation of the lattice. Chemical impurities can be reduced in quantity, but never eliminated completely. With extreme care one might prepare samples with impurities of only one part per billion, as in the purest Ge and Si. If we assume an impurity level of one part per million, and assume further that each impurity resides at the center of a cube containing one million molecules, the impurity molecules are still only about $(10^6)^{1/3}$ or 100 lattice units apart. This is about 250 to 300 Å for most crystals. A 1 mm cube of this purity of, say, sodium chloride would contain about $10^{13}$ impurities.

Defects in the lattice structure are very common as well. Defects are classified either as *point defects,* an error at a single point in the lattice, or as *lattice defects,* a mistake which causes mismatches over a large section of the crystal. Point defects are either *vacancies* or *interstitials.** Frenkel defects and Schottky defects are common pairings of point defects (Figure

*Impurities could be considered to be point defects.

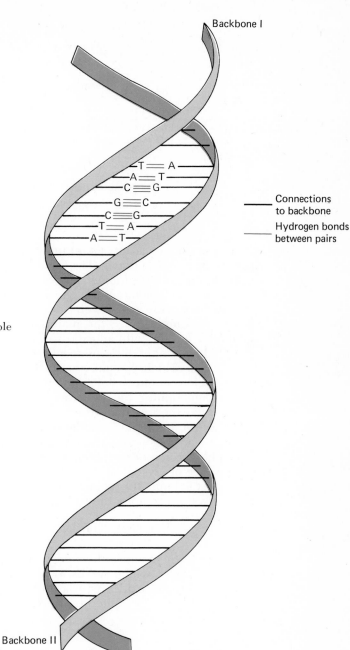

Backbone I

Connections
to backbone

Hydrogen bonds
between pairs

**FIGURE 6.27** Watson-Crick double helix for DNA.

Backbone II

6.28). Frenkel defects in tiny AgBr crystals are thought to be essential to the light sensitivity of photographic film.

Vacancies in an ionic crystal can be of two types, missing positive ions and missing negative ions. Unequal numbers of these two types will alter the stoichiometry of the crystal. The stoichiometries of ionic crystals are variable and are dependent on the manner in which the crystal is prepared. Copper (I) sulfide is one example. The expected formula for this compound is $Cu_2S$, but crystals can be prepared with compositions anywhere between $Cu_{1.7}S$ and $Cu_2S$. That is, as many as 15% of the copper sites in

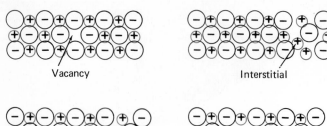

Vacancy                               Interstitial

**FIGURE  6.28**  Point  defects  in
crystals.

Frenkel defect                        Schottky defect
(a vacancy + an interstitial)         (a pair of vacancies)

the lattice can be vacant. Titanium oxide is even less regular. Crystals with
any composition between $Ti_{0.7}O$ and $TiO_{0.7}$ can be prepared. These are
examples of *non-stoichiometric compounds*. Although all ionic compounds
are in fact non-stoichiometric, the deviations are most noticeable in oxides
and sulfides. However, even sodium chloride can be crystallized in a
variety of compositions; the material that results from reacting a limited
amount of sodium with an excess of chlorine is the familiar white salt, but
a limited amount of chlorine reacting with an excess of sodium produces a
blue crystal which is somewhat metallic. Ironically, some of the earliest
evidence for chemical stoichiometry was based on work on non-stoichiometric compounds; the deviations from perfect stoichiometry were attributed to experimental error.

The use of simple formulae such as NaCl or FeO should not be con-
strued as carrying any limitation on the range of composition of a particular
solid phase. Such designations imply only the composition of the solid in
the absence of crystal vacancies.

Nonstoichiometric compounds are not found in gases or in liquid
solutions. For example, the gaseous oxides of nitrogen,

$$NO, NO_2, NO_3, N_2O, N_2O_3, N_2O_4, \text{ and } N_2O_5$$

have exactly the compositions designated. Each of these compounds has a
unique behavior which serves to distinguish it from its neighbors. Compare the nitrogen oxides with a 9.75 g single crystal of sodium chloride,
which could be represented by the formula

$$Na^+_{10^{23}} Cl^-_{10^{23}}$$

If we now remove a single Na atom from the crystal, the formula becomes

$$Na^+_{10^{23}-1} Cl^-_{10^{23}-1} Cl$$

Here the change of one atom produces a (relatively) trivial change in the
composition and in the properties of the crystal.

The most common lattice defects are edge dislocations and screw
dislocations. These are illustrated in Figure 6.29. Lattice defects are important because they are the weak points of the crystal, both mechanically
and chemically. Tiny metal crystals without dislocations are about a thousand times stronger than the usual crystals. Atoms or molecules next to a
defect are not bound to the crystal as strongly as others, and so they are

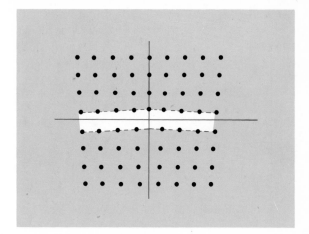

**FIGURE 6.29(a)**  An edge dislocation.

Axis of screw dislocation

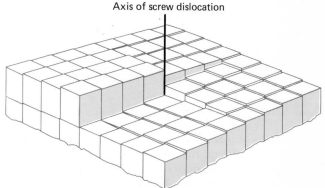

**FIGURE 6.29(b)**  A screw dislocation. The unit cells are represented by small cubes.

more open to attack. Crystal erosion (and growth) usually occurs along screw dislocations.

The positions of the energy levels of crystals are consequences of the energy levels of the molecules, atoms, or ions and of the interactions between the molecules, atoms, or ions. The energy-level patterns are the key to the understanding of a new and vital technology, *solid-state electronics.*

Let us return for a moment to the simplest molecule, $H_2$, and examine it from a different viewpoint. The energy-level pattern for the hydrogen atom can be drawn within a potential energy well (Figure 6.30). The potential energy is $-\dfrac{e^2}{r}$.[*] As two hydrogen atoms are brought together, the changes shown in Figure 6.31 occur. At some separation, say that in part $(b)$, the two atoms begin to interact to a detectable extent, and the degeneracy of the two separated $1s$ orbitals is removed. Two new orbitals are formed, one of lower and one of higher energy. In molecular-orbital language, these two new levels are incipient bonding and antibonding

---

[*]For other atoms the potential energy would be $-Ze^2/r$. Ideally one would use $Z_{\text{eff}}$, but we omit that complication here.

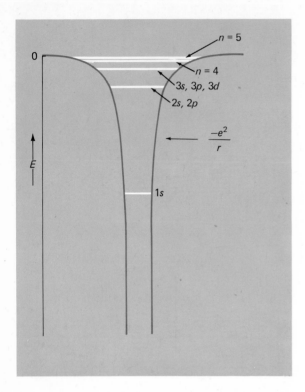

$n = 5$

$0$

$n = 4$

$3s, 3p, 3d$

$2s, 2p$

$E$

$\dfrac{-e^2}{r}$

$1s$

**FIGURE 6.30** Hydrogen atom energy levels. The levels are drawn within a potential energy well.

orbitals. In localized-orbital (or valence-bond) language, these two levels each have one electron localized on each nucleus; the lower level has the two induced dipole moments in-phase, and the upper level has them out-of-phase. The latter description is preferable here. Further reduction of the internuclear separation leads finally to the molecule $H_2$, represented by part $(c)$ of the figure. Here the molecular-orbital description is perfectly adequate.

Which separation represents nature? If the orbitals of the separated units are filled, diagram $(b)$ is appropriate. Two He atoms are held together (albeit weakly) at a separation corresponding to part $(b)$. If the orbitals are only partially filled, the separation is more nearly that of diagram (c). In this latter case a delocalized molecular orbital and a chemical bond are formed. The hypothetical linear $H_4$ molecule can be approached in the same way. Our representation for this molecule is shown in Figure 6.32. Four atomic orbitals combine to form four molecular orbitals which extend over all four nuclei. However, calculations show that the energy difference between the most bonding and the most antibonding orbitals is only slightly larger than this difference in the $H_2$ molecule. In other words, the four levels in linear $H_4$ are packed more closely (in energy) than the two levels in the $H_2$ molecule.

Let us now form a long line of Na atoms, all at the same separation. The appropriate diagram for $N$ atoms is given in Figure 6.33. Each atomic level has split into a band of $N$ levels, each of which can accept two electrons. The width of the bands has increased only slightly, but the density of levels within the band has soared. The bands that are based on the inner-shell atomic orbitals are very narrow and, more significantly, they are filled. The electrons in these levels are localized and contribute little to the properties of the Na crystal. The valence orbitals ($3s$) become a wider band of levels which have electron density spread throughout the crystal. The $N$ levels in

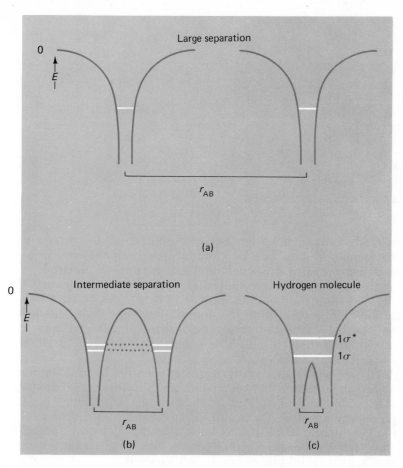

**FIGURE 6.31**   Two hydrogen atoms at various distances. The symbol ——— · · · · ——— represents a single level; an electron is not found in the · · · · region (usually).

this band can accept $2N$ electrons, so the band is only half filled. Electrons in partially filled orbitals that extend throughout the crystal are moved easily by electric fields. Therefore, sodium, a typical metal, is a good conductor of electricity. A similar array of Mg atoms has $2N$ valence electrons, enough to fill the band formed from the $3s$ orbitals. However, magnesium is still a metal because the band width of the $3s$ and $3p$ bands is greater than the $3s$–$3p$ atomic separation, and as a result, the bands overlap. Again there are more delocalized orbitals than valence electrons.

A crystal is a three-dimensional array, yet the qualitative description based on the line of atoms is essentially correct. This is probably true because of the simple crystal structure of metals. Other materials with more complex crystal structures require a more elaborate treatment. This is the case with chemically-bonded crystals, such as silicon. Silicon forms a diamond-like crystal in which each silicon atom is chemically bound to four other atoms located at the corners of a tetrahedron (see Figure 6.23). These chemical bonds are so strong that they invalidate the procedure used to discuss metals.

We can construct a reasonable model for the silicon crystal by bringing together a line of unit cells. In fact, we do not have to bring together the complex unit cell of Figure 6.23; a more primitive unit cell containing only two silicon atoms is adequate. The crucial change is that the silicon atoms

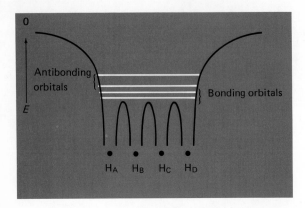

**FIGURE 6.32**   Linear $H_4$ molecule.

have already formed bonding and antibonding orbitals from their $sp^3$-hybrid valence orbitals. The bonding orbitals from $N$ atoms form a band of levels that contains $2N$ levels and can accept $4N$ electrons, and the antibonding orbitals form a second band of the same size (Figure 6.34). The band formed from the bonding orbitals is called the *valence band*; it is exactly filled. The band from the antibonding orbitals is called the *conduction band,* because the empty delocalized levels give metallic conduction when partially filled. The energy difference between the two bands is called the *energy gap.*\*

Metals are materials with no gap between the bands. Materials like diamond, with large gaps between bands, are *insulators,* and the larger the band gap the better the insulator. Application of a huge voltage can provide enough energy for electrons in the valence band to jump the energy gap into the conduction band; this is known as the dielectric breakdown of insulators. Some materials, such as germanium, have narrow enough band gaps that a (relatively) few electrons with only thermal energy reach the conduction band. Because the average kinetic energy of a particle is $3kT/2$, this will happen when the band gap is about $kT$. Such materials are called *intrinsic semiconductors* (Figure 6.35).

---

\*Compare the treatments of sodium and silicon. In sodium the band from $3s$ atomic orbitals has the bonding orbitals at the bottom and the antibonding orbitals at the top. In silicon the formation of strong chemical bonds separates the bonding from the antibonding orbitals so far that they form separate bands. Both qualitatively and quantitatively we must treat the strongest interactions first.

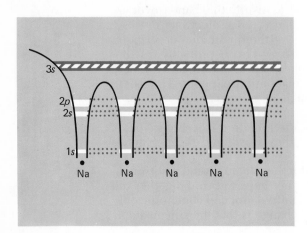

**FIGURE 6.33**   Linear array of Na atoms.

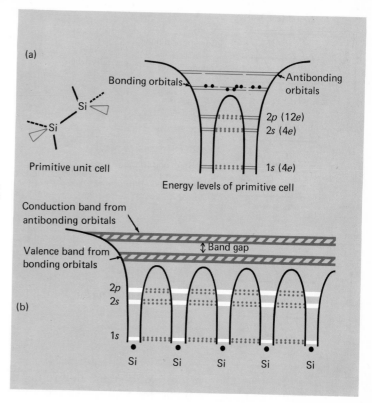

**FIGURE 6.34** Silicon crystal. (a) Primitive cell; (b) linear array of primitive cells.

Calculations show that the widths of the bands, which determine the energy gap and in turn the electrical properties of the crystal, are themselves a function of the strengths of the interactions between the atoms or molecules in the crystal. These interactions can be increased by squeezing the atoms together. The expected conversion of semiconductors and insulators to metals has been accomplished by placing the materials under high pressure (10,000 to 100,000 atm).

Some insulators can be converted to semiconductors by the judicious introduction of impurities (called "doping"). Introduction of a small amount (1 ppm) of phosphorus into a silicon crystal leads to occasional lattice substitutions of P for Si atoms. A phosphorus atom in a silicon crystal adds one extra electron which cannot fit into the valence band. The electron occupies an impurity level which is just below the conduction band, for its energy is lower when it is close to the phosphorus atom, which has a higher nuclear charge.

As with an intrinsic semiconductor, a small investment of thermal energy raises this electron into the conduction band (Figure 6.36). Because the current is carried by a *negative* electron, we call this material an *n*-type semiconductor.

A *p*- (for *positive*) type semiconductor can be made by substitution of an atom with fewer valence electrons than the host atoms. The lower nuclear charge on, say, an aluminum nucleus pushes one level up out of the valence band. Aluminum has one less electron than silicon, so the highest energy level of the valence band is only half filled. Thermal energy can be used to promote a second electron to this level, leaving behind a positive hole (Figure 6.37). The hole can move through the crystal from atom to atom, and so a current flows.

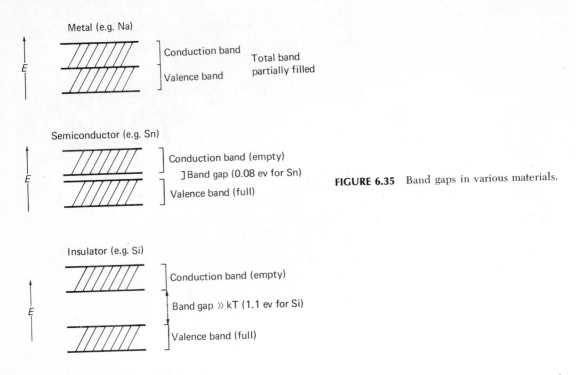

FIGURE 6.35   Band gaps in various materials.

The various solid-state electronic devices, such as transistors and diodes, are made by appropriate combinations of $n$ and $p$ type impurity semiconductors.* Although germanium was initially used, silicon is the favored material today. The proportion and type of impurities in the crystal are critical; the base material must be refined to a very high degree of purity before the careful introduction of the desired impurities. The purification process is expensive. The cost of the host material is immaterial for electronic devices, because such tiny amounts are needed. Photovoltaic cells, which utilize a Si wafer doped on one surface with boron to form a diode, can convert sunlight directly to electric current, but many cells are

---

*See reference 6.7.

Conduction band (empty)

Valence band (full

Insulator

FIGURE 6.36   An $n$-type semiconductor.

] $\sim kT$

Semiconductor

In conducting state

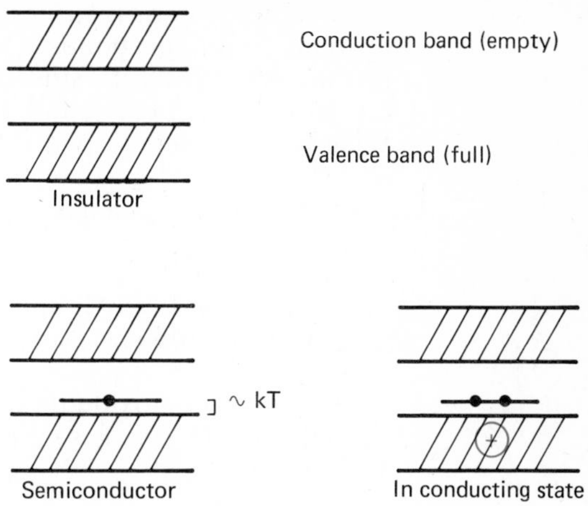

Insulator

**FIGURE 6.37** A *p*-type semiconductor.

Semiconductor    In conducting state

needed to collect much light. The cost is immaterial in the space program where solar batteries are used as power sources in communications satellites. The cost *is* material when one is designing a power system for a poverty-stricken village, and at present the cost of solar batteries is prohibitive for routine application.

---

## PROBLEMS

1. Can lattice vacancies lead to nonstoichiometric compounds in molecular crystals?

2. Does a crystal of nonstoichiometric compound carry a net charge? If not, why not?

3. Use your answer to problem 2 to rationalize why oxides and sulfides are more likely to show large deviations from stoichiometry.

4. Use Figure 6.33 to explain why the minimum energy required to produce electron current in the photoelectric effect experiment is not the ionization energy of the metal.

5. What type of hole will the smaller ion occupy in the following compounds?

$$CsI, \ NH_4F, \ NH_4Br, \ MgS, \ Cs_2O$$

What fraction of the holes will be occupied?

6. Why is a vacuum a better insulator than any material?

7. There are a number of triangular holes in a planar array of closest-packed spheres. How many are there per sphere? What is the largest ion that would fit in this hole?

8. How many ions of each type are there per unit cell in NaCl (see Figure 6.3)?

9. The protein *insulin* forms a molecular crystal (although there may be some hydrogen bonds which help hold it together). The unit cell is a rectangular solid with sides 130 Å, 74.8 Å, and 30.9 Å. The density of the crystal is 1.315 g cm⁻³, and there are six insulin molecules in the unit cell. What is the molecular weight of insulin?

10. Why is $H_2S$ a gas at $-10°C$ when $H_2O$ is a solid at this temperature?

11. What would you expect for the possible structures and properties of silicon carbide (SiC) and boron nitride (BN) (each substance has more than one possible structure in the solid phase)?

12. Silicon dioxide forms hard, high-melting crystals or glass, whereas carbon dioxide is a gas. Would you expect a silicon analog to graphite? Why or why not?

13. Benzene forms a typical molecular crystal. The crystal has numerous electronic absorptions, whereas the molecule has only one. Yet there is no evidence that the excited electrons are free to move through the crystal. Use these facts to sketch a diagram similar to Figure 6.35 for the molecular crystal of benzene.

14. Illustrate by a drawing why an ionic crystal made up of positive and negative ions of equal sizes does not form a crystal based on closest-packing of spheres.

15. The unit cell of KCl is quite similar to that of NaCl (Figure 6.3). The density of KCl at 18°C is 1.989 g cm⁻³. Each edge of the unit cell is 6.291 Å long. Use these data to calculate Avogadro's number.

16. What do you predict for the crystal structure of $BaSO_4$?

17. Use a coin to draw a hexagonal array of spheres like that in Figure 6.12. Label the centers of the spheres with $x$'s, and label the adjacent holes between the spheres with $y$'s and $z$'s, respectively. Each $y$ hole should be surrounded by $z$ holes, and *vice versa*. Now, show by drawing succeeding layers of spheres, that the HCP structure has the packing arrangement $xyxy \ldots$ (or equivalently, $xzxz \ldots$), whereas the FCC structure has the packing arrangement $xyzxyz \ldots$ (or equivalently, $xzyxzy \ldots$).

# References

6.1. "Materials," special issue of Scientific American, September, 1967.
6.2. Runnels, L. K., *Ice*, Scientific American *215*, No. 6, 118 (1966).
6.3. Kornberg, A., *The synthesis of DNA*, Scientific American, *219*, 64 (1968).
6.4. Crick, F., *The Genetic Code*, Scientific American, *215*, No. 4, 55 (1966).
6.5. Pope, M., *Electric Currents in Organic Crystals*, Scientific American, *216*, 86 (1967).
6.6. Brewer, L., *Non-Stoichiometry in Non-Metallic Compounds*. J. Chem. Educ.
6.7. Moore, W., *Seven Solid States* (W. A. Benjamin, New York, 1967).
6.8. Gurney, E. F., *Fundamental Principles of Semiconductors*, Journal of Chemical Education, *46*, 80 (1969).
6.9. Watson, J. D., *The Double Helix: A Personal Account of the Discovery of the Structure of DNA* (Atheneum, New York, 1968).
6.10. Acta Crystallographica. *NOTE*: This is the standard journal for publication of crystal structure studies. Although the text is quite specialized, almost every issue has several interesting figures and photographs.

## CHAPTER 7

# STOICHIOMETRY AND CHEMICAL REACTIONS

"Every student of Science, even if he cannot start his journey where his predecessors left off, can at least travel their beaten track more quickly than they could while they were clearing the way: and so before his race is run he comes to virgin forest and becomes himself a pioneer."

*T. W. Richards\**

## 7.1 INTRODUCTION.

It is usually not difficult to establish whether or not a chemical reaction has taken place between certain chemical species, but it is often difficult to establish exactly what chemical change (reaction) has taken place. One basic problem that confronts the synthetic chemist is to establish the *stoichiometry* of the chemical reaction; what chemical species (if any) are formed (the products) when a given set of chemical species (the reactants) are allowed to interact with one another; and in what relative proportions the products are produced and the reactants consumed. The reaction stoichiometry is described conveniently and completely by means of a *balanced* chemical equation.† The first step toward establishing the stoichiometry of a chemical reaction is the identification of the products of the reaction; there follows a quantitative determination of the amounts of the various products that arise from known amounts of reactants. A balanced chemical equation is then written to agree with the analytical data.

---

\*T. W. Richards, Nobel Lectures. Vol. I (Elsevier Publishing Co., New York, 1966) p. 280–2.

†Actual chemical processes may not be describable by means of a *single* chemical equation; several different reactions may occur simultaneously.

## 7.2 FORMULA (OR MOLECULAR) WEIGHT* AND PERCENTAGE COMPOSITION.

A *mole* is that quantity of a substance whose mass in grams is numerically equal to its formula (or molecular) weight in atomic mass units (amu).† The absolute scale of atomic mass units is established by arbitrarily taking $^{12}C = 12$ (exactly) amu (see Chapter 1).

The formula weight of the compound $K_3Co(CN)_6$ is computed as follows:

| ELEMENT | (atomic weight of element) × (no. of atoms of the element in the formula) = | (contribution of element to the formula weight) |
|---------|---------------------------------------------|----------|
| K | $39.1 \times 3$ = | 117.3 |
| Co | $58.9 \times 1$ = | 58.9 |
| C | $12.0 \times 6$ = | 72.0 |
| N | $14.0 \times 6$ = | 84.0 |
| | TOTAL | 332.2 |

The formula weight of $K_3Co(CN)_6$ is 332.2 amu, and 1 mole of $K_3Co(CN)_6$ has a mass equal to 332.2 g. From the formula weight we can compute the number of moles of substance in any given mass of the substance. For example, 8.76 g of $K_3Co(CN)_6$ is

$$\frac{8.76 \text{ g}}{332.2 \text{ g} \cdot \text{mole}^{-1}} = 0.0264 \text{ mole of } K_3Co(CN)_6$$

The percentage of the total mass that each element comprises in the compound $K_3Co(CN)_6$ is:

$$\% K = \frac{117.3 \times 100}{332.2} = 35.3$$

$$\% Co = \frac{58.9 \times 100}{332.2} = 17.7$$

$$\% C = \frac{72.0 \times 100}{332.2} = 21.7$$

$$\% N = \frac{84.0 \times 100}{332.2} = 25.3$$

The above procedure can be reversed and used to determine the empirical formula of a compound from the observed percentage composition. The empirical formula expresses the combining ratios of the constituent elements of the compound as the smallest possible whole numbers. For example, suppose a certain chemical compound gave the following elemental analysis: 75.91% C; 6.41% H; and 17.72% N. Basing

---

*For the purposes of this section we are taking mass and weight to be synonymous. See, however, problems 15 and 21.

†Or, alternatively, a *mole* is the mass of a substance that contains Avogadro's number of formula units, where Avogadro's number is the number of carbon atoms in exactly 12 g of $^{12}C$ (i.e., $6.022 \times 10^{23}$).

the calculations on a 100.00 g sample for numerical simplicity, we compute the number of moles of C, H, and N in the 100.00 g sample as:

$$\frac{75.91 \text{ g C}}{12.0 \text{ g C/mole of C}} = 6.33 \text{ moles of C}$$

$$\frac{6.41 \text{ g H}}{1.008 \text{ g H/mole of H}} = 6.36 \text{ moles of H}$$

$$\frac{17.72 \text{ g N}}{14.0 \text{ g N/mole of N}} = 1.27 \text{ moles of N}$$

Next we compute the relative numbers of moles:

$$\left.\begin{array}{l}\dfrac{6.33 \text{ moles of C}}{1.27} = 4.98 \;\rightarrow\; 5C \\[2mm] \dfrac{6.36 \text{ moles of H}}{1.27} = 5.01 \;\rightarrow\; 5H \\[2mm] \dfrac{1.27 \text{ moles of N}}{1.27} = 1.00 \;\rightarrow\; 1N\end{array}\right\} C_5H_5N$$

If we now assume that the three elements C, H, and N occur in the compound in the ratio of small whole numbers (the law of simple combining ratios), then we have 5C, 5H, and 1N, or the empirical formula $C_5H_5N$. This represents the simplest possible formula of the compound, but the molecular formula might be some multiple of this, such as $C_{10}H_{10}N_2$ or $C_{15}H_{15}N_3$, both of which have the same percentage composition as $C_5H_5N$.

There are several methods for determining the molecular formula once the empirical formula is known. For example, suppose it were found that a 1.58 g gaseous sample of the above compound in a 0.80 $\ell$ vessel has a pressure of 0.82 atm at 400 K. The apparent molecular weight, $M$, of the compound is then

$$PV = nRT = \frac{w}{M} RT$$

$$M = \frac{wRT}{PV} = \frac{(1.58 \text{ g})(0.0821 \ \ell \cdot \text{atm} \cdot \text{K}^{-1} \cdot \text{mole}^{-1})(400 \text{ K})}{(0.82 \text{ atm})(0.80 \ \ell)} = 79 \text{ g} \cdot \text{mole}^{-1}$$

The formula weight of $C_5H_5N$ is $12.0 \times 5 + 14.0 \times 1 + 5 \times 1 = 79$ g $\cdot$ mole$^{-1}$, and in this case the empirical formula and the molecular formula are the same. If the molecular formula was $C_{10}H_{10}N_2$, then the observed gas pressure in the example given would have been 0.41 atm, which would yield a molecular weight of 158 g $\cdot$ mole$^{-1}$.

## 7.3 STOICHIOMETRY CALCULATIONS.

We must use balanced chemical equations when making stoichiometry calculations. As an example, consider the air oxidation of $Fe_3O_4(s)$ to $Fe_2O_3(s)$. The reaction is

$$Fe_3O_4(s) + O_2(g) \rightarrow Fe_2O_3(s)$$

and the balanced equation is[*]

$$2Fe_3O_4(s) + \frac{1}{2} O_2(g) = 3Fe_2O_3(s)$$

The formula weights of the three compounds are

$$Fe_2O_3 \qquad 159.7 \text{ g} \cdot \text{mole}^{-1}$$

$$Fe_3O_4 \qquad 231.6 \text{ g} \cdot \text{mole}^{-1}$$

$$O_2 \qquad 32.0 \text{ g} \cdot \text{mole}^{-1}$$

Suppose that 25.0 g of $Fe_3O_4(s)$ is oxidized completely by $O_2(g)$ to $Fe_2O_3(s)$. How many grams of $O_2(g)$ are consumed and how many moles of $Fe_2O_3(s)$ are produced? The 25.0 g of $Fe_3O_4(s)$ contains

$$\frac{25.0 \text{ g}}{231.6 \text{ g} \cdot \text{mole}^{-1}} = 0.108 \text{ mole of } Fe_3O_4$$

From the stoichiometry of the balanced equation, we note that half of a mole of $O_2(g)$ is required to oxidize two moles of $Fe_3O_4(s)$, and therefore the number of moles of $O_2$ required to oxidize 0.108 mole of $Fe_3O_4$ is

$$\text{moles of } O_2 \text{ required} = (0.108 \text{ mole } Fe_3O_4) \left\{ \frac{\frac{1}{2} \text{ mole } O_2}{2 \text{ mole } Fe_3O_4} \right\}$$

$$\text{moles of } O_2 \text{ required} = \frac{0.108}{4} = 0.0270 \text{ mole } O_2$$

The number of grams of $O_2$ required is

$$(32.0 \text{ g} \cdot \text{mole}^{-1})(0.0270 \text{ mole}) = 0.864 \text{ g}$$

The number of moles of $Fe_2O_3(s)$ produced is

$$\text{moles of } Fe_2O_3 \text{ produced} = (0.108 \text{ mole } Fe_3O_4) \left\{ \frac{3 \text{ mole } Fe_2O_3}{2 \text{ mole } Fe_3O_4} \right\}$$
$$= 0.162 \text{ mole } Fe_2O_3.$$

As a second example, suppose that 50.0 g of $Fe_3O_4(s)$ is allowed to react with 2.0 g of $O_2(g)$. How many moles of $Fe_2O_3(s)$ are produced? In this case we are given the available amounts of *both* reactants, and consequently we must determine which reactant is *stoichiometrically limiting*. The number of moles of $O_2$ available is

$$\frac{2.0 \text{ g}}{32.0 \text{ g} \cdot \text{mole}^{-1}} = 0.0625 \text{ mole of } O_2$$

---

[*]The balancing coefficients have been chosen such that the total number of Fe and O atoms are the same on the left and right sides of the reaction. Fractional balancing coefficients are encountered frequently in balanced equations, and are no cause for alarm. In the present context, $\frac{1}{2} O_2$ indicates $\frac{1}{2}$ of a *mole* of oxygen.

and the number of moles of $Fe_3O_4$ available is

$$\frac{50.0 \text{ g}}{231.6 \text{ g} \cdot \text{mole}^{-1}} = 0.216 \text{ mole of } Fe_3O_4$$

The number of moles of oxygen required to react with 0.216 moles of $Fe_3O_4$ is

$$(0.216 \text{ mole } Fe_3O_4) \left\{ \frac{\frac{1}{2} \text{ mole } O_2}{2 \text{ mole } Fe_3O_4} \right\} = 0.0540 \text{ mole } O_2$$

Therefore, in this case, $O_2$ is in excess ($0.0625 > 0.0540$) and $Fe_3O_4$ is the stoichiometrically limiting reagent. The number of moles of $Fe_2O_3$ produced is

$$(0.216 \text{ mole } Fe_3O_4) \left\{ \frac{3 \text{ mole } Fe_2O_3}{2 \text{ mole } Fe_3O_4} \right\} = 0.324 \text{ mole } Fe_2O_3$$

The number of moles of $O_2$ consumed is $0.216/4 = 0.0540$ mole, and the number of moles of $O_2$ in excess (i.e., which remain unreacted after all the $Fe_3O_4$ is consumed) is $0.0625 - 0.0540 = 0.0085$ mole.

As a further example, suppose 1.25 g of liquid benzene, $C_6H_6$, is combusted with excess gaseous oxygen in a 0.50 $\ell$ chamber maintained at 300 K. Assuming ideal-gas behavior, calculate the final partial pressure of the $CO_2(g)$ produced, given that the reaction is

$$C_6H_6(\ell) + \frac{15}{2} O_2(g) = 6CO_2(g) + 3H_2O(\ell)$$

The number of moles of benzene consumed is

$$(1.25 \text{ g}/78.0 \text{ g} \cdot \text{mole}^{-1})$$

and, therefore, the number of moles of $CO_2$ produced is

$$\left\{ \frac{1.25}{78.0} \text{ mole } C_6H_6 \right\} \left\{ \frac{6 \text{ mole } CO_2}{1 \text{ mole } C_6H_6} \right\} = 0.096 \text{ mole } CO_2$$

This number of moles of $CO_2(g)$ at 300 K in a 0.50 $\ell$ chamber has a pressure of

$$P = \frac{nRT}{V} = \frac{(0.096 \text{ mole})(0.0821 \ell \cdot \text{atm} \cdot \text{K}^{-1} \cdot \text{mole}^{-1})(300 \text{ K})}{(0.50 \ell)} = 4.7 \text{ atm}$$

It is possible to carry out certain types of stoichiometry calculations without *formally* writing a balanced chemical equation. Such calculations are often carried out in connection with analytical work. By way of example, consider the following problem: The arsenic in a 0.4964 g sample of an arsenic ore was converted quantitatively into $As_2S_5(s)$ that weighed 0.3320 g. Calculate the weight percentage of arsenic in the ore sample. Because the arsenic in the sample is converted quantitatively to $As_2S_5$, the number of grams of *arsenic* in the $As_2S_5(s)$ is the same as was originally present in the ore sample; therefore,

$$(0.3320 \text{ g As}_2S_5) \left\{ \frac{2 \times 74.92 \text{ g As/mole As}_2S_5}{310.15 \text{ g As}_2S_5/\text{mole As}_2S_5} \right\} = 0.1600 \text{ g As}$$

where the term in brackets is the fraction of As in any sample of $As_2S_5$. The percentage of arsenic in the original sample is

$$\left( \frac{0.1600 \text{ g As}}{0.4964 \text{ g of sample}} \right) 100 = 32.2\%$$

Stoichiometry calculations are frequently carried out using data obtained from mixtures of compounds. For example: A 1.257 g sample of an ore yielded 0.3640 g of a mixture of $Fe_2O_3(s)$ and $Al_2O_3(s)$. This mixture of oxides, when heated in the presence of $H_2(g)$ to yield $Fe(s)$ and $Al_2O_3(s)$, underwent a mass *decrease* of 0.0786 g. Calculate the percentage of iron and aluminum in the original sample. The decrease in weight of the oxide mixture obviously arises from the conversion of $Fe_2O_3(s)$ to $Fe(s)$. The reaction is

$$Fe_2O_3(s) + 3H_2(g) = 2Fe(s) + 3H_2O(g)$$

However, because the decrease in sample weight arises from the loss of oxygen by $Fe_2O_3(s)$, the calculation is more direct if we base it on the equation[*]

$$Fe_2O_3(s) = 2Fe(s) + \frac{3}{2} O_2(g)$$

The number of moles of oxygen lost on conversion of $Fe_2O_3$ to $Fe$ is

$$\frac{0.0786 \text{ g}}{32.00 \text{ g} \cdot \text{mole}^{-1}} = 0.00246 \text{ mole O}_2$$

The number of grams of $Fe(s)$ produced on reaction with $H_2$ is

$$(0.00246 \text{ mole O}_2) \left\{ \frac{2 \text{ mole Fe}}{\frac{3}{2} \text{ mole O}_2} \right\} \left( 55.85 \frac{\text{g Fe}}{\text{mole Fe}} \right) = 0.183 \text{ g Fe}$$

The percentage of iron in the original sample is computed to be (0.183 g Fe/1.25 g of sample) $\times$ 100 = 14.6% Fe. The number of grams of $Fe_2O_3(s)$ that was present in the 0.3640 g mixture of oxides is

$$(0.00246 \text{ moles O}_2) \left\{ \frac{1 \text{ mole Fe}_2O_3}{\frac{3}{2} \text{ mole O}_2} \right\} \left\{ 159.70 \frac{\text{g Fe}_2O_3}{\text{mole Fe}_2O_3} \right\} = 0.262 \text{ g Fe}_2O_3$$

The number of grams of $Al_2O_3$ in the oxide mixture is

$$0.364 \text{ g of mixture} - 0.262 \text{ g of Fe}_2O_3 = 0.102 \text{ g Al}_2O_3$$

---

[*]The calculations could also be based on O rather than $O_2$, using the equation $Fe_2O_3(s) = 2Fe(s) + 3O(g)$.

which comprises

$$\frac{0.102 \text{ g Al}_2\text{O}_3}{101.96 \text{ g Al}_2\text{O}_3/\text{mole Al}_2\text{O}_3} = 0.00100 \text{ mole Al}_2\text{O}_3$$

The percentage of Al present in the original ore sample is then

$$\frac{(0.00100 \text{ mole Al}_2\text{O}_3) \left\{ \frac{2 \text{ mole Al}}{1 \text{ mole Al}_2\text{O}_3} \right\} \left\{ 26.98 \frac{\text{g Al}}{\text{mole Al}} \right\}}{(1.257 \text{ g of sample})} \times 100 = 4.30\% \text{ Al}$$

## 7.4 BALANCING CHEMICAL EQUATIONS.

A chemical equation is balanced once the coefficients of all the reactants and products have been chosen such that the total number of atoms of each element involved in the reaction is the same on both sides of the equation, and, in addition, the net electrical charge is the same on both sides of the equation. The first step in the balancing of a chemical equation is to write down the various reactant and product species on their respective sides of the equation.

For the purposes of the discussion to follow, we shall distinguish two types of chemical reactions, namely, those that involve changes in the oxidation states of the elements involved in the reaction (*oxidation-reduction* or "*redox*" reactions), and those that do not involve changes in the oxidation states of the elements involved (*metathetical reactions*).

### a. Metathetical Reactions.

The balancing of metathetical equations is simple and usually can be accomplished by inspection. For example, $AgCl(s)$ has a very low solubility in water but dissolves readily in $Na_2S_2O_3(aq)$ solution of the proper concentration, because of the formation of the soluble $Ag(S_2O_3)_2^{3-}$ complex.[*] The reaction is

$$AgCl(s) + S_2O_3^{2-} (aq) \rightarrow Ag(S_2O_3)_2^{3-} (aq) + Cl^-(aq)$$

which is balanced as follows:

$$AgCl(s) + 2S_2O_3^{2-} (aq) = Ag(S_2O_3)_2^{3-} (aq) + Cl^-(aq).$$

As a second example, mercury (II) oxide, $HgO$, is insoluble in water but reacts with aqueous solutions containing iodide ion, for example, $KI(aq)$ solution. The principal products of the reaction depend on the mole ratio of $I^-$ to $HgO$. At a 1:1 ratio, the product is $HgI^+(aq)$, and at 2:1 it is $HgI_2(s)$, whereas at greater than 2:1 it is $HgI_4^{2-}$ (aq). Taking the 1:1 case first, we have

$$HgO(s) + I^-(aq) \rightarrow HgI^+(aq)$$

---

[*]The abbreviation for aqueous, (aq), is used to indicate species which are dissolved in water. Concentration units will be introduced in the next chapter.

This reaction could be formally balanced by adding $O^{2-}(aq)$ to the right hand side; however $O^{2-}$ is not a stable species in aqueous solution, but undergoes the reaction

$$O^{2-}(aq) + H_2O(\ell) = 2OH^-(aq)$$

The products of the above reaction are, therefore, $HgI^+$ and $OH^-$, and the balanced equation is

$$HgO(s) + I^-(aq) + H_2O(\ell) = HgI^+(aq) + 2OH^-(aq)$$

When $HgI_2(s)$ is the principal mercury-containing product, the balanced equation is

$$HgO(s) + 2I^-(aq) + H_2O(\ell) = HgI_2(s) + 2OH^-(aq)$$

## b.  Oxidation States.

In assigning oxidation states to the various elements in a species or compound, we use the following conventions (or rules) that are to be applied in the order given: [*]
(1) All elements in the elementary state are assigned an oxidation state of zero.
(2) The alkali metals (Li, Na, K, Rb, and Cs) in any compound are assigned an oxidation state of $+1$; the alkaline earth metals (Be, Mg, Ca, Sr, Ba, and Ra), and also Zn and Cd in any compound are assigned an oxidation state of $+2$.
(3) Oxygen is assigned an oxidation state of $-2$ in all of its compounds except the peroxides and superoxides (e.g., $Na_2O_2$ and $KO_2$).
(4) Hydrogen is assigned an oxidation state of $+1$ in all of its compounds except the hydrides (e.g., NaH).
(5) Oxidation states of the other elements in a species usually can be determined by the requirement that the algebraic sum of the oxidation state values for all of the elements in the species must equal the net charge on the species.
These rules[†] have their origin in the atomic structures and electro-negativities of the various elements. However, it is important to realize that the oxidation state does not, in general, correspond to the actual charge on an element in a chemical species. Oxidation states are primarily a convenient bookkeeping device that is useful in balancing chemical reactions.
Some examples of the assignment of oxidation states are:
(a) $Cl_2$: This is a form of the element chlorine, and there is no net charge on the species, and therefore the oxidation state of chlorine in $Cl_2$ is 0.

---

[*]If a conflict arises, then the rule with the lower number takes precedence. In other words, apply the rules in the order given until there is only one element left, which is then assigned an oxidation state consistent with the net-charge condition (rule 5).

[†]The rules, as given, are not sufficient to cover all chemical species, and in such cases it is best to work by analogy (via the periodic chart) with other compounds covered by the rules. In any case, rule 5 must always be satisfied (see problem 20).

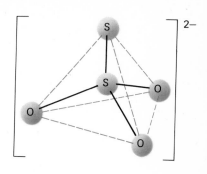

**FIGURE 7.1**   The tetrahedral structure of the thiosulfate ion, $S_2O_3^{2-}$.

(b) $Na_2SO_4$: assigning Na to the $+1$ state, and oxygen to the $-2$ state, we deduce the oxidation state of S to be

$$2(+1) + 4(-2) + 1(x) = 0$$
$$x = +6$$

(c) $SO_2$: assigning oxygen to the $-2$ state, we deduce the oxidation state of sulfur in $SO_2$ as $+4$.

(d) $S_2O_3^{2-}$: assigning oxygen to the $-2$ state, we deduce the average oxidation state of the two sulfur atoms as

$$3(-2) + 2x = -2$$
$$x = +2$$

The $S_2O_3^{2-}$ (thiosulfate) ion has the tetrahedral structure shown in Figure 7.1, and it is evident from the figure that the two sulfur atoms are not equivalent. Rather than regard both sulfur atoms as $+2$, it is preferable to regard the central sulfur atom as $+6$ (by analogy with $SO_4^{2-}$), and the other sulfur atom as $-2$.

(e) NaH: assigning sodium an oxidation state of $+1$, we deduce the oxidation state of hydrogen as $-1$. This compound is a hydride. The oxidation state of hydrogen in hydrides is $-1$.

(f) $Fe_3O_4$: assigning oxygen an oxidation state of $-2$, we deduce an average oxidation state for iron of $+\frac{8}{3}$ (actually, 2Fe at $+3$ and one Fe at $+2$).

## c.   Redox Reactions.

If changes in oxidation states are involved in the reaction, then it is desirable to follow a systematic procedure in balancing the equation, because balancing of oxidation-reduction equations is not as simple as balancing metathetical reactions.

The term *oxidation* denotes a *loss* of electrons (increase in oxidation number), whereas the term *reduction* denotes a *gain* of electrons (decrease in oxidation number). In order for an oxidation-reduction reaction to occur, there must be present an *oxidizing agent* (electron acceptor) and a *reducing agent* (electron donor). In the *ion-electron method* of balancing redox equations, the reaction is separated into two parts (half-reactions), one involving the oxidizing agent and its corresponding reduced species, and the other involving the reducing agent and its corresponding oxidized species. Also, one must know whether the reaction takes place in acidic or basic solution. This is because in *acidic* aqueous solution the principal hydrogen-containing species are $H^+$ and $H_2O$, whereas in *basic* aqueous solution the principal hydrogen-containing species are $OH^-$ and $H_2O$.

Consider the problem of writing a balanced equation for the following reaction:

$$HNO_2(aq) + MnO_4^-(aq) \rightarrow NO_3^-(aq) + Mn^{2+}(aq)$$

The oxidizing agent in this reaction is $MnO_4^-$, and it is reduced by the reducing agent $HNO_2$ to $Mn^{2+}$. The oxidation state of manganese in $MnO_4^-$ is +7, whereas in $Mn^{2+}$ it is +2 (a *gain* of 5 electrons). On the other hand, the oxidation state of nitrogen in $HNO_2$ is +3, whereas in $NO_3^-$ it is +5 (a *loss* of two electrons).

A systematic procedure for balancing redox reactions in acidic solutions is the following:

(1) *Separate the reaction into oxidation and reduction half-reactions*

$$MnO_4^- \rightarrow Mn^{2+} \quad \text{(reduction)}$$
$$HNO_2 \rightarrow NO_3^- \quad \text{(oxidation)}$$

(2) *Balance the two half-reactions with respect to all elements other than hydrogen and oxygen.* The first half-reaction has one Mn on each side, and the second half-reaction has one N on each side. Therefore, in this case, this step is already completed.

(3) *Balance the half-reactions with respect to oxygen using $H_2O$*

$$MnO_4^- \rightarrow Mn^{2+} + 4H_2O$$
$$H_2O + HNO_2 \rightarrow NO_3^-$$

(4) *Balance the half-reactions with respect to hydrogen using $H^+$*

$$8H^+ + MnO_4^- \rightarrow Mn^{2+} + 4H_2O$$
$$H_2O + HNO_2 \rightarrow NO_3^- + 3H^+$$

(5) *Balance the half-reactions with respect to charge using $e^-$ (electrons)*

$$5e^- + 8H^+ + MnO_4^- \rightarrow Mn^{2+} + 4H_2O \tag{7.1}$$

$$H_2O + HNO_2 \rightarrow NO_3^- + 3H^+ + 2e^- \tag{7.2}$$

(6) *Multiply the two half-reactions by the integers that will make the number of electrons supplied by the reducing agent equal to the number of electrons consumed by the oxidizing agent* (conservation of electrons). In the present example each mole of $HNO_2$ gives 2 moles of electrons, whereas each mole of $MnO_4^-$ consumes 5 moles of electrons. Therefore, we multiply the second half-reaction by 5 and the first half-reaction by 2

$$10e^- + 16H^+ + 2MnO_4^- \rightarrow 2Mn^{2+} + 8H_2O \tag{7.3}$$

$$5H_2O + 5HNO_2 \rightarrow 5NO_3^- + 15H^+ + 10e^- \tag{7.4}$$

(7) *Add the two half-reactions and cancel like terms to obtain the final balanced equation.*

$$2MnO_4^-(aq) + 5HNO_2(aq) + H^+(aq) = 2Mn^{2+}(aq) + 5NO_3^-(aq) + 3H_2O(\ell)$$

A half-reaction can always be balanced by itself without reference to a particular accompanying half-reaction. For example, potassium dichromate, $K_2Cr_2O_7$, finds wide use as an oxidizing agent in acidic aqueous solution, where the reduced chromium species is usually the chromic ion, $Cr^{3+}(aq)$. The balanced equation for the half-reaction involving chromium is obtained in the same manner as outlined above:

$$Cr_2O_7^{2-} \rightarrow Cr^{3+}$$

$$Cr_2O_7^{2-} \rightarrow 2Cr^{3+} + 7H_2O$$

$$Cr_2O_7^{2-} + 14H^+ \rightarrow 2Cr^{3+} + 7H_2O$$

$$6e^- + Cr_2O_7^{2-}(aq) + 14H^+(aq) \rightarrow 2Cr^{3+}(aq) + 7H_2O(\ell)$$

The procedure for balancing half-reactions in basic aqueous solution requires an additional step in which any $H^+$ appearing in the half reaction is converted to $H_2O$ using $OH^-$. For example, potassium iodide is sometimes used as a reducing agent in basic aqueous solution, the $I^-$ being oxidized to $IO_3^-$.

$$I^- \rightarrow IO_3^-$$

Following the procedure outlined above, we first balance the half-reaction *as if it occurred* in acidic solution:

$$3H_2O + I^- \rightarrow IO_3^-$$

$$3H_2O + I^- \rightarrow IO_3^- + 6H^+ + 6e^-$$

We now convert the $H^+$ to $H_2O$ by adding $OH^-$ to both sides:

$$3H_2O + I^- + 6OH^- \rightarrow IO_3^- + [6H^+ + 6OH^-] + 6e^-$$

$$3H_2O + I^- + 6OH^- \rightarrow IO_3^- + 6H_2O + 6e^-$$

or

$$I^-(aq) + 6OH^-(aq) \rightarrow IO_3^-(aq) + 3H_2O(\ell) + 6e^-$$

## d. Disproportionation Reactions.

Disproportionation reactions constitute a particularly interesting class of redox reactions. In a disproportionation reaction, a single substance is both oxidized and reduced. An example is the disproportionation of cuprous ion to copper metal and cupric ion in acidic aqueous solution:

$$Cu^+(aq) \rightarrow Cu^{2+}(aq) + Cu(s)$$

The balanced half-reactions are

$$Cu^+ \rightarrow Cu^{2+} + e^- \qquad \text{(oxidation)}$$

$$e^- + Cu^+ \rightarrow Cu(s) \qquad \text{(reduction)}$$

and the balanced equation is

$$2Cu^+(aq) = Cu^{2+}(aq) + Cu(s)$$

## 7.5   NET REACTIONS

Consider the following reactions:

$$CsI(aq) + AgNO_3(aq) = AgI(s) + CsNO_3(aq) \qquad (7.5)$$

$$NaI(aq) + AgClO_4(aq) = AgI(s) + NaClO_4(aq) \qquad (7.6)$$

$$LiI(aq) + AgF(aq) = AgI(s) + LiF(aq) \qquad (7.7)$$

$$\frac{1}{2} CaI_2(aq) + AgNO_3(aq) = AgI(s) + \frac{1}{2} Ca(NO_3)_2(aq) \qquad (7.8)$$

Although these reactions are obviously different, they all involve the precipitation of silver iodide, and if all the electrolytes are strong (that is, exist predominantly as the individual ions in solution), then the net reaction

$$I^-(aq) + Ag^+(aq) = AgI(s) \qquad (7.9)$$

is the same in all cases. For example, $CsI(aq)$, $AgNO_3(aq)$, and $CsNO_3(aq)$ are all strong electrolytes and, therefore, reaction (7.5) can be rewritten as

$$Cs^+(aq) + I^-(aq) + Ag^+(aq) + NO_3^-(aq) = AgI(s) + Cs^+(aq) + NO_3^-(aq)$$

and cancelling like terms on opposite sides of the equality yields reaction (7.9).

If an electrolyte is not "strong," that is, if it exists predominantly as the undissociated species, then the net equation must take this into account. Some of the more common strong and weak electrolytes are listed in Table 7.1. For example, the reactions

$$CaCl_2(aq) + 2AgNO_3(aq) = 2AgCl(s) + Ca(NO_3)_2(aq) \qquad (7.10)$$

$$HgCl_2(aq) + 2AgNO_3(aq) = 2AgCl(s) + Hg(NO_3)_2(aq) \qquad (7.11)$$

can be written as the net reactions

$$2Cl^-(aq) + 2Ag^+(aq) = 2AgCl(s) \qquad (7.10a)$$

**TABLE 7.1**

STRONG AND WEAK ELECTROLYTE RULES

(1) Soluble strong acids: $HCl$, $HBr$, $HI$, $H_2SO_4$, $HNO_3$, $HClO_4$, $HBrO_3$, $RSO_3^- H^+$ (sulfonic acids, e.g., $HOCH_2CH_2SO_3^- H^+$).

(2) Soluble weak acids: bisulfate ($HSO_4^-$), other common inorganic acids, and most organic acids (for example, $CH_3COOH$, $C_6H_5COOH$) ($H_2SiO_3$ insoluble).

(3) Soluble strong bases: $LiOH$, $NaOH$, $TlOH$, $KOH$, $CsOH$, $RbOH$, $Ba(OH)_2$.

(4) Soluble weak bases: $NH_3$, $N_2H_4$, and most organic bases (for example, $CH_3NH_2$, $C_5H_5N$).

(5) Weak electrolytes: all halides, cyanides ($CN^-$), and thiocyanates ($SCN^-$) of $M^{n+}$ ($n \geqq 2$) ions.

and

$$HgCl_2(aq) + 2Ag^+(aq) = 2AgCl(s) + Hg^{2+}(aq) \qquad (7.11a)$$

respectively. Note that $HgCl_2(aq)$ is a weak electrolyte, and thus it appears in the net reaction.

## 7.6  ACID-BASE REACTIONS.

We shall discuss acid-base equilibria in detail in Chapters 11 and 12. It will suffice for the purposes of the present discussion to use the Brönsted acid-base classification. A Brönsted *acid* is defined as a *proton donor*, whereas a Brönsted *base* is defined as a *proton acceptor*. The reaction of an acid with a base yields a *salt*. An aqueous solution is said to be *acidic* if the number of $H^+(aq)$ ions exceeds the number of $OH^-(aq)$ ions, and *basic* if the reverse is true.

The reaction between the strong base $KOH(aq)$ and the strong acid $HCl(aq)$ is

$$KOH(aq) + HCl(aq) = KCl(aq) + H_2O(\ell)$$

or

$$K^+(aq) + OH^-(aq) + H^+(aq) + Cl^-(aq) = K^+(aq) + Cl^-(aq) + H_2O(\ell).$$

Therefore, the net reaction is

$$OH^-(aq) + H^+(aq) = H_2O(\ell)$$

On the other hand, the reaction between the strong base $NaOH(aq)$ and the weak acid $HNO_2(aq)$

$$Na^+(aq) + OH^-(aq) + HNO_2(aq) = H_2O(\ell) + Na^+(aq) + NO_2^-(aq)$$

has the net reaction

$$OH^-(aq) + HNO_2(aq) = H_2O(\ell) + NO_2^-(aq)$$

The acid $HCN(aq)$ is a weak acid and, therefore, the reaction

$$NaCN(aq) + HCl(aq) = HCN(aq) + NaCl(aq)$$

has the net reaction

$$CN^-(aq) + H^+(aq) = HCN(aq)$$

*Salts of weak acids and strong bases* react with water to produce *basic* solutions. For example, the reaction

$$KNO_2(aq) + H_2O(\ell) = KOH(aq) + HNO_2(aq)$$

has the net reaction

$$NO_2^-(aq) + H_2O(\ell) = HNO_2(aq) + OH^-(aq).$$

Because $OH^-(aq)$ is produced in this net reaction, a solution of $KNO_2(aq)$ is basic.

*Salts of weak bases and strong acids* yield *acidic* solutions upon dissolution in water. When $NH_4Cl(s)$ is dissolved in water, we have the net reaction

$$NH_4^+(aq) = NH_3(aq) + H^+(aq)$$

Because $H^+(aq)$ is produced in this net reaction, a solution of $NH_4Cl(aq)$ is acidic.

*Salts formed from weak acids and weak bases* may yield *either acidic or basic* solutions, depending on the relative proton donating and accepting "strengths" of the acids and bases involved. For example, when $NH_4NO_2(s)$ is dissolved in water, the principal reaction that occurs is

$$NH_4^+(aq) + NO_2^-(aq) = NH_3(aq) + HNO_2(aq)$$

It is not possible to state in such a case whether the resulting solution is acidic or basic unless we know the strengths of the two acids, $NH_4^+(aq)$ and $HNO_2(aq)$.

Salts of weak acids that are only slightly soluble in water have a much greater solubility in aqueous solutions of strong acids, owing to the formation of the weak acid in solution. For example, $PbSO_3(s)$ is essentially insoluble in water, but dissolves readily in aqueous solutions of strong acids, because $H_2SO_3(aq)$ is a weak acid:

$$PbSO_3(s) + 2H^+(aq) = Pb^{2+}(aq) + H_2SO_3(aq)$$

On the other hand, $PbSO_4$ is insoluble in water, and also in aqueous solutions of strong acids, because $H_2SO_4$ is a strong acid.

Many cations in aqueous solution are hydrated to a definite extent, for example

$$Be(OH_2)_4^{2+}(aq) \qquad Al(OH_2)_6^{3+}(aq) \qquad Fe(OH_2)_6^{2+}(aq)$$

For certain other cations the extent of hydration is apparently indefinite, for example, $Na^+(aq)$. Anions, as a rough rule, are only weakly (if at all) hydrated. Salts composed of cations involving definite hydration shells and anions of strong acids yield acid solutions on dissolution, because of the loss of $H^+$ from a bound water molecule, for example,

$$Al(OH_2)_6^{3+} = Al(OH_2)_5OH^{2+}(aq) + H^+(aq)$$

A consequence of this type of reaction is the instability of certain salts involving such cations with anions of weak acids. For example, aqueous solutions of $Fe_2(CO_3)_3$ undergo the following decomposition reaction:

$$2Fe(OH_2)_6^{3+}(aq) + 3CO_3^{2-}(aq) = 2Fe(OH)_3 \cdot 3H_2O(s) + 3H_2O + 3CO_2(g)$$

In Tables 7.1 and 7.2 we have set down some solubility rules for salts in aqueous solution, together with some rules for strong and weak acids and

**TABLE 7.2**

SOLUBILITY RULES*

(1) All nitrates ($NO_3^-$), chlorates ($ClO_3^-$), and acetates ($CH_3COO^-$) (except $AgOCOCH_3$ and $Hg_2(OCOCH_3)_2$) are soluble.
(2) All salts of $Li^+$, $Na^+$, $K^+$, $Cs^+$ and $NH_4^+$ are soluble.
(3) All chlorides ($Cl^-$), bromides ($Br^-$), and iodides ($I^-$) (except those of $Ag^+$, $Hg_2^{2+}$, $Tl^+$, and $Pb^{2+}$) are soluble.
(4) All sulfates ($SO_4^{2-}$) (except those of $Ca^{2+}$, $Ba^{2+}$, $Pb^{2+}$, $Ag^+$, and $Hg_2^{2+}$) are soluble.
(5) All sulfides ($S^{2-}$) (except those of $Li^+$, $Na^+$, $K^+$, $Cs^+$, $NH_4^+$, $Mg^{2+}$, $Ca^{2+}$, and $Ba^{2+}$) are insoluble.
(6) All carbonates ($CO_3^{2-}$), phosphates ($PO_4^{3-}$), chromates ($CrO_4^{2-}$), arsenates ($AsO_4^{3-}$), and sulfites ($SO_3^{2-}$) (except those of the alkali metals and $NH_4^+$) are insoluble.

*There are some exceptions that have not been noted in the interests of simplicity; rules apply only to water as the solvent at a solution temperature around 25°C. All compounds are soluble to some extent; however, we shall arbitrarily regard a compound as insoluble if less than 0.01 mole will dissolve in one liter of solution.

bases and strong and weak electrolytes. The rules given in Tables 7.1 and 7.2 are applicable only around 25°C with water as the solvent. Both solubility and acid and base strengths often depend strongly on the solvent and the temperature. For example, $NH_4^+$ is a weak acid in water but a strong acid in $NH_3(\ell)$; HCl is a strong acid in water, but a weak acid in chloroform; LiBr is a strong electrolyte in water, but a weak electrolyte in ether solutions. Generally speaking, ionic-lattice-type solids which dissolve to yield conducting solutions are more soluble in more polar solvents. Molecular-lattice-type solids are, in general, more soluble in non-polar solvents. In other words, *like dissolves like*. A salt, acid, or base is more likely to be weak the less ionic its crystal lattice and the less polar the solvent.

_____ **PROBLEMS**

1. 72.5 g of $CH_4$ is _____ moles, and contains _____ molecules; it will occupy _____ liters at 25°C and 3.50 atm.

2. Compute the percentages by weight of the elements in the following compounds:

(a) $BaSO_4 \cdot 2H_2O$
(b) $NH_4B(C_6H_5)_4$
(c) $C_5H_5NHCl$
(d) $CH_3OCH_2CH_2OCH_2CH_2OCH_3$
(e) $[Co(NH_3)_5OCOCH_3]Cl_2$
(f) $CH_3CHCH_2$
                   |   |
                   O   O
                    \  /
                     C
                     ‖
                     O
(g) $[(CH_3)_2N]_3PO$

(h) $CH_3CH_2CH_2CH_2CH_2OD$

3. The elemental analysis of a compound yielded the following percentage composition by weight:

50.0% F;     21.1% C;     0.886% H;     and     28.1% O.

Determine the empirical formula of the compound.

4. It was found that 4.60 g of the compound whose analysis is reported in problem 3 occupied a volume of 644 cm³ at 100°C and 0.950 atm. Calculate the apparent molecular weight of the compound. What is the chemical formula of the compound?

5. Calculate the maximum number of grams of $Ca_3(PO_4)_2(s)$ that can be obtained from 10.0 g of $P_2O_5(s)$.

6. Determine the oxidation states of the elements marked with an asterisk in the following compounds and ions:

| | | |
|---|---|---|
| (1) $NaCl^*$ | (11) $Fe^*(CN)_6^{3-}$ | (21) $CaC_2^*$ |
| (2) $NaCl^*O$ | (12) $C^*O$ | (22) $C^*H_4$ |
| (3) $NaCl^*O_3$ | (13) $C^*O_2$ | (23) $HC^*HO$ |
| (4) $NaCl^*O_4$ | (14) $Cr^*O_4^{2-}$ | (24) $N^*O_2$ |
| (5) $NaCl^*O_2$ | (15) $Mn^*O_4^{2-}$ | (25) $N^*O_2^-$ |
| (6) $Cl^*O_2$ | (16) $Mn^*O_2$ | (26) $N^*H_3$ |
| (7) $Co_2^*O_3$ | (17) $Mn_2^*O_7$ | (27) $Na_2O^*$ |
| (8) $Co^*O$ | (18) $KO_2^*$ | (28) $Na_2O_2^*$ |
| (9) $Co_3^*O_4$ | (19) $Co^*Cl_4^{2-}$ | (29) $NaO_2^*$ |
| (10) $Sb^*OCl$ | (20) $NiO_2^*H$ | |

7. Under oxygen-free, anhydrous conditions $BaSO_4(s)$ reacts with $P_4(s)$ to yield $BaS(s) + P_4O_{10}(s)$. Calculate the number of grams of $P_4O_{10}(s)$ produced when 15.0 g of $BaSO_4$ is allowed to react completely with excess $P_4(s)$. How many moles of $BaS(s)$ are produced when 15.0 g of $BaSO_4(s)$ is reacted with 2.00 g of $P_4(s)$?

8. A 0.357 g sample of an unknown gas was found to occupy 250 cm³ at 0.350 atm and 35°C. Calculate its apparent molecular weight.

9. Compute the volume occupied by 6.0 g of $CO_2(g)$ at 0°C and 350 Torr.

10. Explain why (a) the most common oxidation state of the Group I elements is +1, (b) the most common oxidation state of the Group II elements is +2, (c) the most common oxidation state of hydrogen is +1, but occasionally hydrogen is assigned a state of −1, and (e) the most common oxidation state of the Group VI elements is −2.

11. Complete and balance the following reactions (if there is no reaction, write N.R.). In those cases where a reaction occurs give the "net reaction" as well as the balanced equation for the "total reaction."

(a) $PbCO_3(s) + HCl(aq) \rightarrow$
(b) $Hg(NO_3)_2(aq) + NaCN(aq) \rightarrow$
(c) $H_3PO_4(aq)$, in excess $+ KOH(aq) \rightarrow$
(d) $H_3PO_4(aq) + KOH(aq)$, in excess $\rightarrow$

(e) $NH_3(g) + HI(g) \rightarrow$
(f) $Fe(NO_3)_3(aq) + KOH(aq) \rightarrow$
(g) $Al(OH)_3(s) + HBr(aq) \rightarrow$
(h) $AgOCOCH_3(s) + HClO_4(aq) \rightarrow$
(i) $CaHPO_4(s) + H_2SO_4(aq) \rightarrow$
(j) $CaCO_3(s) + H_3PO_4(aq) \rightarrow$
(k) $CdS(s) + HBr(aq) \rightarrow$
(l) $NaHCO_3(aq) \rightarrow$
(m) $CH_3COOH(aq) + NH_3(aq) \rightarrow$
(n) $Na_2CO_3(aq) \rightarrow$
(o) $Na_3PO_4(aq) \rightarrow$

12. Given the following list of compounds:

| | | |
|---|---|---|
| (1) $NH_3$ | (5) $PbSO_4$ | (9) $AgOCOCH_3$ |
| (2) $CaCO_3$ | (6) $HNO_2$ | (10) $Na_2SO_3$ |
| (3) KBr | (7) AgI | (11) $NH_4NO_3$ |
| (4) $Cd(OH)_2$ | (8) $CdCl_2$ | (12) $HClO_4$ |

select:
    (a) those which are soluble in water.
    (b) those which are soluble in water and yield basic solutions.
    (c) those which are soluble in water and yield acidic solutions.
    (d) those which are insoluble in water, but soluble in aqueous solutions of strong acids.
    (e) those which are weak acids in water.
    (f) those which are weak electrolytes in water.

13. Consider the following hypothetical chemical reaction:

$$XY_2(s) + ZW_3(g) \rightarrow Z_3Y_2(\ell) + X(s) + W_2(g)$$

The atomic weights of the hypothetical elements are:

$$X = 10.0; \quad Y = 15.0; \quad Z = 20.0; \quad \text{and} \quad W = 30.0.$$

    (a) Balance the equation.
    (b) How many moles of $ZW_3(g)$ are required to produce 3/5 mole of $W_2(g)$?
    (c) How many moles of $X(s)$ can be produced from the reaction of 5.00 g of $XY_2(s)$ with 10.0 g of $ZW_3(g)$?
    (d) How many $cm^3$ of $Z_3Y_2(\ell)$ (density = 2.30 $g/cm^3$) can be formed in the reaction of 0.50 mole of $XY_2(s)$ with excess $ZW_3(g)$?
    (e) How many liters of $W_2(g)$ at 235°C and 2.00 atm can be produced from the reaction of 3.0 moles of $XY_2(s)$ with excess $ZW_3(g)$?

14. A 1.25 g sample of a mixture of NaCl and KCl gave a precipitate of AgCl(s) which weighed 2.50 g. Assuming that *all* the chloride in the mixture was precipitated as AgCl(s), calculate the percentage of sodium and potassium in the mixture. If a second 1.25 g sample of the NaCl + KCl mixture was heated with $H_2SO_4$ and converted to a $Na_2SO_4 + K_2SO_4$ mixture, what would be the weight of the sulfate mixture?

15. Distinguish between the *mass* and the *weight* of an object. How much does an exactly 1 g mass weigh: (a) on your campus; (b) on the surface of the moon?

16. Balance the following oxidation-reduction equations by the ion-electron method

*Acid solution:*

$$CuS(s) + NO_3^-(aq) = Cu^{2+}(aq) + S(s) + NO(g)$$

$$MnO_4^-(aq) + Cl^-(aq) = Mn^{2+}(aq) + Cl_2(g)$$

$$As_2S_5(s) + NO_3^-(aq) = H_3AsO_4(aq) + HSO_4^-(aq) + NO_2(g)$$

$$H_2C_2O_4(aq) + MnO_4^-(aq) = Mn^{2+}(aq) + CO_2(g)$$

$$MnO(s) + PbO_2(s) = MnO_4^-(aq) + Pb^{2+}(aq)$$

*Alkaline solution:*

$$Bi_2O_3(s) + OCl^-(aq) = BiO_3^-(aq) + Cl^-(aq)$$

$$Fe(CN)_6^{3-}(aq) + Cr_2O_3(s) = Fe(CN)_6^{4-}(aq) + CrO_4^{2-}(aq)$$

$$CoCl_2(s) + Na_2O_2(aq) = Co(OH)_3(s) + Cl^-(aq) + Na^+(aq)$$

$$CrI_3(s) + Cl_2(g) = CrO_4^{2-}(aq) + IO_4^-(aq) + Cl^-(aq)$$

$$C_2H_5OH(aq) + I_3^-(aq) = CO_2(g) + HCOO^-(aq) + HCI_3(aq) + I^-(aq)$$

17. A 0.255 g sample of Cu(s) was heated in air in the presence of excess sulfur to produce a copper sulfide phase. The excess sulfur was oxidized to $SO_2(g)$ by air. The final mass of the copper sulfide sample was 0.326 g. Determine the composition of the solid phase.

18. A mixture of AgCl(s) and AgBr(s) weighed 1.060 g. The mixture was heated in an atmosphere of $Cl_2(g)$ which converted it entirely to AgCl. The total weight of AgCl was found to be 0.998 g. If the mixture of AgCl and AgBr came from a 2.000 g sample, what were the percentages of Br and Cl in the original sample?

19. The oxidation states of elements in some compounds cannot be deduced by application of the rules given in the text. Working by analogy with elements in the same groups of the periodic table, assign oxidation states to the elements in the species:

| | | |
|---|---|---|
| (1) FeS | (5) NSF | (9) $CN^-$ |
| (2) $As_2S_3$ | (6) $CCl_4$ | (10) $CH_3NH_2$ |
| (3) $As_2S_5$ | (7) $COCl_2$ | (11) $SCN^-$ |
| (4) BN | (8) $CHCl_3$ | (12) $PF_3$ |

20. Find one or more exceptions to each of the rules given in Table 7.2.

21. The weight of an object immersed in a fluid is not equal to its weight in a vacuum, owing to the buoyancy effect of the fluid. Similarly, if the densities of the object and of the standard weights used to determine

its mass are not the same, then the true mass and the apparent mass are not equal because of the buoyant effect of the air.

(a) Derive an expression for the correction of the apparent mass of an object weighed in air to its true mass, in terms of the observed mass and the densities of the object and the standard masses (weights) used.

(b) Show how the density of an object can be determined by weighing it in air and in a liquid in which it is insoluble.

────────────────────────────────────────────── **Reference**

7.1.  Nash, L. K., *Stoichiometry* (Addison-Wesley Publishing Co., Reading, Mass., 1966).

# CHAPTER 8

# LIQUIDS AND SOLUTIONS

The most complex phase of matter is the liquid phase. Liquids are complex because they possess neither the great regularity of crystals nor the complete randomness of gases. Liquids are intermediate in behavior between solids and gases. X-ray studies show that liquids have short-range order reminiscent of solid lattices, but are devoid of long-range order. A liquid can be thought of as a solid with about 12% of the volume taken up by *mobile* molecular-size holes.†

Liquids are the media for the vast majority of chemical reactions. Most chemical reactions require the intimate mixing of the reactants, and for many materials intimate mixing can occur only when the solid is dissolved in a liquid. The properties of the liquid which is the major component of the solution (the solvent) often are of great importance in determining the course of a reaction. The chemist is concerned especially with the physical and chemical properties of liquids that are important in solutions.

## 8.1 STRUCTURE OF LIQUIDS.

The arrangement of atoms in any material can be deduced from the way in which it scatters x-rays. A crystal scatters x-rays with a regularity that produces a network of spots called the diffraction pattern. The diffraction pattern can be analyzed to give the crystal structure; it can be analyzed in a less complex fashion to give the probability of finding a particular number of atoms a given distance from any arbitrary central atom. A plot of this probability (radial distribution function) versus the distance from a central Na atom in crystalline Na metal is shown as a bar graph in Figure 8.1. Sodium does not happen to crystallize in a closest-packed array, and

---

°P. Debye, *Polar Molecules* (Dover Publications Reprint of the 1928 Edition) p. 7.

†The simple picture is not valid for liquid water, which is about 10% more dense than ice. Ice has a very unusual open structure because of hydrogen bonding.

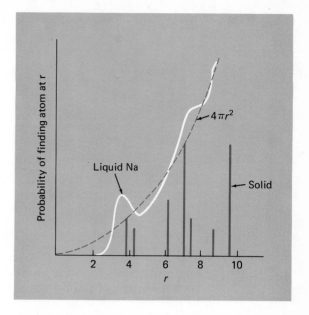

**FIGURE 8.1** Comparison of the radial distribution functions for solid, liquid, and (very dense) gaseous sodium.

each atom is surrounded by only eight nearest neighbors. A bar graph can be used because each neighboring atom is a member of a set of atoms, all of which are the same distance out from the central atom (Figure 8.2).

In a gas the atoms are randomly distributed throughout the volume available, so the probability of finding an atom at the distance $r$ is proportional to the rate at which the volume of a sphere about the central atom increases with increasing $r$. We already know that this is $4\pi r^2$ (see Section 2.6). The continuous curve in Figure 8.1 shows this probability for an extremely dense ideal gas.

A similar analysis of the x-ray scattering pattern from the liquid gives the curve labelled "liquid" in Figure 8.1. The distribution is like that in a crystal close to the central atom, but more like that in a very dense gas at

**FIGURE 8.2** Location of nearest-neighbor Na atoms around a central Na atom in metallic sodium.

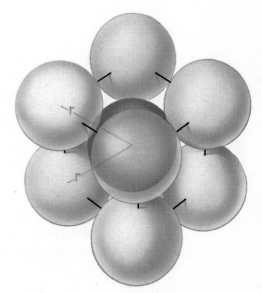

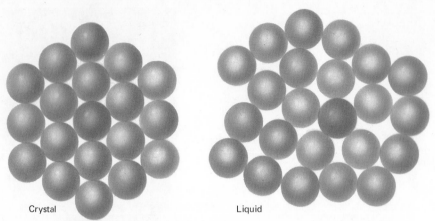

**FIGURE 8.3**   Comparison of order in the crystal and liquid phases.

larger distances from the central atom. For this reason we say that a liquid
is characterized by short-range order and long-range disorder (Figure 8.3).
Liquid sodium has an average of 10.6 atoms close to the central one; that
is closer to the closest-packed result of 12 than to the arrangement in the
Na crystal (8).

The densities of most liquids are within about 12% of those of the
corresponding crystals, and densities of this magnitude can be achieved
only by rather efficient packing of the molecules or atoms. Thus, the liquid
must be almost as regular as the crystal. Water is an interesting, but atypical,
example. The solid (ice) has a very open structure caused by the dominance
of hydrogen-bond formation over packing considerations. In the less
ordered liquid, the packing is of greater relative importance, and the
molecules in water actually are packed closer than in ice. Thus, ice has
lower density than water, and ice floats. The strengths of the attractions
between densely packed molecules combined with the lack of long-range
order in the liquid preclude the existence of a simple equation of state for
liquids. Attempts have been made to treat liquids as very imperfect gases
and as very imperfect crystals. One interesting approach has been the
attempt to derive the properties of liquids from an equation of state for
vacancies in the liquid treated as a gas.

## 8.2   MOLECULAR MOTION IN LIQUIDS.

Tiny particles suspended in a liquid seem to dance erratically when
viewed through a microscope. These particles are under constant bombardment by the molecules in the liquid surrounding them; the movements
observed are caused by momentary imbalances in the number of collisions
against opposing sides of the particle. Study of this motion, called *Brownian
motion*, provides information about collision frequencies and distribution
of molecular velocities in liquids. These studies indicate that the kinetic
energies in the liquid are about the same as those in the gas; for example,
the average kinetic energy of a molecule in the liquid is also $3kT/2$. The
potential energies of molecules in the liquid are lower than for gas-phase
molecules because of the strong intermolecular interactions.* However,
as the temperature is raised, a point is reached where the kinetic energy
of an especially rapidly moving molecule is sufficient for the molecule to

*The potential energy of the molecules in an ideal gas is taken to be zero.

escape the attractions holding it to neighboring molecules, and the energetic molecule escapes from the surface. This occurs quite frequently at the boiling point; thus, the boiling point is one indicator of the strength of the attractive forces between molecules in the liquid.

## 8.3  TYPES OF LIQUIDS.

The forces of adhesion which hold crystals and liquids together are of the same nature, so the classifications for crystals and liquids are the same. An ionic crystal usually melts to an ionic liquid. The ions in this liquid are reasonably mobile, and ionic liquids are fair conductors of electricity. The properties of liquid metals are, except for mechanical strength, very similar to those of solid metals. There is one metal which is a liquid at room temperature and atmospheric pressure (mercury), and the combination of no mechanical strength and good electrical conductivity is very useful in the construction of delicate electrical switching devices. Liquid sodium is used as a coolant in nuclear reactors because it has the flow characteristics of any liquid and the high heat conductivity of a metal, plus a low nuclear reactivity.

Most materials which are liquid at room temperature are held together by induced dipole attractions; these materials form molecular crystals. Like their crystals (molecular crystals), these liquids are electrical insulators and poor heat conductors. The introduction of disorder into chemically bonded crystals, like diamond, requires the rupture of true chemical bonds. These crystals decompose rather than melt.

Hydrogen-bonded crystals usually melt to form hydrogen-bonded liquids. Such liquids usually dissolve appreciable amounts of ionic compounds. In their pure state they are very poor conductors of electricity, but the presence of minute amounts of ionic compounds increases this conductivity greatly. Water is a familiar example of a hydrogen-bonded liquid.

## 8.4  PHYSICAL PROPERTIES OF LIQUIDS.

There are several properties of liquids that have practical importance to chemists. The *viscosity* is a measure of the resistance of a liquid (or, for that matter, a gas) to flow. The viscosity also indicates the maximum speed at which molecules can move through the liquid; it limits the rates of diffusion and hence the rate of some chemical reactions. A representative set of viscosities is given in Table 8.1.

The *dielectric constant* is defined experimentally as the ratio of the capacitance of the liquid to that of air; it is larger than unity for all materials. The dielectric constant ($\epsilon$) modifies Coulomb's law

$$f = \frac{q_1 q_2}{\epsilon r_{12}^2} \tag{8.1}$$

where $q_1$ and $q_2$ are charges, and $r_{12}$ is the distance between the charges. A dielectric constant greater than one reduces the forces between ions. This reduction can be very significant in hydrogen-bonded liquids; for example, water has $\epsilon = 78.5$ at 25°C. This reduction in force between ions makes it possible to separate oppositely charged ions without the investment of large amounts of energy. Therefore, liquids with large dielectric constants ($\epsilon > 40$) are good solvents for ionic compounds.

The dielectric constant is a complex function of the total dipole

**TABLE 8.1**

PHYSICAL PROPERTIES OF SELECTED LIQUIDS AT 25°C

| COMPOUND | FORMULA | VISCOSITY° | DIELECTRIC CONSTANT† |
|---|---|---|---|
| Acetone | $CH_3\overset{\overset{O}{\|\|}}{C}CH_3$ | 0.32 | 20.7 |
| Ammonia | $NH_3$ | | 16.9 |
| Benzene | $C_6H_6$ | 0.61 | 2.27 |
| Carbon tetrachloride | $CCl_4$ | 0.90 | 2.23 |
| Chloroform | $CH_3Cl$ | 0.54 | 4.79 |
| Dichloromethane | $CH_2Cl_2$ | | 9.0 |
| Diethyl ether | $CH_3CH_2-O-CH_2CH_3$ | 0.22 | 4.33 |
| Dimethyl sulfoxide | $CH_3\overset{\overset{O}{\|\|}}{S}CH_3$ | | 48.0 |
| Ethyl alcohol | $CH_3CH_2OH$ | 1.10 | 24.3 |
| Formamide | $H\overset{\overset{O}{\|\|}}{C}-NH_2$ | 3.30 | 106. |
| Hexane | $C_6H_{14}$ | 0.294 | 1.88 |
| Sulfuric acid | $H_2SO_4$ | 20. | |
| Water | $H_2O$ | 0.95 | 78.5 |
| Castor oil | $--$ | 750. | 10.3 |

°Although the numbers are in centipoises, which is the standard viscosity unit, we can best regard them as relative values.

†At 1000 Hz; value is frequency-dependent.

moment of the molecule. It is low ($\epsilon \approx 2$) when the liquid consists of non-polar molecules which have only induced dipole moments, but it is quite large when molecules have large permanent dipole moments; for example,

dimethyl sulfoxide $\left( CH_3-\overset{\overset{O}{\|\|}}{S}-CH_3 \right)$ has $\epsilon = 48$ at 25°C.

The combination of strong hydrogen bonds and the resultant high dielectric constant in a liquid makes possible an important phenomenon, self-ionization or, as it is more commonly called, self-dissociation. The hydrogen bonds in a group of water molecules are so strong that an occasional hydrogen atom can leave its "parent" molecule and join a neighbor (Figure 8.4). This produces an ion pair of a hydroxide ion ($OH^-$) and a hydrogen ion $H(OH_2)_n^+$. Successive proton transfers throughout the extended hydrogen-bonded structure serve to separate these ions; little energy is required for this separation because the dielectric constant of water is high. The ions arising from self-dissociation can never be removed by purification.

A molecule within a liquid is subject to attractive forces in all directions, but a molecule on the surface "feels" attractions only on one side (Figure 8.5). This produces a net inward force on surface molecules which tends to minimize the number of surface molecules, and in turn the area of the surface. This force is given the name *surface tension*. An isolated

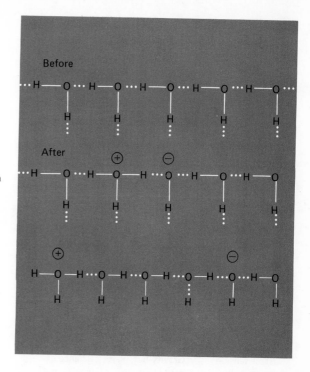

**FIGURE 8.4**   Self-ionization of water by proton transfer.

liquid sample tends to form a spherical drop, for a sphere is the shape which has the smallest surface for a given volume. Any liquid whose molecules attract each other strongly has high surface tension; water is a good example.

A liquid in contact with a solid which it wets does not form spherical drops. Wetting apparently occurs when the attractions between the molecules in the solid and those in the liquid are stronger than those between the liquid molecules themselves. Then each liquid molecule has lower potential energy close to a solid molecule, and a thin film of liquid is formed. Water wets very clean glass, but it forms droplets on dirty glass. The "dirt" usually contains greases and oils; these non-polar compounds do not attract the polar water molecules.

One consequence of surface tension is the formation of a meniscus when a tube is lowered into a liquid. If the liquid wets the tube, as for example a clean glass tube in water, surface tension actually pulls a column

**FIGURE 8.5**   Comparison of forces on surface molecules with the forces on a molecule buried in the liquid.

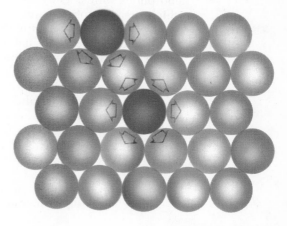

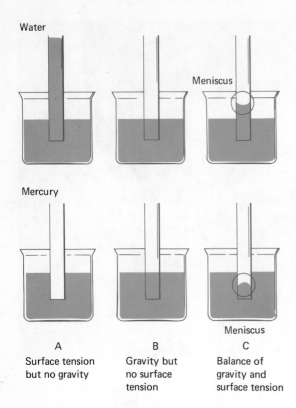

Water

Mercury

| A | B | C |
|---|---|---|
| Surface tension but no gravity | Gravity but no surface tension | Balance of gravity and surface tension |

**FIGURE 8.6** Capillary action. The height, $h$, to which a liquid rises in a capillary is given by $h = 2\sigma/rg\rho$, where $\sigma$ is the surface tension, $r$ is the radius of the capillary, $\rho$ is the density of the liquid, and $g$ is the gravitational acceleration.

of water up the tube until the tendency for smaller surface area is balanced by the weight of the column. The same glass tube, when lowered into mercury (which does not wet glass) depresses the liquid level inside the tube. Here surface tension holds the level down, and the weight of the liquid outside pushes it up (Figure 8.6). The meniscus distorts the readings of manometers and barometers, but this can be corrected for.

## 8.5   AMORPHOUS SOLIDS.

Not all solids are crystalline. When liquids of high viscosity are cooled rapidly, a rigid structure is formed before the molecules have time to orient properly to form a crystal. These materials are called glasses or amorphous solids. They are distinguished by lack of long-range regularity, which means isotropic physical properties, and lack of a sharp melting point.[*] Amorphous solids have liquid-like structures.

A wide variety of materials form non-crystalline solids under the proper conditions, but the most familiar examples are materials with large molecules (e.g., rubbers, plastics, and glasses). Silicon dioxide ($SiO_2$) forms the crystal quartz.[†] However, when liquid $SiO_2$ is cooled quickly, it forms the amorphous solid silica glass (called fused quartz). Fused quartz is very transparent over an unusually wide range of wavelengths; ultraviolet spectrometers always use fused quartz optics.

As the temperature of a glass is raised, the viscosity of the glass gradually decreases until, at high temperatures, the glass flows. This softening

---

[*]Lack of a sharp melting point can indicate that the sample is crystalline, but very impure.
[†]There are actually twenty-three different crystal forms for this substance.

range extends over several hundred degrees, although the visible effects may appear only at the highest temperatures in the range. Glass objects are shaped at high temperatures at which the glass flows; as the glass is cooled to room temperature, it contracts. This contraction (which is uneven) introduces severe strains into an already brittle material. In many ways pure silica glass is the ideal glass; the thermal contraction is very slight, so the resulting silica glass is quite strong. It is relatively immune to chemical attack as well. Unfortunately, silica glass is very difficult to work. The temperatures required are very high ($\sim 1600°C$), and the range of temperatures at which the material has the proper flow characteristics is narrow. The working temperature can be lowered and the working range can be extended by the addition of oxides ($B_2O_3$, $Na_2O$, and $CaO$) to the $SiO_2$.

## 8.6 SOLUTIONS AND CONCENTRATION SCALES.

The major constituent of a solution is called the *solvent*, and the minor constituents are called the solutes. The amount of solute dissolved in a given amount of the solution is referred to as the *concentration*.

There are several different types of concentration scales used to describe the composition of a solution. The most frequently employed scales

| SCALE | DEFINITION | SYMBOL |
|---|---|---|
| molarity | $\dfrac{\text{moles of solute}}{\text{liter of solution}}$ | $M$ |
| mole fraction (of $i$) | $\dfrac{\text{number of moles of the } i\text{th component}}{\text{total number of moles}}$ | $X_i$ |
| molality | $\dfrac{\text{moles of solute}}{\text{kilogram of solvent}}$ | $m$ |
| formality | $\dfrac{\text{moles of solute based on a given chemical formula}}{\text{liter of solution}}$ | $F$ |
| normality | $\dfrac{\text{equivalents of solute based on a given chemical equation}}{\text{liter of solution}}$ | $N$ |

The molality and mole fraction scales have the advantage of being temperature-independent scales, because they are defined in terms of a mass of solute per mass of solvent. The molarity, formality, and normality scales all involve the volume of the solution, and consequently, these scales are temperature dependent. A 1.00 m solution of KCl in water is 1.00 m at all temperatures, but a solution of KCl in water that is 1.00 M at 25°C is not exactly 1.00 M at $t \neq 25°C$.

The formality scale is used occasionally in equilibrium calculations when one wants to distinguish between the *total* concentration of a substance in solution and the *equilibrium* concentrations of the various species involving that substance. For example, an aqueous solution formed by dissolving 0.10 mole of $HgCl_2(s)$ in sufficient 1 M HCl(aq) to make a liter of solution contains the following Hg(II) species: $Hg^{2+}(aq)$, $HgCl^+(aq)$, $HgCl_2(aq)$, $HgCl_3^-(aq)$, and $HgCl_4^{2-}(aq)$, where (representing the concentration of a species by placing parentheses around it)

$$0.10 = (Hg^{2+}) + (HgCl^+) + (HgCl_2) + (HgCl_3^-) + (HgCl_4^{2-})$$

However, such a solution can be regarded formally as containing 0.10 mole of $HgCl_2$ per liter of solution, hence the designation 0.10 F. The actual equilibrium concentrations of the various Hg(II)-containing species can then be designated using the molarity scale. In practice, this convention is not rigorously adhered to, and many chemists use the molarity scale to designate both total and equilibrium concentrations.

The normality scale was designed to simplify certain types of stoichiometry calculations. This scale is of dubious value because of its potential ambiguity, and is in the process of being abandoned by chemists. However, we shall give a few problems involving the normality scale so that the reader will gain the familiarity necessary to deal with it if it is encountered elsewhere.

It is advantageous at this point to develop the interrelationships between the various concentration scales. We shall restrict our discussion to solutions involving only two components, but the extension to more complicated solutions is not difficult. Designating solute quantities by a subscript 2 and solvent quantities by a subscript 1, we have, by definition (where $n_i$ is the number of moles of $i$), the mole fractions

$$X_2 = \frac{n_2}{n_1 + n_2} \qquad X_1 = \frac{n_1}{n_1 + n_2} \tag{8.2}$$

and, by inspection,

$$X_1 + X_2 = 1$$

To get $X_2$ in terms of the molality, we note that if we have a solution containing $n_2$ moles of solute per kilogram of solvent, then (from the definition of molality) $n_2 = m$ and $n_1 = 1000/W_1$, where $W_1$ is the molecular weight of the solvent; therefore,

$$X_2 = \frac{m}{m + \dfrac{1000}{W_1}} \tag{8.3}$$

In the special case when water is the solvent, $W_1 = 18.02$, and

$$X_2 = \frac{m}{m + 55.49}$$

The conversion of molality to molarity (or vice-versa) requires a knowledge of the density of the solution. The number of grams of solvent in a liter of a solution containing a solute with a molality $M$ and a molecular weight $W_2$ is

$$(1000 \text{ ml})\rho - MW_2$$

where $\rho$ is the density of the solution in $g \cdot ml^{-1}$. With the use of this result, we can write

$$\left(\frac{\text{moles of solute}}{\text{kg of solvent}}\right) \times \left(\frac{\text{g solvent}}{\text{liter solution}}\right) \times \left(\frac{1 \text{ kg}}{1000 \text{ g}}\right) = m \, (1000 \, \rho - MW_2)/1000$$

but the left hand side of this expression is also equal to the number of moles of solute per liter of solution, and therefore

$$m = \frac{M}{\rho - \left(\dfrac{MW_2}{1000}\right)} \tag{8.4}$$

As examples of the use of the $m$, $M$, and $X$ scales, and also Equations (8.2), (8.3), and (8.4), consider the following problems:

(a) How many grams of $K_4Fe(CN)_6 \cdot 3H_2O$ are required to prepare 550 ml of a 0.125 M solution at 25°C? From a table of atomic weights we compute the formula weight of potassium hexacyanoferrate (II) trihydrate as 422 g $\cdot$ mole$^{-1}$. Letting the unknown weight be $y$, we have, from the definition of molarity,

$$0.125 \frac{\text{mole}}{\text{liter}} = \frac{(y/422 \text{ g} \cdot \text{mole}^{-1})}{0.550 \text{ liter}}$$

$$y = (0.125 \text{ mole} \cdot \text{liter}^{-1})(0.550 \text{ liter})(422 \text{ g} \cdot \text{mole}^{-1}) = 29.0 \text{ g}$$

To prepare such a solution, we dissolve 29.0 g of $K_4Fe(CN)_6 \cdot 3H_2O$ in water and dilute to 550 ml total volume at 25°C.

(b) In how many grams of water should 29.0 g of $K_4Fe(CN)_6 \cdot 3H_2O$ be dissolved to prepare a 0.125 m solution? Letting the number of kilograms of water required by $y$, we have, from the definition of molality,

$$0.125 \frac{\text{moles}}{\text{kg}} = \frac{(29.0 \text{ g}/422 \text{ g} \cdot \text{mole}^{-1})}{y}$$

$$y = \left(\frac{29.0 \text{ g}}{422 \text{ g} \cdot \text{mole}^{-1}}\right) \left(\frac{1}{0.125 \text{ mole} \cdot \text{kg}^{-1}}\right) = 0.550 \text{ kg}$$

To prepare this solution, we dissolve 29.0 g of $K_4Fe(CN)_6 \cdot 3H_2O$ in 550 g of water. Note that the amounts of water used in cases (a) and (b) are not the same. The density of the 0.125 M solution is 0.984 g $\cdot$ ml$^{-1}$, and the number of grams of water required in case (a) is approximately (550 ml)(0.984 g $\cdot$ ml$^{-1}$) $-$ (0.125 mole $\cdot$ liter$^{-1}$)(422 g $\cdot$ mole$^{-1}$) $=$ 488 g of $H_2O$.

(c) A 100 g sample of $K_2SO_4$ was dissolved in 850 g of water. The density of the solution was found to be 1.09 g $\cdot$ ml$^{-1}$ at 25°C. Calculate the molality, the mole fraction of $K_2SO_4$, and the molarity (at 25°C) of this solution. The formula weight of $K_2SO_4$ is 174.3 g $\cdot$ mole$^{-1}$, and, therefore, the molality of the resulting solution is

$$m = \frac{(100 \text{ g}/174.3 \text{ g} \cdot \text{mole}^{-1})}{(0.850 \text{ kg})} = 0.675 \text{ mole} \cdot \text{kg}^{-1}$$

the mole fraction of $K_2SO_4$ is

$$X_2 = \frac{(100/174.3)}{(850/18.02) + (100/174.3)} = 0.0120$$

and the molarity of the solution is

$$0.675 = \frac{M}{1.09 - \dfrac{M(174.3)}{1000}}$$

$$M = \frac{(0.675)(1.09)}{1 + (0.675)(0.1743)} = 0.658 \text{ mole} \cdot \text{liter}^{-1}$$

(d) An aqueous solution of phosphoric acid is 85 weight per cent $H_3PO_4$ and has a density of 1.70 g·ml$^{-1}$ at 20°C. Calculate the mole fraction $H_3PO_4$, the molality, and the molarity (at 20°C) of this solution. The molecular weight of $H_3PO_4$ is 98 g·mole$^{-1}$, and (arbitrarily) taking 100 g of 85% $H_3PO_4(aq)$, we have for the mole fraction of $H_3PO_4$

$$X_2 = \frac{\text{(moles of } H_3PO_4)}{\text{(moles of } H_3PO_4) + \text{(moles of water)}} = \frac{(85/98)}{(85/98) + (15/18)} = 0.510$$

whereas for the molality we compute

$$X_2 = 0.510 = \frac{m}{55.5 + m}$$

$$m = \frac{(55.5)(0.510)}{(0.495)} = 57.2 \text{ mole/kg}$$

and for the molarity

$$57.2 = \frac{M}{1.70 - \frac{M(98)}{1000}}$$

$$M = \frac{(57.2)(1.70)}{1 + (57.2)(0.098)} = 14.7 \text{ moles·liter}^{-1}$$

(e) How many milliliters of 85% $H_3PO_4(aq)$ are required to prepare 300 ml of 2.0 M $H_3PO_4(aq)$? The *number of moles* of a solute is not changed by addition of solvent, and therefore the product

$$MV = (\text{mole·liter}^{-1})(\text{liter}) = \text{mole}$$

remains unchanged in a dilution. Consequently, we can write for a dilution (where the subscripts $b$ and $a$ refer to the solution *before* and *after* dilution, respectively)

$$M_b V_b = M_a V_a \tag{8.5}$$

In example (d) we found that the molarity of the 85% $H_3PO_4(aq)$ is 14.7 mole·liter$^{-1}$, and thus

$$(14.7 \text{ mole·liter}^{-1})V_b = (2.0 \text{ mole·liter}^{-1})(0.300 \text{ liter})$$

$$V_b = \frac{(2.0)(0.300)}{14.7} = 0.041 \text{ liter}$$

To prepare 300 ml of 2.0 M $H_3PO_4(aq)$ at 20°C from 85% $H_3PO_4$ at 20°C, we dilute 41 ml of the concentrated solution to a final volume of 300 ml using water.[*]

(f) Given that the reaction that occurs when copper metal is dissolved in moderately concentrated aqueous nitric acid is

---

[*]This is not the same as adding 41 ml of the concentrated acid to 259 ml of water because, in general, $V_1 + V_2 \neq V_{total}$.

$$3 \, Cu(s) + 8HNO_3(aq) = 3Cu(NO_3)_2(aq) + 2NO(g) + 4H_2O(\ell)$$

how many milliliters of 6.0 M $HNO_3(aq)$ are required to dissolve 1.50 g of copper? The number of moles of $Cu(s)$ to be dissolved is

$$\frac{1.50 \text{ g}}{63.5 \text{ g} \cdot \text{mole}^{-1}} = 0.0236 \text{ mole}$$

We now compute from the reaction stoichiometry that the number of moles of $HNO_3$ required to react with 0.0236 mole of $Cu(s)$ is

$$(0.0236 \text{ mole Cu}) \left( \frac{8 \text{ mole } HNO_3}{3 \text{ mole Cu}} \right) = 0.0630 \text{ mole } HNO_3$$

The volume of 6.0 M $HNO_3(aq)$ that contains 0.0630 mole of $HNO_3$ is

$$0.0630 \text{ mole} = (6.0 \text{ mole} \cdot \text{liter}^{-1})V$$

$$V = \frac{0.0630}{6.0} = 0.0105 \text{ l} = 10.5 \text{ ml}$$

Hence, 10.5 ml of 6.0 M $HNO_3(aq)$ are required to dissolve 1.50 g of $Cu(s)$.

(g) Given the following reaction:

$$H_3PO_4(aq) + 2KOH(aq) = K_2HPO_4(aq) + 2H_2O(\ell)$$

how many milliliters of 1.35 M $KOH(aq)$ are required to convert 40.0 ml of 2.00 M $H_3PO_4(aq)$ to $K_2HPO_4(aq)$? The number of moles of $H_3PO_4$ reacted is

$$(0.0400 \text{ liter})(2.00 \text{ mole} \cdot \text{liter}^{-1}) = 0.0800 \text{ mole}$$

The number of moles of KOH required to convert 0.0800 mole of $H_3PO_4$ to $K_2HPO_4$ is

$$(0.0800 \text{ mole } H_3PO_4) \left\{ \frac{2 \text{ mole KOH}}{1 \text{ mole } H_3PO_4} \right\} = 0.160 \text{ mole KOH}$$

The volume of 1.35 M KOH that contains 0.160 mole of KOH is

$$0.160 \text{ mole} = (1.35 \text{ mole} \cdot \text{liter}^{-1})V$$

$$V = \frac{0.160}{1.35} = 0.119 \text{ liter} = 119 \text{ ml}$$

## 8.7 NORMALITY AND EQUIVALENT WEIGHT.

The normality of an oxidizing or reducing agent is defined as the number of equivalent weights per liter. *Equivalent weight* is defined (for redox reactions) as that weight in grams of a substance that yields or accepts Avogadro's number of electrons. For example, if the balanced half-reaction for the oxidizing agent is

$$5e^- + MnO_4^- + 8H^+ \rightarrow Mn^{2+} + 4H_2O$$

then each mole of $MnO_4^-$ consumes 5 moles of electrons, and the equivalent weight of $KMnO_4$ is

$$\frac{158.04 \text{ g} \cdot \text{mole}^{-1}}{5 \text{ equiv} \cdot \text{mole}^{-1}} = 31.61 \text{ g} \cdot \text{equiv}^{-1}$$

On the other hand, for the oxidation of $HNO_2$ to $NO_3^-$ we have

$$HNO_2 + H_2O = NO_3^- + 3H^+ + 2e^-$$

The $HNO_2(aq)$ is prepared by acidifying $KNO_2(aq)$. Upon acidification, each mole of $KNO_2$ gives a mole of $HNO_2$, and each mole of $HNO_2$ yields 2 moles of electrons (when oxidized to nitrate, $NO_3^-$). Therefore, the equivalent weight of $KNO_2$ is

$$\frac{85.11 \text{ g} \cdot \text{mole}^{-1}}{2 \text{ equiv} \cdot \text{mole}^{-1}} = 42.56 \text{ g} \cdot \text{equiv}^{-1}$$

It is important to recognize that normality can be ambiguous; for example, with some reducing agents $MnO_4^-(aq)$ is reduced to $MnO_2(s)$ rather than $Mn^{2+}(aq)$:

$$3e^- + MnO_4^-(aq) + 4H^+(aq) \rightarrow MnO_2(s) + 2H_2O$$

$$\text{equivalent weight of } KMnO_4 = \frac{158.04}{3} = 52.68 \text{ g}$$

Therefore, one must specify both the reduced and oxidized forms when giving the normality of a solution; better yet, the ambiguity can be avoided by the use of the molarity scale.

## SI Units for Concentrations.*

The SI units for the various concentration scales are as follows:

(a) *molality* $= \dfrac{\text{moles of solute}}{\text{kilogram of solvent}} = \text{mol} \cdot \text{kg}^{-1}$

(b) *molarity* $= \dfrac{\text{moles of solute}}{\text{liter of solution}} = \text{mol} \cdot \text{dm}^{-3}$

(note that $1l = 10^{-3} \text{ m}^3 = 1 \text{ (decimeter)}^3 = 1 \text{ dm}^3$)

(c) *mole fraction;* mole fraction is a dimensionless quantity. The density of a solution can be expressed in SI units as $\text{g} \cdot \text{cm}^{-3}$ or $\text{kg} \cdot \text{dm}^{-3}$.

---

*See Appendix 4 for a discussion of SI units.

## PROBLEMS

1. Select from Table 8.1 those substances that are likely to contain hydrogen bonds. Do the values of the viscosities or the dielectric constants show any relation to the extent of such hydrogen bonding?

2. Why are the dielectric constants of the liquids $CCl_4$, $CHCl_3$, and $CH_2Cl_2$ in the observed order?

3. Calculate the force between a unipositive and a uninegative charge 5 Å apart in water and in acetone.

4. Would the meniscus correction in a manometer be larger in a small diameter tube or in a large diameter tube?

5. At what pressure is the average separation of atoms in Na gas equal to the nearest neighbor distance in solid Na?

6. Lithium reacts with water to form lithium hydroxide and hydrogen gas. Compute the weight of lithium that is required to prepare 200 ml of 0.030 M LiOH(aq).

7. Calculate the mole fractions of ethanol ($C_2H_5OH$) and water in a solution prepared by mixing 100 g of each substance.

8. How many grams of $NH_4Cl$ are required to prepare (a) 250 ml of a 0.052 M aqueous solution; (b) 250 ml of a 0.052 M solution of $NH_4Cl$ in $NH_3(\ell)$; and (c) approximately 250 ml of a 0.052 m $NH_4Cl$(aq) solution?

9. How would you prepare approximately 1 liter of a 0.275 molal aqueous solution of $KMnO_4$?

10. An aqueous sulfuric acid solution is 48.0 wt % $H_2SO_4$ and has a density of 2.18 g $\cdot$ ml$^{-1}$(20°C). Calculate $X_2$, $m$, and $M$ for this solution. How would you prepare 2.50 liters of 0.500 M $H_2SO_4$(aq) from 48% solution? How many ml of 0.0500 M solution are required to neutralize (a) 50 ml of 0.325 M NaOH, and (b) 50 ml of 0.0125 M $Ca(OH)_2$(aq)?

11. How many milliliters of 2.00 M NaOH(aq) are required to convert 75.0 ml of 1.25 M $H_3PO_4$(aq) to $Na_3PO_4$(aq)?

12. Water reacts with excess phosphorous pentachloride to form hydrogen chloride gas and phosphoric acid. Suppose the evolved hydrogen chloride is to be dissolved in dimethylformamide $\left[ (CH_3)_2 \overset{\displaystyle O \atop \displaystyle \|}{N}CH \right]$. Compute the number of milliliters of water that must be added to excess $PCl_5$ in order to prepare 400 ml of 0.050 M HCl in dimethylformamide.

13. A 26.802 g sample of $Na_2C_2O_4$ was dissolved in water and diluted to exactly 1 liter. A 25.00 ml aliquot of acidified $KMnO_4$(aq) was then titrated with the sodium oxalate solution to give $Mn^{2+}$(aq) and $CO_2$(g). This titration required 35.50 ml of the oxalate solution. Calculate the molarity and normality of the permanganate solution.

14. Compute the molarity and molality range for each of the following commercially available solutions:

| Solution | Weight Per Cent Solute | Specific Gravity |
|---|---|---|
| $NH_3$(aq) | 28–30 | 0.90 |
| HCl(aq) | 37–38 | 1.19 |
| $H_2SO_4$(aq) | 95.5–96.5 | 1.84 |
| $HNO_3$(aq) | 70.0–71.0 | 1.42 |
| $CH_3COOH$(aq) | 99.5–100.0 | 1.05 |

# CHAPTER 9

# PRINCIPLES OF CLASSICAL THERMODYNAMICS

## 9.1  INTRODUCTION.

Modern thermodynamics is comprised of several subdisciplines, the oldest of which is classical thermodynamics. Classical thermodynamics deals solely with the macroscopic (bulk) properties of matter, and the classical theory is formulated without reference to any microscopic (atomic or molecular) theories of matter. The macroscopic approach of classical thermodynamics stands in stark contrast to the microscopic approach to the study of matter employed in the preceding chapters. The broad applicability of the principles of classical thermodynamics is in large measure a consequence of the model-independent nature of its construction. However, a very high price was paid for this independence—namely, classical thermodynamics is incapable of explaining in microscopic terms why a particular system behaves in the observed way. Deeper insights into the behavior of matter are achieved through the unification of the macroscopic and microscopic approaches. The unified approach is employed in statistical thermodynamics. However, the study of statistical thermodynamics requires a familiarity with the principles of classical thermodynamics. Furthermore, it is of great importance to understand clearly the limitations on behavior to which all macroscopic systems, whatever their microscopic nature, are subject.

Classical thermodynamics is an indispensable tool in chemistry. In particular, it enables us to characterize the equilibrium states of systems, to analyze the energetics of chemical reactions, to rule out impossible

---

°G. N. Lewis and M. Randall, *Thermodynamics*, 2nd ed. revised by K. S. Pitzer and L. Brewer (McGraw-Hill Book Co., Inc., New York, 1961).
†Professor C. Y. Nical.

**FIGURE 9.1** *Waterfall*, by Maurits C. Escher. (Reproduced by permission from the collection of C. V. S. Roosevelt, Washington, D.C.).

processes, to interrelate the observable properties of matter, and to test proposed models for actual systems. For these reasons we shall attempt to develop a more than casual acquaintance with some of the basic concepts and principles of thermodynamics.

## 9.2  THERMAL EQUILIBRIUM.  THE TEMPERATURE CONCEPT AND THERMOMETERS.

Although temperature is a familiar concept, the development of the temperature concept is not a simple task. To measure temperature, we measure other quantities that change when the temperature changes, such as *length* (of a column of mercury or alcohol), *volume* or *pressure* (of a gas), *resistance* (of a platinum wire or a thermistor), or *voltage* (of a thermocouple).* We cannot agree on a standard temperature scale without first agreeing on a device that will be used to measure temperature. A temperature-measuring device is called a *thermometer.*

We measure the temperature of a system by placing a thermometer in thermal contact with the system. When the physical properties of the thermometer no longer change with time, the system and the thermometer have reached *thermal equilibrium.* The walls separating the system and the thermometer must conduct heat; walls that conduct heat are called *conducting* or *diathermic* walls. Walls that do not conduct heat are called *insulating* or *adiabatic* walls. Systems separated by adiabatic walls are not (in general) in thermal equilibrium. Thermal equilibrium is the time-independent state achieved by two systems that have been allowed to interact through diathermic walls.

Systems in thermal equilibrium with one another can be said to possess a property that insures their being in thermal equilibrium with one another; this property is called the *temperature. Temperature is the property of a system that determines whether or not the system is in thermal equilibrium with other systems.* A necessary and sufficient condition for two or more systems to be in thermal equilibrium with one another is that they have the same temperature.

All that is necessary to establish a temperature scale is to select a thermometer and to assume a correspondence between the temperature and a physical property of the thermometer. To this end, we pick a gas held at a constant volume as our thermometer, and we take the temperature, $t$, of this system to be a linear function of the pressure of this system:

$$t = kP$$

Two temperatures, $t_1$ and $t_2$, on this linear-$P$ scale are related by an expression of the type

$$\frac{t_1}{t_2} = \frac{P_1}{P_2}$$

Numerical values for temperatures can be obtained from this equation if we choose some easily reproducible fixed point of some convenient system

---

*The frequency of a cricket's chirp can be used to measure temperature; to the number of chirps in 15 seconds, add 37 to obtain the temperature in degrees Fahrenheit. The result is accurate to within ±1°F in the range 50–80°F.

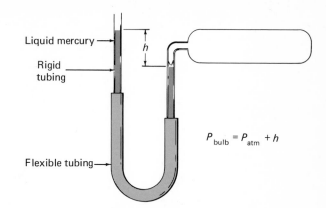

**FIGURE 9.2** Constant-volume gas thermometer.

and arbitrarily assign a value to $t$ for this system. The standard fixed point that has been chosen is the *triple point of water*, for which we take $t_{tpt} = 273.16$ degrees (exactly).* Thus, we have for the temperature, $t$, of some system of interest,

$$t = t_{tpt} \left(\frac{P}{P_{tpt}}\right) = 273.16 \left(\frac{P}{P_{tpt}}\right)$$

where $P_{tpt}$ and $P$ refer to the measured pressures of the gas *inside the thermometer* when the thermometer is thermally equilibrated with the triple-point cell and the system whose temperature is desired, respectively. Diagrams of a constant-volume gas thermometer and a triple-point cell are presented in Figures 9.2 and 9.3, respectively.

In addition to constant-volume gas thermometers, there are many other types of systems that can serve as thermometers, such as a column of liquid of length $L$ at fixed pressure in a uniform capillary:

$$t = 273.16 \left(\frac{L}{L_{tpt}}\right)$$

This is the familiar liquid-in-glass type of thermometer. In principle, any macroscopic system possessing a property that varies monotonically with temperature can be used as a thermometer.

Let us suppose that we have available to us a variety of thermometers, and that we determine, as accurately as possible, the temperature of some system of interest. The results of such measurements are rather disconcerting; a different value of the temperature is found for each different type of thermometer that we use. The best agreement, however, is obtained for the various gas thermometers, and the agreement is better the lower the value of $P_{tpt}$. The difficulty can be overcome by means of the *ideal-gas temperature scale*. We define the ideal-gas temperature, $T$, as follows:

$$T = \lim_{P_{tpt} \to 0} t = 273.16 \left\{ \lim_{P_{tpt} \to 0} \left(\frac{P}{P_{tpt}}\right) \right\} \tag{9.1}$$

---

*This particular value was chosen (1954) to preserve the closest possible numerical agreement with older temperature scales.

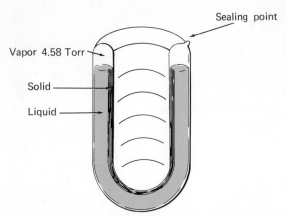

Sealing point

Vapor 4.58 Torr

Solid

Liquid

**FIGURE 9.3**   Cross-sectional view of a triple-point cell. The cell is a double-walled, Dewar-type flask. Water of the highest obtainable purity is placed between the cell walls, and repeatedly degassed (freeze-pump-thaw) to eliminate any dissolved gases. The cell is then carefully sealed. A coolant liquid at a temperature below the triple-point temperature is then introduced into the center of the flask to freeze a portion of the water. The coolant is then removed. When the three phases, ice, liquid water, and water vapor, attain thermal equilibrium, the temperature of the cell is exactly 273.16 degrees, by definition. The value of $P_{tpt}$ (i.e., the pressure of the gas *inside a thermometer bulb* when the thermometer is thermally equilibrated with the triple-point cell) is then obtained by inserting the bulb of a constant-volume gas thermometer (Figure 9.2) into the triple-point cell, allowing the combined system to attain thermal equilibrium, and measuring the pressure of the gas in the constant-volume gas thermometer. This yields $P_{tpt}$ for the particular gas thermometer.

The determination of the ideal-gas temperature of, let us say, the melting point of tin requires an extended series of measurements using progressively smaller amounts of gas (resulting in progressively smaller measured values of $P$ and $P_{tpt}$) in the thermometer. The results obtained are extrapolated to $P_{tpt} = 0$, as depicted in Figure 9.4. This procedure is effective because as the gas pressure becomes progressively lower, the gas temperature and gas pressure becomes more nearly proportional. The proportionality becomes exact in the limit of zero pressure, where the various gases behave ideally.

The temperatures of several points on the ideal-gas temperature scale

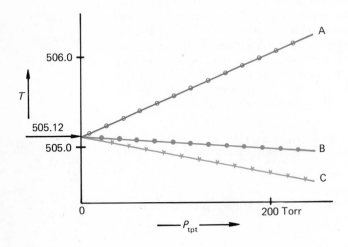

**FIGURE 9.4**   Determination of the 1-atm melting point of tin on the ideal-gas temperature scale. A, B, and C are different gases. As the amount of gas in the constant-volume gas thermometer is reduced, both $P_{tpt}$ and $P$ (at m. pt. of Sn) for the gas in the thermometer bulb decrease. It is the ratio $(P_{m.pt.Sn}/P_{tpt})$ times 273.16 that attains the value of 505.12 K in the limit of zero pressure.

TABLE 9.1

TEMPERATURES OF SOME REFERENCE POINTS ON THE IDEAL-GAS TEMPERATURE SCALE*
(1 atm total pressure)

| BOILING POINTS (K) | | MELTING POINTS (K) | |
|---|---|---|---|
| helium | 4.22 | mercury | 234.29 |
| hydrogen | 20.40 | water | 273.15 |
| nitrogen | 77.35 | tin | 505.12 |
| oxygen | 90.19 | cadmium | 594.26 |
| carbon dioxide | 194.67 | zinc | 692.73 |
| water | 373.15 | antimony | 903.89 |
| mercury | 629.81 | silver | 1235.08 |
| sulfur | 717.82 | gold | 1337.58 |
| | | platinum | 2045.00 |
| | | tungsten | 3660.00 |

*See Metrologia 5, No. 2, April 1969. The effective operating range of the ideal-gas thermometer is 1 to 600 K. Outside this range other thermometers must be used to establish the temperatures of reference points. (See reference cited above and references 9.1 and 9.2 for further details.)

are presented in Table 9.1. Temperatures on the single-fixed-point (Kelvin-Giauque) ideal-gas temperature scale are designated by the unit K (Kelvin).

The systems whose temperatures are given in Table 9.1 can be used to calibrate (at several fixed points) more convenient thermometers, such as resistance thermometers and thermocouple thermometers. These calibrated thermometers can then in turn be used, in conjunction with an empirical equation that accurately describes their non-linear behavior, to make routine temperature measurements rapidly and reliably.*

In addition to the Kelvin-Giauque temperature scale, both the Celsius and Fahrenheit temperature scales are widely used. The magnitude of a degree on the Celsius and Kelvin-Giauque scales is the same. The triple point of water on the Celsius scale is defined as exactly $+0.01°C$, whereas the normal (1 atm) boiling point of water is defined as exactly $100°C$. The Celsius scale has *two fixed points*. The algebraic relationship between the Kelvin-Giauque and Celsius scales is

$$\boxed{t(°C) = T(K) - 273.15}$$

which follows from the fact that when $t = +0.01°C$, $T = 273.16$ K. The Fahrenheit scale also has two fixed points. On this scale the normal freezing and boiling points of water are arbitrarily assigned the values $32°F$ (exactly) and $212°F$ (exactly), respectively. The ratio of the magnitudes of a degree Celsius and a degree Fahrenheit is 5/9, and the relationship between the Celsius and Fahrenheit scales is

$$t(°C) = (5/9) \{t(°F) - 32\}$$

---

*A mercury thermometer calibrated at 0°C and 100°C is about +0.1°C in error at 50.0°C if *linear* behavior is assumed between 0°C and 100°C.

As an example of the application of these equations, we note that the body temperature of a relaxed, healthy human is usually within a degree of 98.6°F, which corresponds to a Celsius temperature of 37.0°C, and a Kelvin temperature of 310.2 K.

## 9.3  THE ENERGY CONCEPT.

Energy, like temperature, is a fundamental concept of an essentially abstract nature. We distinguish various types of energy by the various formulae that we use to compute them. Thus, the potential energy of a mass point in a uniform and constant gravitational field is given by $mgh$, where $m$ is the mass, $g$ is the gravitational acceleration, and $h$ is the distance to a convenient reference point. The translational kinetic energy of a mass is given by $mv^2/2$, where $v$ is the velocity of the mass. One of the earliest formulations of energy conservation was the postulate that, for a mass point subject only to a gravitational force, the sum of kinetic and potential energies is a constant, that is,

$$mgh + mv^2/2 = \text{constant}$$

This equation checks with experiment, provided that the mass is non-magnetic and uncharged. Suppose the mass has a charge $q_1$ and interacts with another mass carrying a charge $q_2$. In this case it is necessary to include the Coulombic energy, $kq_1q_2/r_{12}$ (where $r_{12}$ is the distance between the charge centers and $k$ is a constant) in order to obtain agreement with experimental observations. Thus, for a charged body moving in a constant gravitational field containing another charged body, we have

$$mgh + mv^2/2 + kq_1q_2/r_{12} = \text{constant}$$

If the body emits or absorbs photons of frequency $\nu$, then the above equation must be further modified to

$$mgh + mv^2/2 + kq_1q_2/r_{12} + nh_p\nu = \text{constant}$$

where $h_p$ is Planck's constant and $n$ is the number of photons ($n > 0$ if the photons are absorbed). All available experimental evidence leads to the conclusion that, whatever energy is in fundamental terms, *energy is conserved*, provided only that all the relevant energy terms are included in the sum. It is also clear from experiment that *only changes in energy are measurable*. We are unable to determine the absolute energy of a system. All we can say is that the system now has, for example, more potential and less kinetic energy than it once did. Or, if a system interacts with its surroundings,* then the amount of energy gained (or lost) by the system is exactly equal to the amount of energy lost (or gained) by the surroundings. Because only energy changes are measurable, it is unnecessary to attempt to separate the energy concept from the conservation principle.

In summary, we must require that the total energy (system plus surroundings) always be conserved; we can write down the formulae for calculating the various types of energy given up or taken in by a system,

---

*The "surroundings" are defined as all parts of the universe that can exchange either energy or matter with the system.

and we can state the necessary mathematical requirements of the (total) energy function. This is the best that we can do in providing an answer to the question—what is energy?

Application of the principle of energy conservation to an *isolated system* (i.e., a system that cannot exchange either energy or matter with its surroundings) yields the result that, *for any process whatsoever occurring within the system*,

$$\Delta E_{sys} = E_{final} - E_{initial} = 0 \qquad \text{(isolated system)}$$

where $E_{sys}$ is the total energy of the system. If the system can exchange energy with its surroundings, then we can write, *for any process whatsoever* that the system undergoes,

$$\Delta E_{sys} + \Delta E_{surr} = 0$$

In other words, if a system undergoes some process in which a quantity of energy is transferred to the surroundings, then the energy of the system must decrease by an *exactly* equal amount.

## 9.4 EQUATIONS OF STATE AND REVERSIBLE PROCESSES.

We know from experiment that the state of a definite amount of a gas of fixed composition is completely determined, in the absence of external force fields, by the specification of its pressure and volume. In particular, the temperature, energy, and all of the other thermodynamic properties have definite values. Because a definite mass of gas of fixed composition has only two independent thermodynamic coordinates, there must exist equations of the type $E = E(P,V)$, $P = P(V,T)$, and so forth. Such equations are called *equations of state*.

We take as a postulate based on experimental observation that *an equation of state always exists for defined states of a system*. We do not, however, require that we know the *particular* form of the equation of state for a system, but only that such an equation exists for defined states. (In fact, for most systems of interest we cannot write down such an equation.)

Some simple examples of equations of state follow:

(a) *an ideal gas*

$$PV = nRT$$

where $n$ is equal to the number of moles of gas, and $R$ is the gas constant;

(b) *a van der Waals gas*

$$\left(P + \frac{n^2a}{V^2}\right)(V - nb) = nRT$$

wherein $a$ and $b$ are constants characteristic of the particular gas (see Table 5.5);

(c) *a simple solid*

$$V = V_0 \{1 + \alpha T - \beta(P - 1)\}$$

where $V_0$ is the volume at $T = 0$ and $P = 1$ atm, and $\alpha$ and $\beta$ are constants characteristic of the particular solid.

Equations of state must be obtained from experimental data or from molecular theory. Thermodynamics is completely powerless to generate an equation of state, owing to the model-independent nature of thermodynamic theory. Once the equation of state is available, however, thermodynamic theory can be used to deduce many of the thermodynamic properties of the system.

Equations of state exist only for equilibrium states of a system. An equilibrium state is a state for which there exist no *finite* unbalanced forces either within the system or between the system and its surroundings. A process (change in state) is said to be *reversible* when the states through which the system passes during the process are all equilibrium states. In the analysis of reversible processes we can always employ the equation of state for the substance, whereas for *irreversible* processes we cannot, in general, use the equation of state to analyze the *process*, because the thermodynamic coordinates (e.g., $P$, $T$) do not have defined values during an irreversible process. From an experimental point of view, if we wish some process to take place reversibly, we subject the system of interest to a sequence of small changes in some variable (e.g., pressure), until the desired change is accomplished. The smaller the increment of change, the more nearly reversible the process.

It is important to realize that *the values of changes in the thermodynamic coordinates* ($\Delta P$, $\Delta V$, $\Delta T$, $\Delta E$, . . .) *depend only on the initial and final states of the system*, and not at all on how the given change in state was accomplished. The foregoing is true for all thermodynamic coordinates; that is,

$$\Delta X = X_{\text{final}} - X_{\text{initial}}$$

where $X$ is any thermodynamic coordinate. This means that, if a system undergoes an irreversible process between thermodynamically defined initial and final states, we can use the equation of state in the calculation of the change in any thermodynamic coordinate of the system, because the change is independent of the path. However, in order to use the equation of state, we must devise a reversible process with the same initial and final states as the irreversible process.

## 9.5  WORK AS A MODE OF ENERGY TRANSFER.

When a force, $F$, causes a displacement, $\Delta x = x_2 - x_1$, of a system parallel to the direction of the force, the work done on the system is equal to the area under the curve ($F$ plotted versus $x$) between the initial ($F_1$, $x_1$) and final ($F_2$, $x_2$) points.[*] In the simplest case the force is constant throughout the process. This situation is depicted in Figure 9.5. The area under the curve between the points $x_1$ and $x_2$ is simply $F_1(x_2 - x_1)$, and, therefore, the work done on the system is

$$w = F_1(x_2 - x_1) = F_1 \Delta x$$

---

[*]That is to say,

$$\text{work} = \int_{x_1}^{x_2} F \, dx$$

if the line of action of the force is parallel to the displacement.

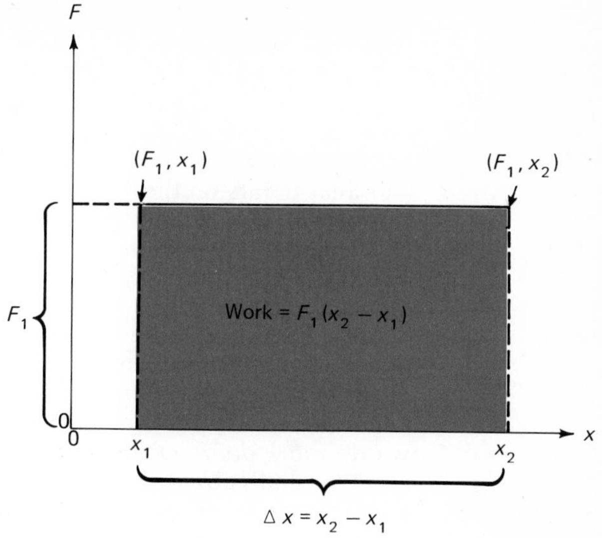

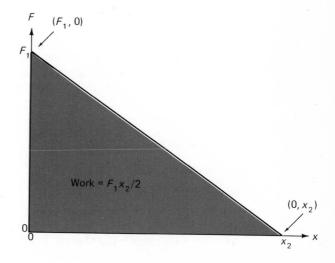

**FIGURE 9.5** Work done on a system that is displaced from $x_1$ to $x_2$ under the action of the constant force $F_1$ acting parallel to the displacement.

The units of work are

$$\text{work} = (\text{force})(\text{displacement})$$

$$(\text{g} \cdot \text{cm} \cdot \text{sec}^{-2})(\text{cm}) = \text{g} \cdot \text{cm}^2 \cdot \text{sec}^{-2} = \text{erg}$$

or

$$(\text{kg} \cdot \text{m} \cdot \text{sec}^{-2})(\text{m}) = \text{kg} \cdot \text{m}^2 \cdot \text{sec}^{-2} = \text{joule} = 10^7 \text{ erg}$$

The dimensions of work are the same as the dimensions of energy.

As an example of a process in which the force is not constant, consider a case for which the initial force is $F_1$, and decreases *linearly* to zero throughout the displacement (i.e., our final point is $F_2 = 0$, $x = x_2$). In the interests of algebraic simplicity we shall take $x_1 = 0$. This situation is depicted graphically in Figure 9.6. The area under the curve is the area

**FIGURE 9.6** Work done on a system that moves from $x_1 = 0$ to $x_2$ under the action of a force, $F$, where

$$F = F_1 \left(1 - \frac{x}{x_2}\right).$$

of the triangle of which the $F$ versus $x$ curve is the hypotenuse; that is, (1/2)(base)(height) or

$$w = F_1 x_2 / 2$$

There are infinitely many paths for moving between any given initial and final points on the $F,x$ plane. For each of the paths, the area under the curve, which is defined by the path, is different. *The value of w depends on the path.* The path *must* be specified before $w$ can be calculated. When a system undergoes a particular change in thermodynamic state, the value of $w$ depends on how the change in state is accomplished. This means that work is not a thermodynamic coordinate, and, therefore, $w$ is not a property of a system. Systems do not possess work. Energy is not work. *Work is a mode of energy transfer.* Our sign convention for $w$ is to take $w$ as positive when the energy transfer is from the surroundings to the system. In other words, *work done on the system is taken as positive.*

We shall now take a specific thermodynamic system, namely, 2.50 moles of helium gas, and calculate the value of $w$ for some simple processes. The helium gas is placed in a cylinder equipped with a piston and immersed in a large ice-water bath. The ice-water bath is thermally insulated with plastic foam. The gas is the system, and the piston, cylinder, and ice-water bath are the surroundings. Let us suppose that, after thermal equilibrium has been achieved between the gas and its surroundings, the pressure of the gas is found to be 3.50 atm. The temperature of the gas is 273 K and, assuming the gas behaves ideally, we compute for the volume of the gas

$$V_1 = \frac{nRT_1}{P_1} = \frac{(2.50 \text{ mole})(0.08205 \, \ell \cdot \text{atm} \cdot \text{K}^{-1} \cdot \text{mole}^{-1})(273 \text{ K})}{(3.50 \text{ atm})} = 16.0 \, \ell$$

Suppose now that the pressure on the gas is suddenly reduced from 3.50 atm to 2.00 atm. This change produces a non-equilibrium situation because the force acting on the inside of the piston is greater than that acting on the outside and, therefore, the piston is driven outward by the expanding gas. The gas will continue to expand under irreversible conditions until it attains a pressure of 2.00 atm at 273 K. The volume of the gas in the final equilibrium state is

$$V_2 = \frac{(2.50 \text{ mole})(0.08205 \, \ell \cdot \text{atm} \cdot \text{K}^{-1} \cdot \text{mole}^{-1})(273 \text{ K})}{(2.00 \text{ atm})} = 28.0 \, \ell$$

Because the gas expands against a constant external pressure, $P_2$, we calculate for the work done by the gas in the irreversible, isothermal expansion ($A$ is the area of the piston)

$$w = F \Delta x = (F/A) A \Delta x = -P \Delta V$$

$$w = -P_2 \Delta V = -P_2 (V_2 - V_1)$$

The minus sign arises because when the gas expands ($\Delta V > 0$) it does work on the surroundings, and, therefore, $w$ is negative according to our convention. In a compression, work is done on the gas ($\Delta V < 0$) and $w = -P \Delta V > 0$. For the process under discussion, we compute for $w$

$$w = -(2.00 \text{ atm})(28.0 \, \ell - 16.0 \, \ell) = -24.0 \, \ell \cdot \text{atm}$$

The energy units, *liter-atmospheres*, are not commonly employed, so we shall convert our result to both joule and calorie units, using ratios of the gas constant in the appropriate units:

$$w = \frac{(-24.0 \; \ell \cdot \text{atm})(8.314 \; \text{J} \cdot \text{K}^{-1} \cdot \text{mole}^{-1})}{(0.08205 \; \ell \cdot \text{atm} \cdot \text{K}^{-1} \cdot \text{mole}^{-1})} = -2.43 \times 10^3 \; \text{J} = -2.43 \; \text{kJ}$$

$$w = \frac{(-24.0 \; \ell \cdot \text{atm})(1.99 \; \text{cal} \cdot \text{K}^{-1} \cdot \text{mole}^{-1})}{(0.08205 \; \ell \cdot \text{atm} \cdot \text{K}^{-1} \cdot \text{mole}^{-1})} = -582 \; \text{cal} = -0.582 \; \text{kcal}$$

Let us now consider the *reversible, isothermal* compression of the above gas, initially in the state (2.00 atm, 273 K), back to the state (3.50 atm, 273 K). In other words, we shall return the gas to its original state along a different path. The work done on an *ideal gas* in a *reversible, isothermal compression* is given by*

$$w = -nRT \ln \left(\frac{V_2}{V_1}\right) \tag{9.2}$$

From this expression we compute

$$w = -(2.50 \; \text{mole})(8.314 \; \text{J} \cdot \text{K}^{-1} \cdot \text{mole}^{-1})(273 \; \text{K})(2.303) \log \left(\frac{16.0}{28.0}\right) = 3.18 \times 10^3 \; \text{J}$$

where we have used the relation $\ln x = 2.303 \log x$. The work done on the gas in the reversible, isothermal compression from 2.00 atm to 3.50 atm ($w = 3.18$ kJ) is *greater* than the work done by the gas in the irreversible, isothermal, isobaric expansion from 3.50 to 2.00 atm ($-w = 2.43$ kJ). If the gas were allowed to expand isothermally along the reversible path, then the work done by the gas would be† $-w = 3.18$ kJ. In Figure 9.7, the work done by an ideal gas that undergoes a reversible, isothermal expansion is compared with the work done by the same gas that undergoes an irreversible, isothermal, isobaric expansion between the same initial and final states. Although it is not obvious from the special case considered above, the work done by a system in passing from a given initial to a given final state is always a *maximum* when the process is carried out reversibly.

The results obtained above show that work is not, in general, conserved. The *closed cycle*

---

*The result can be derived using calculus, because

$$w = -\int_{V_1}^{V_2} P_{\text{ext}} dV$$

where $P_{\text{ext}}$ is the external pressure acting on the gas. For a reversible process $P_{\text{ext}} = P$(of the gas) and, furthermore, $P = nRT/V$, because the gas is ideal. Hence,

$$w = -\int_{V_1}^{V_2} \frac{nRT}{V} dV = -\int_{V_1}^{V_2} nRT \; d \ln V = -nRT \int_{V_1}^{V_2} d \ln V = -nRT \ln \left(\frac{V_2}{V_1}\right)$$

†Note that $V_1$ and $V_2$ for the compression are equal to $V_2$ and $V_1$, respectively, for the expansion, because the gas is returned to its initial state. Thus, as expressed in Equation (9.2), the work done in a *reversible* process has the same *magnitude* in both directions but opposite *sign*.

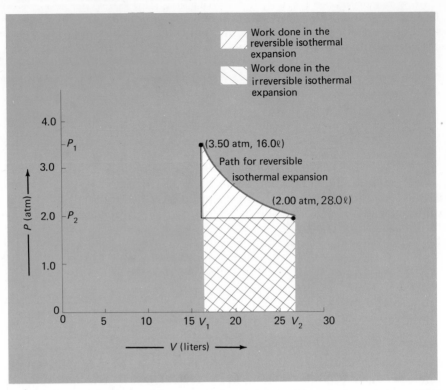

**FIGURE 9.7** Comparison of the amount of work involved in an isothermal, irreversible, isobaric expansion with that involved in a reversible, isothermal expansion. For an irreversible process it is not possible, in general, to specify the path on the $P, V$ plane, although it may be possible (as in this figure) to mark off the area corresponding to the work done.

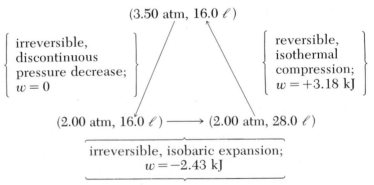

involves a *net input* of energy as work to the system of $3.18 - 2.43 = 0.75$ kJ. This amount of work corresponds to the area enclosed by the solid curves in Figure 9.7. When the gas is returned to its initial state, its total energy must be the same, by the conservation principle

$$\Delta E_{sys} = 0 \text{ (closed cycle)}$$

This means that there must be a *net output* of 0.75 kJ of energy from the system by some mode other than that of work, when the system is taken around the closed cycle.

## 9.6   HEAT AS A MODE OF ENERGY TRANSFER.

Experiments show that the change in state of a system *within an adiabatic enclosure,* produced by the performance of a given amount of work on the system, is independent of the manner in which the work is performed. Combining this observation with the energy conservation principle, we can write

$$\Delta E_{sys} = w \qquad \text{(adiabatic change in state)}$$

where $\Delta E_{sys}$ is the energy change of the system, and $w$ is the amount of work done on the system.

A change in state of a system within a diathermic enclosure can be produced without performing work on it, namely, by raising or lowering the temperature of the surroundings. There must, therefore, exist a mode of energy transfer that does not require the action of unbalanced forces. This alternate mode of energy transfer simply requires a temperature difference between a diathermically enclosed system and its surroundings. This mode of energy transfer is called *heat.* Our sign convention for heat, $q$, is to take heat added to the system as positive. Any process that does not involve the transfer of energy as heat is said to be adiabatic ($q = 0$). For a particular change in state of a system, the value of $q$ depends on how the change in state is accomplished. This means that $q$ is not a property of the system. Systems do not possess heat. Energy is not heat. *Heat is a mode of energy transfer.*

Returning to the above cycle, we recall that there was a net output of 0.75 kJ of energy by a mode other than that of work. Consequently, this energy must have left the system as heat. *Heat and work are the two ways in which energy can be transferred from one system to another system.*

## 9.7   THE FIRST LAW OF THERMODYNAMICS.

Application of the conservation principle to a system undergoing a general process that may involve the transfer of energy as either heat, or work, or a combination of the two, yields

$$\boxed{\Delta E_{sys} = q + w} \tag{9.3}$$

This equation is a somewhat simplified mathematical statement of the first law of thermodynamics. Although the first law is often stated as simply the conservation of energy, the first law also postulates the existence of the thermodynamic coordinate $E$, the *internal energy.* The passage of a system from one defined (i.e., equilibrium) state to another always involves a definite change in $E$, completely independent of the manner in which the change in state is accomplished. For all the different possible paths between two states, $q$ and $w$ will be different, but $\Delta E_{sys}$ remains the same.

*Energy is conserved whether or not the process is reversible.* However, for many irreversible processes neither $q$ nor $w$ for the *actual* process can be calculated, because the system passes through a sequence of non-equilibrium states for which its thermodynamic coordinates are not defined. If the initial and final states are defined, we can always find $\Delta E$ for the process (if we have the equation of state) by devising a reversible path between these states and computing $q$ and $w$ along the reversible path.

The first law requires that the *sum* $q + w$ for the reversible path must be equal to $\Delta E$ for the irreversible path.

The application of the first law requires that all of the relevant work terms be included in the calculation of $w$. That is, we must take into account the various possible ways in which energy can be exchanged as work by the system with the surroundings. In many cases of interest there is only one way. If our system is a gas, liquid, or solid of fixed composition, subject to an external pressure, which does not interact with any external force fields (e.g., magnetic, electric, or gravitational) to any experimentally significant extent, and which is not too finely divided (so that surface effects can be ignored), then we need only consider simple compression (or expansion) work.*

$$\Delta E_{gas} = q + \text{compression work}$$

On the other hand, if our system of interest is an electrochemical cell, then we must consider both compression work and electrical work:

$$\Delta E_{cell} = q + \text{compression work} + \text{electrical work}$$

If a charge of Z coulombs is passed through a cell of constant voltage, $\mathscr{E}$, then the electrical work done on the cell is $-\mathscr{E}Z$ ($1$ volt $\times$ $1$ coulomb$=1$ joule).

## 9.8   HEAT CAPACITY AND THE ENTHALPY FUNCTION. CALORIMETRY.

The heat capacity of a system is the rate at which the temperature of the system increases as a result of the input of energy as heat into the system. A system will have two or more types of heat capacity; the different types are distinguished by the different conditions (e.g., constant pressure or constant volume) under which the energy is transferred.

If a system of fixed composition, whose thermodynamic state is completely determined by the specification of its temperature and volume, undergoes a constant volume process, then no energy can be exchanged as work with the surroundings, and therefore, from the first law, we obtain $\Delta E = q$. We define the heat capacity at constant volume, $C_V$, as

$$C_V = \lim_{\Delta T \to 0} \left( \frac{q_V}{\Delta T} \right) = \lim_{\Delta T \to 0} \left( \frac{\Delta E}{\Delta T} \right)_V \tag{9.4}$$

where $q_V$ is the energy absorbed as heat by the system at fixed $V$ when the temperature of the system is increased by $\Delta T$. The units of heat capacity are either $\text{cal} \cdot \text{deg}^{-1}$ or $\text{joule} \cdot \text{deg}^{-1}$. To determine $C_V$ precisely at a particular temperature, the temperature increment, $\Delta T$, must be kept as small as practicable, because what we measure experimentally is the average value of $C_V$ over the temperature range $\Delta T$. This is the experimental significance of the notation $\lim_{\Delta T \to 0}$ (or, "in the limit as $\Delta T$ goes to zero") in Equation (9.4). The heat capacity of a substance (system) is directly proportional to the mass of the substance. The heat capacity of 2 grams of water is exactly twice the heat capacity of 1 gram of water, at the same conditions.

---

*Unless otherwise stated, we shall henceforth assume that external force fields and surface effects can be ignored in our thermodynamic analysis.

A *calorimeter* is a device used to determine the amount of energy evolved or absorbed as heat in a process. A suitable calorimeter for the study of combustion reactions is the bomb calorimeter. A bomb calorimeter is a heavy-walled, metal container, equipped with appropriate electrical leads, that is charged with the reactants (e.g., sucrose crystals and oxygen gas at about 30 atm). The container is tightly sealed and immersed in a fluid (usually water) containing a sensitive thermometer. The bomb and surrounding fluid are enclosed by adiabatic walls (e.g., plastic foam). Upon ignition, a chemical reaction (combustion of sucrose) takes place (irreversibly) in the closed container. Because the entire setup is essentially isolated by rigid adiabatic walls, we can write

$$\Delta E_{total} = \Delta E_{rctn} + \Delta E_{surr} = 0$$

where $\Delta E_{surr}$ represents the energy change in the container, its component parts, and the surrounding fluid, and $\Delta E_{rctn}$ represents the energy change associated with the reaction. Provided that the temperature change, $\Delta T = T_{final} - T_{initial}$, is kept sufficiently small, the *total* heat capacity at constant volume, $C_V$, can be taken as constant within the required precision, and we have from Equation (9.4)

$$\Delta E_{rctn} = -\Delta E_{surr} = -C_V \Delta T$$

The value of $C_V$ is determined electrically using a resistance heater, and this value together with the measured value of $\Delta T$ for the combustion yields $\Delta E_{rctn}$.

Although bomb calorimetry is an important source of thermo-dynamic data on chemical systems, chemists are primarily interested in processes taking place at constant pressure (e.g., open to the atmosphere). To facilitate the thermodynamic analysis of constant-pressure processes, we define a thermodynamic coordinate, $H$, called the *enthalpy*, as

$$\boxed{H = E + PV} \tag{9.5}$$

The enthalpy change, $\Delta H$, for a process is given by

$$\Delta H = \Delta E + \Delta(PV) = q + w + \Delta(PV)$$

For a constant-pressure process taking place in a system involving only expansion work, $w = -P\Delta V$, and $\Delta(PV) = P_2 V_2 - P_1 V_1$, which is $P\Delta V$ because $P$ is constant. Substituting these results into the above expression yields $\Delta H = q_p$. In other words, the heat absorbed by a system *for which only expansion work is possible*, undergoing a constant pressure process, is equal to the enthalpy change of the system for that process.[*]

---

[*]The result $\Delta H = q_p$ is not applicable to systems involving other than expansion work. For example, for the enthalpy change of an electro-chemical cell undergoing a discharge at constant pressure and voltage, we have

$$\Delta H = \Delta E + \Delta(PV) = q + w + \Delta(PV) = q - P\Delta V - \mathscr{E}Z + P\Delta V$$

or

$$\Delta H = q_p - \mathscr{E}Z$$

The heat capacity at constant pressure of a system involving only expansion work is

$$C_P = \lim_{\Delta T \to 0} \left(\frac{q_p}{\Delta T}\right) = \lim_{\Delta T \to 0} \left(\frac{\Delta H}{\Delta T}\right)_P \tag{9.6}$$

which is analogous to Equation (9.4). The measurement of $C_P$ for a condensed phase whose vapor pressure is negligible at the temperature of interest can be accomplished by loading the sample into a suitable container of known heat capacity, and determining $C_P$ by electrically heating the sample and container at constant total pressure. Subtraction of $C_P$ for the container yields $C_P$ for the sample, because $C_{P,total} = C_{P,sample} + C_{P,container}$.

Some representative values of $\bar{C}_P$ ($=C_P/n$, where $n$ is the number of moles of substance), the molar heat capacity, are presented in Table 9.2. As a simple example of the use of the data in Table 9.2, we compute the enthalpy increase of a sample of 2.00 moles of $H_2O(\ell)$ when its temperature is increased isobarically from 273 K to 373 K:

$$q_p = \Delta H = n\bar{C}_P \, \Delta T = (2.00 \text{ mole})(18.04 \text{ cal} \cdot \text{K}^{-1} \cdot \text{mole}^{-1})(373-273) \text{ K}$$
$$= 3.61 \text{ kcal}$$

The value of $\Delta H$ for an isothermal, isobaric process (e.g., the dissolution of a solid in a liquid, a dilution reaction, oxidations, neutralizations,

**TABLE 9.2**

MOLAR HEAT CAPACITIES AT CONSTANT PRESSURE

| SUBSTANCE | $\bar{C}_P^o$(cal · K$^{-1}$ · mole$^{-1}$) 298 K | 1000 K | SUBSTANCE | $\bar{C}_P^o$(cal · K$^{-1}$ · mole$^{-1}$) 298 K |
|---|---|---|---|---|
| $H_2(g)$ | 6.88 | 7.31 | $C_{10}H_8(s)$ | 39.2 |
| $O_2(g)$ | 7.91 | 8.20 | $Cu(s)$ | 5.86 |
| $N_2(g)$ | 6.97 | 7.72 | $H_2O(\ell)$ | 18.04 |
| $I_2(g)$ | 8.79 | 9.06 | $H_2O(s)$(273 K) | 9.0 |
| $CO_2(g)$ | 8.87 | 12.46 | $I_2(s)$ | 13.14 |
| $H_2O(g)$ | 8.03 | 9.76 | $Hg(\ell)$ | 6.69 |
| $He(g)$ | 4.97 | 4.97 | $C(s)$(diamond) | 1.46 |
| $C(s)$(graphite) | 4.37 | 4.97 | $Br_2(\ell)$ | 18.09 |
| | | | $AgCl(s)$ | 12.14 |
| | | | $SiC(s)$ | 6.42 |
| | | | $CuO(s)$ | 10.11 |
| | | | $Cu_2O(s)$ | 15.21 |
| | | | $AsI_3(s)$ | 25.28 |
| | | | $Bi_2O_3(s)$ | 27.13 |

1 cal = 4.184 J

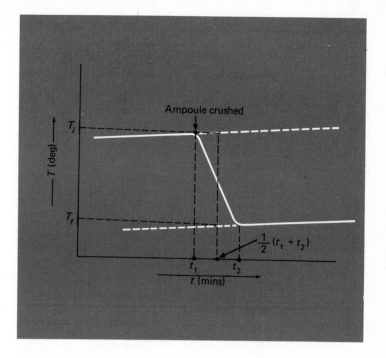

**FIGURE 9.8** Temperature versus time plot for the dissolution of KCl(s) in water. In this case $\Delta T < 0$ because KCl absorbs energy when it dissolves; i.e., $\Delta H$(dissolution) $> 0$. The values of $T_i$ (the initial temperature) and $T_f$ (the final temperature) are obtained from the intersections of the temperature extrapolation with a vertical line drawn from the time $\frac{1}{2}(t_1 + t_2)$.

precipitations, and so forth) can be measured[*] using an adiabatic calorimeter (e.g., a Dewar flask), the contents of which are held at fixed pressure. Let us suppose that the process of interest is the dissolution of KCl(s) in water. A sample of KCl(s) is sealed in a thin-walled glass ampoule, connected to the tip of the shaft of a constant-speed stirrer, and immersed in the water. Because no adiabatic wall is ever perfectly adiabatic, and also, because the stirring action increases the temperature of the system, the temperature of the system is monitored periodically (see Figure 9.8). When the temperature is observed to be increasing linearly with time, the stirrer is depressed to crush the glass ampoule. The calorimeter and contents are adiabatically enclosed, and therefore

$$\Delta H_{total} = q_p = 0 = \Delta H(\text{dissolution}) + \Delta H(\text{calorimeter} + \text{contents})$$

or

$$\Delta H(\text{dissolution}) = -C_{P,total}(T_f - T_i),$$

where the last equality holds if $C_{P,total}$, the heat capacity of the calorimeter and its contents, can be taken as constant over the range $T_f - T_i$. The value of $C_{P,total}$ can be measured using an electrical resistance heater.

---

[*] In any calorimetric investigation of a chemical reaction, it is necessary to ascertain:
(1) the nature of the reaction, that is, what are the reactants and the products;
(2) the extent of the reaction, that is, how much product(s) is formed; and
(3) the reaction rate — if the reaction is too slow, then the value of $\Delta H$ cannot be accurately determined, because of unavoidable heat leaks.

## 9.9    ENTHALPY CHANGES IN CHEMICAL REACTIONS.

In this section our intent is to acquire a familiarity with the magnitudes of the thermal effects that are involved in various types of reactions.

### a.    Heats of Solution, Dilution, and Mixing.

The value of $\Delta \bar{H} (=\Delta H/n)$ for the dissolution of a solid in a liquid (heat of solution), or the dilution of a solution (heat of dilution), or the mixing of two liquids or solutions (heat of mixing), depends on the temperature and total pressure at which the process takes place, and also on the compositions of the solutions involved. The effect of composition is apparent in the data for the dissolution of KCl(s) in water given in Table 9.3. The effect is usually more pronounced for electrolyte solutions than for non-electrolyte solutions, because of the much greater deviations from ideal-solution behavior that occur in electrolyte solutions. If we have a solution formed from the components $A$ and $B$, then the solution will be ideal, if the $A$ to $A$, $B$ to $B$, and $A$ to $B$ interaction forces are all equal. In other words, if the potential energy of an $A$ particle surrounded by $B$ particles is the same as when it is surrounded by $A$ particles, then the $A + B$ solution will be ideal. The heat of mixing of two components that form an ideal solution is zero. Two liquids that form an ideal solution are meta- and para-xylene:

$m$-xylene            $p$-xylene

Ideal-solution behavior is favored by very close structural and chemical similarities, and although not unknown, ideal-solution behavior is rare.

Note that, for a constant-pressure process, $\Delta H = q_p$ for both reversible and irreversible processes. A positive value of $\Delta H$ for a process means that $q_p$ is positive and, therefore, that heat is absorbed by the system if the process is carried out isothermally. If the process is not isothermal, then a positive value of $\Delta H$ will lead to a temperature drop in the system when the reaction occurs. Thus, when KCl(s) is dissolved in water under adiabatic conditions, the temperature of the final solution will be less than the initial temperature of the KCl(s) and water because $\Delta H \simeq +4$ kcal/mole for dissolving KCl in water. To restore the final solution to the initial temperature, we must put in about 4 kcal/(mole KCl) of energy as heat. If $\Delta H > 0$, then the process is said to be *endothermic*, whereas if $\Delta H < 0$, the process is said to be *exothermic*.

**TABLE 9.3**

**HEATS OF SOLUTION, DILUTION, AND MIXING**

**I. Heats of Solution**
(25°C, 1 atm total pressure)

| Process | $\Delta H$ (kcal/mole of solute) |
|---|---|
| $KCl(s) + \infty\, H_2O(\ell) = KCl(\infty\, aq)$* | 4.12 |
| $KCl(s) = KCl(m=1, aq)$† | 3.94 |
| $KCl(s) = KCl(m=4.81,$ satd aq soln$)$‡ | 3.27 |
| $KCl(s) + 55.5\, H_2O(\ell) = KCl(m=1, aq)$** | 4.09 |
| $Tl(s) + \infty\, Hg(\ell) = Tl(\infty\, Hg)$ | 0.73 |
| $AgCl(s) + \infty\, H_2O(\ell) = AgCl(\infty\, H_2O)$ | 15.62 |
| $HCl(g) + \infty\, H_2O(\ell) = HCl(\infty\, H_2O)$ | −17.89 |
| $H_2SO_4(\ell) + \infty\, H_2O(\ell) = H_2SO_4(\infty\, H_2O)$ | −22.77 |
| $NiCl_2(s) + \infty\, H_2O(\ell) = NiCl_2(\infty\, H_2O)$ | −19.8 |
| $NH_4Cl(s) + \infty\, H_2O(\ell) = NH_4Cl(\infty\, H_2O)$ | 3.53 |
| $NH_4NO_3(s) + \infty\, H_2O(\ell) = NH_4NO_3(\infty\, H_2O)$ | 6.14 |
| $AgNO_3(s) + \infty\, H_2O(\ell) = AgNO_3(\infty\, H_2O)$ | 5.40 |

**II. Heats of Dilution**
(25°C, 1 atm total pressure)

| Process | $\Delta H$ (kcal/mole of solute) |
|---|---|
| $NH_4Cl(100\, H_2O) + 900\, H_2O(\ell) = NH_4Cl(1000\, H_2O)$ | −0.06 |
| $H_2SO_4(100\, H_2O) + \infty\, H_2O(\ell) = H_2SO_4(\infty\, H_2O)$ | −5.17 |
| $HCl(100\, H_2O) + \infty\, H_2O(\ell) = HCl(\infty\, H_2O)$ | −0.30 |
| $CuSO_4(50\, H_2O) + \infty\, H_2O(\ell) = CuSO_4(\infty\, H_2O)$ | −1.56 |
| $NH_4NO_3(50\, H_2O) + \infty\, H_2O(\ell) = NH_4NO_3(\infty\, H_2O)$ | +0.31 |

**III. Heats of Mixing††**

| Process | $\Delta H$ (kcal/mole of solute) |
|---|---|
| $CS_2(\ell) + CH_3COCH_3(\ell)$ | 0.349 |
| $C_6H_6(\ell) + C_6H_5CH_3(\ell)$ | 0.011 |
| $n-C_5F_{12}(\ell) + n-C_5H_{12}(\ell)$ | 0.370 |
| $NaCl(1.0\, m, aq) + LiCl(1.0\, m, aq)$ | 0.021 (per Kg of H₂O) |

1 cal = 4.184 J
* Dissolution of the solid in a very large excess of solvent.
† Dissolution of the solid in a very large excess of 1 m KCl(aq).
‡ Dissolution of the solid in a very large excess of 4.81 m KCl(aq).
** Dissolution of the solid in sufficient water to produce a 1 m KCl(aq) solution.
†† Components are mixed in a 1:1 mole ratio.

## b.  Heats of Neutralization and Complexation.

Unusually large $\Delta H$ values for the mixing of two electrolyte solutions often arise when one or more of the product species is only slightly dissociated. By way of example, $\Delta H$ for the mixing of NaOH(aq, 0.001 m) with HCl(aq, 0.001 m) is $-13.3$ kcal/mole at 25°C, whereas $\Delta H$ for the mixing of NaCl(aq, 0.001 m) with HCl(aq, 0.001 m) at 25°C is <0.1 kcal/mole. In fact, the value of $\Delta H$ for the mixing of any dilute aqueous solution of a strong acid with a dilute aqueous solution of a strong base at 25°C is approximately equal to $-13.3$ kcal per mole of water formed, because the major thermal effect arises from the formation of neutral water from $H^+(aq)$ and $OH^-(aq)$ ions:

$$HX(aq, m) + MOH(aq, m) \xrightarrow{\Delta H_1} H_2O(\ell) + MX(aq, m/2)$$

At low $m$, $\Delta H_1$ is approximately equal to $\Delta H$ for the reaction

$$H^+(aq) + OH^-(aq) = H_2O(\ell)$$

provided HX and MOH are completely dissociated electrolytes.

The great difference in $\Delta H$ values for the processes

$$Ba(NO_3)_2(aq,m) + CaCl_2(aq,m) \rightarrow \text{(solution)} \quad \Delta H \approx 0 \text{ kcal per mole } Ba^{2+}$$

$$Hg(NO_3)_2(aq,m) + CaCl_2(aq,m) \rightarrow \text{(solution)} \quad \Delta H \approx -13 \text{ kcal per mole } Hg^{2+}$$

arises from the formation of the weak electrolyte $HgCl_2(aq)$ in the latter case. Some $\Delta H$ values for ionization and complexation reactions are presented in Table 9.4.

**TABLE 9.4**

$\Delta H°$ VALUES FOR SOME IONIZATION AND COMPLEXATION REACTIONS AT 25°C

| REACTION | $\Delta H°(\text{kcal} \cdot \text{mole}^{-1})$ |
|---|---|
| $H_2O(\ell) = H^+(aq) + OH^-(aq)$ | 13.3 |
| $HCOOH(aq) = HCOO^-(aq) + H^+(aq)$ | $-0.03$ |
| $CH_3COOH(aq) = CH_3COO^-(aq) + H^+(aq)$ | $-0.06$ |
| $H_2C_2O_4(aq) = HC_2O_4^-(aq) + H^+(aq)$ | 1.6 |
| $H_3PO_4(aq) = H_2PO_4^-(aq) + H^+(aq)$ | 2.6 |
| $H_2PO_4^-(aq) = HPO_4^{2-}(aq) + H^+(aq)$ | 0.99 |
| $HPO_4^{2-}(aq) = PO_4^{3-}(aq) + H^+(aq)$ | 3.5 |
| $NH_4^+(aq) = NH_3(aq) + H^+(aq)$ | 12.48 |
| $HCl(aq) = Cl^-(aq) + H^+(aq)$ | 0 |
| $NH_4OH(aq) = NH_4^+(aq) + OH^-(aq)$ | 0.86 |
| $HgCl_2(aq) = HgCl^+(aq) + Cl^-(aq)$ | 7.3 |
| $HgCl^+(aq) = Hg^{2+}(aq) + Cl^-(aq)$ | 5.4 |
| $HgI^+(aq) = Hg^{2+}(aq) + I^-(aq)$ | 17.3 |
| $HgI_4^{2-}(aq) = Hg^{2+}(aq) + 4I^-(aq)$ | 44.3 |

1 cal = 4.184 J

## c. Heats of Combustion.

Highly exothermic ($\Delta H \ll 0$, energy evolved as heat) chemical reactions find wide practical use as thermal energy sources. All such reactions can be considered as the reaction of a fuel (e.g., hydrogen, methane, acetylene, or hydrazine) with an oxidizer (e.g., oxygen, fluorine, or nitrogen tetroxide). The very exothermic reaction of hydrogen with oxygen was used to power the second and third stages of the Apollo moon rockets.[*] The reaction of methane ($CH_4$, "natural gas") with oxygen is utilized in the gas stoves and heaters of millions of homes throughout the world, whereas butane ($CH_3CH_2CH_2CH_3$) is used for the same purpose in areas not serviced by gas mains. More exotic fuels such as methanol ($CH_3OH$) and nitromethane ($CH_3NO_2$) are used in some high-performance combustion engines. The reaction between the fuel N,N-dimethylhydrazine ($H_2NN(CH_3)_2$) and nitrogen tetroxide ($N_2O_4$)

$$H_2NN(CH_3)_2(\ell) + 2N_2O_4(\ell) = 3N_2(g) + 2CO_2(g) + 4H_2O(g)$$

was used to power the jet engine on the Apollo 11 lunar escape vehicle.[†] The values of $\Delta H$ for some representative cases are listed in Table 9.5.

The food we eat constitutes the fuel necessary to maintain our body temperature and internal functions, and to provide the energy necessary to move around. If we consume more food than we need for our normal level of activity, then the excess that is not eliminated is stored in the body as fat. One pound of excess weight is roughly equivalent to 3500 kcal of stored energy. It is evident from the first law that *calories do count*. The energy values of some food substances are presented in Table 9.6. The amounts of food that must be consumed per day to maintain a constant body weight for moderately active men and women in the 18–35 year range are:[‡]

| weight (lbs) | sex | kcal per day |
|---|---|---|
| 110 | F | 1,850 |
| 132 | F | 2,150 |
| 132 | M | 2,550 |
| 176 | M | 3,250 |

As examples of fuel consumption by the human body, we note that it requires about 1500 kcal per day for a man, and 1200 kcal per day for a woman, merely to keep the body functioning at close to the zero activity level (lying still in bed), whereas a one hour walk over average terrain consumes 140 kcal. It is also worth noting (last two entries in Table 9.4) that a gram of fat contains about three times the caloric value of a gram of protein.

---

[*] Liquid hydrogen is particularly suitable for a rocket fuel, both because of its very low mass and because of its very exothermic reaction with oxygen (see Table 9.5). It must be stored at around 20 K, however, and it is explosively flammable in the presence of oxygen.

[†] This reaction is particularly suitable for such a purpose because combustion starts spontaneously on mixing—no battery or spark plugs are required.

[‡] These values were prepared by the Food and Nutrition Board of the National Academy of Sciences, National Research Council.

TABLE 9.5

$\Delta H°$ VALUES AT 25°C FOR THE REACTIONS OF SOME COMMON FUELS WITH OXIDIZERS

| REACTION | $\Delta H°$ (kcal) per mole of fuel | per gram of fuel |
|---|---|---|
| $H_2(g) + (1/2)O_2(g) = H_2O(g)$ | −57.102 | −28.32 |
| $H_2(g) + F_2(g) = 2HF(g)$ | −129.6 | −64.29 |
| $CH_4(g) + 2O_2(g) = CO_2(g) + 2H_2O(g)$ | −190.27 | −11.86 |
| $CH_3{-}\overset{\displaystyle CH_3}{\underset{\displaystyle CH_3}{C}}{-}CH_2CHCH_3(\ell) + (25/2)O_2(g) = 8CO_2(g) + 9H_2O(g)$ (with CH₃ on CHCH₃) | −1303.9 | −11.41 |
| $H_2NN(CH_3)_2(\ell) + 2N_2O_4(\ell) = 3N_2(g) + 2CO_2(g) + 4H_2O(g)$ | −418.8 | −6.98 |
| $H{-}C{\equiv}C{-}H(g) + (3/2)O_2(g) = 2CO_2(g) + H_2O(g)$ | −299.22 | −11.49 |
| $C_6H_6(\ell) + (15/2)O_2(g) = 6CO_2(g) + 3H_2O(g)$ | −746.80 | −9.56 |
| $CH_3OH(\ell) + (3/2)O_2(g) = CO_2(g) + 2H_2O(g)$ | −151.13 | −4.72 |
| $CH_3NO_2(\ell) + (3/4)O_2(g) = CO_2(g) + (1/2)N_2(g) + (3/2)H_2O(g)$ | −152.59 | −2.50 |
| $N_2H_4(\ell) + O_2(g) = N_2(g) + 2H_2O(g)$ | −126.30 | −3.94 |
| $Fe(s) + S(s) = FeS(s)$ | −22.72 | −0.41 |
| $Mg(s) + (1/2)O_2(g) = MgO(s)$ | −143.77 | −5.91 |
| $2Li(s) + (1/2)O_2(g) = Li_2O(s)$ | −142.4 | −20.52 |
| *Biological Fuels* | | |
| $C_6H_{12}O_6(s, glucose) + 9O_2(g) = 6CO_2(g) + 6H_2O(\ell)$ | −673. | −3.74 |
| $CH_3CHCOOH(\ell, lactic\ acid) + (9/2)O_2(g) = 3CO_2(g) + 3H_2O(\ell)$ (with OH) | −326. | −3.62 |
| $CH_3(CH_2)_{14}COOH(s, palmitic\ acid) + 24O_2(g) = 16CO_2(g) + 16H_2O(\ell)$ | −2380. | −9.30 |
| $2H_2NCH_2COOH(s, glycine) + \dfrac{13}{2}O_2(g) = 4CO_2(g) + N_2(g) + 5H_2O(\ell)$ | −234.5 | −3.12 |

1 cal = 4.184 J

## d. Standard Enthalpy Changes. The Use of $\Delta \overline{H}_f°$ Values.

Actual calorimetric experiments, carried out at a particular temperature and pressure, yield a $\Delta H$ value appropriate to the experimental temperature and pressure. In order to achieve economies in tabulation, it is
conventional to refer all $\Delta H$ values to a particular pressure. The pressure
agreed upon is one standard atmosphere ($1.013 \times 10^5$ N · m$^{-2}$ = 1.013
bar = 760 Torr). This leads us to the concept of *standard states*, which for
pure or fixed-composition phases are chosen arbitrarily as follows:
  (a) The standard state of a gas is chosen as the ideal gas at 1 atm
pressure and the temperature of interest.
  (b) The standard state for a condensed phase is chosen as the pure

TABLE 9.6

ENERGY VALUES OF SOME COMMONLY CONSUMED SUBSTANCES*

| SUBSTANCE | Cal†/8 oz.‡ | SUBSTANCE | Cal†/8 oz.‡ |
|---|---|---|---|
| lettuce (iceberg) | 28 | vodka, gin, scotch, brandy (80 proof) | 536 |
| asparagus (raw spears) | 33 | ice cream (chocolate, bulk) | 573 |
| carrots (raw) | 78 | bread (white) | 613 |
| milk (skim) | 89 | mozzarella cheese | 632 |
| beer (lager) | 104 | hamburger (broiled) | 652 |
| orange juice | 120 | bread (French) | 658 |
| apples | 121 | frankfurters (cooked) | 690 |
| bananas | 131 | popcorn | 876 |
| cherries (sweet) | 143 | parmesan cheese | 880 |
| milk (whole, 3.5% fat) | 159 | sirloin steak (round bone, meat and fat) | 882 |
| clams (fresh, steamed) | 182 | chocolate (milk base) | 1168 |
| crab (fresh, steamed) | 210 | cashews (roasted, salted) | 1278 |
| champagne (extra dry, 12.5% alcohol) | 232 | potato chips | 1288 |
| cottage cheese | 240 | cream (heavy whipping) | 1532 |
| yogurt (plain "low fat") | 246 | butter | 1625 |
| chicken (broiled, meat only) | 310 | margarine (regular) | 1633 |
| sirloin steak (round bone, no fat) | 472 | mayonnaise | 1665 |
| yogurt (strawberry) | 520 | lard | 1984 |
| ham (broiled) | 532 | | |

*Data (adjusted to a common weight of substance) from "Count Your Calories," Dell Publishing Co., New York, N.Y. (1968), where much more extensive tables can be found.

†Note that the nutritionist's "calorie" is equal to 1 thermochemical kcal. The reasons for this are presumably psychological in nature. There are 1,278,000 thermochemical calories in 8 oz. of cashews. The nutritionist's calorie is distinguished from the thermochemical calorie by writing it with a capital C.

‡Solid or liquid, whichever is applicable. Values given assume complete combustion of the foodstuff to $CO_2(g)$ and $H_2O(\ell)$.

material in a reproducible equilibrium state at a total pressure of 1 atm and the temperature of interest.

We designate $\Delta H$ for an isothermal reaction as $\Delta H°$, the *standard enthalpy change*, when every reactant is consumed and every product is produced in its 1 atm standard state at that temperature. The value of $\Delta H°$ depends only on the temperature. Because the enthalpy of a mole of a substance is defined as $\bar{H} = \bar{E} + P\bar{V}$, our inability to determine an absolute value for $\bar{E}$ means that we cannot determine an absolute value of $\bar{H}$. Absolute enthalpies are not measurable—only enthalpy differences can be determined. However, it is convenient to tabulate enthalpies of substances relative to some arbitrarily chosen zero of enthalpy, because enthalpy data can then be tabulated by compound rather than by reaction. Such a procedure leads to a considerable saving of space, because the number of chemical compounds is far less than the number of possible reactions between these compounds.

A $\Delta H°$ value for a reaction at a particular temperature gives us the difference in enthalpy between the products and reactants in their standard states at that temperature. Mathematically speaking, we have

$$\Delta H_{rctn} = \sum_i \nu_i \bar{H}_i°$$

where the $\nu_i$ values are the balancing coefficients (positive for products and negative for reactants), and $\overline{H}_i^\circ$ is the molar enthalpy of species $i$ in its 1 atm standard state. We can rewrite the above equation as

$$\boxed{\Delta H_{rctn}^\circ = \sum_i \nu_i \Delta \overline{H}_{f,i}^\circ} \tag{9.7}$$

where $\Delta \overline{H}_{f,i}^\circ$ is the *standard enthalpy of formation of species i from the elements*, if we agree to set the values of $\Delta \overline{H}_{f,i}^\circ = 0$ at all $T$ for an arbitrarily chosen physical state (gas, liquid, or solid) of each *element*. For example, the reaction

$$H_2(g) + (1/2)O_2(g) = H_2O(\ell)$$

has $\Delta H_{rctn}^\circ = -68.32$ kcal at 298 K. From the above equations we obtain $\left( \nu_{H_2O} = +1, \ \nu_{H_2} = -1, \text{ and } \nu_{O_2} = -\dfrac{1}{2} \right)$

$$\Delta H_{rctn}^\circ = \overline{H}^\circ [H_2O(\ell)] - (1/2)\overline{H}^\circ [O_2(g)] - \overline{H}^\circ [H_2(g)] = -68.32 \text{ kcal}$$

$$\Delta H_{rctn}^\circ = \Delta \overline{H}_f^\circ [H_2O(\ell)] - (1/2)\Delta \overline{H}_f^\circ [O_2(g)] - \Delta \overline{H}_f^\circ [H_2(g)] = -68.32 \text{ kcal}$$

From these two expressions it can be seen that setting $\Delta \overline{H}_f^\circ = 0$ for the elements $H_2(g)$ and $O_2(g)$ yields, for the standard enthalpy of formation of liquid water from the elements,

$$\Delta \overline{H}_f^\circ [H_2O(\ell)] = \overline{H}^\circ [H_2O(\ell)] - \left(\frac{1}{2}\right) \overline{H}^\circ [O_2(g)] - \overline{H}^\circ [H_2(g)]$$

$$= -68.32 \text{ kcal} \cdot \text{mole}^{-1} \text{ (at 298.15 K)}$$

In assigning the value of $\Delta \overline{H}_f^\circ = 0$ to the elements, it is important to specify what particular state, gas, liquid, or solid (of which there may be several allotropic modifications), is chosen. As long as one is consistent, it makes no difference which form is chosen; however, it is usually advantageous with regard to experimental convenience to set $\Delta \overline{H}_f^\circ = 0$ for that state of each element which is the most stable at 1 atm total pressure and 25°C.

In order to obtain $\Delta \overline{H}_f^\circ$ for a substance at a particular temperature we need only to obtain $\Delta H^\circ$ for a reaction in which all the $\Delta H_f^\circ$ values are known except the one of interest. Conversely, we can calculate $\Delta H^\circ$ for any reaction involving species whose $\Delta \overline{H}_f^\circ$ values are all known. This is possible because $H$ is a state function, and therefore $\Delta H$ is independent of path:*

---

*The path-independent property of $\Delta H$ is often referred to as *Hess's Law*. The path-independent nature of $\Delta H$ is a direct consequence of the fact that $H$ is a state function. Hess's law is derivable from the first law of thermodynamics.

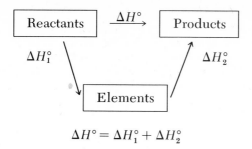

$$\Delta H^\circ = \Delta H_1^\circ + \Delta H_2^\circ$$

Some examples follow:
(a) For the reaction

$$H_2O(\ell) = H_2O(g)$$

$\Delta H_{rctn}^\circ = +10.52$ kcal at 298 K, and therefore

$$10.52 \text{ kcal} = \Delta \overline{H}_f^\circ [H_2O(g)] - \Delta \overline{H}_f^\circ [H_2O(\ell)]$$

$$10.52 \text{ kcal} = \Delta \overline{H}_f^\circ [H_2O(g)] - (-68.32 \text{ kcal})$$

$$\Delta \overline{H}_f^\circ [H_2O(g)] = -57.80 \text{ kcal} \cdot \text{mole}^{-1} \text{ at } 298 \text{ K}$$

(b) For the reaction

$$H_2O_2(g) = H_2O(g) + (1/2)O_2(g)$$

$\Delta H_{rctn}^\circ = -25.27$ kcal at 298 K, and therefore

$$-25.27 \text{ kcal} = \Delta \overline{H}_f^\circ [H_2O(g)] + (1/2) \Delta \overline{H}_f^\circ [O_2(g)] - \Delta \overline{H}_f^\circ [H_2O_2(g)]$$

$$-25.27 \text{ kcal} = -57.80 + 1/2 \, (0) - \Delta \overline{H}_f^\circ [H_2O_2(g)]$$

$$\Delta \overline{H}_f^\circ [H_2O_2(g)] = -32.53 \text{ kcal} \cdot \text{mole}^{-1} \text{ at } 298 \text{ K}$$

(c) For the reaction

$$N_2H_4(\ell) + O_2(g) = N_2(g) + 2H_2O(\ell)$$

$\Delta H_{rctn}^\circ = -148.69$ kcal at 298 K, and therefore

$$-148.69 \text{ kcal} = \Delta \overline{H}_f^\circ [N_2(g)] + 2\Delta \overline{H}_f^\circ [H_2O(\ell)] - \Delta \overline{H}_f^\circ [N_2H_4(\ell)] - \Delta \overline{H}_f^\circ [O_2(g)]$$

$$-148.69 \text{ kcal} = 0 + 2(-68.32) - \Delta \overline{H}_f^\circ [N_2H_4(\ell)] - 0$$

$$\Delta \overline{H}_f^\circ [N_2H_4(\ell)] = 12.05 \text{ kcal} \cdot \text{mole}^{-1} \text{ at } 298 \text{ K}$$

Proceeding in the above manner, one can construct a table of $\Delta \overline{H}_f^\circ$ values at 298 K (Appendix 5), which can in turn be used to calculate $\Delta H_{rctn}^\circ$ values. Such a table can be compiled for any convenient temperature, but

25°C (298.15 K) is the temperature that has been agreed upon for most of the major compilations of thermodynamic data. Such a table of $\Delta \bar{H}_f^\circ$ values eliminates the need to determine $\Delta H^\circ_{rctn}$ experimentally for every conceivable reaction of interest involving species with known $\Delta \bar{H}_f^\circ$ values. In addition, tables of $\Delta \bar{H}_f^\circ$ values enable us to compute and utilize $\Delta \bar{H}^\circ_{rctn}$ values for numerous chemical reactions that are difficult to carry out experimentally in a clean fashion, owing to the simultaneous involvement of numerous competing reactions and other complications, such as extreme reaction conditions. For example, in the petrochemical industry it is desirable to know the values of $\Delta H^\circ_{rctn}$ for reactions involving the conversion of straight-chain hydrocarbons to branched-chain hydrocarbons, because the former are abundant in crude oil, whereas the latter make better gasolines. Knowing the enthalpy change in a reaction gives the chemist vital information regarding suitable reaction conditions for maximizing the yield of the desired products. A reaction that falls into the above category is the conversion of *n*-octane to 2,2,4-trimethylpentane (*iso*-octane)

$$
\text{(1)} \quad CH_3CH_2CH_2CH_2CH_2CH_2CH_2CH_3(\ell) = CH_3\overset{\overset{\displaystyle CH_3}{|}}{\underset{\underset{\displaystyle CH_3}{|}}{C}}CH_2\overset{\overset{\displaystyle CH_3}{|}}{C}HCH_3(\ell)
$$

                n-octane                        iso-octane

This reaction cannot be carried out cleanly, but it is a matter of no great difficulty to carry out the combustion reactions:

$$
\text{(2)} \quad CH_3CH_2CH_2CH_2CH_2CH_2CH_2CH_3(\ell) + (25/2)\,O_2(g)
$$
$$
= 8CO_2(g) + 9H_2O(\ell)
$$

$$
\text{(3)} \quad CH_3\overset{\overset{\displaystyle CH_3}{|}}{\underset{\underset{\displaystyle CH_3}{|}}{C}}CH_2\overset{\overset{\displaystyle CH_3}{|}}{C}HCH_3(\ell) + (25/2)\,O_2(g) = 8CO_2(g) + 9H_2O(\ell)
$$

In this case we can compute $\Delta \bar{H}^\circ_{rctn(1)}$ directly from $\Delta \bar{H}^\circ_{rctn(2)}$ and $\Delta \bar{H}^\circ_{rctn(3)}$

$$
\Delta H^\circ_{rctn(1)} = \Delta H^\circ_{rctn(2)} - \Delta H^\circ_{rctn(3)} = 1.3\ \text{kcal} \quad (298\ \text{K})
$$

because $\Delta H$ is independent of path.

## 9.10   ESTIMATION OF $\Delta H^\circ_{rctn}$ FROM AVERAGE BOND ENERGIES.

An approximate method for the estimation of $\Delta H^\circ$ for a reaction, which is useful when insufficient $\Delta \bar{H}_f^\circ$ data are available, involves the use of bond energies (more precisely, bond enthalpies). For example, $\Delta H^\circ_{298} = 221.14$ kcal for the reaction

$$
H_2O(g) = O(g) + 2H(g)
$$

This reaction involves the breaking of two O—H bonds, and we can compute the *average* O—H "bond energy" in $H_2O$, $\epsilon_{O-H}$, at 298 K as follows:

$$\epsilon_{O-H} = \frac{\Delta H^\circ}{2} = 110.57 \text{ kcal (at 298 K)}$$

It is important to note that 110.57 kcal is *not* the enthalpy change for the reaction

$$H_2O(g) = OH(g) + H(g)$$

For this reaction, $\Delta H^\circ_{298} = 119.95$ kcal, and this is the value of the *bond dissociation energy* of an O—H bond in water. Note also that $\Delta H^\circ_{298} = 221.14 - 119.95 = 101.19$ kcal for the reaction

$$OH(g) = O(g) + H(g)$$

The dissociation energy of a given type of bond, say an O—H bond, depends on the species in which the bond occurs. Thus, the average O—H bond energy in H—O—H is not exactly, but only approximately, equal to the O—H bond dissociation energy in $CH_3OH$. However, as long as we are talking about bonds of a given order (single, double, and so forth), it is true that the bond energy of, for example, an O—H bond does not usually vary by more than about ±10% from one species to another.

By using the available data, one can build up a table of average bond energies. For example, from $\Delta \overline{H}^\circ_{f,298}$ data we can compute $\Delta H^\circ_{298}$ for the reaction.

$$\begin{array}{c} H \qquad\qquad H \\ \diagdown \qquad\quad \diagup \\ O\!-\!O \end{array} \; (g) = 2H(g) + 2O(g)$$

$$\Delta H^\circ_{298} = 2\,\Delta \overline{H}^\circ_{f,298}\,[\,H(g)\,] + 2\,\Delta \overline{H}^\circ_{f,298}\,[\,O(g)\,] - \Delta \overline{H}_{f,298}\,[\,H_2O_2(g)\,]$$

$$\Delta H^\circ_{298} = 2\,(52.095) + 2\,(59.553) - (-32.58) = 255.88 \text{ kcal}$$

$$\Delta H^\circ_{298} = 2\,\epsilon_{O-H} + \epsilon_{O-O} = 255.9 \text{ kcal}$$

and, since $\epsilon_{O-H} = 110.6$ kcal, we compute $\epsilon_{O-O} = 34.7$ kcal.

As an example of the use of bond energies, we shall estimate $\Delta H^\circ_{298}$ for the reaction

$$\begin{array}{cc} H \;\; H \\ | \;\;\; | \\ H\!-\!C\!-\!C\!-\!H(g) = H\;-\!C\!\equiv\!C\!-\!H(g) + 2H\!-\!H(g) \\ | \;\;\; | \\ H \;\; H \end{array}$$

given the following average bond energies: $\epsilon_{C\equiv C} = 199.6$; $\epsilon_{C-H} = 98.7$; $\epsilon_{H-H} = 104.2$; and $\epsilon_{C-C} = 82.6$ kcal. For the reaction we have

$$\Delta H^\circ_{298}(\text{rctn}) = 6\epsilon_{C-H} + \epsilon_{C-C} - 2\epsilon_{C-H} - \epsilon_{C\equiv C} - 2\epsilon_{H-H}$$

$$\Delta H^\circ_{298}(\text{rctn}) = 4\epsilon_{C-H} + \epsilon_{C-C} - \epsilon_{C\equiv C} - 2\epsilon_{H-H}$$

Note that we have to put in the energy necessary to break the bonds that

**TABLE 9.7**

AVERAGE BOND "ENERGIES" (298 K; kcal $\cdot$ mole$^{-1}$)*

$$C(s, graphite) = C(g); \Delta H_{298}^{\circ} = 172 \text{ kcal} \cdot \text{mole}^{-1}$$

| BOND | $\epsilon$ | BOND | $\epsilon$ | BOND | $\epsilon$ |
|---|---|---|---|---|---|
| H—H | 104 | C—C | 83 | C≡N | 210 |
| H—F | 135 | C=C | 147 | N—N | 38 |
| H—Cl | 103 | C≡C | 194 | N=N | 100 |
| H—Br | 88 | C—F | 105 | N≡N | 226 |
| O—O | 33 | C—Cl | 79 | N—H | 93 |
| O=O | 118 | C—Br | 66 | F—F | 37 |
| O—H | 111 | C—S | 62 | Cl—Cl | 58 |
| C—H | 99 | C=S | 114 | Br—Br | 46 |
| C—O | 84 | C—N | 70 | | |
| C=O | 170 | C≡N | 147 | | |

1 cal = 4.184 J

*From S. H. Maron and C. F. Prutton, *Principles of Physical Chemistry* (4th Edition, Macmillan Publishing Co., New York, 1965); p. 151.

are broken (+$\epsilon$), whereas we get back the energy from the bonds that are formed (−$\epsilon$). Using the above data, we compute

$$\Delta H_{298}^{\circ} = 4(98.7) + 82.6 - 199.6 - 2(104.2) = +69.4 \text{ kcal}$$

For comparison purposes, we now compute $\Delta H_{298}^{\circ}$(rctn) from the available $\Delta \overline{H}_{f,298}^{\circ}$ data:

$$\Delta H_{298}^{\circ}(\text{rctn}) = \Delta \overline{H}_{f,298}^{\circ} [\text{C}_2\text{H}_2(g)] + 2\Delta \overline{H}_f^{\circ} [\text{H}_2(g)] - \Delta \overline{H}_{f,298}^{\circ} [\text{C}_2\text{H}_6(g)]$$

$$\Delta H_{298}^{\circ}(\text{rctn}) = 54.194 + 2(0) - (-20.236) = +74.430 \text{ kcal}$$

Thus, the percentage error in the bond-energy estimate of $\Delta H^{\circ}$ is about $5 \times 100/74 = 6.7\%$. A collection of average bond energies is given in Table 9.7.

## 9.11   THE SECOND LAW OF THERMODYNAMICS.

The development of the second law of thermodynamics was hindered by innumerable attempts to make the principle of energy conservation serve the purposes of the entropy concept. Thermodynamics is more than "Energetics." The concept of energy is not sufficient to deduce the second law. In the words of Max Planck,*

"But when we pass from the consideration of the first law of thermodynamics to that of the second, we have to deal with a new fact, and it is evident that no definition, however ingenious, although it contain no contradiction in itself, will ever permit of the deduction of a new fact."

---

*Max Planck, *Thermodynamics* (Dover Reprint, 1945).

The first law of thermodynamics places no restriction on energy transfers other than that of conservation. It is known from experiment, however, that there are restrictions, in addition to conservation, on energy transfers. Some examples of such restrictions are:

(a) If two systems with temperatures $T_1$ and $T_2$ $(T_2 > T_1)$ are placed within an adiabatic enclosure and are allowed to interact through a diathermic wall, then the equilibrium temperature, $T_f$, of the combined system always is found to meet the condition $T_1 \leq T_f \leq T_2$.

(b) Work must be done on a system to effect the transfer of energy as heat from that system to surroundings which are at a temperature higher than that of the system. (You have to pay the power company to run your refrigerator and air conditioner.)

(c) If any system undergoes a process any part of which is irreversible, then the system cannot be restored to its original state without giving rise to a change in state of some part of its surroundings.

(d) Innumerable naturally-occurring (i.e., spontaneous) processes are unidirectional. Any isolated system proceeds spontaneously* towards a state of thermal, mechanical, and chemical equilibrium.

Paper will burn spontaneously in oxygen if ignited, but the reverse process, that is, the spontaneous recombination of the combustion products to the piece of paper and oxygen gas, has never been observed. There is no incantation that will cause a broken egg to reassemble itself spontaneously. Ice at any temperature above 273.15 K at 1 atm always melts spontaneously. Certain occurrences are inevitable. Things are never exactly the same today as they were yesterday. In the words of Omar Khayyam:†

The moving finger writes, and having writ moves on, and all your tears and wit will not suffice to change a word of it.

More subtle, but equally germane, is the observation that when a system that requires the specification of only two thermodynamic co-ordinates to determine its state undergoes a *reversible, adiabatic process*, we are free to pick only one of the thermodynamic coordinates of the final state. If our system is a given number of moles of gas of a fixed composition, then we can choose the final pressure of the gas in a reversible, adiabatic expansion, but once this is done the final temperature is determined, and vice versa. However, if the same system undergoes an *irreversible, adiabatic process* from the same initial state to a state with the same final pressure as in the reversible case, then the final temperature of the gas is not unique, but rather can have any one of the innumerable values depending on the details of the process.

The fact that we can select only the final temperature (or the final pressure) of a system that undergoes a reversible, adiabatic process implies the existence of a thermodynamic coordinate whose value is constant during this process. We shall call this coordinate *entropy* and give it the symbol $S$.

Fixing the temperature and the pressure of the initial state determines $S$, and if $S$ does not change in the reversible, adiabatic process, then when we specify the final pressure of the system, both $S$ and $P$ of the final state are determined. Fixing two of the thermodynamic coordinates of the sys-

---

*This statement carries no connotations regarding the rate of such processes. Spontaneous is not synonymous with immediate.

†Translated by Edward Fitzgerald.

tem fixes the values of all the other coordinates (such as $T$, $E$, $H$, and $V$), because we have only two independent coordinates.

Because $S$ is postulated to be a state function, the value of $\Delta S$ for any system undergoing any particular change in state must be independent of path; that is, the value of $\Delta S$ for any system depends only on the final and initial states

$$\Delta S = S_f - S_i$$

It is sufficient for our purposes to state the second law of thermodynamics in mathematical terms as follows:

(a) For any *adiabatic* process that a system undergoes,

$$\boxed{\Delta S_{sys} \geq 0} \qquad \text{(adiabatic)} \qquad (9.8)$$

where the equality sign holds if the process is *reversible,* and the inequality sign holds if the process is *irreversible.*

(b) For any *isothermal* process that a system undergoes,

$$\boxed{\Delta S_{sys} \geq q_{sys}/T} \qquad \text{(isothermal)} \qquad (9.9)$$

where the equality sign holds if the process is *reversible,* and the inequality sign holds if the process is *irreversible.* As can be seen from Equation (9.9), the dimensions of entropy are energy/deg. The most commonly employed entropy units are cal $\cdot$ K$^{-1}$ (or gibbs;[*] 1 gibbs = 1 cal $\cdot$ K$^{-1}$) and J $\cdot$ K$^{-1}$.

If we place two systems that are in thermal contact within an adiabatic enclosure, and regard one of the systems as functioning as the surroundings for the other system, then for any process taking place within the enclosure we have, using Equations (9.8) and (9.9),

$$\boxed{\Delta S_{total} = \Delta S_{sys} + \Delta S_{surr} \geq 0} \qquad (9.10)$$

where the equality sign applies if the process is reversible, and the inequality sign applies if the process is irreversible.

From Equation (9.10) it follows that *the total entropy is not conserved in an irreversible process. Whenever an irreversible process takes place there is a net entropy production.* All real processes are to some extent irreversible and, therefore, involve entropy production. Whenever a system undergoes an irreversible process, it is impossible to return the system to its initial state without leaving a change in state in some part of the surroundings. It is never possible to put everything back the way it was before the occurrence of an irreversible process. This is the essence of what is meant by an irreversible process.

Any spontaneous occurrence of a change in the state of an adiabatically enclosed system is sufficient to show that the system was not in a state of equilibrium before the spontaneous process took place. A system in an equilibrium state has exhausted its ability to undergo spontaneous processes. The occurrence of a spontaneous change in state leads to entropy

---

[*]Named for the American mathematician and physicist, J. Willard Gibbs.

production and, therefore, *in an equilibrium state the entropy of a system is a maximum* for the given energy and volume. In an equilibrium state there are no unbalanced forces and no temperature gradients.

A consideration of the entropy production associated with naturally occurring processes apparently led Rudolf Clausius to his famous statement of the first two laws of thermodynamics:

> The energy of the Universe is constant;
> the entropy is increasing to a maximum.

The reasoning involved in Clausius's proposition implies that the Universe is proceeding toward an equilibrium state. In such a state, no system could extract energy at the expense of some part of its surroundings and use it to perform work; no work can be done, and all life must cease.* Extrapolation of this reasoning back in time seems to imply that the Universe must have had a beginning. Scientifically speaking, however, it must be said that we really do not understand the operation of the Universe well enough to be certain that the second law is valid throughout the Universe, or whether or not it always was, and always will be, valid. In particular, the above argument assumes that the Universe is an isolated system. In fact, because everything that exists is (by definition) a part of the Universe, the concept of the Universe as an isolated system seems meaningless. According to Einstein, the Universe is finite but unbounded (like the surface of a sphere), and general relativity theory requires that the Universe pulsate — perhaps the second law holds only during an expansion period like the one we are now in. In any event, the second law is here now, and it is a powerful tool for use in the analysis of the behavior of matter that can be applied to macroscopic systems in the real world with essentially complete confidence.†

We shall now illustrate the entropy concept by means of some specific examples of entropy calculations. Suppose that a sample of 2.50 moles of an ideal gas at 300 K expands isothermally into a vacuum from an initial volume of $V_1$ to a final volume of $V_2 = 2V_1$. An ideal gas is a gas whose equation of state is $PV = nRT$, and whose internal energy (for a given mass of gas) depends only on the temperature, $E = E(T)$. During the expansion there is no restraining force on the expanding gas and, therefore, no work is done ($w = 0$). Furthermore, because the process is isothermal, $\Delta E_{gas} = 0$, and, therefore, by the first law $q = 0$; that is, the expansion is also adiabatic. What is the value of $\Delta S$ for the gas? The process is adiabatic and irreversible and, hence, using the second law (Equation (9.8)), we obtain $\Delta S_{gas} > 0$. There is no transfer of energy either as heat or work to the surroundings and, hence, $\Delta S_{surr} = 0$. Entropy is a state function and, therefore, the value of $\Delta S_{gas}$ for a given change in state is independent of the path. For this

---

*There is no immediate cause for alarm. The natural thermal death of the Universe is estimated to be at least ten to twenty billion years off. In any event, no earthlings will be around anyway, because our sun will exhaust itself long before numerous other stars have expired.

†It can be calculated from statistical-thermodynamic theory that the probability of observing a detectable ($\sim.003$ cal $\cdot$ K$^{-1}$) violation of the second law is less than the probability that a band of wild monkeys punching *randomly* on typewriters would type out the complete works of Shakespeare 15 quadrillion ($15 \times 10^{15}$) times in succession without a single error. (Quoted in *The Second Law*, H. A. Bent, New York, Oxford University Press, 1965.) The probability of observing a violation of magnitude $\Delta S$ is given by $e^{-\Delta S/k}$, where $k$ is Boltzmann's constant.

reason $\Delta S_{gas}$ for the irreversible expansion into a vacuum is equal to $\Delta S_{gas}$ for a reversible path *between the same initial and final states.* We allow the gas to expand reversibly at 300 K from $V_1$ to $V_2 = 2V_1$. This reversible process is isothermal, and therefore $\Delta E_{gas} = 0$, because the gas is ideal. From this result, together with the first and second laws, it follows that

$$q_{rev} = -w_{rev} = T\Delta S_{gas}$$

The work done on the gas in the *reversible* expansion is (Section 9.5)

$$w_{rev} = -nRT \ln \left(\frac{V_2}{V_1}\right) = -nRT \ln 2,$$

and therefore

$$\Delta S_{gas} = \frac{q_{rev}}{T} = nR \ln 2$$

$$\Delta S_{gas} = (2.50 \text{ mole})(1.987 \text{ cal} \cdot K^{-1} \cdot \text{mole}^{-1})(2.303)\log 2 = +3.45 \text{ cal} \cdot K^{-1}$$
$$= 3.45 \text{ gibbs}$$

As a second example, suppose that 2.50 moles of a real gas undergo a reversible, isothermal expansion at 500 K from an initial pressure of 50.0 atm to a final pressure of 1.00 atm. Let us also suppose that experimental measurements on the system during the expansion yield the following results for the energy exchanged as work and heat:

$$q = +10,107 \text{ cal} \qquad w = -9,803 \text{ cal}$$

The value of $w$ is negative because work is done by the gas on the surroundings, whereas the positive value of $q$ indicates that the heat flow is from the surroundings to the gas. From the first law we compute for the energy change of the gas

$$\Delta E_{gas} = q_{gas} + w_{gas} = +10,107 - 9,803 = 304 \text{ cal}$$

We also compute from the first law that the energy change in the surroundings is

$$\Delta E_{surr} = -\Delta E_{gas} = -304 \text{ cal}$$

From the second law we compute that the entropy change of the gas is (reversible, isothermal process)

$$\Delta S_{gas} = \frac{q_{gas}}{T} = \frac{10,107 \text{ cal}}{500 \text{ K}} = 20.2 \text{ gibbs}$$

whereas the entropy change for the surroundings is

$$\Delta S_{surr} = \frac{q_{surr}}{T} = \frac{-q_{gas}}{T} = \frac{-10,107 \text{ cal}}{500 \text{ K}} = -20.2 \text{ gibbs}$$

and $\Delta S_{tot} = \Delta S_{gas} + \Delta S_{surr} = 0$, as is required by the second law for a reversible process. If the gas is returned to its initial state, then for the return process, whatever its detailed nature, $\Delta E_{gas} = -304$ cal, $\Delta S_{gas} = -20.2$ gibbs, and $\Delta E_{surr} = +304$ cal. If the return process is reversible, then $\Delta S_{surr} = +20.2$ gibbs, whereas if it is irreversible, then $\Delta S_{surr} > +20.2$ gibbs.*

Suppose now that the 2.50 moles of gas in the state (500 K, 50 atm) expand irreversibly and isothermally into a vacuum to a final pressure of 1.00 atm. The initial and final states of the gas are the same here as in the reversible, isothermal expansion discussed above, and thus

$$\Delta E_{gas} = +304 \text{ cal} \qquad \Delta S_{gas} = +20.2 \text{ cal}$$

However, no work is done by the gas, and therefore

$$\Delta E_{gas} = +304 \text{ cal} = q_{gas}$$

From the second law we know that $\Delta S_{gas} + \Delta S_{surr} > 0$, and thus $\Delta S_{surr} > -20.2$ gibbs. Furthermore, from the second law,

$$\Delta S_{surr} \geq \frac{q_{surr}}{T} = \frac{-q_{gas}}{T} = \frac{-304 \text{ cal}}{500 \text{ K}} = -0.61 \text{ gibbs}$$

Therefore, the net entropy production in the irreversible expansion is

$$\Delta S_{gas} + \Delta S_{surr} \geq +20.2 - 0.61 = +19.6 \text{ gibbs}$$

Note that the reverse of the above process (that is, the spontaneous isothermal contraction of the gas from 1.00 atm to 50.0 atm with $w = 0$) is impossible. Such a process is impossible because it would lead to a value of $\Delta S_{tot} \leq -19.6$ gibbs.† To return the gas to its initial state, work must be done on the gas, because the return process must meet the condition $\Delta S_{tot} \geq 0$, or, in the special case under consideration, $\Delta S_{surr} \geq +20.2$ gibbs.

## 9.12 THE THIRD LAW OF THERMODYNAMICS.

The heat absorbed by a system in a *reversible, isothermal* process is given by the second law as

$$q_{rev} = T\Delta S$$

where $\Delta S$ is the change of entropy of the system. If we cause some process (e.g., a phase transformation or a chemical reaction) to take place under reversible, isothermal conditions, and determine $q_{rev}$ calorimetrically, we can compute $\Delta S$. For example, we can determine the amount of heat input necessary to melt a known mass of ice at 1 atm and 273.15 K, and thereby obtain the entropy difference between liquid and solid water. Many

---

*This result shows that in the irreversible compression to the state (500 K, 50 atm), more heat is given up by the gas to the surroundings than in the reversible case. Since $q_{gas} < -10,107$ cal and $\Delta E_{gas} = -304$, $w_{gas} > +9,803$ cal, and therefore, more work must be done to effect the given change in state in the irreversible process than in the reversible one.

†That such a process is impossible is also obvious from experience (gases do not spontaneously contract to leave an evacuated space), but then the second law is based on experience.

chemical reactions can be made to take place reversibly and isothermally in an electrochemical cell. We can place the cell in a calorimeter and measure $q_{rev}$ for the cell under reversible discharge, thereby obtaining $\Delta S$ for the cell reactions at that temperature.[*]

A consideration of $\Delta S$ values for a wide variety of isothermal processes (including chemical reactions) leads to the observation that, as the temperature at which the isothermal process is carried out decreases, the value of $\Delta S$ for the process decreases, and approaches zero as the temperature approaches zero. These results were noted by Walther Nernst, and he introduced (1906) the general postulate that for any isothermal process

$$\lim_{T \to 0} \Delta S = 0 \qquad\qquad (9.10)$$

The Nernst postulate requires that the entropy change for any process (including all chemical reactions) must attain a value of zero in the limit of zero absolute temperature, that is, $\Delta S = 0$ at $T = 0$.

The entropy change of a substance whose temperature is changed at constant pressure can be obtained from heat capacity data. The results of such experiments show that the entropy of a substance always decreases as the temperature of the substance decreases. These observations led Max Planck (1911) to extend Nernst's ideas by postulating that the entropy for any substance must become equal to zero at absolute zero, that is, $S = 0$ at $T = 0$.

Additional heat capacity measurements on a wide variety of substances showed that both the Nernst and Planck postulates had to be qualified. The qualification amounted to the exclusion from the postulates of those substances (such as glasses, strained solids, solid solutions, and certain crystals with randomly oriented subunits) that retained internal disorders down to the lowest temperatures of the heat capacity measurements.

Gilbert N. Lewis and Merle Randall (1923) put forth a qualified version of Planck's postulate that has withstood the onslaught of experimentation and that constitutes a statement of the third law of thermodynamics.[†] *If the entropy of each element in some crystalline state be taken as zero at the absolute zero of temperature, then every substance has a finite positive entropy; but at the absolute zero of temperature the entropy may become zero, and does so become in the case of perfect crystalline substances.*[‡]

The third law of thermodynamics forms the basis of the *practical absolute entropy scale.* On this scale the entropy of a perfect crystalline sub-

---

[*]There are other methods that also yield $\Delta S$ values for reactions. We shall postpone a discussion of these methods until Chapters 11, 12, and 13.

[†]An alternate statement of the third law is that *absolute zero is unattainable.* The lowest *bulk-lattice* temperature so far obtained in the laboratory is $10^{-3}$ K.

[‡]A "perfect crystalline substance" is one that does not retain any significant internal disorders at the lowest temperature of the heat capacity measurements. In many cases it is possible to correct the measured entropy for internal disorder that persists in the crystal to very low temperatures (see reference 9.9).

stance is equal to zero at the absolute zero of temperature.[*] Also, the entropy of any substance is positive for all $T > 0$, and increases with increasing temperature.

As mentioned above, third-law entropies can be obtained from heat-capacity measurements. In Figure 9.9 we have plotted the calorimetrically determined third-law entropy of oxygen (at 1 atm pressure) as a function of temperature. The vertical jumps on the graph correspond to entropy increments associated with phase transitions (e.g., melting and vaporization). Figure 9.9 illustrates the increasing value of the entropy associated with increasing temperature.

If the third-law entropies of all but one of the substances involved in the reaction are known, then the value of $S°$ for that substance can be computed from the value of $\Delta S°$ and the known third-law entropies for the other substances. In Appendix 5 we have tabulated $\bar{S}°_{298}$ values for some common substances. The choice of standard states is the same as that discussed in Section 9.9d. To illustrate the use of the data in Appendix 5, we have computed $\Delta S°_{298}$ values for some reactions:

1) $CH_4(g) + 2O_2(g) = CO_2(g) + 2H_2O(g)$

$\Delta S°_{298} = \bar{S}°_{298} [CO_2(g)] + 2\bar{S}°_{298} [H_2O(g)] - \bar{S}°_{298} [CH_4(g)] - 2\bar{S}°_{298} [O_2(g)]$

$\Delta S°_{298} = 51.06 + 2(45.104) - 44.492 - 2(49.003) = -1.23$ gibbs

2) $C(s, graphite) + O_2(g) = CO_2(g)$

$\Delta S°_{298} = \bar{S}°_{298} [CO_2(g)] - \bar{S}°_{298} [C(s)] - \bar{S}°_{298} [O_2(g)]$

$\Delta S°_{298} = 51.06 - 1.372 - 49.003 = 0.68$ gibbs

3) $Mg(s) + (1/2)O_2(g) = MgO(s)$

$\Delta S°_{298} = \bar{S}°_{298} [MgO(s)] - \bar{S}°_{298} [Mg(s)] - (1/2)\bar{S}°_{298} [O_2(g)]$

$\Delta S°_{298} = 6.55 - 7.81 - (1/2)(49.003) = -25.76$ gibbs

4) $3C_2H_2(g) = C_6H_6(\ell)$

$\Delta S°_{298} = 41.30 - 3(47.997) = -102.69$ gibbs

5) $H_2O(\ell) = H_2O(g)$

$\Delta S°_{298} = 45.104 - 16.71 = 28.39$ gibbs

6) $I_2(s) = I_2(g)$

$\Delta S°_{298} = 62.28 - 27.76 = 34.52$ gibbs

7) $N_2H_4(\ell) + O_2(g) = N_2(g) + 2H_2O(g)$

$\Delta S°_{298} = 2(45.104) + 45.77 - 28.97 - 49.003 = 58.02$ gibbs

---

[*]Nuclear entropy (that is, entropy arising from randomness in the nuclei) is ignored on the practical absolute entropy, and this is the reason for the designation of this entropy scale as *practical*. This neglect of nuclear entropy is of no consequence in balanced chemical reactions, because the nuclei are conserved in such cases, and therefore there is no net nuclear entropy change in the reaction. Nuclear entropy persists in a crystal down to very low temperatures ($\leq 10^{-3}$ K).

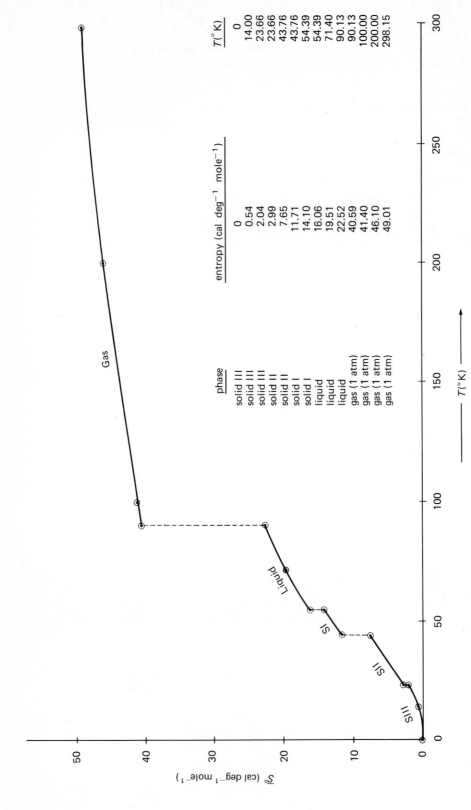

**FIGURE 9.9**   Entropy of oxygen as a function of temperature (at 1 atm total pressure); vertical jumps correspond to phase transitions.

| phase | entropy (cal deg$^{-1}$ mole$^{-1}$) | $T(^\circ$K) |
|---|---|---|
| solid III | 0 | 0 |
| solid III | 0.54 | 14.00 |
| solid III | 2.04 | 23.66 |
| solid II | 2.99 | 23.66 |
| solid II | 7.65 | 43.76 |
| solid I | 11.71 | 43.76 |
| solid I | 14.10 | 54.39 |
| liquid | 16.06 | 54.39 |
| liquid | 19.51 | 71.40 |
| liquid | 22.52 | 90.13 |
| gas (1 atm) | 40.59 | 90.13 |
| gas (1 atm) | 41.40 | 100.00 |
| gas (1 atm) | 46.10 | 200.00 |
| gas (1 atm) | 49.01 | 298.15 |

## 9.13   A DIGRESSION ON THE PHYSICAL INTERPRETATION OF ENTROPY.

The purpose of this section is to give the entropy concept a more intuitive meaning. The first, and most important, point to recognize is that *entropy is not a thing,* any more than is energy. To say something is not a thing, however, is not the same as saying that it has no physical significance.

Entropy was described by the thermodynamicist J. Willard Gibbs as a measure of the "mixed-up-ness" of a system; the more disordered (or randomized) a system is, the higher is its entropy. Things left to themselves proceed to a state of maximum possible disorder (entropy), subject to certain constraints such as fixed energy and volume. Constructions fall apart and living matter, upon death, rots. The existence of life, which spontaneously builds itself up into a more ordered state, on first sight seems to be at variance with the second law and, in fact, some scientists do not think that the second law is applicable to living systems. Any living system, however, builds itself up only at the expense of degrading its environment. All living systems are open systems. Very little, if any, life could exist on the earth if it were not for the sun's radiation continually impinging upon the earth's surface. This spontaneous radiation production by the sun is accompanied by a tremendous entropy production. The second law does not say that the entropy of a system can never spontaneously decrease—what it says is that the entropy of any system (or collection of systems) *in an adiabatic enclosure* can never spontaneously decrease. There is a big difference between a system that is *open* (that is, a system that can exchange energy and matter with its environment), and a system that is adiabatically enclosed (cannot exchange matter or energy as heat) or isolated (cannot exchange either matter or energy). A skeptical scientist placed in an adiabatic enclosure (e.g., a thick-walled, hermetically sealed plastic box) would soon change his mind about the applicability of the second law (correctly formulated) to living systems.

Compared at the same pressure and temperature, liquids are more disordered than solids, and gases are more disordered than liquids. This is because the molecules of a gas have more freedom to move around than those in a liquid, and similarly for the molecules of a liquid as compared to a solid. The melting of a solid is accompanied by an entropy increase of 2 to 9 gibbs per mole for the substance, and the vaporization of a liquid by an increase of 18 to 25 gibbs per mole (see Chapter 10).

As we increase the temperature of a system, keeping the pressure constant, the entropy always increases. As the temperature increases, the particles increase the extent of their motion, and there is a corresponding loss of information as to the precise location of all the particles at any instant—the substance is more disordered. The isothermal expansion of a gas always gives rise to an entropy increase of the gas. The molecules of the gas have greater volume available to move around in, and consequently, the configurational randomness of the gas is greater. An interesting and simple experiment, which demonstrates the above point of regarding entropy increases as increases in randomness or disorder, is the following: take a rubber band (the larger the better) and grasp it firmly with both hands between the thumbs and first fingers with the thumbs close together. Now moisten your upper lip with your tongue, and quickly stretch the rubber band taut,* placing the stretched section to the moistened lip. If this is

---

*Be careful not to stretch the rubber band so tightly that it breaks, because you might hurt yourself.

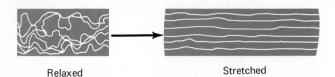

Relaxed                              Stretched

FIGURE 9.10 The relaxed and stretched states of rubber.

done quickly enough, you will note that the rubber band is warm. If you keep the band tightly stretched until it no longer feels warm, then let it relax quickly, and place it to your moistened lip, you will note that the band feels cool. The long polymer molecules in rubber in the relaxed condition are in a very randomly-oriented, kinky-coiled configuration (see Figure 9.10). When the band is stretched, the long polymer molecules become oriented in a more-or-less-parallel configuration.[*] In the spontaneous, adiabatic relaxation of the band, $\Delta S_{band} > 0$, and the temperature of the band decreases, because of the conversion of internal thermal (or kinetic) energy into internal potential energy. If we stretch the band isothermally, then as a consequence of the increased order $\Delta S_{band} = \dfrac{q_{rev}}{T} < 0$; therefore, $q_{rev} < 0$, and energy is given off as heat. If we relax the band isothermally, $\Delta S_{band} = \dfrac{q_{rev}}{T} > 0$, and energy is absorbed as heat.

All macroscopic spontaneous processes can be visualized as mixing processes, of which there are two distinct types (E. A. Guggenheim):

(a) mixing or "spreading" of the particles over positions in space ("configurational randomness"); and

(b) mixing or "spreading" of the available energy of a system over the particles themselves ("thermal randomness").

When two or more substances spontaneously mix to form a homogeneous solution, the entropy of the mixture is higher than the sum of the entropies of the separated components, because the particles in the solution are spread over a larger number of positions in space. The spontaneous process of temperature equalization, which occurs when two bodies at different temperatures are brought into thermal contact, can be visualized as a mixing of the available energy more completely over the combined system. In some cases configurational and thermal randomness factors oppose each other. This situation arises in the case of two immiscible substances like water and oil. Partial (or total) immiscibility occurs when the intermolecular interaction energy between unlike molecules is much different from that between like molecules. In such a case, although complete mixing would lead to increased configurational randomness, it could be accomplished only at the expense of a decrease in thermal randomness of the system. That is to say, some internal kinetic energy would have to be converted to internal potential energy if the particles are to be placed in energetically less favorable positions.

The interpretation of entropy as a measure of disorder, or randomness, or mixed-up-ness of a system can also be applied to chemical reactions (see the $\Delta S°$ values for the reactions on page 265). For reactions involving only pure phases of a particular type (i.e., all solids, all liquids, or all gases)

---

[*] X-ray studies on rubber show that the stretched material is definitely more crystalline (ordered) than the unstretched material. The stretched rubber gives an x-ray diffraction pattern, whereas the unstretched rubber does not.

we expect $\Delta S° > 0$ if the number of moles of products exceeds the number of moles of reactants, and $\Delta S° < 0$ if the reverse is true; similarly, we expect $\Delta S° \simeq 0$ if the number of moles of reactants and products is the same. This is generally, but not always, observed. Some exceptions arise because the prediction is based only on configurational factors. If the reaction involves both gaseous and condensed phases, then we expect $\Delta S° > 0$ if there are more moles of gas among the products than among the reactants, and $\Delta S° < 0$ if the reverse is true. The interpretation of $\Delta S°$ values for homogeneous reactions in solution, or heterogeneous reactions involving solution phases, is not as simple as the above types, because of the involvement of the solvent.

_____ **PROBLEMS**

1. Calculate the work done (in joules, calories, ergs, and $\ell \cdot$ atm) in lifting a 150 pound man 1 mile off the surface of the earth ($g = 980$ cm $\cdot$ sec$^{-2}$). Repeat the calculation (in joules) for lifting the man 1 mile off the surface of the moon ($g_{\text{moon}} = (1/6)g_{\text{earth}}$).

2. Calculate the work done at 300 K when a spring, which is extended to four times its equilibrium length ($L_0$), is suddenly released and allowed to return to its equilibrium length.

3. Why does a piece of metal at 70°F feel cooler to the touch than a piece of cardboard at the same temperature?

4. Suggest two types of thermometers in addition to those mentioned in the text.

5. (a) Derive the expression

$$t(°C) = \frac{5}{9} \{t(°F) - 32\}$$

   (b) What temperature on the Fahrenheit scale corresponds to absolute zero?

6. (a) Derive a relationship between $\Delta H$ and $\Delta E$ in which neither $P$ nor $V$ appears for a chemical reaction involving only ideal gases.

   (b) Derive a general relationship between $\Delta H$ and $\Delta E$ for a reaction carried out at constant pressure.

   (c) Compare the magnitude of the $P\bar{V}$ product for $H_2O(g)$ with that for $H_2O(s)$ and $H_2O(\ell)$ at 273 K and 1 atm (Take $\rho_\ell = 1.00$ and $\rho_s = 0.91$ g $\cdot$ cm$^{-3}$).

7. The Law of Dulong and Petit states that around 300 K and above, the total heat capacity of a solid containing $n$ atoms per formula unit is roughly given by $6.0 \, n$ cal $\cdot$ K$^{-1} \cdot$ mole$^{-1}$. Using this relation, compute $n$ for the solids given in Table 9.2 and compare with the actual $n$ values.

8. Explain how you might use the second law of thermodynamics to distinguish a dream from conscious observation.

9. The solubility of silver chloride in water at 25°C is only $1.7 \times 10^{-5}$ mole · liter$^{-1}$, and therefore $\Delta H$ for the dissolution of AgCl in water cannot be directly measured. Suggest a calorimetric method for the determination of the heat of solution of AgCl in water and show how $\Delta H_{soln}$ can be obtained from your method.

10. Suggest an explanation for the difference in the $\Delta H°$ values for the reactions (gaseous species)

$$H_2O = HO + H \qquad \Delta H°_{298} = 120 \text{ kcal}$$
$$HO = H + O \qquad \Delta H°_{298} = 101 \text{ kcal}$$

(Hint: Consider the bonding in the species HOH and OH.)

11. Using the table of bond energies given in the text, estimate $\Delta H°_{298}$ for the following reactions, and compare (where possible) your results with $\Delta H°_{298}$ calculated from data in Appendix 5:

a) $CO(g) + 2H_2(g) = CH_3OH(g)$

b) $CO(g) + 2H_2(g) = CH_3OH(\ell)$

c) $CH_4(g) + 4Cl_2(g) = CCl_4(g) + 4HCl(g)$

d) $CH_3CH_3(g) + (7/2)O_2(g) = 2CO_2(g) + 3H_2O(g)$

e) $CH_3CH = CH_2(g) + Cl_2(g) = CH_3CHCH_2Cl(g)$
$\qquad\qquad\qquad\qquad\qquad\qquad\qquad\quad |$
$\qquad\qquad\qquad\qquad\qquad\qquad\qquad\quad Cl$

f) $N_2H_4(\ell) + 2H_2O_2(\ell) = N_2(g) + 4H_2O(g)$

g) $CH_4(g) + Cl_2(g) = CH_3Cl(g) + HCl(g)$

h) $CH_4(g) + Cl_2(g) = CH_3Cl(\ell) + HCl(g)$

i) $C_2N_2(g) + O_2(g) = 2CO(g) + N_2(g)$

j) $CH_3CH_3(g) + 2Cl_2(g) + 2F_2(g) = Cl_2CHCHF_2(\ell) + 2HCl(g)$
$\qquad\qquad\qquad\qquad\qquad\qquad\qquad\qquad\qquad + 2HF(g)$

k) $H_2C = CH_2(g) + H_2(g) = CH_3CH_3(g)$

12. Using data in Appendix 5, compute $\Delta H°$ and $\Delta S°$ values for the following reactions at 298 K:

a) $CO(g) + (1/2)O_2(g) = CO_2(g)$

b) $2Na(s) + O_2(g) = Na_2O_2(s)$

c) $0.947 Fe(s) + (1/2)O_2(g) = Fe_{0.947}O(s)$

d) $Hg_2Cl_2(s) + Br_2(\ell) = Hg_2Br_2(s) + Cl_2(g)$

e) $PCl_5(g) + 2NO(g) = 2NOCl(g) + PCl_3(g)$

f) $SO_2(g) + NO_2(g) = NO(g) + SO_3(g)$

g) $C_6H_6(\ell) + (15/2)O_2(g) = 6CO_2(g) + 3H_2O(g)$

h) $3Fe_2O_3(s) = 2Fe_3O_4(s) + (1/2)O_2(g)$

13. Suggest a method for measuring $\bar{H}_{600} - \bar{H}_{300}$ for a solid.

14. Suppose a gas at 600 K and 10 atm expands irreversibly and adiabatically to 1 atm and 300 K, and then is reversibly returned to 600 K and 10 atm. Compute the values of $\Delta S_{gas}$ and $\Delta E_{gas}$ for the complete cycle. What can you say about the values of $\Delta S_{tot}$ and $\Delta E_{tot}$?

15. What is the relation between the amount of energy discharged as heat into the surroundings by a refrigerator, the energy required to run the refrigerator, and the amount of energy removed as heat from the inside of the refrigerator? Comment on the effect of large numbers of air conditioners on the outside air temperature of a city on a hot, windless summer day.

16. Can you devise an isothermal process for which $\Delta S_{sys} < q_{sys}/T$?

17. Calculate the entropy changes in problems 1 and 2. Take $F = (1 \times 10^{-3} \text{ dyne} \cdot \text{cm}^{-1})(L - L_0)$ for the spring.

18. (a) Assuming He(g) behaves ideally, compute $\Delta H$ and $\Delta S$ for 2.4 moles of He(g) that expand isothermally and reversibly from 10.0 atm to 1.0 atm at 300 K. Also compute $\Delta S$ for the surroundings in this process. (b) Suppose 2.4 moles He(g) at 300 K are adiabatically and reversibly compressed from 1.0 atm to 10.0 atm. Compute $\Delta S$ for the gas and $\Delta S$ for the surroundings.

19. Arrange the following reactions in terms of increasing $\Delta S°$ values, and then check your predicted order using data in Appendix 5:

a) $CO(g) + 2H_2(g) = CH_3OH(\ell)$

b) $H_2O_2(\ell) = H_2O(\ell) + (1/2)O_2(g)$

c) $CH_4(g) + 4Cl_2(g) = CCl_4(\ell) + 4HCl(g)$

d) $H_2(g) + (1/2)O_2(g) = H_2O(\ell)$

e) $SF_6(g) = SF_4(g) + F_2(g)$

f) $n\text{-}C_4H_{10}(\ell) + (13/2)O_2(g) = 4CO_2(g) + 5H_2O(g)$

g) $n\text{-}C_4H_{10}(\ell) = i\text{-}C_4H_{10}(\ell)$

h) $2Fe(s) + (3/2)O_2(g) = Fe_2O_3(s)$

i) $3Fe(s) + 2O_2(g) = Fe_3O_4(s)$

20. A current of 1.3 amps flows for 35 sec through a 5.0 ohm resistor with $C_P = 100$ cal $\cdot$ deg$^{-1}$, maintained at 300 K by circulating water. Calculate $\Delta S$ for the resistor and $\Delta S$ for the water. Electrical energy (delivered as heat) is $i^2 Rt$, where $i$ is current, $R$ is resistance, and $t$ is time.

21. Outline how you would go about determining the S—S bond energy in $CH_3$—S—S—$CH_3$.

22. Using the data given in Table 9.4, compute $\Delta H°$ for the reactions

$$CH_3COOH(aq) + OH^-(aq) = CH_3COO^-(aq) + H_2O(\ell)$$

$$H_3PO_4(aq) + OH^-(aq) = H_2PO_4^-(aq) + H_2O(\ell)$$

$$H_2PO_4^-(aq) + OH^-(aq) = HPO_4^{2-}(aq) + H_2O(\ell)$$

Rationalize your results.

23. Suppose we have two identical Hooke's Law springs, one of which is at its equilibrium length and the other of which is held at twice its equilibrium length. Suppose now that each of these springs is dissolved in a given amount of acid. By how much, if at all, will the energy evolved as heat differ for the two springs? (Note that the force required to stretch a Hooke's Law spring is given by $F = k(L - L_0)$, where $k$ is a constant and $L_0$ is the equilibrium length.) Assume that the force constant for the spring is temperature independent. Will the results be different if the force constant is temperature dependent?

24. Suppose 3.50 moles of $CH_4(g)$ are mixed with 20.0 moles of $O_2(g)$ in an adiabatic enclosure at 298 K. A spark is produced in the mixture and the $CH_4$ is completely burned to $CO_2(g)$ and $H_2O(g)$. Assume ideal-gas behavior and compute the final temperature of the gas mixture. See Table 9.2 and Appendix 5 for the necessary data. (Assume constant $C_P$ values.)

25. Assuming He(g) behaves ideally, and has $\bar{C}_P = 4.97$ cal $\cdot$ K$^{-1} \cdot$ mole$^{-1}$, compute $\Delta E$, $\Delta H$, and $\Delta S$ for 2.5 moles of He(g) that undergo the change of state

$$He(g)(1000 \text{ K}, 5 \text{ atm}) \rightarrow He(g)(400 \text{ K}, 0.10 \text{ atm})$$

*irreversibly*. What can you say about $\Delta S$ for the surroundings in this process?

26. Energy is sometimes defined as the "capacity to do work." What do you think of this definition—is it meaningful?

27. Calculate $\Delta S_{tot}$ when a 1 kg mass at 300 K drops freely a distance of 10 meters into a heat bath at 300 K.

28. Can the second law of thermodynamics be used to determine which of two events took place first? If so, under what conditions?

# References

9.1 L. G. Rubin, *Temperature-Concepts Scales and Measurement Techniques* (a Leeds and Northrup Company reprint of Tech. Memo T-538 issued by the Research Division of Raytheon Co.).

9.2 M. W. Zemansky, *Temperatures Very Low and Very High* (Van Nostrand Reinhold Co., New York, 1964).

9.3 P. W. Bridgman, *The Nature of Thermodynamics* (Harper Torchbook Reprint, 1941).

9.4 S. W. Angrist and L. G. Hepler, *Order and Chaos* (Basic Books, Inc., New York, 1967).

9.5 H. A. Bent, *The Second Law* (Oxford University Press, New York, 1965).

9.6 A. L. Lehninger, *Bioenergetics* (W. A. Benjamin, Inc., New York, 1965).

9.7 M. Planck, *Treatise on Thermodynamics*, 3rd ed., 1926 (Dover Publications, Inc., New York, 1945).

9.8 L. K. Nash, *Introduction to Chemical Thermodynamics* (Addison-Wesley Publishing Co., Reading, Mass., 1969).

9.9 P. A. Rock, *Chemical Thermodynamics: Principles and Applications* (The Macmillan Co., Collier-Macmillan Limited, London, 1969).

## CHAPTER 10

# PHASE EQUILIBRIA

There was a sound like that of the gentle closing of a portal as big as the sky, the great door of heaven being closed softly. It was a grand AR-WHOOM.
I opened my eyes— and all the sea was ice-nine.
The moist green earth was a blue-white pearl.

*Kurt Vonnegut, Jr.\**

## 10.1 THE GIBBS ENERGY MINIMUM AS AN EQUILIBRIUM CRITERION.

The processes of primary interest in chemical thermodynamics are those that take place at constant temperature and pressure. To facilitate the thermodynamic analysis of isothermal, isobaric processes, a new thermo-dynamic coordinate, called the *Gibbs energy*,† $G$, has been defined:

$$G = H - TS = E + PV - TS \qquad (10.1)$$

If the system of interest undergoes an isothermal, isobaric change in its thermodynamic state, then

$$\Delta G = \Delta E + P\Delta V - T\Delta S$$

Combining this result with the first law, we obtain

$$\Delta G = q + w + P\Delta V - T\Delta S$$

If only pressure-volume work is involved, then the work, $w$, done on the system is $w = -P\Delta V$, and therefore

$$\Delta G = q - T\Delta S$$

Furthermore, from the second law, $\Delta S \geqq q/T$ (isothermal process), and thus,

$$\Delta G \leqq 0 \qquad (10.2)$$

---

\*K. Vonnegut, Jr., *Cat's Cradle* (Delacorte Press, Dell Publishing Co., Inc., New York, 1971).

† This function is named after the thermodynamicist J. Willard Gibbs.

The equality sign in Equation (10.2) applies if the process is reversible, and the inequality sign applies if the process is spontaneous and irreversible. At equilibrium, no spontaneous process can occur, and therefore our equilibrium criterion for a system of fixed mass, composition, temperature, and pressure is* $\Delta G = 0$. In other words, at constant temperature and pressure, the Gibbs energy is a minimum at equilibrium.

The molar Gibbs energy

$$\bar{G} = G/n = \frac{1}{n} (H - TS) = \bar{H} - T\bar{S} \tag{10.3}$$

of a fixed-composition phase in an equilibrium state is completely determined (in the absence of external force fields and surface effects) by the specification of its temperature and pressure.

## 10.2   THE PHASE RULE.

In order to apply thermodynamic theory to any system we must know the minimum number of thermodynamic variables, including those defining the composition, that must be specified to determine completely an equilibrium state of the system. The solution to this problem, for the general case of any number of phases involving any number of components, was obtained by J. Willard Gibbs and is known as the *Gibbs phase rule*. All that is required for the application of the phase rule is the ability to enumerate the phases and the components of a system.

### a. Phases.

*Each physically or chemically distinct, homogeneous* (i.e., uniform throughout in chemical composition and physical state), *and mechanically separable part of a system is called a phase.* A phase need not be continuous in a physical sense; e.g., a mass of crystalline solid constitutes only one phase no matter how many individual crystals are involved, and likewise with a number of droplets of a particular liquid versus a physically continuous mass of the same liquid. Following are some examples of phase enumeration:

(a) A system composed of a saturated aqueous solution of NaCl (with excess solid present) in a closed, air-free vessel contains the phases
    (1) NaCl(s),
    (2) aqueous solution of NaCl, and
    (3) water vapor.

(b) A system composed of ice in contact with air-saturated water contains the phases
    (1) $H_2O(s)$ saturated with air,
    (2) $H_2O(\ell)$ saturated with air, and
    (3) air saturated with $H_2O(g)$

(c) A liquid solution of ethyl alcohol and water in a sealed vial with a vapor space contains the phases

---

* Equation (10.2) is not applicable if other types of work are involved. For example, with an electrochemical cell operating at fixed $T$, $P$, and voltage, $\mathscr{E}$, we obtain ($w = -P\Delta V - \mathscr{E}Z$)

$$\Delta G \leq -\mathscr{E}Z$$

where Z is the quantity of charge that passes through the cell with voltage $\mathscr{E}$.

      (1) liquid solution of ethanol and water, and
      (2) gaseous solution of ethanol and water.
(d) A mixture of carbon tetrachloride and water in a sealed vial with a vapor space contains the phases
      (1) $CCl_4(\ell)$ saturated with $H_2O$,
      (2) $H_2O(\ell)$ saturated with $CCl_4$, and
      (3) a gaseous solution of $CCl_4$ and $H_2O$.

## b. Components.

The number of *components, C,* of a system is equal to the maximum number of independently variable composition variables within the system. We can compute the number of components from the expression

$$C = N - r \qquad (10.4)$$

where $N$ is the total number of chemically distinct constituents, and $r$ is the number of restrictive conditions (that is, $r$ is the number of independent algebraic relationships among the composition variables). Following are some examples of component enumeration:

    (a) A homogeneous solution of carbon tetrachloride ($CCl_4$) and benzene ($C_6H_6$) contains $C = 2 - 0 = 2$ components.

The significance of the result 2 in this case is that we can vary at will the number of moles of $CCl_4$ in the solution while holding the number of moles of $C_6H_6$ constant, and vice versa.

    (b) In the nitrogen-hydrogen-ammonia system, several possibilities arise:

    (i)   In any gaseous mixture of $N_2$, $H_2$, and $NH_3$ around 25°C (1 phase), where the gases do not react with one another, we have $C = 3 - 0 = 3$.

    (ii)  In any gaseous mixture of $N_2$, $H_2$, and $NH_3$ at high temperature where the equilibrium *

$$N_2(g) + 3H_2(g) = 2NH_3(g) \qquad \text{(1 phase)}$$

        is rapidly established, we have $C = 3 - 1 = 2$. The number of components is reduced by one, because of the reaction equilibrium.

    (iii) In any gaseous system obtained by heating pure $NH_3(g)$ to a temperature where the equilibrium in (ii) is rapidly established, we have in addition the relationship

$$P_{H_2} = 3P_{N_2}$$

        which arises because of the special way in which the system was prepared, and therefore $C = 3 - 2 = 1$.

---

  *As we shall see in Chapter 11, when a chemical reaction attains an equilibrium state, there exists an algebraic relationship among the composition variables of the reactants and products, that is, the number of independent composition variables is reduced by one, for each independent equilibrium reaction.

(c) A solution of acetic acid in water contains the following constituents:[*]

$CH_3COOH(aq)$, $CH_3COO^-(aq)$, $H_2O(\ell)$, $H^+(aq)$, and $OH^-(aq)$   (1 phase)

The restrictive conditions are

$$CH_3COOH(aq) = CH_3COO^-(aq) + H^+(aq)$$

$$H_2O(\ell) = H^+(aq) + OH^-(aq)$$

$$(H^+) = (CH_3COO^-) + (OH^-)$$

and, thus, $C = 5 - 3 = 2$. The first two of these restrictive conditions are the weak-acid-dissociation equilibria, whereas the third (where parentheses around a species represent the concentration of that species) is the *electroneutrality condition*. The electroneutrality condition must always be included among the restrictive conditions for any electrolyte solution, because all electrolyte solutions are electroneutral. The electroneutrality condition expresses the requirement that for any electrolyte solution the amount of positive charge per unit volume must equal the amount of negative charge per unit volume. We know that electrolyte solutions are electroneutral because there is no detectable electric field outside of the solutions.

(d) The thermal decomposition of zinc carbonate leads to the formation of zinc oxide and gaseous carbon dioxide

$$ZnCO_3(s) = ZnO(s) + CO_2(g)   \text{(2 solid phases and a gas phase)}$$

and $C = 3 - 1 = 2$.

In this case $C$ remains equal to 2 even if the system is prepared solely from pure $ZnCO_3(s)$. This is because $ZnO(s)$ and $CO_2(g)$ are formed in different phases, and therefore a single composition variable is inadequate for describing the composition of each of the three phases.

## c. Degrees of Freedom.

Gibbs showed that the number of *independent* thermodynamic coordinates (or degrees of freedom), $F$, that must be specified to determine completely the equilibrium state of a system composed of $\phi$ phases and $C$ components is given by[†]

$$\boxed{F = C - \phi + 2}$$   (The Phase Rule)   (10.5)

Thus, for a one-component system, $F = 3 - \phi$. If there is only one phase (e.g., $H_2O(\ell)$), then $F = 2$, and both the temperature and the pressure must be specified before the state of a pure phase is determined. If there

---

[*]See, however, problem 26.

[†]There are $C + 2$ variables (namely, the $C$ components, the temperature, and the pressure), and there is a necessary restriction on the simultaneous variations of these $C + 2$ variables in each one of the $\phi$ phases of the system. Thus, there are $C + 2 - \phi$ independent variables in all.

are two phases of a single component in equilibrium (e.g., $H_2O(s) = H_2O(\ell)$, or $H_2O(\ell) = H_2O(g)$, or $H_2O(s) = H_2O(g)$), then $F = 1$, and when the temperature is specified the equilibrium value of the pressure is fixed and vice versa. If there are three phases of a single component in equilibrium (e.g., $H_2O(g) = H_2O(\ell) = H_2O(s)$), then $F = 0$, and we cannot vary the equilibrium values of the pressure and the temperature, their values being completely determined by the nature of the substance. Four phases of a single component cannot coexist in equilibrium under any conditions, because such a system has $F = -1$, which is impossible.

## 10.3 PHASE EQUILIBRIA IN ONE-COMPONENT SYSTEMS. GENERAL CONSIDERATIONS.

For any two-phase equilibrium in a one-component system, we deduce from the phase rule that $F = 1 - 2 + 2 = 1$, and therefore there is a unique value of the pressure at each temperature for which the two phases are in equilibrium. For a one-component system, $X$, involving only a single solid phase, there are three different two-phase equilibria possible, namely

(a) $X(s) = X(\ell)$     (melting)
(b) $X(s) = X(g)$     (sublimation)
(c) $X(\ell) = X(g)$     (vaporization)

If we were to construct a plot of $P$ versus $T$, using the equilibrium values of $P$ and $T$ for any of these three equilibria, we would obtain a *phase-equilibrium line*. All points representing states of the system for which the two phases are in equilibrium form the phase-equilibrium line. The $P$ vs. $T$ diagram containing the various phase-equilibrium lines is known as the *phase diagram*. The general appearance of such a diagram for a single component system involving a single solid phase is depicted in Figure 10.1. There are three phase-equilibrium lines on this diagram, namely the *melting-point curve*, the *sublimation-pressure curve*, and the *vapor-pressure curve*. These three curves intersect one another at the triple point.

The *normal melting point* of a solid is defined as the temperature at which the pure solid and liquid are in equilibrium with each other at 1 standard atmosphere total pressure. The *normal boiling point* of a liquid is defined as the temperature at which the equilibrium vapor pressure of the liquid is equal to 1 standard atmosphere. A substance whose triple point lies above 1 atm does not possess a *normal* melting or boiling point, but such a substance does exhibit a *normal* (i.e., 1 atm) *sublimation point*. Substances whose triple points lie below 1 atmosphere do not possess a normal sublimation point.

The melting-point curve (Figure 10.1) is quite steep (the melting point is not much affected by the pressure), and usually has a positive slope. The melting points of most (but not all) substances increase in an approximately linear manner with increasing pressure at an average rate of about $+0.03$ deg atm$^{-1}$. This means that a pressure increase in the range of 30 to 40 atm is required to increase the melting point of most solids by about 1 degree.

The equilibrium sublimation pressure of a solid increases in an approximately exponential manner from 0 K up to the triple point, where the solid-gas equilibrium line terminates abruptly. Above the triple point the solid and gaseous phases cannot coexist in equilibrium with each other.

The equilibrium vapor pressure of the liquid increases in an approxi-

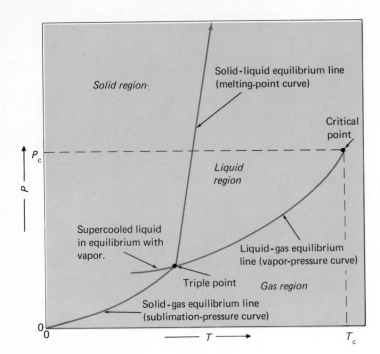

**FIGURE 10.1** Generalized phase diagram for a one-component system involving a single solid phase.

mately exponential manner from the triple point to the critical point, at which point the vapor pressure curve abruptly terminates. The critical point is the point where the molar volumes, molar enthalpies, and molar entropies of the gaseous and liquid phases are equal. A gas above its critical point cannot be liquefied no matter how high the pressure of the gas is made. In fact, above the critical point, there is no longer a distinction between the gas and the liquid, and for this reason substances above their critical points are sometimes designated as *fluids*.

Before proceeding to a more detailed consideration of the various phase equilibrium lines in a phase diagram, we note here that if two phases, $\alpha$ and $\beta$, of a substance are in equilibrium, then, from Equation (10.2), their molar Gibbs energies $\bar{G}_\alpha$ and $\bar{G}_\beta$ must be equal:

$$\Delta \bar{G}_{tr} = \bar{G}_\beta - \bar{G}_\alpha = 0 = (\bar{H}_\beta - T\bar{S}_\beta) - (\bar{H}_\alpha - T\bar{S}_\alpha)$$

where the subscript $tr$ designates the transition from the $\alpha$ to the $\beta$ phase. In other words,

$$\Delta \bar{H}_{tr} = T\Delta \bar{S}_{tr} \qquad \text{(at equilibrium)} \quad (10.6)$$

If $\beta$ is the high-temperature phase, that is, the phase lying to the right of the $P,T$ equilibrium line, then $\Delta \bar{S}_{tr(\alpha \to \beta)}$ is necessarily positive. This is because energy must be added as heat ($q > 0$) to the system to convert the $\alpha$-phase to the $\beta$-phase at any pair of equilibrium values of $P$ and $T$, and from the second law, $\Delta \bar{S}_{tr} = q_{tr}/T > 0$. Also, from Equation (10.6), $\Delta \bar{H}_{tr} = T(q_{tr}/T) = q_{tr} > 0$ for the conversion from the low-temperature to the high-temperature phase.

## a. The Melting-Point Curve.

The melting-point curve gives the pressure at which a substance melts as a function of the temperature. Two points $(P_1, T_1)$ and $(P_2, T_2)$ that lie on the melting-point curve are related approximately by the equation*

$$\frac{P_2 - P_1}{T_2 - T_1} = \frac{\Delta P}{\Delta T} \simeq \frac{\Delta \bar{S}_{fus}}{\Delta \bar{V}_{fus}} \qquad (10.7\text{a})$$

or (solving for $T_2$, the melting point at a total pressure $P_2$)

$$T_2 = T_1 + \left(\frac{\Delta \bar{V}_{fus}}{\Delta \bar{S}_{fus}}\right)(P_2 - P_1) \qquad (10.7\text{b})$$

where the subscript *fus* (fusion) indicates the transition solid → liquid, and the bars denote molar quantities. In Figure 10.2 we have plotted the melting-point curves for benzene and potassium. There is an easily discernible departure from linearity in these melting point curves, which arises from the dependence of $\Delta \bar{S}_{fus}$ and $\Delta \bar{V}_{fus}$ on temperature and pressure. However, the value of $(\Delta \bar{S}_{fus}/\Delta \bar{V}_{fus})$ over a moderate range of pressure ($< 2000$ atm) usually can be taken to be constant without serious error (see Figure 10.2). The value of $\Delta \bar{S}_{fus}$ is necessarily positive,† and thus the criterion for increasing melting point with increasing pressure is $\Delta \bar{V}_{fus} > 0$. In other words, because

$$\Delta \bar{V}_{fus} = M\left(\frac{1}{\rho_\ell} - \frac{1}{\rho_s}\right) \qquad (10.8)$$

(where $M$ is the molecular weight and $\rho_\ell$ and $\rho_s$ are the densities of the liquid and solid phases, respectively) the melting point will increase ($T_2 > T_1$) with increasing pressure ($P_2 - P_1 > 0$), if the solid phase is denser than the liquid at the same temperature, that is, if $\rho_s > \rho_\ell$. Almost all substances meet this condition, with water being the most important exception. An increase in pressure stabilizes the more dense phase relative to the less dense phase, owing to the smaller volume occupied by the denser phase ($V = $ mass/density).

The pressure that appears in Equations (10.7a) and (10.7b) is not the partial pressure of the substance over the solid and liquid phases. The gaseous phase of a substance cannot coexist in equilibrium with both the solid *and* liquid phases of the substance at any pressure other than the triple-point pressure. The pressure exerted on the condensed phases can be produced by enclosing the sample in a piston and cylinder arrangement

---

*An expression of this form is also applicable to solid-solid transformations (that is, equilibrium transformations between two solid phases of a pure substance such as red and white phosphorus, gray and white tin, or rhombic and monoclinic sulfur). In such a case, $\Delta \bar{V}_{fus}$ and $\Delta \bar{S}_{fus}$ should be changed to $\Delta \bar{V}_{tr}$ and $\Delta \bar{S}_{tr}$. Equation (10.7a) is an approximate form of the Clapeyron equation, $dP/dT = \Delta S/\Delta V$, which applies to phase-equilibrium lines.

†For most solids, $\Delta \bar{S}_{fus}$ lies in the range from 2 to 9 gibbs · mole⁻¹; however, much larger values (up to 20 gibbs · mole⁻¹) occasionally are observed for substances that are polymeric in the solid state (e.g., $AlCl_3$).

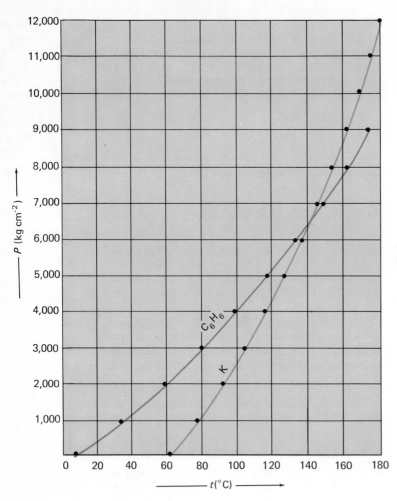

**FIGURE 10.2** Melting-point curves for benzene and potassium (1 kg·cm⁻² = 0.968 atm). Data from reference 10.1.

in which the two condensed phases are located without any vapor space. The pressure could also be provided by a chemically inert, essentially insoluble gas; however, this increases the number of components by one.

## b. The Vapor-Pressure and Sublimation-Pressure Curves.

The dependence of the equilibrium vapor pressure and equilibrium sublimation pressure on the temperature is given approximately in each case by an equation of the form

$$\log P = \frac{-\Delta \bar{H}_{tr}}{2.303\ RT} + \text{constant} \tag{10.9}$$

where $\Delta \bar{H}_{tr}$ (which is necessarily positive for the vaporization and sublimation transitions) is equal to $\Delta \bar{H}_{vap}$ for the equilibrium vapor pressure curve, and is equal to $\Delta H_{sub}$ for the sublimation pressure curve.[*] Equation (10.9) indicates that a plot of $\log P$ versus $1/T$ has a slope[†] at any point equal to $-\Delta \bar{H}_{tr}/2.303\ R$. If $\Delta \bar{H}_{tr}$ is constant, then the $\log P$ versus $1/T$ curve will be linear.

In Figures 10.3 and 10.4 we have plotted $\log P$ versus $1/T$ for some representative vaporization and sublimation reactions. In general, $\Delta \bar{H}_{vap}$ and $\Delta \bar{H}_{sub}$ are functions of $T$ and $P$; however, it is evident from the linearity of the plots in these two figures that $\Delta \bar{H}_{vap}$ and $\Delta \bar{H}_{sub}$ may be taken to be constant over a pressure range of a few atmospheres without serious error.[‡] The approximation $\Delta \bar{H}_{vap}$ = constant breaks down as the critical point is approached, because $\Delta \bar{H}_{vap}$ rapidly approaches zero in the vicinity of the critical point.

Given that $\Delta \bar{H}_{vap}$ is constant to within the desired precision, and using Equation (10.9), we obtain for the relationship between two points $(P_1, T_1)$ and $(P_2, T_2)$ on the equilibrium vapor-pressure curve

$$\log \left(\frac{P_2}{P_1}\right) = \frac{-\Delta \bar{H}_{vap}}{2.30\ R} \left\{\frac{1}{T_2} - \frac{1}{T_1}\right\} = \frac{\Delta \bar{H}_{vap}}{2.30\ R} \left\{\frac{T_2 - T_1}{T_1 T_2}\right\} \qquad (10.10)$$

As an example of the application of this equation, we shall compute $\Delta \bar{H}_{vap}$ for water. The equilibrium vapor pressure of water at 25°C is 23.8 Torr and the normal boiling point is 100°C. Using Equation (10.10), we obtain

$$\log \left(\frac{760\ \text{Torr}}{23.8\ \text{Torr}}\right) = \frac{-\Delta \bar{H}_{vap}}{(2.30)(1.99\ \text{cal} \cdot \text{K}^{-1} \cdot \text{mole}^{-1})} \left\{\frac{1}{373\ \text{K}} - \frac{1}{298\ \text{K}}\right\}$$

or

$$\Delta \bar{H}_{vap} = \frac{-(2.30)(1.99)(1.504)}{(0.00268 - 0.00335)} = 10.3\ \text{kcal} \cdot \text{mole}^{-1}$$

This is the amount of energy in kilocalories that must be added as heat to vaporize 18 grams of water.

It often is necessary in scientific work to know the vapor pressure of a liquid at a particular temperature. If sufficient data are available, we can simply construct a plot of $\log P$ versus $1/T$ and interpolate for the desired result. If, on the other hand, $\Delta \bar{H}_{vap}$ is available, the calculation can be carried out using Equation (10.10), provided that in addition to $\Delta \bar{H}_{vap}$ we know the vapor pressure of the liquid at one temperature; for example, we might have available the normal boiling point of the liquid. In many cases $\Delta \bar{H}_{vap}$ is not known, and all that is available is the vapor pressure at a particular temperature—most often the normal boiling point. In such a case, a rough estimate of the equilibrium vapor pressure can be made using

---

[*]Note that Equation (10.9) can also be written as

$$P = (\text{const}')\exp(-\Delta \bar{H}_{tr}/RT).$$

[†]A linear equation of the form $y = mx + b$ ($m$ and $b$ are constants) has a slope $m = (y_2 - y_1)/(x_2 - x_1)$, where $(x_1, y_1)$ and $(x_2, y_2)$ are any two points on the line. In the above case, $y = \log P$ and $x = 1/T$.

[‡]Note, however, the slight departure from linearity in the $\log P$ versus $1/T$ plot for $H_2O(\ell) = H_2O(g)$ in Figure 10.3. In this case the range of $P_{H_2O}$ given in the figure extends from 0 to 10 atm.

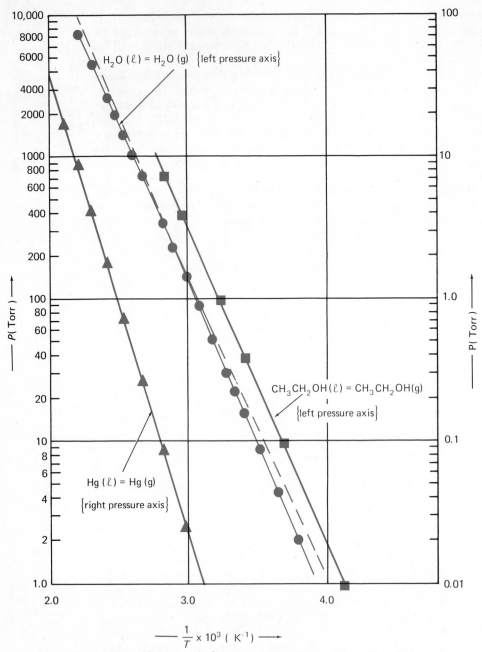

**FIGURE 10.3** Plots of $\log P$ vs. $1/T$ for some vaporization reactions. The plot is made on semi-log paper, which does the job of taking logarithms for us.

*Trouton's rule.* Trouton's rule states that the entropy of vaporization *at the normal boiling* of an "unassociated" liquid is approximately equal to 21 gibbs $\cdot$ mole$^{-1}$; that is, $\Delta \bar{H}_{vap} \simeq 21 \, T_{b.pt.}$. An associated liquid is one involving *specific* intermolecular interactions, such as hydrogen bonding or dimerization, that effectively make the basic molecular unit different in

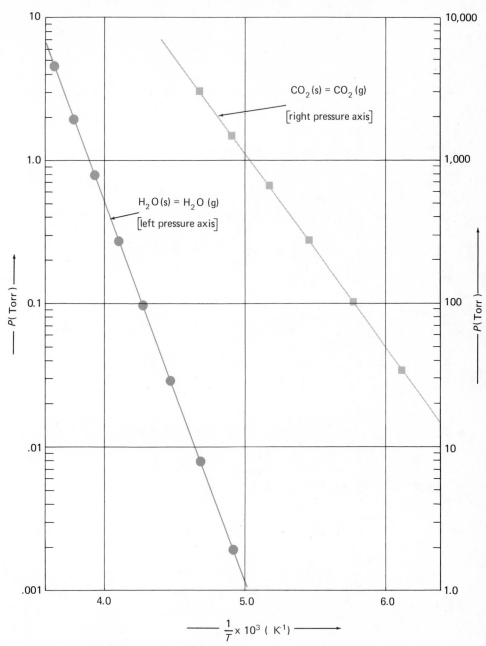

**FIGURE 10.4** Plots of log $P$ vs. $1/T$ for some sublimation reactions.

the liquid and gas phases. A rough idea of the validity of Trouton's rule can be obtained by examining the $\Delta \bar{S}_{vap}$ values in Table 10.1. Triple points, critical points, normal melting points, and normal boiling points, together with values of $\Delta \bar{H}_{vap}$, $\Delta \bar{H}_{fus}$, $\Delta \bar{S}_{vap}$, and $\Delta \bar{S}_{fus}$ for some common substances also can be found in Table 10.1.

**TABLE 10.1**

PHASE EQUILIBRIUM DATA FOR SOME ONE-COMPONENT SYSTEMS.*

| Substance | Triple Point | | Critical Point | | Normal Melting Point | Normal Boiling Point | Melting | | Vaporization | |
| | $T_{tp}$ K | $P_{tp}$ atm | $T_c$ K | $P_c$ atm | $T_m$ K | $T_b$ K | $\Delta \bar{H}_{fus}$ kcal·mol⁻¹ | $\Delta \bar{S}_{fus}$ gibbs·mol⁻¹ | $\Delta \bar{H}_{vap}$ kcal·mol⁻¹ | $\Delta \bar{S}_{vap}$ gibbs·mol⁻¹ |
|---|---|---|---|---|---|---|---|---|---|---|
| $NH_3$ | | | 405.50 | 111.3 | 195.4 | 239.73 | 1.351 | 6.914 | 5.581 | 23.28 |
| $Br_2$ | 216.6 | 5.11 | 584 | 102 | 265.90 | 332.36 | 2.527 | 9.50 | 7.05 | 21.2 |
| $CO_2$ | | | 304.19 | 72.85 | (217.0)ᵇ | (194.7)ᵃ | (1.99)ᵇ | (9.2)ᵇ | (6.03)ᵃ | (30.98)ᵃ |
| $Cl_2$ | | | 417.2 | 76.1 | 172.17 | 239.10 | 1.531 | 8.89 | 4.878 | 20.40 |
| $H_2$ | 13.8 | | 33.23 | 12.80 | 13.84 | 20.26 | 0.028 | 2.0 | 0.215 | 10.6 |
| $^2H_2$ | | | 38.34 | 16.43 | 18.63 | 23.59 | 0.0471 | 2.53 | 0.2939 | 12.46 |
| He | | | 5.20 | 2.26 | (3.5)ᶜ | 4.215 | (0.005)ᶜ | (1.5)ᶜ | 0.02 | 4.7 |
| HCl | | | 324.6 | 81.5 | 158.96 | 188.12 | 0.476 | 2.99 | 3.860 | 20.52 |
| $N_2$ | | | 127.2 | 33.54 | 63.18 | 77.36 | 0.172 | 2.72 | 1.335 | 17.26 |
| $O_2$ | 54.34 | | 154.77 | 50.14 | 54.4 | 90.18 | 0.106 | 1.95 | 1.630 | 18.07 |
| $H_2O$ | 273.16 | 0.0060 | 647.4 | 218.3 | 273.15 | 373.15 | 1.436 | 5.26 | 9.717 | 26.04 |
| $^2H_2O$ | | | 644.1 | 216 | 276.97 | 374.57 | 1.501 | 5.42 | 9.933 | 26.52 |
| $CH_3Cl$ | | | 416.27 | 65.93 | 175.43 | 248.93 | 1.537 | 8.76 | 5.14 | 20.6 |
| $CH_3COOH$ | | | 594.8 | 57.1 | 289.76 | 391.7 | 2.80 | 9.66 | 5.66 | 14.4 |
| $CH_3CH_2OH$ | | | 516 | 63 | 158.6 | 351.5 | 1.200 | 7.57 | 9.25 | 26.32 |
| $CCl_4$ | | | 556.4 | 44.97 | 250.28 | 349.9 | 0.59 | 2.36 | 7.17 | 20.5 |
| $CH_3COCH_3$ | | | 508.7 | 46.6 | 178.2 | 329.7 | 1.36 | 7.64 | 7.23 | 21.9 |
| $CH_3CH_2OCH_2CH_3$ | | | 467.8 | 35.6 | 149.9 | 307.8 | ... | ... | 6.22 | 20.2 |
| $C_6H_6$ | | | 562.7 | 48.6 | 278.7 | 353.3 | 2.37 | 8.49 | 7.37 | 20.9 |
| Li | | | | | 453.7 | 1599 | 0.723 | 1.594 | 30.8 | 19.26 |
| LiBr | | | | | 823 | 1583 | 4.2 | 5.1 | 35.4 | 22.4 |

*Data taken from *American Institute of Physics Handbook*, 2nd edition, D. E. Gray, editor (McGraw-Hill Book Co., New York, 1963), where much more extensive tables can be found.

1 kcal·mol⁻¹ = 4.184 kJ·mol⁻¹

1 gibbs·mol⁻¹ = 4.184 J·K⁻¹·mol⁻¹

(a) Normal sublimation point data.
(b) Melting point at $P = 5.1$ atm.
(c) $P_{tot} = 100$ atm.

# 10.4 PHASE DIAGRAMS FOR ONE-COMPONENT SYSTEMS.

We now proceed to a consideration of the phase diagrams for some one-component systems.

**(a) Water.** The phase diagram for water is presented in two parts in Figures 10.5 and 10.6. Equilibrium values of $P$ and $T$ for vaporization, sublimation, and melting are presented in Table 10.2. Plots of $\log P$ versus $1/T$ for the vaporization and sublimation phase transformations have already been presented in Figures 10.3 and 10.4.

In Figure 10.5, the curve $TC$ is the equilibrium vapor-pressure curve for liquid water. For any given temperature in the range from 273.16 K (triple point, point $T$ in the figure) to 647.2 K (critical point, point $C$ in the figure) there is one, and only one, value of the pressure of gaseous $H_2O$ for which pure liquid water and pure gaseous water can exist in equilibrium with each other ($F = 1 - 2 + 2 = 1$). Point $B$ in Figure 10.5 is the normal boiling point of water. A liquid open to the atmosphere boils when it attains the temperature at which its equilibrium vapor pressure equals atmospheric pressure. When the barometric pressure at sea level is exactly 1 atm, pure water boils at 100.00°C at sea level, whereas at 8000 feet elevation on the same day it boils at 92°C. If we continuously add energy as heat to a sample of water open to the atmosphere, its temperature continues to increase until it boils, at which temperature further input of energy as heat increases the amount of water vaporized, but does not increase the temperature of the liquid. The energy is simply consumed in the evaporation of the water ($540 \text{ cal} \cdot \text{g}^{-1}$). It takes energy to vaporize a liquid at any temperature, and this effect is one of several utilized by the human body to regulate its temperature at about 310 K. When the body becomes overheated it sweats; the evaporation of the water from the surface of the body consumes about 540 calories per gram of water evaporated. This same effect is utilized in evaporative air coolers.

**FIGURE 10.5** Phase diagram for water at $P$ <2000 atm. The pressure axis is not to scale. C, critical point; T, triple point (ice I-liquid-vapor); M, normal melting point; B, normal boiling point. (See Figure 10.6 for $P$ >2000 atm.)

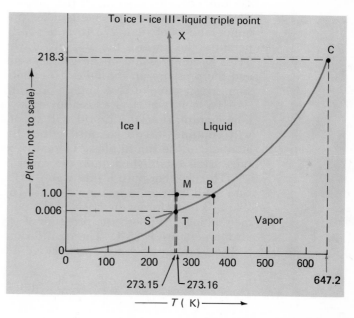

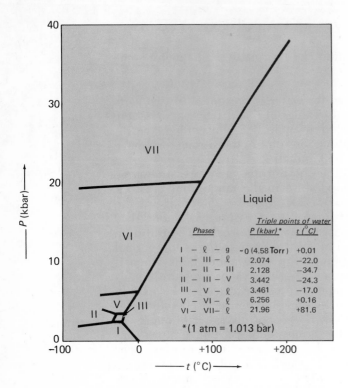

FIGURE 10.6 Phase diagram for water at high pressures, showing the regions of stability of the solid phases of water. (Data from J. W. Stewart, *The World of High Pressure*, D. van Nostrand Pub. Co., Princeton, N.J., 1967).

| Phases | Triple points of water | |
|---|---|---|
|  | P (kbar)* | t (°C) |
| I – ℓ – g | ~0 (4.58 Torr) | +0.01 |
| I – III – ℓ | 2.074 | −22.0 |
| I – II – III | 2.128 | −34.7 |
| II – III – V | 3.442 | −24.3 |
| III – V – ℓ | 3.461 | −17.0 |
| V – VI – ℓ | 6.256 | +0.16 |
| VI – VII – ℓ | 21.96 | +81.6 |

*(1 atm = 1.013 bar)

Suppose we have a sample of $H_2O(g)$ at 100 Torr pressure enclosed in a cylinder that is equipped with a piston and immersed in an 80°C thermostat. As the piston is slowly depressed, the gas pressure steadily increases until it reaches a value of 355.1 Torr (Table 10.2 and Figure 10.5). At this pressure the first liquid forms, and further depression of the piston does not lead to an increase in pressure, but rather to a conversion of more gas to liquid. The conversion of gas to liquid continues with further depression of the piston until the gas phase is *completely* exhausted, at which point the pressure again begins to rise, owing to the compression of the liquid. This same sequence of events takes place for any $H_2O(g)$ sample at any temperature between 273.16 and 647.2 K. However, the lower the temperature, the lower is the pressure at which the liquid first appears. Above 647.2 K, compression of the gas does not lead to separation of the liquid, and the pressure on the fluid rises continuously throughout the compression, irrespective of how large a pressure is applied.

If a sample of $H_2O(\ell)$ is allowed a vapor space and held at a particular temperature, then the liquid will spontaneously begin to evaporate, and will continue to do so until either the equilibrium vapor pressure is attained, or the liquid phase is totally exhausted. If the equilibrium vapor pressure is established, then the vapor phase is said to be saturated.

The fact that pure liquid water has a definite equilibrium vapor pressure at all temperatures between 273.16 and 647.2 K (as does $H_2O(s)$ from 0 to 273.16 K) forms the basis of the percentage relative humidity scale in meteorology. Suppose the pressure of water in the air is 17.5 Torr and the air temperature is 25°C. The equilibrium vapor pressure of water at 25°C is 23.8 Torr, and the per cent relative humidity is

$$\frac{17.5}{23.8} \times 100 = 73\%$$

TABLE 10.2

PHASE EQUILIBRIUM DATA FOR WATER*

| $H_2O(s) = H_2O(g)$ | | $H_2O(\ell) = H_2O(g)$ | | $H_2O(\ell) = H_2O(g)$ | | $H_2O(s,I) = H_2O(\ell)$ | |
|---|---|---|---|---|---|---|---|
| $t(°C)$ | $P(Torr)$ | $t(°C)$ | $P(Torr)$ | $t(°C)$ | $P(atm)$ | $t(°C)$ | $P(atm)$ |
| −98 | 0.000015 | −16 | 1.34 | 100 | 1.00 | 0 | 0 |
| −90 | 0.0007 | −10 | 2.149 | 120.1 | 2.00 | −5 | 590 |
| −80 | 0.0004 | 0 | 4.578 | 152.4 | 5.00 | −10 | 1093 |
| −70 | 0.00194 | 0.01 | 4.580 | 180.5 | 10.00 | −15 | 1539 |
| −60 | 0.00808 | 10 | 9.209 | 213.1 | 20.00 | −20 | 1907 |
| −50 | 0.02955 | 20 | 17.535 | 251.1 | 40.00 | −22 | 2047 |
| −40 | 0.0966 | 25 | 23.756 | 276.5 | 60.00 | | |
| −30 | 0.2859 | 30 | 31.824 | 300.0 | 84.80 | | |
| −20 | 0.776 | 40 | 55.324 | 350.0 | 163.2 | | |
| −10 | 1.950 | 50 | 92.51 | 374.2 | 218.3 | | |
| 0 | 4.579 | 60 | 149.38 | | | | |
| | | 70 | 233.7 | | | | |
| | | 80 | 355.1 | | | | |
| | | 90 | 525.76 | | | | |
| | | 100 | 760.00 | | | | |

*Data on sublimation and vaporization from *CRC Handbook of Chemistry and Physics,* 48th edition, Chemical Rubber Co., Cleveland, Ohio. Data on melting from *The Phase Rule,* 9th edition, by A. Findlay, A. N. Campbell, and N. O. Smith, Dover Publications, Inc.

If the atmospheric temperature in the above example were to decrease to about 20°C (see Table 10.2), then the air would become saturated with water vapor and condensation would occur. Thus, air having 17.5 Torr of water vapor pressure is said to have a *dew point* of 20°C. A related phenomenon is the condensation of moisture from the air onto the surfaces of objects whose temperature is below the dew point of the air. A common case is the formation of water droplets on the outside of a glass containing a cold drink. If the relative humidity of the air is high, then the capacity of the air for taking up additional water vapor is correspondingly low. This is why we become uncomfortable in a high humidity atmosphere; the air in the vicinity of the body surface is readily saturated with water evaporating from the skin surface, and when the air becomes saturated, there is no further net loss of water from the surface, and consequently no cooling.

If proper precautions are taken, pure liquid water can be obtained in equilibrium with water vapor below the triple point. Under such conditions the water is said to be *supercooled* (see curve *ST* in Figure 10.5). The equilibrium vapor pressure of supercooled liquid water is necessarily higher than that of solid water (see curve *OT* in Figure 10.5) at the same temperature, and if an ice crystal is injected into the supercooled liquid, then crystallization of the liquid will occur spontaneously. If the temperature of the system is held constant, then the entire mass of liquid will become solid. If the supercooled liquid is placed in an enclosed thermostated space together with the solid, but not in actual contact with it, the mass of the solid will increase at the expense of the liquid until the entire liquid phase is exhausted.

The curve *OT* in Figure 10.5 is the sublimation-pressure curve of ice. At any temperature in the range from 0 to 273.16 K there is a unique value of the vapor pressure of water for which equilibrium exists between the gas and solid. Suppose we have a sample of $H_2O(g)$ at 0.50 Torr enclosed in a cylinder that is equipped with a piston, the entire apparatus being immersed in a 263 K thermostat. When the piston is depressed slowly and continuously, the gas pressure rises steadily until it reaches a value of 1.95 Torr, at which point ice begins to form and the pressure stops rising until the entire gas phase is exhausted. At this point the pressure once again begins to rise, owing to the compression of the solid. Ice can be evaporated (sublimed) by maintaining the vapor pressure of water over the ice below its equilibrium value. This effect is utilized to keep the "frost" out of one form of "frost-free" refrigerator. Cool, dry air is blown through the freezing compartment and carries out water vapor at a pressure equal to or less than the equilibrium value at the temperature of the freezing compartment.

The curve $TX$ in Figure 10.5 is the melting-point curve of ice, and it gives the temperature at which pure solid and liquid water are in equilibrium at a particular total pressure. An unusual feature of this curve for water is its negative slope ($\rho_{solid} < \rho_{liquid}$). The melting point of ice *decreases with increasing pressure* at the rate of about 0.0075 K·atm$^{-1}$. The melting-point curve in Figure 10.5 was arbitrarily cut off at about 250 atm; it extends to much higher pressures, where it terminates at the ice I-ice III-$H_2O(\ell)$ triple point* (2047 atm, $-22°C$). If we take a piece of ice I at, say, $-15°C$, and squeeze it hard enough (1600 atm), it will melt.†

Solid water phases other than ice I (i.e., ordinary find-it-in-your-freezer ice) are capable of existing in equilibrium with $H_2O(\ell)$ at up to, and evidently beyond,‡ 40,000 atm (see Figure 10.6). The other known solid phases of water (ices II, III, V, VI, VII, and the metastable ice IV, which is not shown in Figure 10.6) all have $\rho_{solid} > \rho_{liquid}$ (when compared at the same $T$ and $P$), and thus exhibit increasing melting points with increasing pressure (see Figure 10.6). Ice VII at 38,000 atm does not melt until its temperature reaches 200°C! Ice V at 6000 atm and 0°C sinks in $H_2O(\ell)$. Below about $-22°C$, ice cannot be liquefied by application of pressure, no matter how great the pressure is made.

The ice I-$H_2O(\ell)$-$H_2O(g)$ triple point of water occurs at 273.16 K and 4.58 Torr. The triple point is the *only* point ($F = 1 - 3 + 2 = 0$) that is common to the vapor-pressure, sublimation-pressure, and melting-point curves. This triple point of water is the single most important fixed point in thermometry (see Chapter 9).

**(b) Carbon Dioxide.**     The phase diagram for carbon dioxide is presented in Figure 10.7, and phase equilibrium data for $CO_2$ are given in Table 10.3. There are several noteworthy differences between this phase diagram and that for water (Figure 10.5). The melting-point curve, $TX$, has a positive slope, and therefore the melting point of $CO_2(s)$ increases with increasing pressure. The solid-liquid-vapor triple point lies above

---

*Note that a system with two or more solid phases has more than one triple point.

†Some other substances that exhibit a decreasing melting point with increasing pressure are the elements gallium, antimony, bismuth, iron, and germanium; carbon as diamond, some alloys containing these elements, and a few other compounds besides water. At very high pressures the number of cases with $\rho_\ell > \rho_s$ increases, because liquids are usually more compressible than solids.

‡There is no evidence of a solid-liquid critical point analogous to that found on the vapor-pressure curve. There may be solid-solid critical points (e.g., for Ce metal), but this has not been unambiguously proven.

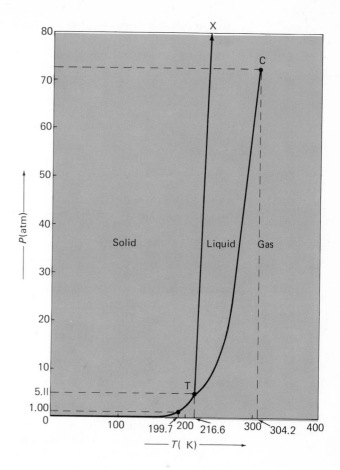

**FIGURE 10.7** Phase diagram for carbon dioxide.

one atmosphere (5.11 atm, 216.6 K), and therefore $CO_2$ does not possess a normal melting or boiling point. Below 5.11 atm of $CO_2(g)$ the solid does not melt as we add energy as heat to it, but rather, it sublimes—thus the designation "Dry Ice." However, as is evident from the phase diagram,

**TABLE 10.3**

PHASE EQUILIBRIUM DATA FOR CARBON DIOXIDE.*

| $CO_2(s) = CO_2(g)$ | | $CO_2(\ell) = CO_2(g)$ | | $CO_2(s) = CO_2(\ell)$ | |
|---|---|---|---|---|---|
| $t(°C)$ | $P(\text{Torr})$ | $t(°C)$ | $P(\text{Torr})$ | $t(°C)$ | $P(\text{Torr})$ |
| −180 | 0.000013 | −50 | 5128 | −56.6 | 5.11 |
| −160 | 0.0059 | −40 | 7548 | | |
| −140 | 0.431 | −30 | 10718 | | |
| −120 | 9.81 | −20 | 14781 | | |
| −100 | 104.81 | −10 | 19872 | | |
| −80 | 672.2 | 0 | 26142 | | |
| −78.2 | 760.0 | 10 | 33763 | | |
| −70 | 1486.1 | 20 | 42959 | | |
| −60 | 3073.1 | 30 | 54086 | | |

*Data from *CRC Handbook of Chemistry and Physics*, 48th edition, Chemical Rubber Co., Cleveland, Ohio.

carbon dioxide does possess a normal (i.e., 1 atm) sublimation point (which water does not) at 194.7 K. This is the approximate temperature of $CO_2(s)$ stored in closed but not tightly sealed containers. This low temperature, coupled with a low price, makes solid $CO_2$ a widely used one-shot refrigerant for maintaining foods well below their freezing temperatures.

If we pump $CO_2(g)$ into a tank whose temperature is held between 216.6 and 304.2 K (the critical temperature), $CO_2(\ell)$ will be formed when the $CO_2(g)$ pressure reaches the equilibrium value. For example, at 296.2 K the pressure of $CO_2(g)$ in a tank cannot be made to exceed the equilibrium value of 60.6 atm, because at this pressure the $CO_2(g)$ is in equilibrium with $CO_2(\ell)$. As long as there is some $CO_2(\ell)$ in the tank, the tank pressure will remain at 60.6 atm provided the temperature is held at 296.2 K.[*] As $CO_2(g)$ is removed from a tank containing $CO_2(\ell)$ and $CO_2(g)$, the gas pressure is maintained constant by evaporation of some of the liquid. If we take a sample of $CO_2(\ell)$ in equilibrium with $CO_2(g)$ and raise its temperature above 304.2 K, the system will go over to one phase. The critical point of $CO_2$ lies at 72.9 atm and 304.2 K, which makes this substance convenient for the observation of critical-point phenomena such as *critical opalescence* (light scattering due to density fluctuations at the critical point).

(c) **Carbon:** The phase diagram for carbon is of particular interest because it defines the conditions of temperature and pressure that are required for the production of diamonds from graphite.[†] Although diamonds are highly prized by many people for embellishing their persons, the industrial market for diamonds far exceeds that for jewels. Diamond is the hardest substance known; it finds wide use as an abrasive and cutting material and where a high resistance to wear is required.

The phase diagram for carbon is presented in Figure 10.8. This phase diagram is quite different in appearance from that for water or carbon dioxide. There are two reasons for this: The first of these is that the pressure axis has been made logarithmic in order to facilitate the plotting of a wide range of pressures ($10^{-3}$ to $10^5$ kbar); the second is that there are three distinct solid phases involved. The ordinary form of solid carbon is graphite ($\rho = 2.25$ g·cm$^{-3}$ at 1 atm, 25°C).

As is evident from the phase diagram, graphite is the thermodynamically stable form of carbon at ordinary temperatures and pressures. The normal sublimation point of graphite is about 4200 K. No other solid exists at higher temperatures than this at one atmosphere. Very pure graphite, when heated (in the absence of oxidizers) to a high temperature ($>1500°C$) and compressed, yields pyrolytic graphite, a material with excellent thermal conductivity along its surface and low thermal conductivity perpendicular to the surface. Pyrolytic graphite, discovered by Thomas Edison in 1896, is used in rocket nose cones, rocket nozzles, and pipe bowls.

The graphite-liquid-vapor triple point occurs at about 100 bar and 4250 K, which is the highest known triple point. The melting-point curve of graphite starts out with a positive slope, but at about 55 kbar and 4900 K the melting point reaches a maximum. At this maximum the densities of

---

[*] This is to be contrasted with such gases as tank $N_2$ or He, both of which are well above their critical points at 23°C.

[†] See Chapter 6 for the structures of diamond and graphite.

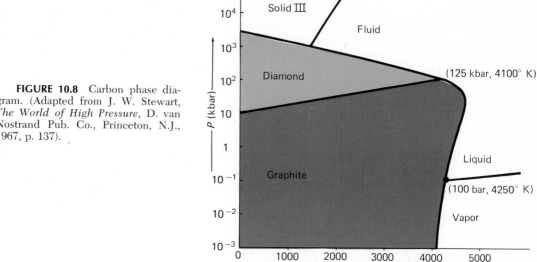

**FIGURE 10.8** Carbon phase diagram. (Adapted from J. W. Stewart, *The World of High Pressure,* D. van Nostrand Pub. Co., Princeton, N.J., 1967, p. 137).

the solid and liquid phases are equal. At still higher pressures the melting point decreases, and the graphite melting-point curve terminates at the graphite-diamond-liquid triple point.

In order to produce diamonds from graphite it is necessary to subject graphite to a temperature and pressure that lie within the region of stability of diamond. At 300 K about 15 kbar of pressure is required. Although it is absolutely necessary to subject the graphite to a $P$ and $T$ where it is thermodynamically unstable with respect to diamond in order to convert the graphite to diamond, thermodynamics gives no guarantee that the transition will actually take place. The *no* of thermodynamics is quite emphatic, but the *yes* is only conditional. This is because equilibrium thermodynamics has nothing to say about the rates of processes. If the rate of conversion is very slow, then the graphite will not be converted to diamond. Many unsuccessful attempts were made to produce diamonds from graphite by subjecting the latter to high pressure, before the first successful synthesis was carried out at the General Electric Research Laboratories in 1955. Pressures around 120 kbar and temperatures as high as 3000 to 4100 K were employed. The high temperatures facilitate the conversion because of the increase in conversion rate with increase in temperature. In these initial experiments the high temperature was maintained for only a short time (0.01 sec), and the conversion was incomplete. If the high temperature is maintained for a longer time, a greater percentage conversion can be obtained. At 2500 K, essentially complete conversion was achieved in a few minutes at 150 kbar. Using an iron or nickel catalyst, appreciable conversion can be obtained at 60 kbar and 1500 K. Once the diamonds are produced, the pressure and temperature are quickly reduced and the carbon is "trapped" in the diamond form, which is metastable at ordinary conditions. The rate of conversion of diamond to graphite at ordinary temperatures is so slow as to be completely negligible. If the diamonds are heated at low pressure to a high temperature ($>1500$ K), then they revert to graphite. It is worth noting that there is no, repeat *no*, way that can be devised to convert graphite to diamond within the region of $T$ and $P$ where graphite is the more stable form, nor is there any guarantee that graphite, subjected to a $T$ and $P$ where diamond is the more stable form, will convert to diamonds. It is not correct, however, to conclude that diamonds cannot be produced at low pressures. If we take some carbon-containing compound,* say, $CH_4(g)$, at conditions where it is unstable with respect to C (diamond) and $H_2(g)$, then a synthesis of diamond is possible. This is not the same, however, as the conversion of *graphite* to diamond. Passing methane gas, at low pressures and about 1300 K, over diamond dust leads to a few per cent increase in the mass of the diamonds.

The region labeled Solid III in Figure 10.8 is the region of stability of another solid phase of carbon, apparently with a cubic-close-packed structure.† These cubic diamonds have been identified in meteorite fragments, indicating that the fragments came from material that was subjected to pressures in excess of $10^3$ kbar, or about 1 million atmospheres.

**(d) Helium:**     The phase diagram for helium of normal isotopic composition (99.99987% $^4$He) is presented in Figure 10.9. This phase diagram is very unusual. Helium does not exhibit a solid-liquid-vapor triple point, and, provided the pressure is maintained below about 29 atm, liquid

---

*Or, for that matter, *gaseous* carbon.
†There is also evidence for a fourth solid phase of carbon with a hexagonal structure. It is not known whether this is a stable phase.

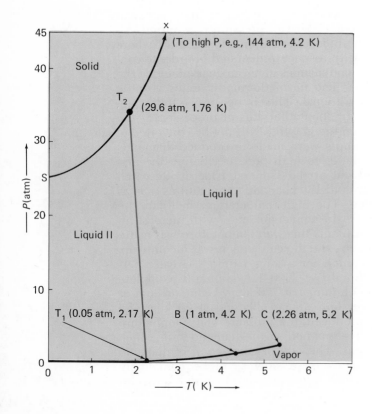

**FIGURE 10.9** The phase diagram for helium: B, normal boiling point; C, critical point; OC, vapor pressure curve; SX, melting curve; $T_1T_2$, the λ-line.

helium does not solidify *no matter how low its temperature*. The normal boiling point at 4.2 K and the critical point at 5.2 K are in both cases the lowest known. The most unusual feature is the existence of two different isotropic liquid phases, designated helium I and helium II. Helium is the only known substance that exhibits such behavior.

Liquid helium I can be produced by running the low temperature gas (precooled with liquid hydrogen) through an adiabatic expansion. At 1 atm pressure of He(g), the liquid helium I forms at 4.2 K. The temperature of the liquid can be still further reduced (to about 1.0 K) by pumping off He(g) with high-speed vacuum pumps (evaporative cooling). During this evaporative process the liquid boils vigorously as its temperature drops, until a temperature of 2.17 K is reached. At this point the vigorous bubbling suddenly stops, and the liquid becomes quiescent. This phenomenon has been traced to the fact that the thermal conductivity of the helium has suddenly increased one-million-fold. That a dielectric liquid should suddenly become a better heat conductor than silver or copper was quite unexpected and baffling to early investigators. Helium II has many other interesting properties. For example, vessels with very tiny holes that hold liquid helium I above 2.17 K suddenly begin to leak like sieves below this temperature. If an empty container, closed at one end, is partially immersed in helium II, such that its open end remains well above the liquid surface, the liquid will flow up and over the separating wall into the empty container until the levels are the same. This is accomplished by means of a surface film 50 to 100 atoms thick ($\sim$ 100 Å) that flows at a rate of about $10^{-4} \text{cm}^3 \cdot \text{sec}^{-1} \cdot \text{cm}^{-1}$. The flow rate depends primarily on the temperature, being zero for $T > 2.17$ K and increasing rapidly to a maximum of about $10^{-4} \text{cm}^3 \cdot \text{sec}^{-1} \cdot \text{cm}^{-1}$ below this temperature. The flow rate into and out of the central vessel is independent of the difference in levels, the path length, and the height of the rim above the surface. If the container is raised again so that the level in the inner vessel is higher than that in the outer, the liquid will flow back out of the inner vessel until the liquid levels are again the same. These unusual effects have been traced to the fact that helium II is essentially non-viscous, and encounters less flow resistance the narrower the orifice through which the liquid has to pass. The unique properties of helium II have led to its designation as a *superfluid*.

The explanation for the properties of helium II is quite complicated and not yet complete; it is clear that the behavior of helium II cannot be explained in classical terms. It has been suggested that when the temperature of liquid helium is brought down to 2.17 K, the He atoms begin to condense in momentum (as opposed to position) space, and it is this condensation in momentum space that gives rise to a quantum (or super-) fluid.

---

## PROBLEMS

1. Ethyl chloride, $CH_3CH_2Cl$, is frequently used as a local surface anesthetic. Discuss the reasons why this compound is suitable for such proposes.

2. Determine the number of phases in each of the following cases and devise a method for *mechanically* separating the phases:

(a) A mixture of sand, sawdust, iron filings, gold dust, and zinc granules.

(b) A mixture of carbon tetrachloride and water in a closed vial with a vapor space.

(c) An aqueous solution saturated with sodium chloride and sucrose, which is in equilibrium with ice and open to the atmosphere.

(d) Clean air.

3. Determine the number of components and the number of degrees of freedom in each case:

(a) Cases (a), (b) and (c) in problem 2.

(b) The system formed in the reduction of zinc oxide with carbon at elevated temperatures that involves the following equilibria:

$$ZnO(s) + C(s) = Zn(g) + CO(g)$$

$$ZnO(s) + CO(g) = Zn(g) + CO_2(g)$$

Does your result depend on how the system is prepared?

4. Using the following data for water, estimate the melting point of ice at 1540 atm and compare your result with that given in Table 10.2. $\Delta \bar{H}_{fus} = 1.44$ kcal, $\rho_{\ell} = 1.00$ g $\cdot$ cm$^{-3}$ and $\rho_s = 0.91$ g $\cdot$ cm$^{-3}$.

5. The equilibrium vapor pressure of ice at $-40°C$ is 0.0966 Torr, whereas at $0°C$ it is 4.58 Torr. Using these data, estimate $\Delta \bar{H}_{sub}$ for water.

6. Suppose an aqueous solution was observed to begin freezing at $1.0°C$. What can you say about the composition of the solid phase?

7. Construct the phase diagram of water *to scale* in the region from 0 to 700 K and from 0 to 250 atm using data in Table 10.3, on both $P$ versus $T$ and log $P$ versus $T$ coordinates.

8. Using the data in Tables 10.2 and 10.3, graphically determine $\Delta \bar{H}_{sub}$ and $\Delta \bar{H}_{fus}$: (a) for H$_2$O; and (b) for CO$_2$. Use your results to compute $\Delta \bar{S}_{vap}$ and $\Delta \bar{S}_{fus}$, and $\Delta \bar{S}_{sub}$ at $P = 1$ atm for H$_2$O and CO$_2$, respectively.

9. Suppose a sample of supercooled water at 263 K in a closed dewar flask is agitated until crystallization begins. What will be the temperature of the water in the flask when crystallization ceases?

10. Discuss the effectiveness of two-phase equilibrium systems as thermostats. What is the heat capacity of a two-phase equilibrium system?

11. What is the minimum amount of data necessary to make a rough sketch of the phase diagram of a substance exhibiting (a) a single solid phase, and (b) two solid phases?

12. The normal boiling point of a certain ether is $260°C$. Using Trouton's rule, estimate its vapor pressure at $23°C$.

13. Suppose the completely immiscible liquids H$_2$O($\ell$) and C$_6$H$_6$($\ell$) are placed together in an air-free flask, which is then sealed. What would be the pressure in the flask at $25°C$? What is the boiling temperature of the system when it is open to the atmosphere?

14. Estimate the temperature of a piece of $CO_2(s)$ exposed to the atmosphere on days when the atmospheric pressure is 730, 760, and 800 Torr, respectively.

15. Sulfur exhibits two solid phases, rhombic and monoclinic sulfur. Make a rough sketch on log $P$ versus $T$ coordinates of the phase diagram given in the following data:

| Triple Point | $t(°C)$ | $P$ |
|---|---|---|
| R – M – g | 95.5°C | 0.01 Torr |
| M – $\ell$ – g | 120°C | 0.025 Torr |
| R – M – $\ell$ | 151°C | 1290 atm |
| R – $\ell$ – g | 113°C | 0.020 Torr |

16. Arrange the following gases in the order of increasing critical temperature (Table 10.1) and explain in qualitative terms the reason(s) for the observed order: $NH_3$, $Br_2$, $CO_2$, $Cl_2$, $H_2$, $D_2$, He, HCl, $N_2$, $O_2$, $H_2O$, $D_2O$, $CH_3Cl$.

17. Determine the maximum *possible* number of triple points for a substance exhibiting (a) two solid phases, a liquid, and a gas phase, and (b) five solid phases, a liquid, and a gas phase.

18. Can the equilibrium vapor pressure of a pure liquid be increased by dissolution of a solute?

19. It is usually a fairly good approximation to take $\Delta \bar{H}_{vap}$ as independent of $P$ over a range of a few atmospheres. Is this approximation also applicable to $\Delta \bar{S}_{vap}$?

20. Will phase separation occur if helium gas at 10 K is subjected to very high pressure?

21. What can you say about the relative magnitudes of $\Delta \bar{H}_{sub}$ and $\Delta \bar{H}_{vap}$ of a substance in the vicinity of the triple point? How does this result manifest itself on the $P$ versus $T$ phase diagram?

22. What can you say about the signs and relative magnitudes of $\Delta S_{tot}$, $\Delta S_{sys}$, and $\Delta S_{surr}$ for the following processes?

$$H_2O(\ell, 263 \text{ K}, 1 \text{ atm}) \rightarrow H_2O(s, 263 \text{ K}, 1 \text{ atm})$$

$$H_2O(\ell, 383 \text{ K}, 1 \text{ atm}) \rightarrow H_2O(g, 383 \text{ K}, 1 \text{ atm})$$

23. Using the van der Waals equation,

$$\left(P + \frac{a}{\bar{V}^2}\right)(\bar{V} - b) = RT$$

plot $P$ versus $\bar{V}$ (the molar volume) at $T$ values of 250, 275, 300, 304.2, 325, and 350 K, for $CO_2$. Take $a = 3.59 \; \ell^2 \cdot \text{atm} \cdot \text{mole}^{-2}$ and $b = 0.0427 \; \ell \cdot \text{mole}^{-1}$. Discuss the significance of your results as regards gas liquification and the critical point.

24. Discuss the significance to life on earth of the fact that the density of ice I is less than that of liquid water.

25. Although it is possible to obtain a metastable supercooled liquid in equilibrium with its vapor, it is not possible to obtain a solid above its melting point (i.e., "superheated" solid) in equilibrium with its liquid. Why?

26. In computing the number of components in water via the expression $C = N - r$, only the species $H_2O(\ell)$, $H^+(aq)$, and $OH^-(aq)$ were considered. Why was it permissible to ignore species such as $H_3O^+$, $H_5O_2^+$, $H_7O_3^+$, $H_9O_4^+$, and other such ions, as well as $(H_2O)_2$, $(H_2O)_3$, and so on? Similarly, why was it permissible to ignore isotopically different species such as $H_2O^{16}$, $H_2O^{17}$, $D_2O^{18}$, and $HDO^{17}$?

27. The ratio of the equilibrium vapor pressure of small spherical drops of radius $r$, $P_r$, to that of the bulk liquid, $P_o$, is given by $P_r/P_o = \exp(2\sigma \bar{V}/rRT)$, where $\sigma$ is the surface tension, $\bar{V}$ is the molar volume, $R$ is the gas constant, and $T$ is the absolute (Kelvin) temperature. Compute $P_r/P_o$ at 25°C for water droplets with radii of 1 mm, $1 \times 10^{-4}$ mm, and $6 \times 10^{-8}$ mm, respectively. Take $\sigma = 80$ dyne $\cdot$ cm$^{-1}$ for water.

28. Discuss the effect of subjecting a living organism to a high vacuum.

29. The equilibrium ratio of gas pressures at two heights, $h$ and 0, $P_h/P_o$, in a column of gas at uniform temperature, subject along its entire length to a uniform gravitational field, is given by $P_h/P_o = \exp(-Mgh/RT)$, where $M$ is the molecular weight, $g$ is the gravitational acceleration, $R$ is the gas constant, and $T$ is the absolute temperature. Taking the molecular weight of air as 29 g $\cdot$ mole$^{-1}$, estimate the boiling point of water at 30,000 feet elevation. Would this be the approximate boiling point of water in an airplane at that elevation?

30. Phosphorus exhibits three solid phases, namely red (violet), white (yellow), and black. The white modification is *metastable at all values of T and P*; the black modification is the stable form at very high pressures, but can be obtained in a metastable (?) state at 1 atm. Given the following data, make a rough sketch of the phase diagram: the red-liquid-gas triple point occurs at 590°C and 43 atm, the white-liquid-gas triple point occurs at 44°C and 0.18 mm, liquid white and supercooled liquid red are the same, and the values of $\Delta P/\Delta T$ for the melting of white and red phosphorus are roughly equal.

31. The equilibrium vapor pressure of solid carbon dioxide is 1.00 atm at −78.3°C. The enthalpy change for the reaction

$$CO_2(s) = CO_2(g)$$

is 6.03 kcal at −78.3°C. The polar caps on Mars are thought to be composed primarily of $CO_2(s)$. Given that the pressure of $CO_2$ in the Martian atmosphere is about 4.0 Torr, estimate the temperature at the surface of the polar caps on Mars.

32. Assuming that one is able to ice skate because of the formation of a thin surface film of water arising from the decrease in the melting point of

ice due to blade pressure, suggest an explanation for the observation that it is impossible to ice skate at very low temperatures (the sensation being somewhat like trying to skate on glass).

_____ References

10.1.  A. Findlay, *Phase Rule*, 9th Edition, revised by A. N. Campbell and N. O. Smith (Dover Publications, 1951).

10.2.  F. D. Ferguson and T. K. Jones, *The Phase Rule* (Butterworths Pub., Washington, 1966).

10.3.  J. W. Stewart, *The World of High Pressure* (Van Nostrand Rinehold Co., New York, 1967).

10.4.  D. Schoenberg, *Superconductivity* (Cambridge University Press, 1965).

10.5.  W. H. Keesom, *Helium* (Elsevier Publishing Co., Amsterdam, 1942).

10.6.  J. V. Sengers and A. L. Sengers, *The Critical Region*, Chem. Eng. News, *46*, 104 (1968).

10.7.  C. Domb, *Thermodynamics of Critical Points*, Physics Today, *21*, 23 (1968).

10.8.  K. Mendelssohn, *The Quest for Absolute Zero* (McGraw-Hill Book Co., New York, 1966).

10.9.  C. A. Knight, *The Freezing of Supercooled Liquids* (Van Nostrand Rinehold Co., New York, 1967).

# CHAPTER 11

# CHEMICAL EQUILIBRIUM
## Part I: Principles

## 11.1  THE CHEMICAL POTENTIAL AND THE ACTIVITY.

The thermodynamic analysis of chemical equilibria is based upon the concept of the *chemical potential*. Just as energy is transferred spontaneously down a temperature gradient as heat, and down a pressure gradient as work, energy is transferred spontaneously down a chemical potential gradient as mass. This analogy is depicted schematically in Figure 11.1. The chemical potential is a kind of chemical pressure that governs the escaping tendency of a substance from one phase to another, or from one region of a particular phase to some other region of that phase. When a system is in a state of equilibrium, no spontaneous processes‡ (including mass transfer) can take place, and, therefore, *at equilibrium the chemical potential of a given component of the system must be the same throughout the system.* If mass transfer is possible between the system and the surroundings (open system), then, at equilibrium, the chemical potential of a given component must be the same in the system and in the surroundings.

We shall designate the chemical potential of the species $i$ by the symbol $\bar{G}_i$ (read as the chemical potential of $i$). The chemical potential of a pure phase in an equilibrium state (in the absence of external force fields

---

*C. M. Guldberg and P. Waage, *Studier over Affiniteten* (Special reprint of Vid-Selsk. Forhandlinger for 1864), Facsimilia Scientifica et Technica Norvegica 15. Translated by Professor H. Hope, University of California at Davis.

†J. Willard Gibbs, *On the Equilibrium of Heterogeneous Substances*, Transactions of the Connecticut Academy, III, October 1875 (reprinted by Dover Publications, New York, 1961 as *The Scientific Papers of J. Willard Gibbs*, Vol. I, Thermodynamics).

‡Recall that by "process" we mean a change in state.

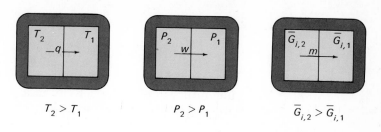

**FIGURE 11.1** The gradient of chemical potential as the driving force for mass transfer. The mass that flows down the chemical potential gradient is the substance $i$.

and surface effects) is completely determined by the specification of the temperature and pressure of the phase. *For a pure phase* the chemical potential is simply the Gibbs energy per mole, that is, $\bar{G}_i = G_i/n_i$ (Chapter 10). On the other hand, the chemical potential of a species in solution depends not only on the temperature and pressure, but also on the *composition* of the solution. The chemical potential of a solution species is a *property of the solution* that depends not only on the nature of the species itself, but also upon the nature and relative amounts of the other species present in the solution. Although the chemical potential of a species depends upon the composition of a phase, it does not depend upon the total amount of the phase. The *chemical potential is an intensive quantity* like temperature and pressure. The units of the chemical potential are energy/mole.

The change in the chemical potential of a species that undergoes a change of state (without passing from one phase to another) under isothermal conditions is conveniently expressed in terms of the logarithm of the ratio of the *activities* of the species in the two states:

$$\bar{G}_i - \bar{G}_i' = RT \ln \frac{a_i}{a_i'} \qquad \text{(isothermal change of state)} \qquad (11.1)$$

where $a_i$ is the activity of $i$ in the final (unprimed) state, and $a_i'$ is the activity of $i$ in the initial (primed) state; $R$ is the gas constant and $T$ is the absolute temperature. The activity may be thought of as the *active* (or *effective*) concentration of a solution species, or the *active* (or *effective*) pressure of a gas. If the substance behaves ideally, then the activity of a solution species is equal to its concentration, and the activity of a gaseous species is equal to its pressure. If the solution or gas is not ideal, then the activity will deviate from the concentration or pressure by an amount that is greater, the greater the deviation from ideality. Like the chemical potential, the activity is a measure of the escaping tendency of a substance from a phase, or from one region of a phase to another region of a phase. At equilibrium, the activity of a substance is uniform throughout a phase and is time-independent.

Because only *differences* in the chemical potential are measureable, it is conventional to refer the chemical potential of a species in a particular phase of the system to an arbitrarily chosen *standard state* at each temperature. We designate the standard state by means of a superscript zero. Thus, the change in the chemical potential of a species that undergoes the change in state

$$\text{standard state} \xrightarrow[\text{no phase change}]{\text{isothermal}} \text{arbitrary state}$$

is given by

$$\bar{G}_i - \bar{G}_i^\circ = RT \ln \left(\frac{a_i}{a_i^\circ}\right)$$

The numerical value of the activity in a particular state is fixed by arbitrarily setting $a_i^\circ$, the activity in the standard state, equal to unity. Thus

$$\boxed{\bar{G}_i - \bar{G}_i^\circ = RT \ln a_i}$$

(11.2a)

Like the chemical potential, the activity of a pure substance is a function of temperature and pressure, and the activity of a species in solution is a function of temperature, pressure, and composition.

If the species of interest undergoes an isothermal change of state that also involves a change in phase, say, from the $\alpha$ to the $\beta$ phase, then the change in the chemical potential is given by

$$\bar{G}_{i\beta} - \bar{G}_{i\alpha} = \bar{G}_{i\beta}^\circ - \bar{G}_{i\alpha}^\circ + RT \ln \frac{a_{i\beta}}{a_{i\alpha}}$$

(11.2b)

If the standard state is the same in both phases (which is not necessarily the case), then $\bar{G}_{i\beta}^\circ = \bar{G}_{i\alpha}^\circ$, and this equation reduces to Equation (11.1).*

The difference in chemical potential between two real states of a substance is a measurable quantity, and consequently its value must be independent of the choice of standard states.† However, the values of $\bar{G}_i^\circ$ and $a_i$ depend on the choice of standard state.

We shall employ the following (arbitrarily chosen) standard states:

(a) For Gases: *the ideal gas at 1 atm and the temperature of interest.* The activity of an ideal gas is equal to the pressure of the gas:

$$a_i = P_i \quad \text{(ideal gas)}$$

(11.3)

If the gas is not ideal, then $a_i \neq P_i$. At sufficiently low pressures, all gases behave ideally, and the approximation of a real gas as an ideal gas is more reliable the lower the pressure of the gas. Because the deviation of $a_i/P_i$ from unity is not large within the low-to-moderate pressure (say, less than 5 atm) range, we shall assume ideal-gas behavior in our thermodynamic analysis of chemical equilibria.‡

(b) For Solids and Liquids: *the pure (fixed-composition) bulk* (i.e., not too finely divided)§ *solid or liquid at 1 atm total pressure and the temperature of interest.* For solids we must add the additional qualifica-

---

*Note that if the standard state chosen for $i$ is the same in the two phases, then at equilibrium $(\bar{G}_{i\beta} = \bar{G}_{i\alpha})$, $a_{i\beta} = a_{i\alpha}$. However, if the standard states are different, then $a_{i\beta} = a_{i\alpha} \cdot \exp\{(\bar{G}_{i\alpha}^\circ - \bar{G}_{i\beta}^\circ)/RT\} = a_{i\alpha} \cdot (\text{constant})$ at a given $T$.

†The system has no way of knowing what state we have chosen as the standard state.

‡This is not a necessary assumption, but it does simplify the treatment considerably without much loss in accuracy.

§It is sufficient for our purposes to take "not too finely divided" to mean particles of average diameter greater than about $10^{-3}$ cm.

tions that the substance not be strained. A strained material has a higher Gibbs energy (and therefore a higher activity) than the same amount of unstrained material at the same temperature and pressure; the excess Gibbs energy is equal to the minimum amount of work that must be done on the unstrained material to produce the given degree of strain. Similarly, the activity of finely-divided particles of a solid, or very small droplets of a liquid, is higher than the bulk material, owing to the additional surface energy of the finely-divided material. Work must be done to produce small droplets from bulk material, and conversely small particles will coalesce spontaneously into larger particles. The activity of a condensed phase varies only very slowly with pressure. Consequently, the activity of a pure condensed phase can be set equal to unity at all pressures and temperatures without serious error, unless the pressure is very high ($>100$ atm).

(c) **For Solution Species:** There are two commonly employed choices of standard state for solution species, namely, the *solvent* and *solute* standard states. The solvent standard state is always the preferred choice for a solution species, but unfortunately it is not always a feasible choice. If the solution is liquid and the thermodynamic properties of a solution species as a pure liquid are known (or can be obtained) at the temperature of the solution, then the solvent standard state is chosen. This standard state is the same as that described for pure liquids under (b). Similarly, if the solution is a solid solution and the necessary data on the pure solid are obtainable at the temperature of interest, then the solvent standard state for the solid is chosen. This standard state is the same as that described for solids under (b) above.

When the foregoing conditions cannot be met, then a solute standard state must be chosen. The solute standard state is the hypothetical, ideal, unit-composition solution at 1 atm total pressure and the temperature of interest. It is sufficient for our purposes to define an ideal solution as one for which the activities of all the components of the solution are equal to their respective concentrations,† that is, $a_i = X_i$, or $a_i = m_i$, or $a_i = M_i$.

For the solvent standard state, the activity of $i$ *in solution* at a particular composition is given by

$$a_i(\text{soln}) \simeq \frac{P_i}{P_i^\circ} \qquad (11.4)$$

where $P_i$ is the equilibrium pressure of gaseous $i$ over the solution, and $P_i^\circ$ is the equilibrium vapor pressure of pure $i$ at the same temperature as the solution. For example, suppose an aqueous salt solution has an equilibrium vapor pressure of water over the solution of 19.5 Torr at 25°C. The equilibrium vapor pressure of pure water at 25°C is 23.5 Torr; therefore, the activity of water in the solution is $a_{\text{H}_2\text{O}} = 19.5/23.5 = 0.83$. Ideality in the gas phase in no way implies ideality in the solution phase in equilibrium with the gas phase.

---

† Because, in general, $c_i \neq m_i \neq X_i$, this means that the value of $a_i$ for a solution species depends on the composition scale chosen, and this is also true for the standard states themselves. Therefore, when a solute standard state is chosen, the value of $\bar{G}_i^\circ$ depends on the composition scale used.

For the activity of a species in solution relative to a solute standard state on the molality composition scale, we have

$$a_{im}(\text{soln}) \simeq \frac{P_i}{P_{im}^*} \qquad \text{(molality)} \qquad (11.5)$$

The quantity $P_{im}^*$ is the pressure of the species $i$ in the gas phase in equilibrium with the solution containing $i$ in the solute standard state. This quantity must be determined experimentally by measurements on the dilute solutions. The establishment of the solute standard state on the molality scale is equivalent to the determination of the value of $P_{im}^*$ at the temperature of interest. The determination of $P_{im}^*$ requires an extended series of measurements at low solute concentrations.

By way of example, consider the choice of standard states for the components of the following solutions:

(a) Liquid solutions containing ethanol ($CH_3CH_2OH$) and water at 25°C. In this case both ethanol and water exist as pure liquids at 25°C and 1 atm total pressure, and we therefore choose a solvent standard state for both components.

(b) Liquid solutions of oxygen in water at 25°C. Pure $O_2(\ell)$ is not obtainable at 25°C and 1 atm; therefore, a solute standard state must be chosen for $O_2$ dissolved in water (e.g., the hypothetical $a_{O_2} = m_{O_2} = 1$ solution at 25°C and 1 atm). We take pure $H_2O(\ell)$ at 1 atm and 25°C as our choice for the standard state (solvent) of water.

(c) Aqueous solutions of NaCl at 20°C. For water we choose the solvent standard state (pure $H_2O(\ell)$ at 20°C and 1 atm total pressure), whereas for NaCl we must choose a solute standard state (the hypothetical $a_{NaCl} = m_{NaCl} = 1$ solution at 20°C and 1 atm).

(d) Silver-gold alloys at 250°C. This alloy is a solid solution.† Because both silver and gold can be obtained as the pure solids at 250°C and 1 atm, we choose a solvent standard state for both substances—the pure bulk unstrained solids at 1 atm and 250°C.

## 11.2 THE ACTIVITY OF SOLUTION SPECIES. RAOULT'S LAW.

Raoult's Law states that the equilibrium vapor pressure of a volatile component over a solution is given by the product of its mole fraction *in the solution* and the equilibrium vapor pressure of the pure liquid at the same temperature as the solution; that is,

$$P_1 = X_1 P_1^\circ \qquad (11.6)$$

For example, suppose we have solution of sucrose in water at 25°C that is 0.500 m in sucrose. The equilibrium vapor pressure of pure water at 25°C is 23.75 Torr. The mole fraction of water in the solution is

$$X_{H_2O} = \frac{(1000/18.016)}{(1000/18.016) + 0.500} = 0.991$$

---

†Not all alloys are solid solutions. Some consist of intermetallic compounds (such as brass), and other alloys (such as steel) are heterogeneous.

and Raoult's Law predicts that the equilibrium vapor pressure of water over the solution is

$$P_{H_2O} = X_{H_2O} P^\circ_{H_2O} = (0.991)(23.75 \text{ Torr}) = 23.53 \text{ Torr}$$

which is equal to the measured value.

In Figure 11.2 we have plotted $P_{H_2O}$ vs $X_{H_2O}$ for aqueous sucrose solutions over the range from 0 to 2 m in sucrose, which corresponds to the range from 1.000 to 0.965 in $X_{H_2O}$. It can be seen in Figure 11.2 that Raoult's Law holds for water in these solutions over the range $1.00 \geq X_{H_2O} \geq 0.987$, which corresponds to sucrose concentrations in the range $0 \leq m \leq 0.7$. Above 0.7 m in sucrose, the equilibrium vapor pressure of water is less than that predicted by Raoult's Law.

If Raoult's Law holds for a component in a solution, then the activity of that component in the solution is equal to its mole fraction:

$$a_i \simeq \frac{P_i}{P^\circ_i} = \frac{X_i P^\circ_i}{P^\circ_i} = X_i$$

If Raoult's Law does not hold, then $a_i \neq X_i$ and the solution is said to be non-ideal.

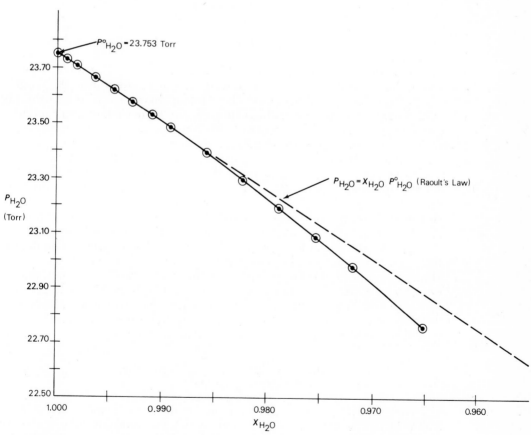

**FIGURE 11.2**   Equilibrium vapor pressure of water over aqueous sucrose solutions at 25°C.

We have no *a priori* way of knowing the range of concentration over which Raoult's Law holds for a volatile component of a solution. However, we can say that $P_1 \to X_1 P_1^\circ$ as $X_1 \to 1$; that is, the major component of a solution approaches Raoult's Law behavior as the mole fraction of the major component approaches unity.

## 11.3 HENRY'S LAW AND DILUTE SOLUTIONS.

An important class of solutions is the dilute solution. As is implied by the name, dilute solutions have one major component (*the solvent*), denoted by a subscript 1, and one (or possibly more than one) minor component (*the solute*) denoted by a subscript 2. A dilute solution is defined thermodynamically as one in which the solvent obeys Raoult's law and the solute obeys *Henry's law*. Henry's law states that the equilibrium vapor pressure of the solute over the solution is directly proportional to its concentration in the solution. For a given solute in a given solvent the proportionality constant, or *Henry's-law constant*, depends on the temperature and the concentration scale chosen. On the molality scale Henry's law takes the form

$$P_2 = P_{2m}^* \, m \tag{11.7}$$

where $P_2$ is the pressure of the solute over the solution, and $P_{2m}^*$ is the (experimentally determined) Henry's-law constant.

If Raoult's law holds for the solvent, then Henry's law holds for the solute and vice versa. There is no *a priori* way to determine the range of composition over which Henry's law holds; this must be established by experiment. Furthermore, the value of the Henry's-law constant for a given solute in a given solvent at a given temperature is a property of the *solution*; it cannot be obtained from measurements on the pure solute, but rather it must be obtained from measurements on dilute solutions.

Consider the determination of the Henry's-law constant for $O_2$ dissolved in water, that is, $P_{O_2,m}^*$. Experimental measurements of the solubility of $O_2$ in water as a function of pressure yield $P_{O_2}$ and $m_{O_2}$. The resulting data are plotted on $P_{O_2}$ versus $m_{O_2}$ coordinates. If the measurements are carried out at sufficiently low pressures, a straight line is obtained; the slope of the straight line is equal to the Henry's law constant. The data for the $O_2 + H_2O$ system are presented in Figure 11.3, which shows that Henry's law holds for $O_2$ in $H_2O$ over the range $0 \leq P_{O_2} < 6$ atm, and $P_{O_2}$ deviates only slightly from Henry's law even at $P_{O_2}$ values as high as 10 atm.

As an example of the use of Henry's law, we shall now calculate the molality of $O_2$ in water in equilibrium with the atmosphere at 25°C. The partial pressure of $O_2(g)$ in the air is 0.20 atm, and thus

$$m_{O_2} = \frac{0.20 \text{ atm}}{773 \text{ atm} \cdot \text{mole}^{-1} \cdot \text{kg } H_2O} = 2.6 \times 10^{-4} \text{ mole/kg } H_2O$$

Henry's-law constants for some common gases in water are given in Table 11.1. The data in Table 11.1 show that gas solubility decreases with increasing temperature. This is invariably the case for gases dissolved in water (but not in all solvents).

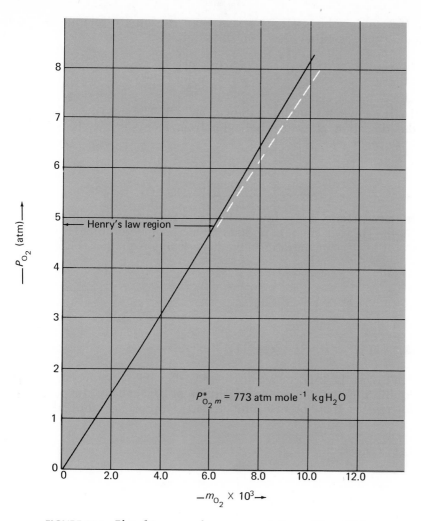

**FIGURE 11.3** Plot of $P_{O_2}$ vs $m_{O_2}$ for aqueous solutions of $O_2$ at 25°C.

**TABLE 11.1**

HENRY'S-LAW CONSTANTS FOR SOME COMMON GASES IN AQUEOUS SOLUTION.

| GAS | $t$ (°C) | $P_{2m}^{*}$ (atm · mole$^{-1}$ · kg $H_2O$) |
|---|---|---|
| $N_2$ | 0 | 950 |
| $N_2$ | 25 | 1610 |
| $H_2$ | 25 | 1310 |
| He | 25 | 2650 |
| $CO_2$ | 0 | 13.2 |
| $CO_2$ | 15 | 22.2 |
| $CO_2$ | 25 | 29.4 |
| $H_2S$ | 25 | 10 |
| $O_2$ | 0 | 453 |
| $O_2$ | 25 | 773 |
| $O_2$ | 40 | 962 |

If a solute obeys Henry's law, then its activity relative to a solute standard state is equal to its concentration:

$$a_2 = \frac{P_2}{P_{2m}^*} = \frac{m_2 P_{2m}^*}{P_{2m}^*} = m_2$$

If Henry's law does not hold, then $a_2/m_2 \neq 1$. The ratio $a_2/m_2$ is a measure of the deviation of solute behavior from Henry's law; it is called the *activity coefficient*, $\gamma_2$

$$\boxed{\gamma_2 = a_2/m_2}$$  (11.8)

Note that if Henry's law holds, then $\gamma_2 = 1$. On the other hand, if we are not in the Henry's law region, but $P_{2m}^*$ has been determined from measurements at lower concentrations, we have (using Equation (11.5))

$$\gamma_2 = \frac{a_2}{m_2} = \frac{P_2}{m_2 P_{2m}^*} \neq 1$$  (11.9)

## 11.4 ELECTROLYTE SOLUTIONS.

In Figure 11.4 we have constructed a plot of $P_{HCl}$ versus $m_{HCl}$ for HCl(g) in equilibrium with an aqueous solution of HCl at 25°C. Inspection of the figure reveals that, in the limit as $m_{HCl}$ approaches zero, the quantity

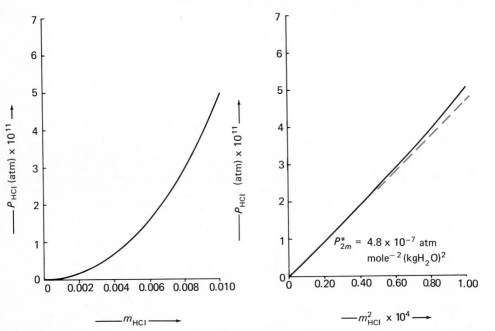

**FIGURE 11.4**  Partial pressure of HCl(g) over HCl(aq) solutions as a function of $m_{HCl}$ at 25°C.

**FIGURE 11.5**  Partial pressure of HCl(g) over HCl(aq) solutions as a function of $m_{HCl}^2$ at 25°C.

$P_{HCl}/m_{HCl}$ approaches zero, rather than $P_{2m}^*$, as expected from Henry's law. In Figure 11.5 we have constructed a plot of $P_{HCl}$ versus $m_{HCl}^2$ using the same data as were used in Figure 11.4. Figure 11.5 shows that, over the range $0 \leqslant m_{HCl}^2 < 0.6 \times 10^{-4}$, Henry's law is obeyed if it is expressed in the form

$$P_{HCl} = P_{HCl,m}^* \, m_{HCl}^2 \qquad (11.10)$$

rather than $P_{HCl} = P_{HCl,m}^* \, m_{HCl}$. The reason why Equation (11.10) holds is that in aqueous solution HCl is a *strong electrolyte*; that is, the solution species are $H^+(aq)$ and $Cl^-(aq)$, rather than *undissociated* HCl(aq). When dealing with strong electrolytes, we can set up a solute standard state by writing the activity of the electrolyte as the *product* of the activities of the individual ions that it yields in solution, and then taking (in the special case of a 1:1 electrolyte) Henry's law in the form $P_2 = P_{2m}^* m^2$. The experimental determination of $P_{2m}^*$ is then sufficient to establish this solute standard state. The activity of HCl(aq) relative to a strong electrolyte solute standard state is then given (on the molality composition scale) by

$$a_{HCl} = a_{H^+} a_{Cl^-} = \frac{P_{HCl}}{P_{HCl,m}^*} \qquad (11.11)$$

and, taking $a_i = m_i \gamma_i$ (Equation (11.9)), we have

$$a_{HCl} = a_{H^+} a_{Cl^-} = m_{H^+} m_{Cl^-} \gamma_{H^+} \gamma_{Cl^-} \qquad (11.12)$$

If HCl is the only significant source of $H^+(aq)$ and $Cl^-(aq)$ in the solution, then $m_{H^+} = m_{Cl^-} = m_{HCl}$. Also, because $\gamma_{H^+}$ and $\gamma_{Cl^-}$ are not separately measurable—only the product $\gamma_{H^+} \gamma_{Cl^-}$ being susceptible to experimental determination—it is convenient to define the *mean ionic activity coefficient*, $\gamma_\pm$, as

$$\gamma_{\pm(HCl)} = (\gamma_{H^+} \gamma_{Cl^-})^{1/2} \qquad (11.13)$$

Substitution of these equations into Equation (11.12) yields

$$a_{HCl} = a_{H^+} a_{Cl^-} = m_{HCl}^2 \gamma_{\pm(HCl)}^2 = \frac{P_2}{P_{2m}^*} \qquad (11.14)$$

In the region where $P_2 = P_{2m}^* \, m_{HCl}^2$, the mean activity coefficient of HCl, $\gamma_{\pm(HCl)}$, is equal to unity.†

In Table 11.2 we have assembled $\gamma_\pm$ data for some common electrolytes in water. The data in Table 11.2 show that strong electrolytes in water deviate significantly from dilute solution behavior even at low concentrations; consequently, it is seldom a good assumption to take $\gamma_\pm = 1$ for an

---

†For a strong electrolyte like $CaCl_2(aq)$ which yields $Ca^{2+} + 2Cl^-$ in water, we have

$$a_{CaCl_2} = a_{Ca^{2+}} a_{Cl^-}^2 = m_{Ca^{2+}} m_{Cl^-}^2 \gamma_{Ca^{2+}} \gamma_{Cl^-}^2 = m_{Ca^{2+}} m_{Cl^-}^2 \gamma_{\pm(CaCl_2)}^3 = \frac{P_2}{P_{2m}^*}$$

where $\gamma_{\pm(CaCl_2)} = (\gamma_{Ca^{2+}} \gamma_{Cl^-}^2)^{1/3}$. Note that the mean ionic activity coefficient is the *geometric mean* of the individual ion activity coefficients.

TABLE 11.2

MEAN ACTIVITY COEFFICIENT DATA* FOR SOME COMMON ELECTROLYTES IN AQUEOUS SOLUTION AT 25°C, 1 atm.

| | MOLALITY | | | | | | |
|---|---|---|---|---|---|---|---|
| ELECTROLYTE | 0.10 | 0.30 | 0.50 | 0.70 | 1.00 | 2.00 | 3.00 |
| HCl(aq) | 0.796 | 0.756 | 0.757 | 0.772 | 0.809 | 1.009 | 1.316 |
| KCl(aq) | 0.769 | 0.687 | 0.649 | 0.626 | 0.603 | 0.572 | 0.568 |
| NaCl(aq) | 0.778 | 0.710 | 0.681 | 0.667 | 0.657 | 0.668 | 0.719 |
| NaOH(aq) | 0.764 | 0.706 | 0.688 | 0.680 | 0.677 | 0.707 | 0.782 |
| LiClO$_4$(aq) | 0.812 | 0.792 | 0.808 | 0.839 | 0.887 | 1.158 | 1.582 |
| KNO$_3$(aq) | 0.739 | 0.614 | 0.545 | 0.496 | 0.443 | 0.333 | 0.269 |
| CaCl$_2$(aq) | 0.518 | 0.455 | 0.448 | 0.460 | 0.500 | 0.792 | 1.483 |
| Na$_2$SO$_4$(aq) | 0.452 | 0.325 | 0.270 | 0.237 | 0.204 | 0.154 | 0.139 |
| H$_2$SO$_4$(aq) | 0.266 | 0.183 | 0.156 | 0.142 | 0.132 | 0.128 | 0.142 |
| K$_3$PO$_4$(aq) | 0.312 | 0.211 | 0.175 | 0.156 | —— | —— | —— |
| LaCl$_3$(aq) | 0.314 | 0.263 | 0.266 | 0.285 | 0.342 | 0.825 | —— |

*Data taken from R. A. Robinson and R. H. Stokes, *Electrolyte Solutions* (Butterworths, London, 1959).

electrolyte, even at quite low concentrations. The data on HCl(aq) in Table 11.3 show that $\gamma_\pm$ deviates from unity by 3.5% even at concentrations as low as 0.001 molal.

The determination of $\gamma_\pm$ for an electrolyte involves an extended series of time-consuming experiments of high precision,* and such data are often not available for particular cases of interest. In certain cases a useful estimate of $\gamma_\pm$ for $m \leq 0.1$ can be made using the Debye-Hückel equation:

$$\log \gamma_\pm = \frac{-A \mid Z_+ Z_- \mid I^{1/2}}{I + I^{1/2}} \qquad (11.15)$$

*In most cases the equilibrium value of $P_2$ is much too small to measure, and other methods must be used to obtain $a_2$. The two principal methods are: (a) vapor pressure measurements on the solvent which give $a_1$ as a function of $m_2$ from which $a_2$ can be computed; and (b) electrochemical cell measurements which yield $a_2$ directly (see Chapter 13).

TABLE 11.3

ACTIVITY AND ACTIVITY COEFFICIENT DATA FOR HCl(aq) AT 25°C, 1 atm

| m | $\gamma_\pm$(obs) | $\gamma_\pm$(calc)° | $a_2$(obs) |
|---|---|---|---|
| 0.0001 | 0.988 | 0.988 | $9.8 \times 10^{-9}$ |
| 0.001 | 0.965 | 0.965 | $9.3 \times 10^{-7}$ |
| 0.01 | 0.904 | 0.899 | $8.17 \times 10^{-5}$ |
| 0.1 | 0.796 | 0.754 | $6.34 \times 10^{-3}$ |
| 1 | 0.809 | 0.555 | 0.655 |
| 10 | 10.44 | 0.409 | $1.09 \times 10^4$ |

*Calculated using the expression

$$\log \gamma_\pm = \frac{-0.511 \, m^{1/2}}{1 + m^{1/2}} \, .$$

where $Z_+$ is the charge on the cation and $Z_-$ is the charge on the anion of the electrolyte of interest, $A$ is a constant for a given solvent at a given temperature and pressure, and $I$ is the ionic strength of the solution. The ionic strength is defined by the equation (G. N. Lewis)

$$\boxed{I \equiv \frac{1}{2} \sum_i m_i Z_i^2}$$ (11.16)

where the sum extends over *all* the ions in solution. For water at 25°C, $A = 0.511$, and for HCl in water $Z_+ = +1$ and $Z_- = -1$; hence,

$$I = \frac{1}{2} \left\{ m_{H^+}(1)^2 + m_{Cl^-}(1)^2 \right\} = m$$

and therefore

$$\log \gamma_\pm = \frac{-0.511\ m^{1/2}}{1 + m^{1/2}}$$

The results obtained for $\gamma_\pm$ using this equation are compared with the experimentally determined values for HCl(aq) in Table 11.3.

As a second example, consider the application of the Debye-Hückel equation to the estimation of $\gamma_\pm$ for $CaCl_2$ in an aqueous solution which is 0.010 m in $CaCl_2$(aq) and 0.050 m in NaCl(aq). The ionic strength of the solution is

$$I = \frac{1}{2} \left\{ m_{Na^+}(1)^2 + m_{Cl^-}(1)^2 + m_{Ca^{2+}}(2)^2 \right\}$$

$$I = \frac{1}{2} (0.050 + 0.070 + 0.040) = 0.080$$

and

$$\log \gamma_\pm = \frac{-0.511\,(2)\,(0.080)^{1/2}}{1 + (0.080)^{1/2}}$$

or $\gamma_\pm = 0.595$.

Equation (11.15) predicts that the value of $\gamma_\pm$ in a given solvent at a given temperature, total pressure, and ionic strength is the same for all strong electrolytes that have the same value for $|Z_+ Z_-|$.

## 11.5 APPLICATION OF THERMODYNAMICS TO CHEMICAL REACTIONS. THE EQUILIBRIUM CONSTANT.

In this section we shall develop several basic thermodynamic relationships that are useful in the analysis of chemical equilibria. The expressions obtained will be used in subsequent sections(a) to predict whether or not a given chemical reaction is spontaneous for a particular set of conditions, and (b) to calculate the equilibrium distribution of species under various conditions.

Consider the generalized balanced chemical equation

$$mM + nN + \ldots = xX + yY + \ldots$$ (11.17)

where the reactants $(M, N, \ldots)$ and products $(X, Y, \ldots)$ can be pure solids, liquids, gases, or solution species, and $m, n, \ldots, x, y, \ldots$ are the balancing coefficients. The chemical potentials of the various reactants and products are given by expressions of the type (Equation (11.2a))

$$\bar{G}_M = \bar{G}_M^\circ + RT \ln a_M$$

The total Gibbs energy change, $\Delta G$, for Equation (11.17) is given by[*]

$$\Delta G = (x\bar{G}_X + y\bar{G}_Y + \ldots) - (m\bar{G}_M + n\bar{G}_N + \ldots) \qquad (11.18)$$

If we restrict our attention to *isothermal* reactions, then the substitution of an expression like that for $\bar{G}_M$ shown above for each reactant and product into Equation (11.18) yields

$$\Delta G = (x\bar{G}_X^\circ + y\bar{G}_Y^\circ + \ldots) - (m\bar{G}_M^\circ + n\bar{G}_N^\circ + \ldots)$$
$$+ RT \{(x \ln a_X + y \ln a_Y + \ldots) - (m \ln a_M + n \ln a_N + \ldots)\} \quad (11.19)$$

Making the identification

$$\Delta G^\circ = (x\bar{G}_X^\circ + y\bar{G}_Y^\circ + \ldots) - (m\bar{G}_M^\circ + n\bar{G}_N^\circ + \ldots) \qquad (11.20)$$

where $\Delta G^\circ$ is the *standard* Gibbs energy change for the reaction, Equation (11.19) can be rewritten as[†]

$$\Delta G = \Delta G^\circ + RT \ln \left\{ \frac{a_X^x a_Y^y \ldots}{a_M^m a_N^n \ldots} \right\} \qquad (11.21)$$

The *activity quotient*, $Q$, is defined as

$$\boxed{Q = \frac{a_X^x a_Y^y \ldots}{a_M^m a_N^n \ldots}} \qquad (11.22)$$

and this definition of $Q$ enables us to write Equation (11.21) in the more compact form

$$\boxed{\Delta G = \Delta G^\circ + RT \ln Q} \qquad \text{(isothermal reaction)} \qquad (11.23)$$

which gives $\Delta G$ in terms of $\Delta G^\circ$ and the activities of the various reactants and products. It is of the utmost importance for the thermodynamic analysis of chemical equilibria to understand clearly the difference between $\Delta G$ and $\Delta G^\circ$. The distinction between these two quantities is outlined diagrammatically in Figure 11.6. In particular, $\Delta G \neq \Delta G^\circ$ unless $Q = 1$. When all of the reactants and products are in their unit activity standard states, then $Q = 1$, and $\Delta G = \Delta G^\circ$. To determine whether or not a given

---

[*]The value of $\Delta G$ for a reaction is independent of path; it depends only on the final states of the products and the initial states of the reactants. This is because G is a state function.

[†] Note that $n \ln a = \ln a^n$; $\ln a_i^n + \ln a_j^m = \ln a_i^n a_j^m$; and $-\ln a^n = \ln \frac{1}{a^n}$.

The Distinction Between $\Delta G$ and $\Delta G°$.

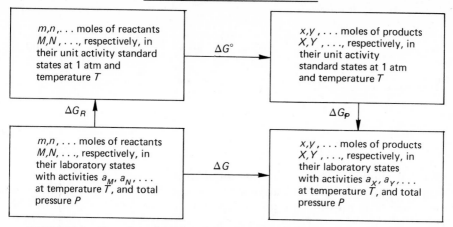

**FIGURE 11.6** The value of $\Delta G$ is independent of path, hence

$$\Delta G = \Delta G_R + \Delta G° + \Delta G_P$$

Furthermore,

$$\Delta G_R = m(\bar{G}_M° - \bar{G}_M) + n(\bar{G}_N° - \bar{G}_N) + \cdots \cdot$$

$$\Delta G_R = -RT \ln a_M^m - RT \ln a_N^n - \cdots \cdot$$

and

$$\Delta G_P = x(\bar{G}_X - \bar{G}_X°) + y(\bar{G}_Y - \bar{G}_Y°) + \cdots \cdot$$

$$\Delta G_P = RT \ln a_X^x + RT \ln a_Y^y + \cdots \cdot$$

Substitution of the expressions for $\Delta G_P$ and $\Delta G_R$ into the above expression for $\Delta G$ yields Equation (11.23).

reaction, with reactants and products at particular activities, is spontaneous under the stated conditions of temperature, pressure, concentrations, and so forth, we insert the values of the activities of the reactants and products into Equation (11.23), together with the value of $\Delta G°$ for the reaction at the given temperature. If the calculated value of $\Delta G$ is less than zero, then the reaction takes place spontaneously* in the direction written under the stated conditions (see Equation (10.2)).

The equilibrium criterion for the reaction is $\Delta G = 0$ (Equation (10.2)), and therefore *at equilibrium* we obtain from Equation (11.23)

$$\Delta G° = -RT \ln Q_{eq} \qquad (11.24)$$

where $Q_{eq}$ is the value of $Q$ *at equilibrium*. Because $\Delta G°$ for a given reaction depends only on the temperature and the choice of standard states, the value of $Q_{eq}$ must depend only on the temperature for a given choice of standard states. In other words, once we have chosen our standard states, the value of $Q_{eq}$ is a constant for a given reaction taking place at a given

*Recall that the designation "spontaneous" carries no implication regarding the rate of the reaction.

temperature. For this reason, $Q_{eq}$ is called the *equilibrium constant*, and is given the symbol $K$:

$$\Delta G° = -RT \ln K \qquad (11.25)$$

The algebraic form of $K$ is the same as that for $Q$, namely

$$K = \frac{a_X^x a_Y^y \cdots}{a_M^m a_N^n \cdots} \qquad (11.26)$$

There is, however, a very important distinction between $K$ and $Q$; the values of the activities that appear in the equilibrium constant expression are those that prevail at equilibrium, and conversely, *only equilibrium values of the activities can be inserted into the expression for K*. On the other hand, non-equilibrium $a_i$ values can be inserted into the $Q$ expression. Once we have chosen the standard states of the reactants and products, the value of $K$ for a reaction is fixed at a given temperature. On the other hand, $Q$ can have innumerable values, depending upon how we make up the system at a given temperature. The difference between $Q$ and $K$ can be emphasized further by combining Equations (11.23) and (11.25):

$$\Delta G = \Delta G° + RT \ln Q = -RT \ln K + RT \ln Q$$

$$\Delta G = RT \ln (Q/K) \qquad (11.27)$$

If $Q > K$ (a number), then $\Delta G > 0$ and the reaction is spontaneous from right-to-left; if $Q < K$, then $\Delta G < 0$ and from the reaction is spontaneous from left-to-right; if $Q = K$, then $\Delta G = 0$ and the reaction is at equilibrium.

The equilibrium constant is a dimensionless number. This is a consequence of the fact that each activity in the $K$ expression is actually of the form[*]

$$a_i = \frac{a_i}{a_i^°_{(std.st.)}} = \frac{a_i}{1}$$

because we have taken the activity in the standard state, $a_i^°_{(std.st.)}$, to be identical to 1. This is not the same, however, as saying that we do not have to worry about units when dealing with equilibrium constants. For example, our choice of standard state for a gas is the ideal gas at a pressure of 1 atm. Our unit of pressure then is the atm, and the pressure of a gas must be taken in atm when calculating its activity. When dealing with solution species, the composition scale used must be the same as in the standard state. If our standard state is the hypothetical ideal 1 molal solution (at 1 atm), then we must calculate $a_i$ from the relation[†] $a_i = \gamma_i m_i$, and not, for example, from the relation $a_i = \gamma_i M_i$. Furthermore, because $\Delta G°$ and $K$ depend on the choice of standard states, when one of these quantities is

---

[*] See Figure 11.6, Equation (11.2), and the discussion preceding Equation (11.2).
[†] Note also that the value of the activity coefficient depends on the composition scale.

calculated from the other using Equation (11.25), the standard states applicable to the resulting quantity are automatically fixed.

For a given choice of standard states, the value of $K$ depends only on the temperature, and in particular, the value of $K$ is independent of the *total* pressure in the reaction system at equilibrium.

We shall now give some specific examples of how the equilibrium constant expression is written under various approximations regarding the activities of the species involved:

(a) The decomposition of calcium carbonate to calcium oxide and carbon dioxide has the following stoichiometry:

$$CaCO_3(s) = CaO(s) + CO_2(g)$$

The most general expression for the equilibrium constant for this reaction is

$$K = \frac{a_{CaO(s)} a_{CO_2(g)}}{a_{CaCO_3(s)}}$$

Unless the total pressure is very high, it is usually a good approximation to take the activity of a pure condensed phase as equal to unity.[*] Thus, assuming $P_{tot}$ does not greatly exceed 1 atm, we can take $a_{CaO(s)} = 1.00$ and $a_{CaCO_3(s)} = 1.00$, without serious error, and therefore

$$K \simeq a_{CO_2(g)}$$

If, in addition, the $CO_2(g)$ behaves ideally (for example, when $P_{CO_2}$ is small), then we have

$$K \simeq P_{CO_2}$$

where $P_{CO_2}$ is the equilibrium pressure of $CO_2(g)$ over the mixture of the two solid phases $CaO(s)$ and $CaCO_3(s)$.

(b) The hydrolysis of the ester ethyl acetate has the following stoichiometry

$$\begin{array}{cccc} H_2O + CH_3CH_2OCOCH_3 = CH_3COOH + CH_3CH_2OH \\ (1) \qquad\qquad (2) \qquad\qquad\quad (3) \qquad\qquad (4) \end{array}$$

For notational simplicity, we shall number the substances involved in the manner indicated. Let us now suppose that the reaction is carried out in aqueous solution:

$$CH_3CH_2OCOCH_3(aq) + H_2O(\ell) = CH_3COOH(aq) + CH_3CH_2OH(aq)$$

The equilibrium constant has the form

$$K = \frac{a_3 a_4}{a_1 a_2}$$

---

[*]See problem 3.

If we choose our standard states as follows:

H$_2$O
    the pure liquid at 1 atm (solvent standard state) (This choice is indicated by writing H$_2$O($\ell$) rather than H$_2$O(aq).)

$\left.\begin{array}{l} \text{CH}_3\text{CH}_2\text{OCOCH}_3 \\ \text{CH}_3\text{COOH} \\ \text{CH}_3\text{CH}_2\text{OH} \end{array}\right\}$
    the hypothetical, ideal $m_i = 1$ solute standard states in H$_2$O at 1 atm

then

$$K = \frac{m_3 m_4}{X_1 m_2} \cdot \frac{\gamma_3 \gamma_4}{\gamma_1 \gamma_2}$$

In dilute aqueous solution, if Henry's law holds for 2, 3, and 4, then $\gamma_2 = \gamma_3 = \gamma_4 = 1.00$, and because Raoult's law holds for H$_2$O (the major component), $\gamma_1 = 1.00$. Thus

$$K \simeq \frac{m_3 m_4}{X_1 m_2}$$

In very dilute solution, $X_1 \simeq 1$ and

$$K \simeq \frac{m_3 m_4}{m_2}$$

Therefore, $K$ can be computed from the equilibrium values of $m_2$, $m_3$, and $m_4$ in dilute aqueous solution. Once the numerical value of $K$ has been correctly determined, it can be equated to $a_3 a_4 / a_1 a_2$, even though the concentrations of the various species involved are such that $\gamma_i \neq 1$. ($K$ is independent of concentration.)

In the above case (as in many others) there are numerous possible combinations of choices of standard states for the species involved in a reaction, and each different set of choices leads to a different numerical value of $K$ for the reaction. Consequently, it is necessary to indicate just what standard states have been chosen in the thermodynamic treatment of the reaction; otherwise, the reported value of the equilibrium constant is meaningless.

The value of the equilibrium constant for a reaction involving one or more solution species for which a solute standard state has been chosen depends on the solvent in which the reaction is run. This is true even if the solvent is neither a reactant or product, but merely serves as the medium in which the reaction is carried out. This is because the value of the standard chemical potential of a species, $\bar{G}_i^\circ$, in a solute standard state is dependent on the Henry's law constant for the solute in the given solvent, and the value of the Henry's law constant for a solute depends on the solvent.

(c) As an example of the treatment of equilibria involving electrolytes, consider the equilibrium between PbCl$_2$(s) and water saturated with lead chloride:

$$\text{PbCl}_2(\text{s}) = \text{Pb}^{2+}(\text{aq}) + 2\text{Cl}^-(\text{aq})$$

The equilibrium constant is given by (with a strong electrolyte solute standard state for $PbCl_2(aq)$, as indicated by the equilibrium)

$$K = \frac{a_{Pb^{2+}(aq)}\, a^2_{Cl^-(aq)}}{a_{PbCl_2(s)}}$$

Around 1 atm total pressure, $a_{PbCl_2(s)} = 1$, and

$$K = a_{Pb^{2+}(aq)}\, a^2_{Cl^-(aq)}$$

Using the relation $a_i = c_i\gamma_i$, we obtain

$$K = (Pb^{2+})\gamma_{Pb2+} \cdot (Cl^-)^2\gamma^2_{Cl-} = (Pb^{2+})(Cl^-)^2\gamma^3_{\pm}$$

Only at extremely low concentrations can we take $\gamma_{\pm} = 1.00$. However, if the ionic strength is held constant, then we can usually assume that $\gamma_{\pm} =$ constant $(\neq 1)$.

(d) As our next example, consider the weak-acid-dissociation equilibrium

$$CH_3COOH(aq) = CH_3COO^-(aq) + H^+(aq)$$

The equilibrium constant is given by (where $X \equiv CH_3COO$)

$$K = \frac{a_{X^-}\, a_{H^+}}{a_{HX}}$$

$$K = \frac{(X^-)\gamma_{X^-} \cdot (H^+)\gamma_{H^+}}{(HX)\gamma_{HX}} = \frac{(X^-)(H^+)\gamma^2_{\pm(HX)}}{(HX)\gamma_{HX}}$$

where $\gamma_{\pm(HX)}$ is the mean *ionic* activity coefficient of $H^+(aq) + CH_3COO^-(aq)$, and $\gamma_{HX}$ is the activity coefficient of the *neutral* species $CH_3COOH(aq)$ $(\gamma_{\pm(HX)} \neq \gamma_{HX})$. In dilute aqueous solution,[*] $\gamma_{HX} \simeq 1$, and

$$K \simeq \frac{(X^-)(H^+)\gamma^2_{\pm}}{(HX)}$$

If the only significant source of $H^+$ and $X^-$ is HX, then $(X^-) = (H^+)$ and

$$K \simeq \frac{(H^+)^2\gamma^2_{\pm}}{(HX)}$$

(e) As our last example, consider the equilibrium between $Ag^+(aq)$ and $Ag(NH_3)_2^+(aq)$ ions in a $AgNO_3(aq)$ solution containing $NH_3(aq)$; that is,

$$Ag^+(aq) + 2NH_3(aq) = Ag(NH_3)_2^+(aq)$$

---

[*]Generally speaking, *neutral* solutes do not deviate as greatly from ideal $(\gamma = 1)$ dilute solution behavior as do electrolytes, and it is usually a satisfactory approximation in all but the most precise work to take $\gamma = 1$ for neutral solutes at moderate to low concentrations.

The equilibrium constant expression is

$$K = \frac{a_{Ag(NH_3)_2^+}}{a_{Ag^+}\, a_{NH_3}^2} = \frac{(Ag(NH_3)_2^+)}{(Ag^+)(NH_3)^2} \cdot \frac{\gamma_{Ag(NH_3)_2^+}}{\gamma_{Ag^+}\, \gamma_{NH_3}^2}$$

It would appear from this expression that there is no way to put the equilibrium constant solely in terms of measurables, because of the presence of the single ion activity coefficient terms $\gamma_{Ag^+}$ and $\gamma_{Ag(NH_3)_2^+}$. This is not the case, however, because we can multiply the numerator and denominator of the $K$ expression by $a_{NO_3^-} = \gamma_{NO_3^-}(NO_3^-)$, to obtain

$$K = \frac{(Ag(NH_3)_2^+)\, \gamma_{\pm[Ag(NH_3)_2 NO_3]}^2}{(Ag^+)(NH_3)^2\, \gamma_{\pm[AgNO_3]}^2\gamma_{NH_3}^2}$$

The necessity for the introduction of the $a_{NO_3^-}$ arises because the reaction was written as a *net* reaction, that is, without the $NO_3^-(aq)$ on the left and right hand sides of the reaction

$$Ag^+(aq) + NO_3^-(aq) + 2NH_3(aq) = Ag(NH_3)_2^+(aq) + NO_3^-(aq)$$

Although it is true that the value of $K$ is independent of the nature of the anion, provided both electrolytes can be treated as strong (for example, K is the same whether the anion is $NO_3^-$, $F^-$, or $ClO_4^-$), it is *not true* that the nature of the anion can be completely ignored in the experimental determination of $K$. This is because, in general, $\gamma_{\pm(AgNO_3)} \neq \gamma_{\pm(AgClO_4)}$ at the same ionic strength unless the Debye-Hückel law holds. Note that if Equation (11.15) does hold for the above solution at equilibrium, then $\gamma_{\pm[Ag(NH_3)_2NO_3]} = \gamma_{\pm(AgNO_3)}$, and (assuming $\gamma_{NH_3} \simeq 1$)

$$K \simeq \frac{(Ag(NH_3)_2^+)}{(Ag^+)(NH_3)^2}$$

It is a common practice in the study of equilibria involving ionic species to attempt to hold the $\gamma_\pm$ terms for the electrolytes involved at a constant value by keeping the ionic strength, $I$, constant using a large excess of a suitable "inert" supporting electrolyte (e.g., $LiClO_4$ or $KNO_3$ for aqueous solution equilibria). Such a procedure is not necessarily effective, because, in general, $\gamma_\pm$ depends not only on the total ionic strength, but also on the nature of the various ions present in the solution. Equation (11.15) is only an *approximation*. The value of the "equilibrium constant" determined in this way is applicable only at the particular value of the ionic strength at which it was determined. To be more specific, we return to the dissociation of the weak acid HX(aq) discussed in example (d) above:

$$K = \frac{(H^+)(X^-)\, \gamma_{\pm(HX)}^2}{(HX)\, \gamma_{HX}}$$

Assuming that $\gamma_{\pm(HX)}$ and $\gamma_{HX}$ are constant at a given total $I$, we have

$$K = \frac{(H^+)(X^-)}{(HX)}(constant)$$

*where the value of (constant) depends on the value of I.* Very frequently the quantity that is determined in electrolyte equilibrium studies is $K_{app}$ (*app* for apparent)

$$K_{app} = \frac{K}{(\text{constant})} = \frac{(H^+)(X^-)}{(HX)} = \text{function of total ionic strength}$$

and not $K$. This distinction is important and, if ignored, can lead to considerable confusion, to say nothing of erroneous results, when using tabulated values of $K_{app}$.

## 11.6   THE TEMPERATURE DEPENDENCE OF THE EQUILIBRIUM CONSTANT.

For a reaction taking place under isothermal conditions, we have from the definition of the Gibbs energy ($G = H - TS$)

$$\Delta G = \Delta H - T\Delta S \tag{11.28a}$$

For the isothermal conversion of the reactants in their respective standard states to products in their respective standard states, Equation (11.28a) becomes

$$\Delta G° = \Delta H° - T\Delta S° \tag{11.28b}$$

The choice of standard states that serves to define $\Delta G°$ also serves to define $\Delta H°$ and $\Delta S°$. Combining Equation (11.28b) with the relation $\Delta G° = -RT \ln K$ yields

$$\boxed{\ln K = 2.303 \log_{10} K = -\frac{\Delta H°}{RT} + \frac{\Delta S°}{R}} \tag{11.29}$$

This equation shows that the *magnitude* of the equilibrium constant at a given temperature is determined by the magnitude of *both* $\Delta H°$ and $\Delta S°$ for the reaction. As the temperature decreases, the value of $\Delta H°$ becomes more important in determining the magnitude of $K$, because of the inverse temperature dependence in the $-\Delta H°/RT$ term; conversely, as the temperature increases, the magnitude of $\Delta S°$ becomes more important in determining the magnitude of $K$ for the same reason. *The equilibrium constant is necessarily a non-zero, positive quantity*

$$K = \exp(-\Delta G°/RT) > 0$$

Equation (11.29) also tells us that a plot of $\log_{10} K$ versus $1/T$ has a slope at any point equal to $-\Delta H°/2.303R$. If the value of $\Delta H°$ is constant over the entire range of temperature covered in the plot, then the resulting curve will be a straight line. If $\Delta H°$ is a function of temperature over the experimental range, then a plot of $\log K$ versus $1/T$ will exhibit curvature. It is still true in such a case that the slope of the $\log K$ versus $1/T$ curve at any $T$ is equal to $-\Delta H°/2.303\ R$ *at that temperature.*

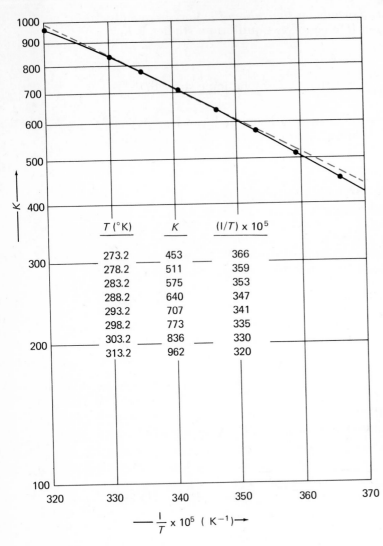

| $T$ (°K) | $K$ | $(1/T) \times 10^5$ |
|---|---|---|
| 273.2 | 453 | 366 |
| 278.2 | 511 | 359 |
| 283.2 | 575 | 353 |
| 288.2 | 640 | 347 |
| 293.2 | 707 | 341 |
| 298.2 | 773 | 335 |
| 303.2 | 836 | 330 |
| 313.2 | 962 | 320 |

**FIGURE 11.7** Plot of $\log_{10} K$ versus $1/T$ for the reaction
$$CO_2(aq) = CO_2(g)$$
The 1 atm ideal gas and the unit molality ideal solution are the standard states chosen for $CO_2(g)$ and $CO_2(aq)$, respectively.

In Figure 11.7 we have constructed a $\log K$ versus $1/T$ plot over the range from 273 to 313 K for the reaction

$$O_2(aq) = O_2(g)$$

$$K = \frac{a_{O_2(g)}}{a_{O_2(aq)}}$$

The curvature of the plot in Figure 11.7 shows that $\Delta H°$ is slightly dependent on temperature over the given range of temperature. From Equation (11.29) we note that at the two temperatures $T_2$ and $T_1$ the respective values of $\log K$ are given by

$$\log K_2 = \frac{-\Delta H_2°}{2.30\, RT_2} + \frac{\Delta S_2°}{2.30\, R}$$

and

$$\log K_1 = \frac{-\Delta H_1^\circ}{2.30\ RT_1} + \frac{\Delta S_1^\circ}{2.30\ R}$$

Subtracting the second of these two expressions from the first, *and assuming that $\Delta H^\circ$ is temperature-independent*, we obtain

$$\log \frac{K_2}{K_1} = \frac{-\Delta H^\circ}{2.30\ R}\left\{\frac{1}{T_2} - \frac{1}{T_1}\right\} = \frac{\Delta H^\circ}{2.30\ R}\left(\frac{T_2 - T_1}{T_1 T_2}\right) \qquad (11.30)$$

where we have taken $\Delta S_1^\circ = \Delta S_2^\circ$, because if $\Delta H^\circ$ is independent of temperature, then $\Delta S^\circ$ is also independent of temperature.

Equation (11.30) tells us that if $\Delta H^\circ > 0$, then $K$ increases with increasing $T$, whereas if $\Delta H^\circ < 0$, then $K$ decreases with increasing $T$. The sign and magnitude of the *temperature dependence* (as opposed to the magnitude at a particular temperature of the equilibrium constant) are governed *solely* by the sign and magnitude of $\Delta H^\circ$ for the reaction.[*]
Solving Equation (11.30) for $\Delta H^\circ$, we obtain

$$\Delta H^\circ = \frac{2.30\ R(\log K_2 - \log K_1)}{\left\{\dfrac{1}{T_1} - \dfrac{1}{T_2}\right\}}$$

Using this expression to analyze the data in Figure 11.7, we compute $\Delta H^\circ$ over the range from 303 to 313 K as

$$\Delta H^\circ = \frac{2.30(1.99\ \text{cal}\cdot\text{K}^{-1}\cdot\text{mol}^{-1})(\log 962 - \log 836)}{(0.00330 - 0.00320)\ \text{K}^{-1}} = 2.6\ \text{kcal}$$

whereas over the range from 273 to 278 K an analogous calculation yields $\Delta H^\circ = 3.4$ kcal. Because $\Delta H^\circ$ is varying only slowly with $T$, an average value of $\Delta H^\circ$ over the range from 273 to 313 K can be obtained from Figure 11.7 by drawing a best-fit straight line through the points (indicated in the figure by the dashed line). Using this line, we compute from the graph

$$\Delta H^\circ = \frac{4.58(\log 982 - \log 438)}{(0.00370 - 0.00320)} = 3.2\ \text{kcal}$$

In certain cases the temperature dependence of $\Delta H^\circ$ is much larger than that found above, and in such cases the assumption that $\Delta H^\circ$ is constant is so poor that it is a totally useless approximation. Such cases arise

---

[*]The *general* equation governing the temperature dependence of $K$ is known as the *van't Hoff equation*:

$$\frac{d \ln K}{dT} = \frac{\Delta H^\circ}{RT^2}$$

from which we deduce that, if $\Delta H^\circ > 0$, the $d \ln K/dT > 0$, and $K$ increases with increasing temperature. Equation (11.30) is obtained from this expression on integration between limits $(K_1, T_1)$ and $(K_2, T_2)$ with the assumption that $\Delta H^\circ$ is constant.

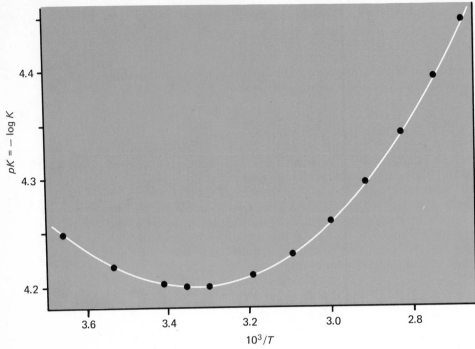

**FIGURE 11.8** Plot of $-\log K$ versus $1/T$ for the aqueous ionization of benzoic acid over the range from 0 to 100°C.

quite frequently for ionization reactions in solution. An example is the ionization of benzoic acid in water:

$$C_6H_5COOH(aq) = C_6H_5COO^-(aq) + H^+(aq)$$

In Figure 11.8 we have plotted $pK = -\log K$ versus $1/T$ for this reaction over the range from 273 to 373 K. In this case, the temperature dependence of $\Delta H°$ is such that over the range from 0 to 100°C the value of $\Delta H°$ changes sign.

## 11.7 THE USE OF $\Delta \bar{G}_f°$ VALUES IN THERMODYNAMIC CALCULATIONS.

In order to achieve the maximum potential usefulness of Gibbs energy data, it is convenient to define the quantity $\Delta G_f°$, the *standard Gibbs energy of formation of a substance from the elements.*

A $\Delta G°$ value for a reaction at a particular temperature gives us the difference in Gibbs energy between the products and reactants in their respective standard states at that temperature, that is (see Equation (11.21)),

$$\Delta G°_{rctn} = \sum_i \nu_i \bar{G}_i° \tag{11.31}$$

where the $\nu_i$ values are the balancing coefficients (positive for products and negative for reactants). In a manner analogous to that used in defining $\Delta \bar{H}_f^\circ$ values (see Section 9.8(d)), we can rewrite Equation (11.31) as

$$\Delta G_{rctn}^\circ = \sum_i \nu_i \Delta \bar{G}_{f,i}^\circ \qquad (11.32)$$

if we agree to set $\Delta G_{f,i}^\circ = 0$, at all $T$, for some arbitrarily chosen state of each element. The form of each element for which we set $\Delta G_f^\circ = 0$ at all $T$ is usually that form which is most stable (i.e., has the lowest molar Gibbs energy) at 1 atm and 25°C.* As long as one is consistent, it makes no difference from a computational point of view which form is chosen; however, there is a considerable experimental advantage to the above choice.

By way of example, consider the reaction

$$H_2(g) + \left(\frac{1}{2}\right) O_2(g) = H_2O(\ell)$$

for which $\Delta G_{rctn}^\circ = \Delta G^\circ = -56.688$ kcal at 298.15 K. Using Equations (11.31) and (11.32), we obtain

$$\Delta G^\circ = \bar{G}^\circ [H_2O(\ell)] - \frac{1}{2} \bar{G}^\circ [O_2(g)] - \bar{G}^\circ [H_2(g)] = -56.688 \text{ kcal}$$

$$\Delta G^\circ = \Delta \bar{G}_f^\circ [H_2O(\ell)] - \frac{1}{2} \Delta \bar{G}_f^\circ [O_2(g)] - \Delta \bar{G}_f^\circ [H_2(g)] = -56.688 \text{ kcal}$$

If we set $\Delta \bar{G}_f^\circ = 0$ for $H_2(g)$ and $O_2(g)$, then the standard Gibbs energy of formation of liquid water at 298.15°K is: $\Delta \bar{G}_f^\circ [H_2O(\ell)] = -56.688$ kcal $\cdot$ mol$^{-1}$.

The procedure for building up a table of $\Delta \bar{G}_f^\circ$ values is strictly analogous to that discussed in Section 9.8(d) for $\Delta \bar{H}_f^\circ$ values. In Appendix 5 we have assembled $\Delta \bar{G}_f^\circ$ values for some common substances. As an example of the use of such data, consider the calculation of $\Delta G^\circ$ and $K$ at 25°C for the reaction

$$H_2(g) + 2AgCl(s) = 2Ag(s) + 2H^+(aq) + 2Cl^-(aq)$$

The value of $\Delta G^\circ$ is given by (Equation (11.32))

$$\Delta G^\circ = 2\Delta \bar{G}_f^\circ [Ag(s)] + 2\Delta \bar{G}_f^\circ [H^+(aq)] + 2\Delta \bar{G}_f^\circ [Cl^-(aq)] - \Delta \bar{G}_f^\circ [H_2(g)]$$
$$- 2\Delta \bar{G}_f^\circ [AgCl(s)]$$

In Appendix 5 we find the following data

| SPECIES | $\Delta \bar{G}_{f,298}^\circ$(kcal $\cdot$ mol$^{-1}$) |
|---|---|
| Ag(s) | 0 (by convention) |
| H$^+$(aq) | 0 (by convention) |
| Cl$^-$(aq) | $-31.372$ |
| H$_2$(g) | 0 (by convention) |
| AgCl(s) | $-26.244$ |

---

*The form of the element for which we have taken $\Delta H_f^\circ = 0$ must be the same form for which we take $\Delta G_f^\circ = 0$ (see Section 9.8(d)).

and thus

$$\Delta G° = 2(-31.372) - 2(-26.244) = -10.256 \text{ kcal}$$

Combining this result with the relation $\Delta G° = -RT \ln K$ yields

$$\log K = \frac{-(-10.256 \times 10^3 \text{ cal})}{(2.3026)(1.9872 \text{ cal} \cdot \text{K}^{-1} \cdot \text{mol}^{-1})(298.15 \text{ K})} = +7.517$$

or $K = 3.29 \times 10^7$ (25°C).

## 11.8   SPONTANEITY IN CHEMICAL REACTIONS.

The value of $\Delta G°$ at 300 K for the reaction

$$Pb(s) + 2AgCl(s) = 2Ag(s) + PbCl_2(s)$$

is $-22.06$ kcal. Therefore, at 300 K

$$\Delta G = \Delta G° + RT \ln Q$$

$$\Delta G = -22.06 \times 10^3 \text{ cal} + (1.99)(300)(2.303)\log \left\{ \frac{a^2_{Ag(s)} \, a_{PbCl_2(s)}}{a_{Pb(s)} \, a^2_{AgCl(s)}} \right\}$$

At 1 atm all the $a_i = 1$, and hence, $\Delta G = -22.06$ kcal. In other words, the reaction is spontaneous ($\Delta G < 0$) in the direction written (left to right) at 1 atm and 300 K. Because the value of $\Delta G$ in this case *does not change* as reactants are converted to products, the reaction will proceed* until whichever of the two reactants is stoichiometrically limiting is *totally* exhausted. Conversely, if $PbCl_2(s)$ and $Ag(s)$ are brought in contact at 1 atm and 300 K, no $Pb(s)$ or $AgCl(s)$ will form ($\Delta G = +22.06$ kcal).

The distinction between $\Delta G$ and $\Delta G°$ is especially important to keep in mind when dealing with reactions that involve a solution phase. For example, consider the reaction

$$CH_3COOH(aq) = H^+(aq) + CH_3COO^-(aq) \quad K_{298} = 1.76 \times 10^{-5}$$
(molality composition scale, solute standard states)

for which, at 25°C, $\Delta G°$ is

$$\Delta G° = -(1.99 \text{ cal} \cdot \text{K}^{-1} \cdot \text{mol}^{-1})(298 \text{ K})(2.303)\log(1.76 \times 10^{-5}) = +5.46 \text{ kcal}$$

In this case (or in any case for that matter) we cannot conclude that, owing to the positive value of $\Delta G°$, no products will form spontaneously when we dissolve $CH_3COOH(\ell)$ in water at 25°C. This is because *it is* $\Delta G$, *and not* $\Delta G°$, *that determines whether or not the indicated process is spontaneous*

---

*The rate of this reaction is expected to be extremely slow because it involves bringing together two solid phases. The reaction rate can be greatly increased, however, by providing a more favorable reaction pathway. For example, the reactants could be mixed in the presence of $H_2O(\ell)$. A small amount of $AgCl(s)$ dissolves in the water and this reacts with $Pb(s)$.

*at a particular temperature and pressure.* In this case, even at 1 atm total pressure, $\Delta G \neq \Delta G°$. The value of $\Delta G$ (in kcal) at 25°C is given by[*]

$$\Delta G = +5.46 + 1.36 \log \left\{ \frac{a_{H^+(aq)} \, a_{CH_3COO^-(aq)}}{a_{CH_3COOH(aq)}} \right\}$$

and with $a_{CH_3COOH} = 0.10$, and $a_{H^+(aq)} \simeq a_{CH_3COO^-} \simeq 0$, the value of $\Delta G$ is very large and *negative.* Therefore, the dissociation of $CH_3COOH(aq)$ take place *spontaneously* with $a_{CH_3COOH}$ decreasing, and $a_{H^+}{}_{(aq)}$ and $a_{CH_3COO^-(aq)}$ increasing, until

$$1.36 \log \frac{a_{H^+(aq)} \, a_{CH_3COO^-(aq)}}{a_{CH_3COOH(aq)}} = -5.46 \text{ kcal}$$

which corresponds to the equilibrium state, $\Delta G = 0$. Because of the positive value of $\Delta G°$, the values of $a_{H^+(aq)}$ and $a_{CH_3COO^-(aq)}$ at equilibrium will be small ($\sim 1.3 \times 10^{-3}$) compared to the equilibrium value of $a_{CH_3COOH(aq)}$ ($\sim 0.10$), but this is not the same thing as saying that because $\Delta G°$ is positive the reaction is not spontaneous in the direction written. Spontaneity at constant temperature and pressure is governed by $\Delta G$, not $\Delta G°$, and the value of $\Delta G$ depends on the actual value of the activities of the reactants and products, whereas the value of $\Delta G°$ for a given reaction depends (for a given choice of standard states) only on the temperature.

## 11.9 A DIGRESSION ON THE SIGNIFICANCE OF THE ACTIVITY.

On first encounter, the activity function comes across as a somewhat mysterious concept, and there is a widespread tendency to avoid a discussion of equilibrium constants in terms of activities at the introductory level. This is unfortunate, because without the activity function the equlibrium constant expression does not have the particularly simple form given in Equation (11.26). More importantly, if we express $K$ in the form of Equation (11.26), but with concentrations (or pressures) in place of activities, then we no longer have a quantity that is a constant, except in the limit of very low concentrations.

The activity function was *invented* in order to preserve the simple algebraic form of $K$, which otherwise prevails only at low concentrations and pressures, over the entire possible range of concentrations and pressures.

The physical basis for the activity function is that the concentration of a solution species (or pressure of a gaseous species) is not, in general, a reliable quantitative measure of the escaping tendency of a species from a phase, because of the existence of strong, specific interactions between the species. By way of example, consider a solution of NaCl in water (Figure 11.9). We expect that the escaping tendency of water from the solution is less than that from pure water simply on the basis that the

---

[*]This problem could just as well be worked with Equation (11.27):

$$\Delta G = 1.36 \log \left\{ \left( \frac{a_{H^+(aq)} \, a_{CH_3COO^-(aq)}}{a_{CH_3COOH(aq)}} \right) \Big/ 1.76 \times 10^{-5} \right\}$$

Surface

Surface

Pure water

NaCl in water

**FIGURE 11.9**

number of water molecules per unit area of solution surface is less than that for pure water. This effect is measured by Raoult's Law, which says that the equilibrium vapor pressure (which measures the escaping tendency) of water over the solution is directly proportional to the mole fraction of water in the solution ($P_{H_2O} = X_{H_2O} P_{H_2O}^\circ$). The greater the concentration of NaCl, the lower $X_{H_2O}$, and thus the lower $P_{H_2O}$. However, Raoult's Law holds for $H_2O$ in a NaCl(aq) solution only at very low concentrations of NaCl. At higher concentrations the equilibrium partial pressure of water over the solution is less than that calculated from Raoult's Law. This is a consequence of the fact that the ions in solution interact electrostatically with the dipoles on water molecules, and thereby bind some of the water molecules to themselves (Figure 11.10). As a consequence, the average energy of the water molecules in the solution is reduced relative to pure water, which leads to a reduced escaping tendency of water (a lower activity). Note that because $a_{H_2O(\ell)} = P_{H_2O}/P_{H_2O}^\circ$, if Raoult's Law holds, then $a_{H_2O(\ell)} = X_{H_2O}$, whereas if $P_{H_2O}/P_{H_2O}^\circ < X_{H_2O}$, then $a_{H_2O(\ell)} < X_{H_2O}$; that is, $a_{H_2O} = \gamma_{H_2O} X_{H_2O}$ ($\gamma_{H_2O} \neq 1$).

Similar considerations apply to the NaCl in the solution. If the solution is very dilute, then Henry's Law holds ($P_{NaCl} = m_{NaCl}^2 P_{NaCl}^*$) and the escaping tendency of NaCl from the solution (i.e., the activity) is equal to

$$a_{NaCl} = \frac{P_{NaCl}}{P_{NaCl}^*} = m_{NaCl}^2$$

However, Henry's Law holds only at very low concentrations because the electrostatic forces between ions are long-range forces. Consequently, $a_{NaCl} < m_{NaCl}^2$ and

$$a_{NaCl} = \frac{P_{NaCl}}{P_{NaCl}^*} = m_{NaCl}^2 \gamma_{\pm(NaCl)}^2$$

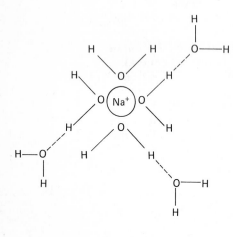

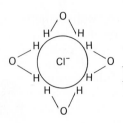

**FIGURE 11.10** Solvated ions. Water molecules above and below the plane of the page are not shown.

The activity can be thought of as an effective concentration of a species. In other words, strong specific interactions between species in solution (which manifest themselves as deviations from either Raoult's Law (solvent) or Henry's Law (solute)) make the species behave as if it were present at a concentration smaller (or larger) than the actual concentration. This effective or active concentration of a species is called the activity.

_____ **PROBLEMS**

1. The equilibrium vapor pressure of water at 25°C over a 2.00 molal solution of $CaCl_2(aq)$ is 20.53 Torr. Calculate the activity and the activity coefficient of water (based on a solvent standard state) in this solution. ($P^\circ_{H_2O} = 23.75$ Torr at 25°C.)

2. The activity coefficients of water at 25°C in 6.00 m and 76.00 m $H_2SO_4(aq)$ are 0.257 and 7.18, respectively. Calculate the activity of water in these solutions and the equilibrium vapor pressure of water over the solutions.

3. The average partial pressure of $CO_2(g)$ in the air is $3 \times 10^{-4}$ atm. Estimate the concentration of $CO_2(aq)$ in water in equilibrium with air at 25°C (see Table 11.1 for additional data).

4. Calculate the activities of the following electrolytes in water at 25°C (see Table 11.2 for additional data).
    (a) 0.30 m NaCl(aq)
    (b) 0.10 m $CaCl_2(aq)$
    (c) 0.70 m $Na_2SO_4(aq)$
    (d) 0.50 m $K_3PO_4(aq)$
    (e) 0.050 m NaCl(aq) + 0.010 m $CaCl_2(aq)$ [Take $\gamma_\pm = 0.75$ for NaCl, 0.58 for $CaCl_2$]

5. Given that $-\log K = 4.221$ at 35°C and $-\log K = 4.241$ at 45°C for the reaction

$$C_6H_5COOH(aq) = H^+(aq) + C_6H_5COO^-(aq)$$

estimate $\Delta G^\circ$, $\Delta H^\circ$, and $\Delta S^\circ$ for this reaction at 40°C. State your assumptions.

6. Given that $\Delta H^\circ = 50.0$ kcal and $\Delta S^\circ = 50$ gibbs for a certain reaction, compute $K$ at 300 K and 1000 K. Assume $\Delta H^\circ$ and $\Delta S^\circ$ are independent of temperature.

7. It is frequently necessary to use water that is free of dissolved oxygen. Two ways in which deoxygenated water is prepared are: (1) vigorous boiling of the water, followed by cooling of the boiled water in the absence of oxygen; and (2) prolonged bubbling of oxygen-free $N_2(g)$ through the water in a narrow-necked flask open to the atmosphere. Discuss why these procedures are effective.

8. Show that: (a) the equilibrium constant for the reverse reaction is the reciprocal of that for the forward reaction; (b) if we add two reactions

to obtain a third reaction, then the equilibrium constant for the resulting reaction is equal to the product of the equilibrium constants for the two reactions that are added.

9. The activity of a condensed phase is given approximately by the expression

$$a \simeq \exp \left\{ \frac{\bar{V}(P-1)}{RT} \right\}$$

where $P$ is the pressure and $\bar{V}$ is the molar volume of the condensed phase. The molar volume of $CaCO_3(s)$ is 36.9 $cm^3 \cdot mol^{-1}$. Compute $a_{CaCO_3}$ at $P = 1, 10, 100,$ and 10,000 atm.

10. Calculate the ionic strengths of each of the following solutions:

(a) 0.25 m $CaCl_2(aq)$
(b) 0.25 m $CaCl_2(aq)$ + 0.15 m $K_2HPO_4(aq)$
(c) 0.10 m $NaCl(aq)$ + 0.20 m $KNO_3(aq)$ + 0.025 m $K_3Fe(CN)_6(aq)$
(d) 0.10 m $CH_3COOH(aq)$ + 0.10 m $HCl(aq)$
(e) 0.050 m $K_2CaFe(CN)_6(aq)$
(f) 0.40 m $NaCl(aq)$ + 0.15 m $KCl(aq)$ + 0.02 m $CaCl_2(aq)$

11. Using data in Table 11.2, calculate $\Delta \bar{G}_{NaCl(aq)}$ at 25°C for the reaction

$$NaCl(aq, 0.100 \text{ m}) = NaCl(aq, 1.00 \text{ m})$$

Compare your result with that obtained assuming $\gamma_{\pm} = 1$.

12. Using data in Table 11.2, plot log $\gamma_{\pm}$ versus $I^{1/2}(1 + I^{1/2})$ for $HCl(aq)$, $LiClO_4(aq)$ and $CaCl_2(aq)$. Also plot on the same graph log $\gamma_{\pm}$ versus $I^{1/2}/(1 + I^{1/2})$ using the Debye-Hückel equation for these three electrolytes.

13. (a) Using data in Appendix 5, compute $\Delta G°$ and $K$ at 25°C for each of the following reactions. (b) Write out the equilibrium constant expression for each of the following reactions assuming: (i) all species are non-ideal and $P \gg 1$ atm; (ii) $P_{tot} \simeq 1$ atm and all gases are ideal; (iii) $P_{tot} \simeq 1$ atm and all $\gamma_i = 1.00$; (iv) $P_{tot} \simeq 1$ atm, but $\gamma_{\pm} \neq 1$ (express $K$ in terms of $m_i$ and $\gamma_{\pm,i}$ where possible:

(a) $CO(g) + 2H_2(g) = CH_3OH(\ell)$
(b) $BaSO_4(s) = Ba^{2+}(aq) + SO_4^{2-}(aq)$
(c) $Cd(s) + Hg_2SO_4(s) = Cd^{2+}(aq) + SO_4^{2-}(aq) + 2Hg(\ell)$
(d) $CH_3COOH(aq) = H^+(aq) + CH_3COO^-(aq)$
(e) $CaCO_3(s) = CaO(s) + CO_2(g)$
(f) $\frac{1}{2} Cl_2(g) + Ag(s) = AgCl(s)$
(g) $H_2(g) + 2AgI(s) = 2H^+(aq) + 2I^-(aq) + 2Ag(s)$
(h) $Br_2(aq) + 2Fe(CN)_6^{4-}(aq) = 2Fe(CN)_6^{3-}(aq) + 2Br^-(aq)$
(i) $C_6H_6(\ell) + \frac{15}{2} O_2(g) = 6CO_2(g) + 3H_2O(g)$

14. (a) Use data in Appendix 5 to calculate $\Delta H°$ at 25°C for the reactions in problem 13. (b) Using the results in (a) and in problem 13, estimate $K$ for the reactions in problem 13 at 50°C.

15. Use the data in Table 11.1 to construct a plot of log $K$ versus $1/T$ for the reaction

$$CO_2(g) = CO_2(aq)$$

and evaluate $\Delta H°$ at 25°C from the plot.

16. For the reaction

$$C(s,graphite) = C(s,diamond)$$

$$\Delta H° = -450 \text{ cal} \qquad \Delta S° = -0.79 \text{ gibbs} \qquad \Delta V° = -1.92 \text{ cm}^3$$

(a) Compute $\Delta G°$ at 298 K
(b) Compute $\Delta G$ at 298 K and 1 atm. Which is the more stable form of carbon at 298 K and 1 atm?
(c) Using the expression for the activity of a solid given in problem 9, compute the pressure at which the two forms are in equilibrium at 298 K and at 500 K.

17. State the assumptions necessary to arrive at the following approximate equilibrium-constant expressions:

(a) $NH_3(aq) + H_2O(\ell) = NH_4^+(aq) + OH^-(aq)$   $K \approx (NH_4^+)(OH^-)/(NH_3)$
(b) $Ag_2SO_4(s) = 2Ag^+(aq) + SO_4^{2-}(aq)$   $K \approx (Ag^+)^2(SO_4^{2-})$
(c) $CO_2(g) + OH^-(aq) = HCO_3^-(aq)$   $K \approx (HCO_3^-)/(OH^-)P_{CO_2}$

18. At 900 K the equilibrium constant for the reaction

$$2SO_2(g) + O_2(g) = 2SO_3(g)$$

is $K = 10$. Suppose we mix the following numbers of moles of the three gases. Predict in which direction the reaction will proceed toward equilibrium.

(a) 2 moles $SO_3$ + 2 moles $SO_2$ + 1 mole $O_2$
(b) 1 mole $SO_3$ + 1/2 mole $SO_2$ + 1/3 mole $O_2$
(c) 2 moles $SO_3$ + 1 mole $SO_2$ + 2/5 mole $O_2$

# CHAPTER 12A

Every system in stable chemical equilibrium subjected to the influence of an external cause that tends to vary either its temperature or its condensation (pressure, concentration, number of molecules per unit volume) as a whole or merely in some of its parts, cannot experience other than those internal modifications which, if they were to be produced alone, would lead to a change of temperature or condensation of opposite sign to that resulting from the external cause.

*Henry Le Chatelier\**

# CHEMICAL EQUILIBRIUM

## Part II. Applications to Homogeneous Equilibria

## INTRODUCTION.

Any system when left undisturbed proceeds spontaneously and inexorably towards an equilibrium state. The rate at which the system approaches an equilibrium state depends on the particular conditions of temperature, pressure, and the concentrations of the various species in the system. There are two basic criteria for the existence of an equilibrium state:

(1) the temperature, the total pressure, and the concentration of each species (within a given phase) must be uniform and time-independent; and

(2) an equilibrium state involving a chemical reaction must be approachable from either direction (left-to-right or right-to-left for the reaction as written) and independent of the direction of approach.†

An equilibrium state, when viewed on the macroscopic level, is a *static* state in which there is no net change with time. However, when viewed on the microscopic level, an equilibrium state is found to be a *dynamic* state in which competing processes (e.g., forward and back reactions) are exactly balanced.

A particularly useful principle, which enables us to make qualitative predictions about the response of an equilibrium system to a perturbation, is **Le Chatelier's Principle:** *If a system in an equilibrium state is subjected*

---

\*Henry Le Chatelier, Comptes Rendus de l'Academie des Sciences, 99, 786 (1884) quoted in J. H. Hildebrand and R. E. Powell, *Principles of Chemistry*, 7th Edition (The Macmillan Co., New York, 1964), pp. 180–1.

†This point is discussed in more detail in Section 15.6; see, in particular, Figure 15.4.

*to a change in conditions that puts the system out of equilibrium, then the direction in which the system proceeds to a new equilibrium state will be such as partially to offset the change in conditions.*

Some examples of the application of Le Chatelier's Principle are as follows:

(1) $\qquad$ $HX(aq) \rightleftharpoons H^+(aq) + X^-(aq) \qquad \Delta H = -1 \text{ kcal}$

(a) If we add more HX, then the reaction proceeds from left to right ($\longrightarrow$) to a new equilibrium state, leading to an increase in the concentration of $H^+(aq)$ and $X^-(aq)$ (as well as HX(aq)).

(b) If we add some NaX, then the reaction proceeds from right to left ($\longleftarrow$), leading to an increase in the concentration of HX and to a decrease in the concentration of $H^+(aq)$.

(c) If we add NaOH, then the reaction proceeds $\longrightarrow$, because of the decrease in the concentration of $H^+(aq)$ $[H^+(aq) + OH^-(aq) \rightarrow H_2O(\ell)]$. The new equilibrium state will have a higher concentration of $X^-(aq)$ and a lower concentration of HX(aq) and $H^+(aq)$ than the original equilibrium state.

(d) If we add water to the equilibrium system (dilution), then the reaction proceeds $\longrightarrow$ to a new equilibrium state, leading to an increase in the *percentage* of HX(aq) that is dissociated into $H^+(aq)$ and $X^-(aq)$. The concentrations of all three species, $H^+(aq)$, $X^-(aq)$, and HX(aq), in the new equilibrium state are less than in the original equilibrium state (because of the dilution effect). However, the concentrations of $H^+(aq)$ and $X^-(aq)$ are greater, and the concentration of HX(aq) is less, than would be estimated on the basis of dilution without net reaction. In other words, on dilution the reaction proceeds in the direction that produces the greater number of moles (left-to-right in the case under consideration, because there are two product species and only one reactant species).

(e) If we increase the temperature of the system, then the reaction proceeds $\longleftarrow$. This is because $\Delta H = q_P = -1 < 0$. The reaction absorbs energy as heat when it proceeds right-to-left, and an increase in temperature leads to an input of energy as heat to the system. In general, if $\Delta H < 0$ for a reaction, then the equilibrium will shift from right to left when the temperature is increased. If $\Delta H > 0$ (energy absorbed as heat), then the equilibrium will shift from left to right when the temperature is increased.

(2) $\qquad$ $AgX(s) \rightleftharpoons Ag^+(aq) + X^-(aq) \qquad \Delta H \simeq +12 \text{ kcal}$

(a) Add $AgNO_3(aq)$: $\longleftarrow$
(b) Add NaX(aq): $\longleftarrow$
(c) Add $H_2O$: $\longrightarrow$ (But the *equilibrium* concentrations of $Ag^+(aq)$ and $X^-(aq)$ are unaffected, provided AgX(s) is in excess).
(d) Add more AgX(s): No effect on the concentrations of $Ag^+(aq)$ and $NO_3^-(aq)$, because the activity of a pure solid is independent of the amount of the solid present. The equilibrium is not disturbed by the addition of AgX(s).
(e) Increase in the temperature: $\longrightarrow$ ($\Delta H = q > 0$)

(3) $\qquad$ $N_2(g) + 3H_2(g) \rightleftharpoons 2NH_3 \qquad \Delta H = -22 \text{ kcal}$

(a) Increase in $P_{N_2}$: $\longrightarrow$
(b) Increase in $P_{H_2}$: $\longrightarrow$
(c) Increase in $P_{NH_3}$: $\longleftarrow$
(d) Increase in the total pressure: $\longrightarrow$ (The reaction proceeds in

the direction of the smaller number of moles, because this partially offsets the perturbation that is due to a compression of the system).

(e) Increase in temperature: ⟵ ($\Delta H$ = q < 0).

The qualitative ideas embodied in Le Chatelier's Principle as applied to chemical reactions can be put on a quantitative basis with the *Law of Mass Action* (Guldberg and Waage). For the generalized reaction

$$mM + nN + \cdots = xX + yY + \cdots$$

we define the activity quotient $Q$ as (Chapter 11)

$$Q = \left\{ \frac{a_X^x \, a_Y^y \, \cdots}{a_M^m \, a_N^n \, \cdots} \right\}$$

where the $a_i$ values are the activities of the various species. The Law of Mass Action states that at equilibrium $Q$ = constant, called the equilibrium constant, $K$:

$$Q_{eq} = K = \left\{ \frac{a_X^x \, a_Y^y \, \cdots}{a_M^m \, a_N^n \, \cdots} \right\}_{eq}$$

The equilibrium constant for a given reaction is a dimensionless number that depends only on the temperature for a given choice of standard states. In particular, the value of $K$ is independent of the individual $a_i$ values that prevail at equilibrium. $Q$ and $K$ have the same algebraic form, but the activities that appear in the $K$ expression are those that prevail at equilibrium; conversely, only equilibrium $a_i$ values can be inserted into the $K$ expression. On the other hand, non-equilibrium $a_i$ values can be inserted into the $Q$ expression.

The criterion for spontaneity in chemical reactions is provided by the ratio $Q/K$:

If $Q/K$ < 1, then the reaction proceeds ⟶

If $Q/K$ > 1, then the reaction proceeds ⟵

If $Q/K$ = 1, then the reaction is at equilibrium.

## 12.1   ACID-BASE EQUILIBRIA IN SOLUTION.

In this chapter we shall analyze numerous equilibrium problems under the assumption that the activities of the various solution species involved are equal to their concentrations. That is, we shall assume that the activity coefficients of the various species are equal to unity. There are several reasons why we choose to make this simplification, even though we know that in many cases serious errors may arise because of it. These reasons are:

(1) The *rigorous* thermodynamic treatment of solution equilibria is difficult, and often leads to cumbersome algebraic expressions that tend to obscure many important and qualitatively useful concepts in equilibrium reasoning.

(2) In many cases we seek only order-of-magnitude answers, and the neglect of activity coefficients is often justifiable in terms of the desired accuracy.

(3) The *first step* in the rigorous thermodynamic treatment of solution equilibria is the calculation of the approximate concentrations of the various species under the assumption that $\gamma_i = 1$.

It is important, nonetheless, to remember that it is only when using activities that the expression for the equilibrium constant is really a constant, and that the use of concentrations in place of activities is often a crude approximation. Once we have obtained some proficiency in calculating the equilibrium compositions of electrolyte solutions, we shall undertake a more rigorous analysis of some simple equilibria in order to obtain a feeling for the magnitudes of the errors involved in the assumption that $\gamma_i = 1$.

Consider the dissociation of the weak acid HB (for example, $HNO_2$, $CH_3COOH$, $NH_4^+$, $H_2PO_4^-$) in aqueous solution[*]

$$HB(aq) = H^+(aq) + B^-(aq)$$

The equilibrium constant for this reaction (often designated as $K_a$) is given by

$$K_a = \frac{a_{H^+(aq)} \, a_{B^-(aq)}}{a_{HB(aq)}} \simeq \frac{(H^+)(B^-)}{(HB)}$$

where we have indicated the concentration of a species by placing parentheses around it.

The equilibrium constant for the reaction of the base $B^-$ (for example, $NO_2^-$, $CH_3COO^-$, $NH_3$, $HPO_4^{2-}$) with water

$$B^-(aq) + H_2O(\ell) = HB(aq) + OH^-(aq)$$

(often designated as $K_b$) is given by

$$K_b = \frac{a_{HB(aq)} \, a_{OH^-(aq)}}{a_{B^-(aq)} \, a_{H_2O(\ell)}} \simeq \frac{(HB)(OH^-)}{(B^-)}$$

where in dilute solution $a_{H_2O} \simeq X_{H_2O} \simeq 1$.

---

[*] It is a very common practice to write this reaction as

$$HB(aq) + H_2O(\ell) = H_3O^+(aq) + B^-(aq)$$

rather than as given above. We shall not follow this practice for two reasons. The first of these is that the reported thermodynamic equilibrium constants for acid dissociation do not refer to the reaction involving $H_2O(\ell)$ as a reactant, but rather refer to the reaction that we have written. The second reason is that the formula $H_3O^+$ (the "hydronium ion") conveys the impression that this species adequately describes the proton in aqueous solution. There is very little *direct* experimental evidence to support the existence of $H_3O^+$ in aqueous solutions. The evidence does, however, suggest that the species $H_9O_4^+(aq)$ is a better representation

of the proton in water than is $H_3O^+$. Fortunately, it is not necessary for thermodynamic purposes to specify the state of the proton in aqueous solution any more definitely than $H^+(aq)$— by which formula we simply mean *the solvated proton in all its equilibrium forms*.

TABLE 12.1

CONJUGATE ACID-BASE PAIRS.

| ACID | BASE | ACID | BASE |
|------|------|------|------|
| $H^+(H_2O)_n$ | $(H_2O)_n$ | $CH_3COOH$ | $CH_3COO^-$ |
| $HSO_4^-$ | $SO_4^{2-}$ | $H_2PO_4^-$ | $HPO_4^{2-}$ |
| $H_3PO_4$ | $H_2PO_4^-$ | $NH_4^+$ | $NH_3$ |
| $Fe(OH_2)_6^{3+}$ | $Fe(OH_2)_5OH^{2+}$ | $HPO_4^{2-}$ | $PO_4^{3-}$ |
| $HNO_2$ | $NO_2^-$ | $H_2O$ | $OH^-$ |

According to the Brönsted acid-base classification system, when an acid HB (proton-donor) loses a proton, a base (proton-acceptor) $B^-$ is formed. The species HB and $B^-$ are called a *conjugate pair*; $B^-$ is the conjugate base of the acid HB, and HB is the conjugate acid of the base $B^-$. Some conjugate acid-base pairs are given in Table 12.1.

Values of $pK_a = -\log K_a$ at 25°C for some common acids in aqueous solution are given in Table 12.2. The larger the value of $K_a$ (the smaller the value of $pK_a$) for an acid, the stronger the acid, and the weaker its conjugate base.

The strongest acid that can exist at appreciable concentrations in aqueous solution is $H^+(aq)$. Any acid whose conjugate base is a weaker base in aqueous solution than water will be a strong (i.e., completely dissociated) acid because the acid will be deprotonated by water. The strong acids in water are $HClO_4$, HCl, HI, HBr, $HBrO_3$, $H_2SO_4$ ($HSO_4^-$ is weak), $HNO_3$, and $RSO_3H$ (sulfonic acids). The strongest base that can exist at appreciable concentrations in aqueous solution is $OH^-(aq)$, because any stronger base, for example, $O^{2-}(aq)$ or $NH_2^-(aq)$, will be protonated by the acid $H_2O$. In liquid ammonia solutions, on the other hand, the strongest acid that can exist in appreciable concentrations is $NH_4^+(amm)$, whereas the strongest base is $NH_2^-(amm)$. In $CH_3CH_2OH$ (ethanol) solutions the strongest acid is $H^+(eth)$, and the strongest base is $CH_3CH_2O^-(eth)$. Acids which are weak in one solvent may be strong in another solvent, and vice versa.

Water itself is a Brönsted acid:

$$H_2O(\ell) = H^+(aq) + OH^-(aq)$$

The equilibrium constant for this reaction is called the *ion-product* or *autoprotolysis* constant of water, and is designated as $K_w$:

$$K_w = \frac{a_{H^+(aq)}\, a_{OH^-(aq)}}{a_{H_2O(\ell)}} \simeq (H^+)(OH^-)$$

(in dilute aqueous solutions $a_{H_2O(\ell)} \simeq X_{H_2O} \simeq 1$). Like most weak-acid ionization constants, $K_w$ changes significantly with temperature:

|  | 0°C | 25°C | 60°C |
|------|------|------|------|
| $K_w$ | $0.12 \times 10^{-14}$ | $1.00 \times 10^{-14}$ | $9.6 \times 10^{-14}$ |

The autoprotolysis constant for the solvent ethanol at 25°C is

$$CH_3CH_2OH(\ell) = H^+(eth) + CH_3CH_2O^-(eth)$$

$$K_e = \frac{a_{H^+(eth)}\, a_{CH_3CH_2O^-(eth)}}{a_{CH_3CH_2OH(\ell)}} = 8 \times 10^{-20} \simeq (H^+)(CH_3CH_2O^-)$$

The equilibrium constants $K_a$ and $K_b$ for a conjugate acid-base pair are related through the solvent autoprotolysis constant, as shown for water:

$$K_a = \frac{a_{H^+}\, a_{B^-}}{a_{HB}} \qquad\qquad K_b = \frac{a_{HB}\, a_{OH^-}}{a_{B^-}\, a_{H_2O}}$$

$$K_a K_b = \frac{a_{H^+}\, a_{B^-}}{a_{HB}} \cdot \frac{a_{HB}\, a_{OH^-}}{a_{B^-}\, a_{H_2O}} = \frac{a_{H^+}\, a_{OH^-}}{a_{H_2O}} = K_w$$

Thus, for a conjugate acid-base pair in water at any temperature,

$$\boxed{K_a K_b = K_w}$$

(12.1)

whereas at 25°C, $K_a K_b = 1.0 \times 10^{-14}$.

A convenient measure of the acidity of a solution is the $pH$,[*]

$$\boxed{pH = -\log a_{H^+} \simeq -\log (H^+)}$$

(12.2)

The $pH$ of pure water at 25°C is computed by starting with

$$K_w = 1.00 \times 10^{-14} \simeq (H^+)(OH^-)$$
$$-\log (1.00 \times 10^{-14}) = 14.00 \simeq -\log \{(H^+)(OH^-)\}$$

and, using the electroneutrality condition in pure water,

$$(H^+) = (OH^-)$$

we obtain

$$14.00 = -\log (H^+)^2 = 2\, pH$$

or $pH = 7.00$. Any aqueous solution for which $(H^+) > (OH^-)$ is said to be *acidic*; any aqueous solution for which $(OH^-) > (H^+)$ is said to be *basic*; and any aqueous solution for which $(H^+) = (OH^-)$ is said to be *neutral*. At 25°C, aqueous solutions with $pH < 7.00$ are acidic, those with $pH > 7.00$ are basic, and those with $pH = 7.00$ are neutral. At 60°C, a neutral aqueous solution has a $pH = \dfrac{1}{2} (-\log 9.6 \times 10^{-14}) = 6.51$.

---

[*]There are certain operational difficulties with this definition, in that it involves the non-measurable, single-ion activity $a_{H^+}$. We shall take up this point in more detail in Chapter 13.

**TABLE 12.2**

$pK_a(\equiv -\log K_a)$ VALUES FOR SOME ACIDS IN WATER AT 25°C
(Molality composition scale, solute standard states)

### I. MONOPROTIC ACIDS

| ACID | | $pK_a$ |
|---|---|---|
| $F_3CCOO\underline{H}$ | trifluoroacetic | −0.23 |
| $\underline{H}IO_3$ | iodic | 0.79 |
| $\underline{H}SO_4^-$ | monohydrogensulfate ion | 1.99 |
| $ClCH_2COO\underline{H}$ | chloracetic | 2.87 |
| $\underline{H}F$ | hydrofluoric | 3.17 |
| $\underline{H}NO_2$ | nitrous | 3.35 |
| $HCOO\underline{H}$ | formic | 3.75 |
| $CH_3\underline{C}HCOO\underline{H}$ <br> $\quad\ \ \mid$ <br> $\quad\ \ OH$ | lactic | 3.86 |
| ⬡—$COO\underline{H}$ | benzoic | 4.20 |
| ⬡—$N\underline{H}_3^+$ | anilinium ion | 4.62 |
| $\underline{H}N_3$ | hydrazoic | 4.72 |
| $CH_3COO\underline{H}$ | acetic | 4.76 |
| ⬡$N{-}\underline{H}^+$ | pyridinium ion | 5.17 |
| $HO\underline{Cl}$ | hypochlorous | 7.53 |
| $HO\underline{B}r$ | hypobromous | 8.60 |
| $\underline{H}BO_2$ | boric | 9.19 |
| $N\underline{H}_4^+$ | ammonium ion | 9.24 |
| $HC\underline{N}$ | hydrocyanic | 9.32 |
| ⬡—$O\underline{H}$ | phenol | 10.00 |
| $HO\underline{I}$ | hypoiodous | 12.3 |

### II. DIPROTIC ACIDS

| ACID | | $pK_{a_1}$ | $pK_{a_2}$ |
|---|---|---|---|
| $\underline{H}_3NN\underline{H}_3^{2+}$ | hydrazinium ion | −0.88 | 8.11 |
| $\underline{\underline{H}}_2CrO_4$ | chromic | −0.08 | 6.45 |
| $HOOCCOO\underline{H}$ | oxalic | 1.27 | 4.27 |
| $\underline{H}_3\overset{+}{N}CH_2COO\underline{H}$ <br> $\ ②\qquad\qquad ①$ | glycinium ion | 2.35 | 9.78 |
| $HOOC\underline{C}H{-}\underline{C}HCOO\underline{H}$ <br> $\qquad\ \mid\quad\ \ \mid$ <br> $\qquad\ OH\ \ OH$ | tartaric | 2.70 | 4.05 |

**TABLE 12.2**—*Continued*

| Structure | Name | $pK_{a_1}$ | $pK_{a_2}$ | | |
|---|---|---|---|---|---|
| 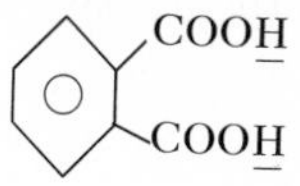 (phthalic structure: benzene ring with two COOH groups) | phthalic | 3.10 | 5.40 | | |
| 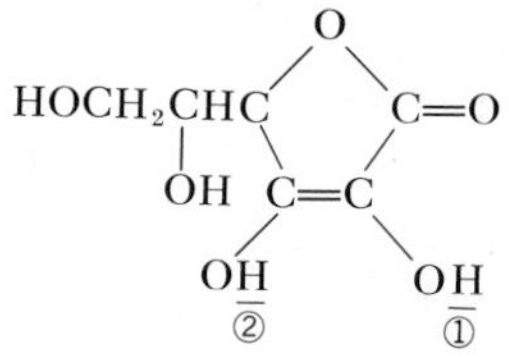 (ascorbic acid structure) | ascorbic (Vitamin C) | 4.17 | 11.57 | | |
| $CO_2(aq)$ | carbonic | 6.35 | 10.33 | | |
| $\underline{\underline{H_2S}}$ | hydrogen sulfide | 7.0 | 12.92 | | |
| $\underline{\underline{H_2O}}(\ell)$ | water | 14.00 | >36 | | |

**III. TRIPROTIC ACIDS**

| ACID | | $pK_{a_1}$ | $pK_{a_2}$ | $pK_{a_3}$ | |
|---|---|---|---|---|---|
| $\underset{③}{HSCH_2}\underset{②}{\underset{|}{\underset{NH_3^+}{CH}}}\underset{①}{COOH}$ | cysteinium ion | 1.96 | 8.54 | 10.51 | |
| $\underline{\underline{H_3PO_4}}$ | phosphoric | 2.15 | 7.21 | 12.36 | |
| $\underline{\underline{H_3AsO_4}}$ | arsenic | 2.22 | 6.96 | 11.40 | |
| (nitrilotriacetic structure) | nitrilotriacetic (NTA) | 2.5 | 2.8 | 10.2 | |
| 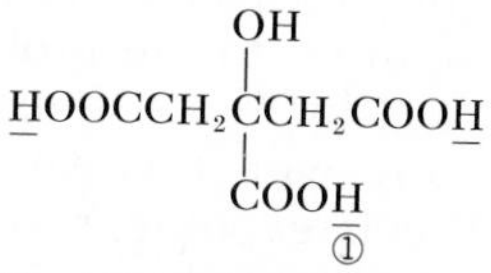 (citric acid structure) | citric | 3.13 | 4.76 | 6.40 | |

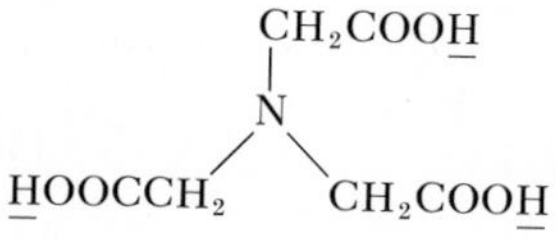

**IV. TETRAPROTIC ACID**

| Structure | Name | $pK_{a_1}$ | $pK_{a_2}$ | $pK_{a_3}$ | $pK_{a_4}$ |
|---|---|---|---|---|---|
| (EDTA structure) | 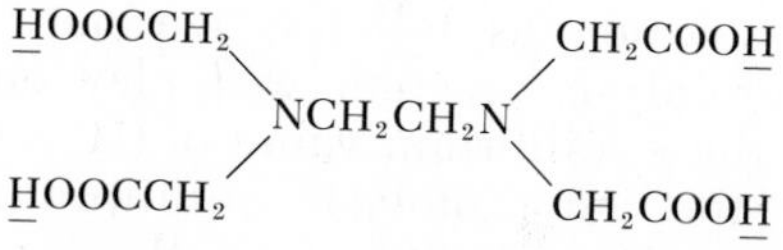 ethylenediaminetetraacetic | 2.18 | 2.73 | 6.20 | 10.0 |

**V. GAS-SOLUTION EQUILIBRIUM CONSTANTS**

$CO_2(g) = CO_2(aq)$    $K = 0.034$

$H_2S(g) = H_2S(aq)$    $K = 0.10$

$NH_3(g) = NH_3(aq)$    $K = 15$

**VI. MISCELLANEOUS EQUILIBRIUM CONSTANTS OF USE IN ACID-BASE CHEMISTRY**

$HF_2^-(aq) = F^-(aq) + HF(aq)$    $K = 0.26$

$H_2O(\ell) + Cr_2O_7^{2-}(aq) = 2HCrO_4^-(aq)$    $K = 0.030$

Consider the problem of calculating the equilibrium concentrations of the various species present in a 0.100 F $HNO_2(aq)$ solution at 25°C and 1 atm, given $K_a = 4.5 \times 10^{-4}$, and $K_w = 1.0 \times 10^{-14}$. There are two *independent* equilibria to be considered,[*] namely

No matter how many equilibria a given species is involved in, there is one and only one equilibrium concentration of that species for a given set of conditions. In other words, the value of (H⁺) in these two equilibrium constant expressions must be the same at equilibrium

$$HNO_2(aq) = H^+(aq) + NO_2^-(aq)$$

$$K_a = 4.5 \times 10^{-4} \simeq \frac{(H^+)(NO_2^-)}{(HNO_2)} \tag{12.3a}$$

and

$$H_2O(\ell) = H^+(aq) + OH^-(aq)$$

$$K_w = 1.0 \times 10^{-14} \simeq (H^+)(OH^-) \tag{12.3b}$$

We have four unknowns in this problem:

$$(H^+), \ (OH^-), \ (HNO_2), \text{ and } (NO_2^-)$$

and consequently, the general solution to this problem requires four *independent* equations involving these unknowns. The equilibrium constant expressions, (12.3a) and (12.3b), provide two equations, and the additional relationships are: a *mass balance* on nitrite

$$0.100 = (NO_2^-) + (HNO_2) \tag{12.3c}$$

which expresses the fact that all the nitrite is in one or the other of the two forms $NO_2^-(aq)$ and $HNO_2(aq)$; and the *electroneutrality condition*

$$(H^+) = (NO_2^-) + (OH^-) \tag{12.3d}$$

which expresses the fact that the solution as a whole must be electrically neutral. Equations (12.3a–d) constitute a formal solution to the problem, and using these four equations we can grind out by brute force the equilibrium concentrations of the four species. However, this can be a very difficult and time-consuming procedure, and it is expedient to seek an approximate solution. The validity of the results obtained can then be tested in the unused equations for the exact solution.

We approach our approximate solution as follows: we note that $K_a \gg K_w$, and, therefore, the autoprotolysis reaction will play only a relatively minor role in determining the equilibrium value of $(H^+)$. Consequently, let us ignore (initially) the autoprotolysis equilibrium. If $HNO_2$ is the only significant source of $H^+(aq)$, then from the reaction stoichiometry and the mass balance equation we have

$$(H^+) \simeq (NO_2^-) \quad \text{and} \quad (HNO_2) = 0.100 - (H^+)$$

respectively. The substitution of these equations into the $K_a$ expression yields

$$4.5 \times 10^{-4} \simeq \frac{(H^+)(NO_2^-)}{(HNO_2)} \simeq \frac{(H^+)^2}{0.100 - (H^+)}$$

---

[*] The equilibrium $NO_2^-(aq) + H_2O(\ell) = HNO_2(aq) + OH^-(aq)$ is not independent because $K_b = K_w/K_a$.

or

$$(H^+)^2 + 4.5 \times 10^{-4}(H^+) - 4.5 \times 10^{-5} \simeq 0$$

Using the quadratic formula,* we obtain

$$(H^+) \simeq \frac{-4.5 \times 10^{-4} \pm \sqrt{(4.5 \times 10^{-4})^2 + 4(4.5 \times 10^{-5})}}{2}$$

and taking the positive root (negative concentrations are *physically impossible*) yields

$$(H^+) \simeq 6.4 \times 10^{-3} \text{ M} \simeq (NO_2^-)$$

from which we compute

$$(HNO_2) \simeq 0.100 - (NO_2^-) \simeq 0.100 - 0.0064 = 0.094 \text{ M}$$

$$(OH^-) \simeq \frac{1.0 \times 10^{-14}}{(H^+)} \simeq \frac{1.0 \times 10^{-14}}{6.4 \times 10^{-3}} = 1.6 \times 10^{-12} \text{ M}$$

Checking the approximate solution in the exact electroneutrality equation, we find

$$(H^+) = (OH^-) + (NO_2^-)$$

$$6.4 \times 10^{-3} = 1.6 \times 10^{-12} + 6.4 \times 10^{-3} \simeq 6.4 \times 10^{-3}$$

Hence, it was a good approximation to ignore the autoprotolysis equilibrium in our calculation. The *pH* of the above solution is

$$pH \simeq -\log (6.4 \times 10^{-3}) = 2.19$$

The percentage of the acid $HNO_2$ that is dissociated in 0.100 F aqueous solution at 25°C is

$$\frac{6.4 \times 10^{-3} \times 100}{0.100} = 6.4\%$$

The most time-consuming part of the above solution involved solving the quadratic equation. It is often possible to obtain a solution more rapidly than was done above by solving a simpler form of the quadratic equation as an approximation. This more approximate solution is as follows: in the expression

$$4.5 \times 10^{-4} \simeq \frac{(H^+)^2}{0.100 - (H^+)}$$

---

*The general solution to the quadratic equation

$$ax^2 + bx + c = 0$$

is

$$x = \frac{-b \pm \sqrt{b^2 - 4ac}}{2a}$$

if $(H^+) \ll 0.100$, then

$$4.5 \times 10^{-4} \simeq \frac{(H^+)^2}{0.100}$$

or

$$(H^+) \simeq (4.5 \times 10^{-5})^{1/2} \simeq 6.7 \times 10^{-3} \text{ M}$$

and $pH = 2.17$. These results are to be compared with the "exact" values of $(H^+) \simeq 6.4 \times 10^{-3}$ M and $pH = 2.19$. We shall arbitrarily adopt the following criterion as to the acceptability of such approximation: we regard the approximation

$$a - x \simeq a$$

as acceptable if $x$ is less than 5% of $a$. In the above case the approximation

$$0.100 - (H^+) \simeq 0.100$$

yields $(H^+) = 6.7 \times 10^{-3}$ M, and

$$\frac{6.7 \times 10^{-3} \times 100}{0.100} = 6.7\%$$

Hence, in this case the exact solution of the quadratic equation in $(H^+)$ is required. Nonetheless, it is advisable to try the more approximate solution first, because in many cases it leads to an error of less than 5% with a considerable savings in time.

*If a salt of a weak acid is added to a solution of the acid, the ionization of the acid is repressed and the pH of the solution increases.* Consider once again the ionization of $HNO_2(aq)$:

$$HNO_2(aq) = H^+(aq) + NO_2^-(aq)$$

$$K_a \simeq \frac{(H^+)(NO_2^-)}{(HNO_2)}$$

If $(NO_2^-)$ is increased, then $(H^+)$ must decrease if $(H^+)(NO_2^-)/(HNO_2)$ is to remain constant, as is required by the equilibrium-constant expression. As an example, let us now compute the concentration of $H^+(aq)$ in a solution that is 0.1000 F in $HNO_2$ and 0.1000 F in $NaNO_2$ at 25°C. The formal solution to this problem is very similar to the previous problem, with the exceptions that: (i) the mass balance on nitrite is

$$0.2000 = (NO_2^-) + (HNO_2)$$

and (ii) the electroneutrality condition is

$$(Na^+) + (H^+) = (NO_2^-) + (OH^-)$$

because of the presence of $Na^+$ from $NaNO_2$. We seek an approximate solution by neglecting the autoprotolysis reaction as a significant source of $H^+(aq)$ relative to $HNO_2(aq)$. The equilibrium concentrations of $HNO_2$, $NO_2^-$, and $H^+$ are obtained from the reaction stoichiometry, together with

If a salt of a weak acid is added to a solution of the acid, the ionization of the acid is repressed and the pH of the solution increases.

the realization that there must be some $H^+(aq)$ arising from $HNO_2(aq)$ (otherwise the $K_a$ expression would not be satisfied)

Species: $\qquad HNO_2(aq) \quad = \quad H^+(aq) \quad + \quad NO_2^-(aq)$
equilibrium concentration: $0.1000 - (H^+) \qquad (H^+) \qquad 0.1000 + (H^+)$

Hence,

$$4.5 \times 10^{-4} \simeq \frac{(H^+)(NO_2^-)}{(HNO_2)} \simeq \frac{(H^+)\{0.1000 + (H^+)\}}{\{0.1000 - (H^+)\}}$$

If $(H^+) \ll 0.1000$, then

$$4.5 \times 10^{-4} \simeq (H^+) \frac{(0.1000)}{(0.1000)} = (H^+)$$

and using this result, we compute

$$(NO_2^-) \simeq 0.1000 + (H^+) \simeq 0.1004 \text{ M}$$

$$(HNO_2) \simeq 0.1000 - (H^+) \simeq 0.0996 \text{ M}$$

$$(OH^-) \simeq \frac{1.0 \times 10^{-14}}{4.5 \times 10^{-4}} = 2.2 \times 10^{-11} \text{ M}$$

Our approximate solution can be checked for reliability in the exact electroneutrality equation (which was not employed in obtaining the approximate solution):

$$(Na^+) + (H^+) = (NO_2^-) + (OH^-)$$

$$0.1000 + 0.0004 = 0.1004 + 2.2 \times 10^{-11}$$

$$0.1004 = 0.1004$$

which justifies our approximate solution. The percentage of the $HNO_2(aq)$ that is dissociated is

$$\frac{4.5 \times 10^{-4} \times 10^2}{0.100} = 0.45\%$$

compared to 6.4% for 0.100 F $HNO_2(aq)$ with no added $NaNO_2(aq)$. The *pH* of the solution is

$$pH \simeq -\log (4.5 \times 10^{-4}) = 3.35$$

which is considerably higher (less acidic solution) than the *pH* of 0.10 F $HNO_2(aq)$ in the absence of added $NaNO_2$, which is 2.19.
   *The salt of a weak acid yields a basic solution on dissolution in water because of the hydrolysis reaction.* For example, consider a 0.100 F aqueous solution of $NaNO_2$. The hydrolysis reaction is

The salt of a weak acid yields a basic solution on dissolution in water because of the hydrolysis reaction.

$$NO_2^-(aq) + H_2O(\ell) = HNO_2(aq) + OH^-(aq)$$

$$K_b = \frac{K_w}{K_a} = \frac{1.00 \times 10^{-14}}{4.5 \times 10^{-4}} = 2.2 \times 10^{-11} \simeq \frac{(HNO_2)(OH^-)}{(NO_2^-)}$$

We also have

$$1.00 \times 10^{-14} \simeq (H^+)(OH^-)$$

$$0.100 = (NO_2^-) + (HNO_2)$$

$$(Na^+) + (H^+) = (OH^-) + (NO_2^-)$$

or, because $NaNO_2(aq)$ is a strong electrolyte $(Na^+) = 0.100$, and

$$0.100 + (H^+) = (OH^-) + (NO_2^-)$$

which gives us four equations and four unknowns. We now seek an approximate solution by neglecting the autoprotolysis reaction as an important source of $OH^-(aq)$. If this approximation is valid, then from the reaction stoichiometry and mass balance equation we have

$$(HNO_2) \simeq (OH^-)$$

$$(NO_2^-) \simeq 0.100 - (OH^-)$$

Substitution of these equations into the $K_b$ expression yields

$$2.2 \times 10^{-11} \simeq \frac{(OH^-)^2}{0.100 - (OH^-)}$$

and, with the additional approximation that $(OH^-) \ll 0.100$, we obtain

$$(OH^-) \simeq (2.2 \times 10^{-12})^{1/2} \simeq 1.5 \times 10^{-6}\ M$$

which verifies the $(OH^-) \ll 0.100$ approximation. The concentrations of the various species at equilibrium are

$$(HNO_2) \simeq (OH^-) \simeq 1.5 \times 10^{-6}\ M$$

$$(NO_2^-) \simeq 0.100\ M$$

$$(H^+) \simeq \frac{1.00 \times 10^{-14}}{1.5 \times 10^{-6}} \simeq 6.7 \times 10^{-9}\ M$$

and

$$pH \simeq -\log(6.7 \times 10^{-9}) = 8.17$$

Many metal ions in aqueous solution are hydrated, and these bound water molecules can ionize to yield acidic solutions.

The solution is basic.

*Many metal ions in aqueous solution are hydrated, and these bound water molecules can ionize to yield acidic solutions.* For example, the equilibrium constant at 25°C for the reaction

$$Fe(OH_2)_6^{3+}(aq) = Fe(OH_2)_5OH^{2+}(aq) + H^+(aq)$$

is $1.0 \times 10^{-3}$. We compute the equilibrium concentrations of the various species in a 0.10 F solution of $Fe(NO_3)_3(aq)$ at 25°C as follows:

$$1.0 \times 10^{-3} \simeq \frac{(Fe(OH_2)_5OH^{2+})(H^+)}{(Fe(OH_2)_6^{3+})} \simeq \frac{(H^+)^2}{0.10 - (H^+)}$$

$$(H^+) \simeq \frac{-1.0 \times 10^{-3} + \sqrt{1.0 \times 10^{-6} + 4.0 \times 10^{-4}}}{2}$$

$$(H^+) \simeq (Fe(OH_2)_5OH^{2+}) \simeq 9.5 \times 10^{-3} \text{ M}$$

$$(Fe(OH_2)_6^{3+}) \simeq 0.09 \text{ M}$$

The *pH* is 2.02, which indicates that the solution is quite acidic. The *pK$_a$* values for some common cations are presented in Table 12.3.

**TABLE 12.3**

*pK$_a$* VALUES FOR SOME COMMON CATIONS IN AQUEOUS SOLUTION AT 25°C.

| CATION | $pK_a$ |
|---|---|
| $Tl^{3+}(aq)$ | 1.2 |
| $Sn^{2+}(aq)$ | 2.0 |
| $Ga^{3+}(aq)$ | 2.62 |
| $Fe(OH_2)_6^{3+}(aq)$ | 3.0 |
| $In^{3+}(aq)$ | 3.7 |
| $Al(OH_2)_6^{3+}(aq)$ | 4.95 |
| $Pb^{2+}(aq)$ | 7.9 |

## 12.2 BUFFERS.

An important property of a solution containing a mixture of a weak acid and its conjugate base is its buffer action. *A buffer is a system that maintains the concentration of* $(H^+)$ *at an approximately constant level when acids or bases are added.* Some representative examples of buffer systems in aqueous solution are:

$$HNO_2 + NaNO_2$$

$$CH_3COOH + NaOCOCH_3$$

$$NH_4Cl + NH_3$$

$$NaHCO_3$$

$$NH_4OCOCH_3$$

$$HCl \text{ (concentrated)}$$
$$NaOH \text{ (concentrated)}$$
} pseudo-buffers

As an example of the analysis of a buffer solution, consider a buffer composed of 0.100 F $NH_4Cl(aq)$ and 0.100 F $NH_3(aq)$ at 25°C ($K_a = 5.5 \times 10^{-10}$ for $NH_4^+(aq)$). The general formulation of the solution to the problem of finding the equilibrium concentrations of all the species in this buffer solution is as follows:

*acid dissociation*: $NH_4^+(aq) = NH_3(aq) + H^+(aq)$

$$K_a = 5.5 \times 10^{-10} \simeq \frac{(NH_3)(H^+)}{(NH_4^+)} \qquad (12.4a)$$

*autoprotolysis*: $H_2O(\ell) = H^+(aq) + OH^-(aq)$

$$K_w = 1.0 \times 10^{-14} \simeq (H^+)(OH^-) \qquad (12.4b)$$

*mass-balance*: $\qquad 0.200 = (NH_4^+) + (NH_3) \qquad (12.4c)$

*electroneutrality*: $(NH_4^+) + (H^+) = (OH^-) + (Cl^-) = (OH^-) + 0.100$ (12.4d)

An approximate solution can be obtained by neglecting the autoprotolysis equilibrium. We then have

$$5.5 \times 10^{-10} \simeq \frac{(NH_3)(H^+)}{(NH_4^+)} \simeq \frac{\{0.100 + (H^+)\}\,(H^+)}{\{0.100 - (H^+)\}}$$

Assuming that $(H^+) \ll 0.100$, we compute $(H^+) \simeq 5.5 \times 10^{-10}$, $(NH_4^+) \simeq$ 0.100 M, and $pH = 9.26$.

    Suppose that we now add to the above solution an amount of KOH(s) that makes the solution 0.010 F in KOH(aq). If we assume (a) that the addition of solid KOH does not significantly change the volume of the solution, and (b) that the neutralization reaction $(NH_4^+ + OH^- \rightarrow NH_3 + H_2O)$ is quantitative, then we obtain the following values for the *formalities** of $NH_4^+(aq)$ and $NH_3(aq)$:

$$[NH_4^+] = 0.100 - 0.010 = 0.090 \text{ F}$$

$$[NH_3] = 0.100 + 0.010 = 0.110 \text{ F}$$

where the square brackets denote formal concentrations. The *equilibrium concentrations* are computed as follows:

$$5.5 \times 10^{-10} \simeq \frac{(H^+)\{0.110 + (H^+)\}}{\{0.090 - (H^+)\}} \simeq \frac{(H^+)(0.110)}{(0.090)}$$

and $(H^+) \simeq 4.5 \times 10^{-10}$ M, $(NH_4^+) \simeq 0.090$ M, and $(NH_3) \simeq 0.110$ M. The *pH* of the solution is $pH = -\log (4.5 \times 10^{-10}) = 9.35$. The change in *pH* upon addition of the KOH(s) is $\Delta(pH) = 9.35 - 9.26 = +0.09$ units. In contrast, consider an unbuffered aqueous solution at 25°C with $pH = 9.26$, to which KOH(s) is added until the solution is 0.010 F in KOH. The initial $(OH^-)$ is

$$(OH^-) \simeq \frac{1.0 \times 10^{-14}}{5.5 \times 10^{-10}} \simeq 2 \times 10^{-5} \text{ M}$$

---

*We shall distinguish formal concentrations from equilibrium concentration by the use of square brackets for the former and parentheses for the latter.

that is, the solution is $2 \times 10^{-5}$ M in, say, KOH(aq). The final $(OH^-)$ is

$$(OH^-) \simeq 2 \times 10^{-5} + 0.010 \simeq 0.010 \text{ M}$$

The final $pH$ is

$$pH \simeq -\log (1.0 \times 10^{-12}) = 12.00$$

The change in $pH$ is $\Delta(pH) = 12.00 - 9.26 = 2.74$, which is about 30 times greater than in the buffered case.

    The change in $pH$ of a buffer solution upon addition of a given amount of strong acid or strong base is less, the higher the buffer concentration. For example, if the $NH_4Cl + NH_3$ buffer is prepared with 1.00 F $NH_4Cl$(aq) and 1.00 F $NH_3$(aq), then addition of enough KOH(s) to make the solution 0.010 F in KOH yields $[NH_4^+] = 0.99$ F and $[NH_3] = 1.01$ F, from which we compute $(H^+) \simeq 5.4 \times 10^{-10}$ M, and $pH = 9.27$. The change in $pH$ is only $9.27 - 9.26 = 0.01$ units. The change in $pH$ of the unbuffered solution in 274 times as great. *For high buffer capacity, use high buffer concentrations.*

    In choosing a buffer it is desirable (in order to maximize the effective $pH$ range of the buffer) to choose a weak acid whose $pK_a \simeq pH$ desired. This follows from the expression

$$K_a \simeq \frac{(H^+)(B^-)}{(HB)}$$

from which we obtain, on taking the negative logarithm of both sides,

$$-\log K_a \simeq -\log (H^+) - \log \frac{(B^-)}{(HB)}$$

or[*]

$$pK_a \simeq pH - \log \frac{(B^-)}{(HB)}$$

From this equation we note that when $(B^-) \simeq (HB)$, then $pK_a \simeq pH$ and the given buffer has the maximum buffer range.

    Any buffer has a limited capacity to maintain an approximately constant $pH$, and, if enough strong acid or base is added, then the buffering action can be swamped. For example, if we add KOH(s) up to 1.0 F KOH(aq) to the $[NH_4Cl] = 0.10$ F, $[NH_3] = 0.10$ F buffer, then $[NH_4^+] \simeq 0$, $[NH_3] \simeq 0.20$ F, $[OH^-] \simeq 0.90$ F, and

$$NH_3(aq) + H_2O(\ell) = NH_4^+(aq) + OH^-(aq)$$

$$K_b = \frac{1.0 \times 10^{-14}}{5.5 \times 10^{-10}} = 1.8 \times 10^{-5} \simeq \frac{(NH_4^+)\{(NH_4^+) + 0.90\}}{\{0.20 - (NH_4^+)\}}$$

---

[*]A more approximate form of this equation, namely

$$pK_a \simeq pH - \log \frac{[B^-]}{[HB]}$$

(where $[B^-]$ and $[HB]$ denote formal concentrations) is known as the Henderson-Hasselbalch equation in the biological sciences. The H.-H. equation involves the assumption that $(H^+)$ and $(OH^-) \ll [B^-]$ and $[HB]$.

from which we compute (assuming $(NH_4^+) \ll 0.20$)

$$(NH_4^+) \simeq \frac{1.8 \times 10^{-5}(0.20)}{(0.90)} = 4 \times 10^{-6} \text{ M}$$

Therefore, $(H^+) \simeq \dfrac{1.0 \times 10^{-14}}{0.90} = 1.1 \times 10^{-14}$, and $pH \simeq 13.96$, or $\Delta(pH) = 13.96 - 9.26 = 4.70$ units. The buffer was simply overwhelmed by that much KOH.

If a buffer is defined simply in terms of its ability to resist changes in $pH$ upon addition of relatively small amounts of acid or base, then we must regard strong bases and acids as buffers. This is shown in Table 12.4, where we have compared the buffering action of a strong acid and a strong base with two conjugate acid-base pair buffers. There is a significant distinction, however, between the two types of buffer. A true buffer differs from a pseudo-buffer in that its $pH$ is nearly unaffected by dilution with the solvent (see Table 12.4).

Buffers are used frequently in analytical chemistry because $pH$ control is essential in causing certain desired reactions to take place quantitatively, and in suppressing others. In chemical kinetics, buffers are used both to maintain constant $pH$ and to study the effect of $(H^+)$ on the reaction rate by using a series of buffers of known $pH$. The desired $pH$, together with the requirement that the buffer components not interfere chemically with the system of interest, determines the choice of buffer. Buffers are usually concentrated solutions, and the neglect of activity coefficients in such concentrated solutions can give rise to serious errors ($>\pm 1$ $pH$ unit) in the calculated $pH$. It is advisable to measure the $pH$ with a calibrated $pH$ meter (Chapter 13) when working with uncharacterized buffers.

Buffer solutions are encountered frequently in living systems, constant $pH$ being required for the maintenance of the delicate balance of the complex sequences of chemical reactions and reaction rates essential to the existence of life. For example, the blood is buffered by a complex mixture of phosphates, carbonates, and proteins, and exhibits a remarkably constant $pH$ in the range 7.3 to 7.5 ($\sim 37°C$). At a $pH < 7.3$ the blood cannot efficiently remove $CO_2$ from the cells, whereas at a $pH > 7.8$ the blood cannot give off $CO_2$ to the lungs. Human blood $pH$ values outside the range 7.0 to 7.8 are incapable of supporting life.

**TABLE 12.4**

| SOLUTION | $pH$ INITIAL | $pH$ AFTER 1 ml OF 0.10 M HCl ADDED TO 100 ml OF THE SOLUTION | $pH$ AFTER 1 ml OF 0.10 M NaOH ADDED TO 100 ml OF THE SOLUTION | $pH$ AFTER 10-FOLD DILUTION WITH WATER |
|---|---|---|---|---|
| 0.10 M HCl(aq) | 1.00 | 1.00 | 1.01 | 2.00 |
| 0.10 M NaOH(aq) | 13.00 | 12.99 | 13.00 | 12.00 |
| {0.10 M CH₃COOH(aq) +0.10 M NaOCOCH₃(aq)} | 4.76 | 4.75 | 4.77 | 4.76 |
| {0.10 M NH₃(aq) +0.10 M NH₄Cl(aq)} | 9.26 | 9.25 | 9.28 | 9.27 |

In the preceding examples of acid-base equilibria we were able to neglect the autoprotolysis equilibrium and to restrict our attention to a single equilibrium. However, it is not always possible to achieve such simplification. For example, let us compute the equilibrium concentrations of the various species present in a $1.0 \times 10^{-5}$ M solution of hydrocyanic acid in water at $25°C$ ($K_a = 6.0 \times 10^{-10}$). The relevant equilibria are

$$HCN(aq) = H^+(aq) + CN^-(aq)$$

$$6.0 \times 10^{-10} \simeq \frac{(H^+)(CN^-)}{(HCN)} \qquad (12.5a)$$

$$H_2O(\ell) = H^+(aq) + OH^-(aq)$$

$$1.0 \times 10^{-14} \simeq (H^+)(OH^-) \qquad (12.5b)$$

In addition, we have the mass-balance and electroneutrality conditions

$$1.0 \times 10^{-5} = (CN^-) + (HCN) \qquad (12.5c)$$

$$(H^+) = (CN^-) + (OH^-) \qquad (12.5d)$$

Assuming, as was done in previous cases, that the autoprotolysis equilibrium can be neglected, we have

$$6.0 \times 10^{-10} \simeq \frac{(H^+)^2}{\{1.0 \times 10^{-5} - (H^+)\}}$$

from which we obtain $(H^+) \simeq 7.7 \times 10^{-8}$ M, or $pH = 7.11$. This answer must be incorrect because, no matter how weak, no acid can partially dissociate into $H^+(aq)$ and $X^-(aq)$ to yield a *basic* solution. The unreliability of the above solution is also indicated by the failure of the numerical solution to check in the electroneutrality condition:

$$7.7 \times 10^{-8} \neq 7.7 \times 10^{-8} + \frac{1.0 \times 10^{-14}}{7.7 \times 10^{-8}} = 2 \times 10^{-7}$$

In this case we must include the autoprotolysis reaction as a significant source of $H^+(aq)$. This is because $HCN(aq)$ is present at a very low concentration.[*] Consequently, both equations (12.5a) and (12.5b) play an important role in determining the equilibrium value of $(H^+)$. Combining equations (12.5d) and (12.5c), we obtain

$$6.0 \times 10^{-10} \simeq \frac{(H^+)(CN^-)}{(HCN)} = \frac{(H^+)\{(H^+)-(OH^-)\}}{\{1.0 \times 10^{-5} - (CN^-)\}}$$

Using equations (12.5b) and (12.5d), we can obtain from this expression an equation involving only $(H^+)$ as an unknown:

$$6.0 \times 10^{-10} \simeq \frac{(H^+)\left\{(H^+) - \dfrac{1.0 \times 10^{-14}}{(H^+)}\right\}}{\left\{1.0 \times 10^{-5} - (H^+) + \dfrac{1.0 \times 10^{-14}}{(H^+)}\right\}}$$

This is a *cubic* equation in $(H^+)$, and the direct solution is tedious, so let us seek an approximate solution. We know that $HCN(aq)$ is a very weak acid because $K_a$ is

---

[*] The concentration of the weak acid at which the autoprotolysis reaction becomes important depends on the value of $K_a$. The smaller the value of $K_a$, the higher the concentration of the acid at which the ionization of water must be taken into account.

so small, and therefore we shall assume that most of the cyanide is present as HCN(aq), that is (from the above $K_a$ expression),

$$(HCN) = 1.0 \times 10^{-5} - (H^+) + \frac{1.0 \times 10^{-14}}{(H^+)} \simeq 1.0 \times 10^{-5}$$

Substitution of this value for (HCN) into the $K_a$ expression yields

$$6.0 \times 10^{-10} \simeq \frac{(H^+) \left\{ (H^+) - \frac{1.0 \times 10^{-14}}{(H^+)} \right\}}{1.0 \times 10^{-5}}$$

and

$$(H^+)^2 \simeq 6.0 \times 10^{-15} + 1.0 \times 10^{-14} = 1.6 \times 10^{-14}$$

from which we compute $(H^+) \simeq 1.26 \times 10^{-7}$ M, and $pH \simeq 6.90$. The reliability of the approximation $(HCN) \simeq 1.0 \times 10^{-5}$ M can be verified using the exact mass-balance equation (12.5c). Because

$$(CN^-) \simeq \frac{6.0 \times 10^{-10}(HCN)}{(H^+)} \simeq \frac{6.0 \times 10^{-10} \times 1.0 \times 10^{-5}}{1.26 \times 10^{-7}} = 4.7 \times 10^{-8} \text{ M}$$

we have

$$1.0 \times 10^{-5} = (CN^-) + (HCN)$$
$$1.0 \times 10^{-5} = 4.7 \times 10^{-8} + 1.0 \times 10^{-5} = 1.0 \times 10^{-5}$$

which justifies our assumption.

## 12.3 MULTIPLE EQUILIBRIA IN ACID-BASE REACTIONS. THE METHOD OF PRINCIPAL EQUILIBRIUM.

The method of principal equilibrium is often useful in solving complex equilibrium problems in systems involving several equilibria. In this method all possible equilibria involving *principal species as reactants* are compared, and if one of the equilibria has an equilibrium constant much larger ($\gtrsim 10^3$) than any of the others, then it is assumed that only this particular equilibrium is of importance in determining the concentrations of the *principal species*. The principal (or major) species are those species whose concentrations are large compared to other (minor) species. Once the concentrations of the principal species have been obtained, the other equilibrium constant expressions are used to compute the concentrations of the species assumed to be present at relatively minor concentrations. If the results of these calculations show that the concentrations of the minor species are small compared to the principal species, and if, in addition, the complete set of results is consistent with the exact mass-balance and electroneutrality equations, then confidence can be placed in the results. Several examples will serve to illustrate the method:

**Example 1:** Calculate the concentrations of all the species in a 0.100 F $H_3PO_4$(aq) solution at 25°C, given that $K_{a1} = 7.1 \times 10^{-3}$, $K_{a2} = 6.2 \times 10^{-8}$, and $K_{a3} = 4.4 \times 10^{-13}$ for $H_3PO_4$(aq).

The relevant equilibria are

(1) $H_3PO_4(aq) \rightleftarrows H^+(aq) + H_2PO_4^-(aq)$  $K_{a1} = 7.1 \times 10^{-3} \simeq \dfrac{(H^+)(H_2PO_4^-)}{(H_3PO_4)}$

(2) $H_2PO_4^-(aq) \rightleftarrows H^+(aq) + HPO_4^{2-}(aq)$  $K_{a2} = 6.2 \times 10^{-8} \simeq \dfrac{(H^+)(HPO_4^{2-})}{(H_2PO_4^-)}$

(3) $HPO_4^{2-}(aq) \rightleftarrows H^+(aq) + PO_4^{3-}(aq)$   $K_{a3} = 4.4 \times 10^{-13} \simeq \dfrac{(H^+)(PO_4^{3-})}{(HPO_4^{2-})}$

(4) $H_2O(\ell) \rightleftarrows H^+(aq) + OH^-(aq)$        $K_w = 1.0 \times 10^{-14} \simeq (H^+)(OH^-)$

A mass balance on "$PO_4$" yields

(5) $0.100 = (H_3PO_4) + (H_2PO_4^-) + (HPO_4^{2-}) + (PO_4^{3-})$

and the electroneutrality condition is

(6) $(H^+) = (H_2PO_4^-) + 2(HPO_4^{2-}) + 3(PO_4^{3-}) + (OH^-)$

This problem involves six unknowns:

$(H_3PO_4)$, $(H_2PO_4^-)$, $(HPO_4^{2-})$, $(PO_4^{3-})$, $(H^+)$, and $(OH^-)$

and, because we have six independent equations, the problem can be solved by brute force. However, the problem can be solved much more rapidly by noting that reaction (1) is the principal equilibrium. Reaction (1) is the principal equilibrium because (a) $K_{a1}$ is much greater than $K_{a2}$, $K_{a3}$, or $K_w$ (i.e., $H_3PO_4(aq)$ is a stronger Brönsted acid than $H_2PO_4^-(aq)$, $HPO_4^{2-}(aq)$ or $H_2O(\ell)$), and (b) reaction (1) also involves the species $H_3PO_4(aq)$, which is expected to be a principal species because the solution was prepared from $H_3PO_4(aq)$. Alternatively, we note that because $H_3PO_4(aq)$ is a weak acid the concentration of $H_2PO_4^-(aq)$ is small relative to the concentration of $H_3PO_4(aq)$, and further (for similar reasons), $(PO_4^{3-}) \ll (HPO_4^{2-}) \ll (H_2PO_4^-)$.

If we assume that reaction (1) is the principal equilibrium, then (ignoring for the moment all other equilibria) we can write

$$H_3PO_4(aq) \rightleftarrows H^+(aq) + H_2PO_4^-(aq)$$
$$0.100 - (H^+) \qquad (H^+) \qquad\quad (H^+)$$

(where we have used the relation $(H^+) \simeq (H_2PO_4^-)$, which follows from the reaction stoichiometry) and

$$7.1 \times 10^{-3} \simeq \frac{(H^+)(H_2PO_4^-)}{(H_3PO_4)} \simeq \frac{(H^+)^2}{0.100 - (H^+)}$$

or

$$(H^+)^2 + 7.1 \times 10^{-3}(H^+) - 7.1 \times 10^{-4} = 0$$

The solution of this quadratic equation is

$$(H^+) \simeq 0.024 \text{ M} \simeq (H_2PO_4^-)$$

and thus

$$(H_3PO_4) = 0.100 - 0.024 = 0.076 \text{ M}$$

The $pH$ of the solution is $pH \simeq -\log(2.4 \times 10^{-2}) \simeq 1.62$.

The concentrations of the (presumed) minor species are now calculated using the above results, together with the other equilibrium constant expressions:

$$(HPO_4^{2-}) \simeq \frac{6.2 \times 10^{-8}(H_2PO_4^-)}{(H^+)} = 6.2 \times 10^{-8} \text{ M}$$

$$(PO_4^{3-}) \simeq \frac{4.4 \times 10^{-13}(HPO_4^{2-})}{(H^+)} = \frac{4.4 \times 10^{-13} \times 6.2 \times 10^{-8}}{2.4 \times 10^{-2}} = 1.1 \times 10^{-18} \text{ M}$$

$$(OH^-) \simeq \frac{1.0 \times 10^{-14}}{2.4 \times 10^{-2}} = 4.2 \times 10^{-13} \text{ M}$$

These results show that the presumed minor species are in fact minor species. The above solution to the problem can be checked using the exact mass-balance and electroneutrality conditions:

*mass-balance check*

$$0.100 = 0.076 + 0.024 + 6.2 \times 10^{-8} + 1.1 \times 10^{-18} = 0.100$$

*electroneutrality check*

$$0.024 = 0.024 + 2(6.2 \times 10^{-8}) + 3(1.1 \times 10^{-18}) + 4.2 \times 10^{-13} = 0.024$$

**Example 2:** Calculate the concentrations of all the species in an aqueous solution that is 0.100 F in $NaH_2PO_4(aq)$ and 0.100 F in $Na_2HPO_4(aq)$.

The relevant equilibria are the same in this example as in Example 1 above. The mass-balance equation is

$$0.200 = (H_3PO_4) + (H_2PO_4^-) + (HPO_4^{2-}) + (PO_4^{3-})$$

and the electroneutrality equation is $[(Na^+) = 0.300 \text{ M}]$

$$0.300 + (H^+) = (H_2PO_4^-) + 2(HPO_4^{2-}) + 3(PO_4^{3-})$$

Which equilibrium is the principal equilibrium? The principal equilibrium depends on how the system is prepared. In this case the system was prepared from $NaH_2PO_4$ and $Na_2HPO_4$; therefore, the prime candidate for the principal equilibrium is reaction (2) in Example 1. Reaction (1) is *not* the principal equilibrium, because it does not contain both of the principal species $H_2PO_4^-(aq)$ and $HPO_4^{2-}(aq)$, and, in addition, it involves only the *minor species* $H_3PO_4(aq)$ *as a reactant*.

If we assume that reaction (2) is the principal equilibrium, then we have

$$H_2PO_4^-(aq) \rightleftarrows H^+(aq) + HPO_4^{2-}(aq)$$

formal concentration}  0.100  0  0.100

equilibrium concentration} 0.100−(H⁺)  (H⁺)  0.100+(H⁺)

and

$$6.2 \times 10^{-8} \simeq \frac{(H^+)(HPO_4^{2-})}{(H_2PO_4^-)} \simeq \frac{(H^+)\{0.100 + (H^+)\}}{\{0.100 - (H^+)\}} \simeq (H^+)$$

$(H^+) = 6.2 \times 10^{-8}$ M $(pH = 7.21)$, and $(H_2PO_4^-) \simeq (HPO_4^{2-}) \simeq 0.100$ M.

The concentrations of the minor species are:

$$(H_3PO_4) \simeq \frac{(H^+)(H_2PO_4^-)}{7.1 \times 10^{-3}} \simeq \frac{6.2 \times 10^{-8} \times 0.100}{7.1 \times 10^{-3}} = 8.7 \times 10^{-7}\, M$$

$$(PO_4^{3-}) \simeq \frac{4.4 \times 10^{-13}(HPO_4^{2-})}{(H^+)} \simeq \frac{4.4 \times 10^{-13} \times 0.100}{6.2 \times 10^{-8}} = 7.1 \times 10^{-7}\, M$$

$$(OH^-) \simeq \frac{1.0 \times 10^{-14}}{6.2 \times 10^{-8}} = 1.6 \times 10^{-7}\, M$$

*mass-balance check*

$$0.200 = 8.7 \times 10^{-7} + 0.100 + 0.100 + 7.1 \times 10^{-7} = 0.200$$

*electroneutrality check*

$$0.300 + 6.2 \times 10^{-8} = 0.300 = 0.100 + 2(0.100) + 3(7.1 \times 10^{-7}) = 0.300$$

A comparison of Examples 1 and 2 shows that the choice of the principal equilibrium requires some care, and the choice involves more than simply comparing $K$ values. No matter how large a $K$ value a reaction has, it is not a candidate for the principal equilibrium unless it involves all of the principal species. Furthermore, it is frequently necessary to combine two (or more) equilibria into one in order to obtain a reaction that involves *all* of the principal species. This point is illustrated in Example 3.

**Example 3:**   Calculate the concentrations of all the species present in a 0.100 F $NH_4CN(aq)$ solution at 25°C, given $K_a = 5.5 \times 10^{-10}$ for $NH_4^+(aq)$, and $K_b' = 1.7 \times 10^{-5}$ for $CN^-(aq)$. The relevant equilibria are

(1) $H_2O(\ell) = H^+(aq) + OH^-(aq)$      $K_w = 1.0 \times 10^{-14} \simeq (H^+)(OH^-)$

(2) $NH_4^+(aq) = H^+(aq) + NH_3(aq)$      $K_a = 5.5 \times 10^{-10} \simeq \dfrac{(H^+)(NH_3)}{(NH_4^+)}$

(3) $CN^-(aq) + H_2O(\ell) = HCN(aq) + OH^-(aq)$

$$K_b' = 1.7 \times 10^{-5} \simeq \frac{(HCN)(OH^-)}{(CN^-)}$$

In addition to these three relationships among the six unknowns, there are also two mass-balance equations and the electroneutrality equation:

(4) $0.100 = (NH_4^+) + (NH_3)$

(5) $0.100 = (CN^-) + (HCN)$

(6) $(NH_4^+) + (H^+) = (CN^-) + (OH^-)$

Equations (1) through (6) constitute the formal solution to the problem, but fortunately, it is not necessary to solve this set of six simultaneous equations directly. Both $NH_4^+(aq)$ and $CN^-(aq)$ are probably principal species because we are dealing with a solution of $NH_4CN(aq)$; therefore, we write a reaction involving both of these species as reactants, namely[*]

(7) $NH_4^+(aq) + CN^-(aq) = NH_3(aq) + HCN(aq)$

This equation results from the addition of equation (3) to equation (2), and subtraction of equation (1) from the result; therefore, $K$ is given by

$$K = \frac{K_a K_b'}{K_w} \simeq \frac{(NH_3)(HCN)}{(NH_4^+)(CN^-)} \simeq \frac{5.5 \times 10^{-10} \times 1.7 \times 10^{-5}}{1.0 \times 10^{-14}} = 0.935$$

Because $K$ is much greater than $K_a$ or $K_b'$, reaction (7) is probably the principal equilibrium in the solution (we cannot be absolutely sure until we determine whether $NH_4^+$, $CN^-$, $NH_3$, and $HCN$ are the principal species). We shall assume that (7) is the principal equilibrium and solve the problem by initially neglecting all other equilibria. From the reaction stoichiometry and mass-balance conditions we obtain

$$NH_4^+(aq) \quad + \quad CN^-(aq) \quad \rightleftarrows NH_3(aq) + HCN(aq)$$

$$0.100 - (NH_3) \quad 0.100 - (HCN) \quad (NH_3) \quad\quad (HCN)$$

but $(NH_3) \simeq (HCN)$ (stoichiometry); hence,

$$0.935 \simeq \frac{(NH_3)(HCN)}{(NH_4^+)(CN^-)} \simeq \frac{(NH_3)^2}{\{0.100 - (NH_3)\}^2}$$

or (taking the square root of both sides)

$$0.967 \simeq \frac{(NH_3)}{0.100 - (NH_3)}$$

from which we compute

$$(NH_3) \simeq (HCN) \simeq 0.049 \text{ M, and } (NH_4^+) \simeq (CN^-) \simeq 0.051 \text{ M}$$

We now check to determine whether or not equation (7) is the principal equilibrium by calculating the concentrations of the (assumed) minor species:

$$(H^+) \simeq \frac{5.5 \times 10^{-10}(NH_4^+)}{(NH_3)} \simeq \frac{5.5 \times 10^{-10} \times 0.051}{0.049} = 5.7 \times 10^{-10} \text{ M}$$

$$(OH^-) \simeq \frac{1.0 \times 10^{-14}}{(H^+)} = 1.8 \times 10^{-5} \text{ M},$$

---

[*]This equilibrium is not independent of equilibria (1), (2), and (3), because this equation can be obtained by adding equations (2) and (3) and subtracting (1).

These results substantiate our assumption that $H^+$ and $OH^-$ are minor species (i.e., have much lower equilibrium concentrations than those appearing in the principal equilibrium). As a further check on our calculated results we use the electroneutrality condition, which requires

$$(NH_4^+) + (H^+) = (OH^-) + (CN^-)$$

Substitution of our calculated concentrations yields

$$0.051 + 5.7 \times 10^{-10} = 1.8 \times 10^{-5} + 0.051$$

or, $0.051 = 0.051$. Thus, we have found a set of equilibrium concentrations which satisfy equations (1) through (6), and therefore constitute a solution to the problem.

**Example 4:** Calculate the concentrations of all the species present in a 0.100 F $NaHCO_3$(aq) solution at 25°C.

If an aqueous solution containing dissolved $CO_2$ is in equilibrium with $CO_2$(g) over the solution, then the concentration of $CO_2$(aq) in the solution is fixed by the value of $P_{CO_2}$ over the solution via the equilibrium

$$CO_2(g) = CO_2(aq)$$

At 25°C,

$$K' = 0.034 \simeq \frac{(CO_2)}{P_{CO_2}}$$

At equilibrium the value of $(CO_2)$ is fixed by the value of $P_{CO_2}$, and this is true no matter how many equilibria in the solution involve $CO_2$(aq).

Part of the $CO_2$ in a solution of carbon dioxide in water is simply physically dissolved as $CO_2$(aq), and part of it is in the form $H_2CO_3$(aq). The equilibrium constant for the reaction

$$CO_2(aq) + H_2O(\ell) = H_2CO_3(aq)$$

at 25°C is

$$K = 2.6 \times 10^{-3} \simeq \frac{(H_2CO_3)}{(CO_2)}$$

which means that 99.74% of the $CO_2$ in solution is present as $CO_2$(aq) and only 0.26% is present as $H_2CO_3$(aq). We shall designate aqueous carbonic acid by the formula $CO_2$(aq) because this is the principal species. For the first and second acid dissociation constants, $K_{a1}$ and $K_{a2}$, of carbonic acid we have* (where the numerical values of the equilibrium constants are for 25°C)

$$CO_2(aq) + H_2O(\ell) = H^+(aq) + HCO_3^-(aq)$$

$$K_{a1} = 4.3 \times 10^{-7} \simeq \frac{(H^+)(HCO_3^-)}{(CO_2)}$$

---

*Note that for the reaction $H_2CO_3$(aq) $= H^+$(aq) $+ HCO_3^-$(aq), we have

$$K_{a1}' = \frac{K_{a1}}{K} = \frac{4.3 \times 10^{-7}}{2.6 \times 10^{-3}} = 1.7 \times 10^{-4} \simeq \frac{(H^+)(HCO_3^-)}{(H_2CO_3)}$$

$$HCO_3^-(aq) = H^+(aq) + CO_3^{2-}(aq)$$

$$K_{a2} = 4.7 \times 10^{-11} \simeq \frac{(H^+)(CO_3^{2-})}{(HCO_3^-)}$$

An aqueous solution of $NaHCO_3$ involves the following unknowns:

$$(HCO_3^-), (CO_3^{2-}), (H^+), (OH^-), (CO_2), (H_2CO_3), (Na^+)$$

If we assume that $Na^+$ is not directly involved in any of the various equilibria, then in a 0.100 F $NaHCO_3$(aq) solution we have $(Na^+) = 0.100$ M. Among the remaining six unknowns we must find six independent algebraic relationships. The relevant relationships are as follows:

(1) $HCO_3^-(aq) = H^+(aq) + CO_3^{2-}(aq)$

$$K_{a2} = 4.7 \times 10^{-11} \simeq \frac{(H^+)(CO_3^{2-})}{(HCO_3^-)}$$

(2) $HCO_3^-(aq) = CO_2(aq) + OH^-(aq)$

$$K_{b1} = \frac{K_w}{K_{a1}} = \frac{1.0 \times 10^{-14}}{4.3 \times 10^{-7}} = 2.3 \times 10^{-8} \simeq \frac{(CO_2)(OH^-)}{(HCO_3^-)}$$

(3) $CO_2(aq) + H_2O(\ell) = H_2CO_3(aq)$

$$K = 2.6 \times 10^{-3} \simeq \frac{(H_2CO_3)}{(CO_2)}$$

(4) $H_2O(\ell) = H^+(aq) + OH^-(aq)$

$$K_w = 1.0 \times 10^{-14} \simeq (H^+)(OH^-)$$

(5) mass balance

$$0.100 = (HCO_3^-) + (CO_3^{2-}) + (CO_2) + (H_2CO_3)$$

(6) electroneutrality

$$(Na^+) + (H^+) = (HCO_3^-) + (OH^-) + 2(CO_3^{2-})$$

Equations (1) through (6) constitute the formal solution to the problem, but rather than attempting a direct solution let us try the principal equilibrium approximation.

Note that $HCO_3^-$(aq) is both an acid and a base; the principal equilibrium involves the transfer of a proton from one $HCO_3^-$(aq) to another $HCO_3^-$(aq).

Although equilibrium (3) has a much larger equilibrium constant than (1), (2), or (4), it is not a candidate for the principal equilibrium because it does not involve the principal species $HCO_3^-$. However, the reaction

$$2HCO_3^-(aq) = CO_3^{2-}(aq) + CO_2(aq) + H_2O(\ell)$$

$$K'' = \frac{K_{a2}}{K_{a1}} = \frac{4.7 \times 10^{-11}}{4.3 \times 10^{-7}} = 1.1 \times 10^{-4} \simeq \frac{(CO_3^{2-})(CO_2)}{(HCO_3^-)^2}$$

is a likely candidate because $K'' \gg K_{a2}$ or $K_{b1}$ and the reaction involves the principal species $HCO_3^-$(aq) as a reactant. Assuming that this is the principal equilibrium, we have from the reaction stoichiometry and mass-balance condition

$$(CO_2) \simeq (CO_3^{2-})$$

$$(HCO_3^-) \simeq 0.100 - 2(CO_3^{2-})$$

Substitution of these expressions into the expression for $K''$ yields

$$1.1 \times 10^{-4} \simeq \frac{(CO_3^{2-})^2}{\{0.100 - 2(CO_3^{2-})\}^2}$$

or, taking the square root of both sides

$$1.05 \times 10^{-2} \simeq \frac{(CO_3^{2-})}{\{0.100 - 2(CO_3^{2-})\}}$$

Solving this equation for $(CO_3^{2-})$ yields

$$(CO_3^{2-}) \simeq 1.0 \times 10^{-3} \, M \simeq (CO_2)$$

$$(HCO_3^-) \simeq 0.100 - 2(1.0 \times 10^{-3}) = 0.098 \, M$$

The concentrations of the minor species are

$$(H^+) \simeq \frac{4.7 \times 10^{-11}(HCO_3^-)}{(CO_3^{2-})} \simeq \frac{4.7 \times 10^{-11} \times 0.098}{1.0 \times 10^{-3}} = 4.6 \times 10^{-9} \, M$$

$$(OH^-) \simeq \frac{1.0 \times 10^{-14}}{(H^+)} \simeq 2.2 \times 10^{-6} \, M$$

$$(H_2CO_3) = 2.6 \times 10^{-3}(CO_2) = 2.6 \times 10^{-3} \times 1.0 \times 10^{-3} = 2.6 \times 10^{-6} \, M.$$

The presumed minor species are in fact the minor species. The $pH$ of the solution is 8.34, which is basic. The solution is basic because the $HCO_3^-(aq)$ can take protons from $H_2O(\ell)$ more readily than it can give them up, that is, $K_{b1} \gg K_{a2}$. As an independent check on our approximate solution, we substitute the above results into the electroneutrality condition:

$$(Na^+) + (H^+) = (HCO_3^-) + (OH^-) + 2(CO_3^{2-})$$

$$0.100 + 4.6 \times 10^{-9} = 0.098 + 2.2 \times 10^{-6} + 2(1.0 \times 10^{-3})$$

$$0.100 = 0.100$$

The average atmospheric pressure of $CO_2(g)$ is about $3 \times 10^{-4}$ atm. Therefore, any aqueous carbonate solution at 25°C which has a concentration of $CO_2(aq)$ greater than

$$(CO_2) \simeq K'P_{CO_2} \simeq 0.034 \times 3 \times 10^{-4} \simeq 1.0 \times 10^{-5} \, M$$

and which is open to the atmosphere, is unstable with respect to loss of $CO_2$ to the atmosphere. In the above 0.100 M $NaHCO_3(aq)$ solution we found $(CO_2) \simeq 1.0 \times 10^{-3}$ M and consequently this solution, if left open to the atmosphere, can undergo the following decomposition reaction:

$$2NaHCO_3(aq) \rightarrow Na_2CO_3(aq) + CO_2(g) + H_2O(\ell)$$

This reaction will proceed until $(CO_2) \simeq 1.0 \times 10^{-5}$ M. Of course, extensive $CO_2$ evolution can be prevented by storing the solution in a tightly closed container with only a small vapor space. The equilibrium partial pressure of $CO_2(g)$ over 0.100 M $NaHCO_3(aq)$ is

$$P_{CO_2} \simeq \frac{1.0 \times 10^{-3}}{0.034} = 0.03 \text{ atm}$$

The loss of $CO_2$ to the atmosphere is the reason why carbonated beverages go flat if not stored in tightly-capped containers.

## 12.4   DISTRIBUTION DIAGRAMS FOR POLYPROTIC ACIDS.

If a polyprotic acid is partially neutralized, the *pH* determines the distribution of species. For example, the major carbon-containing species in the aqueous carbonic acid system are $CO_2(aq)$, $HCO_3^-(aq)$, and $CO_3^{2-}(aq)$. The first and second acid-dissociation equilibrium constants are

$$K_{a1} \simeq \frac{(H^+)(HCO_3^-)}{(CO_2)} \quad \text{and} \quad K_{a2} \simeq \frac{(H^+)(CO_3^{2-})}{(HCO_3^-)}$$

respectively. We now define the variable $\alpha_i$ as the fraction of the total concentration of a particular set of species that a given member of the set comprises. For the carbonic acid system we have (neglecting $H_2CO_3(aq)$ because it is a minor species)

$$\alpha_{CO_2} = \frac{(CO_2)}{(total)} = \frac{(CO_2)}{(CO_2) + (HCO_3^-) + (CO_3^{2-})}$$

$$\alpha_{HCO_3^-} = \frac{(HCO_3^-)}{(total)}$$

$$\alpha_{CO_3^{2-}} = \frac{(CO_3^{2-})}{(total)}$$

where

$$\alpha_{CO_2} + \alpha_{HCO_3^-} + \alpha_{CO_3^{2-}} = 1$$

Using the $K_{a1}$ and $K_{a2}$ expressions, we can rewrite the $\alpha_i$ expressions as

$$\alpha_{CO_2} = \frac{1}{1 + \dfrac{K_{a1}}{(H^+)} + \dfrac{K_{a1}K_{a2}}{(H^+)^2}} = \frac{1}{\beta}$$

$$\alpha_{HCO_3^-} = \frac{K_{a1}}{\beta(H^+)}$$

$$\alpha_{CO_3^{2-}} = \frac{K_{a1}K_{a2}}{\beta(H^+)^2}$$

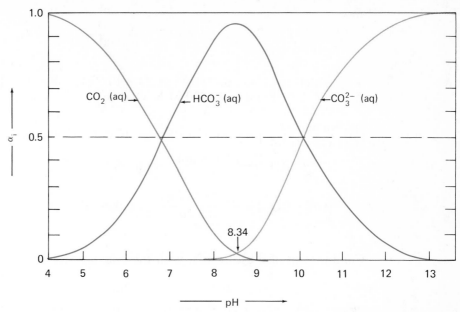

**FIGURE 12.1** Distribution diagram for the aqueous carbonic acid system at 25°C.

The $\alpha_i$ values at a given temperature depend only on the $(H^+)$ concentration of the solution, and in particular, they are independent of the total acid concentration. In Figure 12.1 we have plotted the $\alpha_i$ values for the carbonic acid system as a function of $pH$. This plot is called a *distribution diagram,* and its value is that it enables us to read off the fraction of the total that any of the three species $CO_2(aq)$, $HCO_3^-(aq)$, and $CO_3^{2-}(aq)$ comprises at any $pH$. We note that for $pH < 4$ we have $\alpha_{CO_2} \simeq 1$, whereas for $pH > 13$ we have $\alpha_{CO_3^{2-}} \simeq 1$. The $pH$ of the solution at which the fraction $\alpha_{HCO_3^-}$ is a maximum is 8.34, which is the $pH$ of $NaHCO_3(aq)$. The intersection points at $\alpha_i = 0.50$ correspond to $pK_{a1}$ and $pK_{a2}$.

## 12.5 ACID-BASE TITRATIONS.

Consider a 25.00 ml sample of 0.1000 F $CH_3COOH(aq)$ at 25°C to which 0.1000 M $NaOH(aq)$ is added in controlled amounts by means of a buret. The question of interest here is the manner in which the equilibrium $pH$ of the solution changes upon the stepwise addition of $NaOH(aq)$.

The equilibrium of interest is

$$CH_3COOH(aq) = H^+(aq) + CH_3COO^-(aq)$$

$$K_a = 1.76 \times 10^{-5} \simeq \frac{(H^+)(CH_3COO^-)}{(CH_3COOH)}$$

Before any base is added we have

$$(H^+) \simeq (CH_3COO^-)$$

$$(CH_3COOH) \simeq 0.1000 - (H^+)$$

and

$$1.76 \times 10^{-5} \simeq \frac{(H^+)^2}{0.1000 - (H^+)}$$

from which we compute $(H^+) \simeq 1.33 \times 10^{-3}$ M, and $pH \simeq 2.88$. After the addition of 10.00 ml of 0.1000 M NaOH we have*

$$\left(\begin{array}{l}\text{millimoles of} \\ \text{CH}_3\text{COOH left}\end{array}\right) = (25.00)(0.1000) - (10.00)(0.1000) = 1.50 \text{ mmole}$$

$$[\text{CH}_3\text{COOH}] = \frac{1.50 \text{ mmole}}{35.00 \text{ ml}} = 0.0428 \text{ F}$$

$$[\text{CH}_3\text{COO}^-] = \frac{(10.00)(0.1000)}{(35.00)} = 0.0286 \text{ F}$$

Combination of these results with the reaction stoichiometry yields

$$(\text{CH}_3\text{COO}^-) = 0.0286 + (H^+)$$

$$(\text{CH}_3\text{COOH}) = 0.0428 - (H^+)$$

which leads to

$$1.76 \times 10^{-5} \simeq \frac{(H^+)\,\{0.0286 + (H^+)\}}{\{0.0428 - (H^+)\}}$$

from which we compute $(H^+) \simeq 2.63 \times 10^{-5}$, and $pH \simeq 4.58$. Analogous calculations yield a $pH = 5.35$ after 20.00 ml of base added, and $pH = 6.45$ after 24.50 ml of base added.

At the *equivalence point* (that is, at the point in the titration where the total number of moles of base added is equal to the total number of moles of acid originally present), we have 50.00 ml of 0.0500 F NaOCOCH$_3$(aq):

$$\left(\begin{array}{l}\text{millimoles of CH}_3\text{COOH} \\ \text{initially present}\end{array}\right) = (25.00)(0.1000) = 2.50 \text{ mmole}$$

$$\left(\begin{array}{l}\text{volume in ml of 0.1000 M NaOH} \\ \text{containing 2.50 millimoles}\end{array}\right) = V = \frac{2.50}{M} = \frac{2.50}{0.1000} = 25.00 \text{ ml}$$

$$\left(\begin{array}{l}\text{total volume at the} \\ \text{equivalence point}\end{array}\right)^\dagger = 25.00 + 25.00 = 50.00 \text{ ml}$$

$$[\text{CH}_3\text{COO}^-] = \frac{(25.00)(0.1000)}{(50.00)} = 0.0500 \text{ F}$$

At the equivalence point and beyond, it is more convenient to work with the conjugate-base-protonation equilibrium than with the acid-dissociation equilibrium, because the $K_b$ reaction involves both CH$_3$COO$^-$(aq) and OH$^-$(aq), which are the principal species beyond the equivalence point.

---

*We have assumed for the purposes of computation of *formal* concentrations that the neutralization reaction CH$_3$COOH(aq) + OH$^-$(aq) → CH$_3$COO$^-$(aq) + H$_2$O($\ell$) is complete.
†Assuming additivity of volumes.

For a 0.0500 F $NaOCOCH_3(aq)$ solution, we have

$$CH_3COO^-(aq) + H_2O(\ell) = CH_3COOH(aq) + OH^-(aq)$$

$$K_b = \frac{1.00 \times 10^{-14}}{1.76 \times 10^{-5}} = 5.7 \times 10^{-10} \simeq \frac{(CH_3COOH)(OH^-)}{(CH_3COO^-)}$$

$$5.7 \times 10^{-10} \simeq \frac{(OH^-)^2}{0.0500 - (OH^-)}$$

$$(OH^-) \simeq 5.4 \times 10^{-6} \, M$$

$$pH \simeq -\log \left( \frac{1.0 \times 10^{-14}}{5.4 \times 10^{-6}} \right) = 8.73$$

After 25.50 ml of 0.1000 M NaOH have been added, we have

$$[OH^-] = \frac{(0.50)(0.1000)}{(50.50)} = 9.9 \times 10^{-4} \, F$$

$$[CH_3COO^-] = \frac{2.50}{50.50} = 0.0495 \, F$$

and

$$5.7 \times 10^{-10} \simeq \frac{(CH_3COOH) \{9.9 \times 10^{-4} + (CH_3COOH)\}}{\{0.0495 - (CH_3COOH)\}}$$

$$(CH_3COOH) \simeq 2.9 \times 10^{-8} \, M$$

$$(OH^-) \simeq 9.9 \times 10^{-4} \, M$$

$$pH \simeq -\log \left\{ \frac{1.0 \times 10^{-14}}{9.9 \times 10^{-4}} \right\} = 11.00$$

This result shows that, only 0.50 ml past the equivalence point, the $pH$ of the solution is completely determined by $[OH^-]$, because $(OH^-) \simeq [OH^-]$. After 30.00 ml of base

$$[OH^-] = \frac{(5.00)(0.1000)}{(55.00)} \simeq 9.09 \times 10^{-3} \, M \simeq (OH^-)$$

and $pH = 11.96$. In Figure 12.2 we have plotted the results of the above calculation in the form $pH$ versus ml of 0.1000 M NaOH added (together with the results for the simpler case of 25.00 ml of 0.1000 M HCl(aq) titrated with 0.1000 M NaOH). The important point to be noted in this figure is that the $pH$ of the solution changes very rapidly in the vicinity of the equivalence point. On going from 0 to 24.50 ml of base added, the $pH$ changes by $6.45 - 2.88 = 3.57$ units, whereas over the range from 24.50 to 25.50 ml of base, the $pH$ changes by $11.00 - 6.45 = 5.55$ units. This rapid change in $pH$ in the vicinity of the equivalence point is utilized in acid-base titrations to determine the position of the equivalence in the titration of acids or bases of unknown concentration. The steeper the titration curve in the vicinity of the equivalence point, the more precisely the equivalence point can be established. The slope in this region decreases with decreasing $K_a$. When the acid is very weak, the inflection is very slight, and the equivalence point cannot be precisely established. Inflection points of

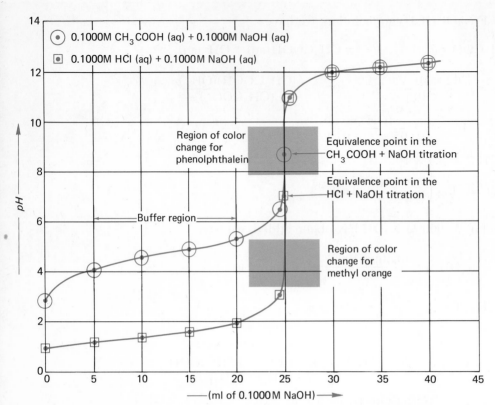

**FIGURE 12.2**   Titration curves for a strong acid (HCl) with a strong base (NaOH), and a weak acid ($CH_3COOH$) with a strong base (NaOH).

titration curves in water for acids with $pK_a > 9$ cannot be determined accurately, because they give almost no inflection. The volume in which the titration is carried out also affects the steepness of the titration curve; the more dilute the solution, the less steep the curve.

Acid-base titrations can conveniently be carried out either potentiometrically using a $H^+$-sensitive glass electrode (see Chapter 13), or by using an acid-base indicator.

An acid-base indicator is a weak organic acid[*], the acid and base forms of which have different colors.

$$HIn(aq) = H^+(aq) + In^-(aq)$$

acid form                            basic form

$$K_I \simeq \frac{(H^+)(In^-)}{(HIn)}$$

---

[*]An example is methyl orange ($pK_I \approx 3.9$), p-benzenesulfonic acid-azo-dimethylaniline.

$$In^- \equiv {}^-O - \overset{\displaystyle O}{\underset{\displaystyle O}{\overset{\|}{\underset{\|}{S}}}} - \langle\!\bigcirc\!\rangle - N{=}N - \langle\!\bigcirc\!\rangle - N \overset{\displaystyle CH_3}{\underset{\displaystyle CH_3}{\big<}} \qquad \text{(yellow)}$$

$$HIn \equiv {}^-O - \overset{\displaystyle O}{\underset{\displaystyle O}{\overset{\|}{\underset{\|}{S}}}} - \langle\!\bigcirc\!\rangle - \underset{\displaystyle H}{N} {-} N{=}\langle\!\bigcirc\!\rangle{=} N^+ \overset{\displaystyle CH_3}{\underset{\displaystyle CH_3}{\big<}} \qquad \text{(red)}$$

The color of the solution depends on the value of the ratio $(HIn)/(In^-)$, which in turn depends on the value of $H^+(aq)$:

$$\frac{(HIn)}{(In^-)} \simeq \frac{(H^+)}{K_I}$$

The color of the solution, therefore, can be taken as a measure of the $(H^+)$, or *pH*. The indicator is chosen such that its $pK_1$ value is within $\pm 1$ *pH* unit of the *pH* at the equivalence point. In the vicinity of the equivalence point $(H^+)$ changes very rapidly and therefore so does $(HIn)/(In^-)$, which leads to a change in the color of the solution in the equivalence-point region. The indicator concentration is made small in order to minimize the amount of titrant necessary to convert it from one form to another.

The equivalence point is what we want to know, but the end point (that is, the point at which our indicator changes color) is what we observe. In choosing an end-point indicator, we want it to indicate an end point as close as possible to the equivalence point, because the two are not necessarily identical. If the equivalence point and end point are not the same within the experimental error, a correction must be applied.

In a titration a polyprotic acid acts as a mixture of monoprotic acids present in equimolar amounts. If the $K_a$ values are sufficiently different $(K_{a1}/K_{a2} \geqslant 10^3)$, and not too large or too small, then there is an inflection point in the curve for each removable proton. And, if present alone in solution, the concentration of a polyprotic acid may be determined by titration to any of these inflection points. Some examples are:

$H_2SO_4$: only $HSO_4^-(aq)$ is a weak acid with $K_a = 1.0 \times 10^{-2}$, and only one inflection point is observed.

$H_3PO_4$: $K_{a1}/K_{a2} = 1.1 \times 10^{+5}$, and $K_{a2}/K_{a3} = 1.4 \times 10^5$, and in this case we can titrate to either of the first two end points (but not the third, because $pK_{a3} = 12.36$).

$CO_2(aq)$: $K_{a1}/K_{a2} = 6000$, and we can titrate to either of the two end points.

## 12.6 GAS-PHASE REACTIONS.

An economically important industrial synthesis of ammonia is based on the reaction

$$N_2(g) + 3H_2(g) = 2NH_3(g)$$

Equilibrium calculations can be used to predict the maximum yield of $NH_3$ as a function of pressure and temperature.

From the data in Appendix 5 we compute $\Delta G^\circ_{298}$ and $\Delta H^\circ_{298}$ for this reaction as

$$\Delta G^\circ_{298} = 2(-3.94) - 1(0) - 3(0) = -7.88 \text{ kcal}$$

$$\Delta H^\circ_{298} = 2(-11.02) - 1(0) - 3(0) = -22.04 \text{ kcal}$$

From $\Delta G^\circ_{298}$ we calculate $K_{298}$ as

$$\log K_{298} = \frac{-\Delta G^\circ_{298}}{2.303\,RT} = \frac{+7880}{2.303 \times 1.987 \times 298.2} = 5.775$$

so that $K_{298} = 5.96 \times 10^5$. Assuming that $\Delta H^\circ$ is independent of $T$ over the range 298–500 K, we have (Van't Hoff equation)

$$\log \frac{K_{500}}{K_{298}} = -\frac{\Delta H^\circ}{2.303\,R}\left\{\frac{1}{500.0} - \frac{1}{298.2}\right\}$$

and

$$\log K_{500} = 5.775 + \frac{22{,}040}{2.303 \times 1.987}\left\{\frac{1}{500.0} - \frac{1}{298.2}\right\} = -0.744$$

so that $K_{500} = 0.180$. The negative value of $\Delta H^\circ$ leads to a decreasing $K$ with temperature. At 500 K we have

$$0.180 = \frac{a^2_{NH_3(g)}}{a_{N_2(g)}\,a^3_{H_2(g)}} \simeq \frac{P^2_{NH_3}}{P_{N_2}\,P^3_{H_2}}$$

where the second equality involves the ideal-gas approximation. This system, when prepared in a purely arbitrary manner, has three degrees of freedom, $F = 2 - 1 + 2 = 3$, and thus, the specification of three variables is necessary to determine the state of the system. Taking $T = 500$ K, there remain only two independent variables. It is obvious from the equilibrium constant expression that fixing any two of the three variables $P_{N_2}$, $P_{NH_3}$, or $P_{H_2}$ fixes the value of the third. For example, if we require that at equilibrium $P_{N_2} = 1.00$ atm and $P_{H_2} = 3.00$ atm, then we compute

$$P_{NH_3} \simeq \{(0.180)(1.00)(3.00)^3\}^{1/2} \simeq 2.21 \text{ atm}$$

for the equilibrium pressure of $NH_3$.

We can express $K$ in terms of the mole fractions $X_{N_2}$, $X_{H_2}$, and $X_{NH_3}$, and the total pressure, $P_{tot}$, using the relation $P_i = X_i P_{tot}$:

$$K = \frac{P^2_{NH_3}}{P_{N_2}\,P^3_{H_2}} = \frac{(X_{NH_3} P_{tot})^2}{(X_{N_2} P_{tot})(X_{H_2} P_{tot})^3} = \frac{X^2_{NH_3}}{X_{N_2} X^3_{H_2} P^2_{tot}}$$

(where $X_{NH_3} + X_{N_2} + X_{H_2} = 1$). A consideration of $K$ expressed in terms of the $X_i$ values and $P_{tot}$ leads to the conclusion that an increase in the total equilibrium pressure increases the yield[*] of $NH_3$. This follows because, if we increase $P_{tot}$ from, say, 1 atm to 2 atm, then the quantity $X^2_{NH_3}/X_{N_2} X^3_{H_2}$ must increase by a factor of 4 in order that the product

$$\frac{X^2_{NH_3}}{X_{N_2} X^3_{H_2}} \cdot \frac{1}{P^2_{tot}}$$

remain constant. For example, suppose that $P_{tot} = 1.00$ atm and $X_{H_2} = 0.10$; then

$$X_{N_2} = 1 - X_{H_2} - X_{NH_3} = 0.90 - X_{NH_3}$$

---

[*] The value of $P_{tot}$ has no effect on the equilibrium $X_i$ values for a gas reaction in which the number of moles of products is the same as the number of moles of reactants.

and

$$0.180 \simeq \frac{X_{NH_3}^2}{(0.90 - X_{NH_3})(0.10)^3(1)^2}$$

from which we compute $X_{NH_3} \simeq 0.012$. If we increase $P_{tot}$ to 2.00 atm, keeping $X_{H_2} = 0.10$, then

$$0.180 \simeq \frac{X_{NH_3}^2}{(0.90 - X_{NH_3})(0.10)^3(2.00)^2}$$

from which we compute $X_{NH_3} \simeq 0.025$, or an increase of roughly a factor of 2 over the previous case.

Suppose we mix $n$ moles of $N_2(g)$ with $h$ moles of $H_2(g)$ and allow the system to come to equilibrium at 500 K. How many moles of $NH_3$ are produced? Referring to the reaction stoichiometry, we note that, if we let the *fraction* of $N_2$ that is consumed be $\alpha$, then

|  | $N_2$ | $+$ | $3H_2$ | $=$ | $2NH_3$ |
|---|---|---|---|---|---|
| initial number of moles: | $n$ | | $h$ | | $0$ |
| equilibrium number of moles: | $n(1-\alpha)$ | | $h-3n\alpha$ | | $2n\alpha$ |
| equilibrium mole fractions: | $\dfrac{n(1-\alpha)}{n+h-2n\alpha}$ | | $\dfrac{h-3n\alpha}{n+h-2n\alpha}$ | | $\dfrac{2n\alpha}{n+h-2n\alpha}$ |

where the equilibrium mole fractions were computed as follows: the *total* number of moles present at equilibrium is

$$n(1-\alpha) + h - 3n\alpha + 2n\alpha = n + h - 2n\alpha$$

and thus

$$X_{N_2} = \frac{n(1-\alpha)}{n+h-2n\alpha}$$

and so forth. Substitution of the equilibrium mole fraction into the equilibrium constant expression yields

$$0.180 \simeq \frac{\{2n\alpha/(n+h-2n\alpha)\}^2}{\left\{\dfrac{n(1-\alpha)}{n+h-2n\alpha}\right\}\left\{\dfrac{h-3n\alpha}{n+h-2n\alpha}\right\}^3 P_{tot}^2}$$

$$0.180 \simeq \frac{(2n\alpha)^2(n+h-2n\alpha)^2}{n(1-\alpha)(h-3n\alpha)^3 P_{tot}^2}$$

For the special case in which $n = 1$ and $h = 3$, we obtain

$$0.180 \simeq \frac{16\alpha^2(2-\alpha)^2}{27(1-\alpha)^4 P_{tot}^2}$$

Taking the square root of both sides, we obtain $0.55\, P_{tot} = \dfrac{\alpha(2-\alpha)}{(1-\alpha)^2}$. This quadratic can be solved for $\alpha$ once $P_{tot}$ is specified. At $P_{tot} = 1.00$ atm we compute $\alpha = 0.20$, and the number of moles of $NH_3$ is $2n\alpha = 2(0.20) = 0.40$.

Ammonia is synthesized from the elements on an industrial scale using the Haber Process. For the reaction

$$N_2(g) + 3H_2(g) = 2NH_3(g)$$

$\Delta H° < 0$, and therefore the equilibrium constant *decreases* with increasing temperature. Thus, the equilibrium mole fraction of ammonia will be greater the lower the temperature. At a given temperature, the equilibrium value of $X_{NH_3}$ will be greater the higher the total pressure. Consequently, thermodynamics predicts high pressure and low temperature for maximizing the yield of $NH_3$. However, the rate of the reaction around room temperature is so slow as to be negligible. Most reactions proceed at a faster rate the higher the temperature. The above reaction must be run at elevated temperatures, even though the equilibrium yield is not as favorable as at lower temperatures, in order to make the reaction proceed at an economically-feasible rate. The industrial process is based on a compromise between kinetic (rate) and thermodynamic (yield) considerations. The actual process is run at 500 to 600°C and $P_{tot} \simeq 500$ atm in the presence of an iron-molybdenum catalyst. Without the catalyst, which increases the rate, the process would not be economically feasible, because the yield is too small at the still higher temperatures that would be required in the absence of the catalyst. The reactant gases must be purified to avoid poisoning the catalyst. The value of $K$ at 500°C (773 K) is $1.50 \times 10^{-5}$. For stoichiometric amounts of nitrogen and hydrogen the following results obtain at equilibrium (500°C):

| $P_{tot}$(atm) | $P_{NH_3}$(atm) | CONVERSION TO $NH_3$ |
|:---:|:---:|:---:|
| 1.00 | $1.26 \times 10^{-3}$ | 0.25% |
| 500.0 | 152 | 46.6% |

Thus, by keeping the pressure high and the temperature at moderate values the process becomes feasible.

An industrial process that involved the synthesis of ammonia from the elements at ordinary temperatures and pressures would quickly make the Haber Process obsolete, because the latter requires fuel to maintain the high temperature, and the use of compressors and expensive, heavy-walled apparatus to maintain the high pressure. The thermodynamics of the reaction is such (larger $K$ at lower $T$) that a low-temperature process is possible. What is needed is the appropriate catalyst. Such catalysts, called *nitrogenases*, are found in nature in the organisms *Clostridium pasteurianum* (anaerobic) and *Azotobacter vinelandii* (aerobic), as well as others. These organisms convert $N_2$ to $NH_3$ in a complex process that involves several successive reductions. The principal steps (which also involve 4 or 5 ATP molecules/2 electrons transferred) are thought to be

$$N \equiv N \xrightarrow{2e^-} H—N = N—H \xrightarrow{2e^-} H_2N—NH_2 \xrightarrow{2e^-} 2NH_3$$

$$\text{(diimide)} \qquad \text{(hydrazine)}$$

The $N_2$-ase enzyme (mol. wt. $\sim$ 270,000) exists in an association-dissociation equilibrium with its component proteins, and is thought to be composed of an iron-molybdenum protein and two different iron proteins. The subunit molecular weights are ~40,000. The specific activity of the $N_2$-ase enzyme is about 80 nanomoles $N_2$ reduced/min/mg of enzyme. The details of the nitrogen-fixation reaction are not well understood, but the following partial mechanism involving the Fe-Mo portion on the enzyme has been proposed[*] (we have written [H] to indicate an unspecified source of hydrogen):

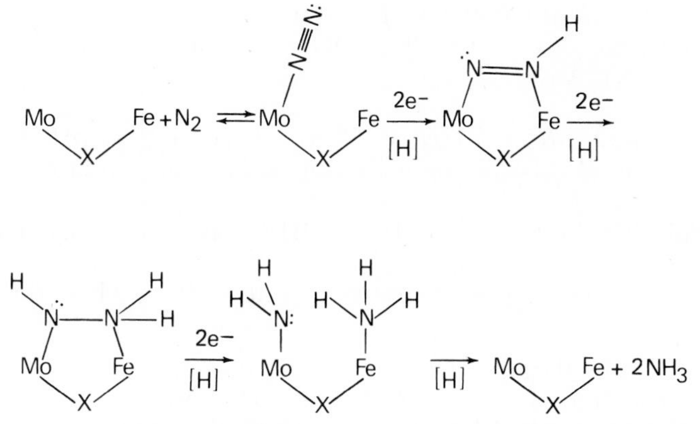

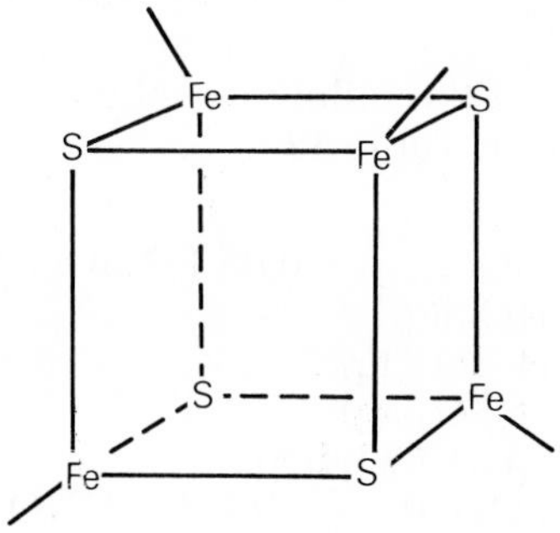

The detailed nature of the electron transfer agent that supplies the electrons used in the reduction is not known. However, this agent is thought to be an iron-sulfur protein which contains an 8-atom Fe-S cubical cluster of the type shown above.

[*]W. F. Hardy, R. C. Burns, and G. W. Parshall, *Bioinorganic Chemistry*, ACS Advances in Chemistry Series #100 (1971).

_____ PROBLEMS [†]

1. Calculate the *pH* of the following solutions:
   (a) 0.010 F HCl(aq)
   (b) 0.0050 F NaOH(aq)
   (c) A solution obtained by mixing equal volumes of the solutions in (a) and (b)

[†] See Table 12.2 for $pK_a$ values.

    (d) 0.16 F HCl(aq) (this is roughly the concentration of HCl in the
        stomach)
    (e) $1.5 \times 10^{-3}$ F $Ca(OH)_2(aq)$
    (f) 0.15 F $CH_3COOH(aq)$
    (g) 0.15 F $NaOCOCH_3(aq)$
    (h) 0.50 F $NH_3(aq)$
    (i) 0.50 F $NH_4Cl(aq)$
    (j) A solution obtained by adding 10 ml of 0.10 F NaOH(aq) to
        100 ml of 0.050 F $CH_3COOH(aq)$
    (k) vinegar (3% by wt. $CH_3COOH$ in $H_2O$)
    (l) 0.010 F $Al(NO_3)_3(aq)$

2. Given that $pK_w = 13.62$ at 37°C, compute the pH of a neutral aqueous solution at this (biologically important) temperature.

3. Given that the $pK_a$ of $ClCH_2COOH(aq)$ is 2.87, compute $K_a$ and $K_b$.

4. Write out the electroneutrality condition for each of the following solutions:

    (a) $CH_3COOH(aq)$         (e) $CO_2(aq)$
    (b) $NH_4Cl(aq)$           (f) $H_2S(aq)$
    (c) $NaCl(aq) + CaCl_2(aq)$   (g) $K_3Fe(CN)_6(aq)$
    (d) $Ca(OH)_2(aq)$        (h) $NaHCO_3(aq)$

5. Calculate the $pH$ of the following solutions at 25°C. The necessary $pK_a$ values can be found in Table 12.2.

    (a) 0.15 F $CH_3COOH(aq)$ + 0.010 F  HCl(aq)
    (b) 0.15 F $CH_3COOH(aq)$ + 0.05 F $NaOCOCH_3(aq)$
    (c) $1.0 \times 10^{-8}$ F HCl(aq)
    (d) $1.0 \times 10^{-4}$ F $H_2SO_4(aq)$
    (e) $1.0 \times 10^{-8}$ F $H_2SO_4(aq)$
    (f) $1.0 \times 10^{-7}$ F $CH_3COOH(aq)$
    (g) 0.100 F $CH_3COOH(aq)$ + 0.100 F $HNO_2(aq)$

6. The human stomach is roughly 0.16 M in HCl(aq). Calculate the number of grams of $Al(OH)_3(s)$ required to neutralize 1.0 ml of stomach acid. Repeat the calculation for $Mg(OH)_2(s)$.

7. The autoprotolysis constant for pure ethanol, $C_2H_5OH$, at 25°C is $8 \times 10^{-20}$. Calculate the $pH$ in a 0.010 F solution of $NaOCH_2CH_3$ (sodium ethoxide) in ethanol at 25°C. Would this solution be considered acidic or basic?

8. A solution is 0.250 F in HX(aq) ($K_a = 2.0 \times 10^{-6}$) and 0.150 F in KX(aq). Calculate the equilibrium concentrations of $K^+$, $X^-$, $H^+$, and $OH^-$.

9. The titration of 25.00 ml of an $NH_3(aq)$ solution required 35.50 ml of 0.1250 M HCl(aq). Calculate the concentrations of $NH_3(aq)$, $NH_4^+(aq)$, $Cl^-(aq)$, $H^+(aq)$, and $OH^-(aq)$ at the equivalence point in this titration. Also calculate the $pH$ of the solution at the equivalence point and give the $pK$ range for suitable two-color acid-base titration end-point indicators.

10. Calculate the $pH$ at the equivalence point in the titration of 25.00 ml of 0.1250 F benzoic acid, $C_6H_5COOH(aq)$, with 0.1065 M KOH(aq).

11. Calculate the *pH* of an aqueous solution of glycine (abbreviated HG) at the isoelectric point. The isoelectric point is the *pH* at which equal numbers of cations and anions exist, i.e., $(H_2G^+) = (G^-)$. The equilibrium constants are:

$$H_2G^+(aq) = HG(aq) + H^+(aq) \qquad K_1 = 4.5 \times 10^{-3}$$
$$HG(aq) = G^-(aq) + H^+(aq) \qquad K_2 = 1.7 \times 10^{-10}$$

12. Calculate the *pH* and the concentrations of all the species in a 0.120 F $H_3AsO_4$(aq) solution at 25°C.

13. Calculate the *pH* of distilled water in equilibrium with atmospheric $CO_2$ at 25°C, given that the approximate pressure of $CO_2$ in the atmosphere is $P_{CO_2} = 3 \times 10^{-4}$ atm.

14. Explain how you would prepare 500 ml of $NH_4Cl$(aq) + $NH_3$(aq) buffer of *pH* = 9.5 from 1.00 F $NH_4Cl$(aq) and 1.00 F $NH_3$(aq). Suppose now that 10.0 ml of 0.10 M HCl(aq) is added to 50 ml of the above buffer. Calculate the change in *pH*.

15. Calculate the concentrations of all the species present at the equivalence point in the titration of 50.0 ml of 0.0450 M aqueous pyridine, $C_5H_5N$(aq), with 0.0650 M HCl.

16. Calculate the concentrations of all the species present in an aqueous solution of $H_2S$ over which the pressure of $H_2S$(g) is 1.0 atm.

17. Construct a distribution diagram for $H_2S$(aq) at 25°C over the *pH* range from 4 to 16.

18. What ratio of volumes of a 0.200 F solution of HA(aq) ($K_a = 1.0 \times 10^{-5}$) and 0.100 F KA(aq) solution should be mixed in order to produce a buffer of *pH* = 4.30?

19. Using data in Appendix 5, compute $K$ at 25°C for the reaction

$$2H_2(g) + CO(g) = CH_3OH(g)$$

Assuming that $a_i = P_i$, solve the following problems:
(a) Calculate the equilibrium values of $P_{CH_3OH}$ and $X_{CH_3OH}$ at 25°C for $P_{tot} = 2.00$ atm, and $P_{H_2} = 1.00$ atm.
(b) Calculate the equilibrium value of $P_{CH_3OH}$ at $P_{H_2} = 0.02$ atm, and $P_{CO} = 0.01$ atm.
(c) Calculate the equilibrium value of $X_{CH_3OH}$ at $X_{CO} = 0.1$, $P_{tot} = 1.00$ atm, and $P_{tot} = 5$ atm.
(d) Compute the number of moles of $CH_3OH$(g) produced at equilibrium when 2.00 moles of $H_2$(g) are mixed with 1.00 mole of CO(g). Take the value of $P_{tot} = 10.0$ atm at equilibrium.

20. To 100.0 ml of 0.137 M $Na_2CO_3$, 80.0 ml of 0.200 M HCl acid is added. Calculate the *pH* of the solution and the equilibrium partial pressure of $CO_2$(g) over the solution.

21. Calculate the $pH$ of a $1.25 \times 10^{-4}$ M solution of aniline, $C_6H_5NH_2$.

22. Calculate the concentrations of all the species present in a 0.100 F $K_2CrO_4(aq)$ solution:

$$CrO_4^{2-}(aq) + H_2O(\ell) = HCrO_4^-(aq) + OH^-(aq) \quad K_b = 3.2 \times 10^{-8}$$

$$2HCrO_4^-(aq) = Cr_2O_7^{2-}(aq) + H_2O(\ell) \qquad\qquad K = 33$$

23. (a) Construct a distribution diagram for $H_3PO_4(aq)$ solutions over the $pH$ range from 1 to 16.
   (b) Construct a titration curve for the titration of 25.0 ml 0.100 F $H_3PO_4$ with 0.100 M HCl(aq). Indicate the buffer regions and equivalence points on the diagram.

24. Calculate the concentrations of all the species present in a 0.10 M solution of $(NH_4)_2HPO_4$.

25. Show, for any salt of the type $BH^+X^-$ (for example, $NH_4OCOCH_3$) for which

$$BH^+(aq) + X^-(aq) = B(aq) + HX(aq)$$

is the principal equilibrium, that $(H^+) \simeq \{K_{a,BH^+} \cdot K_{a,HX}\}^{1/2}$. Use this result to calculate the $pH$ of a $NH_4OCOCH_3(aq)$ solution. Discuss the advantages of the use of such salts as buffers.

26. Show, for any salt of the type MHX (for example, $NaHCO_3$, NaHS, $KHSO_3$) for which

$$2HX^-(aq) = X^{2-}(aq) + H_2X(aq)$$

is the principal equilibrium, that $(H^+) \simeq (K_{a1} \cdot K_{a2})^{1/2}$. Use this result to compute the $pH$ of a $NaHCO_3(aq)$ at 25°C. Discuss the advantages of the use of such salts as buffers.

27. To 100.0 ml of 0.100 M $Na_3PO_4(aq)$, 25.0 ml of 0.600 M HCl was added. Calculate the concentrations of all the species in the resulting solution. Is this solution a buffer?

28. Calculate the concentrations of all the species present in a 0.125 F $NH_4OCOCH_3(aq)$ solution at 25°C.

29. Calculate the concentrations of all the species present in an aqueous solution containing 0.010 mole of $HgCl_2$ per liter. Assume that sufficient $HClO_4(aq)$ has been added to suppress the hydrolysis of $Hg^{2+}(aq)$

$$Hg^{2+}(aq) + Cl^-(aq) = HgCl^+(aq) \qquad K_1 = 5.5 \times 10^6$$

$$HgCl^+(aq) + Cl^-(aq) = HgCl_2(aq) \qquad K_2 = 3.0 \times 10^6$$

$$HgCl_2(aq) + Cl^-(aq) = HgCl_3^-(aq) \qquad K_3 = 7.1$$

$$HgCl_3^-(aq) + Cl^-(aq) = HgCl_4^{2-}(aq) \qquad K_4 = 10.0$$

30. The equilibrium constant for the reaction

$$2SO_2(g) + O_2(g) \rightleftarrows 2SO_3(g)$$

is $K = 10.0$ at 900 K, and $\Delta H° = -29$ kcal.

(a) How will the value of $X_{SO_3}$ at equilibrium be affected by an increase in the total pressure?

(b) How will the value of $X_{SO_3}$ at equilibrium and fixed total pressure be affected by a decrease in the temperature?

(c) Suppose we prepare a gas mixture at 900 K and $P_{tot} = 10$ atm using 0.5 mole of $O_2$, 0.3 mole of $SO_2$, and 2.0 mole of $SO_3$. In which direction will the above reaction proceed (right-to-left or left-to-right)?

(d) Given that at equilibrium $P_{tot} = 5.00$ atm and $P_{O_2} = 0.20$ atm, compute $X_{SO_3}$.

(e) Suppose that 1.00 mole of $SO_3(g)$ is heated to 900 K. Compute the equilibrium partial pressure of $SO_3$ when the total pressure is 5.00 atm.

(f) Compute $K$ for the reaction at 800 K.

31. Given the following equilibrium constants (1000 K)

$$CaCO_3(s) = CaO(s) + CO_2(g) \qquad K_1 = 0.039$$

$$C(s) + CO_2(g) = 2CO(g) \qquad K_2 = 1.9$$

Compute the equilibrium partial pressures of $CO_2(g)$ and $CO(g)$ when $CaCO_3(s)$ and $C(s)$ are heated in a sealed vessel to 1000 K.

32. Calculate the *pH* of a solution prepared by dissolving 9.1 g of $KH_2PO_4(s)$ and 18.9 g of $NaHPO_4(s)$ in water followed by dilution to a final volume of 1.00 liter.

33. Explain how you would prepare 1 liter of $NH_4Cl(aq) + NH_3(aq)$ buffer, *pH* $= 10.1$, from $NH_4Cl(s)$ and 15 M $NH_3(aq)$. Take $[NH_4^+] \sim 0.20$ F.

34. Describe a method that could be used to determine quickly whether or not a solution of unknown composition was buffered or not.

35. Show, for the titration of the weak acid HB with strong base NaOH, that the *pH* at any point on the titration curve is given by

$$pH = pK_a + \log \left\{ \frac{[B^-] + (H^+) - (OH^-)}{[HB] - (H^+) + (OH^-)} \right\}$$

(assume $\gamma_i = 1$). Compare this equation with the Henderson-Hasselbalch equation.

36. In an aqueous $NH_3$ solution, 74% of the $NH_3$ is in the form $NH_3(aq)$. Compute $K$ for the reaction

$$NH_3(aq) + H_2O(\ell) \rightleftarrows NH_4OH(aq)$$

37. Ethylenediaminetetraacetic acid (EDTA) is a tetraprotic acid (see Table 12.2). EDTA is conveniently represented by the abbreviated formula

$H_4Y$. A 0.020 F solution of $Na_2H_2Y(aq)$ is often used to determine the concentrations of $Ca^{2+}(aq)$ and $Mg^{2+}(aq)$ in natural waters. The analysis is based on the formation of octahedral $MY^{2-}$ complexes of the type

(a) Write the mass-balance expression for "Y" in a 0.020 F $Na_2H_2Y(aq)$ solution.

(b) Write the electroneutrality expression for 0.020 F $Na_2H_2Y(aq)$.

(c) Compute the equilibrium constant for the reaction

$$2H_2Y^{2-}(aq) = H_3Y^-(aq) + HY^{3-}(aq)$$

(d) Assume that the reaction in (c) is the principal equilibrium and compute the $pH$ of the solution.

38. Given that human blood is buffered at $pH = 7.4$ and that the partial pressure of $CO_2(g)$ in exhaled air is about 0.04 atm, estimate the concentration of $HCO_3^-(aq)$ in blood at 37°C. Take $K_{a1} = 5 \times 10^{-7}$ and $K_{a2} = 1 \times 10^{-11}$ for $CO_2(aq)$, and $K = 0.02 = (CO_2)/P_{CO_2}$.

# CHAPTER 12B

# CHEMICAL EQUILIBRIUM

*"Until a problem has been logically defined, it cannot be experimentally solved."**

## Part III. Applications to Heterogeneous Equilibria

In Chapter 12A we restricted our application of the principles of chemical equilibrium primarily to homogeneous (one-phase) equilibria. The only type of heterogeneous (multi-phase) equilibrium that we considered in Chapter 12A was that between a gas and a solution containing the dissolved gas. In this chapter we will apply the principles of chemical equilibrium to: (a) equilibria involving a solid and a liquid (solution) phase; (b) two liquid (solution) phases; and (c) a liquid (solution) phase and a gas phase.

If a component is present in two phases, then at equilibrium the chemical potential of that component must be the same in both of the phases. If the chemical potential of a component is the same in both phases, then the ratio of activities of the component in the two phases must be a constant.

A useful concept to keep in mind when analyzing heterogeneous equilibria is that if one of the phases is pure, then the presence of that phase will fix the activity of that component in all other phases with which it is in equilibrium. This is true irrespective of the number of equilibria in which this component participates in the other phases.

## 12.7 SOLUBILITY AND COMPLEX-ION EQUILIBRIA.

The *solubility product*, $K_{sp}$, is the equilibrium constant that characterizes the equilibrium between a solid salt and a solution saturated with

*G. N. Lewis, M. Randall, K. S. Pitzer, and L. Brewer, *Thermodynamics*, 2nd Revised Ed. (McGraw-Hill Pub. Co., New York, 1961) p. 307.

**369**

the ions of which the solid salt is composed. For example, consider the equilibrium

$$AgCl(s) = Ag^+(aq) + Cl^-(aq) \qquad (12.6a)$$

$$K = K_{sp} = \frac{a_{Ag^+(aq)} \cdot a_{Cl^-(aq)}}{a_{AgCl(s)}}$$

where $K_{sp} = 1.8 \times 10^{-10}$ at 25°C (molality composition scale[*]). The value of $K_{sp}$ for a given salt depends on the temperature and solvent. Around 1 atm total pressure, we have $a_{AgCl(s)} = 1$, and

$$K_{sp} = a_{Ag^+(aq)} \cdot a_{Cl^-(aq)} = (m_{Ag^+})(m_{Cl^-})\gamma^2_{\pm(AgCl)}$$

As in Section 12.1, and for the same reasons as cited there, we shall initially neglect activity coefficients (set all $\gamma_i = 1$), and postpone until a later section (12.9) a more rigorous treatment of solubility equilibria. For the present section then, we have for the above equilibrium at 25°C

$$K_{sp} = 1.8 \times 10^{-10} \simeq (Ag^+)(Cl^-)$$

Some other examples (where the particular $K_{sp}$ values refer to a temperature of 25°C) are:

$$Hg_2Cl_2(s) = Hg_2^{2+}(aq) + 2Cl^-(aq) \qquad (12.6b)$$
$$K_{sp} = 1.3 \times 10^{-18} \simeq (Hg_2^{2+})(Cl^-)^2$$

$$Ag_2S(s) = 2Ag^+(aq) + S^{2-}(aq) \qquad (12.6c)$$
$$K_{sp} = 1 \times 10^{-51} \simeq (Ag^+)^2(S^{2-})$$

$$BaSO_4 \cdot 2H_2O(s) = Ba^{2+}(aq) + SO_4^{2-}(aq) + 2H_2O(\ell) \qquad (12.6d)$$
$$K_{sp} = \frac{a_{Ba^{2+}(aq)} \, a_{SO_4^{2-}(aq)} \, a^2_{H_2O(\ell)}}{a_{BaSO_4 \cdot 2H_2O(s)}} = 1.1 \times 10^{-10} \simeq (Ba^{2+})(SO_4^{2-})$$

$$K_2Zn_3[Fe(CN)_6]_2(s) = 2K^+(aq) + 3Zn^{2+}(aq) + 2Fe(CN)_6^{4-}(aq) \qquad (12.6e)$$
$$K_{sp} = 6.7 \times 10^{-43} \simeq (K^+)^2(Zn^{2+})^3(Fe(CN)_6^{4-})^2$$

Solubility-product constants for a number of salts are given in Table 12.5.

## a. Calculation of Salt Solubility from the $K_{sp}$ Expression.

Because the $K_{sp}$ expression describes the equilibrium between a salt and the solution saturated with that salt, this expression can be used to

---

[*]$K_{sp}$ values are usually reported on a molality composition basis. In dilute aqueous solution $\rho \simeq 1.0$ and $M \simeq m$, and the distinction between molality and molarity can usually be ignored. However, this distinction cannot be ignored for solutions whose density differs markedly from unity, for example, in certain non-aqueous solutions. This same distinction should also be kept in mind when working with acid-base equilibria in non-aqueous solutions; $K_a$ values are also usually reported on a molality composition basis.

**TABLE 12.5**

SOLUBILITY PRODUCT CONSTANTS ($K_{sp}$) FOR SALTS IN WATER
(Molality Composition Scale, 25°C)

**HALIDES**

| | | | |
|---|---|---|---|
| $CaF_2(s)$ | $3.9 \times 10^{-11}$ | $PbCl_2(s)$ | $1.0 \times 10^{-4}$ |
| $PbF_2(s)$ | $3.1 \times 10^{-8}$ | $AgBr(s)$ | $8.0 \times 10^{-13}$ |
| $AgCl(s)$ | $1.8 \times 10^{-10}$ | $TlBr(s)$ | $3.5 \times 10^{-6}$ |
| $Hg_2Cl_2(s)$ | $1.3 \times 10^{-18}$ | $CuI(s)$ | $1.1 \times 10^{-12}$ |
| $TlCl(s)$ | $1.7 \times 10^{-4}$ | $PbI_2(s)$ | $1.0 \times 10^{-9}$ |
| | | $AgI(s)$ | $9.8 \times 10^{-17}$ |

**HYDROXIDES**

| | | | |
|---|---|---|---|
| $Al(OH)_3(s)$ | $2 \times 10^{-32}$ | $Cu(OH)_2(s)$ | $3 \times 10^{-20}$ |
| $Cd(OH)_2(s)$ | $3 \times 10^{-14}$ | $Fe(OH)_2(s)$ | $8 \times 10^{-16}$ |
| $Cr(OH)_3(s)$ | $8 \times 10^{-31}$ | $Fe(OH)_3(s)$ | $3 \times 10^{-38}$ |
| $Mg(OH)_2(s)$ | $1.1 \times 10^{-11}$ | $Pb(OH)_2(s)$ | $2 \times 10^{-20}$ |
| $Sn(OH)_2(s)$ | $1 \times 10^{-25}$ | $Zn(OH)_2(s)$ | $1 \times 10^{-16}$ |
| $Ca(OH)_2(s)$ | $8 \times 10^{-6}$ | | |

**SULFIDES**

| | | | |
|---|---|---|---|
| $Ag_2S(s)$ | $1 \times 10^{-51}$ | $CuS(s)$ | $8.7 \times 10^{-36}$ |
| $CdS(s)$ | $1.4 \times 10^{-28}$ | $FeS(s)$ | $1 \times 10^{-19}$ |
| $HgS(s)$ | $8.6 \times 10^{-52}$ | $PbS(s)$ | $8.4 \times 10^{-28}$ |
| $ZnS(s)$ | $1.1 \times 10^{-21}$ | $Tl_2S(s)$ | $7.5 \times 10^{-20}$ |
| $SnS(s)$ | $1.2 \times 10^{-25}$ | | |

**CARBONATES, SULFATES, AND OXALATES**

| | | | |
|---|---|---|---|
| $CaCO_3(s)$ | $4.8 \times 10^{-9}$ | $Ag_2C_2O_4(s)$ | $9 \times 10^{-12}$ |
| $MgCO_3(s)$ | $3 \times 10^{-5}$ | $CaC_2O_4(s)$ | $1.9 \times 10^{-9}$ |
| $PbCO_3(s)$ | $6 \times 10^{-14}$ | $MgC_2O_4(s)$ | $7.1 \times 10^{-7}$ |
| $BaCO_3(s)$ | $5 \times 10^{-9}$ | $Ag_2SO_4(s)$ | $1.7 \times 10^{-5}$ |
| $Ag_2CO_3(s)$ | $6 \times 10^{-12}$ | $CaSO_4(s)$ | $2.4 \times 10^{-5}$ |
| $ZnCO_3(s)$ | $1.5 \times 10^{-11}$ | $Ag_2SO_4(s)$ | $1.2 \times 10^{-5}$ |
| $PbSO_4(s)$ | $1.7 \times 10^{-8}$ | $BaSO_4 \cdot 2H_2O(s)$ | $1.1 \times 10^{-10}$ |
| $Hg_2SO_4(s)$ | $7.1 \times 10^{-7}$ | | |

**MISCELLANEOUS**

| | | | |
|---|---|---|---|
| $Pb_2Fe(CN)_6(s)$ | $9.6 \times 10^{-19}$ | $AgOCOCH_3(s)$ | $4.0 \times 10^{-3}$ |
| $Ag_3Co(CN)_6(s)$ | $8.5 \times 10^{-21}$ | $AgIO_3(s)$ | $3.1 \times 10^{-8}$ |
| $K_2Zn_3[Fe(CN)_6]_2(s)$ | $6.7 \times 10^{-43}$ | $Ag_2CrO_4(s)$ | $1.3 \times 10^{-12}$ |
| $PbCrO_4(s)$ | $2 \times 10^{-14}$ | $Cu(IO_3)_2(s)$ | $7.4 \times 10^{-8}$ |

estimate the solubility of the salt in the solution. Consider the problem of calculating the solubility of $PbI_2(s)$ in water at 25°C. The relevant equilibrium is

$$PbI_2(s) = Pb^{2+}(aq) + 2I^-(aq)$$

and

$$K_{sp} = 1.0 \times 10^{-9} \simeq (Pb^{2+})(I^-)^2$$

A consideration of the reaction stoichiometry leads to the conclusion that the solubility of $PbI_2(s)$ in water is equal to the concentration of $Pb^{2+}(aq)$ (provided that this is the only solution species containing Pb(II)), because each mole of $PbI_2(s)$ that dissolves yields 1 mole of $Pb^{2+}(aq)$ in the solution. The reaction stoichiometry also requires that

$$(I^-) = 2(Pb^{2+})$$

provided that the only source of $I^-(aq)$ and $Pb^{2+}(aq)$ is $PbI_2(s)$. Substitution of this expression into the $K_{sp}$ expression yields

$$1.0 \times 10^{-9} \simeq (Pb^{2+})(2(Pb^{2+}))^2 = 4(Pb^{2+})^3$$

from which we compute $(Pb^{2+}) \simeq (0.25 \times 10^{-9})^{1/3} \simeq 6.3 \times 10^{-4}$ M $\simeq$ solubility of $PbI_2$ in water at 25°C.

The calculation of the solubility of a salt from the $K_{sp}$ expression is only a crude approximation if activity coefficients are neglected. The solubility calculated in the above manner is more reliable the smaller the ionic strength of the solution phase — which, for the special case of the solid in equilibrium with the saturated solution with no added electrolytes present (that affect the $\gamma_i$), means that the calculated solubility is more reliable the smaller it is. Errors arising from neglect of activity coefficients can pale into insignificance relative to the errors that can arise from the failure to consider other equilibria, such as ionic association equilibria, if such equilibria play an important role in determining the total solubility.

## b. The Common-Ion Effect on Salt Solubility.

If the solubility product equilibrium is the only important equilibrium involved, then the solubility of a solid can be significantly suppressed by the addition to the solution of one of the ions involved in the equilibrium. This is known as the *common-ion effect*. For example, consider the solubility of $PbI_2(s)$ in an aqueous solution 0.010 M in $KI(aq)$ at 25°C. From the stoichiometry of the reaction

$$PbI_2(s) = Pb^{2+}(aq) + 2I^-(aq)$$

we have

equilibrium concentration} ——    $(Pb^{2+})$    $0.010 + 2(Pb^{2+})$

where the solubility of $PbI_2(s)$ in the solution is equal to

$$\text{solubility} = (Pb^{2+})$$

Note that each mole of $PbI_2$ that dissolves yields one mole of $Pb^{2+}(aq)$ and *two* moles of $I^-(aq)$. Thus at equilibrium $(I^-)$ is equal to the sum of what was present to begin with [0.010 M] plus twice the solubility of $PbI_2$, that is

$$(I^-) = 0.010 + 2(Pb^{2+})$$

Substitution of this equation into the $K_{sp}$ expression yields

$$1.0 \times 10^{-9} \simeq (Pb^{2+})(I^-)^2 \simeq (Pb^{2+})\{0.010 + 2(Pb^{2+})\}^2$$

Using the trial approximation, $2(Pb^{2+}) \ll 0.01$ (because we expect that the solubility is low), we compute

$$(Pb^{2+}) \simeq \frac{1.0 \times 10^{-9}}{1.0 \times 10^{-4}} = 1.0 \times 10^{-5} \text{ M}$$

which substantiates the approximation. From this result we compute that the solubility of $PbI_2(s)$ in pure water at 25°C is 63($=6.3 \times 10^{-4}/1.0 \times 10^{-5}$) times as great as in 0.010 M KI(aq).

The common-ion effect is easily understood in terms of the $K_{sp}$ expression. Because the equality

$$(Pb^{2+})(I^-)^2 = 1.0 \times 10^{-9}$$

must hold at equilibrium if the solution is saturated with $PbI_2(s)$, an increase in the value of $(I^-)$ necessarily decreases the equilibrium value of $(Pb^{2+})$, and increasing the value of $(Pb^{2+})$ (for example, by adding $Pb(NO_3)_2(aq)$ to the solution) decreases the equilibrium value of $(I^-)$.

The solubility of sparingly soluble electrolytes is also affected by the presence of electrolytes in the solution that do not involve a common ion. For example, the solubility of $PbI_2(s)$ is greater in 1.0 M $NaClO_4(s)$ than in pure water. This is called the *inert-salt effect*, and it has its origin in the effect of the added electrolyte on $\gamma_{\pm}$ for the dissolved salt. The inert-salt effect is negligible *in comparison to* the common-ion effect in determining the solubility of a salt.

A type of problem that is frequently encountered in the laboratory is the following. If two solutions, each of which contains one of the ions of a relatively insoluble salt, are mixed, will the salt precipitate from the mixture, and, if so, how much salt will precipitate?

For example, suppose we mix 50 ml of 0.020 M $Pb(NO_3)_2(aq)$ with 50 ml of 0.020 M $NaI(aq)$, and that we want to calculate the equilibrium concentrations of $(Pb^{2+})$ and $(I^-)$ in the mixture, given that $K_{sp} = 1.0 \times 10^{-9}$ for $PbI_2(s)$. The equilibrium of interest is

$$PbI_2(s) = Pb^{2+}(aq) + 2I^-(aq)$$

and the first question to be answered is: Does *any* $PbI_2(s)$ precipitate? If *not*, then $(Pb^{2+}) = 0.010$ M and $(I^-) = 0.010$ M. These concentrations yield a $Q$ value for the above reaction of

$$Q = (Pb^{2+})(I^-)^2 = (0.010)(0.010)^2 = 1.0 \times 10^{-6}$$

The value of $Q/K$ is

$$Q/K = \frac{1.0 \times 10^{-6}}{1.0 \times 10^{-9}} = 1.0 \times 10^3 > 1$$

Because $Q/K > 1$, the above reaction will proceed from right to left until equilibrium ($Q = K$) is attained. In other words, the precipitation of $PbI_2(s)$ from the mixture is a spontaneous process.

The equilibrium concentrations of $Pb^{2+}(aq)$ and $I^-(aq)$ are computed as follows:

$$\begin{pmatrix} \text{mmoles of } Pb^{2+} \\ \text{in } Pb(NO_3)_2(aq) \\ \text{solution} \end{pmatrix} = \begin{pmatrix} \text{mmoles of } Pb^{2+} \\ \text{in solution at} \\ \text{equilibrium} \end{pmatrix} + \begin{pmatrix} \text{mmoles of } PbI_2(s) \\ \text{precipitated} \end{pmatrix}$$

$$(0.020)(50) = (Pb^{2+})(100) + x$$

$$\begin{pmatrix} \text{mmoles of } I^- \text{ in} \\ NaI(aq) \text{ solution} \end{pmatrix} = \begin{pmatrix} \text{mmoles of } I^- \\ \text{in solution at} \\ \text{equilibrium} \end{pmatrix} + 2 \begin{pmatrix} \text{mmoles of } PbI_2(s) \\ \text{precipitated} \end{pmatrix}$$

$$(0.020)(50) = (I^-)(100) + 2x$$

Combination of these equations with the $K_{sp}$ equation yields

$$1.0 \times 10^{-9} = (Pb^{2+})(I^-)^2 = \left( \frac{1.00 - x}{100} \right) \left( \frac{1.00 - 2x}{100} \right)^2$$

or

$$(1.00 - x)(1.00 - 2x)^2 = 1.0 \times 10^{-3}$$

This equation is a cubic equation in x; the solution can be obtained by trial and error. We know that the value of x lies in the range $0 < x < 0.5$, because the equilibrium concentration of $I^-(aq)$ cannot be negative $[(I^-) = (1.00 - 2x)/100]$. We now guess a value for x, say, 0.48, compute $(1.00 - x)(1.00 - 2x)^2$, and compare it with $1.00 \times 10^{-3}$; we then adjust x until $(1.00 - x)(1.00 - 2x)^2 = 1.0 \times 10^{-3}$. See below.

The trial-and-error solution for the number of mmole of $PbI_2(s)$ formed is $x = 0.478$ mmole. The number of grams of $PbI_2(s)$ formed is $[PbI_2(s)$ contains $461.0$ g $\cdot$ mole$^{-1}]$:

$$0.478 \text{ mmole} \times 0.461 \text{ g} \cdot \text{mmole}^{-1} = 0.220 \text{ g}$$

The equilibrium concentrations of $Pb^{2+}(aq)$ and $I^-(aq)$ are

$$(Pb^{2+}) = \frac{1.00 - 0.478}{100} = 5.2 \times 10^{-3} \text{ M}$$

$$(I^-) = \frac{1.00 - 0.956}{100} = 4.4 \times 10^{-4} \text{ M}$$

| VALUE OF x | $(1.00 - x)(1.00 - 2x)^2$ | COMMENT |
|---|---|---|
| 0.48 | $(0.50)(0.04)^2 = 0.83 \times 10^{-3}$ | x too large |
| 0.47 | $(0.53)(0.06)^2 = 1.9 \times 10^{-3}$ | x too small |
| 0.475 | $(0.525)(0.050)^2 = 1.3 \times 10^{-3}$ | x too small |
| 0.478 | $(0.522)(0.044)^2 = 1.0 \times 10^{-3}$ | x = 0.478 |

## c. Solubility Enhancement of a Solid Arising from the Formation of a Soluble Complex Ion.

In calculating the solubility of a solid from the $K_{sp}$ expression, it is imperative to survey carefully the chemistry of the situation in order to remove the possibility that other potentially important equilibria are overlooked. As an example of a salt-solubility equilibrium involving the formation of a complex ion, consider the problem of calculating the solubility of $CuI(s)$ in 1.00 M $KI(aq)$ at 25°C. The relevant equilibria are

(1)
$$CuI(s) = Cu^+(aq) + I^-(aq)$$

$$K_{sp} = 1.1 \times 10^{-12} \simeq (Cu^+)(I^-)$$

(2)
$$CuI(s) + I^-(aq) = CuI_2^-(aq)$$

$$K = 6.3 \times 10^{-4} \simeq \frac{(CuI_2^-)}{(I^-)}$$

where the second equilibrium involves the formation of the complex ion $CuI_2^-(aq)$. We have four unknowns in this problem, the solubility $s$, $(Cu^+)$, $(I^-)$, and $(CuI_2^-)$. The $K_{sp}$ and $K$ expressions constitute two equations involving these unknowns, and thus, the formal solution to the problem requires two additional equations. These additional equations come from a mass balance on copper and iodide:

for copper

$$s = (Cu^+) + (CuI_2^-)$$

for iodide

$$1.00 + s = (I^-) + 2(CuI_2^-)$$

where the factor of 2 arises because each mole of $CuI_2^-$ contains two moles of iodide. We now have four independent equations and four unknowns, and consequently, an exact solution can be found by brute force. It is more expeditious, however, to use the principal equilibrium approximation method. We note that $K \gg K_{sp}$, and therefore, equilibrium (2) is probably the principal equilibrium—ignoring, for the present, equilibrium (1), we have

$$(I^-) \simeq 1.00 - 2(CuI_2^-)$$

and

$$6.3 \times 10^{-4} \simeq \frac{(CuI_2^-)}{1.00 - 2(CuI_2^-)}$$

from which we compute

$$(CuI_2^-) \simeq 6.3 \times 10^{-4} \text{ M}, \qquad (I^-) \simeq 1.00 \text{ M}$$

The value of $(Cu^+)$ is then computed as

$$(Cu^+) \simeq \frac{1.1 \times 10^{-12}}{(I^-)} \simeq 1.1 \times 10^{-12}$$

which justifies our assumption that $Cu^+$ is a minor species. The solubility of $CuI(s)$ in 1.00 M $KI(aq)$ at 25°C is then

$$s = 1.1 \times 10^{-12} + 6.3 \times 10^{-4} = 6.3 \times 10^{-4} \text{ M}$$

If the solubility were computed without recognizing the importance of equilibrium (2), that is, if we take

$$1.1 \times 10^{-12} \simeq s(1.00 + s) \simeq s$$

then we obtain $s = 1.1 \times 10^{-12}$ M. This result is in error by a factor of $6 \times 10^8$! *Caveat computatus.*

### d. Effect of Particle Size on the Solubility of a Solid.

The solubility of a solid depends, in general, on the particle size. The smaller the crystals, the greater their solubility. Finely-divided particles have a much higher surface Gibbs energy, owing to their larger surface area per unit mass. This higher Gibbs energy for the finely-divided solid gives rise to an activity for the solid phase that is greater than unity, even at 1 atm total pressure. Consider the equilibrium

$$BaSO_4(s) = Ba^{2+}(aq) + SO_4^{2-}(aq)$$

$$K_{sp} = \frac{a_{Ba^{2+}(aq)} \cdot a_{SO_4^{2-}(aq)}}{a_{BaSO_4(s)}} = \frac{(Ba^{2+})(SO_4^{2-})\gamma_{\pm}^2}{a_{BaSO_4(s)}}$$

$$K_{sp} \simeq \frac{(Ba^{2+})(SO_4^{2-})}{a_{BaSO_4(s)}}$$

Because $K_{sp}$ = constant at a given temperature, if $a_{BaSO_4(s)} > 1$, then the product $(Ba^{2+})(SO_4^{2-}) \simeq s^2$, must increase proportionately, relative to the case when $a_{BaSO_4(s)} = 1$, because $K$ is a constant. Crystals of $BaSO_4$ that are $10^{-4}$ cm in diameter are roughly twice as soluble as $BaSO_4$ crystals $10^{-3}$ cm in diameter. The effect of particle size decreases rapidly with increasing particle size, and crystals with diameters $\gtrsim 10^{-3}$ have essentially the same solubility, irrespective of particle size. Because very small particles are less stable than larger particles, a finely-divided solid in contact with a saturated solution will spontaneously grow larger crystals at the expense of the smaller ones.

### e. Solubility Enhancement of Salts of Weak Acids in Acidic Solutions.

Salts of weak acids are more soluble the higher the acidity of the solution, owing to the protonation of the conjugate base of the weak acid.

For example, consider the solubility of silver acetate in aqueous solution. The relevant equilibria are (25°C):

(1)
$$AgOCOCH_3(s) = Ag^+(aq) + CH_3COO^-(aq)$$
$$K_{sp} = 4.0 \times 10^{-3} \simeq (Ag^+)(CH_3COO^-)$$

(2)
$$CH_3COO^-(aq) + H_2O(\ell) = CH_3COOH(aq) + OH^-(aq)$$
$$K_b = 5.7 \times 10^{-10} \simeq \frac{(CH_3COOH)(OH^-)}{(CH_3COO^-)}$$

(3)
$$H_2O(\ell) = H^+(aq) + OH^-(aq)$$
$$K_w = 1.0 \times 10^{-14} \simeq (H^+)(OH^-)$$

We have six unknowns:

$$s, (Ag^+), (CH_3COOH), (CH_3COO^-), (H^+), \text{ and } (OH^-)$$

and, consequently, the general solution to the problem requires three more equations. These are:

(4)    $s = (CH_3COO^-) + (CH_3COOH)$

(5)    $s = (Ag^+)$

(6)    $(H^+) + (Ag^+) = (CH_3COO^-) + (OH^-)$

Because $K_{sp} \gg K_b$, we can try the principal-equilibrium approximation with (1) as the principal equilibrium, in which case

$$(Ag^+) \simeq (CH_3COO^-)$$

and thus

$$s \simeq (Ag^+) \simeq (CH_3COO) \simeq (4.0 \times 10^{-3})^{1/2} = 6.3 \times 10^{-2} \text{ M}$$

Substitution of this result into the $K_b$ expression yields

$$5.7 \times 10^{-10} \simeq \frac{(CH_3COOH)(OH^-)}{6.3 \times 10^{-2}} \simeq \frac{(OH^-)^2}{6.3 \times 10^{-2}}$$
$$(OH^-) \simeq 6.0 \times 10^{-6} \text{ M}$$

and

$$(H^+) \simeq \frac{1.0 \times 10^{-14}}{6.0 \times 10^{-6}} = 1.7 \times 10^{-9}$$

The solution is basic ($pH = 8.77$), because of the protonation reaction. Note that the value of $(CH_3COO^-)$ is fixed by the solubility-product equilibrium.

The use of the principal-equilibrium approximation in the above case is equivalent to the assumption that the protonation of $CH_3COO^-(aq)$ is of negligible importance in determining the total solubility of $CH_3COOAg(s)$. For bases much stronger than $CH_3COO^-(aq)$, for example, $S^{2-}(aq)$, such an approximation may not be adequate. Similarly, if the salt is in equilibrium with an acidified solution, then the above approximation may not be tenable. This can be shown as follows: from equations (1), (2), (4), and (5), we obtain

$$s = (CH_3COO^-) + (CH_3COOH) \simeq \frac{K_{sp}}{(Ag^+)} + \frac{K_b(CH_3COO^-)}{(OH^-)}$$

$$s \simeq \frac{K_{sp}}{s} + \frac{K_b K_{sp}}{s(OH^-)} \simeq \frac{K_{sp}}{s} + \frac{K_b K_{sp}(H^+)}{s K_w}$$

$$s \simeq \frac{K_{sp}}{s} + \frac{K_{sp}(H^+)}{s K_a}$$

or

$$s \simeq \left[ K_{sp} \left\{ 1 + \frac{(H^+)}{K_a} \right\} \right]^{1/2}$$

Substitution of the values of $K_{sp}$ and $K_a$ appropriate to 25°C yields

$$s \simeq \{4.0 \times 10^{-3} + 228(H^+)\}^{1/2}$$

In Figure 12.3 we have plotted $s$ versus $pH$ over the range $1 \leqslant pH \leqslant 14$. Over the range[*] $14 \geqslant pH \geqslant 6$ the value of $s$ is essentially constant at 0.063 M, whereas at $pH < 6$ the solubility rises rapidly, owing to the increased importance of the protonation of $CH_3COO^-$(aq) at low $pH$.

The dramatic influence of $pH$ on the solubility of salts of weak acids is widely used in analytical chemistry to achieve separations of mixtures of cations. This is especially true of separation schemes involving precipitation of hydroxides and sulfides. The separations are achieved by choosing a $pH$ where one or more of the sulfides has a high solubility and the others have a negligible solubility. The effect of $pH$ on solubility is depicted graphically in Figures 12.4(a) and (b) for sulfides and hydroxides, respectively.

## f. Metathetical Reactions.

It is often possible to convert one insoluble salt of a given metal to another insoluble salt of the same metal by treating the insoluble salt with a solution containing a different anion. A related question involves the prediction of which salt will precipitate when a solution containing a given metal ion is mixed with a solution containing two or more different anions with which the metal ion forms insoluble salts. As an example of the analysis of this type of problem, consider the following: Calculate the equilibrium concentrations of $Co(CN)_6^{3-}$(aq), $Cl^-$(aq), and $Ag^+$(aq) that result when excess $Ag_3Co(CN)_6$(s) is treated with 1.00 M KCl(aq) at 25°C. The relevant equilibria are

(1) $\qquad AgCl(s) = Ag^+(aq) + Cl^-(aq)$

$\qquad K_{sp} = 1.0 \times 10^{-10} \simeq (Ag^+)(Cl^-)$

---

[*]The curve is hypothetical at $pH > 10$, owing to the importance of the equilibrium $2AgOCOCH_3(s) + 2OH^-(aq) = Ag_2O(s) + H_2O(\ell) + 2CH_3COO^-(aq)$ at high $pH$.

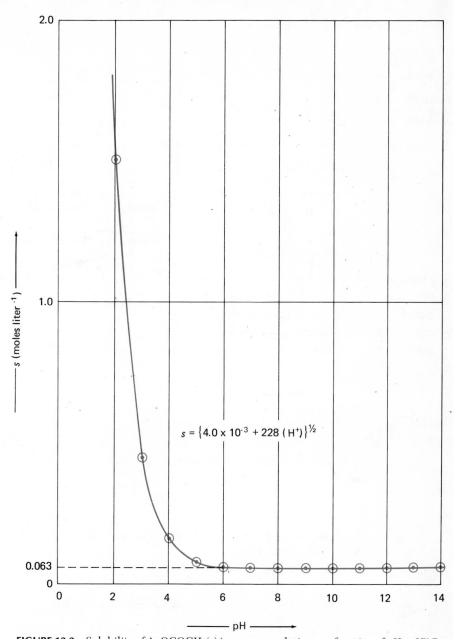

**FIGURE 12.3** Solubility of $AgOCOCH_3(s)$ in aqueous solution as a function of pH at 25°C.

(2) $$Ag_3Co(CN)_6(s) = 3Ag^+(aq) + Co(CN)_6^{3-}(aq)$$

$$K_{sp} = 8.5 \times 10^{-21} \simeq (Ag^+)^3(Co(CN)_6^{3-})$$

Multiplying reaction (1) by 3 and subtracting the result from reaction (2) yields the principal equilibrium

(3) $$Ag_3Co(CN)_6(s) + 3Cl^-(aq) = 3AgCl(s) + Co(CN)_6^{3-}(aq)$$

$$K = \frac{8.5 \times 10^{-21}}{(1.0 \times 10^{-10})^3} = 8.5 \times 10^9 \simeq \frac{(Co(CN)_6^{3-})}{(Cl^-)^3}$$

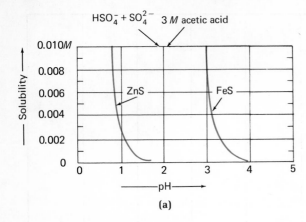

**(a)**

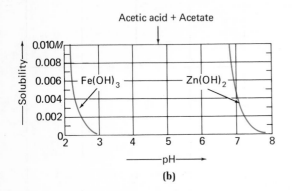

**(b)**

**FIGURE 12.4(a)** Solubility of Zns and FeS in water saturated with $H_2S$ (atm) at various concentrations of $H^+(aq)$. We can separate $Zn^{2+}(aq)$ from $Fe^{2+}(aq)$ by saturating a solution buffered at about $pH = 2$ with $H_2S(g)$. The ZnS precipitates and the $Fe^{2+}(aq)$ remains in solution.

**(b)** Solubility of $Fe(OH)_3$ and $Zn(OH)_2$ at various concentrations of $H^+(aq)$. We can separate $Fe^{3+}(aq)$ from $Zn^{2+}(aq)$ by adjusting the $pH$ of the solution to about $4.8 \pm 1.0$. The iron precipitates as $Fe(OH)_3 \cdot nH_2O$ and the $Zn^{2+}(aq)$ remains in solution.

Figures adapted with minor modification from W. M. Latimer and R. E. Powell, *A Laboratory Course on General Chemistry* (Macmillan Co., 1964), p. 122.

From the reaction stoichiometry, we obtain

$$(Co(CN)_6^{3-}) = \frac{1.00 - (Cl^-)}{3}$$

because only 1/3 of a mole of $Co(CN)_6^{3-}$ is produced for each mole of $Cl^-$ consumed. Thus

$$\frac{\left(\dfrac{1.00 - (Cl^-)}{3}\right)}{(Cl^-)^3} \simeq 8.5 \times 10^9$$

Assuming that $(Cl^-) \ll 1.00$ (because $K$ is so large), we compute $(Cl^-) \simeq 3.4 \times 10^{-4}$ M and $(Co(CN)_6^{3-}) \simeq 0.33$ M. Also

$$(Ag^+) \simeq \frac{1.0 \times 10^{-10}}{3.4 \times 10^{-4}} = 2.9 \times 10^{-7} \text{ M}$$

If we did not specify an *excess* of $Ag_3Co(CN)_6(s)$, and if the initial amount of $Ag_3Co(CN)_6(s)$ were insufficient to yield a solution 0.33 M in $Co(CN)_6^{3-}(aq)$, then at equilibrium none of the $Ag_3Co(CN)_6(s)$ phase would be present. Putting it the other way around, $Ag_3Co(CN)_6(s)$ cannot precipitate from an aqueous solution at 25°C unless the product $(Ag^+)^3(Co(CN)_6^{3-})$ exceeds the value $8.5 \times 10^{-21}$. Conversely, if this product does exceed $8.5 \times 10^{-21}$, then the precipitation of $Ag_3Co(CN)_6$ will proceed* until the value of $(Ag^+)^3(Co(CN)_6^{3-})$ drops to $8.5 \times 10^{-21}$. If we have a solution 1.00 M in $Cl^-$ and 0.33 M in $Co(CN)_6^{3-}$, then we can add $Ag^+$ as $AgNO_3(aq)$ to separate $Cl^-$ from $Co(CN)_6^{3-}$. No $Ag_3Co(CN)_6$ will precipitate until $(Ag^+)$ reaches the value $2.9 \times 10^{-7}$ M, that is, until $(Cl^-)$ is decreased to $3.4 \times 10^{-4}$ M.

It is often possible to dissolve an otherwise insoluble salt by adding a species to the solution that forms a soluble complex with the metal ion. The enhanced solubility of $CuI(s)$ in solutions containing $I^-(aq)$ is an example of this procedure. As a second example, consider the following problem: Calculate the solubility of $AgCl(s)$ at 25°C in a solution that is 0.100 M in $KCl(aq)$ and 0.100 M in $NH_3(aq)$, given the following equilibrium data:

(1)
$$AgCl(s) = Ag^+(aq) + Cl^-(aq)$$

$$K_{sp} = 1.0 \times 10^{-10} \simeq (Ag^+)(Cl^-)$$

(2)
$$Ag(NH_3)_2^+(aq) = Ag^+(aq) + 2NH_3(aq)$$

$$K_2 = 5.0 \times 10^{-8} \simeq \frac{(Ag^+)(NH_3)^2}{(Ag(NH_3)_2^+)}$$

Subtracting reaction (2) from reaction (1), we obtain

(3)
$$AgCl(s) + 2NH_3(aq) = Aq(NH_3)_2^+(aq) + Cl^-(aq)$$

$$K_3 = \frac{1.0 \times 10^{-10}}{5.0 \times 10^{-8}} = 2.0 \times 10^{-3} \simeq \frac{(Ag(NH_3)_2^+)(Cl^-)}{(NH_3)^2}$$

From the stoichiometry of reaction (3), we obtain

$$(Cl^-) \simeq 0.100 + (Ag(NH_3)_2^+)$$

$$(NH_3) \simeq 0.100 - 2(Ag(NH_3)_2^+)$$

Substitution of these equations into the $K_3$ expression yields

$$2.0 \times 10^{-3} \simeq \frac{(Ag(NH_3)_2^+)\{0.100 + (Ag(NH_3)_2^+)\}}{[0.100 - 2(Ag(NH_3)_2^+)]^2}$$

Assuming that $(Ag(NH_3)_2^+) \ll 0.100$ (because $K_3$ is small), we obtain $(Ag(NH_3)_2^+) \simeq 2.0 \times 10^{-4}$ M, and $(Cl^-) \simeq 0.100$ M. The equilibrium concentration of $Ag^+(aq)$ is

$$(Ag^+) \simeq \frac{1.0 \times 10^{-10}}{(Cl^-)} \simeq 1.0 \times 10^{-9} \text{ M}$$

and the solubility of silver chloride in the solution is

$$\text{solubility} = (Ag^+) + (Ag(NH_3)_2^+) = 2 \times 10^{-4} \text{ M}$$

This result shows that the solubility of $AgCl(s)$ in 0.100 M $KCl(aq)$ is increased by a factor of $2 \times 10^5$ in the presence of 0.1 M $NH_3(aq)$. The separa-

---

*This assumes that supersaturation does not occur.

tion of $Ag^+(aq)$ from $Hg_2^{2+}(aq)$ involves the addition of HCl(aq) to precipitate $AgCl(s) + Hg_2Cl_2(s)$. Aqueous $NH_3$ is then added to dissolve the AgCl; the $NH_3$ converts $Hg_2Cl_2(s)$ to $HgNH_2Cl + Hg$:

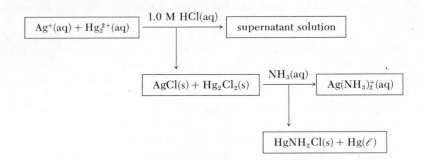

## 12.8 DISTRIBUTION OF A SOLUTE BETWEEN TWO IMMISCIBLE LIQUIDS.

Water and carbon tetrachloride ($CCl_4$) are essentially insoluble in each other; they are immiscible liquids. A flask containing $H_2O$ and $CCl_4$ has two distinct liquid phases: (1) a phase that is predominately* $CCl_4$ ($>99\%$), which is saturated with water; and (2) a phase that is predominately water ($>99\%$), which is saturated with $CCl_4$.

Suppose that we introduce some $I_2(s)$ into a flask containing water and $CCl_4$, and dissolve the iodine. How will the iodine distribute itself between the two phases? The equilibrium of interest is

$$I_2(aq) = I_2(CCl_4)$$

and

$$K = \frac{a_{I_2(CCl_4)}}{a_{I_2(aq)}} \simeq \frac{(I_2)_{CCl_4}}{(I_2)_{H_2O}}$$

and at 25°C we find $K = 50.5$. In other words, the non-polar $I_2$ is preferentially dissolved in the non-polar $CCl_4$ phase. The concentration of $I_2$ in the $CCl_4$ phase is about 50 times as great as in the aqueous phase.

If we have an aqueous phase containing $I_2$ and we shake it up with $CCl_4$, then $I_2$ will pass from the aqueous to the $CCl_4$ phase until the ratio of concentrations is 50.5. This is an example of *solvent extraction*. Extraction of a desired substance from a liquid phase using an immiscible liquid in which the substance is more soluble is a widely used method of achieving separations in synthetic chemistry.

As long as $I_2$ is present in the equilibrium $H_2O + CCl_4$ system, the ratio of concentrations of $I_2$ in the two phases is fixed. This is true irrespective of whether $I_2$ participates in other equilibria in either phase. For example, the complexation reaction

$$I_2(aq) + I^-(aq) = I_3^-(aq)$$

---

*Some liquids are only partially immiscible, e.g., water and ethyl ether. We then speak of the "water-rich" phase and the "ether-rich" phase.

has $K = 770$ at 25°C. Thus, if we have $I_2$ at equilibrium in the two-phase $H_2O + CCl_4$ system, and we introduce some KI into the aqueous phase, then some $I_2$ will pass from the $CCl_4$ to the $H_2O$ phase to form $I_3^-(aq)$. However, at equilibrium, both the complexation equilibrium *and* the distribution equilibrium must be satisfied, and thus $(I_2)_{CCl_4} \simeq 50.5\ (I_2)_{aq}$. In other words, the *total* concentrations of $I_2$ in $CCl_4$ and $H_2O$ are reduced, but the *ratio* of these concentrations remains the same.

Sometimes the substance that is distributed between two immiscible liquids does not have the same form in the two phases. For example, acetic acid in benzene $(C_6H_6)$ exists almost exclusively as the dimer, $(CH_3COOH)_2$, as a result of hydrogen bonding to itself

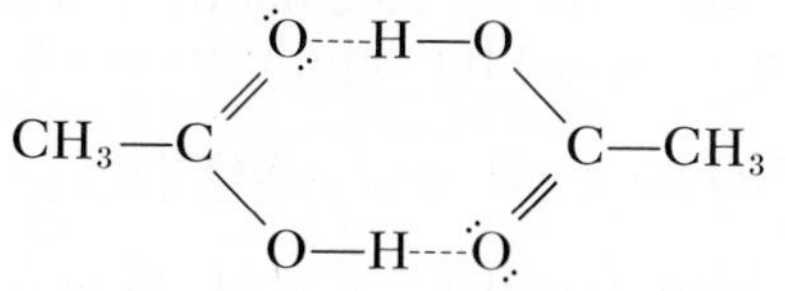

$$CH_3—C \overset{\displaystyle \ddot{O}\text{----}H—O}{\underset{\displaystyle O—H\text{---}\ddot{O}}{\Big\langle\quad\Big\rangle}} C—CH_3$$

and thereby reducing its polarity in the non-polar benzene. Benzene and water are immiscible. The distribution equilibrium for acetic acid in the $H_2O + C_6H_6$ system is

$$2CH_3COOH(aq) = (CH_3COOH)_2(benz)$$

and

$$k \simeq \frac{(CH_3COOH)_{2(benz)}}{(CH_3COOH)^2_{aq}}$$

## 12.9  THE EFFECT OF NON-IDEALITY ON THE EQUILIBRIUM DISTRIBUTION OF SPECIES.

Consider the equilibrium

$$HX(aq) = H^+(aq) + X^-(aq)$$

Using the results obtained in Section 11.2, we can write

$$K_a = \frac{a_{H^+(aq)}\, a_{X^-(aq)}}{a_{HX(aq)}} = \frac{m_{H^+}\, m_{X^-}\, \gamma^2_{\pm(HX)}}{m_{HX}\, \gamma_{HX}} \tag{12.7a}$$

If we assume that $\gamma_{HX} = 1$, then we have

$$K_a \simeq \frac{m_{H^+}\, m_{X^-}\, \gamma^2_{\pm(HX)}}{m_{HX}} \tag{12.7b}$$

If we assume that $\gamma_{\pm(HX)}$ is given by the Debye-Hückel expression, then we obtain

$$K_a \simeq \frac{m_{H^+}\, m_{X^-}}{m_{HX}} \cdot 10^{-1.02\,I^{1/2}/(1+I^{1/2})} \quad (25°C) \tag{12.7c}$$

Lastly, assuming that $\gamma_{\pm(HX)} = 1$ yields

$$K_a \simeq \frac{m_{H^+} m_{X^-}}{m_{HX}} \simeq \frac{(H^+)(X^-)}{(HX)} \tag{12.7d}$$

The exact expression for $K_a$ is given by (12.7a), whereas (12.7b), (12.7c), and (12.7d) are approximations, with the accuracy of the approximation decreasing as we go from (12.7b) to (12.7c) to (12.7d). In Section 12.1 we carried out numerous calculations using approximation (12.7d). In order to obtain a feeling for the magnitudes of the errors that can arise in such calculations, we shall carry out a sample calculation at level (12.7c).

In Section 12.1 we considered the equilibrium

$$HNO_2(aq) = H^+(aq) + NO_2^-(aq)$$

at 25°C (where $K_a = 4.5 \times 10^{-4}$) with 0.100 F $HNO_2(aq)$. Using approximation (12.7d). we found $(H^+) \simeq (NO_2^-) \simeq 6.4 \times 10^{-3}$ M, $HNO_2(aq) \simeq 0.094$ M, and $pH = 2.19$. Using these results to estimate the total ionic strength of the solution, we have (assuming that there are no added electrolytes) $I = 6.4 \times 10^{-3}$. Using this result we obtain

$$4.5 \times 10^{-4} \simeq \frac{(H^+)(NO_2^-)}{(HNO_2)} 10^{-1.02(0.080)/1.08} = 0.84 \frac{(H^+)(NO_2^-)}{(HNO_2)}$$

or

$$5.4 \times 10^{-4} \simeq \frac{(H^+)(NO_2^-)}{(HNO_2)}$$

from which we obtain for 0.100 F $HNO_2(aq)$

$$5.4 \times 10^{-4} \simeq \frac{(H^+)^2}{0.100 - (H^+)}$$

Solving for $(H^+)$ yields

$$(H^+) \simeq (NO_2^-) \simeq 7.1 \times 10^{-3} \text{ M}$$

and

$$(HNO_2) \simeq 0.100 - 0.007 = 0.093 \text{ M}$$

or about[*] a 10% increase in $(H^+)$ and $(NO_2^-)$, and a 1% decrease in $(HNO_2)$. The $pH$ of the solution is computed as $(\gamma_\pm = (0.84)^{1/2})$

$$pH = -\log a_{H^+} \simeq -\log\{(H^+)(\gamma_\pm)\} \simeq -\log\{(7.1 \times 10^{-3})(0.92)\}$$
$$pH \simeq -\log(6.5 \times 10^{-3}) = 2.19$$

---

[*] For a more accurate approximation we should use these results to recompute $I$ and $\gamma_\pm$, and then $(H^+)$, and repeat this process until there is no change in the computed $(H^+)$. The additional change in $(H^+)$ is very slight.

which is just what we obtained on neglecting activity coefficients. This is a consequence of offsetting errors in $(H^+)$ and $\gamma_\pm$ in the calculation of $pH$ in the case where we assumed $\gamma_\pm = 1$. This is a fairly general result; namely, the error in the $pH$ calculated assuming $\gamma_\pm = 1$ is usually much less than the error in the $(H^+)$ calculated assuming $\gamma_\pm = 1$.

Non-ideal solution behavior can have an important effect on the solubility of a salt. As an example, consider the equilibrium

$$AgOCOCH_3(s) = Ag^+(aq) + CH_3COO^-(aq)$$

From Section 11.2 we have (where $X \equiv CH_3COO$)

$$K_{sp} = \frac{a_{Ag^+(aq)}\, a_{X^-(aq)}}{a_{AgX(s)}} = \frac{m_{Ag^+}\, m_{X^-}\, \gamma^2_{\pm(AgX)}}{a_{AgX(s)}} \qquad (12.8a)$$

If $P_{tot} \simeq 1$ atm, and if the AgX(s) phase is unstrained and not too finely divided, then

$$K_{sp} = m_{Ag^+}\, m_{X^-}\, \gamma^2_{\pm(AgX)} \qquad (12.8b)$$

Also (Section 11.2),

$$K_{sp} \simeq m_{Ag^+}\, m_{X^-}\, 10^{-1.02\, I^{1/2}/(1+I^{1/2})} \; (25°C) \qquad (12.8c)$$

$$K_{sp} \simeq m_{Ag^+}\, m_{X^-} \simeq (Ag^+)(X^-) \qquad (12.8d)$$

where (12.8a) and (12.8b) (given the above qualifications) are exact expressions for $K_{sp}$, and (12.8c) and (12.8d) are approximations, with (12.8c) being less approximate than (12.8d). In Section 12.3 we found, using approximation (12.8d), that at 25°C $(K_{sp} = 4.0 \times 10^{-3})$ $(Ag^+) \simeq (X^-) \simeq 6.3 \times 10^{-2}$ M. We now repeat this calculation using approximation (12.8c). Assuming no added electrolytes, the total ionic strength of the solution is $I = 0.063$, and thus

$$4.0 \times 10^{-3} \simeq (Ag^+)(X^-)\, 10^{-1.02(0.25)/1.25}$$

from which we compute

$$(Ag^+) \simeq (X^-) \simeq (6.4 \times 10^{-3})^{1/2} = 0.080 \text{ M}$$

which is 27% greater than the solubility calculated assuming $\gamma_\pm = 1$.

The foregoing calculations are not rigorous, but they show that assuming $\gamma_\pm = 1$ for electrolytes is not a good approximation, and this is especially true for solutions with $I > 0.1$. The errors arising from the assumption that $\gamma_\pm = 1$ at a given ionic strength are greater the higher the value of $Z_+Z_-$ for the electrolyte.

## 12.10  PRESSURE-COMPOSITION AND TEMPERATURE-COMPOSITION DIAGRAMS.

Consider a liquid solution containing two components (which we shall distinguish by the subscripts 1 and 2) that is in equilibrium with a vapor phase containing the same two components. We shall assume for simplicity that the vapor phase is ideal. Given that the vapor phase is ideal, the total pressure over the solution, $P_{tot}$, is given by

$$P_{tot} = P_1 + P_2 \qquad (12.9a)$$

where $P_1$ and $P_2$ are the partial pressures of the two gaseous components. The question that now arises is: How are the total pressure and composition of the gas phase related to the composition of the solution? The answer to this problem is relatively simple if the liquid solution is ideal. In such a case the values of $P_1$ and $P_2$ are given by Raoult's law (Equation (11.6)). The total pressure over a two-component, ideal solution is given by

$$P_{tot} = X_1 P_1^\circ + X_2 P_2^\circ = (1 - X_2) P_1^\circ + X_2 P_2^\circ$$
$$P_{tot} = (P_2^\circ - P_1^\circ) X_2 + P_1^\circ \qquad (12.9b)$$

The composition of the vapor phase in equilibrium with an ideal solution is given by

$$X_2' = \frac{P_2}{P_{tot}} = \frac{X_2 P_2^\circ}{P_{tot}}$$

$$\qquad (12.9c)$$

$$X_1' = \frac{P_1}{P_{tot}} = \frac{X_1 P_1^\circ}{P_{tot}}$$

where the primes denote mole fractions in the vapor. We can express $P_{tot}$ in terms of $X_2'$, $P_2^\circ$, and $P_1^\circ$ by combining equations (12.9b) and (12.9c); the resulting expression is

$$P_{tot} = \frac{P_2^\circ P_1^\circ}{P_2^\circ - (P_2^\circ - P_1^\circ) X_2'} \qquad (12.9d)$$

The use of equations (b) and (d) is depicted graphically in Figure 12.5 for the case of a two-component, ideal solution with $P_1^\circ = 100$ Torr and $P_2^\circ = 200$ Torr.

As an example of solutions exhibiting non-ideality, consider the data on liquid $CCl_4 + CH_3CH_2O\overset{\overset{\displaystyle O}{\|}}{C}CH_3$ (carbon tetrachloride + ethyl acetate) solutions presented in Table 12.6. That the solutions are not ideal is evident from the fact that Raoult's law is not rigorously obeyed, or equivalently, $a_i \neq X_i$.

The pressure-composition data in Table 12.6 are presented graphically in Figure 12.6. A comparison of the pressure-composition curves for this non-ideal system with those in Figure 12.5 for the ideal system reveals several important differences. The curves of $P_{tot}$ over the solution as a function of the mole fraction of component 2 in the solution *and* as a function of the mole fraction of component 2 in the gas phase both exhibit a *maximum*. This maximum[*] is a consequence of the non-ideality of the

---

[*]Minima rather than maxima are sometimes observed in non-ideal solutions. This occurs when $\frac{a_i}{X_i} < 1$ for analogous two-component systems.

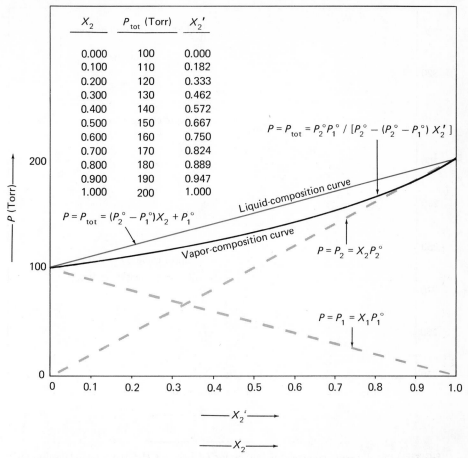

| $X_2$ | $P_{tot}$ (Torr) | $X_2'$ |
|-------|---------|--------|
| 0.000 | 100 | 0.000 |
| 0.100 | 110 | 0.182 |
| 0.200 | 120 | 0.333 |
| 0.300 | 130 | 0.462 |
| 0.400 | 140 | 0.572 |
| 0.500 | 150 | 0.667 |
| 0.600 | 160 | 0.750 |
| 0.700 | 170 | 0.824 |
| 0.800 | 180 | 0.889 |
| 0.900 | 190 | 0.947 |
| 1.000 | 200 | 1.000 |

$$P = P_{tot} = P_2^\circ P_1^\circ / [P_2^\circ - (P_2^\circ - P_1^\circ) X_2']$$

Liquid-composition curve

$$P = P_{tot} = (P_2^\circ - P_1^\circ)X_2 + P_1^\circ$$

Vapor-composition curve

$$P = P_2 = X_2 P_2^\circ$$

$$P = P_1 = X_1 P_1^\circ$$

**FIGURE 12.5**  Pressure-composition diagram for an ideal solution with $P_2^\circ = 200$ Torr and $P_1^\circ = 100$ Torr. Note that the vapor (lower solid curve) is "richer" than the liquid (upper solid curve) ($X_2' > X_2$) in the more volatile component ($P_2^\circ > P_1^\circ$). This is always true for ideal solutions, but it is not necessarily true for non-ideal solutions.

**TABLE 12.6**

DATA ON $CCl_4(2) + CH_3CH_2O\overset{\displaystyle O}{\overset{\displaystyle \|}{C}}CH_3(1)$ SOLUTIONS AT 50°C (pressures in Torr).

| $X_1$ | $X_2$ | $P_1$ | $P_2$ | $P_1/X_1$ | $P_2/X_2$ | $a_1$ | $a_2$ | $a_1/X_1$ | $a_2/X_2$ |
|-------|-------|-------|-------|-----------|-----------|-------|-------|-----------|-----------|
| 0 | 1 | 0 | 306 | — | 306 | 0 | 1.000 | — | 1.00 |
| 0.096 | 0.904 | 34.4 | 277 | 369 | 306 | 0.123 | 0.905 | 1.28 | 1.00 |
| 0.198 | 0.802 | 67.0 | 250 | 338 | 312 | 0.239 | 0.817 | 1.21 | 1.02 |
| 0.425 | 0.575 | 126 | 190 | 296 | 330 | 0.450 | 0.621 | 1.06 | 1.08 |
| 0.684 | 0.316 | 196 | 110 | 287 | 348 | 0.700 | 0.359 | 1.02 | 1.14 |
| 0.806 | 0.194 | 228 | 70.7 | 283 | 364 | 0.814 | 0.231 | 1.01 | 1.19 |
| 0.950 | 0.050 | 266 | 18.6 | 290 | 372 | 0.950 | 0.061 | 1.00 | 1.22 |
| 1 | 0 | 280 | 0 | 280 | — | 1.000 | 0 | 1.00 | — |

Note that $a_1 = P_1/P_1^\circ$ and $a_2 = P_2/P_2^\circ$.

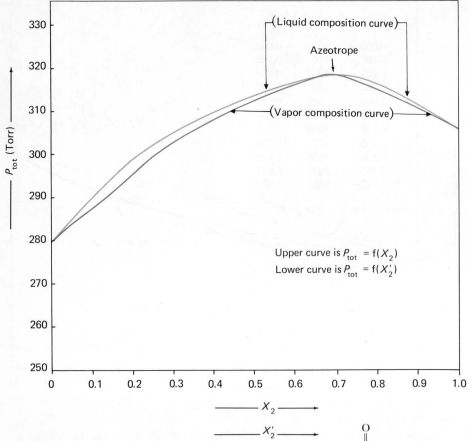

**FIGURE 12.6**   Pressure-composition curves for $CCl_4(2) + CH_3CH_2OCCH_3(1)$ solutions at 50°C.

solution. At the maximum the liquid and vapor compositions are identical ($X_2 = X_2' = 0.67$). A liquid solution whose equilibrium vapor has the same composition as that of the liquid is called an *azeotrope*. An azeotrope distills without change in composition, and in this sense is analogous to a pure liquid. The best known azeotrope occurs in the ethanol + water system at 95% $CH_3CH_2OH$.

In addition to a pressure-composition diagram for a two-component system, we can also construct a temperature-composition diagram. Such diagrams are useful for analyzing distillation processes, which are employed frequently in the laboratory to achieve a separation of the components of a solution. A temperature-composition diagram is a plot of the boiling temperature of the solution versus the composition of the solution *at a fixed total pressure*. If the gas and solution phases are ideal, then equation (b) is applicable. At a fixed total pressure of 1 atm (=760 Torr) this equation yields

$$760 = (P_2° - P_1°)X_2 + P_1°$$

This expression can be used to construct a *T-X* diagram at 1 atm total pressure, if we have available $P_2°$ and $P_1°$ as a function of temperature. To construct the *T-X* diagram we take values of $P_1°$ and $P_2°$ at one temperature and calculate $X_2$; we then repeat the calculation at other temperatures to

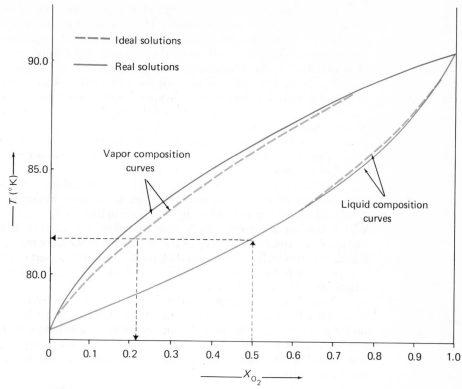

**FIGURE 12.7**   Temperature-composition ($T$-$X$) diagram for the $N_2 + O_2$ system at $P_{tot} =$ 1 atm. A solution with a mole fraction of $O_2$ equal to 0.50 has a boiling point of 81.7 K, and the mole fraction of $O_2$ in the vapor phase in equilibrium with this solution is 0.22.

obtain the boiling temperature as a function of composition. At each temperature we then calculate $X_2'$ (the mole fraction in the vapor phase) from the equation $X_2' = (X_2 P_2^{\circ}/P_{tot})$. In Figure 12.7, we have constructed such a plot for $N_2 + O_2$ solutions assuming ideal-solution behavior. On the same diagram we have also constructed the $T$-$X$ diagram for the $N_2 + O_2$ solutions using *measured* values of $X_2$ and $X_2'$. The $N_2 + O_2$ system deviates only slightly from ideal behavior.

The normal boiling point of pure $N_2(\ell)$ is 77.3 K, and that for pure $O_2(\ell)$ is 90.2 K. The normal boiling points of all possible solutions of $N_2 + O_2$, as well as the compositions of the vapor phase in equilibrium with the liquid, can be read directly off the $T$-$X$ diagram. For example, the boiling point of the solution with $X_{O_2} = 0.50$ is 81.7 K, and the composition of the vapor is $X_{O_2}' = 0.22$. The vapor has a higher mole fraction of the more volatile component ($N_2$) than does the liquid. If some of this vapor is in turn condensed and reboiled, and this process is repeated a sufficient number of times, a separation of the two components can be accomplished. The separation of two components of a solution by distillation is possible, provided the $T$-$X$ diagram has the general form (cigar shape) of that in Figure 12.7.* On the other hand, if the system possesses an azeotrope, then the complete separation of two components by distillation is not possible.

---

*Explain why separation by distillation is easier the greater the difference in boiling points of the two pure liquids.

A fractionating column is a device that carries out the vaporization-condensation cycles automatically. The essential features of such a column are the existence of a temperature gradient along the column ($T$ decreases continuously from the bottom to the top of the column), and the presence of a suitable packing material that provides a large surface area for heat transfer between the rising, high-temperature vapor phase and the descending, low-temperature liquid phase. With a suitably designed column it is possible, in a case like the $N_2 + O_2$ system, to achieve a complete separation of the two components, with the more volatile component issuing from the top of the column at its normal boiling point (assuming $P_{tot} = 1$ atm) and the less volatile component remaining in the distillation flask.

Fractional distillation will not suffice to separate completely the two components of a solution involving an azeotrope. Figure 12.8 shows the general form of the $T$-$X$ diagram for a system like $CCl_4 + CH_3CH_2OCOCH_3$, which has an azeotrope. In a case like that depicted in Figure 12.8, the material issuing from the top of the column has the azeotropic composition; the azeotrope will continue to distill off until only a part of one of the pure components remains in the distillation flask. This material can then be distilled off at its normal boiling point.

There is no requirement that distillations be carried out at 1 atm pressure. Distillations are carried out often at reduced ($<1$ atm) pressure in order to minimize thermal decomposition of a substance. The lower the total pressure over the liquid phase, the lower the temperature at which the liquid boils. Low-temperature distillations are carried out frequently on vacuum lines.

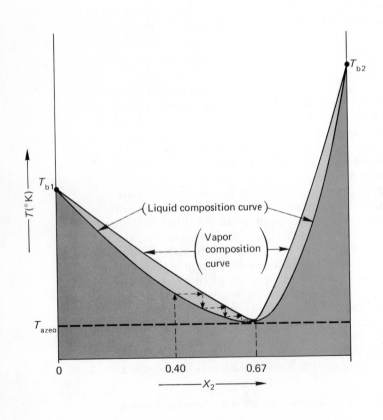

**FIGURE 12.8** Temperature-composition diagram for a system exhibiting a maximum in the $P$-$X$ diagram. A solution with a mole fraction of 0.40 will, upon distillation, yield the azeotrope at $X = 0.67$. The azeotrope will distill over at $T_{azeo}$ until component 2 is exhausted from the liquid phase, at which point pure component 1 will distill over at $T_{b1}$.

## 12.11 COLLIGATIVE PROPERTIES.

Colligative properties can be loosely defined as those properties of a solution that depend only on the number of moles of solute added. The most important colligative properties are:
(1) freezing-point depression;
(2) boiling-point elevation;
(3) osmotic pressure.

The freezing point of a solution is less than that of the pure solvent by an amount which is proportional to the number of moles of solute per kilogram of solvent. The *cryoscopic constant, $K_f$,* is equal to the number of degrees Kelvin by which the freezing point is depressed per mole of (non-dissociating or non-associating) solute per kilogram of solvent (assuming $\gamma_i = 1$). Cryoscopic constants for some common solvents are given in Table 12.7.

As an example, suppose 6.50 g of a solute was dissolved in 600 g of benzene and the measured freezing point of the solution was 4.87°C. In the same apparatus pure benzene was observed to freeze at 5.52°C. Estimate the molecular weight of the solute. Using the cryoscopic constant for benzene we have

$$K_f m = \Delta T = T_0 - T$$

$$\left( 4.99 \; \frac{\text{deg} \cdot \text{kg solvent}}{\text{mole solute}} \right) \left( \frac{6.50 \text{ g}}{\text{M. Wt.}} \right) \left( \frac{1}{0.600 \text{ Kg}} \right) = (5.52 - 4.87) \text{ deg}$$

$$\text{M. Wt.} = \frac{4.99 \times 6.50}{0.600 \times 0.65} = 83 \text{ g} \cdot \text{mole}^{-1}$$

If the solute on dissolution either ionizes or associates, then the freezing point depression will be larger or smaller, respectively, than that predicted from the cryoscopic constant of the solvent. This effect is apparent from the freezing point depressions of 0.010 M aqueous solutions of the substances listed in Table 12.8.

If a *non-volatile solute* is dissolved in a solvent, the boiling point of the solution is higher than that of the pure solvent by an amount which is proportional to the number of moles of solute per kilogram of solvent. The *ebullioscopic constant, $K_b$,* is equal to the number of degrees Kelvin by which the boiling point of the solution is elevated per mole of (non-dis-

TABLE 12.7
CRYOSCOPIC CONSTANTS FOR SELECTED SOLVENTS

| SOLVENT | $K_f$, deg · kg solvent/mole solute |
|---|---|
| $H_2O$ | 1.86 |
| $CH_3CH_2OH$ | 1.20 |
| $C_6H_6$ | 4.99 |
| $CH_3COOH$ | 3.90 |
| camphor | 40.0 |

**TABLE 12.8**

FREEZING-POINT DEPRESSION OF SELECTED COMPOUNDS IN AQUEOUS SOLUTION

| SUBSTANCE | APPROXIMATE FREEZING-POINT DEPRESSION OF 0.010 M AQUEOUS SOLUTION | NUMBER OF SOLUTION SPECIES PRODUCED PER FORMULA UNIT |
|---|---|---|
| $CH_3CH_2OH$ | 0.019 | 1 |
| NaCl | 0.037 | 2 ($Na^+ + Cl^-$) |
| $K_2SO_4$ | 0.056 | 3 ($2K^+ + SO_4^{2-}$) |
| $CH_3COOH$ | 0.020 | 1 |
| $K_3Co(CN)_6$ | 0.075 | 4 ($3K^+ + Co(CN)_6^{3-}$) |
| $Na_4Fe(CN)_6$ | 0.090 | 5 ($4Na^+ + Fe(CN)_6^{4-}$) |

sociating or non-associating) solute per kilogram of solvent (assuming $\gamma_i = 1$). The ebullioscopic constants for water and nitrobenzene are:

| Solvent | deg·kg solvent/mole solute |
|---|---|
| $H_2O$ | 0.52 |
| nitrobenzene | 5.24 |

A 0.20 molal solution of sucrose in water boils at

$$T - T_0 = T - 373.15 = K_b m = 0.52 \times 0.20 = 0.10$$

$$T = 373.25 \text{ K}$$

In order to clarify how the cryoscopic constant arises, we shall now analyze in more detail the equilibrium between a pure solid $A$ and a solution comprised of the solvent $A$ and a solute $B$,

$$A(s) = A(\text{soln}, X_A)$$

For the equilibrium system, $F = 2 - 2 + 2 = 2$, and therefore, if $P_{tot}$ and $X_A$ are specified, then $T$ (the freezing point) is determined. The equilibrium constant, $K$, is given by

$$K = \frac{a_{A(\text{soln})}}{a_{A(s)}}$$

and at 1 atm total pressure

$$K = a_{A(\text{soln})}$$

If the solvent obeys Raoult's law, then $a_A = X_A$ and

$$K = X_{A(\text{soln})}$$

The dependence of $K$ on $T$ is described, as always, by the van't Hoff equation, which (with $\Delta H° = $ constant) yields (Chapter 11)

$$\ln \frac{K_2}{K_1} = -\frac{\Delta H_A°}{R}\left\{\frac{1}{T_2} - \frac{1}{T_1}\right\} = \frac{\Delta \bar{H}_A°}{R}\left(\frac{T_2 - T_1}{T_1 T_2}\right)$$

where $\Delta \bar{H}_A°$ is the molar enthalpy of fusion of pure solid $A$ to the pure liquid. If $T_1 = T_m$, the freezing point of the pure liquid, then $K_1 = 1$, and (given the above assumptions)

$$\ln X_A = -\frac{\Delta \bar{H}_A°}{R}\frac{(T_m - T)}{T_m T}$$

where $T$ is the freezing point of the solution having the mole fraction $X_A$. If the freezing point depression is small, then $T \simeq T_m$ and $(X_A + X_B = 1)$

$$\ln X_A \simeq -\frac{\Delta \bar{H}_A°}{RT_m^2}(T_m - T) \simeq \ln(1 - X_B)$$

If $X_B \ll 1$ (dilute solution), then[*] $\ln(1 - X_B) \simeq -X_B \simeq -\dfrac{1000\,m}{M_A}$, and

$$T_m - T = \left\{\frac{RT_m^2 M_1}{1000\,\Delta \bar{H}_A°}\right\} m = K_f m$$

The quantity in brackets is the cryoscopic constant of the solvent, and it can be used to compute $K_f$ values. From the thermodynamic analysis, it is evident that the reason the freezing point is depressed is that the presence of the solute lowers the chemical potential of the solvent at the normal freezing point below that of the pure solid, and the equality between the chemical potential of pure solid $A$ and that of $A$ in solution is re-established by lowering the temperature.

What happens when the roles of $A$ and $B$ are reversed? That is, consider a solution in which $B$ is the major component ($X_B > X_A$), and suppose that as the temperature is decreased the substance that first separates (freezes) out of the solution is pure $B$. The equilibrium of interest is then

$$B(s) = B(\text{soln}, X_B)$$

A strictly analogous development to that given above yields

$$\ln X_B = -\frac{\Delta \bar{H}_B°}{R}\left\{\frac{1}{T} - \frac{1}{T_{mB}}\right\}$$

where $T_{mB}$ is the normal melting point of pure $B$. This result is to be compared with

$$\ln X_A = -\frac{\Delta \bar{H}_A°}{R}\left\{\frac{1}{T} - \frac{1}{T_{mA}}\right\}$$

which is applicable when $A$ freezes out.

---

[*] For $y \ll 1$, $e^{-y} \simeq 1 - y$. Taking the logarithm of both sides of this expression yields $-y \simeq \ln(1 - y)$.

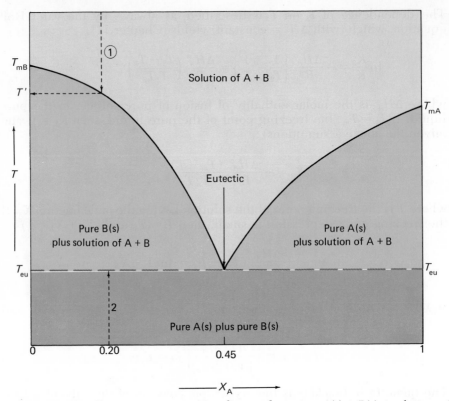

**FIGURE 12.9** Temperature-composition diagram for system A(s) + B(s) + solutions of (A + B) with a eutectic at $X_A = 0.45$.

These two expressions, when plotted on a temperature-composition diagram, yield a result like that depicted in Figure 12.9. The two freezing-point depression curves intersect and terminate at the *eutectic point*, where both pure solids separate out simultaneously from the solution. The phase diagram depicted in Figure 12.9 is particularly instructive in that it allows us to analyze, in a pictorial fashion, phenomena such as mixed melting points, purification by recrystallization, and zone refining. For example, if we have a solution of A + B with $X_A = 0.20$, then as the temperature of the solution is lowered (dashed curve (1) in Figure 12.9), the system remains homogeneous until the freezing-point depression curve is intersected at $T'$. At this temperature pure B(s) begins to separate out of the solution. A further decrease in temperature leads to further separation of B(s) and an enrichment of the solution in A; the temperature and composition of the solution move down along the freezing-point depression curve toward the eutectic. When the eutectic temperature, $T_{eu}$, is reached, both pure B(s) and pure A(s) separate out *simultaneously* from the solution, and with further removal of energy as heat the two solids continue to separate out at $T_{eu}$ until the entire liquid solution is exhausted. Further cooling merely results in a lowering of the temperature of the *heterogeneous* mixture of A(s) + B(s). If we now reheat this heterogeneous mixture of A(s) + B(s), we travel back along exactly the reverse of the path in the cooling experiment. In other words, as the temperature rises (dashed curve (2) in Figure 12.9) there is no phase change until $T_{eu}$ is reached, at which point the eutectic melt is formed. Continued input of

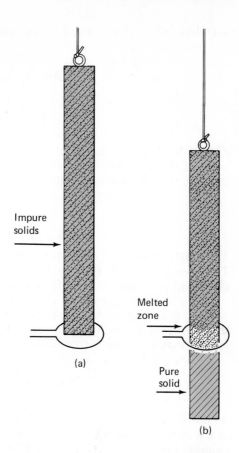

**FIGURE 12.10**  Zone refining.

heat leads to the formation of more eutectic melt at $T_{eu}$ until the $A(s)$ phase is completely exhausted, at which point $T$ begins to rise again as we move up the freezing-point depression curve. The value of $X_{A(soln)}$ decreases as $T$ increases, owing to the dissolution of $B(s)$ in the solution. The last $B(s)$ melts at $T'$. The melting point range is from $T_{eu}$ to $T'$.

An interesting and important application of the thermodynamic principles involved in a phase diagram of the type depicted in Figure 12.9 is *zone refining*. Zone refining is used to produce solids of extremely high purity, especially silicon and germanium for use in transistors. In one version of this method, the solid is placed in a tube which is mounted in the vertical position with an electrical heating loop around the tube (see Figure 12.10). The tube is lowered slowly through the heated loop, which melts the solid in the section of the tube enclosed by the loop. As the tube progresses downward, the pure solid begins to form from the melt of the base of the molten zone. Most of the impurities remain in the molten zone (solution) that is moved upward through the tube. The impurities are thus concentrated in the moving melting zone, and are moved to the top of the tube. The process can be repeated as often as desired to attain the necessary purity. The success of the method depends on the separation of pure solid from the melt. If the formation of solid solutions rather than pure solid is thermodynamically favored, then zone refining will not yield the pure solid.

The thermodynamic analysis of boiling-point elevation is similar (but not identical) to that of the freezing-point depression phenomenon. In this case the equilibrium is

$$A(\text{soln}, X_A) = A(\text{g})$$

$$K = \frac{a_{A(\text{g})}}{a_{A(\text{soln})}} = \frac{P_A}{X_A \gamma_A} = \frac{P_A}{X_A}$$

where the first equality holds if the vapor is ideal, and the last equality holds if the solution is ideal. If the solute is non-volatile, and if we are interested in the normal boiling point, then $P_A = 1.00$ atm and

$$K \simeq \frac{1}{X_A}$$

The smaller the value of $X_A$, the higher the temperature at which $P_A = 1$ atm. The remainder of the analysis, including the identification of the ebullioscopic constant, is left as an exercise.

The *osmotic pressure*, $\Pi$, is defined as the pressure that must be applied to a solution to raise the chemical potential of the solvent to that of the pure solvent at the same temperature. The osmotic pressure is given approximately by the expression

$$\Pi \simeq \frac{-RT}{\bar{V}_A^{\circ}} \ln a_A \simeq \frac{-RT}{\bar{V}_A^{\circ}} \ln X_A \gamma_A \simeq \frac{-RT}{\bar{V}_A^{\circ}} \ln X_A$$

where $\bar{V}_A^{\circ}$ is the molar volume of the solvent. The mole fraction of $H_2O$ in sea water is about 0.974, and assuming* $\gamma_{H_2O} = 1.0$, we compute for the osmotic pressure of sea water at 17°C

$$\Pi = -\frac{(82.06 \text{ cm}^3\text{atm} \cdot \text{K}^{-1} \cdot \text{mol}^{-1})(290 \text{ K})(2.303)}{(18.02 \text{ cm}^3 \cdot \text{mol}^{-1})} \log 0.974 = 34.8 \text{ atm}$$

Suppose that we have a rigid (non-pressure-transmitting), semipermeable membrane that will allow the passage of water, but not of electrolytes (see Figure 12.11). If we apply a pressure $P > \Pi$ to the sea water, then $H_2O$ will pass from the right to the left, giving us pure water from sea water. This process is known as *reverse osmosis*, and is one of the most promising and (potentially) inexpensive ways available for augmenting the world's fresh water supply by tapping the essentially inexhaustible supply of sea water.†

---

*If we have $P_{H_2O}$ for the solution, then $a_{H_2O(\text{soln})} = P_{H_2O}/P_{H_2O}^{\circ}$.
†Portable reverse-osmosis units are commercially available.

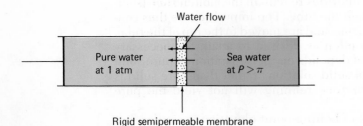

Pure water at 1 atm    Water flow    Sea water at $P > \pi$

Rigid semipermeable membrane

**FIGURE 12.11**   Reverse osmosis.

The osmotic pressure phenomenon is essential for the very existence of cellular life as we know it, in that it acts to keep living cells inflated. Cell walls are semipermeable membranes that allow the passage of water. The activity of water inside the cell is less than that in the intercellular fluids at the same pressure. Therefore, water passes spontaneously from the intercellular fluids into the cell until the osmotic pressure within the cell reaches the value where the equilibrium between inner and outer cell water is established. The osmotic pressure so developed within the cell keeps it inflated. When the osmotic pressure is attained there is no net transfer of water between the inner and outer cellular fluids. Living cells can be ruptured by reducing the solute concentration in the intercellular fluids, and collapsed by increasing the solute concentration in the inter-cellular fluids.

**PROBLEMS***

1. Calculate the solubility of $AgCl(s)$ at 25°C:

   (a) in $H_2O$
   (b) in a solution 0.70 M in $NaCl(aq)$
   (c) in a solution 0.70 M in $NaCl(aq)$ and 0.05 M in $CaCl_2(aq)$
   (d) in a solution 0.10 M in $AgF(aq)$

2. Calculate the solubility of $Hg_2Cl_2(s)$ at 25°C:

   (a) in $H_2O$
   (b) in a solution 0.010 M in $Hg_2(NO_3)_2(aq)$
   (c) in a solution 0.70 M in $CaCl_2(aq)$

3. Calculate the equilibrium constant for the reaction

$$Ag_2SO_4(s) + 2Br^-(aq) = 2AgBr(s) + SO_4^{2-}(aq)$$

4. Estimate the freezing point of sea water, given that the ionic strength of sea water is about 0.67 and that sea water is predominantly composed of monovalent ions ($Cl^-$, $Na^+$, $K^+$) in water.

5. Estimate the concentration of ethylene glycol in radiator water necessary to prevent the formation of ice down to a temperature of $-32°F$.

6. Estimate the boiling point and the osmotic pressure of the solution described in problem 5.

7. Suppose that a given solid is known to be one of three solids, all with the same melting point. Explain how you would determine which of the three solids is identical with the unknown.

8. Consider the following chemical reaction:

$$Ag_2SO_4(s) + 4CN^-(aq) = 2Ag(CN)_2^-(aq) + SO_4^{2-}(aq)$$

*See Table 12.2 for $pK_a$ values and Table 12.5 for $K_{sp}$ values.

Will the solubility of $Ag_2SO_4(s)$ be increased, decreased, or unchanged by:

(a) increasing the concentration of NaCN
(b) decreasing the amount of $Ag_2SO_4(s)$
(c) decreasing the $pH$ of the solution
(d) increasing the $pH$ of the solution
(e) addition of $Ba(NO_3)_2(aq)$
(f) addition of KI(aq)
(g) increasing the temperature ($\Delta H° < 0$)
(h) addition of $KNO_3(aq)$
(i) increasing the total pressure

9. Consider the following chemical reaction:

$$PbI_2(s) + 4OH^-(aq) = Pb(OH)_4^{2-}(aq) + 2I^-(aq)$$

Will the concentration of $Pb(OH)_4^{2-}(aq)$ be increased, decreased or un-affected by:

(a) increasing the concentration of $OH^-(aq)$
(b) addition of $HNO_3(aq)$
(c) decreasing the amount of $PbI_2(s)$
(d) decreasing the concentration of $I^-(aq)$
(e) decreasing the temperature ($\Delta H° < 0$)
(f) increasing the total pressure ($\Delta V° > 0$)
(g) addition of $NaClO_4(aq)$
(h) decreasing the $PbI_2(s)$ particle size

10. The osmotic pressure of an aqueous solution containing 0.250 g of a certain solute was observed to be 5.0 atm at 25°C. Estimate the molecular weight of the solute.

11. Given that $K = 0.05$ for the reaction

$$Zn(OH)_2(s) + 2OH^-(aq) = Zn(OH)_4^{2-}(aq)$$

compute the solubility of $Zn(OH)_2(s)$ in 0.25 M KOH(aq).

12. Calculate the $pH$ at which $Ca(OH)_2(s)$ will begin to precipitate from a solution $3 \times 10^{-2}$ M in $Ca^{2+}(aq)$ at 25°C.

13. Calculate the solubility of $AgOCOCH_3(s)$ in an aqueous solution at 25°C buffered at $pH = 4.0$.

14. How would you distinguish a eutectic from a pure solid?

15. Sharpness of melting point is often used as a criterion of purity for a solid. How do you reconcile this with what is predicted for the melting point range of a solid mixture of $A(s) + B(s)$ with $X_A = 0.999$ by Figure 12.4?

16. Calculate (Table 12.5) the equilibrium constant for the reaction

$$Ag_2SO_4(s) + Ca^{2+}(aq) = CaSO_4(s) + 2Ag^+(aq)$$

Calculate the equilibrium values of $(Ag^+)$, $(Ca^{2+})$, and $(SO_4^{2-})$ when excess $CaSO_4(s)$ is equilibrated with 0.100 M $AgNO_3(aq)$.

17. A 5.35 g sample of aniline, $C_6H_5NH_2$, was dissolved in 250 grams of an organic solvent (freezing point 12.50°C) and the observed freezing point of the solution was 10.75°C. Calculate the cryoscopic constant of the solvent. Estimate the heat of fusion per kilogram, $\Delta H_{kg}$, for the solvent ($R = 1.99$ cal/K mole).

18. Excess $Pb_2Fe(CN)_6(s)$ was equilibrated with an aqueous solution of LiI. The equilibrium concentrations of $I^-(aq)$ and $Fe(CN)_6^{4-}(aq)$ in the aqueous phase were found by chemical analysis to be 0.568 M and 0.108 M, respectively. Estimate the $K_{sp}$ of $PbI_2(s)$.

19. A 50.0 ml sample of 0.100 M NaCl(aq) was mixed with 25.0 ml of 0.250 M $AgNO_3(aq)$ at 25°C. Compute the concentrations of all the species in the aqueous phase at equilibrium.

20. Calculate the $pH$ of an aqueous solution saturated with $Ag_2S(s)$ at 25°C.

21. A 50.0 ml sample of a solution 0.100 M in NaCl and $1.00 \times 10^{-4}$ M in $Na_2S(aq)$ buffered at $pH = 8.0$ is added to 50 ml of a solution 0.050 M in $AgNO_3(aq)$. Compute the concentrations of all the species present at equilibrium.

22. Calculate the solubility of AgBr(s) in a 0.25 M solution of $Na_2S_2O_3(aq)$.

$$Ag^+(aq) + S_2O_3^{2-}(aq) = AgS_2O_3^-(aq) \qquad K_1 = 6.6 \times 10^8$$
$$AgS_2O_3^-(aq) + S_2O_3^{2-}(aq) = Ag(S_2O_3)_2^{3-}(aq) \qquad K_2 = 4.4 \times 10^4$$

23. Calculate the solubility of $PbI_2(s)$ in water at 25°C using the Debye-Hückel equation to estimate $\gamma_\pm$. Repeat the calculation at $I_{tot} = 1.00$.

24. Calculate the concentrations of the various species and the $pH$ of a 0.100 F $CH_3COOH(aq)$ solution at 25°C assuming: (a) $\gamma_\pm = 1$; (b) $\gamma_\pm \neq 1$, no added electrolytes; and, (c) $I_{tot} = 1.00$.

25. A solution 0.30 M in $H^+(aq)$, containing $Ag^+(aq)$, $Cd^{2+}(aq)$, and $Fe^{2+}(aq)$, all at 0.01 M, is saturated with $H_2S(g)$ at 0.50 atm and 25°C. Compute the concentrations of the following species at equilibrium: $Ag^+(aq)$, $Cd^{2+}(aq)$, $Fe^{2+}(aq)$, $S^{2-}(aq)$, $H_2S(aq)$. Discuss the use of sulfide precipitations as a separation method for metal ions.

26. Given the following data (all at 25°C), calculate the equilibrium constants at 25°C for the reactions

$$I_2(s) + I^-(aq) = I_3^-(aq) \tag{1}$$
$$I_2(aq) + I^-(aq) = I_3^-(aq) \tag{2}$$

Solubility of $I_2(s)$ in pure water = $\qquad$ 0.00132 M

Solubility of $I_2(s)$ in 0.100 M $NaNO_3(aq)$ = 0.00130 M

Solubility of $I_2(s)$ in 0.100 M KI(aq) = $\qquad$ 0.05135 M

27. Eggshells, limestone, pearls, oyster shells, coral, and the shells of many other marine organisms are calcium carbonate, $CaCO_3(s)$. The deposition of $CaCO_3(s)$ in various organisms is dependent upon complex biochemical processes occurring in the secreting organs of the animal. A first step in the investigation of any such physiological process is to acquire an understanding of the basic inorganic chemistry involved.

(a) Calculate the equilibrium constant for the reaction

$$Ca^{2+}(aq) + 2HCO_3^-(aq) = CaCO_3(s) + H_2O(\ell) + CO_2(g)$$

(b) The purpose of this problem is to check to see whether or not the chemical composition of the body fluids of, say, a hen are consistent with the precipitation of calcium carbonate. The partial pressure of $CO_2$ in the exhaled air of a hen is estimated as $6 \times 10^{-3}$ atm, the concentration of $Ca^{2+}(aq)$ in the body fluids is about $10^{-3}$ M, and the $pH$ of the body fluids is about 7.5. Show that these data are consistent with the precipitation of $CaCO_3(s)$. Show your calculations and explain your reasoning.

28. Excess $HgI_2(s)$ was equilibrated with a solution initially 0.10 M in $KI(aq)$. Calculate the solubility of $HgI_2(s)$ given

$$HgI_2(s) = Hg^{2+}(aq) + 2I^-(aq) \qquad K_{sp} = 2.0 \times 10^{-28}$$
$$HgI_2(s) + I^-(aq) = HgI_3^-(aq) \qquad K_3 = 1.3 \times 10^{-4}$$
$$HgI_2(s) + 2I^-(aq) = HgI_4^{2-}(aq) \qquad K_4 = 0.79$$

29. Construct a $P$-$X$ diagram at 25°C for a two-component, ideal solution with $P_1° = 250$ Torr, $P_2° = 350$ Torr.

30. Derive equation (12.9d), Section 12.10.

31. Using the data $P_1° = 100$ Torr and $P_2° = 200$ Torr, construct Figure 12.5. What would the figure look like if $P_1° = P_2°$?

32. Plot $P_{tot}$, $P_2$, and $P_1$ versus $X_2$, and $P_{tot}$ versus $X_2$ using data in Table 12.6.

33. Given that the enthalpies of vaporization of components 1 and 2 are $\Delta \bar{H}_1° = 10.0$ kcal mole$^{-1}$ and $\Delta \bar{H}_2° = 15.0$ kcal mole$^{-1}$, construct a $T$-$X$ diagram from the system in problem 29 at $P_{tot} = 760$ Torr.

34. Discuss the use of $NaCl(s)$ to remove (or prevent the formation of) ice on road surfaces. Under what conditions is this an effective procedure? What are the disadvantages?

35. The solubility of the salt $Ag_3X$ in 0.500 M $NaClO_4$ solution was found by analytical methods to be $3.5 \times 10^{-3}$ M. Taking into account activity coefficients, calculate the solubility product of $Ag_3X$.

36. Taking into account activity coefficients, calculate the $pH$ of a solution that is 0.150 M in HCl and 0.250 M in NaCl. Compare your answer with the $pH$ calculated for this solution neglecting activity coefficients.

37. (a) Calculate the solubilities of $ZnS(s)$ and $FeS(s)$ in aqueous solutions buffered at $pH$ values of 0.5, 2.5, and 4.5.
   (b) Calculate the same solubilities in pure water.

38. Calculate the solubilities of $Fe(OH)_3(s)$ and $Zn(OH)_2(s)$ in aqueous solutions buffered at $pH$ values of 2.0, 5.0, and 8.0.

39. Calculate the solubility of CuS in water (a) neglecting hydrolysis of $S^{2-}(aq)$, and (b) including hydrolysis of $S^{2-}(aq)$. In part (b), begin by computing $K$ for the reaction

$$CuS(s) + H_2O(\ell) = Cu^{2+}(aq) + HS^-(aq) + OH^-(aq)$$

Next, show that if $(S^{2-})$ is very small, then $(OH^-) \simeq 1 \times 10^{-7}$ M and $(Cu^{2+}) = 2(HS^-)$. Use these results to compute the solubility and compare with your result in part (a).

_____ **References**

12.1. J. H. Hildebrand and R. E. Powell, *Principles of Chemistry,* 7th Edition (The Macmillan Co., New York, 1964).

12.2. J. N. Butler, *Ionic Equilibrium: A Mathematical Approach* (Addison-Wesley Publishing Co., Reading, Mass., 1964).

12.3. P. A. Rock, *Chemical Thermodynamics: Principles and Applications* (The Macmillan Co., New York, 1969).

The reader interested in the application of thermodynamics to biochemical problems should consult the following introductory texts:

12.4. I. M. Klotz, *Energy Changes in Biochemical Reactions* (Academic Press, New York, 1967).

12.5. A. L. Lehninger, *Bioenergetics* (W. A. Benjamin, Inc., New York, 1965).

# CHAPTER 13

# ELECTROCHEMISTRY

## 13.1 INTRODUCTION.

The investigation of chemical reactions in electrochemical cells offers several advantages over direct equilibrium investigations. The most important advantage to the study of reactions in electrochemical cells is that we, and not nature, get to choose the values of the activities of the various reactants and products. That is to say, if the cell reaction involves a gaseous species, then we can set the pressure of the gas at any convenient value, whereas, if the cell reaction involves a solution species, then we can set the concentration of that species at any convenient value. This is to be contrasted with direct equilibrium studies for which, in general, the equilibrium values of the activities of the reactants and products must be determined by chemical analysis. The analytical problems encountered in the analysis of complex equilibrium reaction mixtures, or the determination of extremely small concentrations, can be formidable.

In principle, any type of reversible chemical reaction can be studied in an electrochemical cell, and certain types of reactions that are very difficult to study by direct equilibrium measurements are easy to study in electrochemical cells. For example, the direct determination of $K_{sp}$ for the reaction

$$\text{AgI(s)} = \text{Ag}^+\text{(aq)} + \text{I}^-\text{(aq)}$$

constitutes a difficult analytical problem because of the low solubility of AgI ($K_{sp} = 8.5 \times 10^{-17}$, $s = 9.2 \times 10^{-9}$ M). It is not a difficult experimental problem, however, to set up an appropriate electrochemical cell to obtain an accurate value for the $K_{sp}$ of AgI(s).

---

*W. H. Nernst, *Nobel Lectures, Chemistry* 1901–1921 (Elsevier Publishing Co., Amsterdam, 1966), pp. 355 and 357.

For a cell the Gibbs phase rule takes the form

$$F = C - \phi + 3$$

as opposed to $F = C - \phi + 2$ for a reaction run in, say, a beaker. The extra degree of freedom arises from the existence of an additional independent, intensive thermodynamic coordinate, namely, the cell voltage. This means that for a reaction involving only pure condensed phases, for example

$$\text{Th(s)} + 2\text{CaO(s)} = \text{ThO}_2\text{(s)} + 2\text{Ca(s)}$$

we have $F = (4-1) - 4 + 3 = 2$ when the reaction is run inside the cell, as opposed to $F = (4-1) - 4 + 2 = 1$ when the reaction is run outside of the cell. Therefore, when this reaction is run in a cell we get to pick *both* the equilibrium temperature and pressure, and this choice of temperature and pressure determines the cell voltage, $\mathcal{E}$. When this reaction is run outside the cell we get to pick the equilibrium temperature *or* the equilibrium pressure, *but not both.* If we choose, say, $P_{eq} = 1$ atm, then there is only one value of $T$ at which equilibrium can be attained, and this equilibrium value of $T$ may be an experimentally inconvenient one.

## 13.2   CONSTRUCTION AND OPERATION OF ELECTROCHEMICAL CELLS.

Consider the chemical reaction

(a)  $\text{Ag(s)} + \text{HI(aq)} = \text{AgI(s)} + \dfrac{1}{2}\,\text{H}_2\text{(g)}$

This complete chemical reaction can be broken down into two half-reactions, the oxidation half-reaction and the reduction half-reaction (Chapter 8):

(b)  $\text{Ag(s)} + \text{I}^-\text{(aq)} = \text{AgI(s)} + \text{e}^-$     *oxidation*

(c)  $\text{H}^+\text{(aq)} + \text{e}^- = \dfrac{1}{2}\,\text{H}_2\text{(g)}$     *reduction*

Each of these half-reactions can be made to take place at one of the electrodes in an electrochemical cell. The *cell diagram* for the cell in which reaction (a) is the cell reaction is

$$\text{Ag(s)} \mid \text{AgI(s)} \mid \text{HI(aq)} \mid \text{H}_2\text{(g, Pt)}$$

where the vertical bars indicate phase boundaries; platinum, Pt, provides the surface on which reaction (c) takes place. This platinum also serves as the electrical connection (electron pathway) from the hydrogen electrode to the external measuring circuit.

A schematic diagram of the cell construction is shown in Figure 13.1. The cell voltage is measured with a potentiometer and a null detector, by finding the voltage that must be placed *in opposition* to the cell voltage to prevent any net current flow. During the operation of the cell as an energy source, the current within the metallic electrodes and external measuring circuit is carried by moving electrons, whereas the current between the

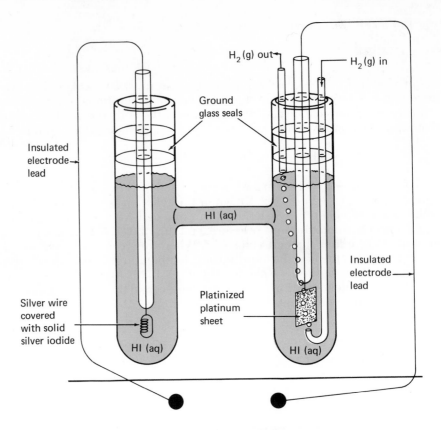

**FIGURE 13.1**    Schematic of the cell: Ag(s) | AgI(s) | HI(aq) | H₂(g, Pt). The HI(aq) solution in the vicinity of the silver-silver iodide electrode is saturated with silver iodide. The HI(aq) solution in the vicinity of the Pt electrode is saturated with hydrogen at a known value of $P_{H_2}$.

two electrodes *within the cell itself* is carried by the ions $H^+$(aq) and $I^-$(aq) moving (in opposite directions) through the solution of HI(aq). In other words, the cell reaction is the source of the energy obtainable from the cell.

In the thermodynamic analysis of cell data, it is conventional to *assume that oxidation takes place at the left electrode* (and, consequently, reduction takes place at the right electrode) *in the cell diagram as written.* (Compare the half-reactions given above with the cell diagram.) This convention is equivalent to the assumption that *negative current* flows through the cell from right to left, and through the external circuit from left to right. Electrons flow spontaneously from a region of negative potential to a region of positive potential, whereas positive current flows spontaneously from a region of positive potential to a region of negative potential. (A negative charge moving in one direction is equivalent to a positive charge moving in the opposite direction.) Electrode polarities serve to define the direction of *external* spontaneous current flow. A negative electrode polarity does not mean that the electrode carries a negative charge. During the spontaneous discharge of a cell, positive ions are produced at the negative electrode and negative *ions* in the solution will move spontaneously toward the negative electrode.

If the *cell reaction* as written proceeds spontaneously from left to right under the conditions prevailing in the cell, then the left electrode will be the negative electrode (and the right electrode the positive electrode), whereas if the cell reaction as written proceeds spontaneously from right to left, then the right electrode will be the negative electrode. The experimentally-determined cell polarity is unambiguous and independent of any arbitrary assumptions about the direction of current in the cell, or the spatial orientation of the cell in the laboratory with respect to the experimenter.

The equation that describes the dependence of the cell voltage on the *activities* of the reactants and products is the Nernst equation. At a given temperature and total pressure, the Gibbs energy change for a reaction taking place in a reversible electrochemical cell is (see Section 10.1)

$$\Delta G = -\mathscr{E}Z$$

where $\mathscr{E}$ is the cell voltage (electromotive force or emf) and $Z$ is the quantity of charge that has passed through the cell. For the occurrence of the cell reaction *as written*, the quantity of charge that has passed through the cell is $Z = nF$, where $F$ is *Faraday's constant*

$$\boxed{F = 96{,}497.2 \ \frac{\text{coulombs}}{\text{mole of electrons}}}$$

and $n$ is the number of moles of electrons transferred from the reducing agent to the oxidizing agent in the balanced cell reaction as written (see Section 8.5). Therefore,

$$\boxed{\Delta G = -nF\mathscr{E}} \tag{13.1}$$

Combining this equation with Equation (11.23)

$$\Delta G = \Delta G° + RT \ln Q \tag{11.23}$$

yields

$$-nF\mathscr{E} = -nF\mathscr{E}° + RT \ln Q$$

or

$$\boxed{\mathscr{E} = \mathscr{E}° - \frac{RT}{nF} \ln Q} \tag{13.2}$$

Equation (13.2) is the **Nernst equation**. In this equation $\mathscr{E}$ is the *observed cell voltage*, and $\mathscr{E}°$ is the *standard cell voltage*, that is, $\mathscr{E}°$ is the voltage that the cell would have if all the reactants and products were in their unit activity standard states (whence $Q = 1$). The observed cell voltage is positive if the left electrode in the cell diagram is the negative electrode, whereas the observed cell voltage is the negative if the right electrode in the cell diagram is the negative electrode.

The standard cell voltage is related to the equilibrium constant for the reaction via the equation

$$\Delta G° = -nF\mathscr{E}° = -RT \ln K$$

$$\boxed{\mathscr{E}^\circ = \frac{RT}{nF} \ln K}$$  (13.3)

At 25°C the Nernst equation becomes ($\mathscr{E}$ in volts)

$$\mathscr{E} = \mathscr{E}^\circ - \frac{0.05916}{n} \log Q$$

because $RT \ln 10/F = 0.05916$ volt·mole of electrons. (Note the change from natural to base 10 logarithm.) Application of the Nernst equation at 25°C to the cell reaction

$$\text{Ag(s)} + \text{HI(aq)} = \text{AgI(s)} + \frac{1}{2} \text{H}_2\text{(g)} \quad (n=1)$$

yields

$$\mathscr{E} = \mathscr{E}^\circ - 0.05916 \log \left\{ \frac{a_{\text{H}_2\text{(g)}}^{1/2} a_{\text{AgI(s)}}}{a_{\text{HI(aq)}} a_{\text{Ag(s)}}} \right\}$$

At a *total* pressure of about 1 atm, $a_{\text{Ag(s)}} = a_{\text{AgI(s)}} = 1.00$, and

$$\mathscr{E} = \mathscr{E}^\circ - 0.05916 \log \left\{ \frac{a_{\text{H}_2\text{(g)}}^{1/2}}{a_{\text{HI(aq)}}} \right\}$$

If the value of $P_{\text{H}_2}$ is about 1 atm or less, then it is a good approximation to take $a_{\text{H}_2\text{(g)}} = P_{\text{H}_2}$; HI(aq) is a strong electrolyte and thus

$$a_{\text{HI(aq)}} = a_{\text{H}+} a_{\text{I}-} = (m_{\text{H}+})(m_{\text{I}-})\gamma^2_{\pm\text{(HI)}} = m^2 \gamma^2_{\pm}$$

where the last equality holds if the only significant source of $\text{H}^+$ and $\text{I}^-$ is HI(aq).[*] Substitution of these results into the above expression yields

$$\mathscr{E} = \mathscr{E}^\circ - 0.05916 \log \left\{ \frac{P_{\text{H}_2}^{1/2}}{m^2 \gamma^2_{\pm}} \right\}$$

This equation gives the dependence of the cell voltage on the partial pressure of $\text{H}_2$, and the concentration of HI(aq). To use this equation to calculate $\mathscr{E}$ we need to know $\mathscr{E}^\circ$ for the cell reaction. The value of $\mathscr{E}^\circ$ can be obtained from cell measurements at very low concentrations, where $\gamma_\pm$ can be estimated from the Debye-Hückel equation, $\log \gamma_\pm = -0.511 m^{1/2}/(1 + m^{1/2})$. The calculations yield $\mathscr{E}^\circ = 0.151$ V (25°C). The value of $\mathscr{E}^\circ$ obtained in this way can then be used, in turn, to calculate the values of $\gamma_\pm$ for HI(aq) from the cell data at higher concentrations, or to compute the cell voltage for particular values of $P_{\text{H}_2}$ and $m$. For example, if $P_{\text{H}_2} = 0.81$ atm and $m = 0.010$, then (assuming $\gamma_\pm \simeq 1$) we compute

$$\mathscr{E} \simeq 0.151 \text{ V} - 0.05916 \log \frac{(0.81)^{1/2}}{(0.01)^2} \simeq -0.083 \text{ V}$$

---

[*]Writing $m_{\text{H}+} m_{\text{I}-} = m^2$ involves the assumption that the solubility of AgI in the cell electrolyte constitutes a negligible source of $\text{I}^-$ compared to HI(aq).

The negative value of $\mathscr{E}$ indicates that the cell reaction is spontaneous in the direction opposite to that written under the above conditions of $m_{HI}$ and $P_{H_2}$.

Suppose that the pressure of $H_2(g)$ is held fixed at 1.00 atm and the above cell is allowed to attain equilibrium by shorting the two electrode leads. What will be the equilibrium value of the concentration of $HI(aq)$?

The relationship between the cell voltage and the value of $Q/K$ is given by the Nernst equation (see Equations (13.2) and (13.3))

$$\mathscr{E} = -\frac{RT}{nF} \ln (Q/K)$$

With the aid of Le Chatelier's principle and the Nernst equation we conclude that:

(a) If $Q/K < 1$,     Then the reaction is spontaneous left-to-right, and       $\mathscr{E} > 0$.

(b) If $Q/K > 1$,     Then the reaction is spontaneous right-to-left, and       $\mathscr{E} < 0$.

(c) If $Q/K = 1$,     Then the reaction is at equilibrium, and       $\mathscr{E} = 0$.

Therefore, we can write the equilibrium condition for the above cell as

$$\mathscr{E} = 0 \simeq 0.151 - 0.0592 \log \left\{ \frac{P_{H_2}^{1/2}}{(H^+)(I^-)} \right\} = 0.151 - 0.0592 \log \left\{ \frac{1.00}{(H^+)(I^-)} \right\}$$

Thus

$$\log (H^+)(I^-) \simeq -\frac{0.151}{0.0592} = -2.551$$

$$(H^+)(I^-) \simeq (H^+)^2 \simeq 28.1 \times 10^{-4}$$

or

$$(H^+) \simeq 0.053 \text{ M}$$

One of the basic reasons for the study of chemical reactions in electrochemical cells is that the cell enables us (a) to choose the value of $Q$ for the reaction, and (b) to determine directly the magnitude of the "driving force" for the reaction (i.e., the value of $\mathscr{E}$) under conditions where $Q \neq K$. We are able to do this for reactions run in cells because the design of the cell forces the reaction to proceed via the external circuit. We control the transfer of electrons from one electrode to the other (i.e., we can make the reaction proceed left-to-right or right-to-left) by means of an adjustable external voltage placed in opposition to the voltage of the cell, and a switch on the potentiometer which is either open (no current flow possible) or closed (current flow possible if $\mathscr{E}_{cell} \neq \mathscr{E}_{external}$). On the other hand, a reaction run in, say, an open beaker proceeds spontaneously and in an uncontrolled manner toward equilibrium.

The design of the cell must be such as to force the reaction to proceed via the transfer of electrons through the external circuit, under the conditions prevailing in the cell. The reaction usually can be forced to proceed via the external circuit by separating the reactant and product species, which would react if they were in direct contact, into the two separate cell compartments. For example, the reaction

$$Cd(s) + Hg_2SO_4(s) = CdSO_4(aq) + 2Hg(\ell)$$

has an $\mathscr{E}° = 1.018$ V at 25°C. Therefore,

$$\mathscr{E} \simeq 1.018 - \frac{0.0592}{2} \log (Cd^{2+})(SO_4^{2-})$$

and this reaction is spontaneous left-to-right ($\mathscr{E} > 0$) for any reasonable concentration of $CdSO_4(aq)$ up to and including a saturated aqueous solution of $CdSO_4(aq)$. Therefore, if we mix $Cd(s)$ with $Hg_2SO_4(s)$ in a beaker in the presence of $H_2O$, then the reaction will proceed spontaneously left-to-right. On the other hand, if we set up the cell

$$Cd(s) \mid CdSO_4(aq) \mid Hg_2SO_4(s) \mid Hg(\ell)$$

$$Cd(s) = Cd^{2+}(aq) + 2e^- \qquad \text{(left electrode)}$$

$$Hg_2SO_4(s) + 2e^- = 2Hg(\ell) + SO_4^{2-}(aq) \qquad \text{(right electrode)}$$

then the spontaneous reaction cannot proceed except via the external circuit because the $Cd(s)$ is not in direct contact with the $Hg_2SO_4(s)$.

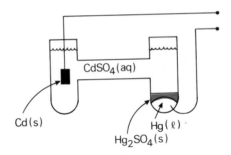

The cell

$$Pb(s) \mid PbSO_4(s) \mid CuSO_4(aq) \mid Cu(s)$$

is *not* operational (it is internally short-circuited) because $Pb(s)$ reacts spontaneously with $CuSO_4(aq)$, and these two phases are in direct contact in the cell.

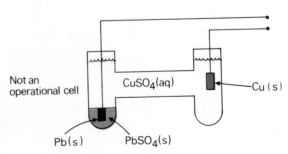

An interesting case of internal short-circuiting occurs in the Ag(s) | AgI(s) | HI(aq) | H₂(g, Pt) cell. The cell reaction

$$Ag(s) + HI(aq) = AgI(s) + \frac{1}{2} H_2(g)$$

is spontaneous left-to-right if the cell voltage is positive. In this cell HI(aq) is in direct contact with Ag(s); consequently, this cell is internally short-circuited for $\mathscr{E} > 0$. However, for $\mathscr{E} < 0$ the reaction is spontaneous right-to-left, and H₂(g) is not in direct contact with AgI(s); therefore, the cell is operational for $\mathscr{E} < 0$, that is, for HI(aq) concentrations less than 0.054 M at $P_{H_2} = 1.00$ atm.

## 13.3 SINGLE ELECTRODE POTENTIALS.

It is not possible to measure the potential of a single electrode by direct thermodynamic methods.* All that can be determined in the laboratory is the difference in potential between two electrodes. It is possible to circumvent this obstacle and set up a table of standard single electrode potentials, if we agree to assign an arbitrarily chosen value to the standard potential of a particular electrode, and then tabulate the values for all other electrodes relative to the arbitrary reference. The major advantage that is gained from tabulating single electrode potentials is that $\mathscr{E}°$ values for complete cells that have not been investigated, but that involve electrodes whose $\mathscr{E}°$ values are known, can be calculated from the single electrode potentials for the two electrodes involved in the cell. The single electrode potential scale is based on the standard hydrogen electrode

$$H^+(aq) + e^- = \frac{1}{2} H_2(g) \tag{13.4}$$

$$\mathscr{E}° = 0 \text{ at all temperatures}$$

The $\mathscr{E}°$ value for a complete cell is equal to the difference between the $\mathscr{E}°$ values for the two electrodes. At this point an ambiguity enters into the problem, because we have two choices. We can take either ($\ell$ = left electrode, $r$ = right electrode)

$$\mathscr{E}°_{cell} = \mathscr{E}°_{ox,\,\ell} - \mathscr{E}°_{ox,r} \quad \text{(standard oxidation potentials)} \tag{13.5}$$

or

$$\boxed{\mathscr{E}°_{cell} = \mathscr{E}°_{red,r} - \mathscr{E}°_{red,\ell}} \quad \begin{array}{l}\text{(standard reduction potentials, or} \\ \text{standard electrode potentials)}\end{array} \tag{13.6}$$

Equations (13.5) and (13.6) show that standard oxidation potentials and standard reduction potentials differ from one another only in sign, that is,

---

*Whether or not single electrode potentials can be determined by extra-thermodynamic methods remains to be seen. No such measurements have been reported, although some suggestions of methods have been made.

$\mathscr{E}_{o.r}^{\circ} = -\mathscr{E}_{red}^{\circ}$. *We shall use the standard electrode (reduction) potential convention in this text.*[*]

From Equation (13.6), and $\mathscr{E}_{cell}^{\circ} = 0.151$ V, we compute the standard electrode potential of the $Ag(s) \mid AgI(s) \mid I^-(aq)$ electrode as

$$\mathscr{E}_{cell}^{\circ} = \mathscr{E}_r^{\circ} - \mathscr{E}_{\ell}^{\circ}$$

$$0.151 = 0 - \mathscr{E}_{\ell}^{\circ}$$

Thus, for the half-reaction

$$AgI(s) + e^- = Ag(s) + I^-(aq)$$

we have

$$\mathscr{E}_{red}^{\circ} = -0.151 \text{ volt } (25°C)$$

By carrying out experiments analogous to those outlined above on other electrodes of interest, a table of standard electrode potentials can be built up (Table 13.1). Once the standard potential of an electrode is known at a particular temperature it can, in turn, be used in cells not involving the hydrogen electrode to determine other standard electrode potentials. For example, at 25°C the measured standard cell voltage for the cell

$$Tl(s) \mid TlI(s) \mid KI(aq) \mid AgI(s) \mid Ag(s)$$

is 0.602 volts. The cell reaction is[†]

$$Tl(s) + AgI(s) = TlI(s) + Ag(s)$$

and

$$\mathscr{E}_{cell}^{\circ} = 0.602 = \mathscr{E}_r^{\circ} - \mathscr{E}_{\ell}^{\circ} = -0.151 - \mathscr{E}_{\ell}^{\circ}$$

or $\mathscr{E}^{\circ} = -0.753$ volt at 25°C for the electrode

$$TlI(s) + e^- = Tl(s) + I^-(aq)$$

An $\mathscr{E}_{cell}^{\circ}$ value can be used to compute the equilibrium constant for the cell reaction, and consequently, to determine whether or not the reaction of interest is spontaneous[‡] in a given direction under the prevailing experimental conditions. In this connection we note that for a reaction taking place at a given temperature and total pressure

$$\Delta G = -nF\mathscr{E} = RT \ln (Q/K)$$

and, if the reaction as written is spontaneous left-to-right, then $\Delta G < 0$, and therefore $\mathscr{E} > 0$. If the reaction is spontaneous right-to-left, then $\Delta G > 0$, and therefore $\mathscr{E} < 0$.

---

[*] To avoid confusion in the meaning of $\mathscr{E}^{\circ}$ values, the I.U.P.A.C. (International Union of Pure and Applied Chemistry) has ruled that standard electrode potentials obtained using Equation (13.5) must be referred to as *standard oxidation potentials*, whereas standard electrode potentials obtained using Equation (13.6) can be referred to as either *standard reduction potentials*, or simply *standard electrode potentials*.

[†] Note that at 1 atm total pressure $\mathscr{E} = \mathscr{E}^{\circ}$ for this cell.

[‡] Recall that *spontaneous* is not synonymous with *immediate*.

As an example of the application of a table of standard electrode potentials, consider the reaction at 25°C between copper metal and an aqueous solution containing silver nitrate:

$$Cu(s) + 2Ag^+(aq) = 2Ag(s) + Cu^{2+}(aq)$$

Will $Cu(s)$ reduce $Ag^+(aq)$ in a solution which is 1.0 M in $AgNO_3(aq)$, and if so, what will be the equilibrium concentrations of $Cu^{2+}(aq)$ and $Ag^+(aq)$ if 1.0 M $AgNO_3(aq)$ is allowed to react with excess $Cu(s)$? From Table 13.1 we obtain the following standard electrode potentials:

(1) $Cu^{2+}(aq) + 2e^- = Cu(s)$ $\qquad \mathscr{E}_1^\circ = +0.337$ volt

(2) $Ag^+(aq) + e^- = Ag(s)$ $\qquad \mathscr{E}_2^\circ = +0.800$ volt

The desired overall reaction is obtained by multiplying the second half-reaction by 2, and then subtracting the first half-reaction from the result. The value of $\mathscr{E}^\circ$ for the whole reaction is[*]

$$\mathscr{E}^\circ = \mathscr{E}_r^\circ - \mathscr{E}_\ell^\circ = \mathscr{E}^\circ(Ag^+ \to Ag) - \mathscr{E}^\circ(Cu^{2+} \to Cu)$$

$$\mathscr{E}^\circ = +0.800 - (+0.337) = +0.463 \text{ V}$$

Substitution of the above $\mathscr{E}^\circ$ value into the Nernst equation yields (at 25°C, and assuming $\gamma_i = 1$)

$$\mathscr{E} \simeq 0.463 - \frac{0.0592}{2} \log \left\{ \frac{(Cu^{2+})}{(Ag^+)^2} \right\}$$

This expression tells us that with $(Cu^{2+}) \sim 0$ and $(Ag^+) \sim 1.0$ M, $\mathscr{E}$ is very large and positive, and therefore the reaction proceeds spontaneously in the direction indicated. Even if the 1.0 M $AgNO_3(aq)$ solution initially contained 2.0 M $Cu^{2+}(aq)$, the reaction would still be spontaneous, because

$$\mathscr{E} \simeq 0.463 - \frac{0.0592}{2} \log \frac{(2.0)}{(1.0)^2} = 0.454 \text{ V} > 0$$

---

[*]Note that the standard electrode potential for the half-reaction

$$2Ag^+(aq) + 2e^- = 2Ag(s)$$

is the same as that for the half-reaction

$$Ag^+(aq) + e^- = Ag(s)$$

Multiplying a half-reaction (or a whole reaction) by some factor does not change $\mathscr{E}^\circ$, because *voltage* (like temperature) *is an intensive quantity*; i.e., it does not depend on the number of moles reacted or produced. Cell voltage is independent of cell size. A two-ton cell of a given type has the same voltage as a one-pounder. However, all other factors being equal, we can obtain 4000 times as much electrical work from the two-ton cell as from the one-pound cell. This is because

$$-w_{max} \text{ (electrical work done by the cell)} = w_{max} \text{ (electrical work done on the cell)} = \Delta G = -nF\mathscr{E}$$

and $\Delta G$ is an extensive quantity; i.e., the value of $\Delta G$ is proportional to the number of moles reacted.

**TABLE 13.1**

STANDARD ELECTRODE POTENTIALS IN VOLTS AT 25°C.

| ACIDIC SOLUTIONS | $\mathscr{E}^{\circ}_{red}$ (V) |
|---|---|
| $Li^{+}(aq) + e^{-} = Li(s)$ | $-3.045$ |
| $K^{+}(aq) + e^{-} = K(s)$ | $-2.925$ |
| $Ca^{2+}(aq) + 2e^{-} = Ca(s)$ | $-2.866$ |
| $Na^{+}(aq) + e^{-} = Na(s)$ | $-2.714$ |
| $Mg^{2+}(aq) + 2e^{-} = Mg(s)$ | $-2.363$ |
| $\frac{1}{2} H_2(g) + e^{-} = H^{-}(aq)$ | $-2.25$ |
| $nH_2O + e^{-} = e^{-}(aq)$ | $(-1.7)$ |
| $Zn^{2+}(aq) + 2e^{-} = Zn(s)$ | $-0.7628$ |
| $Fe^{2+}(aq) + 2e^{-} = Fe(s)$ | $-0.4402$ |
| $Eu^{3+}(aq) + e^{-} = Eu^{2+}(aq)$ | $-0.429$ |
| $Cr^{3+}(aq) + e^{-} = Cr^{2+}(aq)$ | $-0.408$ |
| $Cd^{2+}(aq) + 2e^{-} = Cd(s)$ | $-0.403$ |
| $PbSO_4(s) + 2e^{-} = Pb(s) + SO_4^{2-}(aq)$ | $-0.3588$ |
| $Cd^{2+}(aq) + 2e^{-} = Cd(Hg)(\text{2-phase } 11\% \text{ Cd})$ | $-0.3516$ |
| $Pb^{2+}(aq) + 2e^{-} = Pb(s)$ | $-0.126$ |
| $HgI_4^{2-}(aq) + 2e^{-} = Hg(\ell) + 4I^{-}(aq)$ | $-0.038$ |
| $2H^{+}(aq) + 2e^{-} = H_2(g)$ | $0$ |
| $Cu^{2+}(aq) + e^{-} = Cu^{+}(aq)$ | $+0.153$ |
| $AgCl(s) + e^{-} = Ag(s) + Cl^{-}(aq)$ | $+0.2222$ |
| $Hg_2Cl_2(s) + 2e^{-} = 2Hg(\ell) + 2Cl^{-}(aq)$ | $+0.2680$ |
| $Cu^{2+}(aq) + 2e^{-} = Cu(s)$ | $+0.337$ |
| $Fe(CN)_6^{3-}(aq) + e^{-} = Fe(CN)_6^{4-}(aq)$ | $+0.36$ |
| $Cu^{+}(aq) + e^{-} = Cu(s)$ | $+0.521$ |
| $Hg_2SO_4(s) + 2e^{-} = 2Hg(\ell) + SO_4^{2-}(aq)$ | $+0.6151$ |
| $O_2(g) + 2H^{+}(aq) + 2e^{-} = H_2O_2(aq)$ | $+0.6824$ |
| $Fe^{3+}(aq) + e^{-} = Fe^{2+}(aq)$ | $+0.771$ |
| $Hg_2^{2+}(aq) + 2e^{-} = 2Hg(\ell)$ | $+0.788$ |
| $Ag^{+}(aq) + e^{-} = Ag(s)$ | $+0.7991$ |
| $Pd^{2+}(aq) + 2e^{-} = Pd(s)$ | $+0.987$ |

**TABLE 13.1**

STANDARD ELECTRODE POTENTIALS IN VOLTS AT 25°C. — *Continued*

| ACIDIC SOLUTIONS | $\mathscr{E}^{\circ}_{red}(\mathbf{V})$ |
|---|---|
| $O_2(g) + 4H^+(aq) + 4e^- = 2H_2O(\ell)$ | +1.229 |
| $PbO_2(s) + 4H^+(aq) + 2e^- = Pb^{2+}(aq) + 2H_2O(\ell)$ | +1.455 |
| $PbO_2(s) + SO_4^{2-}(aq) + 4H^+(aq) + 2e^- = PbSO_4(s) + 2H_2O(\ell)$ | +1.682 |
| $H_2O_2(aq) + 2H^+(aq) + 2e^- = 2H_2O(\ell)$ | +1.776 |
| $XeO_3(s) + 6H^+(aq) + 6e^- = Xe(g) + 3H_2O(\ell)$ | +1.8 |
| $Co^{3+}(aq) + e^- = Co^{2+}(aq)$ | +1.808 |
| $S_2O_8^{2-}(aq) + 2e^- = 2SO_4^{2-}(aq)$ | +2.01 |
| $O_3(g) + 2H^+(aq) + 2e^- = O_2(g) + H_2O(\ell)$ | +2.07 |
| $F_2(g) + 2e^- = 2F^-(aq)$ | +2.87 |
| $H_4XeO_6(aq) + 2H^+(aq) + 2e^- = XeO_3(s) + 3H_2O(\ell)$ | +3.0 |
| $F_2(g) + 2H^+(aq) + 2e^- = 2HF(aq)$ | +3.06 |

| BASIC SOLUTIONS | $\mathscr{E}^{\circ}_{red}(\mathbf{V})$ |
|---|---|
| $Ca(OH)_2(s) + 2e^- = Ca(s) + 2OH^-(aq)$ | −3.02 |
| $Mg(OH)_2(s) + 2e^- = Mg(s) + 2OH^-(aq)$ | −2.690 |
| $ZnS(s, wurtzite) + 2e^- = Zn(s) + S^{2-}(aq)$ | −1.405 |
| $2SO_3^{2-}(aq) + 2H_2O + 2e^- = S_2O_4^{2-}(aq) + 4OH^-(aq)$ | −1.12 |
| $Cd(CN)_4^{2-}(aq) + 2e^- = Cd(s) + 4CN^-(aq)$ | −1.028 |
| $2H_2O(\ell) + 2e^- = H_2(g) + 2OH^-(aq)$ | −0.82803 |
| $Cd(OH)_2(s) + 2e^- = Cd(s) + 2OH^-(aq)$ | −0.809 |
| $HgS(s, black) + 2e^- = Hg(\ell) + S^{2-}(aq)$ | −0.69 |
| $Fe(OH)_3(s) + e^- = Fe(OH)_2(s) + OH^-(aq)$ | −0.56 |
| $O_2(g) + e^- = O_2^-(aq)$ | −0.563 |
| $Cu(NH_3)_2^+(aq) + e^- = Cu(s) + 2NH_3(aq)$ | −0.12 |
| $2Cu(OH)_2(s) + 2e^- = Cu_2O(s) + 2OH^-(aq) + H_2O(\ell)$ | −0.080 |
| $HgO(red) + H_2O + 2e^- = Hg(\ell) + 2OH^-(aq)$ | +0.098 |
| $O_2(g) + 2H_2O(\ell) + 4e^- = 4OH^-(aq)$ | +0.401 |
| $2AgO(s) + H_2O(\ell) + 2e^- = Ag_2O(s) + 2OH^-(aq)$ | +0.607 |

Returning to the case where the initial concentration of $Cu^{2+}$ is zero and $(Ag^+) = 1.0$ M, we note that the reaction will proceed until either the $Cu(s)$ is completely consumed, or the reaction attains an equilibrium state. If excess $Cu(s)$ is used, then the reaction will eventually attain equilibrium, in which case $\mathscr{E} = 0$. In other words, if the reaction

$$Cu(s) + 2Ag^+(aq) = Cu^{2+}(aq) + 2Ag(s)$$

were run in an electrochemical cell, then at thermodynamic equilibrium the electrochemical potential drop in the cell, that is, $\mathscr{E}$, is exactly balanced by the difference in the potential of the electrons in the two cell terminal wires. If the circuit is closed, and the cell is allowed to "run down," the cell voltage decreases until $\mathscr{E}_{cell} = 0$, that is, until chemical equilibrium prevails in the cell. Therefore, at equilibrium

$$0 \simeq 0.463 - \frac{0.0592}{2} \log \left\{ \frac{(Cu^{2+})_{eq}}{(Ag^+)^2_{eq}} \right\}$$

From the reaction stoichiometry

$$(Cu^{2+})_{eq} = \frac{(Ag^+)_{initial} - (Ag^+)_{eq}}{2} = \frac{1.0 - (Ag^+)_{eq}}{2}$$

and thus

$$\log \left\{ \frac{1.0 - (Ag^+)_{eq}}{2(Ag^+)^2_{eq}} \right\} \simeq \log K = \frac{2(0.463)}{0.0592} = 15.64$$

or

$$\frac{1.0 - (Ag^+)_{eq}}{2(Ag^+)^2_{eq}} \simeq 4.4 \times 10^{15} = K \simeq \frac{(Cu^{2+})_{eq}}{(Ag^+)^2_{eq}}$$

From this expression, we compute

$$(Ag^+)_{eq} \simeq \left( \frac{0.50}{4.4 \times 10^{15}} \right)^{1/2} = 1.1 \times 10^{-8} \text{ M}$$

and $(Cu^{2+})_{eq} \simeq 0.50$ M. Note that the equilibrium concentration of $Ag^+(aq)$ is very low; this is because $K$ is very large. If the amount of copper metal that was brought into contact with a given volume of 1.0 M $AgNO_3(aq)$ were insufficient to make the solution 0.50 M in $Cu^{2+}(aq)$, then the $Cu(s)$ phase would be *completely* consumed.

    The formation of a slightly soluble solid salt phase makes Ag(I) a much weaker oxidizing agent (more negative standard electrode potential) than $Ag^+(aq)$, and the same is true of complex formation. This effect is apparent in the following $\mathscr{E}°$ values:

| **ELECTRODE** | $\mathscr{E}°_{red}$ **(volt, 25°C)** |
|---|---|
| $Ag^+(aq) + e^- = Ag(s)$ | +0.80 |
| $Ag(NH_3)_2^+(aq) + e^- = Ag(s) + 2NH_3(aq)$ | +0.37 |
| $AgCl(s) + e^- = Ag(s) + Cl^-(aq)$ | +0.22 |
| $Ag_4Fe(CN)_6(s) + 4e^- = Ag(s) + Fe(CN)_6^{4-}(aq)$ | +0.15 |
| $Ag(CN)_2^-(aq) + e^- = Ag(s) + 2CN^-(aq)$ | −0.31 |

In other words, the formation of a complex or insoluble salt of Ag(I) stabilizes this oxidation state relative to Ag metal.

The solubility product of a salt or the equilibrium constant for the complexation of a metal ion can be calculated from the appropriate $\mathscr{E}°$ values. For example, consider the calculation of $K_{sp}$ at 25°C for $Ag_4Fe(CN)_6(s)$, given the following $\mathscr{E}°$ values at 25°C:

(1)  $Ag^+(aq) + e^- = Ag(s)$ $\qquad\qquad\qquad\qquad\qquad \mathscr{E}_1° = +0.800$ V

(2)  $Ag_4Fe(CN)_6(s) + 4e^- = 4Ag(s) + Fe(CN)_6^{4-}(aq)$ $\qquad \mathscr{E}_2° = +0.148$ V

Multiplying reaction (1) by 4 and subtracting the result from reaction (2) yields

$$Ag_4Fe(CN)_6(s) = 4Ag^+(aq) + Fe(CN)_6^{4-}(aq)$$

for which

$$\Delta G° = -4F\,\{(0.148) - (0.800)\} = -RT\ln K_{sp}$$

or

$$\log K_{sp} = \frac{4(-0.652)}{0.05916}$$

and $K_{sp} = 8.6 \times 10^{-45}$. This calculation can be reversed and used to obtain $\mathscr{E}°$ for the silver-silver ferrocyanide electrode, given the $K_{sp}$ for $Ag_4Fe(CN)_6(s)$ and $\mathscr{E}°$ for the $Ag(s)\,|\,Ag^+(aq)$ electrode.

It is sometimes necessary to compute the electrode potential of an electrode from the electrode potentials for other electrodes. Consider the problem of calculating the standard electrode potential at 25°C for the half-reaction

(1)  $Cu^{2+}(aq) + 2e^- = Cu(s)$ $\qquad\qquad\qquad\qquad\qquad \mathscr{E}_1°$

given the $\mathscr{E}°_{red}$ values for the half-reactions

(2)  $Cu^+(aq) + e^- = Cu(s)$ $\qquad\qquad\qquad\qquad\qquad \mathscr{E}_2° = +0.521$ V

(3)  $Cu^{2+}(aq) + e^- = Cu^+(aq)$ $\qquad\qquad\qquad\qquad \mathscr{E}_3° = +0.153$ V

It is evident that half-reaction (1) is obtained by adding half-reactions (2) and (3). However, $\mathscr{E}_1° \neq \mathscr{E}_2° + \mathscr{E}_3°$, because voltage is an intensive quantity and, as such, is not additive. The Gibbs energy change for a reaction (half or whole) is an extensive quantity, and therefore, is additive; thus

$$\Delta G_1° = \Delta G_2° + \Delta G_3°$$

$$-n_1 F\mathscr{E}_1° = -n_2 F\mathscr{E}_2° - n_3 F\mathscr{E}_3°$$

or

$$\mathscr{E}_1° = \frac{1}{n_1}\,(n_2\mathscr{E}_2° + n_3\mathscr{E}_3°) \qquad\qquad\qquad (13.7)$$

Using this expression, we compute

$$\mathscr{E}_1^\circ = \frac{1}{2}\left\{(1)(+0.153) + (1)(+0.521)\right\} = +0.337\ V$$

It is evident from the foregoing that it is not correct simply to add $\mathscr{E}^\circ$ values for half-reactions when calculating an $\mathscr{E}^\circ$ value for a half-reaction from other $\mathscr{E}^\circ$ values. This is to be contrasted with the calculation of $\mathscr{E}^\circ$ *for a complete balanced reaction* from the $\mathscr{E}^\circ$ values for the two half-reactions, where we write

$$\mathscr{E}_{rctn}^\circ = \mathscr{E}_r^\circ - \mathscr{E}_\ell^\circ$$

We can combine the half-reaction $\mathscr{E}_r^\circ$ and $\mathscr{E}_\ell^\circ$ values in this way to obtain an $\mathscr{E}_{cell}^\circ$ value for the complete reaction, because the number of electrons in each half-reaction is made the same before the half-reactions are combined. In other words,

$$\Delta G_{rctn}^\circ = \Delta G_r^\circ - \Delta G_\ell^\circ$$

$$-n_{rctn}F\mathscr{E}_{rctn}^\circ = -n_r F\mathscr{E}_r^\circ + n_\ell F\mathscr{E}_\ell^\circ$$

but $n_{rctn} = n_r = n_\ell$, and therefore

$$\mathscr{E}_{rctn}^\circ = \mathscr{E}_r^\circ - \mathscr{E}_\ell^\circ$$

## 13.4  STABILITY OF OXIDIZING AND REDUCING AGENTS IN WATER.

In this section we shall concern ourselves with the stability of reducing agents and oxidizing agents in aqueous solution.* Powerful reducing agents are capable of liberating hydrogen from aqueous solution, whereas powerful oxidizing agents are capable of liberating oxygen from water. The equilibria in such solvent decomposition reactions are dependent on the hydrogen ion concentration. If the reaction of a reducing or oxidizing agent with water is thermodynamically favorable under the conditions prevailing in the solution, and if the reaction rate is appreciable, then the decomposition of the solution will result. The relevant half-reactions for water are

*These two half reactions and their respective $\mathscr{E}^\circ$ values should be memorized.*

(1)  $2H^+(aq) + 2e^- = H_2(g)$ $\qquad\qquad\qquad \mathscr{E}^\circ = 0$

(2)  $\frac{1}{2}O_2(g) + 2H^+(aq) + 2e^- = H_2O(\ell)$ $\qquad \mathscr{E}^\circ = +1.229\ V$

with the first of importance for reducing agents and the second of importance for oxidizing agents.

As our first example, consider a 1 M aqueous solution of vanadous ion, $V^{2+}(aq)$, with a $pH = 0$. Because

$$V^{3+}(aq) + e^- = V^{2+}(aq) \qquad \mathscr{E}^\circ = -0.24\ V$$

---

*See Section 7.4(c) for definitions of oxidizing and reducing agents.

we have for the reaction

$$V^{2+}(aq) + H^+(aq) = V^{3+}(aq) + \frac{1}{2} H_2(g)$$

$$\mathscr{E}^\circ_{rctn} = 0 - (-0.24) = 0.24 \text{ V}$$

and

$$\mathscr{E} \simeq 0.24 - 0.0592 \log \left\{ \frac{(V^{3+})P_{H_2}^{1/2}}{(V^{2+})(H^+)} \right\}$$

For $(V^{2+}) = 1.0$ M and $(H^+) = 1.0$ M, we obtain

$$\mathscr{E} \simeq 0.24 - 0.0592 \log (V^{3+})P_{H_2}^{1/2}$$

It is evident from this expression that with $(V^{3+}) \sim 0$ and $P_{H_2} \sim 0$, we have $\mathscr{E} > 0$, and therefore, 1 M $V^{2+}(aq)$ is capable of liberating $H_2$ from water at $pH = 0$. The reaction will proceed until either the reactants are exhausted or equilibrium is attained. The condition for equilibrium in the above solution is

$$\mathscr{E} = 0 \simeq 0.24 - 0.0592 \log \left\{ \frac{(V^{3+})P_{H_2}^{1/2}}{(V^{2+})(H^+)} \right\} = 0.24 - 0.0592 \log \left\{ \frac{[1.0 - (V^{2+})]P_{H_2}^{1/2}}{(V^{2+})^2} \right\}$$

from which at $P_{H_2} = 1.0$ atm we compute $(V^{2+}) = 0.0093$ M, and thus $(V^{3+}) = 0.99$ M. If the hydrogen gas is allowed to diffuse continuously out of the reaction vessel, thereby keeping $P_{H_2} \sim 0$, the reaction will go to completion. As an example of the effect of a change in $pH$ on the above equilibrium, we compute[*] that the equilibrium concentration of $V^{2+}(aq)$ is 1.0 M in an aqueous solution *buffered* at $pH = 7.0$, which is initially 1.0 M in $V^{2+}(aq)$, and over which $P_{H_2} = 1.0$ atm.

As an example of the oxidation of water by a strong oxidizing agent, consider a 1.0 M solution of $Co^{3+}(aq)$ with $pH = 0$. From Table 13.1, we obtain

$$Co^{3+}(aq) + e^- = Co^{2+}(aq) \qquad \mathscr{E}^\circ = +1.81 \text{ V}$$

and thus, for the reaction

$$Co^{3+}(aq) + \frac{1}{2} H_2O(\ell) = Co^{2+}(aq) + \frac{1}{4} O_2(g) + H^+(aq)$$

we find that

$$\mathscr{E}^\circ = +1.81 - (1.23) = 0.58 \text{ V}$$

$$\mathscr{E} \simeq 0.58 - 0.0592 \log \left\{ \frac{(Co^{2+})P_{O_2}^{1/4}(H^+)}{(Co^{3+})} \right\}$$

---

[*]These calculations assume that the principal V(III) product is $V^{3+}(aq)$ and not, for example, $VOH^{2+}(aq)$. The details of the calculation are as follows:

$$0 \simeq 0.24 - 0.0592 \log \left\{ \frac{(V^{3+})(1.0)^{1/2}}{[1.0 - (V^{3+})]1.0 \times 10^{-7}} \right\}$$

and $(V^{3+}) \simeq 1 \times 10^{-3}$ M. Actually, $V^{2+}(aq)$ reduces water at $pH = 7$, but the V(III) species formed is not $V^{3+}(aq)$, but $VOH^{2+}(aq)$ and $V(OH)_2^+(aq)$. The above $\mathscr{E}^\circ$ value rules out $V^{3+}(aq)$ as the principal product.

Inserting the concentrations of $H^+(aq)$ and $Co^{3+}(aq)$ given above, and taking $P_{O_2} \sim 0.20$ atm (air), we obtain

$$\mathscr{E} \approx 0.59 - 0.0592 \log (Co^{2+})$$

and $\mathscr{E} > 0$ for $0 \le (Co^{2+}) < 10^{10}$ M. Consequently, 1.0 M $Co^{3+}(aq)$ is capable of oxidizing water with $pH = 0$, and because the reaction rate is appreciable, $Co^{3+}(aq)$ cannot survive at such high concentrations in water.

## 13.5  LATIMER ELECTRODE POTENTIAL DIAGRAMS.

The $\mathscr{E}°$ values for the various half-reactions involving the aqueous species of an element can be summarized conveniently by means of a Latimer electrode potential diagram.* For example, the Latimer electrode potential diagram for copper in acidic aqueous solutions at 25°C is

$$CuO^+(aq) \xrightarrow{\;+1.8\;} Cu^{2+}(aq) \xrightarrow{\;+0.153\;} Cu^+(aq) \xrightarrow{\;+0.521\;} Cu(s)$$
$$\underset{+0.337}{\underline{\hspace{6cm}}}$$

The numbers given are the $\mathscr{E}°_{red}$ values in volts for the half-reactions involving the two copper species connected by the line on which the given $\mathscr{E}°_{red}$ value is placed. For example,

$$CuO^+(aq) + 2H^+(aq) + 2e^- = Cu^{2+}(aq) + H_2O(\ell) \qquad \mathscr{E}° = +1.8 \text{ V}$$

$$Cu^{2+}(aq) + 2e^- = Cu(s) \qquad\qquad\qquad \mathscr{E}° = +0.337 \text{ V}$$

As a simple example of the use of this diagram, consider the following question: Will $Cu^+(aq)$ disproportionate (i.e., undergo a self-oxidation-reduction reaction) in acidic aqueous solution? The disproportionation reaction is

$$2Cu^+(aq) = Cu^{2+}(aq) + Cu(s)$$

From the potential diagram we obtain

$$Cu^{2+}(aq) + e^- = Cu^+(aq) \qquad\qquad \mathscr{E}°_1 = +0.153 \text{ V}$$

$$Cu^+(aq) + e^- = Cu(s) \qquad\qquad \mathscr{E}°_2 = +0.521 \text{ V}$$

and for the disproportionation reaction, we compute

$$\mathscr{E}°_{rctn} = +0.521 - (0.153) = +0.368 \text{ V}$$

---

*The diagram does not provide any new information. What the diagram does is to facilitate the use of $\mathscr{E}°$ values in the analysis of chemical problems involving elements with several accessible oxidation states.

The large positive value of $\mathcal{E}°$ means that $K \gg 1$ (59 mV per power of 10 in $K$ for a reaction with $n = 1$) and the equilibrium lies far to the right. The value of $K$ is

$$\log K = \frac{(1)(0.368)}{(0.0592)} = 6.22$$

$$K = 1.7 \times 10^6 \simeq \frac{(Cu^{2+})_{eq}}{(Cu^+)^2_{eq}}$$

Thus, the disproportionation equilibrium lies far to the right, and therefore, $Cu^+(aq)$ is unstable with respect to disproportionation to $Cu^{2+}(aq)$ and $Cu(s)$. If the solution was initially 1.0 M in $Cu^+(aq)$, then at equilibrium

$$1.7 \times 10^6 \simeq \frac{\left(\dfrac{1.0 - (Cu^+)_{eq}}{2}\right)}{(Cu^+)^2_{eq}}$$

from which we compute $(Cu^+) \simeq 5.4 \times 10^{-4}$ M, and $(Cu^{2+}) \simeq 0.50$ M. We note also from the diagram that 1 M $CuO^+(aq)$ is capable of oxidizing $H_2O(\ell)$ to $O_2(g)$ in 1 M $H^+(aq)$.

Latimer potential diagrams are convenient devices for the analysis of the aqueous solution chemistry of elements exhibiting multiple oxidation states, such as manganese and iodine. The Latimer potential diagrams for manganese in acidic and basic aqueous solutions are given in Figure 13.2.

As examples of the use of these diagrams, consider the problem of predicting the possible reactions in each of the following cases:

(a) $Mn_2(SO_4)_3$ is dissolved in 1 M $H^+(aq)$;
(b) $K_2MnO_4$ is dissolved in 1 M $H^+(aq)$;
(c) $MnSO_4(aq)$ is added to $KMnO_4(aq)$ in 1 M $H^+(aq)$;
(d) $Mn(OH)_2(s)$ is brought in contact with a 1 M $OH^-(aq)$ solution containing $KMnO_4(aq)$;
(e) $Mn(OH)_2(s)$ is treated with an aqueous solution 1 M in $OH^-(aq)$ in the absence of air; and,

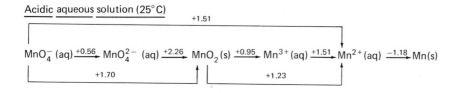

**FIGURE 13.2** Latimer reduction potential diagrams for manganese in aqueous solution at 25°C, values in volts.

(f) $Mn(OH)_2(s)$ is treated with an aqueous solution 1 M in $OH^-(aq)$ in the presence of air.

(a) When $Mn^{3+}(aq)$ is dissolved in 1 M $H^+(aq)$ there are two possible reactions:

(i) disproportionation

$$2H_2O(\ell) + 2Mn^{3+}(aq) = MnO_2(s) + Mn^{2+}(aq) + 4H^+(aq)$$

$$\mathscr{E}° = 1.51 - (0.95) = +0.56 \text{ V}$$

(ii) oxidation of water

$$4Mn^{3+} + 2H_2O(\ell) = 4Mn^{2+}(aq) + O_2(g) + 4H^+(aq)$$

$$\mathscr{E}° = 1.51 - (1.23) = +0.28 \text{ V}$$

Both of these reactions are spontaneous in 1 M $H^+(aq)$ with $(Mn^{3+}) = 1.0$ M; however, the disproportionation reaction predominates because it is favored both thermodynamically and kinetically.[*]
Note that the $MnO_2(s)$ formed in the disproportionation reaction is capable of liberating oxygen gas from a 1 M $H^+(aq)$ solution:

$$2MnO_2(s) + 4H^+(aq) = 2Mn^{2+}(aq) + O_2(g) + 2H_2O(\ell)$$

$$\mathscr{E}° = 1.23 - (1.23) = 0$$

Therefore, $\log K = 0$, and $K = 1$

$$1 \simeq \frac{(Mn^{2+})^2 P_{O_2}}{(H^+)^4}$$

and with $(H^+)_{eq} \simeq 1.0$ M and $P_{O_2} \simeq 0.20$ atm (air), we compute $(Mn^{2+})_{eq} \simeq 2.25$ M. In other words, $MnO_2(s)$ will liberate oxygen from a solution 1 M in $H^+(aq)$ until either $(Mn^{2+})$ reaches 2.25 M, or all the $MnO_2(s)$ is consumed. This reaction is slow.

(b) The aqueous manganate ion, $MnO_4^{2-}(aq)$, is unstable with respect to disproportionation to $MnO_4^-(aq)$ and $MnO_2(s)$ in 1 M $H^+(aq)$

$$3MnO_4^{2-}(aq) + 4H^+(aq) = 2MnO_4^-(aq) + MnO_2(s) + 2H_2O(\ell)$$

$$\mathscr{E}° = 2.26 - (0.56) = 1.70 \text{ V}$$

---

[*]The oxidation of water is usually a slow reaction, because the first step in the reaction mechanism (Chapter 15) usually involves the half-reaction

$$H_2O(\ell) = O(g) + 2H^+(aq) + 2e^- \qquad -\mathscr{E}°_{red} = -2.42 \text{ V}$$

which is thermodynamically unfavorable (large negative $-\mathscr{E}°$ value). In other words, even though the *net* process may be favored thermodynamically, if the detailed mechanism of the reaction involves a step which requires the surmounting of a high Gibbs-energy barrier, then this barrier acts like a bottle-neck to slow down the overall net reaction.

Note also that the products of the disproportionation reaction are themselves unstable in 1 M $H^+(aq)$. Aqueous permanganate ion, $MnO_4^-(aq)$, is capable of oxidizing water:

$$4H^+(aq) + 4MnO_4^-(aq) = 2H_2O(\ell) + 3O_2(g) + 4MnO_2(s)$$
$$\mathscr{E}° = 1.70 - (1.23) = +0.47 \text{ V}$$

The equilibrium constant for this reaction at 25°C is

$$\log K = \frac{12\,(0.47)}{(0.0592)} = 95.3$$

$$K = 2 \times 10^{95} \simeq \frac{P_{O_2}^3}{(H^+)^4(MnO_4^-)^4}$$

The equilibrium concentration of $MnO_4^-(aq)$ for $(H^+) \simeq 1.0$ M and $P_{O_2} \simeq 0.20$ atm is

$$(MnO_4^-)_{eq} \simeq (400 \times 10^{-100})^{1/4} = 4.5 \times 10^{-25} \text{ M}$$

in the presence of $MnO_2(s)$. At $pH = 7.0$ an analogous calculation yields $(MnO_4^-)_{eq} \simeq 4.5 \times 10^{-18}$.

Fortunately, the oxidation of water by permanganate is a slow reaction; otherwise, redox titrations (Section 13.7) with permanganate would be impossible. The reaction is autocatalytic (via $MnO_2(s)$), however, and the older a permanganate solution becomes, the faster it decomposes via the above reaction.

(c) A thermodynamically favorable reaction between $Mn^{2+}(aq)$ and $MnO_4^-(aq)$ in 1 M $H^+(aq)$ solution is

$$3\{2H_2O(\ell) + Mn^{2+}(aq) = MnO_2(s) + 4H^+(aq) + 2e^-\}$$
$$2\{4H^+(aq) + MnO_4^-(aq) + 3e^- = MnO_2(s) + 2H_2O(\ell)\}$$
$$2H_2O(\ell) + 3Mn^{2+}(aq) + 2MnO_4^-(aq) = 5MnO_2(s) + 4H^+(aq)$$
$$\mathscr{E}° = 1.70 - 1.23 = 0.47 \text{ V}$$

(d) A thermodynamically favorable reaction between $Mn(OH)_2(s)$ and $MnO_4^-(aq)$ in 1 M $OH^-(aq)$ is

$$3\{Mn(OH)_2(s) + 2OH^-(aq) = MnO_2(s) + 2H_2O(\ell) + 2e^-\}$$
$$2\{2H_2O(\ell) + MnO_4^-(aq) + 3e^- = MnO_2(s) + 4OH^-(aq)\}$$
$$3Mn(OH)_2(s) + 2MnO_4^-(aq) = 5MnO_2(s) + 2OH^-(aq) + 2H_2O(\ell)$$
$$\mathscr{E}° = 0.59 - (-0.05) = +0.64 \text{ V}$$

Note that $MnO_2(s)$ cannot oxidize water in an aqueous solution with $pH = 14.0$ at 25°C, because this requires an $\mathscr{E}°_{red}$ (half-reaction) more positive than $1.23 + 0.0592 \log (1 \times 10^{-14}) = 0.40$ V.

**TABLE 13.2**

| In 1 M $H^+$(aq) Solution at 25°C we have: | | In 1 M $OH^-$(aq) Solution at 25°C we have: | |
| --- | --- | --- | --- |
| SPECIES | STABILITY | SPECIES | STABILITY |
| Mn(s) | unstable, liberates $H_2$(g) to form $Mn^{2+}$(aq) | Mn(s) | unstable, liberates $H_2$(g) and forms $Mn(OH)_2$(s) |
| $Mn^{2+}$(aq) | stable | $Mn(OH)_2$(s) | unstable, air-oxidized to $MnO_2$(s) |
| $MnO_2$(s) | unstable, oxidizes water | $Mn(OH)_3$(s) | unstable, disproportionates to $MnO_2$(s) and $Mn(OH)_2$(s) |
| $Mn^{3+}$(aq) | unstable, disproportionates to $Mn^{2+}$(aq) and $MnO_2$(s) | $MnO_2$(s) | stable |
| $MnO_4^{2-}$(aq) | unstable, disproportionates to $MnO_4^-$(aq) and $MnO_2$(s) | $MnO_4^{2-}$(aq) | unstable, disproportionates to $MnO_4^-$(aq) and $MnO_2$(s) (less unstable in 1 M $OH^-$(aq) than in 1 M $H^+$(aq)) |
| $Mn^{3+}$(aq) | unstable, oxidizes water (slowly) | $MnO_4^-$(aq) | unstable, oxidizes water to $O_2$(g) and forms $MnO_2$(s) |
| $MnO_4^-$(aq) | unstable, oxidizes water (slowly) | | |

**(e, f)** The reaction between $Mn(OH)_2$(s) and $O_2$(g) in the presence of 1 M $OH^-$(aq) is

$$2H_2O(\ell) + O_2(g) + 4e^- = 4OH^-(aq)$$

$$2\{Mn(OH)_2(s) + 2OH^-(aq) = MnO_2(s) + 2H_2O(\ell) + 2e^-\}$$

$$Mn(OH)_2(s) + O_2(g) = 2H_2O(\ell) + 2MnO_2(s)$$

$$\mathscr{E}° = 0.40 - (-0.05) = +0.45 \text{ V}$$

Note that for $P_{O_2} = 0.20$ atm, we have

$$\mathscr{E} = 0.47 + \frac{0.0592}{4} \log P_{O_2} = 0.48 \text{ V} > 0$$

and therefore, the reaction is spontaneous. In the absence of oxygen gas, that is, with $P_{O_2} \sim 0$, we obtain $\mathscr{E} < 0$, and the reaction does not take place, which, of course, is obvious because with a reactant missing the reaction cannot take place.

In Table 13.2 we have enumerated the stabilities in 1 M $H^+$(aq) and 1 M $OH^-$(aq) of the various manganese species that are given in Figure 13.2.

## 13.6 ELECTRODEPOSITION.

The decomposition of a solution caused by the passage of an electric current through the solution is known as *electrolysis*. The mass, $m$, of a substance that reacts as a result of the passage of Z coulombs of charge through the cell is given by *Faraday's law of electrolysis* as

$$m = \left(\frac{Z}{F}\right)\left(\frac{A}{n}\right) \tag{13.8}$$

In this equation $F$ is the faraday, $A$ is the formula weight of the substance reduced or oxidized, and $n$ is the number of moles of electrons required to reduce or oxidize one formula weight of the substance. Note that $Z/F$ is the number of moles of electrons passed through the external circuit and $A/n$ is the number of grams of the substance reduced or oxidized per mole of electrons.

Combining Equation (13.8) with the definition

$$Z = It \tag{13.9}$$

where $I$ is the current (in amperes), and $t$ is the time (in seconds) that the current $I$ flows, yields

$$m = \frac{ItA}{nF} \tag{13.10}$$

This equation is useful in computing the mass of a material that is reduced or oxidized in a given time owing to the passage of the constant current $I$. If we pass a current of 10 mA for 10 min through a solution of chloroplatinic acid $[H_2PtCl_6]$, the maximum amount of Pt that can be deposited is

$$m = \frac{(10 \times 10^{-3}\ \text{A})(600\ \text{sec})(195.1\ \text{g/mole Pt})}{(4\ \text{mole e}^-/\text{mole Pt})(9.65 \times 10^4\,\text{C/mole e}^-)} = 3.03 \times 10^{-3}\ \text{g}$$

Which reaction occurs at an electrode in an electrolysis depends upon the ionic species present in the solution, upon their concentrations, and also upon the nature of the electrode. In an aqueous solution with inert (platinum or gold) electrodes, the substance that is reduced[*] first will be the one belonging to the couple (half-reaction) with the highest reduction potential under the conditions prevailing in the solution. The substance first oxidized will be the one belonging to the couple which has the lowest reduction potential under the conditions prevailing in the solution.

Consider the problem of computing the minimum voltage that must be applied across two platinum electrodes in a 0.10 M $CdI_2(aq)$ solution to cause the electrodeposition of cadmium metal and the oxidation of iodide ion to solid iodine. As soon as the decomposition begins, the following electrochemical cell is established:

$$\text{Cd(s on Pt)} \mid \text{CdI}_2(\text{aq}) \mid \text{I}_2(\text{s on Pt})$$

The reversible voltage of this cell is computed as follows:

$$
\begin{array}{ll}
\text{Cd}^{2+}(\text{aq}) + 2\text{e}^- = \text{Cd(s)} & \mathscr{E}^\circ_{red} = -0.40\ \text{V} \\
2\text{e}^- + \text{I}_2(\text{s}) = 2\text{I}^-(\text{aq}) & \mathscr{E}^\circ_{red} = +0.54\ \text{V} \\
\end{array}
$$

$$\text{Cd(s)} + \text{I}_2(\text{s}) = \text{Cd}^{2+}(\text{aq}) + 2\text{I}^-(\text{aq})$$

$$\mathscr{E}^\circ_{cell} = 0.54 - (-0.40) = +0.94\ \text{V}$$

$$\mathscr{E} \approx 0.94 - \frac{0.059}{2} \log\,(\text{Cd}^{2+})(\text{I}^-)^2$$

---

[*]Reduction takes place at the *cathode* and oxidation takes place at the *anode* [which may be remembered by the mnemonic R at c (consonants), O at a (vowels)].

For $(Cd^{2+}) \simeq 0.10$ M and $(I^-) \simeq 0.20$ M, we compute from the Nernst equation that

$$\mathscr{E} = 0.94 - \frac{0.059}{2} \log (0.10)(0.20)^2 = 1.01 \text{ V} > 0$$

Note that the cell reaction is spontaneous in a direction opposite to that desired. Consequently, an externally applied voltage greater than 1.01 V is required to reverse the cell reaction.

At thermodynamic equilibrium in the cell there is no *net* reaction occurring in the cell, and thus there is no net current flow through the cell. The passage of current through a cell inevitably involves the development of an *overvoltage* (or *overpotential*). The overvoltage, $\eta$, is defined as

$$\eta = \mathscr{E} - \mathscr{E}_{rev} \qquad (13.11)$$

where $\mathscr{E}$ is the cell voltage during the passage of a particular net current and $\mathscr{E}_{rev}$ is the reversible (Nernst) cell voltage. The magnitude of the overvoltage depends on the magnitude of the electrode current density. The greater the current, the greater is the overvoltage. Overvoltage is a consequence of the need to transfer electrons across metal-solution phase boundaries in the cell.

The magnitude of the overvoltage for a given current density (say, $1 \text{ mAmp} \cdot \text{cm}^{-2} = 10^{-8}$ mole of $e^- \cdot \text{cm}^{-2} \cdot \text{sec}^{-1}$) depends critically on the nature of the electrode reaction and the nature of the electrode surface. Overvoltages associated with electrode reactions involving the evolution of hydrogen or oxygen can be as large as 1 volt at $1 \text{ mAmp} \cdot \text{cm}^{-2}$. Some representative overvoltages for the electrode reaction

$$H^+(aq) + e^- \rightarrow \frac{1}{2} H_2(g)$$

on various metal surfaces at different current densities are:

| $I/A_{sur}$ (Amp $\cdot$ cm$^{-2}$) | Hg($\ell$) | Pt(smooth) | Pt(black) |
|---|---|---|---|
| 0.001 | 0.90 | 0.024 | 0.015 |
| 1.0 | 1.13 | 0.68 | 0.048 |

The formation of $H_2(g)$ at an electrode surface via the reduction of $H^+(aq)$ is thought to proceed through a two-step sequence of the type

(1)   $H^+(soln) + e^-$ (metal) $\rightleftharpoons$ H (surface)

(2)   2H (surface) $\rightarrow H_2(g)$

Either or both of these two steps may act to limit the electrode reaction rate and hence the electrode current. The first step is slow on Hg($\ell$), whereas the second step is slow on platinum. In order to increase the overall rate of these reactions, we must increase the potential of the electrons in the metallic electrode by increasing the voltage applied to the electrode. The increased energy of the electrons in the metal gives rise to an in-

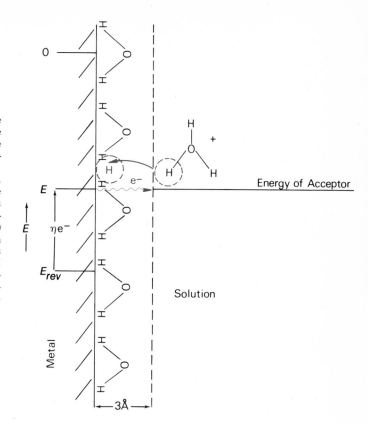

**FIGURE 13.3** The *energy* of the reactive electrons in the metal in the absence of an overvoltage is $E_{rev}$. The energy of the reactive electrons is increased by an amount $\eta e^-$ under the action of the overvoltage $\eta = \mathscr{E} - \mathscr{E}_{rev}$. The overvoltage leads to an increase in the rate of emission of electrons from the metal to $H_3O^+(aq)$. The electrons tunnel (quantum-mechanically) out of the metal to the acceptors in solution. The reduction of $H_3O^+$ is followed by the adsorption of an H atom on the metal surface and the release of an $H_2O$ molecule to the solution. Combination of H atoms on the surface leads to $H_2$ formation.

creased rate of emission of electrons from the metal to the acceptor species in the solution ($H^+(aq)$ in the above case). This situation is depicted schematically in Figure 13.3.

The essential point to be grasped from this discussion is that a non-zero overvoltage is necessary for a net current flow.

Overvoltage phenomena can sometimes be turned to advantage and used to separate metals that would not otherwise be separable electrolytically. For example, with a mercury cathode one can separate iron and zinc from aluminum without appreciable hydrogen evolution:

| HALF-REACTION | $\mathscr{E}$ (~1 M CONCENTRATIONS) | SEQUENCE OF REACTIONS |
|---|---|---|
| $Fe^{3+}(aq) + e^- = Fe^{2+}(aq)$ | $+0.77$ | 1st |
| $Fe^{2+}(aq) + 2e^- = Fe(s)$ | $-0.44$ | 2nd |
| $Zn^{2+}(aq) + 2e^- = Zn(s)$ | $-0.76$ | 3rd |
| $2H^+(aq) + 2e^- = H_2(g)$ | $\sim -1.0$ (overvoltage of hydrogen on mercury) | 4th |
| $Al^{3+}(aq) + 3e^- = Al(s)$ | $-1.66$ | 5th |

The reduction of solution species in acidic aqueous solutions is often accomplished using a *Jones reductor,* which consists of a column of amalgamated zinc granules. Although the evolution of hydrogen gas is thermodynamically favored in the reductor, it does not occur rapidly because of the overvoltage effect.* The Jones reductor can be used, for example, to reduce $Cr^{3+}(aq)$ to $Cr^{2+}(aq)$ by passing the $Cr^{3+}(aq)$ solution through the column and drawing off the $Cr^{2+}(aq)$ (containing $Zn^{2+}(aq)$) at the bottom:†

$$Zn(Hg) + Cr^{3+}(aq) = Cr^{2+}(aq) + Zn^{2+}(aq)$$

$$\mathscr{E}° = -0.41 - (-0.76) = +0.35 \text{ V}$$

The overvoltage phenomenon is another example of the principle that the *no* of thermodynamics is emphatic, but the *yes* of thermodynamics is only conditional. A thermodynamically favorable process may not actually take place because of an unfavorable rate (Chapters 14 and 15).

Because of overvoltage phenomena, reliable computations of decomposition voltages are complicated. One can, however, easily calculate the completeness of separation of two metals by electrolytic means using thermodynamic relations. (When two metals begin to deposit simultaneously at the cathode, their half-cell potentials must be equal.) For the effective electrochemical separation of two metals, their respective $\mathscr{E}°$ values must be 0.2 to 0.3 V apart. Even if this is not the case, it can sometimes be made so by selectively complexing one (or both) of the metals, thereby changing the half-reaction and consequently shifting the $\mathscr{E}°$ value.

## 13.7  OXIDATION-REDUCTION (REDOX) TITRATIONS.

The determination of the amount or concentration of a reducing agent in a sample by titration with a solution of an oxidizing agent of known concentration (or the reverse) is a widely used and important technique in analytical chemistry. The major principles of such titrations are readily understood in terms of the Nernst equation. For example, consider the titration of an acidic aqueous solution of $Fe^{2+}(aq)$ with an acidic aqueous solution of the oxidizing agent ceric ion, $Ce^{4+}(aq)$. At 25°C we have

$$Ce^{4+}(aq) + e^- = Ce^{3+}(aq) \qquad \mathscr{E}°_{red} = +1.65 \text{ V}$$

$$Fe^{3+}(aq) + e^- = Fe^{2+}(aq) \qquad \mathscr{E}°_{red} = +0.77 \text{ V}$$

and the equilibrium constant for the reaction

$$Fe^{2+}(aq) + Ce^{4+}(aq) = Fe^{3+}(aq) + Ce^{3+}(aq)$$

---

*Pure zinc is very unreactive toward $H_2SO_4(aq)$. However, if a piece of zinc in $H_2SO_4(aq)$ is touched with a platinum wire, vigorous evolution of $H_2(g)$ takes place (with dissolution of Zn as $Zn^{2+}$), as long as the zinc is in contact with the platinum. If the platinum wire is removed, the gas evolution stops.

†Notice that the air oxidation of $Cr^{2+}(aq)$ is thermodynamically favored in 1 M $H^+(aq)$. This reaction is rapid, and $Cr^{2+}(aq)$ solutions must be kept under an inert atmosphere (for example, $N_2$). The rapid and quantitative reaction of $Cr^{2+}(aq)$ in acid solutions with oxygen is used to remove traces of oxygen from gases such as nitrogen in "chromous bubblers."

is

$$\log K = \frac{1.65 - 0.77}{0.0592} = 14.86$$

$$K = 7.2 \times 10^{14} \simeq \frac{(Fe^{3+})_{eq}(Ce^{3+})_{eq}}{(Fe^{4+})_{eq}(Ce^{2+})_{eq}}$$

The specific problem of interest is the titration of 25.00 ml of 0.100 M $Fe^{2+}$(aq) with 0.100 M $Ce^{4+}$(aq). The rate of the above reaction is high, and as soon as some of the $Ce^{4+}$(aq) solution is added to the $Fe^{2+}$(aq) solution, the above reaction takes place rapidly and equilibrium is quickly established. Because equilibrium is established, we have

$$\text{``}\mathscr{E}_{cell}\text{''} = 0 = \mathscr{E}_{Ce^{4+}/Ce^{3+}} - \mathscr{E}_{Fe^{3+}/Fe^{2+}}$$

or

$$\mathscr{E}_{Fe^{3+}/Fe^{2+}} = \mathscr{E}_{Ce^{4+}/Ce^{3+}}$$

From this equilibrium condition we obtain

$$+0.77 - 0.059 \log \frac{(Fe^{2+})_{eq}}{(Fe^{3+})_{eq}} = +1.65 - 0.059 \log \frac{(Ce^{3+})_{eq}}{(Ce^{4+})_{eq}}$$

Hence, we can work with either half-reaction to study, for example, the behavior of $\mathscr{E}_{Fe^{3+}/Fe^{2+}}$ as a function of the volume of the $Ce^{4+}$(aq)-containing solution that is added. After 5.00 ml of 0.100 M $Ce^{4+}$(aq) have been added, we have the following:

(25.00 ml)(0.100 M) = 2.50 millimoles of $Fe^{2+}$ initially present

(5.00 ml)(0.100 M) = 0.50 millimoles $Ce^{4+}$ added

$K$ is large, and consequently the reaction proceeds almost to completion. Let the number of millimoles of $Ce^{4+}$ at equilibrium be $x$; then

$$(Ce^{4+})_{eq} = x/30.00 \qquad\qquad (Ce^{3+})_{eq} = (0.50-x)/30.00$$

$$(Fe^{3+})_{eq} = (0.50-x)/30.00 \qquad\qquad (Fe^{2+})_{eq} = (2.50-0.50+x)/30.00$$

Hence

$$\frac{(Fe^{3+})_{eq}}{(Fe^{2+})_{eq}} = \frac{(0.50-x)}{(2.00+x)} \qquad\qquad \frac{(Ce^{3+})_{eq}}{(Ce^{4+})_{eq}} = \frac{0.50-x}{x}$$

$$\left(\frac{0.50-x}{2.00+x}\right)\left(\frac{0.50-x}{x}\right) \simeq \frac{(0.50)^2}{2.00x} \simeq 7.2 \times 10^{14}$$

and $x = 1.7 \times 10^{-16}$ mmole. Therefore,

$$\mathscr{E}_{Fe^{3+}/Fe^{2+}} = +0.77 - 0.059 \log\left(\frac{2.00}{0.50}\right) = +0.73 \text{ V}$$

After 20.0 ml have been added, we have

$$\frac{(Fe^{3+})_{eq}}{(Fe^{2+})_{eq}} = \frac{2.00 - x}{0.50 + x} \qquad \frac{(Ce^{3+})_{eq}}{(Ce^{4+})_{eq}} = \frac{2.00 - x}{x}$$

$$\left(\frac{2.00 - x}{0.50 + x}\right)\left(\frac{2.00 - x}{x}\right) = 7.2 \times 10^{14}$$

and $x = 1.1 \times 10^{-14}$ mmole. Therefore

$$\mathscr{E}_{Fe^{3+}/Fe^{2+}} = +0.77 - 0.059 \log \frac{0.50}{2.00} = +0.81 \text{ V}$$

At the *equivalence point* we have

$$\frac{(Fe^{3+})_{eq}}{(Fe^{2+})_{eq}} = \frac{2.50 - x}{x} \qquad \frac{(Ce^{3+})_{eq}}{(Ce^{4+})_{eq}} = \frac{2.50 - x}{x}$$

$$\frac{(2.50 - x)^2}{x^2} = 7.2 \times 10^{14}$$

and $x = 0.93 \times 10^{-7}$ mmole. Therefore,

$$\mathscr{E}_{Fe^{3+}/Fe^{2+}} = +0.77 - 0.059 \log \frac{0.93 \times 10^{-7}}{2.50} = +1.21 \text{ V}$$

Beyond the equivalence point it is more convenient to work with the $Ce^{4+}/Ce^{3+}$ half-reaction. For example, after 30.0 ml of $Ce^{4+}(aq)$ have been added, we have 2.50 mmole of $Ce^{3+}(aq)$ and 0.50 mmole of $Ce^{4+}(aq)$

$$\mathscr{E}_{Fe^{3+}/Fe^{2+}} = \mathscr{E}_{Ce^{4+}/Ce^{3+}} = +1.65 - 0.059 \log \frac{2.50}{0.50} = +1.61 \text{ V}$$

In Figure 13.4 we have plotted $\mathscr{E}_{Fe^{3+}/Fe^{2+}}$ as a function of the volume of $Ce^{4+}(aq)$ titrant added. It is evident from the figure that there is a very rapid change in $\mathscr{E}_{Fe^{3+}/Fe^{2+}}$ in the vicinity of the equivalence point, and this fact can be used to determine when the equivalence point has been reached. The detection of the equivalence point can be accomplished either potentiometrically, using an indicator electrode* (for example, Pt(s)) and a reference electrode, or with a redox indicator. A suitable indicator for the $Fe^{2+}(aq) + Ce^{4+}(aq)$ titration is the tris($o$-phenanthroline)ferrous ion

$$Fe(o\text{-phen})_3^{3+}(aq) + e^- = Fe(o\text{-phen})^{2+}(aq)$$

blue        red

---

*A suitable cell in which the redox titration can be followed potentiometrically is

$$\text{Pt} \mid Fe^{2+}(aq), Fe^{3+}(aq) \mid\mid KNO_3(4m) \mid\mid KCl(satd) \mid AgCl(s) \mid Ag(s)$$

reference electrode

The $Ce^{4+}(aq)$ solution is added dropwise to the $Fe^{2+}$, $Fe^{3+}$ solution with stirring and the cell voltage is measured after each increment of $Ce^{4+}(aq)$ solution is added. The measured voltage is not equal to $\mathscr{E}_{Fe^{3+}/Fe^{2+}}$, but rather it is equal to $\mathscr{E}_{ref} - \mathscr{E}_{Fe^{3+}/Fe^{2+}} + \mathscr{E}_J$, where $\mathscr{E}_J$ is the liquid junction potential.

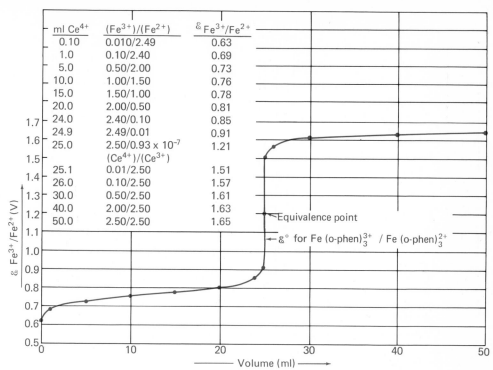

The table shown in the figure:

| ml $Ce^{4+}$ | $(Fe^{3+})/(Fe^{2+})$ | $\mathscr{E}$ $Fe^{3+}/Fe^{2+}$ |
|---|---|---|
| 0.10 | 0.010/2.49 | 0.63 |
| 1.0 | 0.10/2.40 | 0.69 |
| 5.0 | 0.50/2.00 | 0.73 |
| 10.0 | 1.00/1.50 | 0.76 |
| 15.0 | 1.50/1.00 | 0.78 |
| 20.0 | 2.00/0.50 | 0.81 |
| 24.0 | 2.40/0.10 | 0.85 |
| 24.9 | 2.49/0.01 | 0.91 |
| 25.0 | $2.50/0.93 \times 10^{-7}$ | 1.21 |
| | $(Ce^{4+})/(Ce^{3+})$ | |
| 25.1 | 0.01/2.50 | 1.51 |
| 26.0 | 0.10/2.50 | 1.57 |
| 30.0 | 0.50/2.50 | 1.61 |
| 40.0 | 2.00/2.50 | 1.63 |
| 50.0 | 2.50/2.50 | 1.65 |

**FIGURE 13.4** Redox titration curve for the titration of 25.00 ml of 0.100 M $Fe^{2+}$(aq) with 0.100 M $Ce^{4+}$(aq).

for which $\mathscr{E}° = +1.06$ V. When $\mathscr{E}_{Ce^{4+}/Ce^{3+}}$ reaches a value of $+1.06$ V it will oxidize the red form of the indicator to the blue form, which is observed as a color change in the solution.

## 13.8 THE MEASUREMENT OF pH.

The purpose of this section is to explain the way in which emf measurements with suitable electrodes can be used to determine the $pH$ of a solution.

A hydrogen-ion-sensitive glass electrode consists of a silver-silver chloride internal element dipping into an internal solution containing HCl(aq) + NaCl(aq) in a thin-walled bulb of sodium-containing glass (Figure 13.5). The current is carried through the glass membrane primarily by means of sodium ions. A cell involving a glass electrode is

glass electrode

$$Ag(s) \mid AgCl(s) \mid HCl(aq) + NaCl(aq) \mid HCl(aq) \mid Hg_2Cl_2(s) \mid Hg(\ell)$$

glass membrane

The detailed analysis of the operation of the glass electrode is complicated and involves ion-exchange equilibria of the type

$$Na^+(aq) + H^+(glass) = Na^+(glass) + H^+(aq)$$

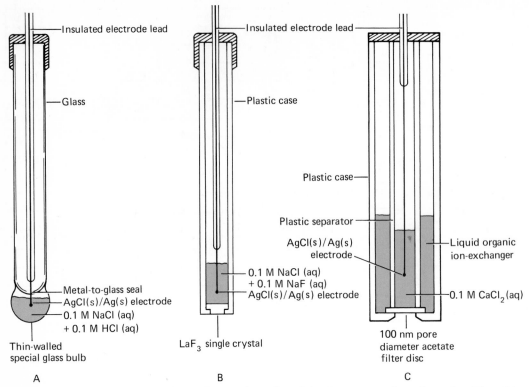

**FIGURE 13.5** Construction of specific-ion electrodes: (A) H⁺-sensitive glass electrode; (B) solid-state F⁻-sensitive electrode; (C) Ca²⁺-sensitive liquid-ion-exchanger electrode.

at the two glass surfaces, and also involves the movement of $Na^+$ ions through the dry glass layer within the glass. For thermodynamic purposes we can ignore these complexities and represent the glass electrode simply as H(glass).* That is, we can represent the above cell as

$$H(glass) \mid HCl(aq) \mid Hg_2Cl_2(s) \mid Hg(\ell)$$

where the glass electrode can be thought of as functioning *like* a conventional hydrogen electrode, that is, $H(glass) = H^+(aq) + e^-$; however, it is important to recognize that this is merely a mnemonic.

The solution whose *pH* is desired is often of unknown composition, and it may or may not contain the chloride ion necessary for the functioning of the calomel reference electrode. In particular, we seldom know the value of the $\gamma_{\pm(HCl)}$ for the solution. For this reason, a saturated calomel reference electrode is used, together with the glass electrode, in a cell of the type

saturated calomel reference<br>
electrode

$$H(glass) \mid \underset{\uparrow}{\text{solution X (containing } H^+(aq))} \overbrace{\underset{\uparrow}{KCl(aq, satd)} \mid Hg_2Cl_2(s) \mid Hg(\ell)}$$

glass membrane          liquid junction

---

*This is not possible in strongly alkaline solutions containing $Na^+(aq)$, where a $Na^+$ correction must be applied to the measured voltage.

This cell involves a liquid junction, that is, a liquid-liquid boundary region containing two interdiffusing electrolyte solutions. The presence of the liquid junction gives rise to a liquid-junction potential, $\mathscr{E}_J$, whose magnitude depends on the composition of solution X. The cell reaction for this cell can be represented as

$$H(glass) + \frac{1}{2} Hg_2Cl_2(s) = Hg(\ell) + H^+(aq, X) + Cl^-(aq, satd\ KCl) \pm ion\ transfer$$

where the "± ion transfer" term enters because of the liquid junction. Application of the Nernst equation to the cell reaction yields

$$\mathscr{E}_X = \mathscr{E}^\circ + \mathscr{E}_{JX} - \frac{RT}{F} \ln a_{H^+(X)} - \frac{RT}{F} \ln a_{Cl^-(satd)}$$

Defining the $pH$ in solution X as

$$pH(X) = -\log a_{H^+(X)}$$

we have

$$pH(X) = F \left( \frac{\mathscr{E}_X - \mathscr{E}^\circ - \mathscr{E}_{JX}}{2.303\ RT} \right) + \log a_{Cl^-(satd)}$$

The operational $pH$ scale is based on the expression

$$pH(X) - pH(S) = -\log \frac{a_{H^+(X)}}{a_{H^+(S)}}$$

The quantity $pH(S)$ is the $pH$ of a standard reference solution whose thermodynamic properties are known in detail and to which a value of $pH(S)$ can be assigned, using a reasonable (but arbitrary) convention for the single-ion activity coefficient (for example, $\gamma_{H^+} = \gamma_{\pm(HCl)}$).

If we make emf measurements on both solutions X and S, using the *same* glass and calomel reference electrodes, then we can write

$$pH(X) = pH(S) + \frac{F(\mathscr{E}_X - \mathscr{E}_S)}{2.303\ RT} - \frac{F(\mathscr{E}_{JX} - \mathscr{E}_{JS})}{2.303\ RT}$$

because $\mathscr{E}^\circ$ and $\log a_{Cl^-(satd)}$ are the same in both measurements. Since $pH(S)$ is known, we can obtain $pH(X)$ from measurements of $\mathscr{E}_X$ and $\mathscr{E}_S$, provided $\mathscr{E}_{JX} - \mathscr{E}_{JS} \simeq 0$. In precise ($\pm 0.02\ pH$ units) work, $pH(S)$ should be within $\pm 2\ pH$ units of $pH(X)$, and the total ionic strength of solution X should be similar to that of solution S in order to insure $\mathscr{E}_{JX} - \mathscr{E}_{JS} \simeq 0$.

The development of the hydrogen-ion-sensitive glass electrode has made the measurement of $pH$ a routine matter, and this fact has greatly facilitated numerous significant scientific discoveries in the chemical and biological sciences. There has been a tremendous upsurge of interest in membrane electrode technology and function in recent years, and special glass electrodes which respond to potassium or sodium ion activities, as well as a more complex type of liquid ion-exchange membrane electrodes, which respond, for example, to the activity of calcium ion, have been developed. A variety of solid state electrodes which respond to the activities of ions such as $F^-(aq)$ and $S^{2-}(aq)$ have also been developed.

The basic idea involved in the development of an ion-specific electrode is to find a membrane (solid, liquid, or glass) that (ideally) is permeable only to the ion of interest. In practice, no ion-selective electrode is ever perfectly selective; if ions $x$ and $y$ can both pass through the membrane, then the electrode voltage usually depends on the activities of ions $x$ and $y$ in the following way:

$$\mathscr{E} = \text{const.} - \frac{RT}{z_x F} \ln\,(a_x + pa_y)$$

where $p$ is the permeability ratio for ions $y$ and $x$ in the membrane $p = P_y/P_x$. For a Nernstian electrode response to ion $x$ in the presence of ion $y$, we require $pa_y \ll a_x$. In other words, a desirable membrane for ion $x$ has $p \ll 1$ for all other ions.

Different types of membrane involve different types of ion transport through the membrane. Liquid ion-exchanger membranes involve the formation of a mobile complex between the ion being measured and a "carrier" (complexing agent). The calcium ion-selective electrode has a liquid ion-exchanger consisting of an organic solvent (e.g., $n$-hexanol) in which a calcium complex of the composition $CaX_2$ is dissolved. The ion $X^-$ is a phosphate diester anion of the type

$$R\!-\!O\!-\!\overset{\displaystyle O}{\overset{\displaystyle \|}{\underset{\displaystyle |}{\underset{\displaystyle O}{P}}}}\!-\!O\!-\!R'$$

A sequence of reactions consistent with the electrode response is (see Figure 13.5; here *ext* refers to the external solution of interest and *int* refers to the internal $CaCl_2(aq)$ solution):

$$Ca^{2+}(ext) + 2X^-(organic) \rightleftharpoons CaX_2(organic)$$

$$CaX_2(organic) \rightleftharpoons Ca^{2+}(int) + 2X^-(organic)$$

$$2AgCl(s) + 2e^- \rightleftharpoons 2Ag(s) + 2Cl^-(int)$$

$$sum: \overline{Ca^{2+}(ext) + 2AgCl(s) + 2e^- \rightleftharpoons Ca^{2+}(int) + 2Cl^-(int) + 2Ag(s)}$$

All of the terms in the net reaction are constant for a given electrode except $Ca^{2+}(ext)$, which is the ion of interest in the external media. Divalent ions such as $Zn^{2+}$ and $Mg^{2+}$ interfere with the response of the electrode to $Ca^{2+}$ by competing with $Ca^{2+}$ for the $X^-(organic)$ carriers.

The fluoride ion-specific electrode has a single crystal of lanthanum fluoride, $LaF_3$, that acts as the ion-selective membrane (Figure 13.5). Lanthanum fluoride is an electrolytic (ionic) conductor. The ionic charge carrier in the crystal lattice is the fluoride ion. The operation of the electrode is consistent with the following reaction sequence:

$$F^-(ext) \rightleftharpoons F^-(LaF_3)$$

$$F^-(LaF_3) \rightleftharpoons F^-(int)$$

$$Ag(s) + Cl^-(int) \rightleftharpoons AgCl(s) + e^-$$

$$sum: \overline{F^-(ext) + Ag(s) + Cl^-(int) \rightleftharpoons AgCl(s) + F^-(int) + e^-}$$

All of the terms in the net reaction are constant for a given $F^-$ electrode except $F^-(ext)$.

Electrodes for measuring gases such as $O_2$ or $CO_2$ are also available. These electrodes utilize a thin Teflon membrane as the semipermeable barrier. The Teflon membrane allows the passage of small molecules like $CO_2$ or $O_2$, but not ions or large neutral molecules. Once the $CO_2$ or $O_2$ has passed through the barrier into an internal solution, its concentration is sensed either by measuring the concomitant change in the $pH$ of a dilute $NaHCO_3(aq)$ solution $(CO_2 + OH^-(aq) \rightarrow HCO_3^-(aq))$ as in the $CO_2$ gas electrode, or by measuring the electrode current arising from the reduction of oxygen on a rhodium coil under a constant, externally-applied voltage:

$$4e^- + O_2(aq) + 2H_2O(\ell) \rightarrow 4OH^-(aq) \qquad \text{(rhodium cathode)}$$

$$4Ag(s) + 4Cl^-(aq) \rightarrow 4AgCl(s) + 4e^- \qquad \text{(silver anode)}$$

The higher the concentration of oxygen, the greater the electrode current. These gas-sensitive electrodes are capable by monitoring $O_2$ and $CO_2$ directly in the gas phase.

## 13.9 BATTERIES.

A battery is simply an electrochemical cell, or a series of identical electrochemical cells, used as a source of power or voltage. The use of batteries is extensive, and undoubtedly will become even more so because of their pollution-free and noise-free operating characteristics. Lightweight battery-powered cars and trucks are already in operation, and hopefully batteries (or fuel cells—see below) will eventually deal the death blow to the infernal, internal combustion engine as a power plant for automobiles.* Batteries are used when mobility, portability, and high reliability are required. For many applications a battery is less expensive and simpler than a plug-in power supply with the same voltage characteristics.

Some common batteries are listed below:

### (a) The Carbon-Zinc Cell (Figure 13.6).

$$^\ominus Zn(s) \mid ZnCl_2 \cdot 2NH_3(s) \mid NH_4Cl(aq) \mid MnO_2(s) \mid Mn_2O_3 \cdot H_2O(s) \mid C^\oplus$$

The principal cell reaction, which is irreversible, is

$$Zn(s) + 2NH_4Cl(aq) + 2MnO_2(s) \rightarrow ZnCl_2 \cdot 2NH_3(s) + Mn_2O_3 \cdot H_2O(s)$$

Actually this equation is a great oversimplification, because the reactions which occur in the cell are many and complex. Their relative importance depends on the rate of current drain and the mode of operation. The chemi-

---

*A number of major scientific and engineering problems remain to be overcome. What is needed is a new lightweight, high-energy-density battery that retains a fair measure of efficiency under high current drain, and that operates with a low to moderate heat production. In principle, none of these difficulties are insuperable, and much research is in progress. The potentially most promising batteries to meet the requirements are of the type

$$Li(s) \mid LiX(\text{non-aqueous solution}) \mid CuX_2(s) \mid Cu(s)$$

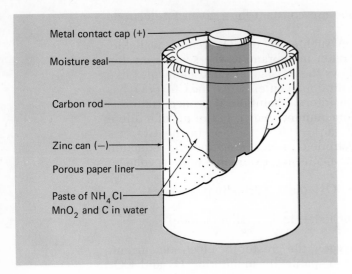

Metal contact cap (+)

Moisture seal

Carbon rod

Zinc can (−)

Porous paper liner

Paste of $NH_4Cl$
$MnO_2$ and C in water

**FIGURE 13.6** The carbon-zinc battery (LeClanche "dry" cell) in the charged condition. The operation of the cell leads to the consumption of $Zn(s)$, $NH_4Cl(aq)$, and $MnO_2(s)$, and the production of $ZnCl_2 \cdot 2NH_3(s)$ and $Mn_2O_3 \cdot H_2O(s)$, as the principal products.

cal reactions occuring in the cell establish a "fresh" voltage of 1.5 to 1.6 V; however, the voltage falls continuously during use, and in addition is especially low during cold weather. The carbon-zinc cell is a so-called *primary* cell, that is, a one-shot, non-rechargeable battery. It is widely used in flashlights, radios, and electronic instruments.[*]

## (b)   The Alkaline-Manganese Battery.

$$^{\ominus}Zn(s) \mid Na_2ZnO_2(aq),\ NaOH(aq) \mid MnO_2(s) \mid Mn_2O_3 \cdot H_2O(s) \mid steel^{\oplus}$$

$$Zn(s) + 2MnO_2(s) + 2NaOH(aq) = Mn_2O_3 \cdot H_2O(s) + Na_2ZnO_2(aq)$$

The construction of the alkaline-manganese battery is similar to that of the carbon-zinc battery. Relative to the carbon-zinc battery (with which it is a competitor), the alkaline-manganese battery has twice the capacity, a longer shelf life, and a steadier voltage under heavy current drain. It has a lower internal resistance, a higher available current, and it can be operated at much lower temperatures (down to −40°C). However, it costs roughly four times as much, primarily because of the more elaborate internal construction necessary to prevent $NaOH(aq)$ leakage.

## (c)   The Mercury Battery (Figure 13.7).

$$^{\ominus}Zn(Hg) \mid ZnO(s) \mid KOH(40\%\ aq) \mid HgO(s) \mid Hg(\ell) \mid steel^{\oplus}$$

$$Zn(Hg) + HgO(s) = ZnO(s) + Hg(\ell)$$

The great advantages of this cell are: a constancy of voltage (1.35 V) during discharge (because there is no change in cell electrolyte composition dur-

---

[*]Attempts should not be made to recharge batteries of this type (or of type (b)), because recharging may lead to an explosion, which results from the build-up of gas pressure in the battery.

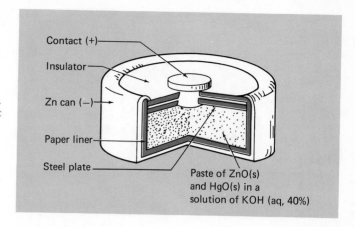

Contact (+)

Insulator

Zn can (−)

Paper liner

Steel plate

Paste of ZnO(s)
and HgO(s) in a
solution of KOH (aq, 40%)

**FIGURE 13.7** The mercury battery. The mercury is formed from HgO(s) at the lower surface of the steel plate.

ing operation); a high capacity (5 to 8 times the carbon-zinc cell); a very long shelf life; and an ability to supply large instantaneous currents.

The mercury battery is used in heart pacemakers, hearing aids, electric watches, and instruments involving transistors (which require a constant voltage). No gas is evolved during discharge, so the mercury battery does not rupture and it is safe to store in instruments (and humans). It costs about five times as much as an alkaline-manganese battery of the same size.

## (d)   The Weston Standard Cell.

There are two types of Weston standard cell, the saturated and unsaturated types:

$^{\ominus}$Cd(Hg) | CdSO$_4 \cdot \frac{8}{3}$ H$_2$O(s) | CdSO$_4$(aq, satd) | Hg$_2$SO$_4$(s) | Hg($\ell$)$^{\oplus}$
2-phase

12.5% Cd

$^{\ominus}$Cd(Hg) | CdSO$_4$(aq) | Hg$_2$SO$_4$(s) | Hg($\ell$)$^{\oplus}$
2-phase

12.5% Cd

The cell reaction for the unsaturated cell is

$$Cd(Hg) + Hg_2SO_4(s) = CdSO_4(aq) + 2Hg(\ell)$$

The voltage of this cell is virtually constant at 1.01875 V (with a CdSO$_4$(aq) concentration corresponding to the saturation solubility at 4°C) over the range from 10 to 30°C, and this cell is widely used as a standard reference voltage in electrochemical work. A group of 10 saturated cells is used to maintain the unit of voltage at the U.S. National Bureau of Standards.

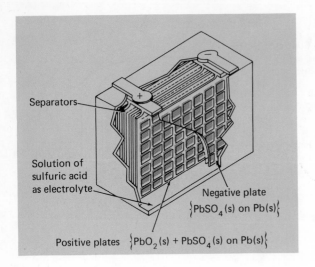

**FIGURE 13.8**   The lead-storage battery.

### (e)   The Lead-Storage Battery (Rechargeable) (Figure 13.8).

$$\ominus Pb(s) \mid PbSO_4(s) \mid H_2SO_4(aq) \mid PbO_2(s) \mid PbSO_4(s) \mid Pb(s)^{\oplus}$$

$$Pb(s) + PbO_2(s) + 2H_2SO_4(aq) = 2PbSO_4(s) + 2H_2O(\ell)$$

When fully charged, this cell develops 2.06 to 2.14 V; in use this drops quickly to 2.01 V, and then drops more slowly owing to dilution of the $H_2SO_4$(aq). The ordinary 12 V automobile battery consists of six of the above cells in series, and is capable of putting out well over 100 Amp for short periods.

### (f)   The Nickel-Cadmium Cell (Rechargeable).

$$\ominus steel \mid Cd(s) \mid Cd(OH)_2(s) \mid KOH(aq), LiOH(aq) \mid NiOOH(s) \mid Ni(OH)_2(s) \mid steel^{\oplus}$$

$$2NiOOH(s) + Cd(s) + 2H_2O(\ell) = 2Ni(OH)_2(s) + Cd(OH)_2(s)$$

This cell, which is a variant of the Edison cell (change Cd to Fe), involves a cell reaction that only consumes water from the aqueous phase, and, as a consequence, the cell voltage is quite steady. The cell develops 1.4 V when fully charged, which drops quickly on use to 1.3 V, and then slowly to 1.0 V. The nickel-cadmium cell is more stable than the lead storage battery, and it can be left inactive (charged or uncharged) for long periods without adverse effects. The development of the completely sealed nickel-cadmium battery has led to its widespread use in innumerable convenience devices—cordless electric shavers, knives, toothbrushes, etc. It is often used in photographic flash units.

In addition to the types of batteries noted above, there are several other important types of cells, such as fuel cells and solar cells. Fuel cells involve the *electrochemical oxidation of a fuel* and offer a major energy utilization advantage over the direct combustion of a fuel.* The Apollo

---

*An electrochemical cell, as opposed to the use of a fuel to power a heat engine, does not require a temperature difference to deliver its available energy as work, and therefore, it does not need to deposit a significant fraction of its available energy as heat in a low-temperature reservoir (see problem 22).

space vehicles derived the bulk of their electrical power from hydrogen-oxygen fuel cells:

$$2H_2(g) + O_2(g) = 2H_2O(\ell)$$

Biochemical fuel cells have been proposed to extract useful work from garbage by electrochemical oxidation. Biological fuel cells running on glucose and oxyhemoglobin have been proposed to power an artificial heart.

Solar cells are solid state photovoltaic cells. A typical solar cell is a silicon wafer 0.5 mm thick and 1 cm² in area which is doped on the light-sensitive surface with a very thin diffused layer of boron. The wafer constitutes a $p-n$ junction[*] and acts as a diode rectifier. Leads are connected to the boron-diffused surface ($p$) and the underlying silicon ($n$). Light striking the boron diffused surface produces "free" electrons and holes. Those electrons with sufficient energy pass across the $p-n$ junction into the pure silicon $n$-type region and then into the external circuit. A typical solar battery consists of eight of the above solar cells in series, and is capable of putting out 14.0 mA at 0.26 V/cm², which corresponds to about 3.4 watts/ft² of active surface.

## 13.10 TRANSMISSION OF ELECTRICAL SIGNALS IN LIVING MATTER.

The origin of electrochemistry can be traced back to the discovery of "animal electricity" by Luigi Galvani in 1791. External evidence for the underlying electrical activity in the body can be obtained from electrodes attached to the skin surface. The electrocardiogram (ECG) is the graphical record of the electrical signals emanating from the heart muscles. These signals are associated with the synchronous contraction of the cells surrounding the ventricles in the heart. The electromyogram (EMG) is the record of electrical signals associated with the operation of the skeletal muscles. The electroencephalogram (EEG) is the record of electrical activity associated with the partially synchronous operation of the billions of neurons in the brain.

A 0.5 sec, 30 mAmp, 60 cycle pulse of ac current through the brain produces a convulsion or fit. Similar currents through the trunk disrupt the synchronous operation of the heart muscle, causing it to stop (ventricular fibrillation). The most dangerous currents for humans are 50-60 cycle ac.

The response of the body to sensory inputs involves the transmission of electrical signals along the sensory nerve fibers to the central nervous system (CNS, brain and spinal cord); the CNS then sends out electrical signals along the motor nerve fibers to the muscles, which lead to muscle contractions. There are different types of sensory receptors (sensors) for each type of sensation. For example, the receptors may be triggered by light (eye), by small pressure changes (ear), by chemicals (nose and mouth), by heat flow (skin), etc.

---

[*]See Chapter 6.

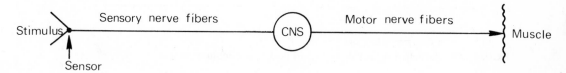

Information is transmitted through the nervous system by means of electrical signals called *action potentials*. An action potential (AP) has a fixed shape, amplitude, and velocity for a given fiber at a particular temperature. The intensity of a stimulus is determined by the number of fibers excited, not by changes in the amplitude or shape of the signal. An action potential is an explosive or regenerative process, a triggered, self-reinforcing event (analogous to the burning of a fuse). The action potential is an all-or-none type of signal with a stimulus *threshold* (the stimulus must exceed a certain minimum value), which protects the system from spontaneous firings triggered by, say, thermal fluctuations.

Human nerves consist of bundles of fibers; each fiber is $10^{-4}$ to $10^{-2}$ mm in diameter and up to 2 meters in length. The action potential is an electrical depolarization wave that moves along the fiber with a velocity of $125 \text{ m} \cdot \text{sec}^{-1}$ ($38°C$, $10^{-2}$ mm diameter fiber), an amplitude of 120 mV, and a width at half height of the order of 1 msec (Figure 13.9).

A nerve fiber or *axon* has a tube-like structure. The interior of the tube (*axoplasm*) is a gel containing water, electrolytes, fibrous proteins, and enzymes. The exterior of the fiber is in contact with the intercellular fluids (blood). The electrical excitability of the axon has been traced to the presence of an excitable membrane (*axolemma*), which is located in the wall of the axon. The axolemma is very thin; the thickness is about 70 Å. Some of the properties of the axolemma are given in Table 13.3. The ionic concentrations and the corresponding transmembrane equilibrium (Nernst) potentials for the various ions are given in Table 13.4. These ionic concentration gradients are the energy source for movement of the action potential along the axon. These gradients are maintained by an

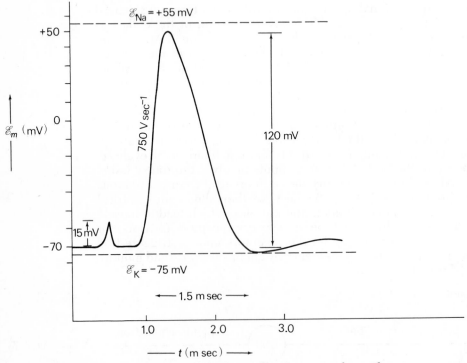

**FIGURE 13.9**  An action potential. The small peak is a stimulus artifact.

**TABLE 13.3**

PROPERTIES OF THE EXCITABLE NERVE AXON MEMBRANE

| | |
|---|---|
| $x_m$ (thickness): | $\sim 70$ Å |
| $c_m$ (capacity): | $1.0 \; \mu F \cdot cm^{-2}$ |
| $\epsilon_m$ (dielectric constant): | $8$–$10$ |
| $r_m \cdot A_m$ (resistance $\times$ area): | $1000$–$3000$ ohm $\cdot$ cm$^2$ (rest) |
| | $25$ ohm $\cdot$ cm$^2$ (excited) |
| | ($10^8$ for phospholipid bilayer) |
| $\rho_m$ (electrical resistivity): | membrane $2 \times 10^9$ ohm $\cdot$ cm |
| | axoplasm 30 ohm $\cdot$ cm |
| | external 20 ohm $\cdot$ cm |
| $(r_m/r_i)^{1/2}$ (length constant): | 6 mm (0.5 mm diam axon) |
| $\tau_m$ (time constant, $r_m \cdot c_m$): | 0.7 msec |
| $D_{K^+,m}$ (diffusion constant): | $\sim 2 \times 10^{-12} \; cm^2 \cdot sec^{-1} \approx 10^{-7} \, D_{K^+ \cdot H_2O}$ |
| $P_{K^+,m}$ (permeability): | $\sim 2 \times 10^{-6}$ cm/sec |
| $\mathscr{E}_m$ (membrane potential, rest): | $-70$ mV (negative inside) |
| $E_m$ (electric field, rest): | $\sim 10^5$ volt $\cdot$ cm$^{-1}$ |

active pumping process (Na-pump) that transfers $Na^+$ from the inside of the axon to the outside, and $K^+$ passes from the outside to the inside during the process. The Na-pump is thought to be a ($Na^+$, $K^+$)-activated ATPase.

If only a single ion could pass through the axolemma, then the trans-membrane potential, $\mathscr{E}_m$, would be equal to the Nernst potential for that ion

$$\mathscr{E}_m = \frac{RT}{z_x F} \ln \left( \frac{a_{x,out}}{a_{x,in}} \right)$$

The membrane potential in the resting state ($\sim -70$ mV, negative inside) is determined primarily by $K^+$, because this is the most permeable ion in the resting state. Depolarizations (outward-directed positive currents that *decrease* $\mathscr{E}_m$) in excess of threshold ($\mathscr{E}_m > -55$ mV) trigger a change in membrane permeability from a value of $P_{Na^+}/P_{K^+} \simeq 1/100$ to $P_{Na^+}/P_{K^+} \simeq 12$:

$$\mathscr{E}_m = \frac{RT}{F} \ln \left\{ \frac{(K^+)_o + (P_{Na^+}/P_{K^+})(Na^+)_o}{(K^+)_i + (P_{Na^+}/P_{K^+})(Na^+)_i} \right\}$$

The change in membrane permeability to $K^+$ and $Na^+$ leads to an inward surge of sodium ions ($\sim 1$ mAmp $\cdot$ cm$^{-2}$ or 4 picomole $Na^+$/cm$^2$/impulse) and a shift in $\mathscr{E}_m$ toward the Nernst potential for $Na^+$ ($+55$ mV). However, the increased $Na^+$ permeability is short-lived and, as $P_{Na^+}$ decreases, $P_{K^+}$ increases leading to an outward $K^+$ current, which restores the trans-membrane potential to its normal resting value. The action potential propagates along the axon by depolarizing (and thereby exciting) the regions adjacent to the excited or active patch.

**TABLE 13.4**

IONIC CONCENTRATION GRADIENTS ACROSS THE NERVE MEMBRANE

|  | inside (mM) | outside (mM) |
|---|---|---|
| (K$^+$) | 400 | 20 |
| (Na$^+$) | 50 | 440 |
| [Ca$^{2+}$]$_{tot}$ | 0.4 | 10 |
| pH | 7.3 | 8.0 |
| (X$^-$) | Cl$^-$ plus organics | mostly Cl$^-$ |

$$\frac{(K^+)_o}{(K^+)_i} \simeq \frac{1}{20} \qquad \frac{(Na^+)_o}{(Na^+)_i} \simeq 9 \qquad \frac{(Ca^{2+})_o}{(Ca^{2+})_i} \sim 10^3 \qquad \frac{(H^+)_o}{(H^+)_i} \sim 0.20$$

$$\mathscr{E}_{K^+} = -75 \text{ mV} \qquad \mathscr{E}_{Na^+} = +55 \text{ mV} \qquad \mathscr{E}_{Ca^{2+}} \sim +90 \text{ mV} \qquad \mathscr{E}_{H^+} \sim -41 \text{ mV}$$

The nature of the processes occurring at the molecular level that give rise to the action potential are not known for certain. These changes may be protein conformational changes, or a change in the activity level of an enzyme that produces a transitory change in the concentrations of membrane carriers for Na$^+$ and K$^+$, or something else.

The action potential travels along the axon until it reaches either a synaptic junction (connection between nerve cells) or a peripheral (neuromuscular) contact. The arrival of an AP at a junction triggers the release of *acetylcholine* .

$$\underset{\text{CH}_3\overset{\displaystyle\text{O}}{\overset{\displaystyle\|}{\text{C}}}\text{OCH}_2\text{CH}_2\overset{+}{\text{N}}(\text{CH}_3)_3}{}$$

from small spherical sacs ($\sim$200 Å in diameter) called *synaptic vesicles*. The acetylcholine diffuses across a gap about 200 Å wide and triggers a depolarization of the post synaptic membrane ($\mathscr{E}_m \to 0$). The acetylcholine is thought to combine with a receptor protein and cause it to undergo a conformational change. The acetylcholine is eventually destroyed by the enzyme acetylcholinesterase (AchE):

$$\text{CH}_3\overset{\displaystyle\text{O}}{\overset{\displaystyle\|}{\text{C}}}\text{OCH}_2\text{CH}_2\overset{+}{\text{N}}(\text{CH}_3)_3 \xrightarrow[\text{H}_2\text{O}]{\text{AchE}} \text{CH}_3\text{COOH} + \text{HOCH}_2\text{CH}_2\overset{+}{\text{N}}(\text{CH}_3)_3$$

## PROBLEMS

1. Concentrated sulfuric acid is often shipped in steel drums. Why is this possible, considering that $\mathscr{E}^{\circ}_{red}$ for Fe$^{2+}$(aq) + 2e$^-$ = Fe(s) is $-0.44$ V?

2. Consider the following electrochemical cells, and in each case write the balanced cell reaction, compute the cell voltage, $\mathscr{E}$ (assuming $\gamma_i = 1$), and in each case indicate the negative electrode:

(a) $Cu(s) \mid CuI(s) \mid KI(0.100 \text{ M, aq}) \mid AgI(s) \mid Ag(s)$

(b) $H_2(g, 0.50 \text{ atm, Pt}) \mid H_2SO_4(1.00 \text{ M, aq}) \mid PbSO_4(s) \mid Pb(s)$

(c) $(Pt) \; Fe^{2+}(aq, 0.10 \text{ M}), \; Fe^{3+}(aq, 0.20 \text{ M}), \; HClO_4(aq, 0.50 \text{ M}) -$
$\mid NaClO_4 \mid KCl(1.0M) \mid Hg_2Cl_2(s) \mid Hg(\ell)$
salt bridge

(d) $H_2(g, 2 \text{ atm, Pt}) \mid HI(0.50 \text{ M, aq}) \mid I_2(s, Pt)$

(e) $Ag(s) \mid Ag_2CrO_4(s) \mid K_2CrO_4(0.150 \text{ M, aq}) \mid PbCrO_4(s) \mid Pb(s)$

(f) $(Au) \; V^{2+}aq(0.10 \text{ M}), \; V^{3+}(aq, 0.20 \text{ M}), \; HClO_4(0.25) \text{ M} \mid NaClO_4 \mid -$
salt bridge

$$K_2SO_4(0.20 \text{ M}) \mid Hg_2SO_4(s) \mid Hg(\ell)$$

3. Given the following data (25°C) ($\mathscr{E}^\circ_{red}$ values in volts)

$$XO_2(s) \xrightarrow{+1.00} X^{3+}(aq) \xrightarrow{-0.60} X^{2+}(aq) \xrightarrow{+0.20} X(s)$$

predict what would happen if:

(a) $XCl_2$ (soluble) were treated with water;

(b) $X(s)$ (metal) were treated with $Fe(NO_3)_3(aq)$ in 1 M $H^+(aq)$.

$$Fe^{3+}(aq) + e^- = Fe^{2+}(aq) \qquad \mathscr{E}^\circ_{red} = +0.77 \text{ V}$$

(c) Calculate the equilibrium constant for the reaction

$$X(s) + 2X^{3+}(aq) = 3X^{2+}(aq)$$

(d) Calculate the equilibrium constant for the reaction

$$XO_2(s) = X(s) + O_2(g)$$

(e) Calculate $\mathscr{E}^\circ_{red}$ for

$$X^{3+}(aq) + 3e^- = X(s)$$

(f) Calculate $\mathscr{E}^\circ_{red}$ for

$$X(OH)_2(s) + 2e^- = X(s) + 2OH^-(aq)$$

given

$$K_{sp} = 1 \times 10^{-18} \text{ for } X(OH)_2(s)$$

(g) Given $\mathscr{E}^\circ_{red} = -0.50 \text{ V}$ for

$$X_2Y(s) + 4e^- = 2X(s) + Y^{4-}(aq)$$

calculate the $K_{sp}$ of $X_2Y(s)$.

(h) Air is bubbled through a 0.01 M solution of $X^{3+}(aq)$ which is 0.20 M in $H^+(aq)$; calculate the equilibrium concentration of $X^{3+}(aq)$.

4. A mercury cathode can often be used to separate two or more metals whose $\mathscr{E}°$ (half-cell) values differ by less than 0.2 V, if the metal with the higher $\mathscr{E}°_{red}$ value is soluble in mercury and the other metals are not. Explain this using the Nernst equation, and find a practical example.

5. Compute the $pH$ at which $H_2(g)$ begins to evolve in an electrolysis involving a 0.10 M solution of the metal A ($A^{2+}(aq) + 2e^- = A(s)$, $\mathscr{E}°_{red} = -0.25$ V). Take the hydrogen overvoltage as 1.00 V.

6. Calculate the voltage of the $Fe^{2+}/Fe^{3+}$ half-reaction at the equivalence point in the titration of 25.00 ml of 0.100 M $Fe^{2+}(aq)$ with 0.100 M $Na_2Cr_2O_7(aq)$.

7. One of the major disadvantages of the lead storage battery as a power source for an automobile is its high mass. What would be a desirable anode metal in a battery from the point of view of low mass and low cost?

8. Consider the following Latimer electrode potential diagram for plutonium (values given in volts) in acid solution:

$$PuO_2^{2+}(aq) \xrightarrow{+0.93} PuO_2^+(aq) \xrightarrow{+1.15} Pu^{4+}(aq) \xrightarrow{+0.97} Pu^{3+}(aq) \xrightarrow{-2.07} Pu(s)$$

with an upper bracket ( ) spanning $PuO_2^{2+}$ to $Pu^{4+}$, a lower bracket ( ) spanning $PuO_2^+$ to $Pu^{3+}$, and a lower bracket spanning $Pu^{4+}$ to $Pu(s)$.

(a) Compute the standard reduction potentials for the couples indicated by ( ) and insert your answers in the parentheses.
(b) Draw circles around those species present in the diagram that are unstable with respect to disproportionation.
(c) Will Pu(s) liberate $H_2(g)$ from a neutral solution? (Show your reasoning.)
(d) Will $Pu^{4+}(aq)$ liberate $O_2(g)$ from an aqueous solution 1 M in $H^+(aq)$? (Show your reasoning.)
(e) Compute the equilibrium constant for the reaction:

$$2H_2O(\ell) + 2Pu^{4+}(aq) = Pu^{3+}(aq) + PuO_2^+(aq) + 4H^+(aq)$$

9. (a) The standard electrode potential of the $Cu^{2+}(aq) \mid Cu(s)$ electrode at 25°C is

$$Cu^{2+}(aq) + 2e^- = Cu(s) \qquad \mathscr{E}° = +0.337 \text{ V}$$

and the solubility product of copper ferrocyanide, $Cu_2Fe(CN)_6$, is $1.2 \times 10^{-16}$. Compute the standard electrode potential of the $Cu_2Fe(CN)_6(s) \mid Cu(s)$ electrode

$$Cu_2Fe(CN)_6(s) + 4e^- = 2Cu(s) + Fe(CN)_6^{4-}(aq)$$

(b) The standard electrode potential of the $Ag^+(aq) \mid Ag(s)$ electrode at 25°C is

$$Ag^+(aq) + e^- = Ag(s) \qquad \mathscr{E}° = +0.800 \text{ V}$$

and the solubility product of silver cobalticyanide, $Ag_3Co(CN)_6$, is $8.5 \times 10^{-21}$ (25°C). Compute the standard electrode potential of the $Co(CN)_6^{3-}(aq) \mid Ag_3Co(CN)_6(s) \mid Ag(s)$ electrode

$$Ag_3Co(CN)_6(s) + 3e^- = 3Ag(s) + Co(CN)_6^{3-}(aq)$$

10. Consider the following Latimer electrode potential diagram for americium ($\mathscr{E}°$ values given in volts) in acid solution at 25°C:

$$AmO_2^{2+}(aq) \xrightarrow{+1.64} AmO_2^+(aq) \xrightarrow{+1.26} Am^{4+}(aq) \xrightarrow{+2.18} Am^{3+}(aq) \xrightarrow{-2.32} Am(s)$$

(a) Compute the standard electrode potentials for the couples indicated by (        ) and insert your answers in the parentheses.
(b) Draw circles around those species present in the diagram that are unstable with respect to disproportionation.
(c) Will $AmO_2^{2+}(aq)$ liberate $O_2(g)$ from an aqueous solution 1 M in $H^+(aq)$? (Show your reasoning.)
(d) Will $Am(s)$ liberate $H_2(g)$ from a neutral aqueous solution? (Show your reasoning.)
(e) Compute the equilibrium constant for the reaction:

$$4H^+(aq) + 2AmO_2^+(aq) = AmO_2^{2+}(aq) + Am^{4+}(aq) + 6H_2O(\ell)$$

11. If the reaction

$$M(s) + R^{2+}(aq) = M^{2+}(aq) + R(s)$$

is to be quantitative to 1 part per 1000, how much difference must there be in the $\mathscr{E}°$ values for the two half-reactions?

12. Construct a titration curve for the titration of 25.00 ml of 0.100 M $KI_3(aq)$ with 0.050 M $Na_2S_2O_3(aq)$.

13. Give the cell diagrams for the electrochemical cells in which the following chemical reactions occur:

(a) $PdI_2(s) + Zn(s) = ZnI_2(aq) + Pd(s)$
(b) $H_2(g) + Cl_2(g) = 2HCl(aq)$
(c) $Cd(s) + Hg_2SO_4(s) = CdSO_4(aq) + 2Hg(\ell)$
(d) $K_4Fe(CN)_6(aq) + \frac{1}{2}Hg_2Cl_2(s) = K_3Fe(CN)_6(aq) + KCl(aq) + Hg(\ell)$

14. In the electrolytic separation of the metal A from the metal B ($\mathscr{E}°_{red,B} > \mathscr{E}°_{red,A}$) from a solution initially 0.125 M in $A^+(g)$ and 0.200 M in $B^{2+}(aq)$, by how many volts must $\mathscr{E}°_{B^{2+}\mid B}$ exceed $\mathscr{E}°_{A^+\mid A}$ in order that the separation of A from B can be considered quantitative (i.e., that no more than 0.01% of the $A^+$, initially present, remains in solution when B begins to deposit)?

15. The standard electrode potentials in volts for the various iodine couples in 1 M $H^+(aq)$ and 1 M $OH^-(aq)$ at 25°C are:
*in 1 M $H^+(aq)$:*

$$H_5IO_6(aq) \xrightarrow{+1.7} IO_3^-(aq) \xrightarrow{+1.14} HIO(aq) \xrightarrow{+1.45} I_3^-(aq) \xrightarrow{+0.54} I^-(aq)$$

with $+1.09$ spanning $IO_3^-$ to $I_3^-$, $+1.20$ spanning $IO_3^-$ to $HIO$, and $+0.54$ spanning $HIO$ to $I^-$.

*in 1 M $OH^-(aq)$:*

$$H_3IO_6^{2-}(aq) \xrightarrow{+0.7} IO_3^-(aq) \xrightarrow{+0.14} IO^-(aq) \xrightarrow{+0.45} I_3^-(aq) \xrightarrow{+0.54} I^-(aq)$$

with $+0.21$ spanning $IO_3^-$ to $I_3^-$, $+0.49$ spanning $IO^-$ to $I^-$, and the lower span.

(a) Using the $\mathscr{E}°$ values given, determine which iodine species are stable in 1 M $H^+(aq)$ and in 1 M $OH^-(aq)$.
(b) Write the equation for the reaction which occurs when $I_3^-(aq)$ is added to 1 M $OH^-(aq)$. Compute the equilibrium constant for the reaction.
(c) Write the equation for the reaction which occurs when NaIO is dissolved in 1 M $OH^-(aq)$. Calculate the equilibrium constant for the reaction.
(d) If 8.0 moles of $I^-(aq)$ are added to 1.0 mole of $MnO_4^-(aq)$ in 1 M $H^+(aq)$, what are the products of the reaction (see Figure 13.2 for data on manganese species)?
(e) If fewer than 3.0 moles of $I^-(aq)$ are added to 1.0 mole of $MnO_4^-(aq)$ in 1 M $H^+(aq)$, what are the products of the reaction?

16. Design a self-sustaining "chromous bubbler," that is, one which can continuously remove $O_2(g)$ from a mixture without the need for *external* replenishment of $Cr^{2+}(aq)$ in the bubbler. Write the reactions that occur in the bubbler when $O_2$ is introduced.

17. (a) Write the balanced half-reactions that occur at the anode and cathode for each of the batteries given in Section 13.9.
(b) Apply the Nernst equation to the cell reaction in cases (d) and (e); estimate $\mathscr{E}$ for the fully charged cells.
(c) Suggest a simple explanation (in terms of the nature of the cell reaction) for the fact that the voltage of the nickel-cadium cell drops only very *slowly* on use.

18. Suggest an explanation for why the reaction

$$Pb(s) + H_2SO_4(aq) = PbSO_4(s) + H_2(g)$$

does not occur spontaneously to an appreciable extent in the lead-storage battery.

19. Suppose the reaction of a cell is

$$Zn(s) + Cu^{2+}(aq) = Cu(s) + Zn^{2+}(aq)$$

which of the following changes will lead to an increase, a decrease, or have no effect on the cell voltage?

    (a)  an increase in $(Cu^{2+})$

    (b)  an increase in $(Zn^{2+})$

    (c)  an increase in the size of the $Zn(s)$ electrode

    (d)  an increase in the temperature $(\Delta S_{rctn} < 0)$

20.  When a cell operates reversibly at fixed temperature and pressure, is the heat exchanged with the surroundings given by $\Delta H$ or $T\Delta S$?

21.  Several new batteries have been proposed involving alkali metal anodes and organic solvents for the electrolyte phase, for example:

$$Li(s) \,|\, LiX(soln) \,|\, CuX_2(s) \,|\, Cu(s)$$

List some of the major advantages, relative to a lead-storage battery, that an operational cell of this type would possess.

22.  Consider the use of hydrogen as a fuel (a) in a heat engine operating between 1000 K and 300 K, and (b) in an electrochemical cell operating at 300 K. The maximum work obtainable from a reaction run in an electrochemical cell is $w_{el} = -\Delta G$, whereas for a heat engine

$$w = -\Delta H \left( \frac{T_2 - T_1}{T_2} \right) \qquad\qquad (T_2 > T_1)$$

Use data in Appendix 5. Assume $P_{H_2} = P_{O_2} = 10$ atm in the fuel cell. Discuss the significance of the two values of $w$.

---

**References**

13.1.  W. M. Latimer, *The Oxidation States of the Elements and Their Potentials in Aqueous Solutions,* Second Edition (Prentice-Hall Inc., New York, 1952).

13.2.  G. Kortüm, *Treatise on Electrochemistry* (Elsevier Pub. Co., New York, 1965).

13.3.  R. A. Durst (ed.), *Ion-Specific Electrodes* (U.S. Government Printing Office, Washington, D.C., 1969), NBS Special Publication 314.

13.4.  D. J. G. Ives and G. J. Janz (eds.), *Reference Electrodes* (Academic Press, New York, 1961).

13.5.  B. Katz, *Nerve, Muscle, and Synapse* (McGraw-Hill, New York, 1966).

# CHAPTER 14

On the basis of the latter hypothesis, however, one can develop the equation, which describes the [rate of the] chemical reaction [i.e., the acid-catalyzed decomposition of sucrose in aqueous solution], in the following way:

$dZ$ is the sugar loss in the time-element $dt$, and one accepts the same to be determined by the equation:

$$-\frac{dZ}{dt} = MZS$$

where $M$ is [the concentration of added acid and $S$ is a constant] . . .

The above equation gives on integration ($Z = Z_0$ at $t = 0$):

$$Z = Z_0 e^{-MSt}$$

L. Wilhelmy (1850)*

# CHEMICAL KINETICS

## Part I. The Reaction-Rate Law

## 14.1  GENERAL CONSIDERATIONS.

The subject matter of chemical kinetics includes the methods for determining at what speed a reaction proceeds (the reaction rate), how a reaction proceeds (the reaction mechanism), and why it takes place at the observed rate according to the observed mechanism(s) (reaction-rate theory).

The reason chemical kinetics is such an important area of chemistry is that much of chemistry, including that involved in living organisms (e.g., you), involves metastable substances—that is, substances that are not at equilibrium with respect to all possible processes. An understanding of the factors that govern the rates and mechanisms of chemical reactions gives us the means to exert a certain degree of control over these reactions. That is to say, we can often choose conditions such that a certain reaction does or does not take place at an appreciable rate. A less pragmatic reason for the study of reaction kinetics is that the determination and prediction of reaction pathways and reaction rates is an intrinsically interesting, challenging, creative, and productive way to spend a part of one's time.

You will recall that equilibrium thermodynamics can answer the question of whether or not a certain reaction *can* occur, but it has nothing whatsoever to say about whether or not a possible reaction actually *will* occur. In addition, given that a possible reaction does take place, equilibrium thermodynamics cannot tell us anything about the rate at which, or the pathway(s) by which, the reaction proceeds to the equilibrium state.

---

*Ludwig Wilhelmy, *Annalen der physik*, *81* (1850), 413–428 and 499–526; quote from p. 418.

Important clues to the reaction mechanism can often be obtained by studying the influence of various factors on the reaction rate, among which are:

(1) *concentration* (of reactants, products, or other added substances whose presence does not affect the reaction stoichiometry but, nonetheless, affects the reaction rate—for example, catalysts and inhibitors);

(2) *temperature;* and

(3) *nature of the reaction medium* (e.g., the type of solvent, the effect of added electrolytes, electric and magnetic fields).

Of all the factors that affect the rate of a reaction, the effect of *concentration,* as expressed by the rate law, has the greatest significance in deciding questions about the reaction mechanism. Therefore, we must understand what a rate law is, what it implies about the mechanism, and what the limitations are on the information that it yields.

## 14.2 THE REACTION RATE LAW.

A balanced chemical equation indicates only what species are consumed and what species are produced in what proportions; that is, it gives the stoichiometry of the reaction. The reaction stoichiometry does not, in general, disclose the pathway by which the reaction occurs. Although a chemical reaction may occur in a single step (in which case it is called an *elementary reaction* or *elementary process,* and the reaction pathway corresponds to that given by the reaction stoichiometry), most reactions take place in a sequence of steps, the sum total of which corresponds to the overall reaction stoichiometry. In such cases there is no completely unambiguous *a priori* method of selecting the actual reaction pathway from among the many stoichiometrically acceptable possibilities.

An example of a reaction that occurs in a single elementary step is the gas-phase dimerization of nitrogen dioxide:

$$2NO_2(g) = N_2O_4(g)$$

The dependence of the rate of formation of the dimer, expressed as moles $\cdot$ liter$^{-1} \cdot$ sec$^{-1}$ (or, more briefly, $M \cdot$ sec$^{-1}$),* on the concentration of gaseous nitrogen dioxide indicates that the forward reaction proceeds via a single elementary reaction

$$NO_2(g) + NO_2(g) \rightarrow N_2O_4(g)$$

Reaction-rate data on the gas-phase reaction between hydrogen and bromine

$$H_2(g) + Br_2(g) = 2HBr(g)$$

indicates that this reaction does not proceed via a single elementary process, but rather proceeds (at least in the very early stages of the reaction) via the following *chain mechanism*†:

---

*These are the preferred units for the reaction rate, although the rate is sometimes expressed as $M \cdot$ min$^{-1}$, or when dealing with gas reactions as Torr $\cdot$ min$^{-1}$, atm $\cdot$ min$^{-1}$, Torr $\cdot$ sec$^{-1}$, or atm $\cdot$ sec$^{-1}$.

†Further into the reaction (i.e., as the HBr concentration builds up) it is necessary to include the chain inhibition step $H + HBr \rightarrow H_2 + Br$ in the reaction mechanism.

$$Br_2 \rightleftharpoons 2Br$$

       initiation of the chain

       termination of the chain

$$Br + H_2 \rightarrow HBr + H$$

$$H + Br_2 \rightarrow HBr + Br$$

       propagation of the chain

$$H_2 + Br_2 = 2HBr \qquad\qquad \text{sum}$$

The data obtained in a kinetic investigation of the decomposition of nitrous acid in acidic aqueous solutions

$$3HNO_2(aq) = H^+(aq) + NO_3^-(aq) + 2NO(g) + H_2O(\ell)$$

are consistent with the following reaction mechanism (i.e., sequence of elementary processes):

(a) $2HNO_2 \rightleftharpoons H_2O + NO + NO_2$     rapid equilibrium

(b) $NO_2 + NO_2 \rightarrow NO^+ + NO_3^-$     slow (rate determining)

(c) $NO^+ + H_2O \rightarrow HNO_2 + H^+$     rapid

Multiplying step (a) by two (because step (b) requires two $NO_2$ species), and adding to the result steps (b) and (c), yields the net reaction stoichiometry. When a reaction mechanism involves two or more steps (elementary processes) and the rate of one of the steps is considerably less than that for any of the other steps, then the slow step is said to be *rate determining*. In other words, the overall rate of a reaction, the mechanism of which involves a sequence of elementary processes, cannot exceed the rate of the slowest step in the sequence. Not all reaction mechanisms involve a *single* rate-determining step. In some cases two or more of the elementary processes involved control the overall reaction rate, and in such cases we speak of the rate-determining steps.

    The relationship between the rate of the reaction and the concentrations of the species that affect the reaction rate is called the *rate law*. After the stoichiometry of the reaction has been established, the usual first step in a kinetic investigation is the determination of the rate law, because the rate law provides the single most important clue to the mechanism of a reaction. The rate law is basically a generalization from experiment, and can be regarded as valid only for the range of conditions (concentration, temperature, nature of the reaction medium, etc.) for which it has been tested. Additional experimentation may prove the need for a modification of the rate law, but if the initial rate law has been formulated properly it will always be contained as a special case in the more general form of the rate law.

## 14.3   FIRST-ORDER KINETICS.

    In order to illustrate the analysis of kinetic data for a reaction that involves first-order kinetics, we shall analyze the gas-phase decomposition of nitrogen pentoxide:

$$N_2O_5(g) = 2NO_2(g) + (1/2) O_2(g)$$

What is done is to take a sample of $N_2O_5$ prepared, say, at $0°C$, where its rate of decomposition is quite slow, and raise its temperature to, say, $45°C$, where the decomposition proceeds at a conveniently measurable rate. The partial pressure of unreacted $N_2O_5$ as a function of time (starting from time zero) can then be obtained either from determinations of the total pressure in the reaction vessel at various times,[*] or by means of spectroscopic techniques that can be employed to obtain the concentration of $N_2O_5$ directly. Let the initial partial pressure (i.e., at $t=0$) of $N_2O_5$ be $P_0$, and let its partial pressure at time $t$ be $P_t$. Utilizing the reaction stoichiometry,[†] the partial pressures of $N_2O_5$, $NO_2$, and $O_2$ at time $t$ are then expressible as

$$P_{N_2O_5} = P_t$$

$$P_{NO_2} = 2(P_0 - P_t)$$

$$P_{O_2} = (P_0 - P_t)/2$$

The total pressure in the reaction vessel at time $t$, $P_{tot}(t)$, is then

$$P_{tot}(t) = P_{N_2O_5} + P_{NO_2} + P_{O_2} = 5P_0/2 - 3P_t/2$$

Therefore,

$$P_t = (5P_0 - 2P_{tot})/3$$

This expression is used to compute the partial pressure of unreacted $N_2O_5$ at any time $t$ from a knowledge of the initial partial pressure of $N_2O_5$, $P_0$ (assuming that this was the only gas present to begin with), and of the total pressure in the reaction system. In Figure 14.1 we have plotted $P_t$ versus $t$ for the decomposition of $N_2O_5$ at $45°C$, with $P_0 = 247.3$ Torr. The rate of disappearance of $N_2O_5$ (that is, $R_{D,N_2O_5}$, where the subscript $D$ stands for disappearance[‡]) at any particular time, $t$, is equal to the negative of the slope[§] of $P_t$ versus $t$ curve at that time:

$$R_{D,N_2O_5}(\text{at } t) = -\text{slope}(\text{at } t) = -\left\{\frac{\text{vertical distance}}{\text{horizontal distance}}\right\} = -\left\{\frac{P_{t_2} - P_{t_1}}{t_2 - t_1}\right\} (\text{at } t)$$

In other words, if we want the slope of the $P_t$ versus $t$ curve at a *particular* $P_t$ (or $t$), we draw a tangent to the curve at that point, obtain the vertical and horizontal increments of the straight line tangent from the graph, and divide the former by the latter. From Figure 14.1 we obtain the data in

---

[*]This method works only if the reaction involves a change in the number of moles of gas on going from reactants to products. Can you explain why this is so?

[†]The analysis of the pressure data as a function of time for the $N_2O_5$ decomposition is complicated by the existence of the equilibrium $2NO_2(g) \rightleftharpoons N_2O_4(g)$. We shall assume in the following treatment that this complication can be ignored.

[‡]$R_{D,N_2O_5} = -dP_{N_2O_5}/dt = -dP_t/dt$.

[§]The slope of a linear equation of the type $y = mx + b$, where $m$ and $b$ are constants, is (where $x_1, y_1$ and $x_2, y_2$ are two points on the line)

$$\frac{y_2 - y_1}{x_2 - x_1} = m.$$

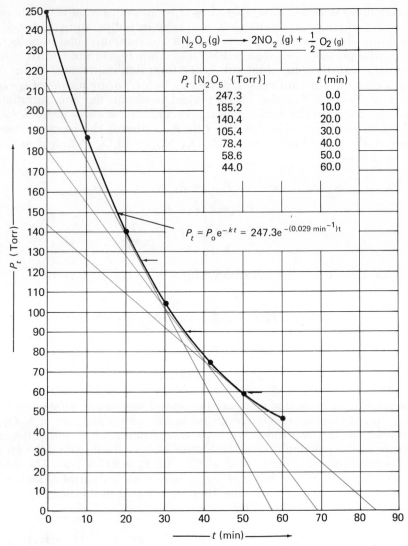

**FIGURE 14.1**   Plot of the pressure of $N_2O_5$ as a function of time for the gas-phase decomposition of $N_2O_5$ ($P_o = 247.3$ Torr, $T = 318K$).

Table 14.1. These results show that the rate of disappearance of $N_2O_5$ is directly proportional to the partial pressure of unreacted $N_2O_5$; that is, the reaction rate law is

$$R_{D,N_2O_5} = kP_{N_2O_5} = kP_t$$

where $k$ is the reaction *rate constant* or specific rate (i.e., the rate when $P_t$ is unity). The rate constant is the proportionality constant between the reaction rate and the concentrations (or partial pressures) upon which the rate depends. The value of the rate constant for this reaction at 45°C is

$$k = 0.029 \text{ min}^{-1} = (0.029 \text{ min}^{-1}) \left( \frac{1 \text{ min}}{60 \text{ sec}} \right) = 4.8 \times 10^{-4} \text{ sec}^{-1}$$

**TABLE 14.1**

RATE CONSTANT DETERMINATION

| $t$(min) | $P_{N_2O_5}=P_t$(Torr) | $R_{D,N_2O_5}=-$slope(Torr·min$^{-1}$) | $R_{D,N_2O_5}/P_{N_2O_5}$(min$^{-1}$)$=k$ |
|---|---|---|---|
| 23.5 | 127.0 | $-\left(\dfrac{214-0}{0-58}\right)=3.69$ | $\left(\dfrac{3.69}{127.0}\right)=0.029$ |
| 35.5 | 90.0 | $-\left(\dfrac{181-0}{0-69.5}\right)=2.60$ | $\left(\dfrac{2.60}{90.0}\right)=0.029$ |
| 50.0 | 58.6 | $-\left(\dfrac{144-0}{0-83.5}\right)=1.72$ | $\left(\dfrac{1.72}{58.6}\right)=0.029$ |

The equation for the curve in Figure 14.1 is

$$P_t = 247.3\, e^{-(0.029\ \text{min}^{-1})t} = P_0 e^{-kt}$$

For example, at $t = 10.0$ min we compute

$$P_t = 247.3\, e^{-0.29} = 247.3 \times 10^{-0.126} = 247.3 \times 0.748 = 185\ \text{Torr}$$

The rate of decomposition of $N_2O_5$ is said to be *first order* in $N_2O_5$, because the reaction rate is proportional to the first power of the $N_2O_5$ concentration.* Note that if the rate law were *not* first order in $N_2O_5$, then $R_{D,N_2O_5}/P_{N_2O_5} \neq$ constant.

The expression

$$P_t = P_0 e^{-kt} \tag{14.1}$$

or its equivalent in terms of concentrations ($P \propto C$)

$$\boxed{C_t = C_0 e^{-kt}} \tag{14.2}$$

is a general result for a first-order-decay process.† Therefore, we can use this expression to determine whether or not a reaction is first order. Taking the natural logarithm of both sides of Equation (14.1) yields

$$\ln P_t = \ln(P_0 e^{-kt}) = \ln P_0 + \ln e^{-kt} = \ln P_0 - kt$$

---

*Note that since $P_{N_2O_5} V = n_{N_2O_5} RT$,

$$P_{N_2O_5} = \left(\frac{n_{N_2O_5}}{V}\right) RT = C_{N_2O_5} RT$$

and the pressure of a gas is directly proportional to its concentration, provided that the gas behaves ideally.

† If $-dx/dt = kx$, then $d \ln x = -kdt$, and $\int_{X_0}^{X_t} d \ln x = -\int_0^t kdt$; therefore, $X_t = X_0 e^{-kt}$.

and converting to base 10 (i.e., common) logarithms we obtain ($\ln 10 = 2.3026 \cong 2.30$)

$$\log P_t = \log P_0 - \left(\frac{k}{2.30}\right) t \qquad (14.3)$$

Equation (14.3) shows that a plot of $\log P_t$ versus $t$ (or $\log C_t$ versus $t$) will be linear (i.e., a straight line) with a slope equal to $-k/2.30$, if the reaction is first order. And conversely (as we shall see subsequently), if the reaction is not first order, then such a plot will not be linear (provided the data are taken sufficiently far into the course of the reaction, say, at least until the reaction is about 75% complete). Such a $\log P_t$ versus $t$ plot can most conveniently be made on semi-log graph paper, which is scaled so that it does the job of taking logarithms for us. In Figure 14.2 we have constructed just such a plot for the $N_2O_5$ decomposition, using the same data as were employed to construct Figure 14.1. It does not take a sharp eye to see that the curve obtained is indeed straight, and, therefore, that the decomposition is first order in $N_2O_5$. From the slope of the line we find the rate constant to be $0.029$ min$^{-1}$, just as was obtained previously. Because semi-log plots are very convenient for determining whether or not the reaction is first order, we shall use them in place of the more tedious method employed in Figure 14.1. Note, however, that in order to compute the slope of a line plotted on

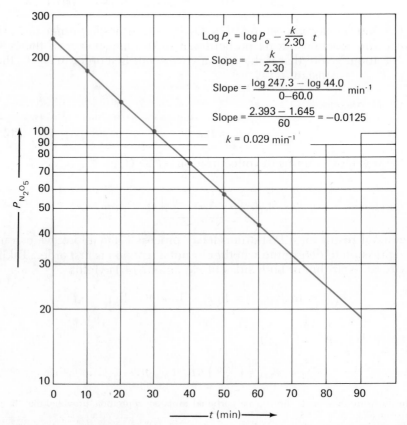

**FIGURE 14.2**    Plot of $\log P_t$ versus $t$ for the decomposition of $N_2O_5$ (see Figure 14.1 for data).

semi-log paper it is necessary to look up the logarithms of the ordinate values used to compute the slope (Figure 14.2).

The *stoichiometry* of the $N_2O_5$ decomposition reaction requires that

Rate of appearance of $NO_2 = 2 \times$ Rate of disappearance of $N_2O_5$

Rate of appearance of $O_2 = (1/2) \times$ Rate of disappearance of $N_2O_5$

or, more briefly (where the subscripts $A$ and $D$ stand for *appearance* and *disappearance*, respectively)

$$R_{A,NO_2} = 2R_{D,N_2O_5} = 4R_{A,O_2}$$
$$R_{A,O_2} = R_{D,N_2O_5}/2 = R_{A,NO_2}/4$$

Because

(a)
$$P_{t(N_2O_5)} = P_0 e^{-kt}$$

we also obtain from the reaction stoichiometry

(b)
$$P_{t(NO_2)} = 2(P_0 - P_t)_{N_2O_5} = 2P_0(1 - e^{-kt})$$

and

(c)
$$P_{t(O_2)} = \frac{1}{2}(P_0 - P_t)_{N_2O_5} = (P_0/2)(1 - e^{-kt})$$

In Figure 14.3 we have sketched equations (a), (b), and (c) on the assumption that the reaction is *unidirectional* (i.e., that the equilibrium state lies far to the right). In other words, we have assumed that the reverse reaction $(2NO_2 + (1/2)O_2 \rightarrow N_2O_5)$ can be neglected, at least in the early stages of the forward reaction. We shall confine ourselves in this chapter to reactions that are essentially unidirectional in the direction indicated, postponing until the next chapter the consideration of reacting systems close to equilibrium. In the latter cases the reverse (or back) reaction cannot be ignored in the kinetic analysis.

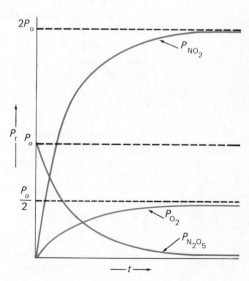

**FIGURE 14.3** Variation of the partial pressures of $N_2O_5$, $NO_2$, and $O_2$ in the decomposition of $N_2O_5$, starting with pure $N_2O_5$ at $P_0$.

A useful concept in the analysis of kinetic data is the *reaction half-life*. The half-life of a reactant is defined as the time required for its concentration to decrease by a factor of 2. For a first-order disappearance we have

$$C_t = C_0 e^{-kt}$$

and, taking $C_t = C_0/2$ when $t = t_{1/2}$, we obtain

$$\frac{1}{2} = e^{-kt_{1/2}}$$

or

$$\ln 2 = 0.693 = kt_{1/2}$$

and, therefore

$$t_{1/2} = \frac{0.693}{k} \tag{14.4}$$

Equation (14.4) shows that *the half-life for a first-order reaction is independent of the initial concentration.* The time required for the disappearance of one-half of the unreacted material is independent of the extent of reaction. For the first-order decomposition of $N_2O_5$ at 45°C we have

$$t_{1/2} = \frac{0.693}{4.8 \times 10^{-4} \text{ sec}^{-1}} = 1.44 \times 10^3 \text{ sec}$$

or, in other words, with $P_0 = 247.3$ Torr, after one-half-life ($1.44 \times 10^3$ sec) the pressure of $N_2O_5$ remaining is 123.7 Torr, after two half-lives ($2.88 \times 10^3$ sec) it is 61.8 Torr, and so on. After $n$ half-lives, the fraction of the original material still remaining is $(1/2)^n$. If we are investigating the kinetics of a reaction and we find that the time required for one half the material present to react is independent of the initial concentration, then we have established that the reaction is first order, and we can calculate the rate constant from the observed half-life.

In 1949, W. F. Libby [Science *109*, 227 (1949)] offered the hypothesis that radioactive carbon-14 ($^{14}C$) is produced in the upper atmosphere through the capture of cosmic ray neutrons by $^{14}N$ nuclei

$$^{14}_{7}N + ^{1}_{0}n \rightarrow ^{14}_{6}C + ^{1}_{1}H$$

The concentration of $^{14}C$ does not increase indefinitely, but remains at an approximately constant level of 12.5 counts per minute per gram (cpm/g) of carbon, because the $^{14}_{6}C$ nucleus is unstable and eventually emits a beta particle:[*]

$$^{14}_{6}C \rightarrow ^{14}_{7}N + ^{0}_{-1}e$$

---

[*] See Chapter 21. The concentration of a radioactive substance is proportional to the number of radioactive disintegrations per unit time per unit mass.

The decay of any radioactive isotope follows first-order kinetics, with the first-order rate constant for the decay process being independent of temperature, pressure, or molecular environment of the radioactive nucleus. The half-life for the decay of radioactive $^{14}C$ is 5720 years, and therefore, the first-order rate constant is (from Equation (14.4)) $k = 0.693/5720$ yr $= 1.21 \times 10^{-4}$ yr$^{-1}$. As long as living matter remains alive, its $^{14}C$ activity (obtained, for example, through the uptake of atmospheric carbon dioxide) remains constant at 12.5 cpm/g. Upon death the uptake of $^{14}C$ stops, and the $^{14}C$ level begins to decline. Thus, 5720 years after the death of an organism, the $^{14}C$ level in the remains (assuming they can be found) is about 6.2 cpm/g of C. This is a very significant observation, because it allows us to date the remains from formerly living matter by determining their $^{14}C$ level. For example, a sample of wood from an Egyptian pharaoh's tomb* has a $^{14}C$ activity of 7.04 cpm/g of carbon. Because the number of counts per minute per gram is directly proportional to the concentration, Equation (14.2) is applicable, and we estimate the age of the wood as

$$t = \frac{2.30}{k} \log \frac{C_0}{C_t} = \frac{2.30}{1.21 \times 10^{-4} \text{ yr}^{-1}} \log \left(\frac{12.5}{7.04}\right) = 4.80 \times 10^3 \text{ yr}$$

Radiocarbon dating is of major importance in archeology. If an old fire site is found, the carbon-containing remains can be used to determine the approximate age of the site.

The age of the earth can be determined by isotopic analysis of lead in minerals containing uranium or thorium isotopes. The uranium isotopes $^{238}U$ and $^{235}U$ decay to $^{206}Pb$ and $^{207}Pb$, respectively, with the following half-lives:

$$^{238}U \rightarrow \, ^{206}Pb \qquad\qquad t_{1/2} = 4.5 \times 10^9 \text{ yr}$$

$$^{235}U \rightarrow \, ^{207}Pb \qquad\qquad t_{1/2} = 7.1 \times 10^8 \text{ yr}$$

$$\text{none} \rightarrow \, ^{204}Pb$$

Because no uranium (or thorium) isotope decays to $^{204}Pb$, we can use the $^{204}Pb$ content (if any is present), together with the known natural abundance of lead of normal isotopic composition to correct for any $^{206}Pb$ or $^{207}Pb$ that was initially present. Using mass spectral data on $^{207}Pb/^{206}Pb$ ratios in rocks [A. O. Nier, Amer. Scient. 54, 359 (1966)], the present best estimate of the age of the earth is $5 \times 10^9$ yr, or about one half-life for the decay of $^{238}U$ to $^{206}Pb$.† We can use this data to construct an earth-existence clock upon which 1 "hour" $= 1 \times 10^9$ years = 1 billion years (Figure 14.4). It is presently about 5 o'clock, give or take a few minutes. Since the oldest known human skeletal fragments (*Kenyapithecus africanus*, recently unearthed by L. S. B. Leakey in Africa) are about $2 \times 10^7$ years old,‡ we compute that man has been around for about

$$\left(\frac{2 \times 10^7}{1 \times 10^9} \text{ hr}\right) \left(\frac{60 \text{ min}}{\text{hr}}\right) = 1.2 \text{ min}$$

---

*We could just as well take a piece of the pharaoh and determine its $^{14}C$ activity.

†A lower limit to the age of the universe can be obtained by isotopic analysis of meteorites.

‡Age estimated by both sedimentation and radiochemical (potassium-argon) dating methods.

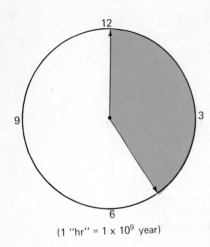

12

9        3        **FIGURE 14.4**   Earth-existence clock.

6

$(1 \text{ "hr"} = 1 \times 10^9 \text{ year})$

of the elapsed time on our earth-existence clock. This, by the way, is only about 1/10 of the time that the overadapted and small-brained dinosaurs lasted.

## 14.4   PSEUDO-FIRST-ORDER KINETICS.

Even though a rate law may involve the concentrations of several species, it is usually possible to simplify the kinetic analysis by flooding the system with all but one of the species. By "flooding" we mean the use of a sufficiently large excess of reactant such that its concentration does not change appreciably during the course of a kinetic run, and therefore, it can be taken as constant.

As an example of the flooding procedure, consider the oxidation of iodide ion by peroxidisulfate in acidic aqueous solutions:

$$S_2O_8^{2-}(aq) + 3I^-(aq) = I_3^-(aq) + 2SO_4^{2-}(aq)$$

It is reasonable to assume (but it is not necessarily so) that the rate law for the forward reaction is of the form

$$R_{D,S_2O_8^{2-}} = (1/3)R_{D,I}^- = k(S_2O_8^{2-})^n(I^-)^m$$

where $n$ and $m$ are constants whose values are to be determined by experiment. If $(I^-) \gg (S_2O_8^{2-})$, then we have, for any given run,

$$R_{D,S_2O_8^{2-}} = k(I^-)^m(S_2O_8^{2-})^n = k_{ps}(S_2O_8^{2-})^n$$

where we have written $k_{ps}$ ($ps =$ pseudo) for $k(I^-)^m$, because $(I^-)$ is in large excess and, therefore, is essentially constant in a particular run. Kinetic data for this reaction are presented in Figure 14.5. The figure shows that the reaction is first order in $(S_2O_8^{2-})$ (i.e., $n = 1$), because $\log C_t$ versus $t$ is linear. The reaction rate is said to be *pseudo-first-order*, however, because $t_{1/2}$ for the disappearance of $S_2O_8^{2-}$, although constant in *any given run* where $(I^-) \gg (S_2O_8^{2-})$, depends on the total iodide ion concentration. From these linear $C_t$ versus $t$ curves we can calculate a pseudo-first-

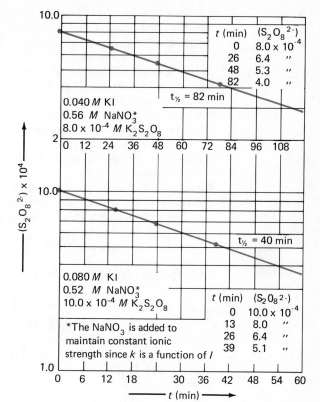

**FIGURE 14.5** Kinetic data (338K) on the reaction

$$S_2O_8^{2-} + 3I^- = I_3^- + 2SO_4^{2-}$$

Experiments at longer times confirm the first-order dependence of rate on $(S_2O_8^{2-})$ found here over one half-life.

order rate constant for each run, and then extract the dependence of the reaction rate on the iodide concentration from these pseudo-first-order rate constants. From run (1) with $(I^-)_0 = 0.040$ and $(S_2O_8^{2-})_0 = 8.0 \times 10^{-4}$ M, we have

$$k_{ps} = \frac{0.693}{t_{1/2}} = \frac{0.693}{82 \text{ min}} = 8.5 \times 10^{-3} \text{ min}^{-1} = k(I^-)^m = k(0.040)^m$$

For run (2) with $(I^-)_0 = 0.080$ and $(S_2O_8^{2-})_0 = 1.0 \times 10^{-3}$ M, we have

$$k_{ps} = \frac{0.693}{40 \text{ min}} = 1.74 \times 10^{-2} \text{ min}^{-1} = k(I^-)^m = k(0.080)^m$$

The correct value of $m$ is that which makes $k$ a constant. For $m = 1$ we find

run (1) $k = 8.5 \times 10^{-3} \text{ min}^{-1}/0.040 \text{ M} = 0.21 \text{ M}^{-1} \cdot \text{min}^{-1}$

run (2) $k = 1.7 \times 10^{-2} \text{ min}^{-1}/0.080 \text{ M} = 0.22 \text{ M}^{-1} \cdot \text{min}^{-1}$

and these numbers are the same within experimental error ($\sim \pm 5\%$). Therefore, we conclude that $m = 1$.

The data in Figure 14.5 show that the rate law for the iodide-peroxidisulfate reaction is

$$R_{D,S_2O_8^{2-}} = (1/3)R_D(I^-) = k(I^-)(S_2O_8^{2-})$$

The reaction is said to be first order in iodide ion, first order in peroxidisulfate ion, and second order overall.[*] This result was established by observing that in the presence of excess iodide ion the reaction rate is proportional to the first power of the peroxidisulfate concentration, and that the half-life of this pseudo-first-order reaction is inversely proportional to the concentration of iodide ion. It is possible that we do not have the complete rate law, because in both experiments the $(H^+)$ was the same. The hydrogen ion concentration is constant throughout the decomposition, even if small, because it is not consumed or produced in the net reaction. If there is an $(H^+)$ dependence it is hidden in $k$, and can only be uncovered by running the reaction at two or more different values of $(H^+)$.

It is instructive to verify our assumption that the concentration of $I^-$ was made sufficiently large relative to the concentration of $S_2O_8^{2-}$ that it did not change appreciably relative to the change in the $(S_2O_8^{2-})$.

For example, for run (2) we have

$$(I^-)_0 = 8.0 \times 10^{-2} \qquad (S_2O_8^{2-})_0 = 10 \times 10^{-4}$$

and after one half-life for $S_2O_8^{2-}$

$$(S_2O_8^{2-}) = (1/2)(10 \times 10^{-4}) = 5 \times 10^{-4} \qquad (50\% \text{ reacted})$$

$$(I^-) = 8.0 \times 10^{-2} - (3/2)(10 \times 10^{-4}) = 7.9 \times 10^{-2} \qquad (1.3\% \text{ reacted})$$

We could just as well have used the same procedure the other way around, that is, with $(S_2O_8^{2-}) \gg (I^-)$, and observed the disappearance of $I^-$ (or the appearance of $I_3^-$) with time.[†]

As a second example of the use of the flooding technique to obtain the rate law, consider the data of Table 14.2 on the reaction between iodide ion and hypochlorite ion in alkaline aqueous solution to yield chloride and hypoiodite ions:

$$I^-(aq) + OCl^-(aq) = IO^-(aq) + Cl^-(aq)$$

Analysis of data: From runs (1) and (2) we conclude that the reaction is first order in $(OCl^-)$, because $t_{1/2}$ (which must refer to $OCl^-$, because this is the limiting reagent) for the disappearance of $OCl^-$ is independent of the initial $OCl^-$ concentration. Runs (1) and (3) show that doubling the $(I^-)$ increases the value of $k_{ps}$ for the disappearance of $OCl^-$ by a factor of 2, and, therefore, the reaction is first order in $(I^-)$ (i.e., hypochlorite is consumed twice as fast when the $(I^-)$ is doubled). From runs (3) and (4) we note that *decreasing* $(OH^-)$ by a factor of 2 *increases* the value of $k_{ps}$ for the disappearance of $OCl^-$ by a factor of 2 (i.e., hypochlorite is consumed more rapidly at lower $(OH^-)$), and therefore, there must be an *inverse* first-order dependence on $(OH^-)$. Note also that the above conclusions are consistent with the data obtained in runs (4) and (5), because halving *both* $(OH^-)$ and $(I^-)$ leaves $k_{ps}$ unchanged. If the above conclusions are correct, then dividing and multiplying the pseudo-first-order rate constant for the

---

[*]The overall order for a reaction whose rate law is of the form

$$\text{Rate} = k(A)^a(B)^b(C)^c \dots$$

is $a + b + c \dots$.

[†]Since $I_3^-(aq)$ is a colored species, its concentration can be conveniently monitored spectrophotometrically.

**TABLE 14.2**

DETERMINATION OF PSEUDO-FIRST-ORDER RATE CONSTANT

| (all concentrations in moles/liter, $T = 298$ K) | | | | | $k_{ps}(\text{sec}^{-1}) = \dfrac{0.693}{t_{1/2}}$ |
|---|---|---|---|---|---|
| Run | $(OH^-)$ | $(I^-)$ | $(OCl^-)$ | $t_{1/2}(\text{sec})$ | |
| 1 | 1.00 | $1.0 \times 10^{-3}$ | $1.0 \times 10^{-5}$ | 11.6 | 0.060 |
| 2 | 1.00 | $1.0 \times 10^{-3}$ | $0.5 \times 10^{-5}$ | 11.5 | 0.060 |
| 3 | 1.00 | $2.0 \times 10^{-3}$ | $1.0 \times 10^{-5}$ | 5.8 | 0.120 |
| 4 | 0.50 | $2.0 \times 10^{-3}$ | $2.0 \times 10^{-5}$ | 2.8 | 0.247 |
| 5 | 0.25 | $1.0 \times 10^{-3}$ | $1.0 \times 10^{-5}$ | 2.9 | 0.240 |

disappearance of $OCl^-$ by the appropriate values of $(I^-)$ and $(OH^-)$, respectively, should yield the same number (the rate constant) for each run. That this is indeed the case is shown by the calculations in Table 14.3. The rate law for the reaction is, therefore,

$$R_{D,I^-} = k \frac{(I^-)(OCl^-)}{(OH^-)}$$

with $k = 60$ sec$^{-1}$ at 25°C. The rate law is first order in $(I^-)$ and $(OCl^-)$, and inverse first order in $(OH^-)$.

## Method of Initial Rates.

The method of initial rates can be used to find the rate law and the rate constant for a reaction. In this method we observe the time required to produce a known, relatively small, quantity of product under conditions such that the concentrations of the reactants do not change appreciably ($<1$ or $2\%$) during the observed time interval. The effect of changes in concentrations of reactants (and possibly of a catalyst) on the rate are de-

**TABLE 14.3**

CALCULATION OF RATE CONSTANT

| Run | $k_{ps}(\text{sec}^{-1})$ | $k_{ps}/(I^-)$ | $k_{ps}(OH^-)/(I^-) = k(\text{sec}^{-1})$ |
|---|---|---|---|
| 1 | 0.060 | 60 | 60 |
| 2 | 0.060 | 60 | 60 |
| 3 | 0.12 | 60 | 60 |
| 4 | 0.247 | 123 | 61 |
| 5 | 0.24 | 240 | 60 |

termined by observing the changes (if any) in the time required to produce the given quantity of product.

For example, suppose we wish to determine the rate law for the reaction

$$3I^-(aq) + S_2O_8^{2-}(aq) \rightarrow I_3^-(aq) + 2SO_4^{2-}(aq) \tag{14.5}$$

Tri-iodide ion reacts with starch to produce an intense blue color. However, in solutions containing both starch indicator and thiosulfate ion, $S_2O_3^{2-}(aq)$, any $I_3^-(aq)$ formed via reaction (14.5) will be preferentially consumed by the reaction

$$I_3^-(aq) + 2S_2O_3^{2-}(aq) \rightarrow 3I^-(aq) + S_4O_6^{2-}(aq) \tag{14.6}$$

The rate of this reaction is very fast compared to reaction (14.5). Only when sufficient $I_3^-(aq)$ has been produced to use up (essentially) all of the $S_2O_3^{2-}(aq)$ will the blue $I_3^-$-starch complex be formed. In other words, a solution containing $I^-(aq)$, $S_2O_8^{2-}(aq)$, $S_2O_3^{2-}(aq)$, starch will remain colorless for a time and then suddenly turn blue; the appearance of the blue color serves as a signal that all of the $S_2O_3^{2-}(aq)$ has been used up. A knowledge of the time required to consume a known quantity of $S_2O_3^{2-}(aq)$ gives us the reaction rate for reaction (14.5) under the prevailing conditions in the solution.

Let us assume that the rate law for reaction (14.5) is of the form

$$R_{A,I_3^-} = k(I^-)^m(S_2O_8^{2-})^n$$

For an extent of reaction that is sufficiently small that $(I^-)$ and $(S_2O_8^{2-})$ can be regarded as constant (and equal to their initial values), the value of $R_{A,I_3^-}$ must also be (essentially) constant and equal to

$$R_{A,I_3^-} \simeq \frac{\Delta(I_3^-)}{\Delta t} = \frac{(I_3^-)_t - 0}{t - 0} = \frac{(I_3^-)_t}{t} = k(I^-)_0^m(S_2O_8^{2-})_0^n$$

From the stoichiometry of reaction (14.6) we have

$$R_{D,I_3^-} = (1/2)R_{D,S_2O_3^{2-}} \simeq -\frac{(1/2)\Delta(S_2O_3^{2-})}{\Delta t} = -\frac{(1/2)\{0 - (S_2O_3^{2-})_0\}}{t - 0}$$

The rate at which $I_3^-(aq)$ is consumed in reaction (14.6) is equal to the rate at which $I_3^-(aq)$ is produced in reaction (14.5); therefore,

$$\frac{1/2(S_2O_3^{2-})_0}{t} \simeq k(I^-)_0^m(S_2O_8^{2-})_0^n$$

where

$$(S_2O_3^{2-})_0 \ll (I^-)_0 \text{ or } (S_2O_8^{2-})_0$$

Suppose that the data of Table 14.4 were obtained in the initial-rate study. Because $(S_2O_3^{2-})_0$ is the same for all runs we can write for any two runs $a$ and $b$

$$\frac{t_a}{t_b} = \frac{(I^-)_{0,b}^m \cdot (S_2O_8^{2-})_{0,b}^n}{(I^-)_{0,a}^m \cdot (S_2O_8^{2-})_{0,a}^n}$$

**TABLE 14.4**

DATA FROM INITIAL-RATE STUDY

| Run | $(S_2O_3^{2-})_0 = 1.00 \times 10^{-3}$ M at 25°C | | |
| | $(I^-)_0$, M | $(S_2O_8^{2-})_0$, M | $t$(sec) |
| --- | --- | --- | --- |
| 1 | 0.100 | 0.100 | 14.0 |
| 2 | 0.050 | 0.100 | 28.2 |
| 3 | 0.100 | 0.050 | 28.1 |
| 4 | 0.050 | 0.050 | 56.3 |
| 5 | 0.200 | 0.100 | 7.1 |

For runs 1 and 2 we have

$$\frac{t_1}{t_2} = \frac{14.0}{28.2} = \frac{(I^-)_{0,2}^m}{(I^-)_{0,1}^m} = \frac{(0.050)^m}{(0.100)^m} = (1/2)^m$$

or $m = 1$. Similarly, from runs 1 and 3 we find $n = 1$, and thus

$$\frac{(1/2)(S_2O_3^{2-})}{t} \simeq k(I^-)(S_2O_8^{2-})$$

A value of $k$ can be computed for each run and the results averaged. For run 1

$$k = \frac{(1/2)(1.00 \times 10^{-3})/14.0 \text{ sec}}{(0.10)(0.10)} = 3.6 \times 10^{-3} \text{ M}^{-1} \cdot \text{sec}^{-1}$$

whereas for run 2

$$k = \frac{(1/2)(1.00 \times 10^{-3})/28.2}{(0.050)(0.100)} = 3.6 \times 10^{-3} \text{ M}^{-1} \cdot \text{sec}^{-1}$$

and so forth.

The method of initial rates provides a fairly rapid and simple procedure for the determination of reaction rate laws and rate constants. The principal disadvantage of the method is that it *may* lead to a rate law which is valid only in the very early stages of the reaction.

## 14.5  SECOND-ORDER KINETICS.

In order to illustrate the analysis of kinetic data for a reaction that involves second-order kinetics, consider the data in Table 14.5 for the gas-phase decomposition of nitrogen dioxide to yield nitric oxide and oxygen:

$$NO_2(g) = NO(g) + (1/2)O_2(g)$$

**TABLE 14.5**

SECOND-ORDER KINETIC DATA; T = 600 K

| $t$(sec) | $(NO_2)(moles \cdot liter^{-1})$ |
|---|---|
| 0 | $5.0 \times 10^{-2}$ |
| 10 | $4.0 \times 10^{-2}$ |
| 20 | $3.3 \times 10^{-2}$ |
| 30 | $2.85 \times 10^{-2}$ |
| 40 | $2.5 \times 10^{-2}$ |
| 50 | $2.2 \times 10^{-2}$ |

We first test these data to see whether the reaction is first order, and it is clear from the non-linearity of the log $C_t$ versus $t$ plot (Figure 14.6) that the reaction is not first order in $NO_2$. In fact, the reaction is second order in $NO_2$. This can be verified by constructing a $C_t$ versus $t$ plot as was done in Figure 14.1, and showing that the reaction rate (equal to the negative of the slope of the $C_t$ versus $t$ curve) at any time, divided by the *square* of the nitric oxide concentration in the reaction system at that time, is a constant; that is,

$$\frac{R_{D,NO_2}(at\ t)}{(NO_2)^2(at\ t)} = \text{constant (if second order in } NO_2)$$

In Figure 14.7 we have constructed just such a plot, using the above data, and at $t = 10$ sec, $C_t = 4.0 \times 10^{-2}$ M, we compute the rate of disappearance of $NO_2$ from the negative of the slope of the curve at this point, as

$$R_{D,NO_2(t=10)} = -\frac{(4.80 - 2.00) \times 10^{-2}\ M}{(0 - 34.5)\ sec} = 8.12 \times 10^{-4}\ M \cdot sec^{-1}$$

and

$$\frac{R_{D,NO_2(t=10)}}{(NO_2)^2_{(t=10)}} = \frac{8.12 \times 10^{-4}\ M \cdot sec^{-1}}{(4.0 \times 10^{-2})^2\ M^2} = 0.50\ M^{-1} \cdot sec^{-1}$$

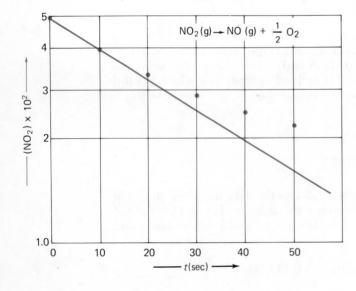

$$NO_2\ (g) \rightarrow NO\ (g) + \frac{1}{2}\ O_2$$

**FIGURE 14.6**   Demonstration that the $NO_2$ decomposition reaction is not first order in $NO_2$.

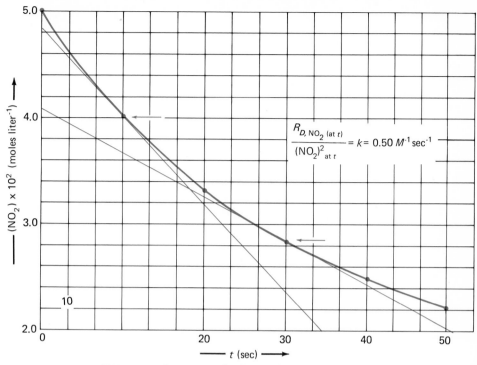

**FIGURE 14.7** $C_t$ versus $t$ plot for the decomposition of $NO_2$:

$$NO_2(g) = NO(g) + \frac{1}{2} O_2(g)$$

whereas at $t = 30$ sec, $C_t = 2.85 \times 10^{-2}$ M, we compute

$$R_{D,NO_2(t=30)} = -\frac{(4.07 - 2.00) \times 10^{-2} \text{ M}}{(0 - 50.7) \text{ sec}} = 4.08 \times 10^{-4} \text{ M} \cdot \text{sec}^{-1}$$

and

$$\frac{R_{D,NO_2(t=30)}}{(NO_2)^2_{(t=30)}} = \frac{4.08 \times 10^{-4} \text{ M} \cdot \text{sec}^{-1}}{(2.85 \times 10^{-2})^2 \text{ M}^2} = 0.50 \text{ M}^{-1} \cdot \text{sec}^{-1}$$

From these results it follows that the rate law for the forward reaction is

$$R_{D,NO_2} = k(NO_2)^2$$

with $k = 0.50$ M$^{-1} \cdot$ sec$^{-1}$ at $T = 600$ K.

Just as with first-order reactions, there is a more convenient graphical technique than the one just outlined for determining whether or not a reaction is second order. The equation for the curve in Figure 14.7 is [*]

$$\frac{1}{C_t} = \frac{1}{5.0 \times 10^{-2} \text{ M}} + (0.50 \text{ M}^{-1} \cdot \text{sec}^{-1})t$$

---

[*] If $-\dfrac{dx}{dt} = kx^2$, then

$$-\int_{x_0}^{x_t} x^{-2} dx = \int_0^t k \, dt, \text{ and } \frac{1}{X_t} = \frac{1}{X_0} + kt.$$

This result can be verified by calculating several $C_t$ values corresponding to various times and comparing the results with the values of $C_t$ that can be read directly off the graph. In the general case of a second-order disappearance, this equation takes the form

$$\frac{1}{C_t} = \frac{1}{C_0} + kt \qquad\qquad (14.7)$$

where $C_0$ is the concentration at $t = 0$ and $k$ is the second-order rate constant. Equation (14.7) says that a plot of $1/C_t$ versus $t$ for a second-order reaction should be linear with a slope equal to the second-order rate constant. Using the same data as were used to construct Figure 14.7, we have plotted $1/C_t$ versus $t$ in Figure 14.8. The linearity of the curve clearly demonstrates the second-order dependence of the reaction rate on the concentration of nitrogen dioxide.

We can obtain an expression for the half-life of a second-order reaction from Equation (14.7). Setting $t = t_{1/2}$ when $C_t = C_0/2$, we obtain

$$\frac{2}{C_0} = \frac{1}{C_0} + kt_{1/2}$$

or

$$t_{1/2} = \frac{1}{kC_0} \qquad\qquad (14.8)$$

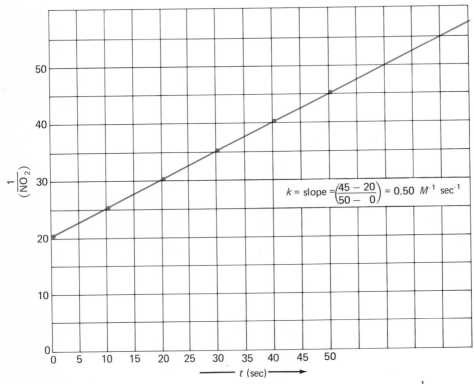

$$k = \text{slope} = \left(\frac{45 - 20}{50 - 0}\right) = 0.50 \ M^{-1} \ \text{sec}^{-1}$$

**FIGURE 14.8**   Plot of $1/C_t$ versus $t$ for the decomposition of $NO_2(g) \rightarrow NO(g) + \frac{1}{2} O_2(g)$ showing the second-order dependence of reaction rate on $(NO_2)$.

For the $NO_2$ decomposition with $C_0 = 5.0 \times 10^{-2}$ M, $k = 0.50$ M$^{-1} \cdot$ sec$^{-1}$, we compute

$$t_{1/2} = \frac{1}{(5.0 \times 10^{-2} \text{ M})(0.50 \text{ M}^{-1} \cdot \text{sec}^{-1})} = 40 \text{ sec}$$

Equation (14.8) shows that the half-life of a second-order reaction is inversely proportional to the initial concentration of the reactant. This means that the smaller the initial concentration of reactant, the longer it takes for half of it to react. This is to be contrasted with a first-order reaction, for which the time required for half of the reactant to disappear is independent of the initial concentration. The reason for this difference is that in a simple first-order-decay process, for example, radioactive decay, the occurrence of the reactive event does not require that two particles come together, and one metastable nucleus of a given type is just as likely to undergo decay in a given time as any other. The *fraction* of the particles that react in a given time is the same no matter what the total number of reactive particles. On the other hand, the reactive event in a second-order reaction requires the *simultaneous participation of two particles* (collision), and therefore, the fraction of molecules that react in a given time will be less the smaller the total number of particles per unit volume that are present to begin with.

Another example of a reaction involving a second-order dependence in the rate law is the oxidation of iodide ion by chlorate ion in aqueous acidic solutions:

$$ClO_3^-(aq) + 9I^-(aq) + 6H^+(aq) = 3I_3^-(aq) + Cl^-(aq) + 3H_2O(\ell)$$

The rate law for this reaction, determined in solutions of constant ionic strength with (H$^+$) held constant by means of a suitable buffer, is

$$R_{D,ClO_3^-} = k'(ClO_3^-)(I^-)$$

Changing the value of (H$^+$) by changing buffers at a fixed temperature leads to change in the magnitude of $k$, with the observed $k$ increasing with increasing (H$^+$). Therefore, $k'$ is a pseudo-second-order rate constant. Let us assume that

$$k' = k(H^+)^n$$

then

$$\log k' = \log k + n \log(H^+)$$

or

$$\log k' = \log k - n \, pH$$

If our assumption is correct, then a plot of the logarithm of the pseudo-second-order rate constant $k'$ versus $pH$ will have a slope of $-n$ and an intercept of $\log k$. Such a plot for this reaction has a slope of $-2.0$ and, therefore, $n = 2$. Hence, the rate law, with the dependence on (H$^+$) explicitly expressed, is

$$R_{D,ClO_3^-} = k(ClO_3^-)(I^-)(H^+)^2$$

The rate law is fourth-order overall, and the units of $k$ are M$^{-3} \cdot$ sec$^{-1}$.

## 14.6  TREATMENT OF KINETIC DATA FOR A REACTION INVOLVING A CATALYST.

Although we shall define a catalyst in mechanistic terms in the next chapter, it will serve our purposes adequately at this point to define a *catalyst* as a substance that does not appear as a reactant in the net chemical reaction but which nonetheless increases the rate of the reaction when its concentration is increased* (all other factors being the same).

As an example of a reaction involving a catalyst (in this case $H^+(aq)$), consider the kinetics of the hydrolysis of sucrose (table sugar) in aqueous solution at 25°C:

$$C_{12}H_{22}O_{11}(aq) + H_2O(\ell) = C_6H_{12}O_6(aq) + C_6H_{12}O_6(aq)$$

(Sucrose)                    (glucose)        (fructose)

From a plot of log(*sucrose*) versus $t$ it was found that the rate law is first order in the sucrose concentration, and further experiments gave the data in Table 14.6. The first point to notice here is that $H^+(aq)$ is not consumed or produced in the net reaction, and consequently its concentration remains constant through the course of the reaction. It is also clear from these data that increasing the $(H^+)$ increases the reaction rate (increases the value of $k_{ps}$). The value of $k_{ps}$ for the disappearance of sucrose is proportional to the $H^+(aq)$ concentration, and therefore the rate law must also be first order in $(H^+)$. Because $(H^+)$ is constant in any given run, we can calculate the pseudo-first-order rate constants, and from these results the second-order rate constant, $k$, can be computed. From these data, then, we conclude that the rate law for the reaction is

$$R_{D,sucrose} = k(sucrose)(H^+)$$

with $k = 5.7 \times 10^2$ $M^{-1} \cdot sec^{-1}$ at 25°C. The possibility still remains that $k$ has hidden in it a dependence on $(H_2O)$. Whether or not this is actually the case cannot be deduced from the data given, because the $(H_2O)$ is constant at 55.5 M in all the runs.†

---

*An inhibitor, on the other hand, slows down a reaction rate when its concentration is increased.

†Can you think of a way in which this uncertainty could be resolved?

**TABLE 14.6**

KINETIC DATA ON SUCROSE HYDROLYSIS

| Run | $(C_{12}H_{22}O_{11})_0$,M | $(H^+)$,M | $k_{ps}(sec^{-1})$ | $k = k_{ps}/(H^+)$,$M^{-1} \cdot sec^{-1}$ |
|-----|------|------|------|------|
| 1 | 0.10 | $1.0 \times 10^{-7}$ | $5.7 \times 10^{-5}$ | $5.7 \times 10^2$ |
| 2 | 0.20 | $1.5 \times 10^{-4}$ | $8.5 \times 10^{-2}$ | $5.7 \times 10^2$ |
| 3 | 0.25 | $2.0 \times 10^{-5}$ | $1.15 \times 10^{-2}$ | $5.7 \times 10^2$ |

An interesting special type of catalysis is *autocatalysis*, that is, the catalysis of a reaction by a product of the stoichiometric reaction. For example, in the reaction

$$A = B + C$$

if either B or C appears in the numerator of the rate law, as in

$$R_{D,A} = k(A)(B)$$

then the reaction is said to be *autocatalytic*. Autocatalysis is often found for reactions that produce $H^+(aq)$, and is also found in ester hydrolysis and in numerous biological reactions.

Suppose we have the above-mentioned reaction stoichiometry and rate law, with the initial concentration of A equal to $(A)_0$ and of B equal to $(B)_0$. In such a case, the concentration of B changes with time in the manner shown[*] in Figure 14.9. Note the S-shape of the curve. Several other features are also worth noting: (a) if there were no initial B to get the reaction started, it would not go at all unless it could find a different (less favorable energetically and, therefore, slower) pathway; (b) as the reaction gets underway it goes with ever increasing rate (Rate = slope) until $(A)_t \simeq (1/2)(A)_0$, when it begins to slow down again because A is depleted. The net reaction shuts down when A is essentially exhausted (assuming that the equilibrium lies far to the right).

Autocatalytic rate behavior is found in the growth of populations.[†] The leveling-off of a population is a result of the fact that the food supply is

---

[*] See Reference 15.8.

[†] Owing to space limitations, we are unable to discuss here the various interesting mechanisms for this type of process.

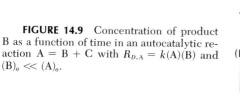

**FIGURE 14.9**  Concentration of product B as a function of time in an autocatalytic reaction $A = B + C$ with $R_{D,A} = k(A)(B)$ and $(B)_0 \ll (A)_0$.

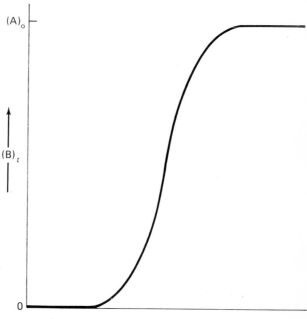

limited. At the present time the earth's human population is growing with ever increasing rate; 1/20 of the people who have *ever* lived on the earth are now alive.

## 14.7  ZERO-ORDER KINETICS.

Consider the adsorption of molecules on a catalytic surface, such as finely-divided platinum or an enzyme (high-molecular-weight, colloidal protein that acts as a catalyst for a biological reaction). Let us suppose that on a given amount of surface area of the catalytic material there are a fixed number of "active sites" (i.e., places where the reaction can occur). The total number of active sites is directly proportional to the total surface area. If the concentration of reactant molecules in the vicinity of the surface is such that essentially every active site has a molecule sitting on it, then the rate of decomposition of the substance on the surface will be constant, and independent of the concentration of the substance in the surrounding medium.* In such a case the reaction rate is controlled solely by how fast the adsorbed molecules undergo reaction on the surface, and $R_{D,X} = k$. This situation is observed, for example, in the catalytic decomposition of $NH_3$ on tungsten. If the rate of consumption of reactant is constant with time, then a plot of $C_t$ versus $t$ will be linear, with a slope equal to the zero-order rate constant for the process (the units are the same as those of $R_D$, i.e., $M \cdot sec^{-1}$), and a $t = 0$ intercept of $C_0$. In other words, for a zero-order reaction,

$$\boxed{C_t = C_0 - kt} \tag{14.9}$$

The half-life of a zero-order reaction is given by

$$\frac{C_0}{2} = C_0 - kt_{1/2}$$

$$t_{1/2} = \frac{C_0}{2k} \tag{14.10}$$

For a reaction proceeding at a constant rate, the time required for one half of the material originally present to react is directly proportional to the initial concentration.

As a simple illustration of the treatment of data for a reaction whose rate law is zero-order in a reactant, consider the homogeneous reaction

$$CH_3CCH_3(aq) + Br_2(aq) = CH_3CCH_2Br(aq) + H^+(aq) + Br^-(aq)$$
$$\overset{\|}{O} \qquad\qquad\qquad\quad \overset{\|}{O}$$

for which the following data were obtained in a run at 25°C with $(CH_3COCH_3) = 1.00$ M and $(H^+) = 0.333$ M;

---

*Think of a large crowd trying to get into a football stadium. Because there is a limit to the number of people that can pass through the turnstiles per second, the rate of filling of the stadium is a constant when people have to wait in line to pass through the turnstiles. (Note this is not the case before game time, when the rate of filling of the stadium is directly proportional to the concentration of people, say the number per square meter, moving toward the turnstiles.)

| $(Br_2), M$ | $t(\text{min})$ |
|---|---|
| $8.0 \times 10^{-3}$ | 0.0 |
| $7.2 \times 10^{-3}$ | 1.0 |
| $6.5 \times 10^{-3}$ | 2.0 |
| $5.7 \times 10^{-3}$ | 3.0 |
| $4.9 \times 10^{-3}$ | 4.0 |
| $4.2 \times 10^{-3}$ | 5.0 |

Because acetone, $CH_3\overset{\overset{\displaystyle O}{\|}}{C}CH_3$, and $H^+(aq)$ are present in excess, their concentrations (to a very good approximation) are constant throughout the run. The above data have been plotted in the form $C_t$ versus $t$ in Figure 14.10. The linearity of the plot shows that with acetone and $H^+(aq)$ in large excess, the rate law for the reaction is (from the curve, $R_{D,Br_2} = -\text{slope} = $ constant)

$$R_{D,Br_2} = k'$$

with $k' = 1.27 \times 10^{-3}$ M $\cdot$ sec$^{-1}$ when $(H^+) = 0.333$ M and $(CH_3COCH_3) = 1.00$ M. The fact that the reaction rate is independent of the concentration

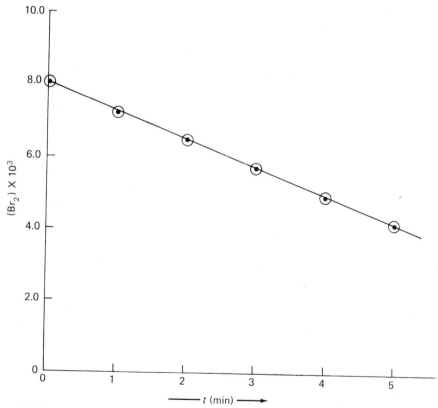

**FIGURE 14.10** Concentration of $Br_2(aq)$ as a function of time for the reaction

$$CH_3COCH_3(aq) + Br_2(aq) = CH_3COCH_2Br(aq) + H^+(aq) + Br^-(aq)$$

for a kinetic run with $(CH_3COCH_3) = 1.00$ M, and $(H^+) = 0.333$ M at 298 K.

of $Br_2$ means that $Br_2$ gets involved in the reaction mechanism after the rate-determining step. Further experiments with different concentrations of acetone and hydrogen ion show that the complete rate law for this reaction is

$$R_{D,Br_2} = k(CH_3COCH_3)(H^+)$$

where $k = 3.75 \times 10^{-3} \, M^{-1} \cdot sec^{-1}$.

## 14.8    MULTITERM RATE LAWS.

The decomposition of peroxidisulfate ion in acidic aqueous solutions has the following stoichiometry:

$$S_2O_8^{2-}(aq) + H_2O(\ell) = 2SO_4^{2-}(aq) + (1/2)O_2(g) + 2H^+(aq)$$

The rate law for the decomposition is first order in $(S_2O_8^{2-})$, and the pseudo-first-order half-lives for the disappearance of $S_2O_8^{2-}$ at various perchloric acid concentrations (at $T = 323$ K) are shown in Table 14.7. These data show that the rate of disappearance of $S_2O_8^{2-}$ is greater the higher the $H^+(aq)$ concentration. However, the rate law cannot simply be

$$R_{D,S_2O_8^{2-}} = k(S_2O_8^{2-})(H^+)$$

because for pure water at 50°C, $(H^+) \simeq 2.3 \times 10^{-7}$ M and $k_{ps} = 0.60 \times 10^{-2}$ min$^{-1}$, whereas for $(H^+) = 1.0 \times 10^{-1}$ M we find $k_{ps} = 4.0 \times 10^{-2}$ min$^{-1}$; that is, approximately an $10^6$-fold increase in $(H^+)$ increases $k_{ps}$ by less than a factor of 10. On the other hand, at high acid concentration, the pseudo-first-order rate constant for the disappearance of $S_2O_8^{2-}$ is roughly proportional to the $(H^+)$. Consequently, the rate law must contain at least two terms, one proportional to $(H^+)$ and one independent of $(H^+)$; that is,

$$R_{D,S_2O_8^{2-}} = k_1(S_2O_8^{2-}) + k_2(S_2O_8^{2-})(H^+) = \{k_1 + k_2(H^+)\}(S_2O_8^{2-}) = k_{ps}(S_2O_8^{2-})$$

In Figure 14.11 we have plotted $k_{ps}$ versus $(H^+)$, and the linearity of the plot confirms the above analysis. From the $(H^+) = 0$ intercept we find $k_1 = 0.60 \times 10^{-2}$ min$^{-1}$ (first-order rate constant for the acid-independent path), and from the slope we find $k_2 = 0.34 \, M^{-1} \cdot min^{-1}$ (second-order rate constant for the acid-dependent path). The two reaction pathways contribute equally to the total reaction rate when

$$k_2(H^+)(S_2O_8^{2-}) = k_1(S_2O_8^{2-})$$

**TABLE 14.7**

| $HClO_4(mole \cdot liter^{-1})$ | $t_{1/2}(min)$ | $k_{ps} = 0.693/t_{1/2}(min^{-1})$ |
|---|---|---|
| 0 (pure water, $(H^+)$ = 2.3 × 10$^{-7}$ M) | 115 | 0.60 × 10$^{-2}$ |
| 0.10 | 17.3 | 4.0 × 10$^{-2}$ |
| 0.20 | 9.0 | 7.7 × 10$^{-2}$ |
| 0.40 | 5.0 | 13.9 × 10$^{-2}$ |

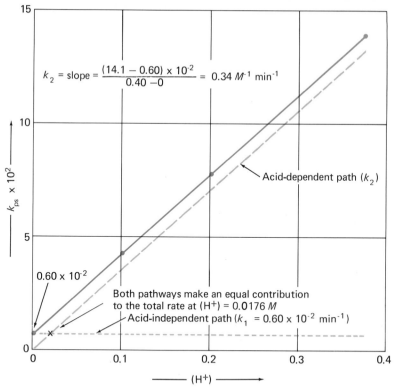

**FIGURE 14.11**   Decomposition of $S_2O_8^{2-}$ in aqueous acidic solutions. Analysis of the pseudo-first-order rate constants showing the existence of a two-term rate law ($T = 323K$).

or

$$(H^+) = \frac{k_1}{k_2} = \frac{0.60 \times 10^{-2}}{0.34} = 1.76 \times 10^{-2} \ M(50°C)$$

Multiterm rate laws are common in solution kinetics. They are a consequence of the existence of two (or more) distinct reaction pathways (mechanisms) by which the reactants can go to products. In the case above we have a reaction pathway involving $H^+(aq)$ before or in the rate-determining step, and a second pathway that does not involve $H^+(aq)$ before or in the rate-determining step.

## 14.9   THE EFFECT OF TEMPERATURE ON REACTION RATES.

All other factors being equal, increasing the temperature at which a reaction is run almost invariably leads to an increase in the rate of reaction; the rate constant increases with increasing temperature. In 1889, S. Arrhenius discovered that the temperature dependence of many reaction-rate constants could be described by an equation of the form

$$k = Ae^{-E_a/RT} \qquad (14.11)$$

For a given reaction, $A$ (the *frequency factor*) and $E_a$ (the *activation energy*) are fairly insensitive to temperature. We shall regard $A$ and $E_a$ as completely independent of temperature in the discussion to follow. The activation energy and the frequency factor provide useful information concerning the reaction mechanism. We shall concern ourselves here with the methods of obtaining $E_a$ and $A$ from experimental data, and postpone until the next chapter a discussion of their significance in terms of the reaction mechanism. From Equation (14.11) we obtain

$$\ln k = \ln A - E_a/RT$$

or

$$\log k = \log A - E_a/2.30\ RT \tag{14.12}$$

Equation (14.12) shows that if the temperature dependence of a rate constant is adequately described by Arrhenius's equation, then a plot of $\log k$ versus $1/T$ (Arrhenius plot) should be linear, with a slope equal to

$$\text{slope} = -E_a/2.30\ R = -E_a/2.30 \times 1.99\ \text{cal} \cdot \text{K}^{-1}\ \text{mole}^{-1} = -E_a/4.58$$

Thus, the slope of the Arrhenius plot multiplied by $-4.58$ yields $E_a$ in $\text{cal} \cdot \text{mole}^{-1}$. The intercept of the plot at $1/T = 0$ yields $\log A$. The $A$ factor can be obtained much more conveniently, however, once we have $E_a$, by substituting a value of $k$ (that falls on the $\log k$ versus $1/T$ curve) and its corresponding value of $T$ into Equation (14.11) or (14.12).

In order to illustrate how $E_a$ is obtained, consider the rate-constant data given in Figure 14.12 for the reaction

$$N_2O_5(g) = 2NO_2(g) + (1/2)O_2(g)$$

From the linearity of the plot[*] in Figure 14.12, we conclude that Equation (14.11) is applicable, and from the slope of the curve $E_a$ is found to be 24.8 $\text{kcal} \cdot \text{mol}^{-1}$.

The rate constants at two different temperatures are given by Equation (14.12) as

$$\log k_2 = \log A - E_a/4.58\ T_2$$

$$\log k_1 = \log A - E_a/4.58\ T_1$$

Subtraction of the second equation from the first yields

$$\log \frac{k_2}{k_1} = -\frac{E_a}{4.58}\left\{\frac{1}{T_2} - \frac{1}{T_1}\right\} = \frac{E_a}{4.58}\left\{\frac{T_2 - T_1}{T_1 T_2}\right\} \tag{14.13}$$

This equation shows that if we have $E_a$ and $k_1$ at $T_1$, we can compute $k_2$ at $T_2$. A useful rough estimate for the influence of temperature on the rate constant is this: around room temperature (300 K), the rate constants for most reactions increase by a factor of about 1.5 to 3.0 for a 10 degree increase in temperature.[†] The value of $E_a$ corresponding to $k_2/k_1 = 2$ for $\Delta T = 305 - 295 = 10$ is computed from Equation (14.13) as

---

[*] Note that $k$ values for *at least three temperatures* are required to determine whether or not $\log k \propto 1/T$, because we can always draw a straight line between any two points.

[†] Factors less than one and greater than thirty are sometimes observed.

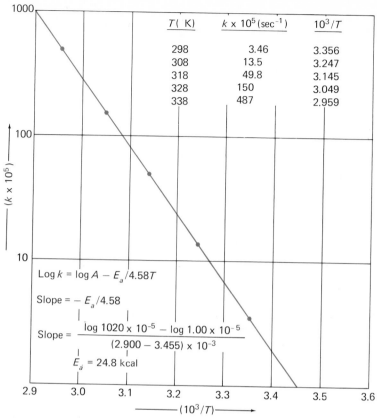

| $T(\text{K})$ | $k \times 10^5 (\text{sec}^{-1})$ | $10^3/T$ |
|---|---|---|
| 298 | 3.46 | 3.356 |
| 308 | 13.5 | 3.247 |
| 318 | 49.8 | 3.145 |
| 328 | 150 | 3.049 |
| 338 | 487 | 2.959 |

$\text{Log } k = \log A - E_a/4.58T$

$\text{Slope} = -E_a/4.58$

$\text{Slope} = \dfrac{\log 1020 \times 10^{-5} - \log 1.00 \times 10^{-5}}{(2.900 - 3.455) \times 10^{-3}}$

$E_a = 24.8 \text{ kcal}$

**FIGURE 14.12**  Determination of the Arrhenius activation energy $E_a$ from the temperature dependence of the reaction-rate constant for the reaction $N_2O_5(g) = 2NO_2(g) + \frac{1}{2} O_2(g)$.

$$\log 2 = \frac{E_a}{4.58} \left\{ \left( \frac{305 - 295}{305 \times 295} \right) \right\}$$

Solving this expression for $E_a$ yields $E_a = 12.5 \text{ kcal} \cdot \text{mol}^{-1}$. A similar calculation shows that for $k_2/k_1 = 3$ and $\Delta T = 10 = 305 - 295$, $E_a = 19.7$ $\text{kcal} \cdot \text{mol}^{-1}$. Note that the larger $E_a$, the greater the increase in $k$ for a given $\Delta T$.

_____ **PROBLEMS**

1. Calculate the time required for the concentration to decrease to 1/10 of its initial value, for a simple first-order reaction with a rate constant $k = 10 \text{ sec}^{-1}$. Repeat the calculation for a second-order reaction with $k = 10 \text{ M}^{-1} \cdot \text{sec}^{-1}$.

2. Show that for a first-order reaction, the time required for 99.9% of the reaction to take place is ten times that required for 50% of the reaction to take place.

3. Assuming that the loss of ability to recall learned material is a first-order process with a half-life of 35 days, compute the number of days required to forget 90% of the material that *you* have learned in preparation for an exam. (Assume constant temperature and that no further reference is made to the previously learned material during this period). Will all students remember the same amount of material after this period (given the above data and assumptions)?

4. Charcoal from the Lascaux Cave in France (the cave with the remarkable prehistoric paintings) was found to have a $^{14}C$ content equal to 14.5% of that in living matter. Given that $t_{1/2}$ for $^{14}C$ is 5720 years, estimate the age of the charcoal.

5. Given the following data on the decomposition of $N_2O_5$

$$N_2O_5(soln) = 2NO_2(soln) + (1/2)O_2(g)$$

at 45°C in carbon tetrachloride ($CCl_4$), determine the rate law and compute the rate constant. Compare your results with the gas-phase decomposition results for $N_2O_5$ given in the text.

| $t$(sec) | $C_{N_2O_5}$, M | $t$(sec) | $C_{N_2O_5}$, M |
|---|---|---|---|
| 0 | 2.33 | 1198 | 1.11 |
| 184 | 2.08 | 1877 | 0.72 |
| 319 | 1.91 | 2315 | 0.55 |
| 526 | 1.67 | 3144 | 0.34 |
| 867 | 1.36 | | |

6. Given the following rate-constant data on the gas-phase decomposition of $NO_2$:

$$NO_2(g) = NO(g) + (1/2)O_2(g)$$

| | 603.5 K | 627.0 K | 656.0 K |
|---|---|---|---|
| $k(M^{-1} \cdot sec^{-1})$ | 0.71 | 1.81 | 4.74 |

Determine the Arrhenius activation energy and the Arrhenius $A$ factor. Use your results to estimate $k$ at 400 K.

7. Assuming that the practical lower limit, owing to counting uncertainties, on the usefulness of the radiocarbon-dating technique is about 1 cpm/g carbon, what is the oldest object that can be dated reliably by this method?

8. The half-life for the decay of $^{238}U$ to $^{206}Pb$ is $4.5 \times 10^9$ years. Estimate the age of a mineral sample in which the $^{206}Pb/^{238}U$ atom ratio (corrected for Pb impurities present at the time of formation) was found to be 0.306.

9. An investigation of the reaction

$$H_2O_2(aq) + 3I^-(aq) + 2H^+(aq) = 2H_2O(\ell) + I_3^-(aq)$$

in buffered aqueous acidic solutions shows that at fixed $(H^+)$ the rate law is

$$R_{D,H_2O_2} = k'(I^-)(H_2O_2)$$

By carrying out runs at different $(H^+)$ values (using buffers), the following data was obtained:

| $(H^+)$ | $k'$ |
|---------|------|
| 0.00100 | $1.17 \times 10^{-2}$ |
| 0.0100  | $1.33 \times 10^{-2}$ |
| 0.0500  | $2.03 \times 10^{-2}$ |
| 0.100   | $2.90 \times 10^{-2}$ |

By plotting $k'$ versus $(H^+)$, show that

$$k' = k_1 + k_2(H^+)$$

and determine $k_1$ and $k_2$. Suppose this reaction were investigated only at $(H^+) < 10^{-3}$ M; would the second reaction pathway be discovered?

10. Given that the major reaction that occurs when an egg is boiled in water is the denaturation of the egg protein (a first-order reaction with $E_a \sim 10$ kcal), calculate how much time is required to cook an egg at 92°C (the boiling point of water at 8000 ft elevation) to the same extent as an egg cooked for 3 minutes at 100°C (the boiling point of pure water at sea level).

11. The denaturation of the virus that causes the rabbit disease *Myxomatosis* is a first-order reaction with rate constants of 0.031 min$^{-1}$ (50.0°C), 0.134 min$^{-1}$ (53.0°C), and 2.0 min$^{-1}$ (60.0°C). Estimate the Arrhenius activation energy for this reaction and compute the value of $t_{1/2}$ at 75.0, 50.0, and 25.0°C. Can you suggest an explanation for the magnitude for the calculated $E_a$? Many rate constants for biologically important reactions have $E_a$ values of the order of 100 kcal. Comment on this with regard to the fact that many living organisms are confined to a rather narrow temperature range of existence.

12. Radioactive potassium 40 decays by emission of a positron to give argon 40

$$_{19}^{40}K \rightarrow _{18}^{40}Ar + _{+1}^{0}\beta$$

with a half-life of $1 \times 10^9$ years. The ages of certain potassium-containing rock samples can be estimated by determining the amount of argon 40 trapped in the rock. Discuss how such experiments could be carried out, and estimate the age-range of rocks that can be dated by this method.

13. The uncoiling of DNA (deoxyribonucleic acid), the molecule that carries the genetic code, is a first-order process with an estimated half-life of 2 minutes at 50°C. Discuss the importance to life of temperature regulation in the human body.

14. Given that the population of the earth is presently doubling every 35 years, estimate the year in which the average land surface space allotment per person on the earth reaches 1 square foot. Take the 1972 popula-

tion of the earth as $3.6 \times 10^9$ people, the circumference of the earth as $2.5 \times 10^4$ mile, and the land surface as 25% of the total earth surface area.

15. Given the following data at 100°C for the reaction

$$A(g) = B(g) + C(g)$$

| $t$(sec) | 0 | 600 | 1200 | 2400 | 3600 |
|----------|------|-----|------|------|------|
| $P_A$(Torr) | 348 | 247 | 185 | 105 | 58 |

find the rate law and calculate the rate constant.

16. Consider the reaction

$$Cr(OH_2)_6^{3+}(aq) + SCN^-(aq) = (H_2O)_5CrNCS^{2+}(aq) + H_2O(\ell)$$

for which the following kinetic data were obtained at 25°C:

| | | (**BUFFERED SOLUTIONS**) | |
|---|---|---|---|
| $(Cr^{3+})$ | $(SCN^-)$ | $(H^+)$ | $t_{1/2}$(sec) |
| $1.0 \times 10^{-4}$ | 0.10 | 1.00 | $3.65 \times 10^6$ |
| $1.0 \times 10^{-3}$ | 0.10 | 1.00 | $3.68 \times 10^6$ |
| $1.5 \times 10^{-3}$ | 0.20 | 1.00 | $1.83 \times 10^6$ |
| $1.5 \times 10^{-3}$ | 0.50 | 1.00 | $7.31 \times 10^5$ |
| $1.0 \times 10^{-3}$ | 0.50 | 0.010 | $5.29 \times 10^5$ |
| $1.0 \times 10^{-3}$ | 0.50 | 0.0010 | $1.53 \times 10^5$ |

(a) Find the rate law and the rate constants.
(b) Calculate the $(H^+)$ at which the two pathways contribute equally to the total reaction rate.

17. Given the following data for the reaction

$$A + B = C$$

taken for a run where the initial concentration of $B$ was 0.100 M find the rate law and the rate constant.

| $t$(sec) | 0 | 2.0 | 3.0 | 4.0 | 5.0 | 6.0 | 8.0 | 12.0 |
|----------|------|------|------|------|------|------|------|------|
| $(A) \times 10^3$ | 2.00 | 1.69 | 1.58 | 1.47 | 1.38 | 1.31 | 1.17 | 0.97 |

Note: It was further observed that $t_{1/2}$ for the disappearance of $A$ was independent of the concentration of $B$.

18. The reaction between bromate ion and iodide ion in acidic aqueous solutions has the stoichiometry:

$$BrO_3^- + 9I^- + 6H^+ = 3I_3^- + Br^- + 3H_2O$$

Given the following kinetic data at 25°C, determine the rate law and the rate constant for the forward reaction.

INITIAL CONCENTRATIONS (moles/liter)

| $(I^-)_0$ | $(BrO_3^-)_0$ | $(H^+)_0$ | $t_{1/2}(sec)$ |
|---|---|---|---|
| $1.00 \times 10^{-3}$ | 0.100 | 0.100 | $3.48 \times 10^3$ |
| $2.00 \times 10^{-3}$ | 0.200 | 0.100 | $1.73 \times 10^3$ |
| $1.00 \times 10^{-3}$ | 0.200 | 0.100 | $1.75 \times 10^3$ |
| $1.00 \times 10^{-3}$ | 0.100 | 0.200 | $8.7 \times 10^2$ |
| $1.50 \times 10^{-3}$ | 0.100 | 0.300 | $3.9 \times 10^2$ |

19. Consider the reaction

$$A + 2B = C + D$$

for which the following kinetic data were obtained at 25°C:

| RUN 1 | | RUN 2 | |
|---|---|---|---|
| $(B)_0 = 2.00$ M | $(A)_0 = 2.50 \times 10^{-3}$ M | $(B)_0 = 0.500$ M | $(A)_0 = 2.50 \times 10^{-3}$ M |
| $t(sec)$ | $(A) \times 10^3$ | $t(sec)$ | $(A) \times 10^3$ |
| 0 | 2.50 | 0 | 2.50 |
| 15 | 2.33 | 100 | 2.22 |
| 60 | 1.92 | 200 | 2.00 |
| 125 | 1.54 | 400 | 1.67 |
| 250 | 1.11 | 500 | 1.18 |

Find the rate law and the rate constant for the forward reaction.

20. The kinetics of the decomposition of phosphine at 950 K was followed by measuring the total pressure in the system as a function of time:

$$4PH_3(g) = P_4(g) + 6H_2(g)$$

The following data were obtained in a run where the reaction chamber contained only pure phosphine at the start of the reaction.

| $t(min)$ | 0.0 | 40.0 | 80.0 | 120.0 |
|---|---|---|---|---|
| $P_{tot}(mm)$ | 100.0 | 150.0 | 166.7 | 172.2 |

(a) Utilize the stoichiometry of the reaction to derive a general relation between $P_t$, $P_0$ and $P_{tot}$ where $P_t$ is the partial pressure of unreacted phosphine at time $t$, $P_0$ is the initial partial pressure of phosphine, and $P_{tot}$ is the total pressure in the reaction chamber.

(b) With the aid of the equation derived in (a) and the data given, find the rate law and calculate the rate constant.

21. Quenching a reaction proceeding in solution by dilution with solvent at the temperature of the solution is effective only for the reactions of certain orders. Explain.

22. Given the following data at 25°C on the reaction

$$A + 2B + C = D + E$$

(where X represents an added substance):

| RUN | $(A)_0$ | $(B)_0$ | $(C)_0$ | $(X)$ | $t_{1/2}(\text{sec})$ |
|---|---|---|---|---|---|
| 1 | $1.00 \times 10^{-2}$ | $5.00 \times 10^{-2}$ | $1.00 \times 10^{-5}$ | $2.00 \times 10^{-2}$ | 160 |
| 2 | $1.00 \times 10^{-2}$ | $5.00 \times 10^{-2}$ | $5.50 \times 10^{-5}$ | $2.00 \times 10^{-2}$ | 159 |
| 3 | $1.00 \times 10^{-2}$ | $1.00 \times 10^{-2}$ | $1.00 \times 10^{-5}$ | $2.00 \times 10^{-2}$ | 161 |
| 4 | $1.00 \times 10^{-2}$ | $1.00 \times 10^{-2}$ | $1.00 \times 10^{-5}$ | $6.00 \times 10^{-2}$ | 53 |
| 5 | $3.00 \times 10^{-2}$ | $5.00 \times 10^{-2}$ | $2.00 \times 10^{-5}$ | $2.00 \times 10^{-2}$ | 18 |

(a) Determine the rate law for the forward reaction.
(b) Compute the rate constant for the forward reaction.

23. Given the following kinetic data for the reaction:

$$3A(aq) + 2B(aq) + C(aq) = 2D(s) + 3S(g)$$

| $(A)_0, M$ | $(B)_0, M$ | $(C)_0, M$ | REACTION RATE |
|---|---|---|---|
| 0.20 | 0.40 | 0.10 | $y$ |
| 0.40 | 0.20 | 0.20 | $16y$ |
| 0.20 | 0.20 | 0.20 | $4y$ |
| 0.40 | 0.40 | 0.10 | $4y$ |

Determine the rate law for the forward reaction. Suppose that the reaction is found to be catalyzed by X(aq), and the reaction rate is found to be directly proportional to (X). Can this fact together with the above data be used to determine the rate law for the catalyzed reaction?

24. The equation that governs the dependence of a rate constant on the total pressure, $P$, at a given temperature is

$$\log \frac{k_P}{k_1} = -\frac{\Delta \bar{V}_a}{RT} (P - 1)$$

where $\Delta \bar{V}_a$ is the activation volume. Typical values of $\Delta \bar{V}_a$ lie in the range $0 \pm 5$ cm$^3 \cdot$ mole$^{-1}$. Biological reactions, such as protein denaturation, often have much larger $\Delta \bar{V}_a$ values. Calculate values of $k_P / k_1$ for two different reactions, one with $\Delta \bar{V}_a = 5$ cm$^3$ and the other with $\Delta \bar{V}_a = 185$ cm$^3$, at $P$ values of 5, 50, and 500 atm.

25. The lower limits to the half-life of a reactant that can be determined by conventional and rapid mixing techniques are 10 sec and 0.1 sec, respectively. Compute the corresponding upper limits on determinable first- and second-order rate constants by the two methods. What significant difference is there between the first- and second-order cases?

## Reference

See the references given at the end of Chapter 15.

## CHAPTER 15

# CHEMICAL KINETICS

## Part II. The Reaction Mechanism

## 15.1 GENERAL CONSIDERATIONS.

Having discussed in Chapter 14 how the rate law is obtained from kinetic data, we shall now discuss what various forms of rate law imply about the reaction mechanism. The first point to recognize is that going from a given mechanism to its corresponding rate law is a much simpler process than the reverse. The reason for this is that the process

$$mechanism \rightarrow rate\ law$$

is basically a mathematical one, whereas the reverse process

$$rate\ law \rightarrow mechanism$$

frequently involves the utilization of extensive chemical knowledge and requires many explicit and implicit chemical judgments. In other words, the more you know about the chemistry of the system of interest, the better your chances for coming up with a plausible mechanism.

Whatever the detailed nature of a proposed mechanism, it must be consistent with the observed rate law and reaction stoichiometry, together with any other experimental facts that have a bearing on the mechanism. The available kinetic evidence may be insufficient to distinguish between two or more possible reaction mechanisms, and further experimentation may be necessary in such cases to distinguish among the various possibili-

---

\*H. Taube, J. Chem. Educ., 36, 451 (1959).   **479**

ties. In fact, in many cases alternate mechanisms exist that cannot be distinguished experimentally on the basis of kinetic evidence alone. Even if only one totally consistent mechanism can be found, there always remains the possibility that another unthought-of mechanism exists that also satisfies the observed facts.

## 15.2   THE ACTIVATED COMPLEX.

Let us assume, in the interests of simplicity, that we are dealing with a reaction whose forward rate law is of the form

$$Rate = kC_xC_y$$

The most general interpretation of such a rate law is that in order for the reaction to occur, some critical events must take place simultaneously in species $x$ and $y$ with a frequency proportional to the product of the concentrations of $x$ and $y$. The collection of species involved in the *simultaneous* occurence of these critical events is called the *activated complex*. The concept of the activated complex does not necessarily imply *close* contact between the reactive species $x$ and $y$.

The *activated-complex postulate* enables us to interpret the quantity $E_a$ (the activation energy) as the *minimum* additional amount of internal, electronic, potential energy that must be pumped into the interacting molecules (e.g., by the conversion of some of the translational kinetic energy of two colliding particles into molecular vibrational energy) in order to bring about the reaction. If this critical amount of energy is available, then, and only then, can the reactive event take place.[*]

*An activated complex is a completely unstable species; it is always in the act of decomposing* to products, to reactants, or to intermediates, and these are the only types of reaction that it can undergo. In particular, the activated complex cannot be intercepted by another reactant species and thereby be converted to something else. The lifetime of an activated complex is estimated to be about the time required for one molecular vibration ($10^{-12}$ to $10^{-13}$ sec).

An activated complex must be carefully distinguished from an *intermediate*. An intermediate is a metastable species formed on the way to the products, but which does not appear as one of the final products. An intermediate requires further activation for reaction and is, in principle at least (and frequently in practice), detectable by physical methods, interceptable by other reactants, and isolable.

We can map the course of a simple, one-step chemical reaction, say

$$A + BC \rightarrow AB + C$$

on an electronic *potential energy surface*. The potential energy surface for the interaction of the three atoms A, B, and C is conveniently constructed as a contour map of $R_{BC}$ versus $R_{AB}$. The contour lines on this map (Figure 15.1) are equipotential energy lines. In Figure 15.1 there is a valley in the potential energy surface for $R_{BC} = R_{BC}^0$ and $R_{AB}$ large (i.e., for A far from BC), owing to the existence of the stable diatomic molecule BC. In addi-

---

[*]Quantum mechanics tells us that this cannot be *rigorously* true, owing to the finite probability of tunneling through the energy barrier. It remains, however, an excellent approximation for most reactions.

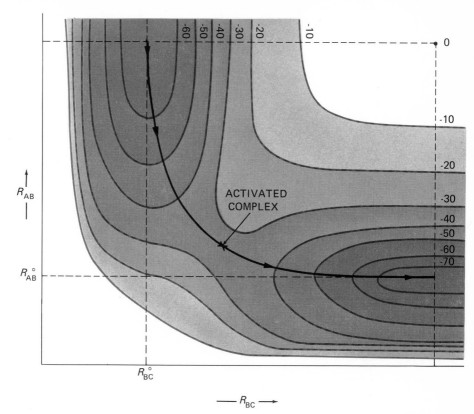

**FIGURE 15.1** Contour map (darker color indicates more negative energies) of the electronic potential energy surface for the reaction $A + BC \rightarrow AB + C$. The numbers given on the equipotential lines are in kcal·mole$^{-1}$; the distances are in Ångstroms. $R_{BC}^{\circ}$ and $R_{AB}^{\circ}$ are the equilibrium internuclear distances of the diatomic molecules BC and AB, respectively. An equipotential line is a line of constant electronic potential energy. The heavy line indicates the path of the reaction.

tion, there is a second valley for $R_{AB} = R_{AB}^{0}$ and $R_{BC}$ large (i.e., for C far from AB), owing to the existence of the stable diatomic molecule AB. For all three particles close together, there is a high point (point **x**) over which the particles must pass as a unit on going from $A + BC$ to $AB + C$. In other words, we have two valleys separated by a mountain pass. The point 0 in the figure is higher than the top of the pass, because at 0 the three particles are all far removed from one another, and their mutual potential energy of interaction is essentially zero.

The atoms must go over a hump (like a mountain pass) on the potential energy surface to get from reactants to products. The course of the reaction is indicated by the dark line (reaction coordinate) on the surface. This line is sketched out (side profile) in Figure 15.2. In Figure 15.2, $E_{af}$ is the activation energy for the forward reaction, $E_{ar}$ is the activation energy for the reverse reaction, and $\Delta E = E_{af} - E_{ar}$ is the overall internal energy change for the reaction. The activated complex is postulated to be the same for the forward and reverse reactions. This *postulate* is known as the *principle of microscopic reversibility*.

The basic supposition in the collision theory of reaction rates is that molecules must collide in order to react. Suppose we have a gas mixture composed of two types of molecules A and BC. At $P_A = P_{BC} = 1$ atm and

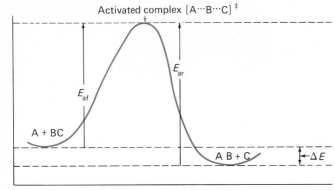

**FIGURE 15.2** Electronic potential energy profile along the reaction coordinate for the reaction $A + BC \rightarrow AB + C$.

300 K, the bimolecular collision frequency $Z_{A,BC}$ is of the order of $10^{33}$ collisions $\cdot sec^{-1} \cdot \ell^{-1}$. If every collision led to reaction, then the initial reaction rate would be of the order of $10^{10}$ mole $\cdot \ell^{-1} \cdot sec^{-1}$, which corresponds to a rate constant of $10^{12}$ $M^{-1} \cdot sec^{-1}$. This is a very large rate constant. A few second-order, gas-phase reactions (e.g., radical-radical recombinations like $H + C_6H_5$) have second-order rate constants as large as $10^{12}$ at 300 K, but most such reactions have much smaller rate constants. The inescapable conclusion is that not all molecular collisions between reactive species lead to reaction. In most collisions the colliding species survive intact.

It is known from the kinetic theory of gases that the *fraction* of bimolecular collisions that involve molecules with a total translational kinetic energy of $E$ or greater is $e^{-E/RT}$. If we interpret $E$ as the activation energy, then we can write for the rate constant

$$k = pZe^{-E/RT}$$

where $Z$ is the collision frequency at unit concentrations of A and BC, and $p$ is the *fraction* of bimolecular collisions that take place with the relative orientation required for reaction. (For example, in the reaction $A + BC \rightarrow AB + C$, the A molecule must collide with the BC molecule on the B side in order to form AB; collisions on the C side do not lead to reaction.) In other words,

$$k = \left(\begin{array}{c}\text{fraction of collisions}\\ \text{with the required}\\ \text{orientation}\end{array}\right) \times \left(\begin{array}{c}\text{collision}\\ \text{frequency}\end{array}\right) \times \left(\begin{array}{c}\text{fraction of collisions}\\ \text{with the required}\\ \text{energy}\end{array}\right)$$

The effect of the $e^{-E/RT}$ term on the magnitude of $k$, and hence on the reaction rate, can be very large. For example, when $E = 10$ kcal $\cdot mol^{-1}$, $e^{-E/RT} = 10^{-7.2}$ at 300 K.

Most bimolecular reaction-rate constants are much smaller than those calculated from the bimolecular collision frequency, because most reactions have a non-zero activation energy. The collision frequency simply represents an upper limit to the reaction rate. Some examples of $p$, $Z$, and $E_a$ are given in Table 15.1.

The form of the collision-theory expression for the rate constant serves as a theoretical justification for the Arrhenius equation. However, it is

**TABLE 15.1**

SOME BIMOLECULAR REACTIONS

| REACTION | $T(K)$ | $Z \times 10^{-13}$ $(\text{cm}^3 \cdot \text{mol}^{-1} \cdot \text{sec}^{-1})$ | $p$ | $E_a(\text{kcal})$ |
|---|---|---|---|---|
| $ClO + ClO \rightarrow Cl_2 + O_2$ | 375 | 2.6 | 0.0022 | 0.0 |
| $NO + O_3 \rightarrow NO_2 + O_2$ | 215 | 6.3 | 0.017 | 2.5 |
| $NO_2 + F_2 \rightarrow NO_2F + F$ | 320 | 5.9 | 0.027 | 10.4 |
| $NO + Cl_2 \rightarrow NOCl + Cl$ | 600 | 9.3 | 0.043 | 20.3 |
| $NO_2 + CO \rightarrow NO + CO_2$ | 665 | 7.4 | 0.16 | 31.6 |

worth noting that simple collision theory does not tell us how to compute $p$ and $E_a$ from molecular parameters.

If the reaction $A + BC \rightarrow AB + C$ involves the formation of an intermediate, then the potential-energy surface in such a case will have three valleys and two passes, and, therefore, two activated complexes. The reaction coordinate diagram for such a case is depicted in Figure 15.3. In this case the two activated complexes differ in which bond is in the process of being formed or broken.

The foregoing description of the activated complex leads to the conclusion that when a reaction mechanism involves a single rate-determining step (one step much slower than all the others), the form of the rate law defines the composition of the activated complex, except for species of undefined order (e.g., solvent molecules). The rate law can be used to establish the composition of the activated complex in essentially the same way that an equilibrium constant expression can be used to establish the formula of a species in equilibrium with other species of known composition. In other words, we assume an equilibrium between the activated complex and the reactant species in the rate-determining step. This "equilibrium" is not a true thermodynamic equilibrium, because the species concentrations are changing with time. Such an equilibrium postulate is tenable (provided the reaction is not extremely fast), because if the inter-

**FIGURE 15.3** Electronic potential energy profile along the reaction coordinate for a reaction involving an intermediate. The intermediate is the species ABC.

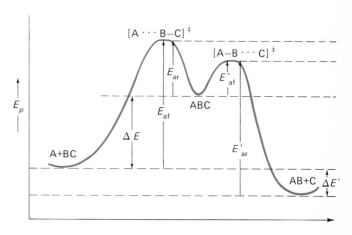

"Reaction Coordinate"

acting species do not possess enough energy to pass over the activation barrier, then they separate and return to reactants. In addition, the "equilibrium" concentration of activated complex is presumably always very small, owing to its short lifetime. The "equilibrium" between the reactants and the activated complex is assumed to be very rapidly established.

## 15.3 MECHANISM TO RATE LAW.

An elementary reaction (or process) is defined as a chemical reaction that takes place in a single step. All elementary processes are chemical reactions, but most chemical reactions are not elementary processes. Elementary processes are classified in terms of the number of species that appear as reactants in the process. A *unimolecular* elementary process is one in which only one molecule is involved as a reactant; some examples are

$$N_2O_5^* \rightarrow NO_2 + NO_3$$

(where * indicates an energetically excited species)

$$CO(v=2) \rightarrow CO(v=1) + h\nu$$

(where $v$ denotes the vibrational quantum number), and

$$^{14}_6C \rightarrow {}^{14}_7N + {}^{\ 0}_{-1}e$$

A *bimolecular* elementary process involves two reacting species; some examples are

$$NO_2 + Ar \rightarrow NO + O + Ar$$

$$NO_2 + NO_2 \rightarrow NO + NO_3$$

$$NO_2 + F_2 \rightarrow O_2NF + F$$

$$N_2O_5 + Ar \rightarrow N_2O_5^* + Ar$$

A *termolecular* elementary process involves three particles as reactants; some examples are

$$H + H + N_2 \rightarrow H_2 + N_2^*$$

$$O + NO + Ar \rightarrow NO_2 + Ar$$

$$I + I + H_2 \rightarrow 2HI$$

The gas-phase recombination of two *atoms* to yield a diatomic molecule will always lead to redissociation to atoms unless another species carries away part of the energy released on bond formation.

Because the rate expression for an elementary process is predictable from the stoichiometry, we have:

(a) *unimolecular elementary process*

$$A \xrightarrow{k} \text{products}$$

$$R_{D,A} = k(A)$$

(b) *biomolecular elementary process*

$$A + B \overset{k}{\rightarrow} \text{products}$$
$$R_{D,A} = R_{D,B} = k(A)(B)$$

(c) *termolecular elementary process*

$$A + B + C \overset{k}{\rightarrow} \text{products}$$
$$R_{D,A} = k(A)(B)(C)$$

or

$$A + A + B \overset{k}{\rightarrow} \text{products}$$
$$R_{D,A} = 2k(A)^2(B) = 2R_{D,B}$$

For any elementary process, the molecularity and order are the same; however, not all first-order reactions are elementary unimolecular processes, not all second-order reactions are elementary bimolecular processes, and so forth. This is one of the reasons why it is much more difficult to go from rate law to mechanism than from mechanism to rate law; it also is one of the reasons why a completely unambiguous choice of mechanism is not always possible.

In the general case, a reaction mechanism consists of a sequence of elementary processes. It is this sequence that we seek to uncover when we set out to determine a reaction mechanism. Consider the hypothetical reaction

$$A + 2B = C + D$$

and let us suppose that the mechanism of this reaction is

(1) $$A + B \rightarrow E$$

(2) $$E + B \rightarrow C + D$$

The products C and D result from a sequence of two elementary processes, which involve the intermediate E. *The products can be formed no faster than the rate of the slowest (rate-determining) step in the sequence.* Therefore, when one step in the sequence is much slower than any of the others, the overall reaction rate will be equal to the rate of that step.

If a mechanism involves a single rate-determining step, then the rate law gives the composition of the activated complex for that step. This is because the reaction rate is proportional to the concentration of the activated complex, and the concentration of the activated complex is in turn proportional to the product of the concentrations of the species that combine to form the activated complex.

Referring to the above mechanism, consider the following two distinct possibilities:

**Case I.** Step (1) is rate-determining; then

(1) $$A + B \overset{k_1}{\rightarrow} E \qquad \text{slow}$$

(2) $\qquad E + B \rightarrow C + D \qquad$ fast

$$R_{D,A} = k_1(A)(B) = k_{obs}(A)(B)$$

Note that as the intermediate E is produced in Step (1), it is rapidly consumed by B in step (2). In effect, B has to wait around for E to form.

**Case II.** Step (1) is a rapid equilibrium and step (2) is rate determining; then[*]

(1) $\qquad A + B \rightleftharpoons E \qquad$ rapid equilibrium, $K_1 = \dfrac{(E)}{(A)(B)}$

(2) $\qquad E + B \overset{k_2}{\rightarrow} C + D \qquad$ slow

and thus

$$R_{D,A} = k_2(E)(B) = k_2 K_1(A)(B)^2 = k_{obs}(A)(B)^2$$

We are permitted to set (E) equal to $K_1(A)(B)$ because the equilibrium is rapidly established in the reaction mixture, and therefore the equilibrium condition holds. The observed rate constant, $k_{obs}$, in this case is equal to $k_2 K_1$.

Let us now consider some actual cases. If the reaction

$$H_2(g) + I_2(g) = 2HI(g)$$

proceeds via the mechanism[†]

$$I_2 \rightleftharpoons 2I \qquad \text{rapid equilibrium, } K_1 = \dfrac{(I)^2}{(I_2)}$$

$$2I + H_2 \overset{k_2}{\rightarrow} 2HI \quad \text{slow}$$

then from the rate-determining step we have for the rate law of the forward reaction

$$R_{D,H_2} = k_2(H_2)(I)^2 = k_2 K_1(H_2)(I_2) = k_{obs}(H_2)(I_2)$$

The composition of the activated complex is $H_2 I_2$.

As a second example, consider the reaction

$$H_2(g) + Br_2(g) = 2HBr(g)$$

A simplified mechanism that is consistent with kinetic data for this reaction (at least in the early stages of the reaction) is

$$Br_2 \rightleftharpoons 2Br \qquad \text{rapid equilibrium, } K = \dfrac{(Br)^2}{(Br_2)}$$

---

[*]We shall denote rate constants by lower-case $k$'s, and equilibrium constants by capital $K$'s.

[†]A mechanism involving I atoms has been used because I atoms have been shown to be an intermediate in this reaction (J. H. Sullivan, J. Chem. Phys., *46*, 73 (1967)).

$$\text{Br} + \text{H}_2 \xrightarrow{k_2} \text{HBr} + \text{H} \quad \text{slow}$$

$$\text{H} + \text{Br}_2 \xrightarrow{k_3} \text{HBr} + \text{Br} \quad \text{fast}$$

From this mechanism we deduce the following rate law:

$$R_{D,\text{Br}_2} = k_2(\text{H}_2)(\text{Br}) = k_2 K^{1/2}(\text{H}_2)(\text{Br}_2)^{1/2} = k_{obs}(\text{H}_2)(\text{Br}_2)^{1/2}$$

The composition of the activated complex is $\text{H}_2\text{Br}^{\ddagger}$ $(= \text{H}_2 + (1/2)\,\text{Br}_2)$. This mechanism requires that the observed rate constant, $k_{obs}$, be equal to $k_2 K^{1/2}$.

As a third example, consider the decomposition of nitrous acid in acidic aqueous solutions:

$$3\text{HNO}_2(\text{aq}) = \text{H}^+(\text{aq}) + \text{NO}_3^-(\text{aq}) + 2\text{NO}(\text{aq}) + \text{H}_2\text{O}(\ell)$$

A mechanism consistent with the known experimental facts for this reaction is

$$2\text{HNO}_2 \rightleftharpoons \text{NO} + \text{NO}_2 + \text{H}_2\text{O} \quad \text{rapid equilibrium,} \quad K_1 = \frac{(\text{NO})(\text{NO}_2)(\text{H}_2\text{O})}{(\text{HNO}_2)^2}$$

$$\text{NO}_2 + \text{NO}_2 \xrightarrow{k_2} \text{NO}^+ + \text{NO}_3^- \quad \text{slow}$$

$$\text{NO}^+ + \text{H}_2\text{O} \xrightarrow{k_3} \text{HNO}_2 + \text{H}^+ \quad \text{fast}$$

In terms of this mechanism, the activated complex is $\text{N}_2\text{O}_4^{\ddagger}$, and the rate law is

$$R_{A,\text{NO}_3^-} = k_2(\text{NO}_2)^2 = \frac{k_2 K_1^2(\text{HNO}_2)^4}{(\text{H}_2\text{O})^2(\text{NO})^2} = k_{obs}\frac{(\text{HNO}_2)^4}{(\text{NO})^2}$$

The composition of the activated complex is* $4\text{HNO}_2 - 2\text{NO} = \text{H}_4\text{N}_2\text{O}_6^{\ddagger}$. And this, less two water (solvent) molecules, is $\text{N}_2\text{O}_4^{\ddagger}$ $(= \text{H}_4\text{N}_2\text{O}_6 - 2\text{H}_2\text{O})$.

Once a mechanism has been proposed, it is usually not a difficult matter to find its corresponding rate law, and the requirement that the mechanism yield the observed rate law is sufficient in many cases to eliminate a large number of otherwise chemically reasonable mechanisms.

## 15.4 RATE LAW TO MECHANISM.

We shall attack the general problem of going from the experimental rate law to the reaction mechanism by means of examples:

(a) The reaction of nitrogen dioxide with fluorine

$$2\text{NO}_2(\text{g}) + \text{F}_2(\text{g}) = 2\text{NO}_2\text{F}$$

is known to follow the rate law

$$R_{D,\text{F}_2} = k_{obs}(\text{NO}_2)(\text{F}_2)$$

---

*Note that we subtract the atoms in the denominator from those in the numerator of the observed rate law in order to obtain the composition of the activated complex (Problem 9).

From this rate law we deduce that the composition of the activated complex is $NO_2F_2\ddagger$. The above rate law will result from a mechanism that involves bringing together one $NO_2$ and one $F_2$ molecule in a rate-determining, bimolecular, elementary process. This step must be followed by another (relatively rapid) step that consumes a second $NO_2$ molecule, as is required by the reaction stoichiometry. A possible mechanism is

$$NO_2 + F_2 \overset{k_1}{\rightarrow} NO_2F + F \quad \text{slow}$$

$$F + NO_2 \overset{k_2}{\rightarrow} NO_2F \quad \text{fast}$$

where an F atom has been used as an intermediate. This mechanism yields the correct rate law

$$R_{D,F_2} = k_1(NO_2)(F_2)$$

The mechanism requires that $k_2 \gg k_1$. This can be checked by producing F atoms (say photochemically) and determining their rate of reaction with $NO_2$. These experiments show that $k_2 \gg k_1$, and therefore provide corroborative evidence for the proposed mechanism.

(b) The reaction

$$NO_2(g) + CO(g) = CO_2(g) + NO(g)$$

has the following rate law ($T < 500$ K):

$$R_{D,NO_2} = k_{obs}(NO_2)^2$$

The activated complex is $N_2O_4\ddagger$, and a possible mechanism is

$$NO_2 + NO_2 \overset{k_1}{\rightarrow} NO + NO_3 \quad \text{slow}$$

$$NO_3 + CO \overset{k_2}{\rightarrow} CO_2 + NO_2 \quad \text{fast}$$

$$R_{D,NO_2} = k_1(NO_2)^2$$

Note that $NO_3$ is postulated as an intermediate in this mechanism.

(c) The decomposition of ozone to oxygen

$$2O_3(g) = 3O_2(g)$$

has the rate law

$$R_{D,O_3} = \frac{k_{obs}(O_3)^2}{(O_2)}$$

The activated complex is $2O_3 - O_2 = O_4\ddagger$. A possible mechanism is

$$O_3 \rightleftharpoons O_2 + O \quad \text{rapid equilibrium,} \quad K_1 = \frac{(O_2)(O)}{(O_3)}$$

$$O_3 + O \overset{k_2}{\rightarrow} 2O_2 \quad \text{slow}$$

$$R_{D,O_3} = 2k_2(O_3)(O) = \frac{2k_2K_1(O_3)^2}{(O_2)}$$

The factor of two arises in the rate law because each occurrence of the rate-determining step removes *two* $O_3$ molecules from the reaction mixture (one as $O_3$, and the other split into O and $O_2$ in the equilibrium step).

(d) The rate law for the reaction between hypochlorite and iodide ions in alkaline aqueous solution

$$OCl^-(aq) + I^-(aq) = IO^-(aq) + Cl^-(aq)$$

is

$$R_{D,OCl^-} = \frac{k_{obs}(I^-)(OCl^-)}{(OH^-)}$$

Because we cannot subtract a $OH^-$ from $I^- + OCl^-$, $k$ *must* have a dependence on $(H_2O)$ hidden in it; the assumption that $k = k'(H_2O)$ yields

$$OCl^- + I^- + H_2O - OH^- = [HOClI^-]^{\ddagger}$$

for the composition of the activated complex. A possible mechanism is

$$OCl^- + H_2O \rightleftharpoons HOCl + OH^- \quad \text{rapid equilibrium, } K_1 = \frac{(OH^-)(HOCl)}{(OCl^-)(H_2O)}$$

$$HOCl + I^- \xrightarrow{k_2} HOI + Cl^- \quad \text{slow}$$

$$HOI + OH^- \xrightarrow{k_3} OI^- + H_2O \quad \text{fast}$$

$$R_{D,OCl^-} = k_2(I^-)(HOCl) = k_2 K_1 \frac{(H_2O)(I^-)(OCl^-)}{(OH^-)}$$

At 25°C, $k_{obs} = 60 \text{ sec}^{-1}$, and $K_1 = 2.9 \times 10^{-7}$ (with pure water as the standard state for $H_2O$). Hence,

$$k_2(H_2O) = \frac{60 \text{ sec}^{-1}}{2.9 \times 10^{-7}} = 2.1 \times 10^8 \text{ sec}^{-1}$$

and if we take $(H_2O) = 55.5$ M, then $k_2 = 3.8 \times 10^6 \text{ M}^{-1} \text{ sec}^{-1}$.

In working from the rate law to the mechanism of a reaction, especially when dealing with reactions in solution, it should be recognized that there may be two or more kinetically equivalent forms of the rate law, owing to the existence of one or more rapid (relative to the rate of the slow step) equilibria in the solution. For example, the decomposition of aqueous ammonium nitrite

$$NH_4^+(aq) + NO_2^-(aq) = N_2(g) + 2H_2O(\ell)$$

follows the rate law

$$R_{A,N_2} = k_{obs}(NH_4^+)(HNO_2)$$

However, the equilibria

$$NH_4^+(aq) \rightleftharpoons H^+(aq) + NH_3(aq)$$

$$HNO_2(aq) \rightleftharpoons H^+(aq) + NO_2^-(aq)$$

are known to be rapidly established relative to the rate of decomposition of $NH_4NO_2(aq)$, and the rate law could be just as correctly formulated as $k'(H^+)(NH_3)(HNO_2)$, or $k''(H^+)^2(NH_3)(NO_2^-)$, or $k'''(NH_4^+)(NO_2^-)(H^+)$.

In all four cases we obtain the same formula for the activated complex $[NH_5NO_2^+]^{\ddagger}$, or that complex less a water molecule, $[NH_3NO^+]^{\ddagger}$. The second of these is believed to be the more reasonable because it can be formed by the plausible sequence

$$HNO_2 + H^+ \rightleftharpoons NO^+ + H_2O \qquad rapid$$

$$NO^+ + NH_3 \rightleftharpoons [NH_3NO^+]^{\ddagger} \rightarrow N_2 + H^+ + H_2O \qquad slow$$

In any case, there is no way to distinguish this mechanism from one involving $NH_4^+$ and $HNO_2$

$$NH_4^+ + HNO_2 \rightarrow N_2 + 2H_2O + H^+ \qquad slow$$

*on the basis of kinetic data alone.* As a consequence, there is no point in arguing, solely on the basis of kinetic data, whether the rate-determining step is between $NO^+$ and $NH_3$ or $NH_4^+$ and $HNO_2$. The reason the first mechanism is preferred over the second is that the activated complex

has the two nitrogen atoms properly positioned to form $N_2$, and, in addition, there is independent evidence for the intermediate $NO^+$ in the reaction.

## 15.5   CATALYSIS.

*A catalyst is a substance that enables the formation of an activated complex of lower activation energy than is possible with the reactants in the absence of the catalyst.* In other words, a catalyst provides a new reaction pathway. A catalyst *does not appear as a reactant* in the net reaction stoichiometry, but it does appear in the rate law.

The reaction between aqueous ceric and thallous ions in acidic aqueous solutions

$$2Ce^{4+}(aq) + Tl^+(aq) = 2Ce^{3+}(aq) + Tl^{3+}(aq)$$

is quite sluggish at room temperature, even though the reaction is thermodynamically favored ($K_{298} = 1.5 \times 10^{12}$). The slowness of this reaction has been ascribed to the fact that the $+2$ state of either cerium or thallium is too unstable to serve as an intermediate. Therefore, the reaction requires a three-body encounter ($2Ce^{4+} + Tl^+$) in the rate-determining step. Three-body encounters are only about $10^{-5}$ times as probable as two-body encounters. Reactions requiring three-body encounters are, therefore, much slower than those of the same activation energy requiring only a two-body encounter. Manganese (II) ion catalyzes the above reaction, because it enables the reaction to proceed via a sequence of two-body encounters, namely

$$Ce^{4+} + Mn^{2+} \rightarrow Ce^{3+} + Mn^{3+}$$

$$Ce^{4+} + Mn^{3+} \rightarrow Ce^{3+} + Mn^{4+}$$

$$Mn^{4+} + Tl^{+} \rightarrow Tl^{3+} + Mn^{2+}$$

A great variety of reactions in solution are catalyzed by acids and bases. Such reactions include hydrolyses of various types, halogenations, hydrations, rearrangements, and polymerizations. Acid catalysis most commonly results when the protonation of a reactant renders it more susceptible to attack by a *nucleophile* (electron donor). An example is the acid-catalyzed hydration of the olefin 2-methylpropene

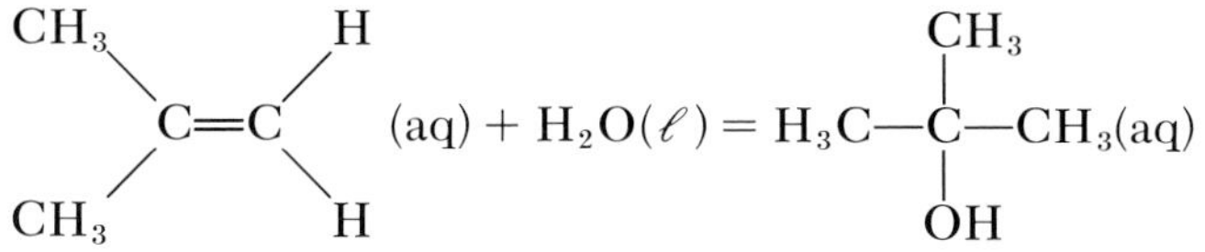

The rate law for this reaction is

$$R_{D,\text{olefin}} = k_{obs}(\text{olefin})(H^+)$$

and a plausible mechanism for this reaction is

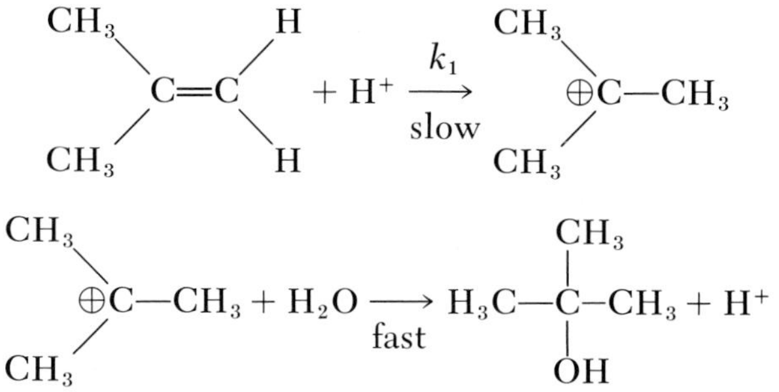

Base catalysis frequently involves removal of a proton by a base, leaving a species more susceptible to attack by an *electrophile* (electron acceptor). An example is the base-catalyzed iodination of acetone, which is catalyzed not only by $OH^-$, but also by other bases such as $CH_3COO^-$:

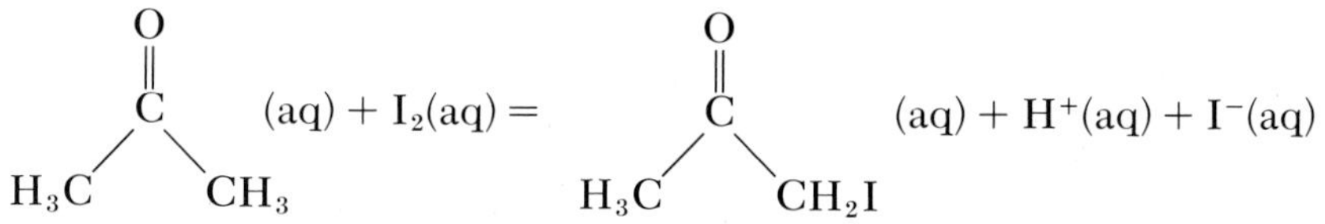

$$R_{D,\text{acetone}} = k_{obs}(\text{acetone})(CH_3COO^-) + k'_{obs}(\text{acetone})(OH^-)$$

A plausible mechanism for the acetate-catalyzed pathway of this reaction is

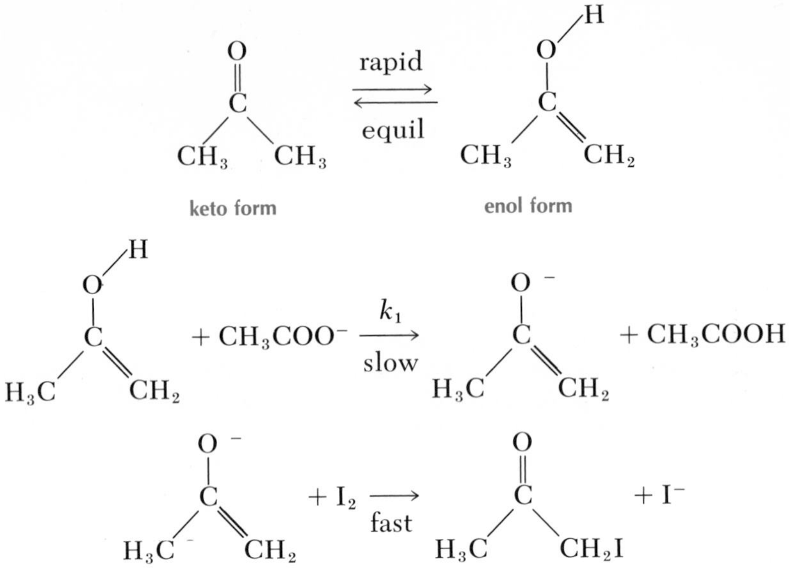

keto form          enol form

$$CH_3COOH \rightleftharpoons CH_3COO^- + H^+$$
rapid

## 15.6 THE REVERSIBILITY OF CHEMICAL REACTIONS.

Consider the reaction

$$CH_3COOH(aq) = H^+(aq) + CH_3COO^-(aq)$$

The equilibrium constant at 25°C is $1.76 \times 10^{-5}$ on the molarity scale. Hence, if we have an aqueous solution of acetic acid for which

$$1.00 \times 10^{-4} = (CH_3COOH) + (CH_3COO^-)$$

then

$$1.76 \times 10^{-5} = \frac{x^2}{c-x} = \frac{x^2}{1.00 \times 10^{-4} - x}$$

Solving the quadratic for $x$ yields $x = 0.38 \times 10^{-4}$ M. The pseudo-first-order rate constant for the dissociation of acetic acid in water at 25°C is $7.9 \times 10^5$ sec$^{-1}$, and, therefore, the half-life for the disappearance of $CH_3COOH(aq)$ is about 0.88 $\mu$sec (1 $\mu$sec $= 10^{-6}$ sec $= 1$ microsecond). The change in the concentrations of the various species with time for a solution initially $1.00 \times 10^{-4}$ M in undissociated acetic acid is presented in Figure 15.4(a). The concentrations of the various species become constant with time when the equilibrium state is achieved. In Figure 15.4(b) we have plotted out the concentrations of the various species as a function of time for the re- verse reaction, where we start with $1.00 \times 10^{-4}$ M NaOCOCH$_3$(aq) and $1.00 \times 10^{-4}$ M HCl(aq). Note that the equilibrium state is the same in both

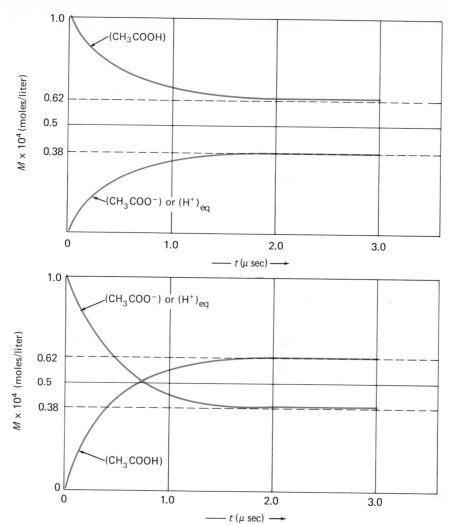

**FIGURE 15.4** (a) The dissociation of $CH_3COOH$ in water at 25°C. $(CH_3COOH)_o = 1.00 \times 10^{-4}$, $(CH_3COO^-)_o \simeq (H^+(aq))_o \simeq 0$, $(CH_3COOH)_{eq} = 0.62 \times 10^{-4}$ M, $(CH_3COO^-)_{eq} = (H^+)_{eq} = 0.38 \times 10^{-4}$ M.

(b) The association of acetate and hydrogen ions in water at 25°C. $(CH_3COO^-)_o = 1.00 \times 10^{-4}$ (from sodium acetate), $(HCl)_o = 1.00 \times 10^{-4}$, $(CH_3COOH)_o = 0$, $(CH_3COOH)_{eq} = 0.62 \times 10^{-4}$ M, $(CH_3COO^-)_{eq} = (H^+)_{eq} = 0.38 \times 10^{-4}$ M.

cases.* Figures 15.4(a) and 15.4(b) serve to illustrate an important facet of a true equilibrium namely, that the same final state is attained starting from either side of the equilibrium.

For the dissociation reaction we have

$$R_{D,HX} = k_1(HX)$$

---

*This is not exactly true because in running the reverse reaction we end up with a solution that also contains $1.00 \times 10^{-4}$ NaCl(aq). The final solution would be the same if the forward reaction were run in the presence of $1.00 \times 10^{-4}$ M NaCl.

whereas for the reverse reaction we have

$$R_{A,HX} = k_{-1}(H^+)(X^-)$$

The net reaction rate at any time is

$$R_{HX,net} = R_{D,HX} - R_{A,HX} = k_1(HX) - k_{-1}(H^+)(X^-)$$

At equilibrium the concentrations of the various species show no further change with time, and, therefore, $R_{D,HX} = R_{A,HX}$. From this result we obtain[*]

$$\frac{(H^+)(X^-)}{(HX)} = \frac{k_1}{k_{-1}} \simeq K$$

The above reaction is an elementary process, that is, it occurs in a single step; the activated complex $(CH_3COOH + nH_2O)$ is the same for the forward and reverse reaction (microscopic reversibility).

At equilibrium the reaction does not cease, but rather a state is achieved for which the rate of disappearance of any reactant via the forward reaction is exactly counterbalanced by the rate of appearance of this same reactant via the reverse reaction. An equilibrium state of a system is a dynamic rather than a static one. That the equilibrium state is indeed dynamic was convincingly demonstrated experimentally by J. N. Wilson and R. G. Dickinson [J. Am. Chem. Soc., 59, 1358 (1937)].[†] These investigators studied the rate of the reaction

$$H_3AsO_4(aq) + 3I^-(aq) + 2H^+(aq) = H_3AsO_3(aq) + I_3^-(aq) + H_2O(\ell)$$

in acidic aqueous solutions by preparing the system with the previously determined *equilibrium* concentrations of $H_3AsO_3$, $I_3^-$, $H_3AsO_4$, $I^-$ and $H^+$. The $H_3AsO_3$ was prepared from radioactive arsenic-76, whereas the $H_3AsO_4$ was prepared from non-radioactive arsenic isotopes. The rate of the reaction was followed by determining the rate of disappearance of $^{76}As$ in $H_3AsO_3$. For any given run the concentrations of the various species remain constant (because they are the equilibrium concentrations), but the $^{76}As$ was found to distribute itself uniformly (i.e., its percentage became the same) over all the arsenic-containing species that were in equilibrium with one another. The rate constant for the forward reaction obtained by this technique agreed within experimental error with that obtained from conventional non-equilibrium kinetic data.

The principle of microscopic reversibility can be used to derive a relationship between the forward and reverse rates of a reaction. Let our reaction be

$$A + 2B = E + F$$

and suppose the rate law for the forward reaction is

$$R_{D,A} = k_f(A)(B)(X)$$

---

[*]This is not rigorously correct as presented here, because the expression given for $K$ contains concentrations, whereas the thermodynamically defined $K$ for the reaction involves activities.

[†]Quoted in J. Hildebrand and R. E. Powell, *Principles of Chemistry*, 7th Edition (Macmillan Co., New York, 1964), p. 179.

where X is a catalyst. Now, the activated complex for the forward reaction is composed of those atoms contained in $A + B + X$, and by microscopic reversibility, the activated complex for the reverse reaction must be composed of those same atoms. From the reaction stoichiometry we note that the only other way these atoms can be obtained is

$$A + B + X = E + F + X - B$$

and therefore, the rate law for the reverse reaction must be

$$R_{A,A} = k_r(E)(F)(X)(B)^{-1} = k_r \frac{(E)(F)(X)}{(B)}$$

This result shows that a catalyst for the forward reaction must also be a catalyst for the reverse reaction, as is known from experiment. A catalyst does not affect the equilibrium constant for a reaction, because it cancels out when we equate $R_{D,A}$ to $R_{A,A}$. The expression $k_r(E)(F)(X)/(B)$ can be factored as follows:

$$k_r \frac{(E)(F)(X)}{(B)} = k_f(A)(B)(X) \cdot \frac{(E)(F)}{(A)(B)^2} \cdot \frac{1}{(k_f/k_r)}$$

but

$$\frac{(E)(F)}{(A)(B)^2} = Q = \text{concentration quotient}$$

and

$$k_f/k_r = K = \text{equilibrium constant}$$

hence

$$\boxed{\text{Reverse Rate} = \text{Forward Rate}\left(\frac{Q}{K}\right)} \qquad (15.1)$$

Both $Q$ and $K$ refer to the complete balanced reaction that occurs when one activated complex is formed and decomposes to products. Equation (15.1) shows us how to obtain the rate law and the numerical value of the rate constant for the reverse reaction from the rate law and numerical value of the rate constant for the forward reaction, together with the equilibrium constant for the reaction.

Applying Equation (15.1) to the previously considered acetic acid dissociation, we have

$$R_{D,HX} = k_1(HX)$$
$$R_{D,HX} = (7.9 \times 10^5 \text{ sec}^{-1})\,(HX)\ (\text{at } 25°C)$$

Therefore

$$R_{A,HX} = (7.9 \times 10^5 \text{ sec}^{-1})(HX) \cdot \frac{(H^+)(X^-)}{(HX)} \cdot \frac{1}{1.76 \times 10^{-5} \text{ M}}$$

$$R_{A,HX} = (4.5 \times 10^{10} \text{ M}^{-1} \cdot \text{sec}^{-1})(H^+)(X^-) = k_{-1}(H^+)(X^-)$$

If we have a complex reaction mechanism, that is, one involving more than one step, the principle of microscopic reversibility requires that the forward and reverse rates of each elementary reaction be equal at equilibrium; the mechanism of the reverse reaction is exactly the reverse of the mechanism of the forward reaction.

If the rate law for the reaction involves two or more terms, then the principle of microscopic reversibility requires the calculation of the reverse rate law from each of the terms of the forward rate law taken separately. The complete reverse rate law is then obtained by adding together the reverse rate laws for each of the separate pathways.

## 15.7 STEADY-STATE KINETICS. MECHANISMS WITH MORE THAN ONE RATE-DETERMINING STEP.

Rate laws of the general form (P = product)

$$R_{A,P} = \frac{k(A)}{1 + k'(B)} \tag{15.2}$$

involving a two-term denominator, arise whenever an intermediate is formed that can either react to give products (productive step) or revert to the original reactants (non-productive step). In such a case

$$k(A) = \text{rate of formation of the intermediate}$$

$$k'(B) = \frac{\text{rate of non-productive step}}{\text{rate of the productive step}}$$

Some examples of reactions with rate laws of this type follow.

### (a)  The Chlorine-Hydrogen Peroxide Reaction.

The reaction between chlorine and hydrogen peroxide in acidic aqueous solution has the stoichiometry

$$Cl_2(aq) + H_2O_2(aq) = 2H^+(aq) + 2Cl^-(aq) + O_2(g)$$

The rate law for the forward reaction is

$$R_{A,O_2} = \frac{k(Cl_2)(H_2O_2)}{1 + k'(H^+)(Cl^-)}$$

What is the mechanism of this reaction? We note that when $k'(H^+)(Cl^-) \ll 1$ (say in the very early stages of the reaction), the apparent rate law is

$$R_{A,O_2} = k(Cl_2)(H_2O_2)$$

and, therefore, the activated complex has the composition $H_2O_2Cl_2^\ddagger$. On the other hand, when $k'(H^+)(Cl^-) \gg 1$ (say in the presence of added excess of HCl) the apparent rate law is

$$R_{A,O_2} = \frac{k}{k'} \frac{(Cl_2)(H_2O_2)}{(H^+)(Cl^-)} = k'' \frac{(Cl_2)(H_2O_2)}{(H^+)(Cl^-)}$$

and, therefore, the activated complex has the composition $HO_2Cl^‡$. Given these results, we devise a mechanism that involves the formation of the activated complexes $H_2O_2Cl_2^‡$ and $HO_2Cl^‡$ in successive steps:

$$Cl_2 + H_2O_2 \underset{k_{-1}}{\overset{k_1}{\rightleftharpoons}} HO_2Cl + H^+ + Cl^- \qquad \text{slow}$$

$$HO_2Cl \overset{k_2}{\rightarrow} O_2 + H^+ + Cl^- \qquad \text{slow}$$

In terms of this mechanism, the rate of appearance of oxygen is given by

$$R_{A,O_2} = k_2(HO_2Cl)$$

*The above mechanism involves three steps of comparable rate.* This mechanism leads to the observed rate law if we assume a *steady state* for the intermediate $HO_2Cl$, that is, if we assume

$$R_{A,HO_2Cl} = R_{D,HO_2Cl}$$

or, in other words,

$$R_{net,HO_2Cl} = 0$$

The validity of the steady state approximation does not require that the net rate of change of the intermediate concentration with time be exactly zero, but rather it requires only that

$$R_{net,intermediate} \simeq 0$$

In other words, we require that the rate of change of the intermediate concentration with time be much smaller than the rate of appearance or disappearance of any reactant or product. This condition will automatically be satisfied if the intermediate concentration remains very low relative to the concentrations of the other species in the reaction.

Applying the steady-state condition to the above reaction yields

$$R_{A,HO_2Cl} = R_{D,HO_2Cl}$$
$$k_1(Cl_2)(H_2O_2) = k_2(HO_2Cl) + k_{-1}(HO_2Cl)(H^+)(Cl^-)$$

Solving for the concentration of the intermediate, $HO_2Cl$, we obtain

$$(HO_2Cl) = \frac{k_1(Cl_2)(H_2O_2)}{k_2 + k_{-1}(H^+)(Cl^-)}$$

and, therefore,

$$R_{A,O_2} = \frac{k_2k_1(Cl_2)(H_2O_2)}{k_2 + k_{-1}(H^+)(Cl^-)} = \frac{k_1(Cl_2)(H_2O_2)}{1 + \dfrac{k_{-1}}{k_2}(H^+)(Cl^-)}$$

We can obtain $k_1$ from runs with $\dfrac{k_{-1}}{k_2}(H^+)(Cl^-) \ll 1$, and we can obtain $(k_2k_1/k_{-1})$ from runs with $\dfrac{k_{-1}}{k_2}(H^+)(Cl^-) \gg 1$.

## (b)   Enzyme Kinetics.

Many reactions in biologically important systems are catalyzed by *enzymes.*[*] For example, the hydrolysis of urea

$$H_2N \diagdown C(=O) \diagup NH_2(aq) + H_2O(\ell) = 2NH_3(aq) + CO_2(g)$$

is catalyzed by the enzyme *urease*. The pseudo-first-order half-life for urea in the uncatalyzed hydrolysis at 25°C is about $10^9$ sec (32 yr), whereas the half-life at the same conditions for the urea-urease (substrate-enzyme) complex is $10^{-4}$ sec. The actual rate of the catalyzed reaction is not, however, $10^{13}$-fold greater than the uncatalyzed reaction, because only a small fraction of the urea is complexed by the enzyme under the reaction conditions. Nonetheless, enzymes are often very effective and specific catalysts for many reactions. Enzymes have one or more active sites, conveniently thought of as cavities on the enzyme surface, at which the reaction can take place when the substrate finds its way into the cavity. After the reaction takes place the products leave the active site, and a new substrate molecule is free to enter.

Many enzyme-catalyzed reactions can be described by the following mechanism (E = enzyme, S = substrate, ES = enzyme-substrate complex, P = product(s)):

(1) $$E + S \underset{k_{-1}}{\overset{k_1}{\rightleftarrows}} ES$$

(2) $$ES \overset{k_2}{\rightarrow} E + P$$

If all three elementary processes proceed at comparable rates (i.e., (1) is not a rapid equilibrium), and a steady state condition is imposed on the intermediate ES, then we have

$$R_{A,ES} = R_{D,ES}$$
$$k_1(E)(S) = k_{-1}(ES) + k_2(ES)$$

From the steady-state condition we obtain

$$(ES)(k_2 + k_{-1}) = k_1(E)(S)$$

The mass-balance condition on the enzyme is

$$(E)_{tot} = (E) + (ES)$$

Therefore,

$$(ES)(k_2 + k_{-1}) = k_1(S) \{(E)_{tot} - (ES)\}$$

---

[*]Enzymes are usually high-molecular-weight, colloidal proteins.

or

$$(ES) = \frac{k_1(S)(E)_{tot}}{k_2 + k_{-1} + k_1(S)}$$

Substitution of this expression for (ES) into the rate law yields

$$R_{A,P} = R_{D,S} = k_2(ES) = \frac{k_1 k_2(S)(E)_{tot}}{k_2 + k_{-1} + k_1(S)} = \frac{k_2(S)(E)_{tot}}{\left(\frac{k_2 + k_{-1}}{k_1}\right) + (S)}$$

In enzyme kinetics studies, one usually determines only the initial reaction rate, and because

$$(S)_0 = (S) + (P) + (ES)$$

we have $(S)_0 \simeq (S)$, if $(ES) \ll (S)$, and $R_{A,P} = R_{A,P}(\text{initial})$. Substitution of these equations into the above rate law yields the *Michaelis-Menten equation*[*]

$$R_{A,P}(\text{initial}) = R_{D,S}(\text{initial}) \simeq \frac{k_2(S)_0(E)_{tot}}{\left(\frac{k_2 + k_{-1}}{k_1}\right) + (S)_0} = \frac{k_2(S)_0(E)_{tot}}{K_m + (S)_0} \qquad (15.3)$$

If we now plot the initial rate versus $(S)_0$ for a fixed value of $(E)_{tot}$, we obtain a curve like that in Figure 15.5. The rate is first order in substrate at low $(S)_0$, and zero order at high $(S)_0$.[†] The maximum initial rate for a given value of $(E)_{tot}$ is $k_2(E)_{tot}$. If $k_2 \ll k_{-1}$, and $k_{-1}/k_1 \gg (S)$, that is, if $K = \frac{k_1}{k_{-1}} = (ES)/(E)(S)$ is small, then

$$R_{D,S} = \frac{k_2 k_1}{k_{-1}} (E)_{tot}(S)$$

---

[*]The quantity $K_m$ is called the *Michaelis constant*, and it is equal to the value of (S) at which $Rate = \frac{1}{2} k_2(E)_{tot}$. The rate constant $k_2$ is often called the *turnover number*.

[†]Compare with Figure 14.11.

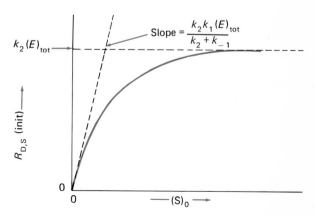

**FIGURE 15.5** Plot of initial rate versus initial substrate concentration for a typical enzyme-catalyzed reaction.

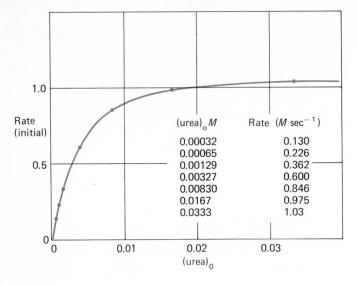

| (urea)$_0$ $M$ | Rate ($M$ sec$^{-1}$) |
|---|---|
| 0.00032 | 0.130 |
| 0.00065 | 0.226 |
| 0.00129 | 0.362 |
| 0.00327 | 0.600 |
| 0.00830 | 0.846 |
| 0.0167 | 0.975 |
| 0.0333 | 1.03 |

**FIGURE 15.6**  Urease-catalyzed hydrolysis of urea with $(E)_{tot} = $ constant. (Adapted from Reference 15.2.)

which corresponds to the special case of a rapid pre-equilibrium between enzyme and substrate.

In Figures 15.6 and 15.7 we have plotted the data[*] of G. B. Kistiakowsky and A. J. Rosenberg, JACS, **74**, 502 (1952), on the urease-catalyzed hydrolysis of urea at a fixed total enzyme concentration. The rates are initial rates at the given initial urea concentration. Figure 15.7 is a reciprocal plot of the data. By taking the reciprocal of both sides of Equation (15.3) we obtain

$$\frac{1}{R_{A,P}(\text{init})} = \frac{1}{k_2(E)_{tot}} + \frac{K_m}{k_2(E)_{tot}(S)_0} \tag{15.3a}$$

From this equation we see that a plot of $1/R(\text{init})$ versus $1/(S)_0$ should be linear with an intercept of $1/k_2(E)_{tot}$, and slope of $K_m/k_2(E)_{tot}$, if the Michaelis-Menten mechanism is consistent with the experimental facts.[†]

## 15.8  MECHANISM TO RATE LAW. A SUMMARY OF WORKING RULES.

Much of the material in the foregoing sections can be conveniently summarized in a set of "working rules" that can serve as aids in arriving at a plausible mechanism from the observed rate law and reaction stoichiometry. Chemists have used these working rules for quite some time, but they were only recently set down in one place.[‡] We have used all of the principles that underlie these rules in the preceding sections of this chapter. The working rules stated below constitute a slightly modified

---

[*]Quoted in Reference 15.2.
[†]Not all enzyme-catalyzed reactions follow Michaelis-Menten kinetics.
[‡]By J. O. Edwards, E. F. Greene, and John Ross. See Reference 15.5.

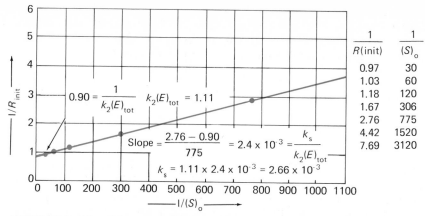

| | $\dfrac{1}{R(\text{init})}$ | $\dfrac{1}{(S)_o}$ |
|---|---|---|
| | 0.97 | 30 |
| | 1.03 | 60 |
| | 1.18 | 120 |
| | 1.67 | 306 |
| | 2.76 | 775 |
| | 4.42 | 1520 |
| | 7.69 | 3120 |

**FIGURE 15.7**   Plot of the inverse of the initial rate versus the inverse initial substrate concentration for the urease-catalyzed hydrolysis of urea. (Adapted from Reference 15.2.)

version (along with a few additions — Rules VII, VIII, and X) of those put forth by Edwards, Greene, and Ross:

I. The chemical composition* and electric charge of the activated complex for the rate-determining step can be obtained from the experimental rate law.

II. The chemical composition of the activated complex deduced from the empirical rate law is ambiguous in the sense that we do not know whether or not a species whose concentration cannot be (or was not) varied (e.g., the solvent) appears in the rate law.

III. When the overall order of the rate law is greater than three, the reaction mechanism probably involves one or more rapidly established equilibria (and, therefore, one or more intermediates) prior to the rate-determining step.

IV. Inverse orders in the rate law indicate rapidly established equilibria prior to the rate-determining step.

V. If the stoichiometric coefficient on the left-hand side (i.e., the reactant side) exceeds the species order in the rate law, there are one or more intermediates involved in the mechanism after the rate-determining step.

VI. Whenever a rate law contains species with non-integral orders, there are intermediates involved in the reaction mechanism. (Non-integral orders in homogeneous reactions indicate that the reaction mechanism involves fragmentation of the non-integral order species).

VII. Multiterm rate laws arise when there is more than one reaction mechanism (pathway) involved. The number of different reaction pathways is equal to the number of terms in the rate law.

VIII. A two-term denominator in the rate law arises when the mechanism involves more than one activated complex, and implies the existence of a steady-state intermediate.

IX. The first guess as to the composition and structure of postulated intermediates should be based on our knowledge of the composition and structure of known species.

---

*See, however, Rule II.

X. The rate law provides no information about the nature of the steps involved in the reaction mechanism that take place after the rate-determining step. In addition, there is no way to distinguish kinetically-equivalent forms of the rate law on the basis of kinetic data alone.

## 15.9    SPECIAL TOPICS IN REACTION KINETICS.

### (a)    Isotopic Labeling.

Useful clues to the nature of a reaction mechanism can often be obtained by means of isotopic-labeling techniques. If we label a particular atom in a species by means of an isotopic substitution, then we can frequently determine just which chemical bonds are broken (or formed) in the mechanism by determining the location of the label in the reaction products. By way of example, consider the following classic study of Polanyi and Szabo on the hydrolysis of the ester $n$-amyl acetate:

$$\underset{\substack{\text{\scriptsize $n$-amyl acetate}}}{\text{CH}_3\text{CH}_2\text{CH}_2\text{CH}_2\overset{\displaystyle \text{H}}{\underset{\displaystyle \text{H}}{\text{C}}}\text{—O—}\overset{\displaystyle \text{O}}{\text{C}}\text{—CH}_3} + {}^{18}\text{OH}^- = \text{CH}_3\text{CH}_2\text{CH}_2\text{CH}_2\text{CH}_2\text{OH} + {}^-\text{O}^{18}\text{—}\overset{\displaystyle \text{O}}{\text{C}}\text{—CH}_3$$

($n$-amyl alcohol)

By using oxygen-18 enriched water and determining that the isotopic enrichment appeared in the acetate ion but not in the $n$-amyl alcohol, it was established that carbon-oxygen bond 2, rather than 1, is cleaved in the reaction mechanism.

In addition to numerous studies of the above type using various isotopic substitutions (e.g., ${}^2\text{H}$ for ${}^1\text{H}$, ${}^{15}\text{N}$ for ${}^{14}\text{N}$, etc.), radioactive carbon-14 has been widely used as a tracer in both chemical and biochemical studies. In particular, ${}^{14}\text{C}$ labeled compounds have been widely used in tracing out reaction sequences in living matter.

### (b)    Nucleophilic Displacement Reactions. Stereochemical Considerations.

Reactions of the general type

$$X + R\text{—}Y \rightarrow X\text{—}R + Y,$$

where X and Y are electron-pair donors (Lewis bases), are often found to proceed via a simple, bimolecular, nucleophilic displacement mechanism. For example,

$$\text{OH}^-(\text{aq}) + \text{H}_2\text{N}\text{—Cl} \overset{k}{\rightarrow} \text{HO}\text{—NH}_2(\text{aq}) + \text{Cl}^-(\text{aq})$$

$$R_{D,\text{NH}_2\text{Cl}} = k(\text{OH}^-)(\text{NH}_2\text{Cl})$$

In reactions of this type X is called the *nucleophile*, Y the *leaving group*, and R—Y the *substrate*. As in the example given, X and Y are often negatively charged species.

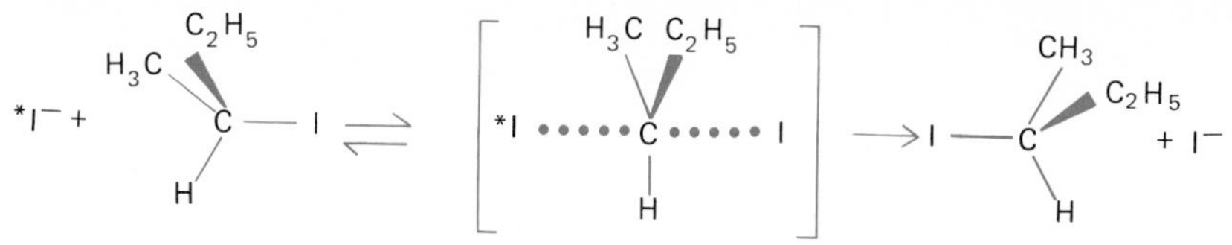

**FIGURE 15.8** Inverse of optical activity in a nucleophilic displacement at an asymmetric center (Walden Inversion). The asymmetric center in this case is the carbon atom with four *different* groups attached to it. All such carbon atoms are asymmetric.

An important characteristic of nucleophilic displacement reactions is the inversion of the optical activity[*] of the substrate when the reaction takes place at an asymmetric center in an optically active substrate. As an example, consider the following isotope-exchange reaction:

$$^*I^- + CH_3CH_2\overset{\overset{\displaystyle CH_3}{|}}{C}HI = CH_3CH_2\overset{\overset{\displaystyle CH_3}{|}}{C}HI^* + I^-$$

$$R_{D,RI} = k(^*I^-)(RI)$$

The observed reversal in the sign of the optical activity of 2-iodobutane is explained by invoking a bimolar, nucleophilic displacement mechanism with the activated complex having the structure indicated in Figure 15.8. The Walden-inversion (or "backside attack") mechanism is somewhat analogous to turning an umbrella inside out.

Inversion of configuration is sufficiently characteristic of bimolecular, nucleophilic displacement reactions that inversion of the geometrical configuration of the substrate, as determined by measurements on optically active compounds, is taken to be sufficient proof of the operation of the bimolecular, nucleophilic displacement mechanism. If racemization (i.e., formation of equal amounts of the two optical isomers) rather than inversion takes place, then this indicates that the leaving group is not displaced by the incoming group in a *concerted* process, but rather, the leaving group leaves before the incoming group attacks. Once the leaving group has departed we have a planar intermediate that can be attacked from either side; hence racemization occurs. Retention of configuration can occur when the nucleophile attacks the substrate on the side of the leaving group.

## (c) Diffusion-Controlled Reactions.

Collisions between neutral gaseous species usually involve only a very short encounter time ($\sim 10^{-13}$ sec), and the colliding particles either react within this time or they separate. In the liquid phase, however, the situation is significantly different. For example, two particles which diffuse together in water remain in the near vicinity of each other for

---

[*]Molecules lacking a plane or center of symmetry have the property of existing in two isomeric forms that are mirror images of one another. The two isomers rotate the plane of plane-polarized light by equal amounts in opposite directions.

**TABLE 15.2**

DATA† FOR SOME DIFFUSION-CONTROLLED REACTIONS IN WATER

| REACTION | $k_{obs}(M^{-1} \cdot sec^{-1}, 25°C)$ |
|---|---|
| $H^+(aq) + OH^-(aq) \rightarrow H_2O(\ell)$ | $1.4 \times 10^{11}$ |
| $H^+(aq) + {}^-OCOCH_3(aq) \rightarrow CH_3COOH(aq)$ | $4.5 \times 10^{10}$ |
| $H^+(aq) + NH_3(aq) \rightarrow NH_4^+(aq)$ | $4.3 \times 10^{10}$ |
| $Cu^{2+}(aq) + e^-(aq)^* \rightarrow Cu^+(aq)$ | $3.3 \times 10^{10}$ |
| $Fe(CN)_6^{3-}(aq) + e^-(aq)^* \rightarrow Fe(CN)_6^{4-}(aq)$ | $3.0 \times 10^9$ |

$^*e^-(aq)$ represents the hydrated electron.

†Data taken from Reference 15.5, J. Halpern, *Some Aspects of Chemical Dynamics in Solution.*

$10^{-11}$ to $10^{-12}$ sec, owing to the restraining effect (or *cage-effect*) exerted by the surrounding solvent. This is sufficient time for the species to undergo something on the order of 10 to 100 vibrations. This has important consequences for solution-phase reactions that have small activation energies. The rate constants for such reactions in solution approach the diffusion-controlled values; that is, essentially every diffusional encounter results in reaction, and the reaction rate constant is governed solely by how fast the reacting particles diffuse together.

Typical values of $k_{diff}$, for monovalent ions in water at 25°C, are of the order of $10^9$ to $10^{10}$ $M^{-1} \cdot sec^{-1}$. A few examples of diffusion-controlled reactions are presented in Table 15.2. The rate constant for the first reaction in the table has the largest value known for a bimolecular reaction in solution.

As a corollary to the foregoing, we note that if a postulated mechanism for a solution reaction requires a bimolecular rate constant significantly greater than the diffusion-controlled value, then it is time to start looking for a new mechanism.

## (d)  The Study of Fast Reactions: Relaxation Techniques.

If a reaction proceeds at a rate which is of such a magnitude that the reaction is essentially complete in a time comparable to the time of mixing of the reactants, then it is evident that a direct determination of the reaction rate law and rate constant cannot be made by conventional kinetic techniques.

In the investigation of a reaction by relaxation methods (M. Eigen), a reaction system *at equilibrium* is perturbed by a pulse of a very short duration ($10^{-4}$ sec or less), and the kinetics of the reaction are followed by monitoring the rate of approach of the reaction system to the new equilibrium state. The pulse may be of electrical current (Joule heating), which produces a temperature increase (*T*-jump) and, therefore, a measurable shift in the equilibrium distribution of species (provided $\Delta H° \neq 0$ for the reaction, and *K* is neither very small nor very large). Alternatively, one may use an electric field pulse (*E*-jump) which produces an increase in the extent of dissociation of a weak electrolyte (Wien effect), or a pressure pulse ($P$-jump), which also produces a shift in the equilibrium distribution of species (provided $\Delta V \neq 0$ for the reaction). The concentrations as a function of time can be followed spectrophotometrically, or conductometrically, and the results displayed on an oscilloscope screen from which a permanent record can be photographed.

Let the system of interest be a reaction of the type

$$A \underset{k_2}{\overset{k_1}{\rightleftharpoons}} B^+ \; + \; C^-$$

$$a-x \quad b+x \quad c+x$$

where $a$, $b$, and $c$ are the concentrations of A, $B^+$, and $C^-$, respectively, that were added in the first place, and $a-x$, $b+x$, and $c+x$ are the concentrations of these species at some time $t$. Assuming that the forward and back reactions are elementary processes, we have for the net reaction rate at any time

$$\text{Rate} = k_1(a-x) - k_2(b+x)(c+x)$$

and at equilibrium

$$0 = k_1(a-x_0) - k_2(b+x_0)(c+x_0)$$

We now define $\Delta x$, the perturbation from the equilibrium state, as

$$\Delta x = x - x_0$$

where $\Delta x \ll x$, $x_0$, $a$, $b$, $c$. Substitution of $x = \Delta x + x_0$ into the rate expression yields

$$\text{Rate} = k_1(a-x_0-\Delta x) - k_2(b+x_0+\Delta x)(c+x_0+\Delta x)$$

$$\text{Rate} = k_1(a - x_0) - k_1(\Delta x) - k_2(b + x_0)(c + x_0) - k_2(b + x_0)(\Delta x)$$
$$-k_2(c+x_0)(\Delta x) - k_2(\Delta x)^2$$

$$\text{Rate} = - \{k_1 + k_2(b+c+2x_0)\}\Delta x$$

In obtaining this result we have utilized the equilibrium condition, and, in addition, have discarded the term in $(\Delta x)^2$, because it is very small compared with all the other terms. We now define the *relaxation time*, $\tau$, as

$$\frac{1}{\tau} = k_1 + k_2(b+c+2x_0) = k_1 + k_2(b_{eq} + c_{eq})$$

where $\tau$ is the time required for $\Delta x$ to fall to $1/e$ of its initial value. Because the rate of approach to the equilibrium state is first order in the displacement, $\Delta x$, from equilibrium,[*] we have

$$\Delta x = \Delta x_0 e^{-t/\tau} \tag{15.4}$$

where $\Delta x_0$ is the value of $\Delta x$ at $t = 0$ (i.e., immediately after the perturbation). Note that when $t = \tau$, $\Delta x = \Delta x_0/e$.

---

[*]That is, $-d\Delta x/dt = \Delta x/\tau$, from which Equation (15.4) follows. Note the similarity in form between Equation (15.4) and the usual first-order-decay equation $C_t = C_0 e^{-kt}$.

The relationship between the relaxation time, $\tau$, and the rate constants and equilibrium concentrations is unique for a given reaction mechanism and reaction type. For example, in the above case, if the proposed mechanism is correct, then a plot of $1/\tau$ versus $b_{eq} + c_{eq}$ will be a *straight line* with intercept $k_1$ and slope $k_2$. Of course, we cannot distinguish by these methods between various mechanisms that yield the same form of the rate law.

Using the E-jump method and following the relaxation conductometrically, Eigen and DeMaeyer (1955) found for the reaction

$$H_2O(aq) \underset{k_2}{\overset{k_1}{\rightleftharpoons}} H^+(aq) + OH^-(aq)$$

that $k_2 = 1.4 \times 10^{11}$ $M^{-1}$ $sec^{-1}$ at 25°C. In neutral aqueous solution at 25°C, $(H^+) = (OH^-) = 1.0 \times 10^{-7}$ M, and therefore

$$t_{1/2} = \frac{1}{1.4 \times 10^{11} \times 1.0 \times 10^{-7}} = 70 \ \mu sec$$

Furthermore,

$$K = \frac{(H^+)(OH^-)}{(H_2O)} = \frac{k_1}{k_2} = 1.8 \times 10^{-16} \ \text{(M units at 25°C for } H^+, OH^-, \text{ and } H_2O)$$

Thus

$$k_1 = 1.8 \times 10^{-16} \text{ M} \times 1.4 \times 10^{11} \text{ M}^{-1} \cdot sec^{-1} = 2.5 \times 10^{-5} \ sec^{-1}$$

and

$$t_{1/2} = 0.693/2.5 \times 10^{-5} \ sec^{-1} = 2.8 \times 10^4 \ sec = 8.0 \ \text{hrs!}$$

In other words, the dissociation of $H_2O$ to yield $H^+$ and $OH^-$ is a slow reaction. For the reaction

$$CH_3COOH(aq) \underset{k_2}{\overset{k_1}{\rightleftharpoons}} H^+(aq) + CH_3COO^-(aq)$$

at 25°C Eigen and DeMaeyer found $k_2 = 4.5 \times 10^{10}$ $M^{-1} \cdot sec^{-1}$ and, therefore,

$$k_1 = 1.76 \times 10^{-5} \times 4.5 \times 10^{10} = 7.9 \times 10^5 \ sec^{-1}$$

Note that in contrast to $H_2O$, the dissociation of $CH_3COOH(aq)$ is very rapid.

## (e)  Reactions in Crossed Molecular Beams.

Molecular beams can be produced by allowing a gas to effuse through a small hole into an evacuated space. The beam can be collimated by means of a second small hole positioned in line with the first, which permits passage of only those gas molecules moving in a particular direction. This collimated molecular beam can then be run through a velocity selector (see Chapter 5) which selects only those molecules in the beam whose velocities lie in a narrow range.

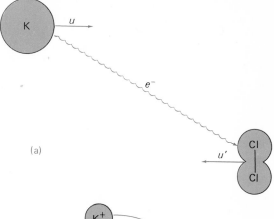

(a)

**FIGURE 15.9** The harpoon mechanism (Polanyi).

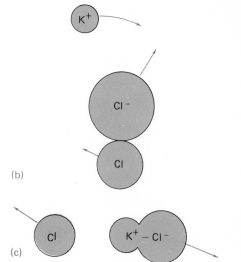

(b)

(c)

Chemical reactions can be carried out using crossed molecular beams. As an example, the dynamics of the reaction

$$K(g) + Cl_2(g) \rightarrow KCl(g) + Cl(g)$$

has been investigated by firing a $K(g)$ beam of known velocity and concentration at a beam of $Cl_2(g)$ of known velocity and concentration. Very detailed information about the nature of the dynamics of chemical reactions on the molecular scale can be obtained from such studies, and much active research is being carried out in this area at the present time.[*] Molecular beam results on the above reaction are quite interesting. The effective collision diameter is large ($\gtrsim 150 Å^2$), indicating that collision between the particles in the usual hard-sphere sense is not required for the reaction to take place. The results are consistent with the long-range transfer of an electron from K to $Cl_2$, followed by the stripping of a chloride ion from $Cl_2^-$ by $K^+$. This sequence of processes has been called the *harpoon* mechanism (see Figure 15.9).

[*] See Reference 15.5, E. F. Green and A. Kuppermann, *Chemical Reaction Cross Sections and Rate Constants.*

## PROBLEMS

1. Given that

$$A + B \underset{k_{-1}}{\overset{k_1}{\rightleftharpoons}} C + D$$

$$C + E \overset{k_2}{\rightarrow} F$$

what relationships among the magnitudes of the various rates will lead to the following rate laws?

$$R_{A,F} = k \frac{(A)(B)(E)}{(D)}$$

$$R_{A,F} = k'(A)(B)$$

$$R_{A,F} = \frac{k(A)(B)}{1 + k'(D)/(E)}$$

2. If the chemical reaction, $A + B = C$, has $\Delta E = 15$ kcal, what is the minimum value that the activation energy for the forward reaction can have?

3. Explain qualitatively why it is desirable to limit postulated mechanistic steps to elementary processes of molecularity three or less.

4. The forward rate law for the reaction

$$3HNO_2(aq) = H^+(aq) + NO_3^-(aq) + 2NO(g) + H_2O(\ell)$$

is

$$R_{A,NO_3^-} = k(HNO_2)^4/(NO)^2$$

Find the rate law for the reverse reaction.

5. The rate law for the reaction

$$Fe^{2+}(aq) + HNO_2(aq) + H^+(aq) = Fe^{3+}(aq) + NO(g) + H_2O(\ell)$$

is reported to be

$$R_{D,\ Fe^{2+}} = k(Fe^{2+})(HNO_2) + k'(Fe^{2+})(HNO_2)(H^+) + k'' \frac{(Fe^{2+})(HNO_2)^2}{(NO)}$$

(a) By how many independent pathways does the reaction proceed?
(b) Give the formula of the activated complex for each of the pathways.
(c) Select from the following set of mechanisms the one that corresponds to each pathway. In each case show how the observed rate constant is related to the rate constant(s) of the elementary processes in the given mechanism.

*Mechanism 1:*

$$HNO_2 + H^+ \underset{k_{-1}}{\overset{k_1}{\rightleftharpoons}} H_2O + NO^+ \qquad \text{(rapid equilibrium)}$$

$$Fe^{2+} + NO^+ \overset{k_2}{\rightarrow} Fe^{3+} + NO \qquad \text{(slow)}$$

*Mechanism 2:*

$$2HNO_2 \underset{k_{-3}}{\overset{k_3}{\rightleftharpoons}} NO + NO_2 + H_2O \qquad \text{(rapid equilibrium)}$$

$$Fe^{2+} + NO_2 \overset{k_4}{\rightarrow} Fe^{3+} + NO_2^- \qquad \text{(slow)}$$

$$NO_2^- + H^+ \overset{k_5}{\rightarrow} HNO_2 \qquad \text{(fast)}$$

*Mechanism 3:*

$$HNO_2 \overset{k_6}{\rightarrow} NO^+ + OH^- \qquad \text{(slow)}$$

$$Fe^{2+} + NO^+ \overset{k_7}{\rightarrow} Fe^{3+} + NO \qquad \text{(fast)}$$

$$OH^- + H^+ \overset{k_8}{\rightarrow} H_2O \qquad \text{(fast)}$$

*Mechanism 4:*

$$Fe^{2+} + HNO_2 \overset{k_9}{\rightarrow} Fe^{3+} + OH^- + NO \qquad \text{(slow)}$$

$$OH^- + H^+ \overset{k_{10}}{\rightarrow} H_2O \qquad \text{(fast)}$$

(d) Find the overall rate law for the reverse reaction.

6. The rate law for the reaction

$$H_3AsO_4(aq) + 3I^-(aq) + 2H^+(aq) = H_3AsO_3(aq) + H_2O(\ell) + I_3^-(aq)$$

is

$$R_{D,H_3AsO_4} = k_f(H_3AsO_4)(H^+)(I^-)$$

with $k_f = 6.3 \times 10^{-3} \ M^{-2} \cdot sec^{-1}$ and $K = 6.3$ at $0°C$. Find the rate law for the reverse reaction and evaluate the rate constant at $0°C$.

7. The reaction

$$2NO_2(g) = 2NO(g) + O_2(g)$$

has the forward rate law

$$R_{D,NO_2} = k_{obs}(NO_2)^2$$

Propose a mechanism consistent with this rate law.

8. Show that if a catalyst could be found that changed the value of the equilibrium constant for a given reaction, then the first law of thermodynamics would be invalidated.

9. Show that we must *subtract* the atoms in the denominator from those in the numerator of the observed rate law in order to obtain the composition of the activated complex.

10. Suppose that the proposed mechanism for a reaction yields the result that the observed rate constant $k_{obs}$ is equal to

$$k_{obs} = \frac{k_1^2 k_2}{k_{-1}^2}$$

where $k_1$, $k_2$, and $k_{-1}$ are rate constants for elementary processes in the mechanism. How is the observed activation energy related to the activation energies of these elementary processes?

11. Suppose that a substance can react in two different ways

$$M \overset{k_1}{\rightarrow} N$$

$$M \overset{k_2}{\rightarrow} P$$

and it was found by analysis that $(N)/(P) = 5.5$. If $k_1 = 0.015$ sec$^{-1}$, what is $k_2$?

12. Consider the following reaction scheme

$$M \rightleftharpoons N \qquad \text{rapid equilibrium } K_{eq}$$

(slow, 1st order) $k_1$ $\downarrow$ $\qquad$ $\downarrow$ $k_2$ $\qquad$ (slow, 1st order)

$$P_1 \qquad P_2$$

How is the product ratio $P_1/P_2$ related to $k_1$, $k_2$, and $K_{eq}$?

13. The dehydration of *t*-butanol in the gas phase

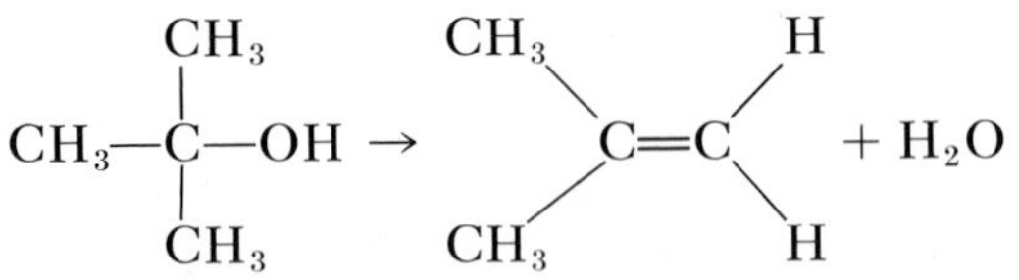

$$CH_3-\underset{\underset{CH_3}{|}}{\overset{\overset{CH_3}{|}}{C}}-OH \rightarrow \underset{CH_3}{\overset{CH_3}{C}}=\underset{H}{\overset{H}{C}} + H_2O$$

is catalyzed by hydrogen bromide. The rate law for the reaction is

$$R_{D,ROH} = k_1(ROH) + k_2(ROH)(HBr)$$

Given that $k_1 = 4.8 \times 10^{11} \cdot e^{-65,500/RT}$ sec$^{-1}$ and $k_2 = 9.2 \times 10^9 \cdot e^{-30,400/RT}$ M$^{-1}$ sec$^{-1}$, determine the concentration of (HBr) at 500°C for which the catalyzed and uncatalyzed pathways contribute equally to the total rate. Compute the half life for *t*-butanol at 500°C in the absence of HBr and in the presence of $1.0 \times 10^{-3}$ M HBr.

14. The rate law for the reaction

$$H_2(g) + I_2(g) \rightarrow 2HI(g)$$

is

$$R_{D,I_2} = k_{obs}(H_2)(I_2)$$

Write down two mechanisms in addition to the one given in the text that are consistent with this rate law. Try to devise experiments that would distinguish between these mechanisms.

15. The rate constants for many enzyme-catalyzed reactions show a parabolic dependence on the temperature; the rate constant increases with temperature in the low range, and then at higher temperatures it decreases with increasing temperature. Suggest an explanation for this behavior.

16. The reaction

$$2NO(g) + O_2(g) = 2NO_2(g)$$

is faster at lower temperatures than at higher temperatures (indicating that the mechanism probably involves at least two steps). The observed rate law is

$$R_{A,NO_2} = k(NO)^2(O_2)$$

Propose a mechanism for the forward reaction that involves only bimolecular steps. Explain what condition is imposed on the activation energies of the elementary processes you have proposed by the observed decrease in $k$ with increasing temperature.

17. Nitramide, $O_2NNH_2$, decomposes in aqueous solution to yield $N_2O$ and $H_2O$

$$O_2NNH_2(aq) = N_2O(g) + H_2O(\ell)$$

The observed rate law is

$$R_{A,N_2O} = k \frac{(O_2NNH_2)}{(H^+)}$$

It was also observed that $O_2NNH^-$ is an intermediate in the decomposition. Propose a reaction mechanism consistent with the observed rate law.

18. The denaturation of a certain virus is a first-order process with an activation energy of 140 kcal/mole. The half-life of the reaction at 29.6°C (98.6°F) is 4.5 hours. Compute the half-life at 32.5°C (105.0°F).

19. The following kinetic data were obtained by Bodenstein on the reaction

$$2HI(g) \underset{k_2}{\overset{k_1}{\rightleftharpoons}} H_2(g) + I_2(g)$$

| $T(K)$ | $k_1 \times 10^4 (M^{-1} \cdot sec^{-1})$ | $k_2 \times 10^2 (M^{-1} \cdot sec^{-1})$ |
|--------|-------------------------------------------|-------------------------------------------|
| 647    | 0.858                                     | 0.522                                     |
| 666    | 2.19                                      | 1.41                                      |
| 683    | 5.11                                      | 2.46                                      |
| 700    | 11.7                                      | 6.42                                      |
| 716    | 25.0                                      | 14.0                                      |
| 781    | 395.                                      | 133.5                                     |

Using these data, compute $E_{af}$, $E_{ar}$, $A_f$, $A_r$, $\Delta H°$, and $\Delta S°$ for this reaction. Also evaluate $K$ for the reaction at 650 and 750 K.

20. Can the presence of a catalyst affect the equilibrium concentrations of the reactants and products? In your answer consider two cases: (a) $a_i = c_i$; and (b) $a_i = \gamma_i c_i$.

21. Suppose the gas-phase reaction

$$A(g) \rightarrow products$$

has the following mechanism:

$$A + A \underset{k_{-1}}{\overset{k_1}{\rightleftarrows}} A^* + A$$

$$A^* \overset{k_2}{\rightarrow} products$$

where $A^*$ represents a vibrationally excited A molecule. Apply a steady-state condition to $A^*$ and find the rate law for the reaction. What does this mechanism predict for the rate law: (a) when $k_{-1}(A)/k_2 \ll 1$; and (b) when $k_{-1}(A)/k_2 \gg 1$?

22. Suggest an explanation for the fact that the rate law for the gas-phase decomposition of a large molecule (e.g., $N_2O_5$, $C_6H_5CH_3$, $C_2H_5OOC_2H_5$) is often first-order, whereas with small molecules (e.g., $NO_2$, HI) the rate law is second-order.

23. An electronically excited mercury atom can either fluoresce (emit a photon of energy $h\nu$) or lose its excitation energy by collision with another species, e.g.,

$$Hg^* \overset{k_1}{\rightarrow} Hg + h\nu$$

$$Hg^* + Ar \overset{k_2}{\rightarrow} Hg + Ar^*$$

These reactions are elementary processes. Find an expression for the fraction of atoms lost by fluorescence at a given pressure of argon.

24. The rate law for the hydrolysis of diphenylchloromethane

$$RCl + H_2O = ROH + H^+ + Cl^-$$

$\{R \equiv (C_6H_5)_2CH\}$ is

$$R_{A,ROH} = \frac{k(RCl)}{1 + k'(Cl^-)}$$

Propose a mechanism consistent with this rate law.

25. The rate law for the reaction between mercurous and thallic ions in acidic aqueous solutions

$$Hg_2^{2+}(aq) + Tl^{3+}(aq) = 2Hg^{2+}(aq) + Tl^+(aq)$$

is

$$R_{D,Hg_2^{2+}} = k\frac{(Hg_2^{2+})(Tl^{3+})}{(Hg^{2+})}$$

Propose a mechanism consistent with this rate law.

26. The rate law for the reaction between ferrous and thallic ions in acidic aqueous solution

$$2Fe^{2+}(aq) + Tl^{3+}(aq) = Tl^+(aq) + 2Fe^{3+}(aq)$$

is

$$R_{D,Fe^{2+}} = \frac{k(Fe^{2+})(Tl^{3+})}{1 + k'(Fe^{3+})/(Fe^{2+})}$$

Propose a mechanism consistent with this rate law. What would the observed rate law be in the presence of an added excess of $Fe^{3+}$?

27. Consider the systems

(1)
$$A + B \underset{k_2}{\overset{k_1}{\rightleftarrows}} C + D$$

(2)
$$2A \underset{k_2}{\overset{k_1}{\rightleftarrows}} C + D$$

and find the relationship between the relaxation time, $\tau$, and the rate constants and equilibrium concentrations.

28. The rate law for the reaction

$$HCOOH(aq) + Tl^{3+}(aq) = 2H^+(aq) + CO_2(aq) + Tl^+(aq)$$

is

$$R_{D,Tl^{3+}} = k_{obs}(HCOO^-)(Tl^{3+})$$

Given also that the equilibrium

$$Tl^{3+} + HCOO^- \rightleftharpoons TlOOCH^{2+} \qquad K$$

is rapidly established compared with the rate of the above reaction, show that the rate law can also be written in the forms

$$R_{D,Tl^{3+}} = k'_{obs}(TlOOCH^{2+}) = \frac{k''_{obs}(HCOO^-)(Tl^{3+})_0}{1 + K(HCOO^-)}$$

where $(Tl^{3+})_0 = (Tl^{3+}) + (TlOOCH^{2+})$. Suppose that the experimenter were unaware of the above equilibrium between $Tl^{3+}$ and $HCOO^-$; what erroneous conclusion might he draw from the rate law expressed in terms of $(HCOO^-)$ and $(Tl^{3+})$?

29. The rate law for the reaction

$$(NC)_5CoOH_2^{2-}(aq) + I^-(aq) = (NC)_5CoI^{3-} + H_2O(\ell)$$

is

$$R_{D,I^-} = \frac{k\{(NC)_5CoOH_2^{2-}\}(I^-)}{1 + k'(I^-)}$$

Propose a mechanism consistent with this rate law. (Hint: postulate $Co(CN)_5^{2-}(aq)$ as an intermediate.)

30. The oxidation of $H_2(aq)$ by $Fe^{3+}(aq)$

$$H_2(aq) + 2Fe^{3+}(aq) = 2Fe^{2+}(aq) + 2H^+(aq)$$

is catalyzed by $Cu^{2+}(aq)$

$$R_{D,H_2} = \frac{k(H_2)(Cu^{2+})^2}{k'(H^+) + k''(Cu^{2+})}$$

Propose a mechanism consistent with this rate law.

31. When aqueous solutions of sodium pentacyanonitrosoferrate (II) are made alkaline with $NaOH(aq)$, the following reactions take place:

(1)   $(NC)_5FeNO^{2-}(aq) + 2OH^-(aq) = (NC)_5FeNO_2^{4-}(aq) + H_2O(\ell)$

(2)   $(NC)_5FeNO_2^{4-}(aq) + H_2O(\ell) = (NC)_5FeOH_2^{3-}(aq) + NO_2^-(aq)$

All three iron-containing species are present at equilibrium. The equilibrium constants for these reactions at 25°C are $K_1 = 1.5 \times 10^6$ and $K_2 = 3.0 \times 10^{-4}$. The rate laws for the forward reactions in the two cases (each obtained under conditions where the other reaction and the reverse rate of the reaction in question could be ignored) are, respectively,

$$R_{D,(NC)_5FeNO^{2-}} = k_f\{(NC)_5FeNO^{2-}\}(OH^-) \qquad k_f(25°C) = 0.55 \ M^{-1} \cdot sec^{-1}$$

$$R_{D,(NC)_5FeNO_2^{4-}} = k'_f\{(NC)_5FeNO_2^{4-}\} \qquad k'_f(25°C) = 7.8 \times 10^{-3} \ sec^{-1}$$

Find the reverse rate laws and rate constants for reactions (1) and (2). Propose mechanisms consistent with the observed rate laws. Explain how you would adjust the experimental conditions such that the kinetics of the two reactions given in this problem could be studied independently.

32. Can Equation (15.1) be used to find the rate law for the reverse reaction if the forward reaction involves a steady-state intermediate?

33. The transition-state theory expression for the reaction rate constant is (Eyring)

$$k = \frac{k_B T}{h} K^{\ddagger} = \frac{k_B T}{h} e^{\Delta S^{\ddagger}/R} e^{-\Delta H^{\ddagger}/RT}$$

where $k_B = 1.38 \times 10^{-16}$ erg·K$^{-1}$, $h = 6.62 \times 10^{-27}$ erg·sec, and $K^{\ddagger}$ s the quasi-thermodynamic equilibrium constant for the reaction in which one mole of the activated complex is formed as sole product. Use the data given in Figure 14.12 to construct a plot of $\log(k/T)$ versus $1/T$. Obtain $\Delta H^{\ddagger}$ from the plot and use it to compute $\Delta S^{\ddagger}$.

34. Compute the rate constant at 300 K for each of the reactions given in Table 15.1. Also compute $\Delta S^{\ddagger}$ for each reaction and compare $p$ with $e^{\Delta S^{\ddagger}/R}$ (see problem 33).

──────────────────────────────────────────────── **References**

15.1. H. Taube, *The Role of Kinetics in Teaching Inorganic Chemistry*, J. Chem. Educ., **36**, 451 (1959).
15.2. E. L. King, *How Chemical Reactions Occur* (W. A. Benjamin, Inc., New York, 1964). This book is an introduction to chemical kinetics and reaction mechanisms written for freshmen chemistry students.
15.3. H. Eyring and E. M. Eyring, *Modern Chemical Kinetics* (Reinhold Publishing Corp., New York, 1965). Some special topics in the field of reaction-rate chemistry are sketched out for undergraduates with an interest in chemistry. (More theoretically oriented than Reference 15.2.)
15.4. G. M. Harris, *Chemical Kinetics* (D. C. Heath and Co., Boston, 1966). This book is useful as a bridge between the material presented in the text in hand and the more advanced treatises on the subject.
15.5. Advisory Council on College Chemistry, Serial Publication Number 37. Reprinted from June, 1968, edition of the Journal of Chemical Education. This reprint includes the following articles:
　　H. Taube, *Mechanisms of Oxidation Reduction Reactions.*
　　J. E. Finholt, *The Temperature-Jump Method for the Study of Fast Reactions.*
　　R. Parsons, *Electrochemical Dynamics.*
　　J. O. Edwards, *Bimolecular Nucleophilic Displacement Reactions.*
　　J. O. Edwards, E. F. Greene, and John Ross, *From Stoichiometry and Rate Law to Mechanism.*
　　J. Halpern, *Some Aspects of Chemical Dynamics in Solution.*
　　E. F. Greene and A. Kuppermann, *Chemical Reaction Cross Sections and Rate Constants* (a fairly advanced treatment).
15.6. W. C. Gardiner, Jr., *Rates and Mechanisms of Chemical Reactions* (W. A. Benjamin, Inc., New York, 1969). This is the most advanced of the general references listed here.
15.7. A. A. Frost and R. G. Pearson, *Kinetics and Mechanism*, 2nd Ed. (John Wiley and Sons, Inc., New York, 1961).

## CHAPTER 16

"The first step in applying the scientific method consists in being curious about the world, about the facts that have been found by observation and experiment. In our science these are the facts of descriptive chemistry."

*Linus Pauling\**

# THE A-GROUP METALS

## 16.1 INTRODUCTION.

Most of the elements are metals (see Figure 17.1). Metals are solid elements in which the number of mobile (conduction) electrons is of the same order of magnitude as the number of atoms. Metals are characterized by high thermal and electrical conductivity and reflectance. Metals are malleable, ductile, strong, and tough. The oxides of metals are basic, and the halides and hydrides are ionic and high-melting.

In this chapter we shall confine our discussion to the chemistry of the Group IA, IIA, IIIA, IVA, and VA metals. The chemistry of the transition metals is discussed in Chapter 19.

## 16.2 THE ALKALI (GROUP IA) METALS.

The Group IA metals (Table 16.1) are called the alkali metals because their hydroxides are all soluble strong bases or *alkalies*. These metals (see Table 16.1) all have a single valence electron in an *s*-orbital outside a noble-gas core. As a consequence, the only oxidation states of any chemical importance are 0 and +1, and the chemistry of these elements is the simplest of any group of elements in the periodic table.

The effect of increasing mass and size on the chemical and physical properties is particularly clear for this group of elements (Table 16.1). The ionization potentials, photoelectric work functions, melting points, boiling points, heat capacities, and the enthalpies of fusion, vaporization, and dehydration all decrease with increasing atomic number for the group. On the other hand, the densities and ionic mobilities increase as we go down the group.

---

\*Linus Pauling, *General Chemistry*, 3rd Ed. (W. H. Freeman and Co., San Francisco, 1970) p. 13.

**TABLE 16.1**

| element | atomic number | atomic weight ($^{12}C = 12$) | principal isotopes (% natural abundance) [half-life] | ionization potential of M(g) in eV | photoelectric work function of M(s) in eV | density (g·cm⁻³ at 20°C) | melting point (°C) | boiling point (°C) | hardness (Moh's scale) | $\Delta H_{fus}$(cal·g⁻¹) | $\Delta H_{eup}$(cal·g⁻¹) | $C_p$(cal·g⁻¹·deg⁻¹ at 0°C) | $-\mathscr{E}^{\circ}_{red}$ {M⁺(aq) + e⁻ = M(s)} at 25°C | 6-coordinate crystal radius of M⁺ in Å | hydrated radius of M⁺(aq)(est) in Å | $-\Delta H_{hyd}$ of M⁺(g), kcal·mole⁻¹ | ionic mobilities of M⁺(aq) at 18°C |
|---|---|---|---|---|---|---|---|---|---|---|---|---|---|---|---|---|---|
| Li lithium | 3 | 6.94 | 6(7.42) 7(92.58) | 5.39 | 2.35 | 0.534 | 181 | 1347 | 0.6 | 103 | 4680 | 0.784 | 3.045 | 0.60 | 3.40 | 124 | 33.5 |
| Na sodium | 11 | 22.9898 | 23(100) 22[2.58y] 24[15.0h] | 5.14 | 2.28 | 0.97 | 97.7 | 892 | 0.4 | 27.5 | 1005 | 0.29 | 2.714 | 0.95 | 2.76 | 97 | 43.5 |
| K potassium | 19 | 39.10 | 39(93.10) 40(0.0118) [1.3×10⁹y] 41(6.88) | 4.34 | 2.24 | 0.86 | 63.6 | 774 | 0.5 | 14.6 | 496 | 0.18 | 2.925 | 1.33 | 2.32 | 77 | 64.6 |
| Rb rubidium | 37 | 85.468 | 85(72.15) 87(27.85) [4.7×10¹⁰y] | 4.18 | 2.12 | 1.53 | 39.0 | 696 | | 6.1 | 212 | 0.080 | 2.925 | 1.48 | 2.28 | 72 | 67.5 |
| Cs cesium | 55 | 132.9055 | 133(100) 135[2.0×10⁶y] 137[30y] | 3.89 | 1.96 | 1.88 | 28.5 | 670 | | 3.8 | 146 | 0.048 | 2.923 | 1.69 | 2.28 | 62 | 68 |
| Fr francium | 87 | | 223[22m] 212[19m] | | | | | | | | | | | | | | |

The ionization potentials show that the electropositive character of the alkali metals increases with increasing atomic number. The metals are extremely reactive (powerful reducing agents), and the reactivity of the metals toward essentially all chemical reagents increases with increasing electropositive character. For example, when treated with water at 25°C lithium reacts slowly, sodium reacts vigorously, potassium inflames, and rubidium and cesium react explosively. Because of their high reactivity, the Group IA elements are never found free in nature. The most common preparation of the alkali metals is by the high-temperature electrolysis of the appropriate molten chloride, for example

$$NaCl(\ell) \xrightarrow{\text{electrolysis}} Na(\ell) + \frac{1}{2} Cl_2(g)$$

Most compounds of the Group IA elements are ionic; however, numerous covalent compounds (volatile and soluble in non-polar solvents) of lithium are known, and even lithium chloride is soluble in alcohol. Furthermore, numerous alkyllithium organometallic compounds can be prepared via reactions of the type

$$C_2H_5Cl \left(\frac{\text{benzene}}{\text{solvent}}\right) + 2Li(s) \rightarrow C_2H_5Li \left(\frac{\text{benzene}}{\text{solvent}}\right) + LiCl(s)$$

At room temperature the compounds $C_2H_5Li$ (ethyl lithium) and $C_6H_5Li$ (phenyl lithium) are colorless solids that exhibit a high volatility and are readily soluble in nonpolar solvents. The anomalous behavior of lithium relative to the other Group IA elements (and its similarity in certain respects to magnesium) is a direct consequence of the small size and high polarizing power of $Li^+$. The high polarizing power results in extensive solvation of $Li^+$ in solution (see $M^+(aq)$ radii in Table 16.1), and an enhanced tendency towards formation of covalent bonds. The organometallic compounds of the other alkali metals are mainly ionic, and, because of their lower solubility, are not used much in chemical synthesis.

The alkali metals themselves have the interesting property of being readily soluble in liquid ammonia to yield solutions that are deep blue when dilute. These solutions conduct current *electrolytically*,[*] and the main current carrier in the solution is thought to be the *solvated electron*. Reversible equilibria of the type

$$Na(soln) = Na^+(soln) + e^-(soln)$$

$$2e^-(soln) = e_2^{2-}(soln)$$

have been postulated to explain the available data on such metal-ammonia solutions. When concentrated by evaporation, the solutions have a bronze color and behave like a liquid metal. Alkali metal-ammonia solutions are used in organic syntheses as strong reducing agents. On long standing, or more rapidly in the presence of a transition-metal catalyst (for example, $Fe_2O_3$), the solutions decompose to yield the *amide* salt, $MNH_2$, and hydrogen gas.

Lithium is the least dense of all the elements that are solids at 25°C and 1 atm. Its density (0.53 g·cm$^{-3}$) is about half that of water at 25°C, and

---

[*] That is, like a salt solution as opposed to a metallic (electronic) conductor.

it floats in oils and ethers. Lithium can be cut with a sharp knife. When freshly cut it has a bright silvery appearance, but it rapidly becomes black in moist air. Lithium has a very wide liquid range (1138°C) and this fact, coupled with its very high heat capacity as a liquid (1.0 cal·g$^{-1}$·deg$^{-1}$), its low viscosity, and its low density, gives liquid lithium considerable promise as a high-temperature heat-transfer medium in nuclear power plants. Liquid lithium is the most corrosive material known. It reacts readily, and often violently, with all but a few materials, namely, tungsten, molybdenum, and low-carbon stainless steel. For example, if a sample of lithium is melted in a glass container, it reacts spontaneously with the glass to produce a hole in the container; the reaction is accompanied by the emission of an intense, greenish-white light.

Solid lithium does not react with thoroughly dried oxygen gas below 100°C, but at higher temperatures $Li_2O$ is readily formed. Near the melting point the metal ignites in oxygen, and the formation of $Li_2O$ is accompanied by a bright red flame. This is in contrast to the other alkali metals, which on reaction with $O_2(g)$ at atmospheric pressure yield $Na_2O_2$, $KO_2$, $RbO_2$, and $CsO_2$, respectively.[*] Lithium also reacts readily with nitrogen at room temperature in the presence of a trace of water to form the black-red, hygroscopic nitride, $Li_3N$. Lithium nitride reacts with water to yield $LiOH(aq)$ and $NH_3(aq)$. The other alkali metals are inert to nitrogen under these conditions, but do react at higher temperatures to yield the nitrides. Because of its high reactivity, lithium is best handled in an atmosphere of argon or helium.

The alkali metals all react readily with the halogens to form water-soluble (excepting LiF) metal-halide salts of the type MX, for example

$$2Na(s) + Cl_2(g) \rightarrow 2NaCl(s)$$

The reactions are both exothermic and chemiluminescent. The oceans contain huge amounts of sodium and potassium chlorides; sea water from the Atlantic or Pacific Ocean is about 2.8% NaCl and 0.08% KCl. The alkali-metal hydroxides are prepared either by electrolysis of the aqueous solution of the chloride salt or by reaction of the metal amalgam with water.

At elevated temperatures the alkali metals react directly with hydrogen to form *hydrides*, for example

$$2Li(s) + H_2(g) \xrightarrow{500-800°C} 2LiH(s)$$

The alkali-metal hydrides are ionic compounds which contain H$^-$ ions, as shown by the fact that electrolysis of the molten salt yields hydrogen at the anode. The hydrides are strong reducing agents and are used in synthetic work. They react readily with water to yield hydrogen gas

$$LiH(s) + H_2O(\ell) \rightarrow LiOH(aq) + H_2(g)$$

In the presence of a trace of water, lithium reacts directly with $CO_2(g)$, both combining with and reducing the gas:

$$4Li(s) + 2CO_2(g) \xrightarrow{\text{trace}\atop H_2O} Li_2O(s) + Li_2CO_3(s) + C(s)$$

---

[*]The ions $O^{2-}$, $O_2^{2-}$, and $O_2^-$ are called *oxide*, *peroxide*, and *superoxide*, respectively (see Table 18.8).

In contrast to the other alkali-metal carbonates, $Li_2CO_3(s)$ is only slightly soluble in water. Lithium carbonate has the interesting property of being an effective drug for the treatment of the manic phase of manic-depressive psychosis.*

Commercially available sodium bicarbonate ($NaHCO_3$, "baking soda") and sodium carbonate are prepared by the *Solvay process* in which $CO_2$ gas is bubbled through an ammoniacal sodium chloride solution

$$NaCl(aq) + NH_3(aq) + CO_2(g) + H_2O(\ell) \xrightarrow{15°C} NaHCO_3(s) + NH_4Cl(aq)$$

$$2NaHCO_3(s) \xrightarrow{80°C} Na_2CO_3(s) + H_2O(g) + CO_2(g)$$

The chemical success of the process depends upon the relatively low solubility of $NaHCO_3$ in water at temperatures of 15°C and below. The commercial success of the process depends upon the recovery of ammonia by the reaction

$$2NH_4Cl(aq) + Ca(OH)_2(s) \rightarrow 2NH_3(g) + CaCl_2(aq) + H_2O(\ell)$$

The calcium hydroxide and carbon dioxide used in the process are both obtained from limestone:

$$CaCO_3(s) \xrightarrow{80°C} CaO(s) + CO_2(g)$$
$$\Big\downarrow H_2O(\ell)$$
$$Ca(OH)_2(s)$$

The alkali metals react directly with all the non-metals, excepting the noble gases. The reaction with carbon yields carbides, such as $Na_2C_2(s)$, which upon treatment with water evolve acetylene gas, $C_2H_2$.

The high content of argon in the earth's atmosphere ($\sim 1\%$) arises from the decay of radioactive $^{40}K$ ($t_{1/2} = 16.1 \times 10^8$ yr) in potassium-containing rocks. Both rubidium and cesium are naturally radioactive elements, and all of the known isotopes of francium are radioactive and short-lived.

The alkali metals have characteristic and well-known flame spectra. The emission lines responsible for the observed spectra are:

---

*See Am. Jour. Psychiatry, *125* [4], Oct. 1968.

| ELEMENT | PRINCIPAL EMISSION LINES (nm) IN THE VISIBLE REGION | | | FLAME COLOR |
|---------|--------------|-------------|-------------|-------------|
| Li | 670 red | 620 orange | | carmine |
| Na | 590 yellow | | | yellow |
| K | 768 red | 404 violet | | violet |
| Rb | 780 red | 420 blue | 358 violet | blue-red |
| Cs | 457 blue | 388 violet | | blue |

The emission lines of lithium and sodium are prominent in sunlight. The lines for K, Rb, and Cs do not appear to an appreciable extent in sunlight, because the surface temperature of the sun ($\sim 6000°C$) is high enough to ionize these atoms.

## 16.3 THE ALKALINE EARTH (GROUP IIA) METALS.

The alkaline earth or Group IIA metals (Table 16.2) all have two valence electrons in an $s$-orbital outside a noble-gas core. As a consequence, the only oxidation states of any chemical importance are 0 and +2, and the chemistry of these elements is relatively simple.

The chemistry of beryllium is significantly different from that of the other Group IIA elements, because of the very strong polarizing power of the small $Be^{2+}$ ion. The chemistry of beryllium is in some respects similar to that of aluminum. There are no crystalline compounds or solutions involving $Be^{2+}$ as such. All Be(II) compounds involve appreciable covalent bonding. The packing in crystals is such that Be(II) is almost always four-coordinate.

The other Group IIA metals have larger sizes and lower ionization potentials, making them more electropositive. As a consequence, the ionic nature of the compounds of the alkaline earths increases down through the group.

Beryllium metal is steel grey in color, light, hard and high-melting. The principal commercial preparations are by electrolysis of the halide and by reduction of $BeF_2$ with Mg. Beryllium and its compounds have a sweet taste and are *extremely poisonous*. The metal does not react with either air (other than to form a thin protective layer of BeO) or concentrated $HNO_3$(aq). Because of its resistance to corrosion and its light weight and high strength, beryllium is often used in instruments (for example, gyroscopes). The combination of physical strength, corrosion resistance, and a very low cross-section for the capture of neutrons makes beryllium an excellent material for the construction of nuclear reactor cores (see Chapter 22). Copper-beryllium alloys are used to make high-quality, non-corroding springs and durable electrical contacts.

Some of the more common reactions of Be are diagrammed in Figure 16.1. Aqueous solutions of Be(II) salts are acidic, owing to the hydrolysis of $Be(OH_2)_4^{2+}$(aq). The strong tendency of Be(II) toward covalent bond formation is evidenced by the formation of covalent organometallics such as $(CH_3)_2Be$ (dimethyl beryllium), which is dimeric in the gas phase (see Figure 16.2).

Magnesium also exhibits a considerable tendency toward covalent bond formation, and its chemistry also differs significantly from the remaining Group IIA metals. However, the differences are not as great as those exhibited by beryllium. Covalently bonded organomagnesium halide compounds of the type RMgX (for example, $C_6H_5MgBr$ and $C_2H_5MgBr$), which are called *Grignard reagents*, are prepared by the direct reaction of magnesium with an alkyl halide under anhydrous conditions in a donor solvent such as ether; for example,

$$C_2H_5Br(ether) + Mg(s) \rightarrow C_2H_5MgBr(ether)$$

Grignard compounds are extensively used in synthetic organic chemistry.[*] Crystallographic studies of the compound $C_6H_5MgBr \cdot 2(C_2H_5)_2O$ show

---

[*]See Chapter 21.

TABLE 16.2

| element | atomic number | atomic weight ($^{12}C = 12$) | principal isotopes (% natural abundance) [half-life] | first and second ionization potentials of M(g) in eV | photoelectric work function of M(s) in eV | density (g·cm⁻³ at 20°C) | melting point (°C) | boiling point (°C) | $\Delta H_{fus}$ (kcal·mole⁻¹) | $\Delta H_{evap}$ (kcal·mole⁻¹) | $-\mathscr{E}°_{red}$ {$M^{2+}$(aq)+2e⁻=M(s)} at 25°C, (volt) | $-\Delta H_{hyd}$ of $M^{2+}$(g)(kcal·mole⁻¹) | charge-to-radius ratio of $M^{2+}$(g) | 6-coordinate crystal radius of $M^{2+}$(g) in Å |
|---|---|---|---|---|---|---|---|---|---|---|---|---|---|---|
| Be beryllium | 4 | 9.01218 | 9(100) 10[2.7 × 10⁶ y] 7[53.6d] | 9.32 18.21 | 3.92 | 1.848 | 1278 | 2970 | 2.34 | — | 1.847 | — | 6.5 | (0.31) |
| Mg magnesium | 12 | 24.305 | 24(78.70) 25(10.13) 26(11.17) 28[21.3h] | 7.64 15.03 | 3.68 | 1.738 | 651 | 1107 | 2.2 | 31.5 | 2.363 | 460 | 3.3 | 0.65 |

| | | | | | | | | | | | | | |
|---|---|---|---|---|---|---|---|---|---|---|---|---|---|
| Ca calcium | 20 | 40.08 | 40(96.97) 42(0.64) 43(0.145) 44(2.06) 46(0.0033) 48(0.18)[$>2 \times 10^{16}$ y] | 6.11 11.87 | 2.71 | 1.55 | 845 | 1487 | 2.2 | 38.6 | 2.866 | 395 | 1.8 | 0.94 |
| Sr strontium | 38 | 87.62 | 84(0.56) 86(9.86) 87(7.02) 88(82.56) 90[28y] | 5.69 10.98 | 2.74 | 2.54 | 769 | 1384 | 2.2 | 33.6 | 2.888 | 355 | 1.2 | 1.10 |
| Ba barium | 56 | 137.34 | 130(0.101) 132(0.097) 134(2.42) 135(6.59) 136(7.81) 137(11.32) 138(71.66) 140[12.8d] 137[7.5y] | 5.21 9.95 | 2.48 | 3.51 | 725 | 1140 | 1.8 | 35.7 | 2.906 | 305 | 1.0 | 1.29 |
| Ra radium | 88 | 226 | 226[1622y] 228[6.7y] 225[14.8d] | 5.28 10.10 | – | 6.0 | 700 | 1740 | – | – | 2.916 | – | 0.7 | 1.50 |

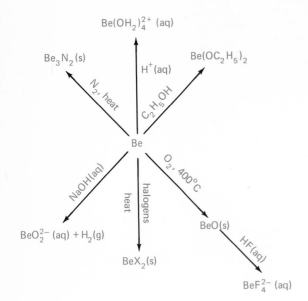

**FIGURE 16.1**   Some reactions of beryllium.

that the magnesium is approximately tetrahedrally coordinated (Figure 16.3). Dialkyl magnesium organometallics of type $(C_2H_5)_2Mg$ are also known. The organometallics of the other Group IIA metals (Ca, Sr, Ba, and Ra) are mainly ionic and are not of much synthetic or commercial importance.

Magnesium metal is prepared by electrolysis of the fused chloride. The metal is light-weight, silvery-white, and fairly tough. Light-weight magnesium alloys (especially Mg/Al) are widely used in airplane and missile construction. The bulk metal is unreactive toward oxygen, but when finely divided it burns in oxygen with a dazzling white flame. Some of the more common reactions of magnesium are diagrammed in Figure 16.4.

The $M^{2+}$ ionic radii of the Group IIA elements are smaller than the $M^{1+}$ ionic radii of the alkali metals of the same period. The metals are highly electropositive, and the heavier members (Ca, Sr, Ba, Ra) exhibit ionic bonding in their compounds. The radii of the $M^{2+}$(aq) ions in Group IIA are greatest for those with the smallest crystallographic radii, as is found for $M^+$(aq) ions in Group IA. The $M^{2+}$(aq) ions of Ca, Sr, Ba, and Ra are not appreciably dissociated into $MOH^+$(aq) and $H^+$(aq).

The Group IIA metals are prepared by electrolysis of the appropriate molten chloride. The heavier Group IIA metals are slightly soluble in liquid ammonia. The resulting solutions are blue in color, and definite ammoniates of the type $Ca(NH_3)_6(s)$ can be obtained. The metals Ca, Sr, Ba, and Ra are silvery-white in appearance when freshly cut, but tarnish readily in air.

Calcium is an essential constituent of plants, bones, teeth, and shells,

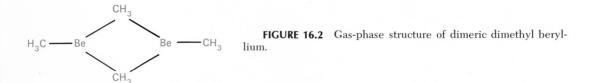

**FIGURE 16.2**   Gas-phase structure of dimeric dimethyl beryllium.

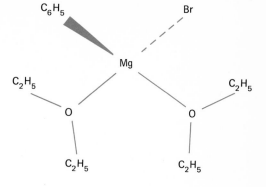

**FIGURE 16.3**  Coordination of magnesium in the Grignard reagent $C_6H_5MgBr \cdot 2(C_2H_5)_2O$.

and is involved as $Ca^{2+}$(aq), and complexes thereof, in numerous biochemical phenomena such as muscle contraction and nerve excitation. Plaster of Paris is $CaSO_4 \cdot \frac{1}{2}H_2O$, which takes up water to form gypsum, $CaSO_4 \cdot 2H_2O$. Asbestos is primarily $CaMg_3(SiO_3)_4$.

Strontium salts are much used in signal flames and pyrotechnics (they produce a brilliant red flame). Strontium-90 salts constitute a major health hazard.[*] The chemistry of barium is very similar to that of calcium, except that barium salts are all poisonous. All of the isotopes of radium are radioactive. The longest-lived isotope of radium is $^{226}Ra$, which is formed from $^{238}U$. Some of the more common reactions of the elements Ca, Sr, Ba, and Ra are diagrammed in Figure 16.5.

## 16.4  THE GROUP IIIA METALS.

The Group IIIA metals, aluminum (Al), gallium (Ga), indium (In), and thallium (Tl) (see Table 16.3), exhibit two important trends that are also

[*] See Chapter 22.

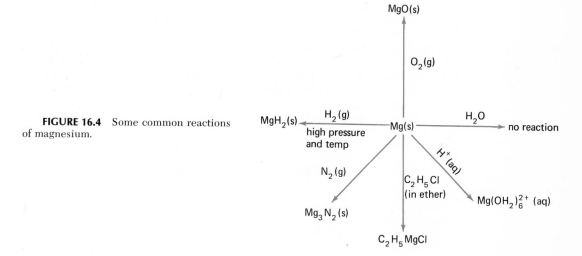

**FIGURE 16.4**  Some common reactions of magnesium.

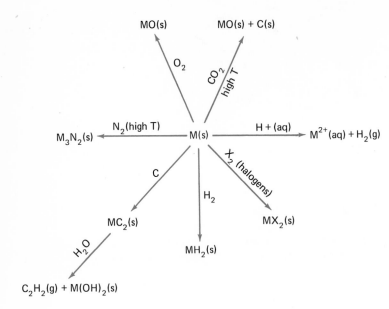

**FIGURE 16.5** Some common reactions of the Group IIA elements Ca, Sr, Ba, and Ra.

found to varying degrees in several other groups in the periodic table. The two trends that are found on descending Group IIIA are:

(1) a decrease in the stability of a higher oxidation state relative to a lower oxidation state (the M(III) state relative to the M(I) state in the case of the Group IIIA metals); and

(2) an increase in the metallic character (for example, in the basicity of oxides) for identical oxidation states.

Aluminum is the most abundant metallic element in the earth's mantle. In addition to its widespread occurrence in silicate minerals, aluminum also is found in enormous deposits of *bauxite*, AlOOH. These deposits are the chief source of the element. Aluminum metal is obtained from bauxite by dissolving the ore in $NaOH(aq)$ and filtering the resulting solution to eliminate $Fe_2O_3 \cdot xH_2O(s)$. The hydrated oxide is dissolved in acid and then reprecipitated by bubbling carbon dioxide through the solution. The precipitate is dissolved in a molten salt mixture composed of cryolite, $Na_3AlF_6$, together with $CaF_2$ and $NaF$, and the metal is obtained from the melt by electrolysis at 800 to 1000°C (Hall Process). The other Group IIIA metals are also obtained by electrolysis of the appropriate molten halide salt, or by electrolysis of aqueous solutions of their salts.

Aluminum is a light, soft,* electropositive metal, but it resists corrosion because of the formation of a tough, adherent, protective layer of the oxide, $Al_2O_3$ (alumina).† Gallium, indium, and thallium are soft, silvery-white metals. Thallium readily tarnishes (black) in air. Indium is soft enough to find use as a metallic O-ring material in metal high-vacuum fittings. Gallium has the widest liquid range (2041°C) of any known substance, but has no particular use.‡

---

*Structural alloys of aluminum (for aircraft and automobiles) contain silicon, copper, magnesium, and other metals to increase the tensile strength and stiffness of the aluminum.

†The electropositive character of Al can readily be demonstrated by rubbing a drop of mercury on the aluminum. The mercury breaks through the oxide layer, exposing unprotected aluminum amalgam to the air, which then reacts to form $Al_2O_3$.

‡Gallium has frequently been proposed as a manometer fluid, but it has the disadvantage of adhering to a glass surface.

**TABLE 16.3**

| element | atomic number | atomic weight | principal isotopes (% natural abundance) [half-life] | ground state electronic configuration of M(g) | ionization potentials in eV | electronegativity (Pauling) | melting point (°C) | boiling point (°C) | crystal radius Å covalent / ionic | photoelectric work function (eV) | density (g·cm⁻³) | $\mathscr{E}^{\circ}_{red}$ {M³⁺(aq) + 3e⁻ = M(s)} at 25°C |
|---|---|---|---|---|---|---|---|---|---|---|---|---|
| Al aluminum | 13 | 26.9815 | 27(100) 26[7.4 × 10⁵ y] | (Ne) 3s²3p | 5.98 18.82 28.44 119.96 | 1.5 | 660.37 | 2467 | 0.51 (Al³⁺) | 4.08 | 2.702 (cubic) | −1.66 |
| Ga gallium | 31 | 69.72 | 69(60.4) 71(39.6) 67[78h] | (Ar) 3d¹⁰4s²4p | 6.00 20.43 30.6 63.8 | 1.6 | 29.8 | 2070 | 0.62 (Ga³⁺) | (4.1) | 5.904 (ortho-rhomb) | −0.53 |
| In indium | 49 | 114.82 | 113(4.28) 115(95.72) [6 × 10¹⁴ y] | (Kr) 4d¹⁰5s²5p | 5.79 18.79 27.9 57.8 | 1.7 | 156.63 | 2000 | 0.81 (In³⁺) | — | 7.30 | −0.342 |
| Tl thallium | 81 | 204.37 | 203(29.50) 205(70.50) 204[3.9y] | (Xe) 4f¹⁴5d¹⁰-6s²6p | 6.11 20.32 29.7 50.5 | 1.8 | 304 | 1457 | 1.44 0.95 (Tl⁺) (Tl³⁺) | 3.68 | 11.85 | +0.72 |
| Sn tin | 50 | 118.69 | 116(14.30) 117(7.61) 118(24.03) 119(8.58) 120(32.85) | (Kr) 4d¹⁰5s²5p² | 7.332 14.63 30.6 39.6 | 1.8 | 231.89 | 2260 | 1.10 (Sn²⁺) 1.40 Sn(IV) | 4.4 | 7.28 (white) 5.75 (gray) | — |
| Pb lead | 82 | 207.2 | 204(1.48) [1.4 × 10¹⁹ y] 206(23.6) 207(22.6) 208(52.3) | (Xe) 4f¹⁴5d¹⁰-4s²6p² | 7.415 15.03 32.0 42.3 | 1.8 | 327.50 | 1744 | 1.21 (Pb²⁺) 1.54 Pb(IV) | 4.06 | 11.34 | — |
| Bi bismuth | 83 | 208.98 | 209(100) 208[75 × 10⁵ y] | (Xe) 4f¹⁴5d¹⁰-4s²6p³ | | 1.9 | 271.44 | 1560 | 1.08 (Bi³⁺) 1.52 Bi(III) | 4.23 | 9.80 (rhomb) | — |

Al and Ga dissolve in both strong aqueous acids and bases (that is, they are *amphoteric*):

$$Al(s) + 3H^+(aq) \rightarrow Al^{3+}(aq) + \frac{3}{2} H_2(g)$$

$$Al(s) + 3H_2O(\ell) + OH^-(aq) \rightarrow Al(OH)_4^-(aq) + \frac{3}{2} H_2(g)$$

aluminate ion

There is some spectroscopic evidence that suggests that the aluminate ion is $AlO_2^-$(aq), rather than $Al(OH)_4^-$(eq)

The latter reaction is utilized in certain types of commercially available preparations (that contain NaOH(s) + Al granules) for unplugging stopped-up sink drains* by gas evolution and grease dissolution. The hydroxides and oxides of aluminum and gallium are also amphoteric:

$$Al(OH)_3(s) + OH^-(aq) \rightarrow Al(OH)_4^-(aq)$$

$$Al(OH)_3(s) + 3H^+(aq) \rightarrow Al^{3+}(aq) + 3H_2O(\ell)$$

Alumina is an extremely stable compound. This stability is evidenced by the ability of aluminum to reduce many metallic oxides to the corresponding metals in the *thermite* process. For example,†

$$2Al(s) + Cr_2O_3(s) \rightarrow Al_2O_3(s) + 2Cr(s) \qquad \Delta H°_{298} = -126 \text{ kcal}$$

Indium and thallium react with aqueous solutions of strong acids, but they are unaffected by strong bases. The oxides and hydroxides of In and Tl are not amphoteric; they are basic. The trivalent state, M(III), is important for all of the group IIIA metals. However, the monovalent state becomes increasingly stable as one goes down the group, and Tl(I) is an important oxidation state in the chemistry of thallium.

Compounds of the Group IIIA metals exhibit both ionic and covalent bonding; however, ionic bonding is somewhat favored. All four metals react with halogens to form compounds with the empirical formula $MX_3$ (for example, $AlCl_3$), but thallium also forms a white monochloride, TlCl, which has properties similar to those of silver chloride. The $MX_3$ fluorides are ionic, whereas the chlorides, bromides, and iodides‡ are low-melting compounds which are dimeric in the vapor state. The halide-bridge structure of the dimer $Al_2Cl_6$ is shown in Figure 16.6.

The $MX_3$ halides (and the $MH_3$ hydrides and $MR_3$ trialkyls as well) are electron-pair acceptors (Lewis acids) that react with electron-pair donors (Lewis bases) to form tetrahedrally coordinated species, such as $AlCl_4^-$ and $(CH_3)_3NAlCl_3$.

As noted previously, the metals dissolve in strong aqueous acids to yield the M(III) oxidation state. The $M^{3+}$(aq) ions of aluminum, gallium, and indium are well-defined cationic species in strongly acidic solutions. The $Al^{3+}$(aq) ion is definitely hexacoordinated, $Al(OH_2)_6^{3+}$(aq), and $Ga^{3+}$(aq) and $In^{3+}$(aq) probably are hexacoordinated species. The $M^{3+}$(aq) ions are Brönsted acids and undergo acid-dissociation reactions of the type

$$Al(OH_2)_6^{3+}(aq) = Al(OH_2)OH^-(aq) + H^+(aq) \qquad K = 1.12 \times 10^{-5} \text{ (25°C)}$$

---

*These preparations evolve $H_2(g)$, which forms explosive mixtures with air.

†This reaction is very dangerous because of the very high temperatures attained.

‡Tl(III)$I_3$ does not exist, because Tl(III) oxidizes $I^-$ to $I_2$. The compound $TlI_3$ is actually thallium (I) tri-iodide.

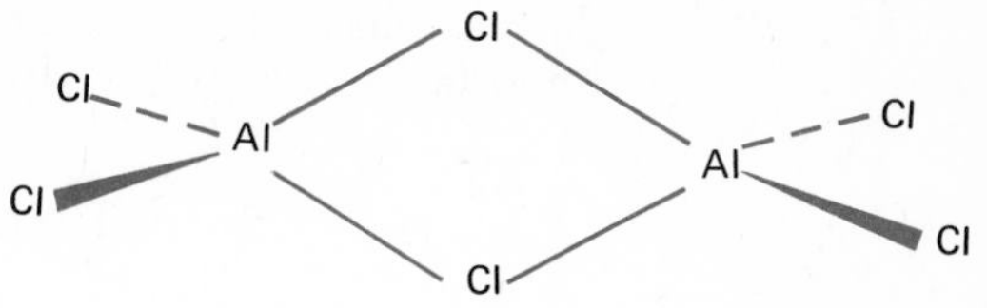

**FIGURE 16.6** The structure of $Al_2Cl_6(g)$.

The dissociation is sufficiently extensive that Group IIIA $M^{3+}$(aq) salts of weak acids spontaneously decompose in water. Slow addition of NaOH(aq) to solutions containing $M^{3+}$(aq) yields the insoluble hydroxides $M(OH)_3$(s), or hydrated oxides, $M_2O_3 \cdot xH_2O$(s). Dehydration of the hydroxides or hydrous oxides yields the oxides $Al_2O_3$ (white), $Ga_2O_3$ (white), $In_2O_3$ (yellow), and $Tl_2O_3$(s) (brown-black). Thallium (III) oxide decomposes above 100°C:

$$Tl_2O_3(s) = Tl_2O(s) + O_2(g)$$

The salts $LiAlH_4$ and $LiGaH_4$, which contain the tetrahedral hydride ions $MH_4^-$, can be prepared from the respective halides and LiH:

$$4LiH(soln) + AlCl_3(soln) \xrightarrow{(C_2H_5)_2O} LiAlH_4(soln) + 3LiCl(s)$$

These hydrides are useful reducing agents in numerous aprotic solvents, but are violently decomposed by water, and occasionally ignite or explode on contact with air.

The organometallic compounds of aluminum are the best known and most important of the Group IIIA organometallics. These compounds can be prepared via the reaction of $AlCl_3$ with the appropriate Grignard reagent:

$$AlCl_3 + RMgCl \rightarrow RAlCl_2, R_2AlCl, R_3Al$$

However, the most important aluminum alkyls (triethyl and tri-isopropyl aluminum) are made by the Zeigler process, in which aluminum metal is heated with hydrogen and an olefin under pressure:

$$6\ C_2H_4 + 3H_2 + 2Al \rightarrow 2(C_2H_5)_3Al$$

Many of the trialkyl aluminum compounds are dimeric, with structures analogous to that of $Al_2Cl_6$ (see Figure 16.6). However, most of the organometallic compounds of Ga, In, and Tl do not show any appreciable tendency to dimerize, although the hydrides are known to do so. Thallium organometallic compounds of the type $(C_2H_5)_2TlCl$ undergo ionization reactions of the type

$$(C_2H_5)_2TlCl \rightleftharpoons (C_2H_5)_2Tl^+ + Cl^-$$

in polar solvents. The ion $(C_2H_5)_2Tl^+$ has a linear C—Tl—C structure analogous to $(C_2H_5)_2Hg$, with which it is isoelectronic.

The M(I) oxidation state is known for all of the Group IIIA metals in the gas phase. However, only in the case of thallium is this state of major importance in solution. In water at 25°C,

$$2e^- + Tl^{3+}(aq) = Tl^+(aq) \qquad \mathscr{E}^\circ_{red} = +1.25 \text{ V}$$

The thallous ion, $Tl^+$, is in some respects similar to $Ag^+$, and in other respects it is similar to $K^+$ and $Rb^+$. For example, the nitrate and fluoride salts are water soluble, whereas the chloride, bromide, chromate, and sulfide salts are insoluble in water. Both TlCl and AgCl darken on exposure to light, whereas the hydroxide TlOH(aq) is a moderately soluble strong base. Thallous compounds are extremely poisonous, and even trace amounts can cause complete loss of body hair.

## 16.5   TIN AND LEAD: THE GROUP IVA METALS.*

Tin and lead are usually obtained from their oxides by reduction with carbon. The resulting impure metals are dissolved in an aqueous solution of a strong acid and are recovered in higher purity from the solution by electrolysis. Some properties of tin and lead are given in Table 16.3. The principal lead ore is *galena*, PbS, and the principal tin ore is *cassiterite*, $SnO_2$.

There are five different solid phases of tin, namely grey (or cubic) tin, white (or tetragonal) tin, rhombic tin, superconducting tin, and a high-pressure, body-centered, tetragonal tin. At 1 atm pressure grey tin is stable with respect to white tin at $t < 13.2°C$, whereas the reverse is true above this temperature. The two forms can coexist in equilibrium with each other at 13.2°C and 1 atm. The transition is sluggish (but not negligibly slow) in either direction around the transition temperature, and both forms can be obtained in the metastable condition. Many fine museum pieces and old organ pipes made of white tin have been ruined because their temperatures were allowed to drop below 13°C for an appreciable period of time; the change in crystal structure disrupts the surface of the metal sufficiently to allow fragments to spall off.

At 1 atm white tin is stable with respect to rhombic tin at $t < 202.8°C$, whereas the reverse is true above this temperature. White tin is malleable, whereas rhombic tin is brittle, and the phenomenon of *tin embrittlement* that occurs when tin is heated above 202.8°C, and then cooled quickly, results from the conversion to the rhombic or brittle phase.

The most unusual phase of tin is the superconducting phase. The phenomenon of superconductivity was discovered by Kammerlingh Onnes in 1911. At that time Onnes was interested in the nature of the temperature dependence of the resistivity of metals at low temperatures. The resistance of a sample of grey tin decreases smoothly with decreasing temperature down to the lowest attainable temperatures. The resistance of a sample of white tin at 1 atm also decreases smoothly with decreasing temperature until a temperature of 3.73°K is reached, at which temperature the resistance drops suddenly and discontinuously to zero. Below 3.73°K white tin is a perfect conductor (a superconductor) of electricity, offering no

---

*See Chapter 17 for the Group IVA non-metals.

resistance to the flow of electrons. As long as the metal is maintained in the superconducting state the current will flow indefinitely in a superconducting loop.*

The principal oxidation states of tin and lead are M(0), M(II) and M(IV). Some important half-reactions for lead and tin in water at 25°C are:

| | $\mathscr{E}^{\circ}_{red}$(volts) |
|---|---|
| $Pb^{2+}(aq) + 2e^- = Pb(s)$ | $-0.126$ |
| $PbO_2(s) + 4H^+(aq) + 2e^- = Pb^{2+}(aq) + 2H_2O$ | $+1.46$ |
| $Sn^{2+}(aq) + 2e^- = Sn(s)$ | $-0.136$ |
| $Sn^{4+}(aq) + 2e^- = Sn^{2+}(aq)$ | $+0.15$ |
| $SnCl_6^{2-}(aq) + 2e^- = SnCl_3^-(aq) + 3Cl^-(aq)$ | $0.0$ |

Aqueous solutions of Sn(IV) and Sn(II)† contain predominantly hydroxy-tin species. The hydroxy-free ions exist in appreciable concentrations only in concentrated solutions of strong acids; even then these ions usually occur as complex ions, such as $SnCl_6^{2-}(aq)$. The ion $Pb^{4+}$ is of no importance as such, and the chemistry of Pb(IV) is limited to the solid state. On the other hand, lead has an extensive chemistry as the $Pb^{2+}(aq)$ ion. The nitrate, acetate, and fluoride of $Pb^{2+}$ are soluble in water, whereas the chromate, sulfate, chloride, bromide, and iodide are insoluble.

The following oxides and sulfides of tin and lead are known: SnO, $SnO_2$, SnS, $SnS_2$, PbO, $PbO_2$, $Pb_3O_4$ ("red lead"), and PbS. The oxides SnO, $SnO_2$, and PbO and the hydroxides $Sn(OH)_2$ and $Pb(OH)_2$ are amphoteric:

$$Pb(OH)_2(s) + 2H^+(aq) = Pb^{2+}(aq) + 2H_2O(\ell)$$

$$Pb(OH)_2(s) + 2OH^-(aq) = Pb(OH)_4^{2-}(aq)$$

As a consequence, water-insoluble lead (II) salts (except PbS) are readily soluble in strongly basic solutions. The sulfides SnS and $SnS_2$ both dissolve in $Na_2S(aq)$ solutions to yield $SnS_3^{2-}(aq)$.

The tetrahalides of the type $SnCl_4$ (tetrahedral) are low-boiling, covalent compounds. Although all four tetrahalides of tin are known, only the lead halides $PbF_4$ and $PbCl_4$ are known. This is because Pb(IV) is reduced by $Br^-$ and $I^-$ to Pb(II). Even in the case of $PbCl_4$ the equilibrium

$$PbCl_4 = PbCl_2 + Cl_2$$

is of some importance, especially at elevated temperatures.

Tin and lead form numerous organometallic compounds, for example, $(C_2H_5)_3SnCl$, $(C_2H_5)_2SnCl_2$, $(C_2H_5)_2Sn$, and $(C_2H_5)_4Pb$ (tetraethyl lead).

---

*Although it is not difficult to demonstrate experimentally that a very large quantity is indeed large, or a very small quantity is indeed small, it is impossible to prove experimentally that a quantity is infinite or exactly zero. The present record for the maintenance of a persistent current in a superconductor is two years (M.I.T.). The experiment ended when a transport strike cut off the supply of liquid helium used to maintain the low temperature. From these experiments it was concluded that the resistivity of a superconductor is less than $10^{-58}$ ohm·cm. For comparison, the resistivity of a "good" conductor like Ag or Cu is about $10^{-6}$ ohm·cm. Lead also exhibits superconductivity at $T \leqslant 7.2°K$.

† The species Sn(II)(aq) is probably primarily the cyclic ion $Sn_3(OH)_4^{2+}(aq)$.

Tetraethyl lead is widely used as an anti-knock additive in gasolines. The $(C_2H_5)_4Pb$ increases the octane number by promoting a smooth burning of the fuel. However, because of the high toxicity of lead compounds, the exhaust fumes from engines utilizing such leaded fuels constitute a serious health menace.

## 16.6  BISMUTH: THE GROUP VA METAL.

Some of the properties of bismuth are in Table 16.3. Bismuth is obtained from the yellow oxide $Bi_2O_3$ by reduction with carbon. The hard and brittle metal is used in numerous alloys, for such applications as in type-metal and in automatic sprinkler systems to combat fires. In the latter, the bismuth alloy is used as the sprinkler plug. It melts at about 100°C, and thereby allows water to escape from the sprinkler. The principal oxidation states of bismuth are Bi(0) and Bi(III), with Bi(V) being of lesser importance. The oxide $Bi_2O_3$ is soluble in aqueous solutions of strong acids. The "bismuthyl" ion $BiO^+(aq)$ (or $Bi(OH)_2^+$) is of some importance in aqueous solution and presumably occurs in the compounds $BiOCl$ and $BiOOH$. Sodium bismuthate, $NaBiO_3$, is a powerful oxidizing agent.

## PROBLEMS

1. Complete and balance the following reactions:

   a) $Mg(s) + BeF_2(s) \rightarrow$

   b) $Na(s) + H_2O(\ell) \rightarrow$

   c) $Cs(s) + NH_3(\ell) \xrightarrow{Fe_2O_3}$

   d) $SrCO_3(s) \xrightarrow{heat}$

   e) $NaOH(aq) + CO_2(g) \rightarrow$

   f) $K(s) + O_2(g) \rightarrow$

   g) $KH(s) + H_2O(\ell) \rightarrow$

   h) $CaC_2(s) + H_2O(\ell) \rightarrow$

   i) $Ca(s) + LiCl(s) \xrightarrow{heat}$

   j) $Li(s) + N_2(g) \xrightarrow{trace \atop H_2O}$

   k) $^{40}K \rightarrow$

   l) $BeF_2(g) + H_2(g) \xrightarrow{heat}$

   m) $Li_3N(s) + H_2O(\ell) \rightarrow$

2. Suggest a possible qualitative explanation for the observation that the standard reduction potential for $Li^+(aq) + e^- \rightarrow Li(s)$ is more negative than that for the other alkali metals, even though the ionization potential of Li is the highest for any alkali metal.

3. Lithium metal is often cleaned by treatment with ethanol, $C_2H_5OH$. Write the principal chemical reaction that takes place when $Li(s)$ is treated with ethanol.

4. The equilibrium constant at 25°C for the reaction

$$Ca(OH)_2(s) = Ca^{2+}(aq) + 2OH^-(aq)$$

is $8 \times 10^{-6}$. Compute the pH of an aqueous solution saturated with $Ca(OH)_2$.

5. Using the ionization potentials given in Table 16.2, estimate $\Delta E$ in kcal for the reaction

$$2M^+(g) = M(g) + M^{2+}(g)$$

where M = Be, Mg, Ca, Sr, and Ba. How do you rationalize your results in connection with the known stable oxidation states of the Group IIA metals in solution?

6. What is the composition of the black coating that forms on the surface of freshly cut thallium which is exposed to air?

7. Suggest a qualitative explanation for the absence of the M(II) oxidation state for the Group IIIA metals.

8. What are the oxidation states of the elements in $Pb_3O_4$?

9. Compute the $pH$ of an aqueous solution at 25°C for which $(TlOH^{2+}) = (Tl^{3+})$ $\{pK_a = 1.2$ for $Tl^{3+}(aq)\}$.

10. The $pK_a$ for $Sn^{2+}(aq)$ is about 2, whereas the $pK_a$ for $Pb^{2+}(aq)$ is about 8. Suggest a qualitative explanation for this difference.

11. Discuss the bonding in $Al_2Cl_6$ and $Be_2(CH_3)_4$.

12. Complete and balance the following reactions (if no reaction occurs, write N.R.):

a) $AlOOH(s) + OH^-(aq) \rightarrow$
b) $Ga(s) + HNO_3(aq) \rightarrow$
c) $Al(OH)_4^-(aq) + CO_2(g) \rightarrow$
d) $Ga(s) + KOH(aq) \rightarrow$
e) $In(s) + HNO_3(aq) \rightarrow$
f) $C_2H_5Cl(\ell) + Na/Pb$ alloy $\rightarrow$

g) $Al(s) + HNO_3$ (conc., aq) $\rightarrow$
h) $Ga_2O_3(s) + NaOH(aq) \rightarrow$
i) $Ga_2O_3(s) + HNO_3(aq) \rightarrow$
j) $In_2O_3(s) + HNO_3(aq) \rightarrow$

k) $In_2O_3(s) + NaOH(aq) \rightarrow$
l) $LiAlH_4(s) + H_2O(\ell) \rightarrow$
m) $LiH + GaCl_3 \xrightarrow{(C_2H_5)_2O}$
n) $PbSO_4(s) + CsOH(aq) \rightarrow$
o) $Sn(OH)_2(s) + KOH(aq) \rightarrow$
p) $SnS(s) + S^{2-}(aq) \rightarrow$
q) $Bi_2O_3(s) + C(s) \xrightarrow{heat}$
r) $Sn(s) + Cl_2(g) \rightarrow$
s) $Pb(OH)_4^-(aq) + OCl^-(aq) \rightarrow$
t) $MgH_2(s) + H_2O(\ell) \rightarrow$

13. Propose structures for the following species:

$AlF_6^{3-}$         $Sn_3(OH)_4^{2+}$         $SnCl_2$         $PbF_4$         $Al_2(C_2H_5)_6$

_____ **References**

16.1. W. M. Latimer and J. H. Hildebrand, Reference Book of Inorganic Chemistry, 3rd edition (The Macmillan Co., New York, 1951).
16.2. F. A. Cotton and G. Wilkinson, Advanced Inorganic Chemistry, 2nd edition (Interscience Publishers, John Wiley and Sons, New York, 1966).

# CHAPTER 17

"Science is the lodestar of practice, without it the latter easily loses itself in the murky and infinite ranks of possibility."

J. N. Fuchs (1844)*

# THE METALLOIDS

## 17.1 INTRODUCTION.

The elements characterized as metalloids exhibit chemical behavior intermediate between those of the metals and the non-metals. The metalloids are the elements† B, Si, Ge, As, Sb, Te, Po, and At (see Figure 17.1). The metalloids fall between the metals and the non-metals in the Periodic Table; they are neither as electropositive as the metals, nor as electronegative as the non-metals. The metalloids are compared and contrasted with the metals and non-metals in Table 17.1.

We shall confine our discussion of the metalloids to the most important members of this class of elements, namely B, Si, Ge, As, and Sb (Table 17.2). Most metalloid-oxygen bonds are particularly strong, and reactions involving the formation of oxides of the metalloids from the elements generally have large negative $\Delta \bar{H}_f^\circ$ and $\Delta \bar{G}_f^\circ$ values (that is, the reactions are highly exothermic), and the oxides are very stable with respect to the metalloid and oxygen. For example, consider the following reactions:

|  | $\Delta G^\circ_{298}$(kcal) | $\Delta H^\circ_{298}$(kcal) |
|---|---|---|
| $Si(s) + O_2(g) = SiO_2(s)$ | $-264.8$ | $-217.7$ |
| $2B(s) + \dfrac{3}{2} O_2(g) = B_2O_3(s)$ | $-304.2$ | $-285.3$ |

The hydrides and halides of the metalloids generally react readily with basic reagents (water, alcohols, and amines); such reactions lead to the

---

*Wilhelm Prandtl, *Johann Nepomuk Fuchs*, Selected Readings in The History of Chemistry (reprinted from J. Chem. Educ., compiled by A. J. Ihde and W. F. Kieffer).
†The elements P, Bi, and Se are sometimes classed with the metalloids.

## PERIODIC TABLE

| Group | Ia | IIa | IIIb | IVb | Vb | VIb | VIIb | VIII | | | Ib | IIb | IIIa | IVa | Va | VIa | VIIa | 0 |
|---|---|---|---|---|---|---|---|---|---|---|---|---|---|---|---|---|---|---|
| 1st period | 1 H | | | | | | | | | | | | | | | | | 2 He |
| 2nd period | 3 Li | 4 Be | | | | | | | | | | | 5 B | 6 C | 7 N | 8 O | 9 F | 10 Ne |
| 3rd period | 11 Na | 12 Mg | | | | | | | | | | | 13 Al | 14 Si | 15 P | 16 S | 17 Cl | 18 A |
| 4th period | 19 K | 20 Ca | 21 Sc | 22 Ti | 23 V | 24 Cr | 25 Mn | 26 Fe | 27 Co | 28 Ni | 29 Cu | 30 Zn | 31 Ga | 32 Ge | 33 As | 34 Se | 35 Br | 36 Kr |
| 5th period | 37 Rb | 38 Sr | 39 Y | 40 Zr | 41 Nb | 42 Mo | 43 Tc | 44 Ru | 45 Rh | 46 Pd | 47 Ag | 48 Cd | 49 In | 50 Sn | 51 Sb | 52 Te | 53 I | 54 Xe |
| 6th period | 55 Cs | 56 Ba | 57* La | 72 Hf | 73 Ta | 74 W | 75 Re | 76 Os | 77 Ir | 78 Pt | 79 Au | 80 Hg | 81 Tl | 82 Pb | 83 Bi | 84 Po | 85 At | 86 Rn |
| 7th period | 87 Fr | 88 Ra | 89 Ac | | | | | | | | | | | | | | | |

\* Lanthanide
  series

**FIGURE 17.1** Placement of metals, metalloids, and nonmetals in the Periodic System. Metalloids are identified by shaded squares. Reprinted by permission of the publisher, from *The Metalloids* by Eugene G. Rochow (Lexington, Mass.: D. C. Heath and Company, 1966).

**TABLE 17.1**

| METALS | METALLOIDS | NON-METALS |
|---|---|---|
| $0.7 \leqslant$ electronegativity $< 1.8$ | $1.8 \leqslant$ electronegativity $< 2.2$ | $2.2 \leqslant$ electronegativity $\leqslant 4.0$ |
| form basic oxides | form weakly acidic oxides | form acidic oxides |
| high electrical and thermal conductance ("free" electrons present in solids) | intermediate electrical and thermal conductance ("semi-conductors," few free electrons) | insulators (essentially no free electrons) |
| high reflectance (owing to presence of "free" electrons) | intermediate reflectance | low reflectance |
| resistance increases with increasing temperature | resistance decreases with increasing temperature | resistance rather insensitive to temperature |
| mostly malleable and ductile, strong and tough | all brittle; deform elastically to the shatter point | ——— |
| form organometallic compounds | form organometallic compounds | ——— |
| form non-volatile and high-melting (ionic) halides and hydrides | form volatile and low-melting (covalent) halides and hydrides | form volatile and low-melting (covalent) halides and hydrides |

**TABLE 17.2**

| element | atomic number | atomic weight | principal isotopes (% natural abundance) half-life | ground state electronic configuration of M(g) | ionization potentials (in eV) | electronegativity (Pauling) | melting point (°C) | boiling point (°C) | ionic crystal radius / covalent | resistivity (ohm-cm at 25°C) | photoelectric work function (eV) | density (g·cm⁻³) |
|---|---|---|---|---|---|---|---|---|---|---|---|---|
| B boron | 5 | 10.81 | 10(19.6)<br>11(80.4) | (He) $2s^2 2p$ | 8.30<br>25.15<br>37.92<br>259.30 | 2.0 | 2200 (α-rhom) | 2530 | 0.20 ($B^{3+}$)<br>1.17 | $2 \times 10^6$ | 4.5 | 2.3 (α-rhom) |
| Si silicon | 14 | 28.086 | 28(92.21)<br>29(4.70)<br>30(3.09)<br>31[2.62h]<br>32[700y] | (Ne) $3s^2 3p^2$ | 8.15<br>16.34<br>33.46<br>45.13 | 1.8 | 1420 (cryst) | 2193 | 0.41 ($Si^{4+}$)<br>1.22 | 40 | 4.5 | 2.4 (cryst) |
| Ge germanium | 32 | 72.59 | 70(20.52)<br>72(27.43)<br>73(7.76)<br>74(36.54)<br>76(7.76)<br>68[280d] | (Ar)$3d^{10}4s^2 4p^2$ | 7.88<br>15.43<br>34.23<br>45.7 | 1.8 | 937.4 | 2830 | 0.53 ($Ge^{4+}$) | (20) | 4.6 | 5.3 (grey-white) |
| As arsenic | 33 | 74.9216 | 75(100)<br>73[76d]<br>74[18d]<br>76[26.5h] | (Ar)$3d^{10}4s^2 4p^3$ | 10.5<br>20.1<br>28.0<br>49.9<br>62.5 | 2.0 | 814 | 613 (subl) | 0.58 ($As^{3+}$)<br>1.21 | $33 \times 10^{-6}$ | 4.9 | 5.7 (grey) |
| Sb antimony | 51 | 121.75 | 121(57.25)<br>123(42.75)<br>125[2.0y] | (Kr)$4d^{10}5s^2 5p^3$ | 8.5<br>(18)<br>24.7<br>44.0<br>55.5 | 1.9 | 630.74 (silvery) | 1440 | 0.92 ($Sb^{3+}$)<br>1.41 | $39 \times 10^{-6}$ | 4.3 | 6.7 (silvery) |
| Te tellurium | 52 | 127.60 | 120(0.89)<br>122(2.46)<br>123(0.87)<br>124(4.61)<br>125(6.99)<br>126(18.71)<br>128(31.79)<br>130(34.48) | (Kr)$4d^{10}5s^2 5p^4$ | 8.96<br>30.5<br>37.7<br>60.0<br>(72) | 2.1 | 450 (silver-white) | 1390 | 2.22 ($Te^{2-}$)<br>1.27 | 0.44 | 4.4 | 6.25 (silver-white) |

formation of bonds between the metalloid and the basic atom (such as O or N) of the reagent. For example,

$$BCl_3(g) + 3CH_3OH(g) \rightarrow B(OCH_3)_3(g) + 3HCl(g)$$

$$SiCl_4(g) + 8CH_3NH_2(g) \rightarrow Si(NHCH_3)_4(g) + 4CH_3NH_3Cl(s)$$

Boron and silicon hydrides react violently with water, evolving hydrogen and forming the hydroxy compounds $B(OH)_3$ and $Si(OH)_4$, which are weak acids.

All of the metalloids react *directly* with simple organic halides such as $CH_3Cl$ and $C_6H_5Cl$ to form organometallic compounds. The organometallic compounds of the metalloids also undergo organometallic replacement reactions:

*direct reaction*

$$2CH_3Cl(g) + Ge(s) \xrightarrow[\text{Cu catalyst}]{320°C} (CH_3)_2GeCl_2(g)$$

*replacement reaction*

$$(CH_3)_2SiCl_2(soln) + 2C_4H_9Li(soln) \rightarrow (C_4H_9)_2Si(CH_3)_2(soln) + 2LiCl(s)$$

## 17.2 BORON.

Boron is a rare element, but it occurs in nature in concentrated deposits of soluble borates such as borax, $Na_2B_4O_7 \cdot 10H_2O$. Several different allotropic modifications of the element are known, all of which are extremely difficult to obtain in high purity. The most dense form ($\rho = 9.3$ $g \cdot cm^{-3}$), the $\alpha$-rhombohedral, is prepared by the pyrolysis of $BI_3$ on a tungsten filament. The structure of $\alpha$-rhombohedral boron is comprised of close-packed $B_{12}$ icosohedra (Figure 17.2).

Simple $B^{3+}$ ions are of no real importance, as such, in boron chemistry, owing to the extremely high polarizing power of B(III) (see the ionization potentials for boron in Table 17.2). Consequently, B(III) is always covalently bonded. All monomeric three-coordinate boron compounds are planar with X–B–X bond angles of 120° ($sp^2$ hybridization, see Figure 17.3).

**FIGURE 17.2** The $B_{12}$ icosohedron. There are 20 triangular faces and 12 apices, with a boron atom situated at each one of the latter.

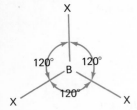

**FIGURE 17.3**   Structure of 3-coordinate boron compounds.

The boron halides (Table 17.3) all have the structure indicated in Figure 17.3. The boron halides are *electron-deficient* compounds (incomplete octets), and possess a vacant $p$-orbital* perpendicular to the plane of the molecule. This vacant $p$-orbital can (and does) act as an electron acceptor, thereby making the boron halides *Lewis acids*, which are capable of reacting with electron donors (*Lewis bases*).

A *Lewis acid* is defined as an *electron acceptor*, and a *Lewis base* is defined as an *electron donor*. The Lewis acid-base classification is a more general one than the Brönsted acid-base classification. Note that $H^+$ is both a Lewis and a Brönsted acid, and $H_2O$ and $NH_3$ are both Lewis and Brönsted bases. However, $BF_3$ is a Lewis acid, but obviously not a Brönsted acid.

The reactions of boron halides with a sufficiently strong Lewis base give rise to four-coordinate boron compounds in which the coordination around boron is tetrahedral ($sp^3$ hybridization), or nearly so. Some examples of reactions in which the boron halides act as Lewis acids are

$$F^- + BF_3 \rightarrow BF_4^-$$

$$(C_2H_5)_2O + BF_3 \rightarrow (C_2H_5)_2OBF_3$$

$$(CH_3)_3N + BCl_3 \rightarrow (CH_3)_3NBCl_3$$

The relative strengths of the boron halides as Lewis acids are:

$$BBr_3 \gg BCl_3 > BF_3$$

The boron halides are formed directly from the elements; they react with water to form boric acid, $B(OH)_3$, and the halogen acid. For example,

$$BCl_3(g) + 3H_2O(\ell) \rightarrow B(OH)_3(s) + 3H^+(aq) + 3Cl^-(aq)$$

---

*This is an over-simplification, owing to the possible importance of $p_\pi(B) + p_\pi(X)$ bonding in these compounds.

**TABLE 17.3**

**THE BORON HALIDES**

| **COMPOUND** | m. pt. (°C) | b. pt. (°C) | $\Delta \bar{G}^\circ_{f,298}$(kcal·mole$^{-1}$) | $\Delta \bar{H}^\circ_{f,298}$(kcal·mole$^{-1}$) |
|---|---|---|---|---|
| $BF_3$ | −127 | −101 | −267.77 | −271.75 |
| $BCl_3$ | −107 | 12.5 | − 92.91 | − 96.50 |
| $BBr_3$ | − 46 | 90.1 | − 55.56 | − 49.15 |

Boric acid is a moderately soluble weak acid in water with $pK = 9.0$ at 25°C:

$$B(OH)_3(aq) + H_2O(\ell) = B(OH)_4^-(aq) + H^+(aq)$$

Boric acid is used in some mouth and eye washes.

Boron forms a large number of hydrides. The boron hydrides are volatile, spontaneously flammable in air, and easily hydrolyzed. The boron hydrides have interesting structures involving multicenter bonds (see Chapter 4).

The simplest boron hydride that can be synthesized in appreciable quantities is diborane, $B_2H_6$, which is prepared by reacting sodium borohydride,* $NaBH_4$, either with $BF_3$ in ether solution, or with $H_2SO_4(aq)$:

$$3NaBH_4(s) + 4BF_3(soln) = 3NaBF_4(s) + 2B_2H_6(aq)$$

$$2NaBH_4(aq) + H_2SO_4(aq) = B_2H_6(g) + 2H_2(g) + Na_2SO_4(aq)$$

The reaction of $B_2H_6$ with water is slow enough to enable $B_2H_6$ to escape from the solution as the gas. Borane, $BH_3$, is unstable with respect to diborane:

$$2BH_3(g) = B_2H_6(g) \qquad K \sim 10^5 \text{ at } 25°C; \Delta H^\circ_{298} \simeq -35 \text{ kcal}$$

However, it is possible to make compounds, such as $(CH_3)_3PBH_3$, which involve the $BH_3$ unit. Diborane is used to prepare all of the higher boranes (for example, $B_4H_{10}$, $B_5H_9$, $B_6H_{12}$, and $B_{10}H_{14}$) and borohydride anions (for example, $B_3H_8^-$ and $B_{12}H_{12}^{2-}$):

$$B_2H_6(g) + H_2 \xrightarrow{\text{heat}} \text{higher boranes}$$

$$B_2H_6(g) + NaBH_4(s) \xrightarrow{\text{heat}} \text{sodium salts of higher borohydride anions}$$

Structures of some higher boranes and borohydride anions are presented in Figure 17.4.

The compound boron nitride, BN, exists in two different modifications. The low-density form is a slippery white powder with a layered structure very similar to that of graphite (see Figure 17.5a). The separation between layers is 3.38 Å. The similarity in structure between graphite and boron nitride suggested the possibility that high pressure might yield a more dense form of BN analogous to diamond. This is the case; the high pressure form of BN has the cubic diamond structure and is called *borazon*. Borazon has a hardness of 9.8, which is exceeded only by that of diamond at 10.0. The extreme hardness of borazon, coupled with its resistance to oxidation (in which respect it is superior to diamond), makes borazon an ideal material for high-speed cutting tools. Another interesting boron-nitrogen compound is borazole, $B_3N_3H_6$, which is structurally (but not chemically) similar to benzene (Figure 17.5b).

---

*The ion $BH_4^-$ (tetrahedral) is the simplest boron hydride *anion*.

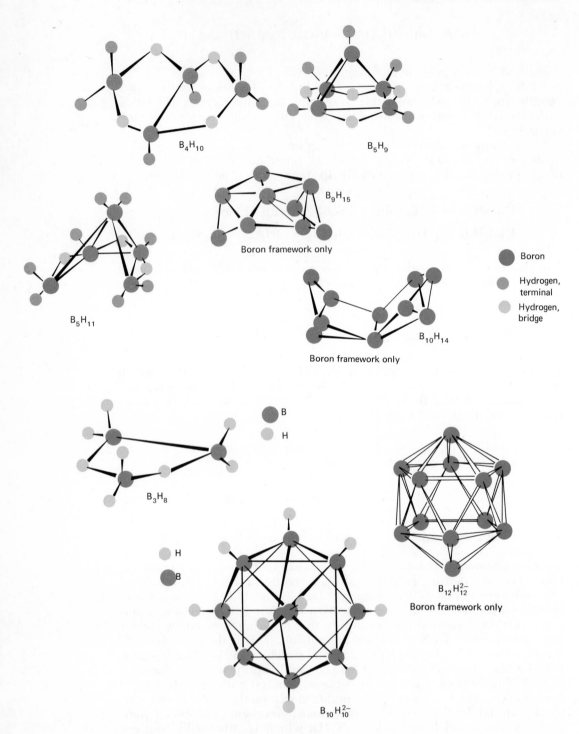

**FIGURE 17.4** **(a)** Structures of some higher boranes. **(b)** Structures of some higher borohydrides.

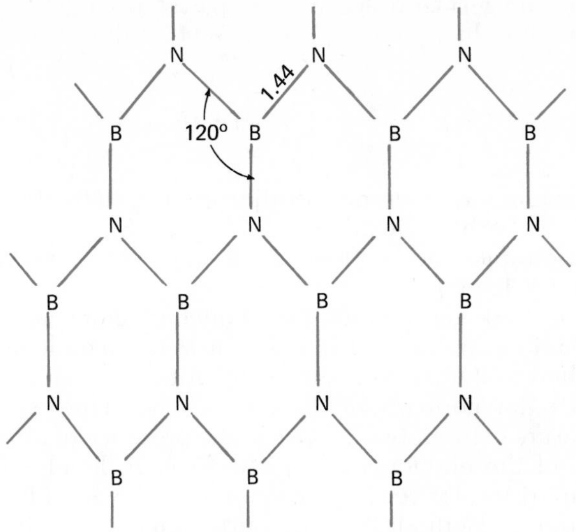

(a)  Structure of layers in low density BN.

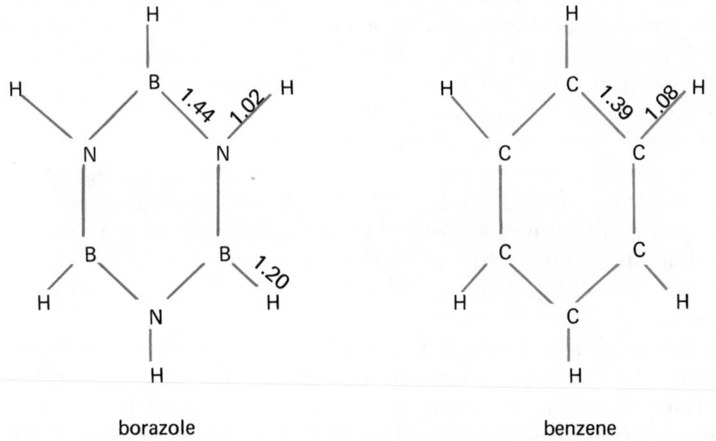

borazole                                                benzene

(b)  Comparison of borazole and benzene structures.

**FIGURE 17.5**    Structures of some boron-nitrogen compounds (distances in Ångstrom units).

# 17.3    SILICON AND GERMANIUM.

## a. The Elements.

Silicon is second only to oxygen in abundance in the earth's mantle. Eighty-seven weight per cent of the mantle is made up of silicon compounds. Silicon itself comprises approximately 28 weight per cent of the mantle. Germanium, on the other hand, amounts to only about 0.001% of the mantle; it is a relatively rare element.

Elemental silicon and germanium can be prepared in about 98% purity by the high-temperature reduction of $SiO_2$ and $GeO_2$ with carbon:

$$SiO_2(s) + C(s) \xrightarrow[\text{(air-free)}]{3000°C} Si(\ell) + CO_2(g)$$

At significantly lower temperatures this equilibrium lies to the left, and at 1200°C silicon burns in carbon dioxide to yield silica, $SiO_2$, and carbon. The most common allotropic modifications of the elements Si and Ge have the diamond structure (Chapter 6).

Elemental silicon and germanium are semiconductors at ordinary temperatures. Germanium and silicon are *intrinsic* semiconductors; the gap between the filled (valence) and empty (conduction) bands is considerably less at 25°C for germanium than for silicon. The specific resistances of Si and Ge are extremely sensitive to the presence of impurities, and the preparation of the elements at sufficiently high levels of purity ($< 1$ part in $10^9$ of impurities, for certain applications) is a very difficult and expensive process. In one method, the 98% pure element is converted to $SiCl_4$ by reaction with chlorine. The silicon tetrachloride is purified by repeated fractional distillation. The purified $SiCl_4$ is then reduced with high purity magnesium metal. The resulting magnesium chloride is then sublimed, leaving behind the elemental silicon. The silicon is next fused to single crystals and repeatedly zone refined to the desired purity. The preparation of $n$-type and $p$-type impurity semi-conductors is accomplished by adding minute (0.0005 to 0.001%) quantities of P (or Sb), or B (or In), respectively, to the "superpure" element.[*]

The most important oxidation states of silicon are Si(0) and Si(IV).[†] These are also the most important oxidation states of germanium, although some Ge(II) compounds (such as GeO and $GeCl_2$) are known. The $Si^{4+}$ ion is of no real importance, as such, because of its extremely high polarizing power. The stereochemistry of Si(IV) is mostly tetrahedral, although a few compounds involving octahedrally coordinated Si(IV) (such as $H_2SiF_6$) are known.

Elementary silicon is a rather unreactive substance, but it can be made to undergo several types of reactions under appropriate conditions. Some of the more common reactions of Si are diagrammed in Figure 17.6. The reactions of germanium are similar to those of silicon, but germanium is somewhat more reactive than Si. For example, germanium dissolves slowly in concentrated nitric and sulfuric acids, which oxidize it to the slightly soluble $GeO_2$. Silicon cannot be dissolved this way, but it will dissolve in a mixture of $HNO_3$ and HF(aq).

### b.  Some Reactions of Si and Ge Compounds.

The silicon halides are low-boiling, highly-reactive compounds (Table 17.4). They all react rapidly and quantitatively with water, for example:

$$SiCl_4(g) + 2H_2O(\ell) \rightarrow SiO_2(s) + 4HCl(aq)$$

---

[*]See Chapter 6 for a discussion of semiconductors.

[†]The only important compound of Si(II) is the monoxide SiO, which is more volatile than $SiO_2$, and which is deposited from the vapor phase on glass lenses (for cameras and binoculars) to form a non-reflective coating.

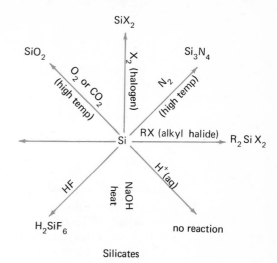

**FIGURE 17.6** Some of the reactions of silicon.

The germanium halides are also readily hydrolyzed, but mixed hydroxo-halides can be isolated from concentrated HX(aq) solutions. Silicon tetra-chloride reacts with lithium aluminum hydride, $LiAlH_4$, to yield mono-silane, $SiH_4$. The silanes $(Si_nH_{2n+2})$ are spontaneously flammable in air, and are strong reducing agents. On the other hand, the germanes $(Ge_nH_{2n+2})$ are much less reactive than the silanes. For example, mono-germane, $GeH_4$, can be prepared by the reaction of $NaBH_4(aq)$ with a suspension of $GeO_2(s)$.

The decidedly amphoteric behavior of the metalloids is exemplified by the reactions of $SiCl_4$ with the base $CH_3NH_2$ and the acid $CH_3COOH$:

$$SiCl_4 + 4CH_3NH_2 \rightarrow Si(NHCH_3)_4 + 4HCl$$

$$SiCl_4 + 4CH_3COOH \rightarrow Si(OCOCH_3)_4 + 4HCl$$

## c.  The Silicates.

Silicates occur widely in nature; they also can be obtained from the fusion of alkali metal carbonates with silica at temperatures around 1300°C. The fusion reaction involves the loss of carbon dioxide. When the

TABLE 17.4

THE SILICON HALIDES

| COMPOUND | m. pt. (°C) | b. pt. (°C) | $\Delta \bar{G}_f^\circ$ (kcal mole⁻¹) | $\Delta \bar{H}_f^\circ$ (kcal mole⁻¹) |
|---|---|---|---|---|
| $SiF_4$ | −77 | —— | −375.88 | −385.98 |
| $SiCl_4$ | −70 | 57.6 | −147.47 | −157.03 |
| $SiBr_4$ | 5 | 153 | −103.2 | − 99.3 |
| $SiI_4$ | 120 | 290 | —— | —— |

$\Delta \bar{G}_f^\circ$ and $\Delta \bar{H}_f^\circ$ refer to the gas at 298 K.

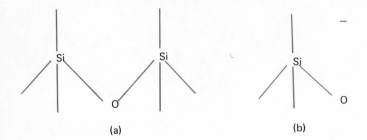

(a)                                              (b)

**FIGURE 17.7** The two types of silicon-oxygen bonds.

alkali content is high, the resulting silicates are water soluble, but with low alkali content the silicates obtained are insoluble in water, because of the formation of large polymeric anions. The ubiquitous silicates are to be found in pottery, cement, bricks, tile, porcelain, glass, vitreous enamel, asbestos, mica, and clay.

In contrast to carbon, silicon does not form double bonds to oxygen. There are two types of silicon-to-oxygen bonding, one of which involves an Si—O—Si linkage, and the other of which involves the Si—O bond (Figure 17.7). The simplest silicate anion is the tetrahedral *orthosilicate* ion, $SiO_4^{2-}$, which is found in the minerals *zircon*, $ZrSiO_4$, and *willemite*, $Zn_2SiO_4$, and in *water glass*, $Na_2SiO_4$(aq). The pyrosilicate ion, $Si_2O_7^{6-}$, has the structure shown in Figure 17.8a. The *pyroxenes* are silicates that involve long, straight-chain-silicate polyanions in which the *meta-silicate unit*, $SiO_3^{2-}$ (Figure 17.8b), is the chain segment. The minerals $MgSiO_3$ and $LiAl(SiO_3)_2$ (*spodumene*) are pyroxenes. In addition to linear chains, both cross-linked and cyclic-chain silicate ions are known. For example, the mineral *beryl*, $Be_3Al_2Si_6O_{18}$, contains the cyclic anion $Si_6O_{18}^{12-}$ (Figure 17.8c). In addition to simple chain and cyclic silicate ions, there are double-stranded, cross-linked silicate chains of composition $(Si_4O_{11}^{6-})_n$ (the *amphiboles*, Figure 17.8d). Asbestos is an amphibole; the fibrous nature of asbestos is a consequence of the underlying silicate chain structure.

The minerals mica and talc contain "infinite," hexagonal-structured, two-dimensional sheets of composition $(Si_2O_5^{2-})$ (Figure 17.8e). The ease with which micas can be fractured into thin sheets is a direct consequence of the silicate sheet structure. The layer separation is typically about 20 Å, with the layers being held together by small divalent ions such as $Mg^{2+}$. The compound β-quartz, $SiO_2$, has an infinite, *three-dimensional*, cross-linked structure in which each oxygen atom is bonded to two silicon atoms (Figure 7.11).

The so-called *framework minerals* contain $SiO_4$ and $AlO_4$ tetrahedra, with each oxygen shared between two tetrahedra. Feldspars, zeolites, and ultramarines are examples of aluminosilicates of this type. These minerals contain cavities of various sizes into which small molecules of appropriate sizes can penetrate, and also possess the property of being able to act as cation exchangers, owing to the presence of loosely-held cations located in certain cavities.

Zeolites have the general composition $M_{x/n}(AlO_2)_x(SiO_2)_y \cdot zH_2O$, where $M^{n+}$ can be $Na^+$, $K^+$, or $Ca^{2+}$, and $z$ is highly variable. Note that the number of cations is independent of $y$; this is because silicon is Si(IV), whereas aluminum is Al(III). *Molecular sieves* are synthetic zeolites, the most common variety of which has the composition $Na_{12}(AlO_2)_{12}(SiO_2)_{12} \cdot 27H_2O$. The compound can be dehydrated by heating and then used in the anhydrous state to remove traces of water from organic solvents such as hexane and ethers. Absolute ethanol can be obtained from 95% ethanol by treating the latter with molecular sieves. Because of the specific spacings in their structure, molecular sieves can also be used to separate straight-chain from branched-chain hydrocarbons.

## d. Glasses and Some Other Useful Silicon-Containing Substances.

The oxides of silicon, germanium, and boron ($SiO_2$, $GeO_2$, and $B_2O_3$) can easily be obtained in the *vitreous* (glassy) state by simply melting the crystalline oxides and chilling the melts. All three oxides can be fused with each other and with the oxides $Li_2O$, $Na_2O$, $K_2O$, $CaO$, $MgO$, $BaO$, $ZnO$, and $Al_2O_3$, as well as certain other substances, to make various types of *glasses*. Glasses have no long-range order, but rather consist of a dis-ordered array of polymeric chains, sheets, or three-dimensional units. Although glasses need not contain any silicon, the most common glasses are silicate glasses. The composition of a silicate glass is usually expressed in parts per hundred of $SiO_2$, as in the following examples:

|  | CaO | Na$_2$O | K$_2$O | Al$_2$O$_3$ | B$_2$O$_3$ |
|---|---|---|---|---|---|
| window glass | 10.5 | 12.5 | 3.8 | 0.3 | 0 |
| laboratory glass | 0 | 4.6 | 0.5 | 1.5 | 12.6 |

Laboratory glassware has a much lower coefficient of thermal expansion than window glass and, as a consequence, is better able to withstand thermal shock. The dissolution of glass in hydrofluoric acid is a conse-quence of the stability of the $SiF_6^{2-}$ ion:

$$SiO_2(s) + 6HF(aq) \rightarrow H_2SiF_6(s) + 4H_2O(\ell)$$

Glass is not soluble in the other halogen acids.

The following are some other silicon-containing substances of con-siderable technological importance:

(1) *Porcelain* is heterogeneous and is much richer in $Al_2O_3$ than are the glasses described above. It has a higher strength (owing to its heterogeneity), and has more chemical resistance than glass.

(2) *Earthenware* is a more porous material than porcelain, primarily because it is fired at a lower temperature.

(3) *Portland cement* is rich in $CaO$, $MgO$, and $Al_2O_3$, as well as $SiO_2$, and is obtained by firing a mixture of limestone, dolomite, and silicate clay. The resulting anhydrous material is then ground to a powder. Crystalline hydrates are formed upon addition of water.

(4) *Silicones* are polymeric organosilicon oxides with the basic com-position $R_2SiO$ (R is a hydrocarbon radical) (Figure 17.9). The methyl silicones are obtained via the hydrolysis of $(CH_3)_2SiCl_2$ to $\{(CH_3)_2SiO\}_4$, which is in turn heated with a trace of KOH to yield straight-chained, elastomeric silicone polymers. Other forms of methyl silicones are widely used as lubricating agents, and also to provide water repellency, although they will allow the passage of air and water vapor.

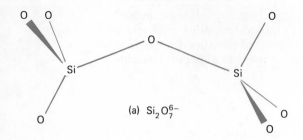

(a) $Si_2O_7^{6-}$

(b) the pyroxenes $(SiO_3^{2-})_n$

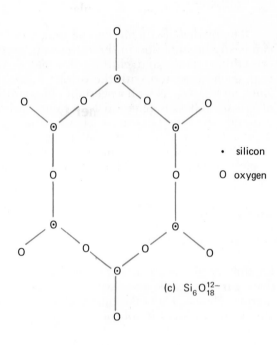

• silicon

O oxygen

(c) $Si_6O_{18}^{12-}$

← Cross links →

• Silicon

O Oxygen

(d) $(Si_4O_{11}^{6-})_n$

Double-stranded,
cross-linked chain

**FIGURE 17.8** Structures of some silicate ions. Note that an O over a dot represents an oxygen atom *above* a silicon atom (dot).

(*Figure 17.8 continued on facing page.*)

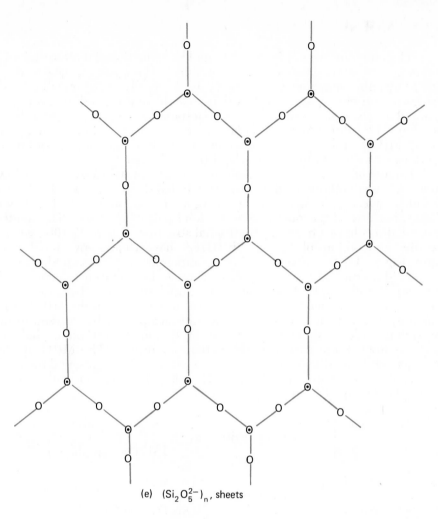

(e)  $(Si_2O_5^{2-})_n$, sheets

**FIGURE 17.8**  (continued)

(5) *Silicone rubber* is obtained from the straight-chained silicones by adding powered silicon dioxide as filler and heating with a trace of oxidizing agent, which produces cross-linkages of the type —Si—CH$_2$—CH$_2$—Si—. The principal advantages of silicone rubber are its wide operating range ($-80°C$ to $+400°C$) and superior resistance to oxygen and other reagents.

**FIGURE 17.9**  A straight-chain silicone.

## 17.4  ARSENIC AND ANTIMONY.

The elements As and Sb are obtained by reduction of their oxides with hydrogen or carbon at elevated temperatures. The elements have a silver-grey, metallic appearance. Antimony is decidedly more metallic in its physical and chemical properties than arsenic, and is used in several important alloys. Arsenic sublimes to the tetrahedral $As_4$ species; $Sb_4$ also is known. The most important oxidation states of arsenic and antimony are $M(0)$, $M(III)$, and $M(V)$. The *free* ions $M^{3+}$ and $M^{5+}$ are of little or no importance, as such, in the chemistry of these elements.

Elemental As and Sb react directly with the halogens to form low-boiling, covalently-bonded tri- and penta-halides (Table 17.5). The trihalides have a pyrimidal structure (Figure 17.10a) and can act as Lewis bases, because of the presence of the lone pair of electrons. The neutral pentahalides have a trigonal-bipyrimidal structure (Figure 17.10b), whereas the halide anions of the type $Sb(III)F_5^{2-}$ have a square-pyrimidal structure (Figure 17.10c). Five-coordinate compounds of $As(V)$ and $Sb(V)$ have trigonal-bipyrimidal type structures, whereas five-coordinate compounds of $As(III)$ and $Sb(III)$ have square-pyrimidal structures, because of the necessity of accommodating the extra pair of electrons in the $M(III)$ compounds. The arsenic and antimony halides are readily hydrolyzed to the hydrated oxides, although in the case of antimony insoluble oxyhalides, such as SbOCl, are formed under certain conditions. The $M(III)$ oxides can also be formed by reaction of oxygen with the elements, and the $M(V)$ oxides can be obtained from the $M(III)$ oxides by oxidation with concentrated $HNO_3(aq)$. The oxides have the compositions

$$\left.\begin{array}{r}As_4O_6\\[4pt]Sb_4O_6\end{array}\right\}\text{—sometimes written as} \rightarrow \left\{\begin{array}{ll}As_2O_3 & \text{acidic}\\[4pt]Sb_2O_3 & \text{amphoteric}\end{array}\right.$$

$$\left.\begin{array}{r}As_4O_{10}\\[4pt]Sb_4O_{10}\end{array}\right\}\text{—often written as} \rightarrow \left\{\begin{array}{ll}As_2O_5 & \text{acidic}\\[4pt]Sb_2O_5 & \text{acidic}\end{array}\right.$$

and the structures are indicated in Figure 17.11. The acidic oxides $Sb_2O_5$, $As_2O_5$, and $As_2O_3$ react with bases such as NaOH(aq) to yield antimonates

TABLE 17.5

HALIDES OF ARSENIC AND ANTIMONY

| Compound | m. pt. (°C) | b. pt. (°C) | $\Delta \bar{G}_f^\circ$(kcal · mole$^{-1}$) | $\Delta \bar{H}_f^\circ$(kcal · mole$^{-1}$) |
|---|---|---|---|---|
| $AsF_3$ | −8.5 | 63 | −216.46 | −220.04 |
| $AsF_5$ | −80 | −53 | —— | —— |
| $AsCl_3$ | −18 | 130 | −58.77 | −61.80 |
| $AsBr_3$ | 32.8 | 221 | −38. | −31. |
| $SbF_3$ | 292 | sublimes | —— | —— |
| $SbF_5$ | 7.0 | 149 | —— | —— |
| $SbCl_3$ | 73.4 | 223 | −72.0 | −75.0 |
| $SbCl_5$ | 2.8 | 140 | −79.91 | −94.25 |
| $SbBr_3$ | 96.6 | 280 | −53.5 | −46.5 |

The quantities $\Delta \bar{G}_f^\circ$ and $\Delta \bar{H}_f^\circ$ refer to the gaseous compounds at 25°C.

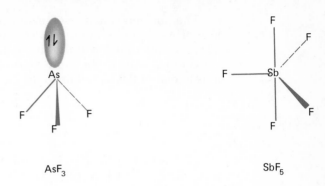

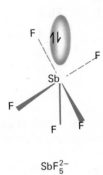

**FIGURE 17.10** Structures of some arsenic and antimony halides.

$(SbO_4^{3-})$, arsenates $(AsO_4^{3-})$, and arsenites $(AsO_3^{3-})$, respectively. The oxide $Sb_2O_3$ is amphoteric:

$$Sb_2O_3(s) + 6NaOH(aq) \longrightarrow 2Na_3SbO_3(aq) + 3H_2O(\ell)$$
$$Sb_2O_3(s) + 3H_2SO_4(aq) \longrightarrow Sb_2(SO_4)_3(s) + 3H_2O(\ell)$$

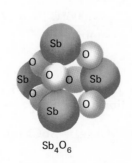
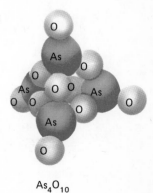

**FIGURE 17.11** Structures of the oxides $Sb_4O_6$ and $As_4O_{10}$.

The poisonous nature of arsenic compounds is well known, the lethal dose of $As_2O_3$ being approximately 0.1 gram. However, small amounts of arsenic compounds have the interesting property of promoting the growth of red blood cells in the bone marrow, and the human body normally contains a total of 0.007 g of arsenic.

## PROBLEMS

1. The boron halides have the following relative strengths as Lewis acids: $BBr_3 \gg BCl_3 > BF_3$. What would you predict for the relative order on steric and electronegativity grounds?

2. Explain why borax is used in several commercially available soaps.

3. Propose a structure for the ion $H_4B_4O_9^{2-}$. (Hint: the structure has two 4-coordinate boron atoms and two 3-coordinate boron atoms.)

4. Identify the $Si_4O_{11}^{6-}$ and $Si_2O_5^{2-}$ units in Figures 17.8d and 17.8e, respectively.

5. Draw the structures of the following species:

(a) BN(cubic)
(b) $\beta$-quartz
(c) $\{(CH_3)_2SiO\}_4$
(d) $As_4$
(e) $H_2Te$

(f) $AsCl_3$
(g) $SbCl_5$
(h) $As_2S_3$
(i) $Sb_2S_5$
(j) $H_4B_3O_7^-$

6. Suggest an explanation for the fact that heterogeneity often confers additional strength to materials.

7. Complete and balance the following reactions:

(a) $C_6H_5Cl + Si \xrightarrow{\text{heat}}$

(b) $Sb + O_2 \rightarrow$

(c) $NH_3 + BF_3 \rightarrow$

(d) $(CH_3)_3N + B_2H_6 \rightarrow$

(e) $BF_3 + H_2O \rightarrow$

(f) $BI_3 \xrightarrow{\text{heat}}$

(g) $Si + Cl_2 \rightarrow$

(h) $GeO_2 + C \xrightarrow[\text{(air free)}]{3000°C}$

(i) $SiCl_4 + Mg \xrightarrow{\text{heat}}$

(j) $LiAlH_4 + SiCl_4 \rightarrow$

(k) $SiH_4 + H_2O \rightarrow$

(l) $(CH_3)_2SiCl_2 + H_2O \xrightarrow[\text{KOH}]{\text{heat}}$

(m) $SbF_3 + F_2 \rightarrow$

(n) $SbF_3 + NaF \rightarrow$

(o) $Sb_2O_5 + KOH \rightarrow$

(p) $SbH_3 \xrightarrow{\text{heat}}$

(q) $Zn(s) + H_3AsO_4(aq) \rightarrow$

8. Using the data in Table 17.5, compute the equilibrium constants at 25°C for the reactions

$$SbCl_5(g) = SbCl_3(g) + Cl_2(g)$$

$$\text{As}(s) + \frac{3}{2}\,\text{F}_2(g) = \text{AsF}_3(g)$$

$$\text{AsCl}_3(g) + \frac{3}{2}\,\text{F}_2(g) = \text{AsF}_3(g) + \frac{3}{2}\,\text{Cl}_2(g)$$

9. The non-volatile oxide $Sb_2O_3$ decreases the fire hazard of organic plastics by extinguishing the flame from an ignited piece of plastic. However, $Sb_2O_3$ functions this way only when chlorinated hydrocarbons of high chlorine content also are incorporated into the plastic. Can you offer an explanation for this behavior?

_____ References

17.1. E. G. Rochow, *The Metalloids*, (D. C. Heath and Co., Boston, 1966).

17.2. F. A. Cotton and G. Wilkinson, *Advanced Inorganic Chemistry*, 2nd edition (Interscience Publishers, John Wiley and Sons, New York, 1966).

# CHAPTER 18

# THE NON-METALS

## 18.1  NITROGEN (GROUP VA).

### a.  General Aspects.

The ground-state electronic configuration of $N(g)$ is $1s^2 2s^2 2p^3$ (Table 18.1). Nitrogen has only four valence orbitals ($2s, 2p_x, 2p_y$ and $2p_z$), and, as a consequence, the maximum number of bonds that can be formed to nitrogen is four. Multiple bonding to nitrogen is quite common, and this factor, coupled with the wide variety of possible oxidation states (nine in all), makes the chemistry of nitrogen extensive and complex.

Although compounds of nitrogen in all nine possible formal oxidation states from $-3$ to $+5$ (or better, $-III$ to $V$) are known (see Table 18.2), the bonding to nitrogen is usually covalent, and, with the exception of the *nitride ion*, $N^{3-}$, nitrogen ions are of little or no importance as such in nitrogen chemistry.

The thermodynamic properties at 298 K of some common nitrogen compounds are given in Table 18.3.

Under ordinary conditions, elemental nitrogen occurs as a diatomic gas. Nitrogen is the principal component of air (78 mole per cent). Diatomic nitrogen is an inert species † because of the very high $N \equiv N$ bond energy:

$$N_2(g) = 2N(g) \qquad \Delta H^\circ_{298} = 226 \text{ kcal}$$

---

*Nobel Lectures, Chemistry (1901–1921) (Elsevier Publishing Co., Amsterdam, 1966) pp. 326–327.

† This is to be contrasted with the highly reactive species $N(g)$ (*atomic nitrogen*) produced when an electric discharge is passed through $N_2(g)$.

**TABLE 18.1**

| element | atomic number | atomic mass | principal isotopes (% natural abundance) [half-life] | ground state electronic configuration of monatomic gaseous element (eV) | electronegativity | melting point (°C) | boiling point (°C) |
|---|---|---|---|---|---|---|---|
| N nitrogen | 7 | 14.0067 | 14(99.63) 15(0.37) 13[10m] | $1s^2 2s^2 2p^3$ | 3.0 | (of $N_2$) $-210.0$ | (of $N_2$) $-195.8$ |
| P phosphorus | 15 | 30.9738 | 31(100) | $(Ne)3s^2 3p^3$ | 2.2 | —— (see text) | —— |
| O oxygen | 8 | 15.9994 | 16(99.759) 17(0.037) 18(0.204) 15[2.03m] | $1s^2 2s^2 2p^4$ | 3.4 | (of $O_2$) $-218.4$ | (of $O_2$) $-183.0$ |
| S sulfur | 10 | 32.06 | 32(95.0) 33(0.76) 34(4.22) 36(0.014) 35[86.7d] | $(Ne)3s^2 3p^4$ | 2.6 | 119 (mono-clinic) | 444.6 |

**TABLE 18.2**

THE FORMAL OXIDATION STATES OF NITROGEN

| OXIDATION STATE | COMPOUND(S) |
|---|---|
| −III | $NH_3$, $Li_3N$, $NH_4Cl$, $NaNH_2$, $Li_2NH$ |
| −II | $N_2H_4$ |
| −I | $H_2NOH$, $NH_2Cl$ |
| 0 | $N_2$ |
| I | $N_2O$, $NHF_2$, $N_2F_2$ |
| II | $NO$, $N_2F_4$ |
| III | $HNO_2$, $NF_3$, $NOCl$, $N_2O_3$, $NOClO_4$ |
| IV | $NO_2$ |
| V | $N_2O_5$, $HNO_3$ |

TABLE 18.3

THERMODYNAMIC PROPERTIES OF NITROGEN COMPOUNDS AT 298 K

(excerpted from *Selected Values of Chemical Thermodynamic Properties*, by D. D. Wagman, W. H. Evans, V. B. Parker, I. Halow, S. M. Bailey, and R. H. Schumm, N. B. S. Technical Note 270-3, 1968, U.S. Government Printing Office) ($\Delta \bar{G}_f^\circ$ and $\Delta \bar{H}_f^\circ$ in kcal mole$^{-1}$, $\bar{S}^\circ$ in gibbs mole$^{-1}$; melting and boiling points (in °C) gathered from several literature sources).

| SPECIES | $\Delta \bar{H}_f^\circ$ | $\Delta \bar{G}_f^\circ$ | $\bar{S}^\circ$ | m.pt. | b.pt. | NAME |
|---|---|---|---|---|---|---|
| $N_2(g)$ | 0 | 0 | 45.77 | −210.0 | −195.8 | nitrogen |
| $NO(g)$ | 21.57 | 20.69 | 50.35 | − | − | nitric oxide |
| $NO_2(g)$ | 7.93 | 12.26 | 57.35 | − | − | nitrogen dioxide |
| $NO_2^-(aq)$ | −25.0 | −8.9 | 33.5 | − | − | nitrite ion |
| $NO_3^-(aq)$ | −49.56 | −26.61 | 35.0 | − | − | nitrate ion |
| $N_2O(g)$ | 19.61 | 24.90 | 52.52 | −98.8 | −88.5 | nitrous oxide |
| $N_2O_3(g)$ | 20.01 | 33.32 | 74.61 | −102 | 3.5 | dinitrogen trioxide |
| $N_2O_4(g)$ | 2.19 | 23.38 | 50.0 | −11 | 21.3 | dinitrogen tetroxide |
| $N_2O_5(g)$ | 2.7 | 27.5 | 85.0 | 30 | 47 | dinitrogen pentoxide |
| $NH_3(g)$ | −11.02 | −3.94 | 45.97 | −33.35 | −77.8 | ammonia |
| $NH_4^+(aq)$ | −31.67 | −18.97 | 27.1 | − | − | ammonium ion |
| $N_2H_4(g)$ | 22.80 | 38.07 | 56.97 | 2 | 114 | hydrazine |
| $HN_3(aq)$ | 62.16 | 76.9 | 34.9 | − | − | hydrazoic acid |
| $HNO_2(aq)$ | −28.5 | −13.3 | 36.5 | − | − | nitrous acid |
| $NF_3(g)$ | −29.8 | −19.9 | 62.29 | − | −129 | nitrogen trifluoride |
| $NOCl(g)$ | 12.36 | 15.79 | 65.52 | −62 | −6 | nitrosyl chloride |

Nitrogen will react directly with highly electropositive elements, such as Li and Mg, to form nitrides ($Li_3N$ and $Mg_3N_2$).

Nitrogen is converted to nitrogen compounds (fixed) by root-nodule bacteria of certain plants (notably legumes). The chemistry of this complex biochemical process, which has been reported to involve a molybdenum-nitrogen complex within the enzyme *nitrogenase*, is not well understood at present.

The industrial fixation of nitrogen is accomplished by a combination of the Haber and Ostwald processes. In the Haber process nitrogen is converted to ammonia by reaction with hydrogen at elevated temperatures (400 to 550°C) and pressures ($10^2$ to $10^3$ atm) using an $Fe/Fe_2O_3$ catalyst:

$$N_2(g) + 3H_2(g) = 2NH_3(g) \qquad \Delta G_{298}^\circ = -3.94 \text{ kcal}; \ \Delta G_{800}^\circ = +16.7 \text{ kcal}$$

Because of the negative value of $\Delta H^\circ$ for the reaction ($\Delta H_{298}^\circ = -11.0$ kcal), the equilibrium lies farther to the right at lower temperatures than at higher temperatures. However, the reaction rate is too slow at, say, 25°C to be economically feasible. Consequently, the process is carried out at 400 to 550°C, in spite of the lower yield. High pressures are utilized to increase the *percentage* conversion to ammonia.

In the Ostwald process, ammonia is converted to nitric acid. The ammonia is first converted to nitric oxide by reaction with oxygen in the presence of a Pt/Rh catalyst at 750–900°C:

$$4NH_3(g) + 5O_2(g) = 4NO(g) + 6H_2O(g)$$
$$\Delta G_{298}^\circ = -229.3; \ \Delta H_{298}^\circ = -216.4 \text{ kcal}$$

The nitric oxide produced is then reacted with oxygen to yield nitrogen dioxide:

$$2NO(g) + O_2(g) = 2NO_2(g) \qquad \Delta G^\circ_{298} = -16.86 \text{ kcal}$$

The gaseous nitric oxide is then dissolved in water to yield nitric acid:

$$3NO_2(g) + H_2O(\ell) = 2HNO_3(aq) + NO(g) \qquad \Delta G^\circ_{298} = -12.62 \text{ kcal}$$

## b.  Nitrogen Hydrides.

Ammonia, $NH_3$, has a pyramidal structure (Figure 18.1(a)). Ammonia and alkyl amines (such as $CH_3NH_2$, $(CH_3)_2NH$ and $(CH_3)_3N$), which also have a pyrimidal structure, are Lewis bases, owing to the presence of the lone pair of electrons on the nitrogen. The ammonium ion, $NH_4^+$, has a tetrahedral structure (Figure 18.1(b)). In aqueous solution the ammonium ion is a weak Brönsted acid,

$$NH_4^+(aq) = H^+(aq) + NH_3(aq) \qquad pK_a = 9.25 \ (25°C)$$

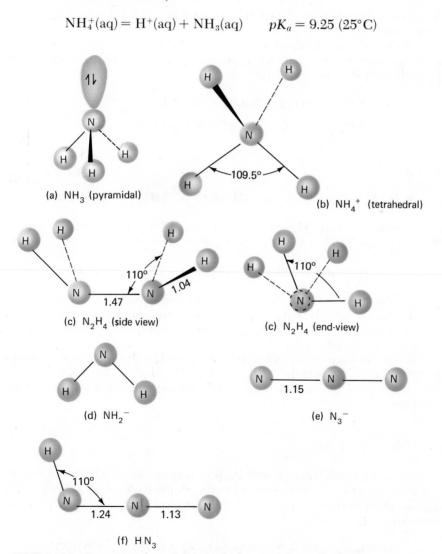

(a) $NH_3$ (pyramidal)

(b) $NH_4^+$ (tetrahedral)

(c) $N_2H_4$ (side view)

(c) $N_2H_4$ (end-view)

(d) $NH_2^-$

(e) $N_3^-$

(f) $HN_3$

**FIGURE 18.1**  Structures of some nitrogen-hydrogen compounds (distances in angstrom units).

whereas ammonia is a weak Brönsted base

$$NH_3(aq) + H_2O(\ell) = NH_4^+(aq) + OH^-(aq) \qquad pK_b = 4.75 \ (25°C)$$

The ammonium ion is decomposed in the presence of strong bases:

$$NH_4^+(aq) + OH^-(aq) \rightarrow NH_3(g) + H_2O(\ell)$$

Various ammonium salts decompose on heating; for example:

$$NH_4Cl(s) = NH_3(g) + HCl(g)$$

liquid ammonium nitrate and aqueous solutions of ammonium nitrate decompose on heating to nitrous oxide and water

$$NH_4NO_3(aq) \rightarrow N_2O(g) + 2H_2O(\ell)$$

Ammonium perchlorate, $NH_4ClO_4$, is used as a solid fuel in various types of missiles.

The compound hydrazine, $N_2H_4$, can be prepared by the reaction of aqueous ammonia with sodium hypochlorite in the presence of a sequestering-gel-catalyst (Raschig synthesis):

$$NH_3 + NaOCl = NaOH + NH_2Cl$$

$$NH_2Cl + NH_3 + NaOH = N_2H_4 + NaCl + H_2$$

Hydrazine and its various derivatives, such as $(CH_3)_2NNH_2$, are used as rocket fuels

$$N_2H_4(\ell) + O_2(g) = N_2(g) + 2H_2O(\ell) \qquad \Delta H_{298}^\circ = -149 \text{ kcal}$$

Hydrazoic acid, $HN_3$, can be prepared by the reaction between hydrazine and nitrous acid in acidic aqueous solution:

$$N_2H_5^+(aq) + HNO_2(aq) = HN_3(aq) + H^+(aq) + 2H_2O(\ell)$$

In aqueous solution $HN_3$ is a weak acid, and a very powerful reducing agent. Liquid $HN_3$ is a dangerously explosive, colorless substance. The lead and mercury salts of hydrazoic acid (azides), $Pb(N_3)_2$ and $Hg(N_3)_2$, are used in detonation caps.

Hydroxylamine, $NH_2OH$, can be prepared by the reduction of nitrates or nitrites either electrolytically or with $SO_2$. Hydroxylamine is a weaker base than ammonia, and a fairly good oxidizing agent in acidic aqueous solution.

## c. Nitrogen Oxides.

Nitrogen forms several low-boiling oxides, all of which are thermodynamically unstable with respect to the elements $N_2$ and $O_2$ (Table 18.3). The reaction between copper and concentrated nitric acid yields nitrogen monoxide (nitric oxide):

$$8HNO_3(aq) + 3Cu(s) = 3Cu(NO_3)_2(aq) + 2NO(g) + 4H_2O(\ell)$$

Nitrogen monoxide is a colorless, electron-deficient, paramagnetic gas[*] that is dimeric in the solid state (Figure 18.2(a)). It reacts with oxygen to form the red-brown, electron-deficient nitrogen dioxide (Figure 18.2(c), and the Ostwald process).

In the gas phase, nitrogen dioxide dimerizes to form dinitrogen tetroxide (Figure 18.2(d)):

$$2NO_2(g) = N_2O_4(g) \qquad \Delta G^{\circ}_{298} = -1.14 \text{ kcal}$$

Condensation of the equilibrium mixture of $NO_2$ and $N_2O_4$ yields a liquid that is almost entirely $N_2O_4$. The reaction between NO and $NO_2$ yields dinitrogen trioxide,

$$NO(g) + NO_2(g) = N_2O_3(g) \qquad \Delta G^{\circ}_{298} = +0.37 \text{ kcal}; \ \Delta G^{\circ}_{273} = -0.45 \text{ kcal}$$

the conversion to $N_2O_3$ being more favorable the lower the temperature.

---

[*]See Chapter 4 for a discussion of the bonding in NO and $NO^+$.

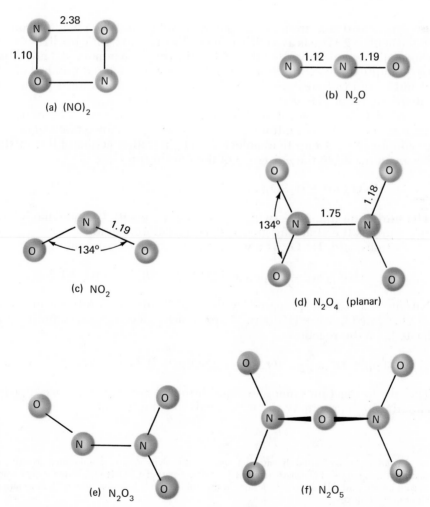

**FIGURE 18.2** Structures of the nitrogen oxides (distances in angstrom units).

The preparation of dinitrogen oxide (nitrous oxide), $N_2O$ (Figure 18.2(b)), via the decomposition of $NH_4NO_3$ has already been mentioned. Nitrous oxide is unique among the nitrogen oxides in that it is non-poisonous. It finds use as a mild anesthetic (laughing gas), and as a propellent in aerosol "whipped cream" cans.

Dinitrogen pentoxide, $N_2O_5$ (Figure 18.2(f), can be prepared via the dehydration of nitric acid,

$$2HNO_3(\ell) + P_2O_5(s) = 2HPO_3(\ell) + N_2O_5(g)$$

and conversely, $N_2O_5$ is the anhydride of nitric acid:

$$N_2O_5(g) + H_2O(\ell) = 2HNO_3(\ell)$$

Solid $N_2O_5$ is an ionic compound, $NO_2^+ NO_3^-$.

## d.  Oxy-acids of Nitrogen.

Nitric acid is a strong acid in aqueous solution. When hot, the concentrated acid ($>2$ M) acts as a fairly powerful and rapid-acting oxidizing agent, especially when mixed with HCl.* However, when cold and dilute, $HNO_3(aq)$ exhibits little, if any, oxidative action. A large number of salts of nitric acid (nitrates) are known, most of which are water soluble. The nitrate ion has a planar structure (Figure 18.3(b)).

A mixture of concentrated nitric and sulfuric acids is widely used in organic chemistry as a nitrating agent (that is, a substance that can be used to add a —$NO_2$ group to a molecule). The nitrating action of this mixture has been traced to the existence of the equilibrium

$$HNO_3(aq) + H_2SO_4(aq) = NO_2^+(aq) + HSO_4^-(aq) + H_2O(\ell)$$

The nitronium ion, $NO_2^+$ (which is isoelectronic with $CO_2$ and has a linear structure), is the attacking group in the nitration reaction.

Nitrous acid, $HNO_2$, is a weak acid in aqueous solution:

$$HNO_2(aq) = H^+(aq) + NO_2^-(aq) \qquad K_{298} = 6.0 \times 10^{-6}$$

Nitrous acid can be prepared by the acidification of a nitrite salt, for example, $NaNO_2(aq) + HCl(aq)$. Nitrous acid decomposes rapidly when heated, via the reaction

$$3HNO_2(aq) = H^+(aq) + NO_3^-(aq) + 2NO(g) + H_2O(\ell) \qquad \Delta G_{298}^\circ = -3.2 \text{ kcal}$$

The above reaction comes to equilibrium if the $HNO_2(aq)$ is kept in a closed container; otherwise, it proceeds to completion.

---

*Aqua regia (so called because it readily dissolves noble metals such as gold and platinum) is composed of 3 parts of 12 M HCl plus 1 part of 12 M $HNO_3$. Aqua regia is known to contain $Cl_2$ and ClNO, and its action on the noble metals is in part due to the formation of chloride complexes.

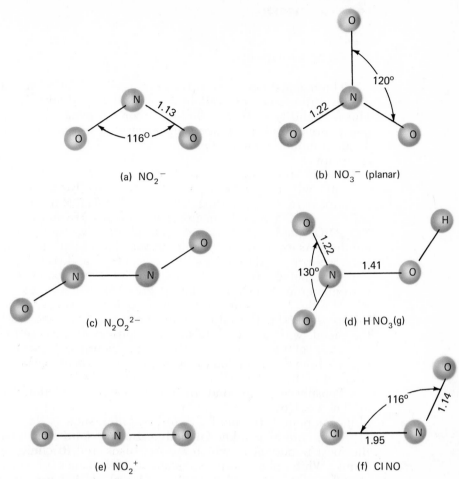

**FIGURE 18.3** Structures of some oxyanions and oxyacids of nitrogen (distances in angstrom units).

## e. Nitrogen Halides.

Nitrogen forms a variety of halogen compounds, all of which are poisonous, and many of which are dangerously explosive. Some examples are: $H_2NCl$ (chloramine), $HNCl_2$ (dichloramine), $NCl_3$ (trichloramine), $NF_3$, $N_2F_2$, $N_2F_4$, and $NFCl_2$. The reaction between NO and a halogen yields nitrosyl halides, XNO

$$2NO(g) + Cl_2(g) = 2ClNO(g)$$

The nitrosyl halides (see Figure 18.3(f)) are reactive compounds that act as powerful oxidizing agents. Their aqueous solutions behave like mixtures of $HNO_2$ and HX, owing to the occurrence of the decomposition reactions

$$XNO(aq) + H_2O(\ell) = HNO_2(aq) + H^+(aq) + X^-(aq)$$

For example, when X = Cl, $\Delta G_{298}^\circ = -38$ kcal for this reaction.

## 18.2  PHOSPHORUS (GROUP VA).

### a.    General Aspects.

The major differences in the chemistries of nitrogen and phosphorus can be attributed to the availability of low-lying $3d$-orbitals on phosphorus. The availability of these $3d$-orbitals for participation in the bonding to phosphorus permits valence shell expansion beyond the octet, and coordination numbers higher than four are found in a number of phosphorus compounds.

Phosphorus readily forms $d_\pi(P) + p_\pi(X)$ bonds of weak to moderate strength, whereas nitrogen, which has no low-lying $d$-orbitals, does not participate in such bonding. On the other hand, nitrogen forms strong $p_\pi(N) + p_\pi(X)$ bonds in many of its compounds and ions (for example, $NO_2$, $NO_2^-$, and $NO_3^-$), whereas phosphorus does not participate in such bonding. As examples of the importance of $d$-orbitals in phosphorus chemistry, we can cite the species $PF_5$, $PF_6^-$, $PO_4^{3-}$, and $HPO_3^{2-}$, which have no nitrogen analogs. In addition, $PX_3$ donors (Lewis bases) form complexes with numerous transition metals in lower oxidation states. The stabilizing effect of the $PX_3$ donors on the lower oxidation states of transition metal ions is ascribed to the ability of phosphorus to form both $\sigma$- and $\pi$-bonds to the metal, with the $d$-orbitals on phosphorus (which are involved in the formation of the $\pi$-bonds to the metal) acting as acceptors of electron density from the metal ion (*back donation*, or $\pi$-*acceptor* bonding), thereby stabilizing the lower oxidation state of the metal.

Phosphorus compounds in all nine possible oxidation states ($-$III to V) are known. However, phosphorus, like nitrogen, is covalently bound in all of its compounds (except for the *phosphides*, such as $Na_3P$).

There are eleven known modifications of solid elemental phosphorus, the most important of which are the black, red (triclinic), and white ($\alpha$) forms. White phosphorus spontaneously inflames in air, and must be stored under water, whereas the red and black forms are stable in air. White phosphorus is unstable with respect to red phosphorus at 25°C and 1 atm. Black phosphorus is the least reactive of all the known phosphorus allotropes. When white phosphorus is heated at about 400°C for several hours, red phosphorus is obtained, whereas when white phosphorus is heated to 220 to 370°C for about eight days in the presence of a mercury catalyst, black phosphorus is obtained. Black phosphorus also can be obtained at high pressures from red or white phosphorus. Vaporization of phosphorus yields $P_4(g)$, which has a tetrahedral structure with a P-P bond distance of 2.21 Å. At higher temperatures, $P_4(g)$ dissociates into $P_2(g)$.

Phosphorus does not occur in nature as the free element, but it is widely distributed in various minerals. The most important mineral source of phosphorus is fluorapatite, $Ca_{10}F_2(PO_4)_6$. Phosphorus is obtained from phosphate rock (as $P_4(g)$) by roasting the crushed rock with silica and carbon.

### b.    Phosphorus Compounds.

The thermodynamic properties of some common phosphorus compounds are given in Table 18.4. Rapid reversible equilibrium is seldom attainable in redox reactions involving phosphorus or its compounds. Consequently, $\Delta \bar{G}_f^\circ$ data cannot be obtained from equilibrium studies on such reactions.

**TABLE 18.4**

THERMODYNAMIC PROPERTIES OF SOME PHOSPHORUS COMPOUNDS AT 298 K
(see Table 18.3 for reference; values of $\Delta \bar{G}_f^\circ$ and $\Delta \bar{H}_f^\circ$ in kcal·mole$^{-1}$; $\bar{S}^\circ$ values in gibbs·mole$^{-1}$).

| SUBSTANCE | $\Delta \bar{H}_f^\circ$ | $\Delta \bar{G}_f^\circ$ | $\bar{S}^\circ$ | m.pt.(°C) | b.pt.(°C) | NAME |
|---|---|---|---|---|---|---|
| P($\alpha$, white) | 0 | 0 | 9.82 | 44.1 | 280 | white phosphorus |
| P(red, triclinic) | −4.2 | −2.9 | 5.45 | − | 416(sub) | red phosphorus |
| P(g) | 75.20 | 66.51 | 38.978 | − | − | − |
| P$_4$(g) | 14.08 | 5.85 | 66.89 | − | − | − |
| PO$_4^{3-}$(aq) | −305.3 | −243.5 | −53. | − | − | (ortho)phosphate ion |
| P$_2$O$_7^{4-}$(aq) | −542.8 | −458.7 | −28. | − | − | pyrophosphate ion |
| P$_4$O$_6$(s) | −392.0 | − | − | 23.8 | 175.4 | phosphorus trioxide |
| P$_4$O$_{10}$ (s, hexagonal) | −713.2 | −644.8 | 54.70 | − | 360(sub) | phosphorus pentoxide |
| PH$_3$(g) | 1.3 | 3.2 | 50.22 | −133. | −87.7 | phosphine |
| HPO$_4^{2-}$(aq) | −308.83 | −260.34 | −8.0 | − | − | monohydrogen phosphate ion |
| H$_2$PO$_4^-$(aq) | −309.82 | −260.17 | 21.6 | − | − | dihydrogenphosphate ion |
| H$_3$PO$_4$(s) | −305.7 | −267.5 | 26.41 | 42.4 | − | (ortho)phosphoric acid |
| H$_3$PO$_4$(aq, undiss) | −307.92 | −273.10 | 37.8 | − | − | (ortho)phosphoric acid |
| PCl$_3$($\ell$) | −76.4 | −65.1 | 51.9 | −112. | 76 | phosphorus trichloride |
| PCl$_3$(g) | −68.6 | −64.0 | 74.49 | −112. | 76 | phosphorus trichloride |
| PCl$_5$(g) | −89.6 | −72.9 | 87.11 | − | 162(sub) | phosphorus pentachloride |
| POCl$_3$(g) | −133.48 | −122.60 | 77.76 | 2 | 105.3 | phosphorus oxychloride |

Phosphorus reacts directly with electropositive metals such as sodium to form phosphides. Hydrolysis of phosphides yields phosphine:

$$Na_3P(s) + 3H_2O(\ell) = PH_3(g) + 3NaOH(aq)$$

Phosphine is not at all basic, and differs markedly in this respect from NH$_3$. Like AsH$_3$, PH$_3$ is extremely poisonous. Phosphine has a pyramidal structure with an H—P—H angle of 93.7°.

Phosphorus reacts directly with the halogens to form halides with the compositions PX$_3$ and PX$_5$; with excess phosphorus PX$_3$ is formed, whereas with excess halogen PX$_5$ is formed. Halide ions of the type PX$_4^+$ and PX$_6^-$ are also known. In the gas phase, equilibria of the type

$$PCl_5(g) = PCl_3(g) + Cl_2(g) \qquad \Delta G_{298}^\circ = +8.9 \text{ kcal}; \ \Delta H_{298}^\circ = +21.0 \text{ kcal}$$

become important at elevated temperatures. Structures of the phosphorus halides, together with a simplified description of the bonding in each case, are given in Table 18.5.

In the *solid* state, PCl$_5$ is actually (PCl$_4^+$)(PCl$_6^-$), whereas PBr$_5$ is actually (PBr$_4^+$)Br$^-$, the difference presumably being due to the larger size of bromine, which makes PBr$_6^-$ unstable. The phosphorus trihalides are Lewis bases, whereas the pentahalides are strong Lewis acids:

$$PCl_3 + BCl_3 \rightarrow Cl_3PBCl_3$$

$$PF_5 + F^- \rightarrow PF_6^-$$

TABLE 18.5

| Species | Structure | Bonding |
|---------|-----------|---------|
| $PCl_3$ | pyramid | $p^3$ to $sp^3$ with a lone-pair on P |
| $PCl_4^+$ | tetrahedron | $sp^3$ |
| $PCl_5$ | trigonal-bipyramid | $sp^3d$ |
| $PCl_6^-$ | octahedron | $sp^3d^2$ |
| $OPCl_3$ | distorted tetrahedron | $sp^3 + \{p_\pi(O) + d_\pi(P)\}$ |
| $OP(OH)_3$ | distorted tetrahedron | $sp^3 + \{p_\pi(O) + d_\pi(P)\}$ |

The trihalides react directly with oxygen to yield phosphorus oxyhalides; for example:

$$2PCl_3 + O_2 \rightarrow 2Cl_3PO$$

The phosphorus halides and oxyhalides are rapidly hydrolyzed by water:

$$PCl_3 + 3H_2O = H_3PO_3 + 3HCl$$

$$PCl_5 + 4H_2O = H_3PO_4 + 5HCl$$

The principal oxides of phosphorus are $P_4O_6$ and $P_4O_{10}$, both of which are acidic. These oxides are formed readily by the reaction of elemental white phosphorus with oxygen, the product being $P_4O_6$ with phosphorus in excess, and $P_4O_{10}$ with oxygen in excess. The structures of $P_4O_6$ and $P_4O_{10}$ are similar to those indicated for $Sb_4O_6$ and $As_4O_{10}$, respectively, in Figure 17.11. In $P_4O_{10}$, the P—O bond length in the P—O—P linkage is 1.62 Å, whereas the P—O bond length in the apical PO groups is only 1.39 Å. The shorter bond length is probably indicative of considerable double-bond character in the apical P—O bonds.

Phosphorus pentoxide (which is really $P_4O_{10}$) is a powdery white solid with a normal sublimation point of 360°C. Below 100°C, $P_4O_{10}$ is one of the most effective drying (or dehydrating) agents known; the equilibrium vapor pressure of water at 25°C over the two-phase system $H_3PO_4(s) + P_4O_{10}(s)$ is less than $2 \times 10^{-17}$ atm. Phosphorus pentoxide is capable of dehydrating $HNO_3$ (to $N_2O_5$) and $H_2SO_4$ (to $SO_3$). This is all the more surprising when it is realized that concentrated sulfuric acid is itself an effective dehydrating agent.

Reaction of the oxides $P_4O_6$ and $P_4O_{10}$ with water produces the phosphorus oxyacids *phosphorous acid* ($H_3PO_3$) and *orthophosphoric acid* ($H_3PO_4$), respectively:

$$P_4O_6(s) + 6H_2O(\ell) \xrightarrow{0°C} 4H_3PO_3(aq)$$

$$P_4O_{10}(s) + 6H_2O(\ell) \xrightarrow{0°C} 4H_3PO_4(aq)$$

Dissolution of $P_4$ in NaOH(aq) yields *hypophosphorous acid* ($H_3PO_2$):

$$P_4(g) + 4NaOH(aq) + 4H_2O(\ell) \rightleftharpoons 4NaH_2PO_2(aq) + 2H_2(g)$$

$$NaH_2PO_2(aq) + HCl(aq) = H_3PO_2(aq) + NaCl(aq)$$

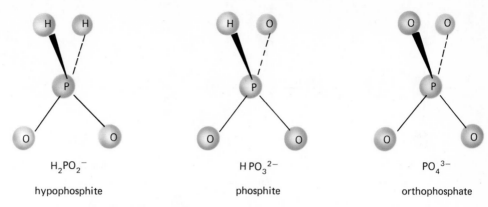

$H_2PO_2^-$

hypophosphite

$HPO_3^{2-}$

phosphite

$PO_4^{3-}$

orthophosphate

**FIGURE 18.4** Structures of some simple phosphorus oxyanions.

All phosphorus oxyacids have P—O—H groups, and some phosphorus oxyacids (such as $H_3PO_3$ and $H_3PO_2$) also have P—H groups. Only the hydrogen bound to oxygen is ionizable. The structures of the completely ionized anions of the phosphorus oxyacids $H_3PO_4$, $H_2(HPO_3)$, and $H(H_2PO_2)$ are given in Figure 18.4. These acids are weak Brönsted acids in aqueous solution, as is indicated by the data in Table 18.6.

In addition to the simple phosphorus oxyanions $PO_4^{3-}$, $HPO_3^{2-}$, and $H_2PO_2^-$, there are a large number of condensed (or poly-) phosphate oxyanions known. The polyphosphate oxyanions involve three distinguishable types of phosphate groups, namely, *chain terminal, chain segment,* and *chain branching* groups (Figure 18.5(a)). The *linear chain* polyphosphates have the general formula $\{P_nO_{3n+1}\}^{(n+2)-}$, the simplest example of which is dipolyphosphate, $P_2O_7^{4-}$ (Figure 18.5(b)). The *cyclic chain* polyphosphates (*metaphosphates*) have the general formula $\{P_nO_{3n}\}^{n-}$, the simplest example of which is trimetaphosphate (Figure 18.5(c)).

Phosphates find wide use in detergents, and also in fertilizers, especially in combination with nitrogen species, for example, $(NH_4)_2HPO_4$. All naturally occurring phosphorus minerals are (ortho)phosphates. Ordinary tooth enamel is *hydroxylapatite*, $Ca_{10}(OH)_2(PO_4)_6$. If fluoride ion (at low concentrations) is incorporated into the diets of children, then a substantial amount of the tooth enamel formed is *fluorapatite*, $Ca_{10}F_2(PO_4)_6(s)$, which is considerably harder and is less soluble in aqueous acids than hydroxylapatite. Consequently, fluorapatite is considerably more resistant to tooth decay than is hydroxylapatite, which

**TABLE 18.6**

$pK_a$ VALUES FOR SOME PHOSPHORUS OXYACIDS IN WATER AT 25°C

| ACID | $pK_{a1}$ | $pK_{a2}$ | $pK_{a3}$ |
|---|---|---|---|
| $H_3PO_4$(aq) | 2.2 | 7.1 | 12.4 |
| $H_2(HPO_3)$ | 1.8 | ? | — |
| $H(H_2PO_2)$ | 1.2 | — | — |

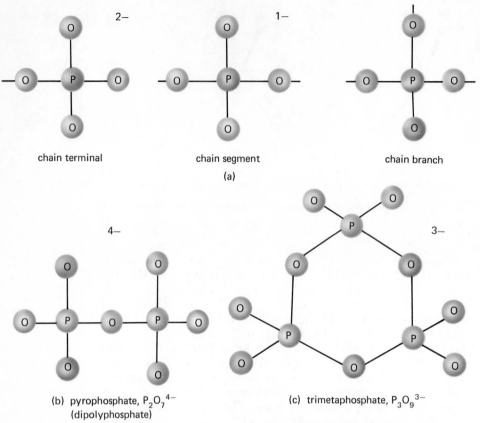

chain terminal          chain segment          chain branch

(a)

(b) pyrophosphate, $P_2O_7^{4-}$
(dipolyphosphate)

(c) trimetaphosphate, $P_3O_9^{3-}$

**FIGURE 18.5** Structures of some simple polyphosphates.

contains a hydroxide ion and hence is more sensitive to the action of food acids.

Synthetic esters of phosphorous acid (such as phosphite triesters, $P(OR)_3$, where R is an "organic" group) and phosphoric acid (such as $OP(OR)_3$) find use as water-soluble and short-lived insecticides. The extremely important life-giving molecules DNA and RNA are polymeric phosphate diesters (see Chapter 21), and the life-taking nerve gases are also esters of phosphoric acid.

## 18.3   OXYGEN (GROUP VIA).

The ground state electronic configuration of O(g) is $1s^2 2s^2 2p^4$. Oxygen has only four valence shell orbitals ($2s$, $2p_x$, $2p_y$, and $2p_z$) available for bond formation, and, as a consequence, the maximum number of bonds that can be formed to oxygen is four. However, because the oxygen atom is only two electrons short of the inert-gas configuration, the most commonly encountered number of bonds to oxygen is two (either two single bonds, or one double bond). Some examples of the types of bonding to oxygen are given in Figure 18.6.

Oxygen and its compounds make up over 49% of the weight of the earth's crust, and about 90% of the weight of the oceans is oxygen. Air is 21% (volume) $O_2$, and enormous quantities of oxygen are produced daily

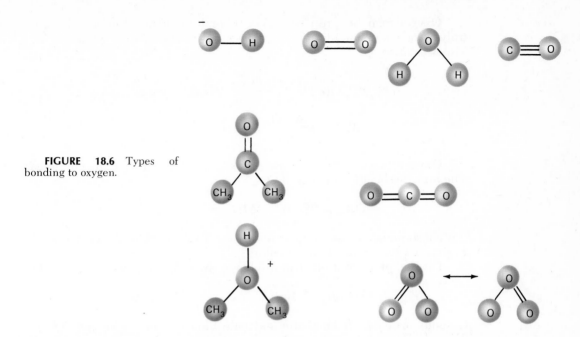

**FIGURE 18.6** Types of bonding to oxygen.

by plants. Oxygen forms compounds with all the elements except He, Ne, and (possibly) Ar. Oxygen reacts directly with all the elements except the halogens, some of the noble metals (such as Pt and Au), and the noble gases.

The better known compounds of oxygen with the elements other than hydrogen are discussed in connection with those elements. The thermodynamic properties of oxygen and oxygen-hydrogen species are given in Table 18.7.

**TABLE 18.7**

THERMODYNAMIC PROPERTIES OF OXYGEN AND OXYGEN-HYDROGEN SPECIES AT 298 K
(see Table 18.3 for reference; $\Delta \bar{G}_f^\circ$ and $\Delta \bar{H}_f^\circ$ values in kcal·mole$^{-1}$, $\bar{S}^\circ$ values in gibbs·mole$^{-1}$; melting and boiling points in °C).

| SUBSTANCE | $\Delta \bar{H}_f^\circ$ | $\Delta \bar{G}_f^\circ$ | $\bar{S}^\circ$ | m.pt. | b.pt. | NAME |
|---|---|---|---|---|---|---|
| O(g) | 59.553 | 55.389 | 38.467 | – | – | atomic oxygen |
| $O_2$(g) | 0 | 0 | 49.003 | −218.4 | −183.0 | oxygen |
| $O_3$(g) | 34.1 | 34.0 | 57.08 | −193 | −112 | ozone |
| H(g) | 52.095 | 48.581 | 27.391 | – | – | atomic hydrogen |
| $H_2$(g) | 0 | 0 | 31.208 | −259.1 | −252.5 | hydrogen |
| $D_2$(g) | 0 | 0 | 34.620 | −254.6 | −249.7 | deuterium |
| $H_2O(\ell)$ | −68.315 | −56.687 | 16.71 | 0.0 | 100.0 | water |
| $D_2O(\ell)$ | −70.411 | −58.195 | 18.15 | 3.82 | 101.4 | heavy water |
| $H_2O$(g) | −57.796 | −54.634 | 45.103 | 0.0 | 100.0 | water |
| $H_2O_2$(g) | −45.69 | −32.05 | 34.4 | – | – | hydrogen peroxide |
| $H_2O_2(\ell)$ | −44.88 | −28.78 | 26.2 | −0.9 | 152.0 | hydrogen peroxide |
| $HO_2^-$(aq) | −38.32 | −16.1 | 5.7 | – | – | biperoxide ion |
| OH(g) | 9.31 | 8.18 | 43.890 | – | – | hydroxyl radical |
| $OH^-$(aq) | −54.970 | −37.594 | −2.57 | – | – | hydroxide ion |

Oxygen can be prepared on a laboratory scale either by the electrolysis of water, or by heating potassium chlorate in the presence of a $MnO_2$ catalyst:

$$2H_2O(\ell) \xrightarrow{\text{electrolysis}} 2H_2(g) + O_2(g)$$

$$KClO_3(s) \xrightarrow[\text{heat}]{MnO_2} KCl(s) + \frac{3}{2} O_2(g)$$

Oxygen is a colorless, odorless, tasteless, diatomic, paramagnetic gas. The high dissociation energy of $O_2$

$$O_2(g) = 2O(g) \qquad \Delta H^\circ_{298} = 119.1 \text{ kcal}$$

is a consequence of the double bond (see Table 18.8) in the $O_2$ molecule. Liquid oxygen is pale blue in color.

Passage of a silent electric discharge through $O_2(g)$ yields *ozone*,° $O_3$:

$$3O_2(g) = 2O_3(g) \qquad \Delta G^\circ_{298} = +78.0 \text{ kcal}$$

Gaseous ozone is diamagnetic and pale blue in color, whereas $O_3(\ell)$ is deep purple, and $O_3(s)$ is black-violet. Ozone can also be produced by the action of ultraviolet light on gaseous $O_2$. This reaction produces a layer of ozone in the upper atmosphere (25 km elevation) that acts as a screen to remove short-wavelength ultraviolet radiation from sunlight. Without this protective ozone layer (the total amount of which is equivalent to a spherical layer 3 mm thick at STP) life, as we know it, would not be possible on the earth. The structure of ozone is given in Figure 18.7(a). The O—O bond distance in ozone (1.28 Å) is close to the O—O bond distance in oxygen (1.21 Å), and considerably shorter than the O—O single bond distance in hydrogen peroxide (1.49 Å), indicating considerable double-bond character in the bonds of ozone.

The most important hydrogen oxide is water, the structure of which is

---

°Ozone is often produced in small quantities in the vicinity of a leaky transformer or a sparking electric motor, and the reader may be familiar with its characteristic electric odor.

**TABLE 18.8**

MOLECULAR ORBITAL DESCRIPTIONS OF THE BONDING IN
SOME DIATOMIC OXYGEN SPECIES

| SPECIES | NUMBER OF VALENCE ELECTRONS | ELECTRONIC CONFIGURATION (Valence Shell Only) | NET NUMBER OF BONDING ELECTRONS | BOND LENGTH (Å) | BOND ENERGY (kcal) | MAGNETIC BEHAVIOR |
|---|---|---|---|---|---|---|
| $O_2^+$ | 11 | $(2\sigma)^2(2\sigma^\circ)^2(3\sigma)^2(1\pi)^4(1\pi^\circ)^1$ | 5 | 1.12 | 153 | paramagnetic |
| $O_2$ | 12 | $(2\sigma)^2(2\sigma^\circ)^2(3\sigma)^2(1\pi)^4(1\pi^\circ)^2$ | 4 | 1.21 | 119 | paramagnetic |
| $O_2^-$ | 13 | $(2\sigma)^2(2\sigma^\circ)^2(3\sigma)^2(1\pi)^4(1\pi^\circ)^3$ | 3 | 1.26 | <90 | paramagnetic |
| $O_2^{2-}$ | 14 | $(2\sigma)^2(2\sigma^\circ)^2(3\sigma)^2(1\pi)^4(1\pi^\circ)^4$ | 2 | 1.49 | <40 | diamagnetic |

**FIGURE 18.7**  Structures of ozone, water, and hydrogen peroxide (distances in angstrom units).

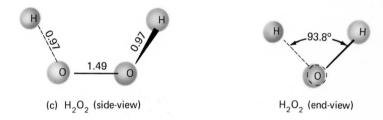

given in Figure 18.7(b). The bonding in water is discussed in Chapters 4 and 6, and the phase behavior of water is discussed in Chapter 10.

Hydrogen peroxide, $H_2O_2$, can be prepared on a laboratory scale by dissolution of $Na_2O_2$, or $CaO_2$, in acidic aqueous solution, or by reaction of organic peroxides with water. The structure of $H_2O_2(g)$ is given in Figure 18.7(c), and the bonding in the peroxide ion is given in Table 18.8. Both hydrogen peroxide and ozone are strong oxidizing agents in both acidic and basic solutions.

The decomposition of hydrogen peroxide in aqueous solution

$$2H_2O_2(aq) = 2H_2O(\ell) + O_2(g) \qquad \Delta G^\circ_{298} = -50.4 \text{ kcal}$$

is very slow at 25°C in the absence of a catalyst. A number of solids (for example, platinum black), various cations (for example, $Fe^{2+}$ and $Fe^{3+}$), HBr and $Br_2$, and the enzyme *catalase* catalyze the evolution of $O_2(g)$ from $H_2O_2$.

There are various oxygen anions known, namely, $O^{2-}$ (oxide), $O_2^-$ (superoxide), and $O_2^{2-}$ (peroxide) (see Table 18.8). In aqueous solution these species are of little importance as such, the predominant related species being $OH^-$ (hydroxide), $HO_2$ (perhydroxyl), and $HO_2^-$ (biperoxide), respectively, as is evident from the following data:

$$O^{2-}(aq) + H_2O(\ell) = 2OH^-(aq) \qquad \Delta G^\circ_{298} < -30 \text{ kcal}$$

$$O_2^-(aq) + H^+(aq) = HO_2(aq) \qquad \Delta G^\circ_{298} \approx -29 \text{ kcal}$$

$$O_2^{2-}(aq) + H^+(aq) = HO_2^-(aq) \qquad \Delta G^\circ_{298} < -30 \text{ kcal}$$

$$HO_2^-(aq) + H^+(aq) = H_2O_2(aq) \qquad \Delta G^\circ_{298} \approx -16 \text{ kcal}$$

The reduction of oxygen in aprotic solvents proceeds via a one-electron step involving $O_2^-$:

$$O_2 + e^- \rightarrow O_2^-$$

whereas the reduction of oxygen in water generally proceeds via a two-electron step:

$$O_2(g) + H_2O(\ell) + 2e^- \rightarrow HO_2^-(aq) + OH^-(aq) \qquad \mathscr{E}^\circ_{red} = -0.076$$

$$O_2(g) + 2H^+(aq) + 2e^- \rightarrow H_2O_2(aq) \qquad \mathscr{E}^\circ_{red} = +0.682$$

These potentials account for the slowness of numerous direct oxidations by oxygen in aqueous solution. If the overall reaction must go through peroxide (or $H_2O_2$), then the potentials of the above half-reactions often make this the rate-determining step.*

## 18.4   SULFUR (GROUP VIA).

### a.   General Aspects.

Sulfur is a pale-yellow, water-insoluble, brittle solid. Sulfur, referred to in Genesis as *brimstone,* is found in nature as the free element, in metal sulfides and sulfates, and as $H_2S$ and $SO_2$. Sulfur is also found in meteorites. The free element is obtained from its subsurface deposits by the *Frasch process.* In this process the sulfur is melted with high-pressure steam and brought to the surface as the liquid. Elemental sulfur is used in black gunpowder, in the vulcanization of rubber,† as a fungicide (especially on grape vines), and (most importantly) in the manufacture of sulfuric acid. A large number of naturally occurring metal sulfides are known (Table 18.9(a)), some of which are important ores. There are also a large number of sulfate minerals known, some of which are listed in Table 18.9(b).

The two principal solid modifications of sulfur are *rhombic* ($\rho = 2.07$ g·cm$^{-3}$) sulfur and *monoclinic* ($\rho = 1.96$ g·cm$^{-3}$) sulfur. At 1 atm the rhombic form is more stable below 95.5°C, whereas between 95.5° and the melting point (119°C), the monoclinic form is the more stable. When heated rapidly through the transition temperature, the metastable rhombic form melts at 112.8°C. Sulfur precipitated from $Na_2S_2O_3$ in the dark separates as bright orange, unstable crystals of $S_6$, but rhombic sulfur, monoclinic sulfur, liquid sulfur, and solutions of sulfur in carbon disulfide ($CS_2$) all contain $S_8$ molecular units in the crown configuration (see Figure 18.8(a)).

Liquid sulfur has the unusual property of becoming more viscous as its temperature is increased. The viscosity increases over the range from 159 to 200°C, but above 200°C the viscosity begins to decrease again, and continues to do so up to the boiling point at 444°C. This increase in viscosity has been traced to the breaking up of the $S_8$ rings into chains, which then polymerize. The average chain length is about $7 \times 10^5$ sulfur atoms per chain at 200°C.

Gaseous sulfur contains $S_8$ and $S_2$ molecules, and possibly also traces of $S_6$ and $S_4$. As the temperature increases, the mole fraction of $S_8$ decreases rapidly, and that of $S_2$ increases rapidly

$$S_8(g) = 4S_2(g) \qquad \Delta G^\circ_{298} = 53.97 \text{ kcal}; \ \Delta H^\circ_{298} = 98.72 \text{ kcal}$$

---

*A reaction with a very positive $\Delta G^\circ$ will most likely have a very high activation energy, and consequently a slow rate.

†In this process the sulfur reacts with unsaturated hydrocarbon polymer chains to form sulfur cross-linkages between the chains.

TABLE 18.9

| (a) SOME SULFIDE MINERALS | | | |
|---|---|---|---|
| $FeS_2$ | pyrite (fool's gold) | $Sb_2S_3$ | stibnite |
| $PbS$ | galena | $As_2S_3$ | orpiment |
| $ZnS$ | sphalerite | $AsS$ | realgar |
| $HgS$ | cinnabar | $CdS$ | greenockite |
| $CaS$ | oldhamite | $CuFeS_2$ | chalcopyrite |
| $PtS$ | braggite | $Ag_2S$ | argentite |
| $Bi_2S_3$ | bismuthinite | $AgS$ | acanthite |
| $MnS$ | albandite | $FeAsS$ | arsenopyrite |
| $Cu_2S$ | chalcocite | $Cu_3AsS_4$ | enargite |
| (b) SOME SULFATE MINERALS | | | |
| $CaSO_4 \cdot 2H_2O$ | gypsum | $SrSO_4$ | celestite |
| $MgSO_4 \cdot 7H_2O$ | epsomite (epsom salts) | $ZnSO_4 \cdot 7H_2O$ | goslarite |
| $CaSO_4$ | barite | $(NH_4)_2SO_4$ | mascagnite |
| $K_2SO_4$ | arcanite | $FeSO_4 \cdot 7H_2O$ | melanterite |

because of the large positive value of $\Delta H°$. The molecule $S_2$, like $O_2$, is paramagnetic, and rapid condensation of $S_2(g)$ yields a highly colored, unstable solid modification of sulfur containing $S_2$ molecules.

The major differences between the chemical behavior of sulfur and oxygen, like those between phosphorus and nitrogen, are a consequence of the presence of valence-shell $3d$-orbitals on sulfur. These $d$-orbitals allow sulfur to assume coordination numbers greater than four, and to participate in $d_\pi(S) + p_\pi(X)$ bonding. The thermodynamic properties of some sulfur compounds are given in Table 18.10.

## b.  Some Reactions of Sulfur.

Sulfur reacts directly with many metals and non-metals (including the halogens). For example, sulfur reacts at 25°C with sodium and mercury:

$$S(s) + 2Na(s) = Na_2S(s)$$

$$S(s) + Hg(\ell) = HgS(s)$$

Most metal sulfides (except those of the alkali metals) are insoluble in water. Numerous heavy-metal sulfides are non-stoichiometric and exhibit variable metal-to-sulfur ratios.

Treatment of most metal sulfides with aqueous solutions of strong acids yields the foul-smelling, poisonous* hydrogen sulfide gas, $H_2S$.

---

*Many sulfur compounds have unpleasant odors. Especially notorious in this aspect is benzyl mercaptan, $C_6H_5CH_2SH$. Methyl mercaptan, $CH_3SH$, is added in small quantities to natural gas to give it an easily detectable odor, and thereby give warning of gas leaks. It is worth noting that $H_2S$ is more poisonous than HCN. At high concentrations $H_2S$ cannot be detected by smell.

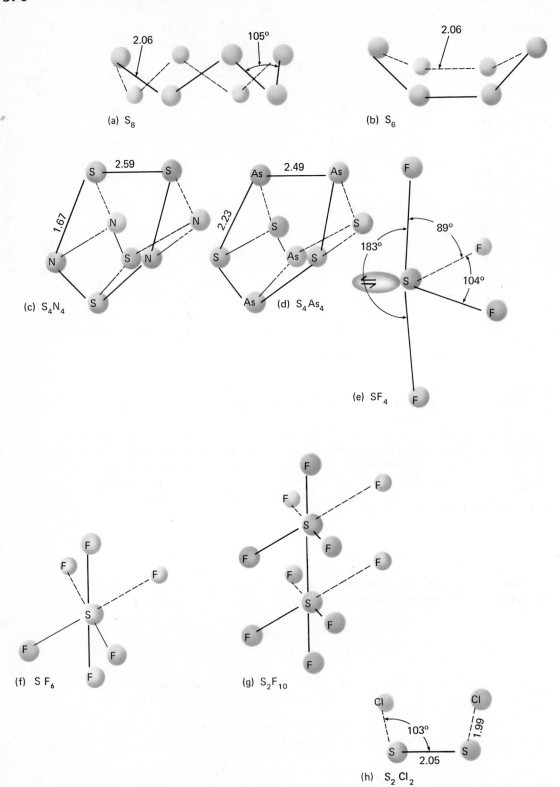

**FIGURE 18.8** Structures of sulfur and some sulfur compounds (bond distances in angstrom units).

TABLE 18.10

THERMODYNAMIC PROPERTIES OF SOME SULFUR COMPOUNDS

(see Table 18.3 for reference; $\Delta \bar{G}_f^\circ$ and $\Delta \bar{H}_f^\circ$ values in kcal $\cdot$ mole$^{-1}$, $\bar{S}^\circ$ values in gibbs $\cdot$ mole$^{-1}$).

| SPECIES | $\Delta \bar{H}_f^\circ$ | $\Delta \bar{G}_f^\circ$ | $\bar{S}^\circ$ | NAME |
|---|---|---|---|---|
| $S(s, rhomb)$ | 0 | 0 | 7.60 | sulfur |
| $S^{2-}(aq)$ | 7.9 | 20.5 | $-3.5$ | sulfide ion |
| $S_2^{2-}(aq)$ | 7.2 | 19.0 | 6.8 | disulfide ion |
| $S_3^{2-}(aq)$ | 6.2 | 17.6 | 15.8 | trisulfide ion |
| $S_8(g)$ | 24.45 | 11.87 | 102.98 | — |
| $SO_2(g)$ | $-70.944$ | $-71.748$ | 59.30 | sulfur dioxide |
| $SO_3(g)$ | $-94.58$ | $-88.69$ | 61.34 | sulfur trioxide |
| $SO_3^{2-}(aq)$ | $-151.9$ | $-116.3$ | $-7.$ | sulfite ion |
| $SO_4^{2-}(aq)$ | $-217.3$ | $-177.97$ | 4.8 | sulfate ion |
| $S_2O_4^{2-}(aq)$ | $-180.1$ | $-143.5$ | 22. | dithionate ion |
| $S_2O_8^{2-}(aq)$ | $-320.0$ | $-265.4$ | 59.3 | peroxodisulfate ion |
| $HS^-(aq)$ | $-4.2$ | 2.88 | 15.0 | bisulfide ion |
| $H_2S(g)$ | $-4.93$ | $-8.02$ | 49.16 | hydrogen sulfide |
| $H_2S(aq)$ | $-9.5$ | $-6.66$ | 29. | hydrogen sulfide |
| $HSO_3^-(aq)$ | $-149.67$ | $-126.15$ | 33.4 | bisulfite ion |
| $HSO_4^-(aq)$ | $-212.08$ | $-180.69$ | 31.5 | bisulfate ion |
| $H_2SO_3(aq)$ | $-145.51$ | $-128.56$ | 55.5 | sulfurous acid |
| $SF_4(g)$ | $-185.2$ | $-174.8$ | 69.77 | sulfur tetrafluoride |
| $SF_6(g)$ | $-289.0$ | $-264.2$ | 69.72 | sulfur hexafluoride |
| $SOCl_2(g)$ | $-50.8$ | $-47.4$ | 74.01 | thionyl chloride |

Hydrogen sulfide gas at 1 atm pressure and 25°C has a solubility in water of 0.1 M, and acts as a weak dibasic acid. Sulfide ion is a very strong base in aqueous solution, and consequently most water-insoluble metal sulfides are readily soluble in strongly acidic aqueous solutions (an exception is HgS).

Sulfur dissolves in aqueous sulfide solutions

$$S(s) + S^{2-}(aq) = S_2^{2-}(aq) \qquad \Delta G_{298}^\circ = -0.5 \text{ kcal}$$

and is soluble in hot aqueous alkali to yield polysulfides such as $S_2^{2-}$, $S_3^{2-}$, $S_4^{2-}$, and so forth.

Sulfur reacts with ammonia in carbon tetrachloride to yield orange-yellow $S_4N_4$, which has the structure shown in Figure 18.8(c). The structure of the mineral *realgar*, $As_4S_4$ (Figure 18.8(d), makes an interesting contrast to that of $S_4N_4$.

The reaction between sulfur and excess fluorine yields $SF_6$ as the major product, with traces of $SF_4$ and $S_2F_{10}$ also being produced. Sulfur hexafluoride has an octahedral structure (Figure 18.8(f)). The bonding can be described as involving $sp^3d^2$ hybridization on sulfur. Although very unstable thermodynamically with respect to hydrolysis,

$$SF_6(g) + 4H_2O(\ell) = 2H^+(aq) + SO_4^{2-}(aq) + 6HF(aq) \qquad \Delta G_{298}^\circ = -112.8 \text{ kcal}$$

$SF_6$ is unaffected by water or by hot alkalies or acids. It is a surprisingly inert compound. There are two sulfur chlorides formed in the reaction between sulfur and chlorine

$$2S(\ell, \text{ in excess}) + Cl_2(g) \rightarrow S_2Cl_2(\ell)$$

$$S(\ell) + Cl_2(g, \text{ in excess}) \xrightarrow[\text{catalyst}]{FeCl_3} SCl_2(\ell)$$

The compound $S_2Cl_2$ (Figure 18.8(h)) is an orange liquid with a foul odor at 25°C. Sulfur dichloride is unstable

$$2SCl_2(\ell) \rightarrow S_2Cl_2(\ell) + Cl_2(g)$$

Reaction of $SCl_2$ with NaF in acetonitrile solvent yields the highly reactive compound sulfur tetrafluoride. For example, $SF_4$ reacts rapidly with water

$$SF_4(g) + 2H_2O(\ell) \rightarrow SO_2(aq) + 4HF(aq) \qquad \Delta G^\circ_{298} = -67.5 \text{ kcal}$$

and is a good fluorinating agent. The sulfur halide $SF_5Cl$ can be prepared by the chlorination of $SF_4$:

$$SF_4 + Cl_2 + CsF \rightarrow SF_5Cl + CsCl$$

Unlike the symmetrical $SF_6$, $SF_5Cl$ is a fairly reactive compound. For example, it can be reduced by $H_2$:

$$2SF_5Cl(g) + H_2(g) \xrightarrow{h\nu} S_2F_{10}(g) + 2HCl(g)$$

The sulfur halides are all volatile compounds (see Table 18.11) that are readily soluble in non-polar solvents.

### c.  The Sulfur Oxy-Acids.

The combustion of sulfur in air yields $SO_2$:

$$S(s) + O_2(g) = SO_2(g) \qquad \Delta G^\circ_{298} = -71.1 \text{ kcal}; \ \Delta H^\circ_{298} = -70.9 \text{ kcal}$$

**TABLE 18.11**

MELTING AND BOILING POINTS OF SOME SULFUR-HALOGEN COMPOUNDS

| COMPOUND | m.pt. (°C) | b.pt. (°C) |
|---|---|---|
| $SF_4$ | −121 | −40 |
| $SF_6$ | − 65 | −51 |
| $S_2F_{10}$ | − 53 | 29 |
| $SF_5Cl$ | − 69 | −19.1 |
| $S_2Cl_2$ | − 80 | 130 |
| $SCl_2$ | − 78 | 59 (decomposes) |

The sulfur dioxide molecule is bent (Figure 18.9(a)) and the short S—O distance implies considerable multiple bonding of the type

$$\ddot{S} \qquad \ddot{S}$$
$$\underset{O}{\diagup}\underset{O}{\diagdown} \quad \leftrightarrow \quad \underset{O}{\diagdown}\underset{O}{\diagup}$$

$(p_\pi(S) + p_\pi(O)$ and $d_\pi(S) + p_\pi(O))$. The dissolution of $SO_2$ in water yields sulfurous acid, $H_2SO_3(aq)$:

$$SO_2(g) + H_2O(\ell) = H_2SO_3(aq) \qquad \Delta G^\circ_{298} = -0.12 \text{ kcal}$$

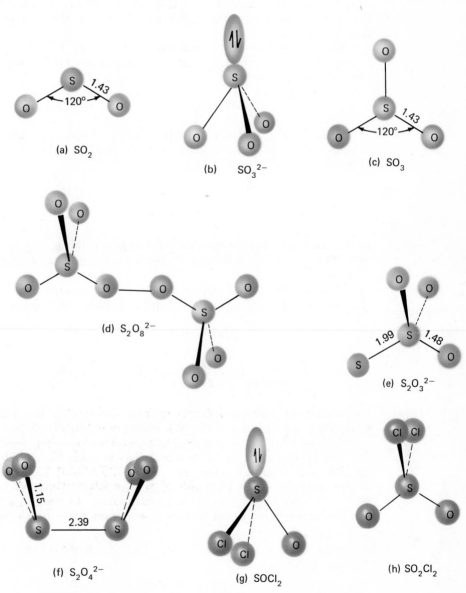

**FIGURE 18.9** Structures of some sulfur-oxygen species (bond distances in angstrom units).

**TABLE 18.12**

$pK_a$ VALUES AT 25°C IN WATER FOR SOME SULFUR OXO-ACIDS

| ACID | $pK_{a1}$ | $pK_{a2}$ |
|---|---|---|
| $H_2SO_3(aq)$ | 1.90 | 7.25 |
| $H_2SO_4(aq)$ | —— | 1.90 |
| $H_2S_2O_4(aq)$ | 0.35 | 2.46 |

Aqueous sulfurous acid is a weak diprotic acid (Table 18.12). The sulfite ion, $SO_3^{2-}$, has a pyrimidal structure, owing to the lone-pair on sulfur (Figure 18.9(b)).

Alkali metal bisulfites can be prepared by bubbling $SO_2$ into an aqueous solution of the carbonate. When alkali-metal bisulfites are heated, *pyrosulfites* are obtained:

$$2NaHSO_3(s) \rightarrow Na_2S_2O_5(s) + H_2O(g)$$

Sulfuric acid, the most widely used and important sulfur compound, is made by the catalytic oxidation of $SO_2$ to $SO_3$, followed by dissolution of $SO_3$ in water: *

$$SO_2(g) + \frac{1}{2}O_2(g) = SO_3(g) \qquad \Delta G^\circ_{298} = -16.94 \text{ kcal}$$
$$SO_3(g) + H_2O(\ell) = H_2SO_4(aq) \qquad \Delta G^\circ_{298} = -32.59 \text{ kcal}$$

In the *lead-chamber process* the oxidation is catalyzed by oxides of nitrogen (NO and $NO_2$), whereas in the *contact process* a platinum gauze catalyst is used.

Gaseous $SO_3$ has a planar structure with a short S—O bond length (Figure 18.9(c)), indicating considerable double bond character:

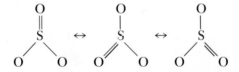

Sulfur trioxide is a strong Lewis acid, and forms acid-base adducts such as $C_5H_5NSO_3$, provided that the $SO_3$ does not oxidize the base.

Sulfuric acid is a strong acid in aqueous solution. However, the bisulfate ion, $HSO_4^-(aq)$, is a much weaker acid (Table 18.12). The sulfate ion has a regular tetrahedral structure. At 25°C, 1 M $H_2SO_4(aq)$ is a poor oxidizing agent; however, the oxidizing power of $H_2SO_4(aq)$ increases rapidly with concentration and temperature, and hot, concentrated sulfuric acid is a potent oxidizing agent that chars paper and dissolves copper.

---

*In the actual plant process, $SO_3$ is first dissolved in $H_2SO_4$, which is then diluted with water:

$$SO_3(g) + H_2SO_4(\ell) = H_2S_2O_7(\ell)$$
$$H_2S_2O_7(\ell) + H_2O(\ell) = 2H_2SO_4(\ell)$$

When alkali metal bisulfates are heated to around 200°C, *pyrosulfates* are obtained,

$$2NaHSO_4(s) \xrightarrow{200°C} Na_2S_2O_7(s) + H_2O(g)$$

and at still higher temperatures, pyrosulfates decompose to sulfates and $SO_3$.

Electrochemical oxidation of sulfates yields peroxodisulfate, $S_2O_8^{2-}(aq)$, the structure of which is given in Figure 18.9(d). Peroxodisulfate has a peroxide linkage, and, as a consequence, it is a powerful oxidizing agent:

$$S_2O_8^{2-}(aq) + 2e^- = 2SO_4^{2-}(aq) \quad \mathscr{E}^°_{red} = +2.01 \text{ V } (25°C)$$

A solution of $H_2S_2O_8$ in water yields hydrogen peroxide:

$$H_2S_2O_8 + 2H_2O = H_2O_2 + 2H_2SO_4$$

Because $H_2S_2O_8$ can easily be made by the electrolysis of cold concentrated $H_2SO_4(aq)$, this provides another route to hydrogen peroxide. Peroxodisulfate oxidations are generally slow, but they are catalyzed by Ag(I), which is oxidized by $S_2O_8^{2-}$ to Ag(III); the Ag(III) then oxidizes the reductant in a fast, two-electron step.

Sulfur dissolves in alkaline aqueous solutions containing sulfite ion to form the thiosulfate ion:

$$S(s) + SO_3^{2-}(aq) = S_2O_3^{2-}(aq) \quad \Delta G^°_{298} = -8.4 \text{ kcal}$$

Sodium thiosulfate is used in analytical chemistry for the volumetric determination of iodine:

$$2S_2O_3^{2-}(aq) + I_3^-(aq) = 2S_4O_6^{2-}(aq) + 3I^-(aq)$$

Solutions of sodium thiosulfate dissolve otherwise insoluble silver salts, such as the chloride and bromide:

$$AgX(s) + 2S_2O_3^{2-}(aq) = Ag(S_2O_3)_2^{3-}(aq) + X^-(aq)$$

Aqueous sodium thiosulfate solution is used as a fixing agent in the photographic film development process, wherein it dissolves the silver bromide on the film that was not photochemically reduced to silver.

The thiosulfate ion is unstable in the presence of strong acids and decomposes to $SO_2(aq)$ and sulfur. The sulfur prepared in this way is metastable and consists of $S_6$ rings (Figure 18.8(b)) rather than $S_8$ rings.

Electrochemical reduction of $HSO_3^-(aq)$ yields dithionite ion, $S_2O_4^{2-}(aq)$ (Figure 18.9(f)). In basic solution dithionite ion is a strong reducing agent:

$$2SO_3^{2-}(aq) + 2H_2O(\ell) + 2e^- = S_2O_4^{2-}(aq) + 4OH^-(aq)$$

$$\mathscr{E}^°_{red} = -1.12 \text{ V } (25°C)$$

primarily because of the weak S—S bond.

Sulfur oxy-halides of the type $SOCl_2$ (thionyl chloride), and $SO_2Cl_2$ (sulfuryl chloride) (Figures 18.9(g) and (h)) can be prepared from the dioxide:

$$SO_2 + PCl_5 \rightarrow SOCl_2 + POCl_3$$

$$SO_2 + Cl_2 \rightarrow SO_2Cl_2$$

These compounds are highly reactive and find use in organic chemistry as chlorinating agents. They are violently decomposed by water.

## 18.5   THE HALOGENS (GROUP VIIA: F, Cl, Br, I, At).

### a.   General Aspects.

Fluorine is a poisonous, pale greenish-yellow, diatomic gas (boiling point $-180°C$). Fluorine is the most reactive of all the elements at $25°C$. It reacts *directly,* and often vigorously, with all of the elements except helium and neon. The extremely corrosive nature of fluorine is evidenced by its reactions with glass (when traces of water are present), ceramics, and carbon. Even water burns in fluorine, the reaction being accompanied by a brilliant flame.

The high reactivity of fluorine is due in part to the weakness of the F—F bond (38 kcal), coupled with the high strength of fluorine bonds to other elements (see Table 18.13). Fluorine is the most electronegative of all the elements. The weakness of the bond in $F_2$ is thought to be a consequence of the high electronegativity of fluorine, coupled with the presence of strong lone-pair repulsions in the $F_2$ molecule.

The principal natural sources of fluorine are the minerals *fluorspar* $(CaF_2)$, *cryolite* $(Na_3AlF_6)$, and *fluorapatite* $(Ca_{10}F_2(PO_4)_6)$. Because of its high reactivity, the free element is difficult to obtain. Elemental fluorine is

TABLE 18.13

HALOGEN SINGLE-BOND ENERGIES (kcal $\cdot$ mole$^{-1}$)

|        | F   | Cl  | Br  | I   |
|--------|-----|-----|-----|-----|
| B      | 154 | 109 | —   | —   |
| C      | 116 | 81  | 68  | 52  |
| N      | 65  | 46  | —   | —   |
| O      | 45  | 45  | 48  | —   |
| Si     | 135 | 91  | 74  | 56  |
| P      | 117 | 78  | 63  | 44  |
| S      | 68  | 61  | 52  | —   |
| H      | 136 | 103 | 88  | 71  |

prepared by the electrolysis of a KF + HF melt in a vessel of copper or nickel, or by electrolysis of a solution of KF in anhydrous HF($\ell$) at 25°C in steel vessels. The metals are protected by fluoride coatings.

Fluorine is used in large quantities to make the gaseous fluorides $^{238}UF_6$ and $^{235}UF_6$, which are then separated in a diffusion process as part of the preparation of enriched uranium for reactor fuel elements. Hydrogen fluoride, HF, is used as an acid catalyst in petroleum refining, in the production of fluorocarbons for plastics and spray cans, and also to etch the glass of light bulbs. Various soluble metal fluorides are used as anti-caries agents, and some transition-metal fluorides are used as welding fluxes. Numerous organofluorine compounds involving C—F bonds are known. For example, *Teflon* is a fluorocarbon polymer ($-CF_2-$)$_n$.

Because of the high strength of C—F bonds (116 kcal·mole$^{-1}$), highly fluorinated organic compounds are usually not oxidizable by oxygen (the O—F bond energy is only 45 kcal·mole$^{-1}$). Numerous non-polymeric organofluorine compounds have low boiling points, and those which are odorless, non-toxic, and non-flammable are used as refrigerants (for example, $CCl_2F_2$ (Freon-12), and $C_2Cl_2F_4$ (Freon-114)).

Fluorine is one of the most powerful oxidizing agents known:

$$F_2(g) + 2e^- = 2F^-(aq) \qquad \mathscr{E}^\circ_{red} = +2.87 \text{ V } (25°C)$$

This high oxidizing power of fluorine leads to high oxidation states for elements combined with fluorine, for example, $SF_6$, $AgF_2$, $PtF_6$, and $IF_7$.

The ground state electronic configuration of F(g) is $1s^2 2s^2 2p^5$, and thus fluorine can complete its octet either by acquiring an additional electron to form the fluorine ion, F$^-$, or by forming a single bond to another atom,[*] X—F. The only known oxidation states of fluorine are −1 and 0. Fluorine compounds with electropositive elements such as sodium and calcium are primarily ionic (for example, NaF and $CaF_2$), whereas fluorine compounds with the more electronegative elements, such as oxygen, chlorine, and carbon, are primarily covalent (for example, $OF_2$, $ClF_3$, and $CCl_2F_2$). There is some evidence that fluorine behaves as a $\pi$-donor, and thereby participates in multiple bonding in some of its compounds.

The high electronegativity of fluorine has important consequences in addition to those noted above. For example, in contrast to $NH_3$, $NF_3$ is not at all basic, and $F_3CCOOH$ is a considerably stronger acid than $CH_3COOH$, the respective acid dissociation constants being 0.59 and $1.76 \times 10^{-5}$ in water at 25°C.

Like fluorine, the other halogens, chlorine (Cl), bromine (Br), iodine (I), and astatine (At), also have atomic ground state electronic configurations of the type (see Table 18.14)

(inert gas core) $ns^2 np^5$

They can complete their octets by acquiring an additional electron to form the respective halide ions, X$^-$, or by forming a single bond to another atom. The halogens can also act as bridging groups (for example, in $Al_2Cl_6$). In contrast to fluorine, the halogens chlorine, bromine, and iodine exhibit higher oxidation states. However, positive oxidation states for these elements are known only in oxy-halogen and interhalogen compounds.

---

[*]There are a few compounds known that involve bridging fluorine atoms, X—F—X. Halide bridges are a rare occurrence in the case of fluorine, but are more common for the other halogens.

**TABLE 18.14**

| element | atomic number | atomic mass | principal isotopes (% natural abundance) [half-life] | electronic structure of X(g) | principal oxidation states | melting point (°C) (of $X_2$) | boiling point (°C) (of $X_2$) | $\mathscr{E}^{\circ}_{red}$ {$X_2 + 2e^- = 2X^-$(aq)} (volts, 25°C) | radius of $X^-$ $\frac{ionic}{covalent}$ (in Å) | electronegativity (Pauling) | $\Delta H_{298}$ in kcal; $X_2(g) = 2X(g)$ | $\Delta H_{298}$ in kcal $2HX(g) = H_2(g) + X_2(g)$ | ionization of potential of X(g) (in eV) | electron affinity of X(g) (in kcal) |
|---|---|---|---|---|---|---|---|---|---|---|---|---|---|---|
| F fluorine | 9 | 18.9984 | 19(100) 18[1.87h] | (He)$2s^2 2p^5$ | −1, 0, | −219.6 | −188.1 | +2.87 | 1.36 / 0.64 | 4.0 | 37.8 | 129.6 | 17.3 | 80 |
| Cl chlorine | 17 | 35.453 | 35(75.53) 37(24.47) 36[$3 \times 10^5$y] | (Ne)$3s^2 3p^5$ | −1, 0, +1, +3 +4, +5 +7 | −101.0 | −34.6 | +1.36 | 1.81 / 0.99 | 3.2 | 58.2 | 44.2 | 13.0 | 83 |
| Br bromine | 35 | 79.904 | 79(50.54) 81(49.46) 77[58h] | (Ar)$3d^{10}$ – $4s^2 4p^5$ | −1, 0, +1, +5 | −7.2 | 58.8 | +1.07 | 1.96 / 1.14 | 3.0 | 53.5 | 17.4 | 11.8 | 78 |
| I iodine | 53 | 126.9045 | 127(100) 131[8.05d] | (Kr)$4d^{10}$ – $5s^2 5p^5$ | −1, 0, +1, +5, +7 | 113.5 | 184.4 | +0.54 | 2.22 / 1.33 | 2.7 | 51.1 | −12.7 | 10.6 | 71 |
| At astatine | 85 | (210) | 210[8.3h] 211[7.5h] | (Xe)$4f^{14}$ – $5d^{10} 6s^2 6p^5$ | — | — | — | — | — | — | — | — | — | — |

The electronegativities and oxidizing powers of the halogens decrease going down the group. All of the halogens are good oxidizing agents, and none of them is found as the free element in nature, their occurrence being limited to halide salts.

Chlorine is found in nature in the form of chlorides, the principal sources being sea water and salt lakes. The important mineral sources are *rock salt* (NaCl), *sylvite* (KCl), and *carnallite* ($KMgCl_3 \cdot 6H_2O$). Chlorine, a pale-green, poisonous, diatomic gas, is obtained via the electrolysis of brine:

$$2H_2O(\ell) + 2NaCl(aq) \xrightarrow{\text{electrolysis}} 2NaOH(aq) + H_2(g) + Cl_2(g)$$

or by the electrolysis of molten NaCl in the manufacture of sodium:

$$2NaCl(\ell) \xrightarrow{\text{electrolysis}} 2Na(\ell) + Cl_2(g)$$

The principal uses of chlorine are: (a) for purification of water (chlorination); (b) as a bleaching agent in the textile and paper industries; and (c) in the synthesis of plastics, dyes, pharmaceuticals, and insecticides such as DDT and 2,4-D.

Bromine is obtained by the passage of chlorine gas through bromide-containing brines which have been adjusted to a pH of 3.5. The resulting bromine is swept out of the brine in a current of air.

Bromine, $Br_2$, is a dense (2.9 $g \cdot cm^{-3}$), red-brown liquid with a pungent odor. Bromine gas and solutions of $Br_2$ in carbon disulfide and carbon tetrachloride are red in color. The principal uses of bromine are the synthesis of ethylene dibromide (which is used as a lead scavenger in anti-knock gasolines), and in the preparation of AgBr (which is used in photography). Bromine is also used as a fumigant, and in the synthesis of flame-proofing agents, dyes, and pharmaceuticals.

Iodine is found as iodide ion in sea water, from which it is assimilated and thereby concentrated by seaweeds. Minerals containing both $I^-$ and $IO_3^-$ (iodate) are also known (for example, *iodyrite*, AgI). Free iodine, $I_2$, can be obtained by air oxidation of acidic aqueous solutions containing iodide, or by electrolysis of aqueous iodide solutions. Elemental iodine is a black solid with a slight metallic luster. Iodine vapor and solutions of $I_2$ in $CS_2$ and $CCl_4$ are a beautiful purple in color, whereas solutions of $I_2$ in water and alcohols are brown, because of the formation of charge-transfer complexes between $I_2$ and these (donor) solvents. Very low concentrations of $I_2(aq)$ can be detected by the brilliant blue color of the molecular complex formed between $I_3^-$ and starch.

Iodine is necessary for the proper functioning of the thyroid gland, and the absence of iodine in the diet causes *goiter*. The radioactive isotope $^{131}I$ ($t_{1/2} = 8$ days) is used in the treatment of thyroid gland disorders.

There are no known naturally occurring isotopes of astatine. All of the known isotopes of At are radioactive and are prepared via nuclear reactions, for example, $^{209}Bi(\alpha, 2n)^{211}At$ (Chapter 22).

## b.  The Hydrogen Halides.

All of the halogens form hydrogen halides with the composition HX. With the exception of HI, these compounds are readily prepared by treating an ionic halide with concentrated sulfuric acid, for example:

$$2NaCl(s) + H_2SO_4(aq, conc) \rightarrow Na_2SO_4(s) + 2HCl(g)$$

This reaction depends on the fact the HCl is less soluble in water and more volatile than $H_2SO_4$. Hydrogen iodide can be prepared via the hydrolysis of $PI_3$, or via the reaction between hydrogen and iodine. The bond energies of the hydrogen halides decrease on descending the group (Table 18.13). The bond in HF is surprisingly strong (136 kcal·mole$^{-1}$).

In water HCl, HBr, and HI are strong acids, whereas hydrogen fluoride is a weak acid:

$$HF(aq) = H^+(aq) + F^-(aq) \qquad K_{298} = 7.2 \times 10^{-4}$$

presumably because of the very strong HF bond. Fluorine forms the strongest hydrogen bonds, and this property is exhibited both in the formation of the bifluoride ion*

$$HF(aq) + F^-(aq) = HF_2^-(aq) \qquad K_{298} = 5.1$$

and in the existence of the polymeric hydrogen-bonded species $(HF)_2$ and $(HF)_6$ in hydrogen fluoride.

## c.    Halogen-Oxygen Compounds.

When fluorine gas is bubbled rapidly through aqueous alkali, the pale-yellow gas $OF_2$ is formed (Figure 18.10(a)). Oxygen difluoride (boiling point $-145°C$) is a potent oxidizing and fluorinating agent that reacts explosively with the other halogens. The oxy-fluoride $O_2F_2$ also is known. Dioxygen difluoride is an orange-yellow solid that melts at $100°C$ and has a structure similar to that of $H_2O_2$ with F replacing H (Figure 18.10(b)). Dioxygen difluoride reacts with explosive violence on contact with numerous substances.

The halogens Cl, Br, and I form a variety of oxides,* only a few of which have been well characterized. We shall restrict our discussion to the chlorine oxides. When chlorine is allowed to react with freshly prepared, moist mercuric oxide, dichlorine oxide (Figure 18.10(c)) is produced:

$$2Cl_2(g) + 2HgO(s) \rightarrow HgCl_2 \cdot HgO(s) + Cl_2O(g)$$

When potassium chlorate, $KClO_3$, is treated with concentrated sulfuric acid in the presence of the reducing agent oxalic acid, $H_2C_2O_4$, gaseous chlorine dioxide, $ClO_2$ (Figure 18.10(d)), is produced. Chlorine dioxide is a yellow, paramagnetic, violently explosive gas which is also a powerful oxidizing agent. Chlorine dioxide disproportionates in basic aqueous solution into chlorite, $ClO_2^-$, and chlorate, $ClO_3^-$, ions:

$$2ClO_2(aq) + 2OH^-(aq) = ClO_2^-(aq) + ClO_3^-(aq) + H_2O(\ell)$$

This reaction is of some industrial importance, because it is used to make the *bleaching powder*, $Ba(ClO_2)_2$, which precipitates from the reaction mixture upon addition of a solution containing $Ba^{2+}(aq)$.

---

*In the solid state the bifluoride ion is linear and symmetric, F–H–F$^-$.

†Compounds involving oxygen and fluorine are called fluorides, because the electronegativity of fluorine is greater than that of oxygen, whereas compounds involving oxygen and the halogens Cl, Br, and I are called oxides because the electronegativity of oxygen is greater than that of Cl, Br, and I.

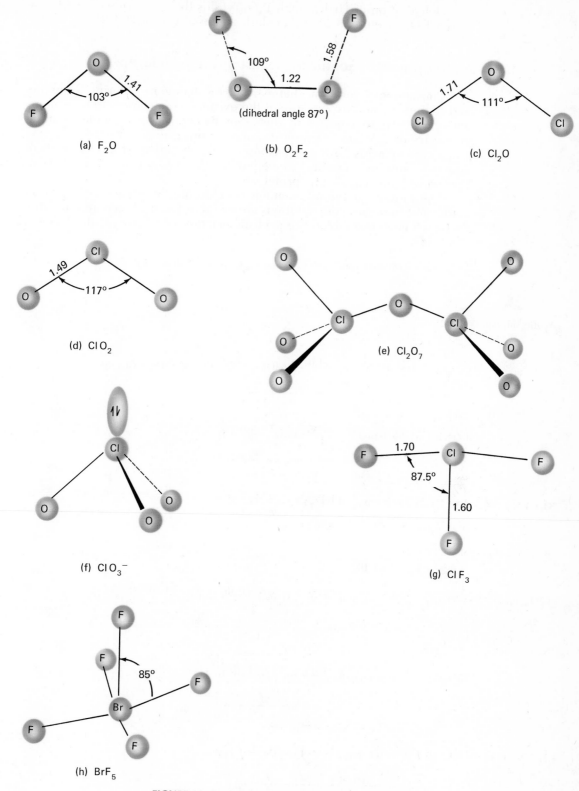

FIGURE 18.10 Structures of some halogen compounds.

Dehydration of $HClO_4$ with $P_4O_{10}$ yields the anhydride of perchloric acid, $Cl_2O_7$ (Figure 18.10(e)), a colorless, unstable, oily liquid.

## d.  The Aqueous Solution Chemistry of the Halogens.

The aqueous solution chemistry of the halogens chlorine, bromine, and iodine is complex, because of the existence of a variety of oxy-anions and oxy-acids. In contrast to the other halogens, the aqueous solution chemistry of fluorine is simple because fluorine does not form any oxy-anions or oxy-acids.* Although complex, the chemistry of the halogens in aqueous solution can be unravelled by the systematic application of thermodynamic and kinetic principles.

The Latimer potential diagrams for chlorine, bromine, and iodine in acidic and basic aqueous solutions are given in Figure 18.11, and in Table 18.15 we have assembled the acid dissociation constants for the halogen oxy-acids.

---

* Some covalent hypofluorites are known, however, such as $CF_3OF$.

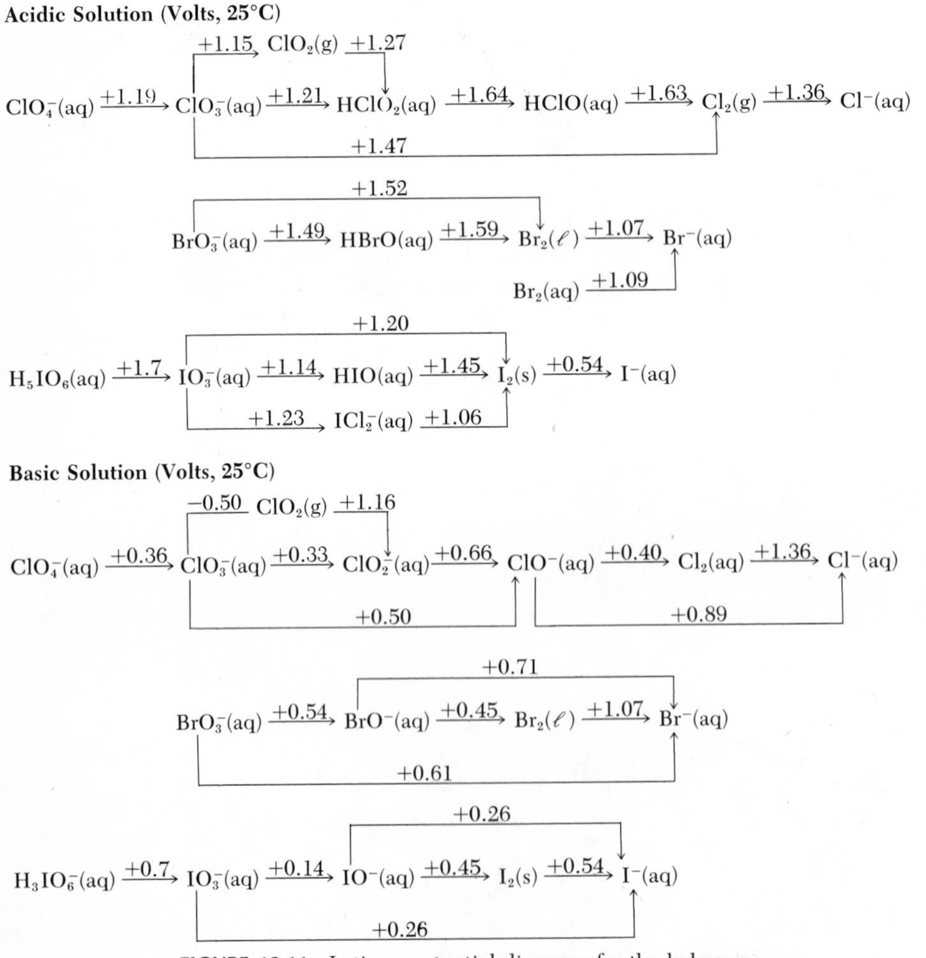

FIGURE 18.11   Latimer potential diagrams for the halogens.

**TABLE 18.15**

ACID DISSOCIATION CONSTANTS FOR THE HALOGEN OXY-ACIDS IN WATER AT 25°C

| CHLORINE | | BROMINE | | IODINE | |
|---|---|---|---|---|---|
| HClO | $3.2 \times 10^{-8}$ | HBrO | $2.1 \times 10^{-9}$ | HIO | $1 \times 10^{-11}$ |
| $HClO_2$ | $1.1 \times 10^{-2}$ | – | – | – | – |
| $HClO_3$ | strong | $HBrO_3$ | strong | $HIO_3$ | 0.16 |
| $HClO_4$ | strong | – | – | $HIO_4$ | ... |
| | | | | $H_5IO_6$ | $5.1 \times 10^{-4}$ |
| | | | | $H_4IO_6^-$ | $2 \times 10^{-7}$ |
| | | | | $H_3IO_6^-$ | $1 \times 10^{-15}$ |
| | | | | $H_4I_2O_9$ | ... |

In Table 18.16 we have indicated the halogen species that are unstable with respect to disproportionation. Of those species which are unstable, some disproportionate very rapidly (for example, $I_2$ in base), whereas others disproportionate more slowly (for example, $ClO^-(aq)$ in base). These latter species can undergo numerous other reactions in addition to disproportionation.

The solubilities of $Cl_2(g)$, $Br_2(\ell)$, and $I_2(s)$ in water at 25°C are 0.091, 0.21, and 0.0013 M, respectively. Upon dissolution in water, these halogens undergo a disproportionation reaction of the type

$$Cl_2(aq) + H_2O(\ell) = H^+(aq) + Cl^-(aq) + HOCl(aq)$$

The equilibrium constants for this type of reaction at 25°C are: $K_{Cl} = 4.2 \times 10^{-4}$, $K_{Br} = 7.2 \times 10^{-9}$, and $K_I = 2.0 \times 10^{-13}$. Thus, an aqueous solution saturated with $Cl_2$ at 1 atm and 25°C has a concentration of free $Cl_2$ equal to about 0.061 M, and $(H^+) = (HOCl) = (Cl^-) = 0.030$ M, whereas for $Br_2$ and $I_2$ the concentrations of free halogen are approximately equal to their solubilities, and the values of $(X^-)$ are $1.2 \times 10^{-3}$ M and $6.4 \times 10^{-6}$ M, respectively. On the other hand, in 1 M $OH^-(aq)$ the above reaction becomes

$$Cl_2(aq) + 2OH^-(aq) = Cl^-(aq) + OCl^-(aq) + H_2O(\ell)$$

and the equilibrium constants are: $K_{Cl} = 13 \times 10^{32}$, $K_{Br} = 3.4 \times 10^{28}$, and $K_I = 2 \times 10^{26}$. In basic solution the concentration of free halogen at equilibrium is very small, and the disproportionation to $X^-$ and $OX^-$ is essentially complete.

As the potentials in Figure 18.11 indicate, both $I^-(aq)$ and $Br^-(aq)$ are unstable with respect to air oxidation in acidic solutions, whereas $Cl^-(aq)$ is stable.

The hypohalous acids can be prepared via the reaction of an aqueous solution of the appropriate halogen with mercuric oxide; for example:

$$2Br_2(aq) + 2HgO(s) + H_2O(\ell) = HgO \cdot HgBr_2(s) + 2HOBr(aq)$$

The hypohalite ions, $XO^-$, undergo disproportionation reactions of the type

$$2ClO^-(aq) = 2Cl^-(aq) + ClO_3^-(aq)$$

**TABLE 18.16**

THERMODYNAMIC AND KINETIC STABILITY OF HALOGEN SPECIES IN
AQUEOUS SOLUTION* AT 25°C

| OXIDATION STATE | CHLORINE | | BROMINE | | IODINE | |
|---|---|---|---|---|---|---|
| | 1 M OH⁻(aq) | 1 M H⁺(aq) | 1 M OH⁻(aq) | 1 M H⁺(aq) | 1 M OH⁻(aq) | 1 M H⁺(aq) |
| −1 | $Cl^-$ | $Cl^-$ | $Br^-$ | $Br^-$ | $I^-$ | $I^-$ |
| 0 | $[Cl_2]$ | $(Cl_2)$ | $[Br_2]$ | $(Br_2)$ | $[I_2 \text{ or } I_3^-]$ | $I_2 \text{ or } I_3^-$ |
| +1 | $(ClO^-)$ | $HOCl$ | $(BrO^-)$ | $HOBr$ | $[IO^-]$ | $(\underline{HOI})$ |
| +3 | $(ClO_2^-)$ | $(HClO_2)$ | − | − | − | − |
| +4 | $[ClO_2]$ | $(ClO_2)$ | − | − | − | − |
| +5 | $(\underline{ClO_3^-})$ | $(ClO_3^-)$ | $BrO_3^-$ | $BrO_3^-$ | $IO_3^-$ | $HIO_3$ |
| +7 | $ClO_4^-$ | $ClO_4^-$ | − | − | $H_3IO_6^{2-}$ | $H_5IO_6$ |

*[ ] disproportionates very rapidly.
( ) disproportionates slowly.
_____ unstable with respect to air oxidation.

for which the equilibrium constants are: $K_{Cl} = 3 \times 10^{26}$, $K_{Br} = 3 \times 10^{11}$, and $K_I = 4 \times 10^{23}$. These values show that the hypohalite ions are all unstable in 1 M OH⁻(aq). The disproportionation of hypochlorite is slow at temperatures below 25°C, and ClO⁻(aq) can be prepared via the reaction

$$Cl_2(aq) + 2OH^-(aq) \xrightarrow{0-10°C} Cl^-(aq) + ClO^-(aq) + H_2O(\ell)$$

On the other hand, the disproportionation of hypobromite is more rapid than that of ClO⁻, and at temperatures in the range between 50 and 80°C the following reaction takes place rapidly and quantitatively:

$$3Br_2(aq) + 6OH^-(aq) = 5Br^-(aq) + BrO_3^-(aq) + 3H_2O(\ell)$$

Around 0°C the rate of disproportionation of BrO⁻ is slow enough that hypobromite can be obtained in basic aqueous solutions by dissolving $Br_2(\ell)$ in NaOH(aq) at 0°C. The disproportionation of hypoiodite is very fast in aqueous solution, and, consequently, IO⁻(aq) has only a transient existence in basic solution.

The chlorite ion, $ClO_2^-$, is thermodynamically unstable with respect to disproportionation into $ClO_3^-$ and Cl⁻ in basic solution, but the reaction is slow. On the other hand, chlorous acid, $HClO_2$, rapidly disproportionates in acid solution:

$$4HClO_2(aq) \rightarrow 2ClO_2(aq) + ClO_3^-(aq) + Cl^-(aq) + 2H^+(aq) + H_2O(\ell)$$

The *halates*, $ClO_3^-$, $BrO_3^-$, and $IO_3^-$, can be prepared by reaction of the appropriate halogen with $HNO_3$(aq, conc), or $H_2O_2$(aq), or by electrolysis of the chlorides. Bromate and iodate are stable in both acidic and basic solution, whereas $ClO_3^-$(aq) is unstable with respect to disproportionation into $Cl^-(aq)$ and $ClO_4^-(aq)$. The latter reaction is slow. The halate ions have a pyramidal structure (Figure 18.10(f)), owing to the presence of the lone-pair electrons on the halogen.

The *perhalates*, $ClO_4^-$, and $IO_4^-$, can be prepared via the electrochemical oxidation of $ClO_3^-(aq)$ and $IO_3^-(aq)$, respectively. The perhalate ions have a tetrahedral structure. Perchloric acid, when hot and concentrated, is a powerful oxidizing agent that is dangerously explosive in the presence of organic matter.

In addition to the (meta)periodate ion, $IO_4^-$, there are several important hydrates such as $H_2IO_5^-(IO_4^- \cdot OH_2)$ and $H_4IO_6^-(IO_4^- \cdot 2OH_2)$. In strongly acidic solutions, the principal I(VII) species is *paraperiodic acid*,

$$H^+(aq) + IO_4^-(aq) + 2H_2O(\ell) = H_5IO_6(aq) \qquad K_{298} = 50$$

The periodic acids are powerful oxidizing agents (being capable of oxidizing manganous ion to permanganate in acidic solution) that act smoothly and rapidly on many organic compounds, and thus find use in synthetic work.

### e. The Interhalogens.

Interhalogen compounds are substances that involve two (or more) different halogens, but nothing else. The neutral interhalogens are generally low-boiling, diamagnetic compounds, readily soluble in non-polar solvents. The known diatomic interhalogens are ClF, BrCl, IBr, ICl, and BrF. These compounds can be prepared by direct combination of the elements mixed in stoichiometric amounts under the appropriate conditions. For example,

$$Cl_2(g) + F_2(g) \xrightarrow[250°C]{Cu(s)} 2ClF(g)$$

Bromine fluoride is the most stable of the diatomic interhalogens *with respect to the elements*, $\Delta \bar{G}_{f,298}^{\circ}\{BrF(g)\} = -26.1$ kcal, whereas bromine chloride is the least stable, $\Delta \bar{G}_{f,298}^{\circ}\{BrCl(g)\} = -0.23$ kcal. Unlike the other interhalogen compounds, BrF readily disproportionates:

$$3BrF(g) \rightarrow Br_2(g) + BrF_3(g)$$

Bromine trifluoride, like $ClF_3$ and $ICl_3$, is a T-shaped molecule. The structure of $ClF_3$ is given in Figure 18.10(g). The structure can be rationalized as involving $sp^3d$ (trigonal-bipyramidal) hybridization on Cl. There are 28 valence electrons in all, six of which are involved in the three Cl-F $\sigma$-bonds, 18 in the nine lone-pairs on the fluorines, and four in two lone-pairs on Cl. One of the lone-pairs on Cl lies in the plane of the molecule, and the other is perpendicular to the molecular plane.

The interhalogen compounds $BrF_5$ and $IF_5$ have a square-pyramidal structure. The structure of $BrF_5$ is shown in Figure 18.10(h). In this molecule the Br atom lies *below* the plane formed by the four F atoms, which lie at the corners of a square. The structure can be rationalized as involving $sp^3d^2$ (octahedral) hybridization on Br. There are 42 valence electrons in all, 10 of which are involved in the five Br-F bonds, 30 of which comprise the 15 lone-pairs on the fluorines, and two of which are involved in a lone-pair on Br. The presence of the lone-pair on Br is thought to be responsible for the fact that the F atoms lie above the Br atoms, the distortion being ascribed to lone-pair repulsions (see Chapter 4 for a discussion of VSEPR Theory).

In general, the polyatomic interhalogens can be prepared from the elements using stoichiometric amounts of the appropriate elements.

There are several *polyatomic halide ions* which have been characterized, the best known of which is the linear tri-iodide ion:

$$I_2(aq) + I^-(aq) = I_3^-(aq) \qquad K_{298} = 1.4 \times 10^2$$

The polyiodide ions $I_5^-$ (V-structure) and $I_7^-$ are also known. Interhalogen triatomic ions, such as $ICl_2^-$, $ClBr_2^-$, and $ClF_2^-$ (all linear) are also known:

$$ICl(aq) + Cl^-(aq) = ICl_2^-(aq) \qquad K_{298} = 1.7 \times 10^2$$

Interhalogen polyhalide anions of the type $ICl_4^-$ (square planar) and $BrF_6^-$ (octahedral) can be prepared via reactions of the type

$$CsCl + ICl_3 \rightarrow CsICl_4$$

$$CsF + BF_5 \rightarrow CsBrF_6$$

## 18.6   THE NOBLE GASES
### (GROUP VIIIA: He, Ne, Ar, Kr, Xe, Rn).

The elements helium (He), neon (Ne), argon (Ar), krypton (Kr), xenon (Xe), and radon (Rn) form a group which has been called the rare gases, the inert gases, and the noble gases. Helium and argon are not rare, and the heaviest four members of the group are not inert. Only the adjective "noble" has proved to be sufficiently ambiguous to survive recent discoveries about the elements of this group.

All of these elements have complete outer $s$ and $p$ subshells. The octet rule implies that these elements should be chemically inert; the only attractions involving atoms of the noble gases should be induced dipole-induced dipole attractions. As usual, the empty $d$ orbitals cannot be ignored, and the heavier elements in this group are known to form compounds. However, the physical properties of these elements can be rationalized by invoking only weak induced-dipole attractions.

All noble gases are monatomic gases at STP. The crystals of the elements are classic examples of molecular crystals; this fact is indicated by the observation that the atomic energy levels in the crystals are the same as for the gases.

The existence of very weak attractive forces between atoms of the noble gases is proved by the physical constants listed in Table 18.17. There would be no solid or liquid phases without attractive forces, but the very low melting points of the solids, the very low boiling points of the liquids, and the very low solubilities in water indicate that all attractive forces involving atoms of noble gases are very weak. For comparison, some physical properties of several non-polar compounds which are isoelectronic with various noble gases are included in the table.

Air is about 1% argon by volume. Helium is found in natural gas deposits. Neon, krypton, and xenon are too rare and expensive for routine usage. Commercially they are used in the discharge lamps called neon

TABLE 18.17

PHYSICAL PROPERTIES OF NOBLE GASES AND THEIR ISOELECTRONIC ANALOGS.

| SPECIES | TRIPLE POINT | BOILING POINT (1 atm) | WATER SOLUBILITY (273 K)(cm³ of gas/kg water) |
|---|---|---|---|
| He | none | 4.2 K | 9.78 |
| $H_2$ | | 20 K | 21.4 |
| Ne | 24 K | 27 K | 14.0 |
| $CH_4$ | | 112 K | |
| Ar | 84 K | 87 K | 52.4 |
| $SiH_4$ | | 161 K | |
| Kr | 116 K | 120 K | 99.1 |
| $GeH_4$ | | 187 K | |
| Xe | 161 K | 165 K | 203.2 |
| Rn | 202 K | 211 K | – |

lights. Most materials are highly reactive towards nitrogen and oxygen at elevated temperatures; the simplest way to exclude the atmospheric gases from a particular volume in order to prevent such reactions is to use an unreactive gas such as argon or helium. Heliarc welding of aluminum is an example of this application; the molten aluminum, which would form $Al_2O_3$ with any available oxygen, is bathed in a stream of helium gas. The protected metal flows together to form a strong joint. Significant amounts of liquid helium are used as a refrigerant in low temperature research.

## a.   Noble Gas Compounds.

Years ago Linus Pauling, among others, pointed out that it should be possible to synthesize compounds of the noble gases. His predictions about the stabilities of xenon fluorides and chlorides led scientists into unsuccessful attempts at the synthesis of these compounds. Inhibited by this failure, chemists let the whole subject of noble gas compounds lie dormant until N. Bartlett in 1962 carried out the reaction

$$O_2 + PtF_6 \rightarrow O_2^+(PtF_6^-)$$

Bartlett reasoned that the analogous reaction

$$Xe + PtF_6 \rightarrow Xe^+(PtF_6^-)$$

should occur as well, for the first ionization energies of $O_2$ and Xe are quite similar. He found that the reaction did occur, although the structure and formal charges implied by the formula $Xe^+(PtF_6^-)$ are still open to question.

Bartlett's discovery opened the floodgates, and within a year many xenon compounds were synthesized. Xenon fluorides ($XeF_2$, $XeF_4$, and $XeF_6$) were made simply by heating the elements Xe and $F_2$ at high pressure in an apparatus made from nickel. This was essentially a repetition of the earliest attempts; the crucial difference was the use of a catalytic nickel container.

TABLE 18.18

SOME COMPOUNDS WHICH CONTAIN NOBLE GAS ATOMS.

| | |
|---|---|
| $KrF_2$ | $XeOF_4$ |
| $KrF_4$ | $XeO_3$ |
| $XeF_2$ | $XeO_4$ |
| $XeF_4$ | $Na_4XeO_6$ |
| $XeF_6$ | $RnF_x$ ($x$ unknown, probably 2, 4, and 6) |

Simple compounds of the noble gases are listed in Table 18.18. Only xenon forms a variety of compounds. Radon might well form as many compounds as does xenon, but the high radioactivity of radon hampers detailed studies with this element. Perhaps more compounds of krypton (and of the lighter noble gases) will be prepared when the proper reagents and catalysts are tried. It does seem clear that the chemistry of the heavier noble gases will always be more complex than the chemistry of the light noble gases.

The most surprising characteristic of the xenon compounds is their stability. For example, although $XeF_4$ is hydrolyzed in water to give HF, some of the Xe is not reduced to the element. Instead the reaction is

$$6XeF_4(s) + 12H_2O(\ell) = 2XeO_3(s) + 4Xe(g) + 3O_2(g) + 24HF(aq)$$

The compound $XeO_3$ is quite stable in water solution; however, it decomposes violently when dry.

## b.  Bonding and Structure in the Xenon Fluorides.

We shall restrict our discussion to the compounds $XeF_2$, $XeF_4$, and $XeF_6$. Xenon difluoride, $XeF_2$, is linear. This structure can be rationalized as follows: there are two electron pairs in bonding orbitals and three electron pairs in nonbonding orbitals about the central xenon atom. The usual geometry for five orbitals around a central core is a trigonal-bipyramid. Because nonbonding orbitals are less compact than bonding orbitals, the more favorable trigonal-bipyramid is that which maximizes the distance between the nonbonding orbitals. The nonbonding orbitals are the orbitals in the plane (120° apart), and the bonding orbitals oppose each other (bonding to nonbonding angle is 90°).

In the tetrafluoride there is a sixth electron pair. As usual, six orbitals form an octahedron. The distance between the two nonbonding orbitals is maximized by placing them on opposite sides of the octahedron. The remaining four bonding orbitals lie in one plane, and thus the square-planar structure of $XeF_4$ is understandable.

The $XeF_6$ molecule is a problem. In this molecule there should be seven electron pairs (six bonding and one nonbonding) around the central Xe atom. The experimental studies of this structure are incomplete, but VSEPR theory predicts a distorted octahedral structure.

These compounds and the other compounds of noble gases serve to demonstrate some of the strengths and weaknesses of quantum chemistry.

The initial prediction about the stability of xenon fluorides was correct, but apparently no one found it convincing enough to continue efforts at synthesis after the first failure. Quantum chemistry has a credibility gap. Once the fluorides were synthesized, the structures of $XeF_2$ and $XeF_4$ were immediately predicted, and only later verified. However, in the hard case, $XeF_6$, quantum chemistry has not been very helpful yet. At this time quantum chemistry is most useful as a context into which experimental facts can be fitted. For example, the octahedral geometry of, say, the complex $FeF_6^{-3}$ tells us that $XeF_4$ should be square planar. However, this occurs only after we interpret the geometry of $FeF_6^{-3}$ in terms of orbitals and electron repulsions. When we are faced with an extremely rare situation, for example, seven electron pairs about one atom, the results available from quantum chemistry are inconclusive.

_____ **PROBLEMS**

*I. Nitrogen*

1. Devise a laboratory-scale preparation of $ND_3$, using $D_2O$ as a source of deuterium.

2. Explain why it is very dangerous to mix an ammonia-type cleanser with a chlorine bleach.

3. Complete and balance the following reactions:

(a) $NO_2^-(aq) + Na(Hg) \rightarrow$
(b) $Pt(s) + HNO_3(aq, conc) + HCl(aq, conc) \rightarrow$
(c) $NO_2(g) + H_2O(\ell) \rightarrow$
(d) $NaNH_2(s) + D_2O(\ell) \rightarrow$
(e) $NO(g) + F_2(g) \rightarrow$
(f) $N_2O_3(g) + H_2O(\ell) \rightarrow$
(g) $NH_4NO_3(aq) \xrightarrow{heat}$

4. Bottles of concentrated $HNO_3(aq)$ are frequently observed to go from colorless to yellow on exposure to light. Write the chemical reaction that is responsible for this color change.

5. Using data in Table 18.3, calculate the equilibrium constant at 25°C for the reaction

$$3NO(g) = N_2O(g) + NO_2(g)$$

6. When solid ammonium dichromate is heated, a spectacular reaction takes place in which a fluffy green solid is produced and gas is evolved, the reaction being accompanied by a bright orange flame. The gas evolved is colorless, odorless, and inert. Give the chemical reaction that takes place when $(NH_4)_2Cr_2O_7$ is heated. (Hint: the reaction is the same in the absence and presence of air.)

7. Give the structures for the following species:

$NCl_3$    BrNO    $(CH_3)_3N$    $N_2F_2$    $NH_2$

8. Ammonium perchlorate, $NH_4ClO_4$, is used as a solid fuel in missiles. Give the principal reaction that takes place when $NH_4ClO_4$ reacts as a fuel.

9. Discuss the bonding in ClNO and $N_2O_2^{2-}$ from a molecular orbital point of view.

10. Show the appropriate combinations of $p_z$ orbitals on nitrogen and oxygen in $NO_3^-$ and $NO_2^-$ used to form molecular orbitals in these ions. Indicate which combinations of the $p_z$ orbitals yield bonding, non-bonding, and antibonding orbitals.

*II. Phosphorus.*
1. Devise a one-step laboratory preparation of DCl, using $D_2O$ as a source of deuterium.

2. Complete and balance the following equations:

(a)  $P + Cl_2(excess) \rightarrow$

(b)  $P(excess) + Cl_2 \rightarrow$

(c)  $PCl_5 + CaF_2 \xrightarrow{350°C}$

(d)  $PF_5 + NaF \xrightarrow{heat}$

(e)  $PBr_3 + O_2 \rightarrow$

(f)  $PBr_3(g) + H_2O(\ell) \rightarrow$

(g)  $PBr_5(s) + H_2O(\ell) \rightarrow$

(h)  $PCl_3 + AsF_3 \xrightarrow{heat}$

(i)  $HNO_3(\ell) + P_4O_{10}(s) \rightarrow$

(j)  $H_2SO_4(\ell) + P_4O_{10}(s) \rightarrow$

(k)  $POCl_3(\ell) + H_2O(\ell) \rightarrow$

3. Draw structures for the following species:

$PH_4^+$    $P_2Cl_4$    $OPF_3$    $PBr_3$    $PBr_4^+$    $PBr_5$    $PF_6^-$

$P_4$    $P_3O_{10}^{5-}$    $P_4O_{12}^{4-}$    $P_4O_6$    $P_4O_{10}$

4. The reaction between phosphorus and sulfur yields the following sulfides under various conditions: $P_4S_3$, $P_4S_5$, $P_4S_7$, and $P_4S_{10}$. Using the structures of the phosphorus oxides as a guide, suggest structures for these sulfides.

*III. Oxygen and Sulfur.*
1. Organic ethers such as ethyl ether, $CH_3CH_2OCH_2CH_3$, on exposure to oxygen form dangerously explosive compounds, which are good oxidizing agents. Give the formula of the compound formed on exposure of ethyl ether to oxygen. Suppose that you suspected that this compound had formed in a bottle of ethyl ether. How would you dispose of the potentially explosive mixture?

2. Complete and balance the following reactions:

(a)  $S(s) + NaOH(aq) \xrightarrow{80°C}$

(b)  $S + NH_3 \xrightarrow[solvent]{CCl_4}$

(c)  $SCl_2 + NaF \xrightarrow[solvent]{CH_3CN}$

(d)  $HSO_3^-(aq) + SO_2(g) \rightarrow$

(e)  $SO_2(g) + Na_2CO_3(aq) \rightarrow$

(f)  $SO_3(g) + Na_2CO_3(aq) \rightarrow$

(g)  $KHSO_4(s) \xrightarrow{heat}$

(h)  $Na_2S_2O_7(s) \xrightarrow{heat}$

(i)  $SOCl_2(\ell) + H_2O(\ell) \rightarrow$

(j)  $SO_2Cl_2(\ell) + H_2O(\ell) \rightarrow$

(k)  $Cu(s) + H_2SO_4(aq, 18M) \xrightarrow{heat}$

(l)  $SO_3^{2-}(aq) + O_2(g) \rightarrow$

(m) $Fe^{2+}(aq) + H^+(aq) + O_2(g) \rightarrow$

(n)  $HSO_3Cl(\ell) + H_2O(\ell) \rightarrow$

3. Suggest a qualitative explanation for the inertness of $SF_6$ as compared to $SF_4$.

4. Draw structures for the following species:

| | | | | |
|---|---|---|---|---|
| $H_2S$ | $S_3^{2-}$ | $S_4^{2-}$ | $CH_3SH$ | $SF_5Cl$ |
| $SCl_2$ | $S_2O$ | $CS_2$ | $SCl_4$ | $S_4N_4F_4$ |
| $S_2O_5^{2-}$ | $S_2O_7^{2-}$ | $S_4O_6^{2-}$ | $HOSO_2ONO$ | $HSO_3Cl$ |

5. There are two known isomers of $S_2F_2$ with very different S—S bond lengths. Draw structures for the two isomers, and discuss the bonding in the two molecules.

6. Discuss the bonding in the species:

$$O_3 \quad H_2O_2 \quad OH \quad HO_2 \quad HO_2^- \quad O_3^-$$

7. Using data in Table 18.10, calculate the equilibrium constant at 25°C for reaction

$$SF_4(g) + F_2(g) = SF_6(g)$$

8. Estimate the fraction of $S_8$ and $S_2$ in gaseous sulfur at the normal boiling point of liquid sulfur (see Table 18.10 for the necessary data).

*IV. The Halogens.*

1. Complete and balance the following reactions (if there is no reaction, write N.R.):

(a)  $F_2O(g) + Cl_2(g) \rightarrow$

(b)  $ClO_3^-(aq) + H^+(aq) + H_2C_2O_4(aq) \rightarrow$

(c)  $ClO_2(aq) \xrightarrow{H^+(aq)}$

(d)  $Cl_2O(g) \xrightarrow{spark}$

(e)  $Cl_2(aq) + HNO_3(aq, 12M) \rightarrow$

(f)  $Br_2(aq) + OH^-(aq) \xrightarrow{50°C}$

(g)  $Br_2(aq) + OH^-(aq) \xrightarrow{0°C}$

(h)  $Br^-(aq) + Cl_2(g) \rightarrow$

(i)  $I^-(aq) + Br_2(aq) \rightarrow$

(j)  $Cl_2(g) + OH^-(aq) \xrightarrow{0°C}$

(k)  $ClO_2^-(aq) + H^+(aq) \xrightarrow{50°C}$

(l)  $I_2(s) + Cl_2(aq) \rightarrow$

(m) $Cl_2(g) + F_2(excess) \rightarrow$

(n)  $I^-(aq) + H^+(aq) + O_2(g) \rightarrow$

(o)  $F_2(g) + H_2O(g) \xrightarrow{heat}$

(p)  $HI(aq) + H_2SO_4(aq, conc) \rightarrow$

2. Draw structures for the following species:

$H_2F_2$    $H_6F_6$    $Br_2O$    $BrO_2$    $Br_2O_7$    $FClO_3$

3. Discuss the bonding in $ClO_2$, $Cl_2O$, and $Cl_2O_7$ from a molecular orbital point of view.

4. The molecule $IF_7$ has a pentagonal bipyramidal structure. Draw the structure and discuss the bonding from a valence-bond point of view. (Hint: use $sp^3d^3$ hybridization on I.)

5. Household bleach is a 5% or 6% solution of sodium hypochlorite. How would you prepare such a solution? How much would it cost you to make one gallon of "bleach"?

*V. The Noble Gases.*
1. List the formal charges on xenon for the compounds listed in Table 18.2. Compare this list with the known oxidation states of iodine.

2. In basic solution, $XeO_3$ disproportionates with the release of xenon gas and the formation of perxenates. Write a balanced reaction for the formation of sodium perxenate, $Na_4XeO_6$.

3. What structures might you expect for the ions $I_3^-$ and $BrF_4^-$? The valence shells are isoelectronic with some noble gas compounds.

4. All known compounds of noble gases contain the heavier noble gases. Why are the heavier noble gases most reactive? Do you think isolable compounds of He and Ne will ever be found? Why, or why not?

5. It is easy to prepare the $HF_2^-$ ion in pure HF. How does this fact affect your arguments used in Problem 4?

## References

18.1. See References 16.1 and 16.2.
18.2. D. M. Yost and H. Russell, *Systematic Inorganic Chemistry (of the 5th and 6th Group Elements)* (Prentice-Hall, New York, 1946).
18.3. W. M. Latimer, *The Oxidation States of the Elements and Their Potentials in Aqueous Solutions,* Second Edition (Prentice-Hall, New York, 1952).
18.4. H. H. Claassen, *The Noble Gases* (D. C. Heath, Boston, Mass., 1966).

## CHAPTER 19

# COORDINATION CHEMISTRY OF THE TRANSITION METALS

"From an examination of the behavior of the [cobalt(III) ammine halide] compounds . . . I concluded that the ammonia molecules . . . must all be directly linked with the metal atom. Further I discovered that the ammonia molecules can be replaced by [halide ions] which enter into direct bond with the metal atom.

"The number of atoms which can be directly linked with an atom forming the centre of a complex molecule . . . [is] called the coordination number . . . ."

*Alfred Werner**

## 19.1 INTRODUCTION.

The transition metals can be divided into two groups:
(a) $3d$, $4d$, and $5d$ transition series,
   $3d$—elements 21–30,
   $4d$—elements 39–48,
   $5d$—elements 57, 72–80; and
(b) $4f$ and $5f$ transition series,
   $4f$—elements 58–71 (the *lanthanides*),
   $5f$—elements 90–103 (the *actinides*).

There are ten elements in each $d$-series and fourteen elements in each $f$-series. Within each series the designated set of orbitals is being filled, one electron at a time, as we move from left to right across the series. For example, the ground states of the neutral, gaseous metal atoms of the $3d$-series have the electron configurations:

|  | 21<br>Sc | 22<br>Ti | 23<br>V | 24<br>Cr | 25<br>Mn |
|---|---|---|---|---|---|
| (Argon core) + | $3d^14s^2$ | $3d^24s^2$ | $3d^34s^2$ | $3d^54s^1$ | $3d^54s^2$ |

|  | 26<br>Fe | 27<br>Co | 28<br>Ni | 29<br>Cu | 30<br>Zn |
|---|---|---|---|---|---|
| (Argon core) + | $3d^64s^2$ | $3d^74s^2$ | $3d^84s^2$ | $3d^{10}4s^1$ | $3d^{10}4s^2$ |

---

*Nobel Lectures, Chemistry* (Elsevier Publishing Co., Amsterdam, 1966) pp. 258–259.

The *Pauli Exclusion Principle* places a restriction on how the electrons fill up the available orbitals; no two electrons in a given atom can have the same four quantum numbers. For a principal quantum number of 3, we have the following possibilities:

for $n = 3$, $\ell$ can take the values $0(s)$, $1(p)$, and $2(d)$;
for $\ell = 2$, $m_\ell$ can take the values $0, \pm 1, \pm 2$;

for any value of $m_\ell$, $m_s$ can take the values $+\dfrac{1}{2}$ or $-\dfrac{1}{2}$.

Therefore, we have five $3d$-orbitals (and five $4d$-, and five $5d$-orbitals as well, for $n = 4$ and $n = 5$, respectively), each of which can accommodate two electrons $\left(\text{with spins} +\dfrac{1}{2} \text{ and } -\dfrac{1}{2}\right)$, for a total of ten electrons in all. This is the reason why we have 10 members for each of the $d$ ($3d, 4d$, and $5d$) transition series.

The order of filling of the $3d$-orbitals is not completely regular; the irregularities in the $3d$ series are Cr with the configuration $3d^5 4s^1$, rather than $3d^4 4s^2$, and Cu with $3d^{10} 4s^1$, rather than $3d^9 4s^2$. These irregularities have been ascribed to the "extra" stability of the half-filled ($d^5$) and completely filled ($d^{10}$) $d$-orbitals (Figure 19.1).

The $d$-transition series elements are characterized by:

(a) incompletely filled $d$-orbitals;
(b) strong horizontal relationships in the periodic table;
(c) variable oxidation states;
(d) colored compounds;
(e) unpaired electrons; and
(f) variable coordination numbers.

The existence of multiple oxidation states, and the absorption of energy in the visible region, is a direct consequence of small energy differences between electronic energy states in the transition-metal atoms.

The variety of possible oxidation states gives rise to a great deal of interesting chemistry for these elements. In addition to their tremendous technological importance (e.g., iron in steel, titanium in jet aircraft and ships, copper in heating, refrigeration and electrical devices), transition-metal compounds are essential to the existence of life, being found, for example, in hemoglobin (iron), cytochrome $c$ (iron), and vitamin B-12 (cobalt).

## 19.2   TRANSITION-METAL COMPOUNDS.

The first synthetic transition-metal compound was prepared by the artist Dreisbach in 1691. Dreisbach made sodium ferrocyanide by heating animal refuse with aqueous sodium carbonate in an iron pot. We shall not pursue the details of this obviously complicated reaction any further, but, rather, shall merely note that not much further progress in transition-metal chemistry occurred until A. Werner and C. Jørgensen undertook a systematic study of the preparation and properties of transition-metal compounds. During the latter part of the nineteenth century and the early part of the twentieth century, the work of these two great chemists triggered an explosion of activity in this field that has continued to the present time.

Much of the early work was carried out with complexes of Pt(IV), Pt(II), Pd(II), Co(III), Cr(III), and Ru(III), because the inertness of the complexes of these ions permitted a wide variety of complexes to be iso-

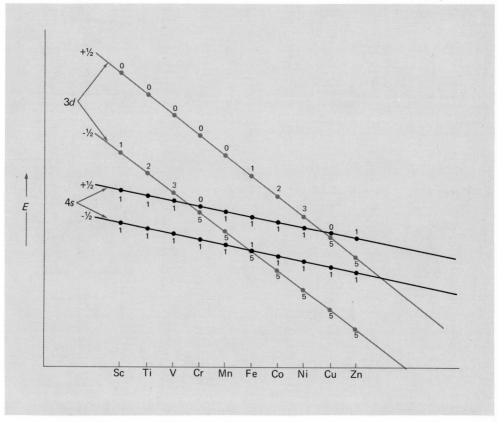

**FIGURE 19.1** Relative energies of the 4s- and 3d-orbitals for *neutral atoms* as a function of nuclear charge. The numbers given represent the number of electrons in the particular level for the element indicated. (After R. Rich, "Periodic Correlations", W. A. Benjamin, Inc., 1965, New York, N.Y., p. 9). For $M^{n+}$ where $n=2$, the 3d level lies below the 4s for all the elements in the series.

lated as pure compounds. As an example, consider Pt(II) complexes involving $NH_3$ and $Cl^-$ as *ligands*.* It was found that compounds having the following stoichiometries could be prepared:

* This is the generic term for the groups that surround the central metal ion in the complex.

| COMPOSITION | MOLES OF AgCl pptd. PER MOLE OF Pt(II) WHEN EXCESS $AgNO_3$ ADDED: | CONDUCTANCE OF AQUEOUS SOLUTION ROUGHLY EQUAL TO THE SAME CONCENTRATION OF: |
|---|---|---|
| (1) $PtCl_2 \cdot 4NH_3$ | 2.00 | $BaCl_2$ |
| (2) $PtCl_2 \cdot 3NH_3$ | 1.00 | KCl |
| (3) $PtCl_2 \cdot 2NH_3$ } isomers | 0 | non-electrolyte |
| (4) $PtCl_2 \cdot 2NH_3$ | 0 | non-electrolyte |
| (5) $KPtCl_3 \cdot NH_3$ | 0 | KCl |
| (6) $K_2PtCl_4$ | 0 | $K_2SO_4$ |

1)   $Pt\ Cl_2 \cdot 4NH_3 \equiv Pt\ (NH_3)_4\ Cl_2$

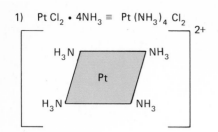

4)*   $Pt\ Cl_2 \cdot 2NH_3 \equiv$ *trans* $- Pt(NH_3)_2Cl_2$

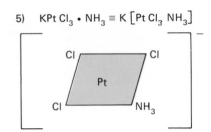

2)   $Pt\ Cl_2 \cdot 3NH_3 \equiv \left[Pt\ (NH_3)_3\ Cl\right]\ Cl$

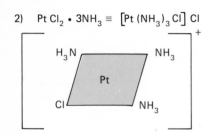

5)   $KPt\ Cl_3 \cdot NH_3 \equiv K\left[Pt\ Cl_3\ NH_3\right]$

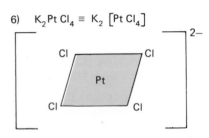

3)*   $Pt\ Cl_2 \cdot 2NH_3 \equiv$ *cis* $- Pt\ (NH_3)_2\ Cl_2$

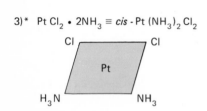

6)   $K_2Pt\ Cl_4 \equiv K_2\left[Pt\ Cl_4\right]$

**FIGURE 19.2**   Compounds of Pt(II) with chloride ion and ammonia.
*Compounds 3 and 4 are geometrical isomers.

Werner correctly concluded from data of this type that there must be a primary coordination sphere involving four groups in a square array around the metal ion. The structures of the complex ions in these salts that were proposed by Werner are as shown in Figure 19.2.

For Pt(IV) complexes with $NH_3$ and $Cl^-$ as ligands, the results shown at the top of the following page were obtained.

These results were explained by assuming an octahedral configuration of ligands around the metal ion. The structures of the complex ions in these salts that were proposed by Werner are shown in Figure 19.3. Werner's proposed structures for both the Pt(II) and Pt(IV) complexes have been confirmed by x-ray diffraction.

Chemical compounds or ions of certain symmetry types exhibit the phenomenon of optical activity. Optically active compounds possess the ability to rotate the plane of polarization of polarized light. The extent of rotation depends on the compound and the frequency of the polarized light used. At a given frequency, where an optically active compound interacts with the polarized light, one of the pair of optical isomers* rotates the plane

---

*Optical isomers always come in pairs.

| Composition | Moles of AgCl pptd. Per Mole of Pt(IV) When Excess $AgNO_3$ Added: | Conductance of Aqueous Solution Roughly Equal to the Same Concentration of: |
|---|---|---|
| (1) $PtCl_4 \cdot 6NH_3$ | 4.00 | $2BaCl_2$ |
| (2) $PtCl_4 \cdot 5NH_3$ | 3.00 | $K_3PO_4$ |
| (3) $PtCl_4 \cdot 4NH_3$ ⎫ isomers | 2.00 | $BaCl_2$ |
| (4) $PtCl_4 \cdot 4NH_3$ ⎭ | 2.00 | $BaCl_2$ |
| (5) $PtCl_4 \cdot 3NH_3$ ⎫ isomers | 1.00 | $KCl$ |
| (6) $PtCl_4 \cdot 3NH_3$ ⎭ | 1.00 | $KCl$ |
| (7) $PtCl_4 \cdot 2NH_3$ ⎫ isomers | 0 | non-electrolyte |
| (8) $PtCl_4 \cdot 2NH_3$ ⎭ | 0 | non-electrolyte |
| (9) $KPtCl_5 \cdot NH_3$ | 0 | $KCl$ |
| (10) $K_2PtCl_6$ | 0 | $K_2SO_4$ |

of polarization in the (+) or positive sense, whereas the other rotates it an equal amount in the (−) or opposite sense. If a molecule possesses a plane or a center of symmetry, then it cannot be optically active. If it does not, then it is (almost certainly) optically active.[*]

Let us consider some examples. The compound $PtCl_2(NH_3)_2$ possesses two geometrical isomers, namely, *cis* and *trans* (Figure 19.2, numbers (3) and (4)). Neither of these geometrical isomers is optically active, however, because they both possess a plane of symmetry[†], namely, the plane in which the Pt(II) and the four ligands all lie. There are also two geometrical isomers (*cis* and *trans*) of the ion $Co(NH_3)_4Cl_2^+$ (Figure 19.4), but neither of these isomers is optically active because they both possess a plane of symmetry. The ion $Co(en)_2Cl_2^+$ also has two geometrical isomers (Figure 19.5) and of these two, one (the *cis*) exhibits optical activity. The two optical isomers of the *cis* complex can be resolved (i.e., separated). Note that the mirror images of the *cis* isomer are not superimposable on one another. These left-and-right-handed *cis* isomers are optical isomers. They have identical thermodynamic and physical properties (except for the direction in which they rotate the plane of polarization of polarized light), but they may differ in their kinetic behavior when reacting with other species that require a specific orientation of the ion before reaction can occur. The thermodynamic properties (such as solubility) of compounds resulting from the *combination* of the isomers with, say, an optically active anion may also differ significantly, owing to packing differences in the solids. This fact is utilized to achieve a separation of a *racemic* mixture (i.e., one containing equal amounts of the two isomers) by fractional crystallization techniques. If *cis*-$Co(en)_2Cl_2^+$ is synthesized via a reaction involving a non-stereospecific pathway,[‡] then a racemic mixture will

---

[*]The general rule is that a compound is optically active if it lacks a rotation-reflection symmetry axis.

[†]Actually these compounds each possess *two* planes of symmetry.

[‡]Many naturally occurring compounds exhibit optical activity. Nature's synthetic pathways are often stereospecific. In fact, the other optical isomer is frequently found to be biologically inactive.

1) $Pt\,Cl_4 \cdot 6NH_3 \equiv \left[Pt\,(NH_3)_6\right]Cl_4$

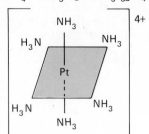

2) $Pt\,Cl_4 \cdot 5NH_3 \equiv \left[Pt\,Cl\,(NH_3)_5\right]Cl_3$

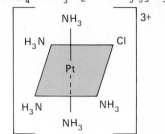

3) $Pt\,Cl_4 \cdot 4NH_3 \equiv cis$ - $\left[Pt\,Cl_2\,(NH_3)_4\right]Cl_2$

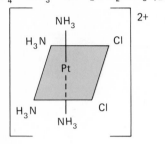

4) $Pt\,Cl_4 \cdot 4NH_3 \equiv trans$ - $\left[Pt\,Cl_2\,(NH_3)_4\right]Cl_2$

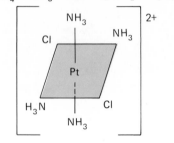

5) $Pt\,Cl_4 \cdot 3NH_3 \equiv cis,\,cis$ - $\left[Pt\,Cl_3\,(NH_3)_3\right]Cl$

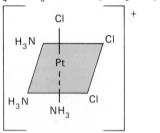

6) $Pt\,Cl_4 \cdot 3NH_3 \equiv cis,\,trans$ - $\left[Pt\,Cl_3\,(NH_3)_3\right]Cl$

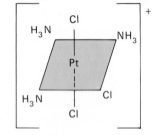

**FIGURE 19.3**  Compounds of Pt(IV) with chloride ion and ammonia *(continued on facing page)*.

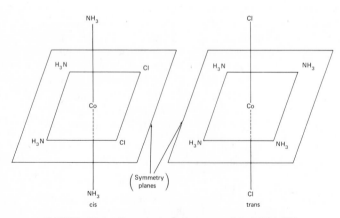

**FIGURE 19.4**  Geometrical  isomers of $Co(NH_3)_4Cl_2^+$.

7)  $Pt\,Cl_4 \cdot 2NH_3 \equiv cis-\left[Pt\,Cl_4\,(NH_3)_2\right]$

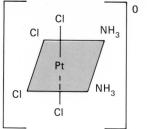

8)  $Pt\,Cl_4 \cdot 2NH_3 \equiv trans-\left[Pt\,Cl_4\,(NH_3)_2\right]$

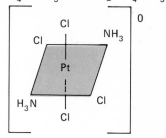

9)  $K\,Pt\,Cl_5 \cdot NH_3 \equiv K\left[Pt\,Cl_5\,(NH_3)\right]$

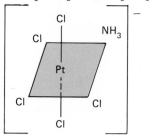

10)  $K_2\,Pt\,Cl_6 = K_2\left[Pt\,Cl_6\right]$

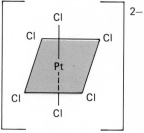

**FIGURE 19.3**   Continued.

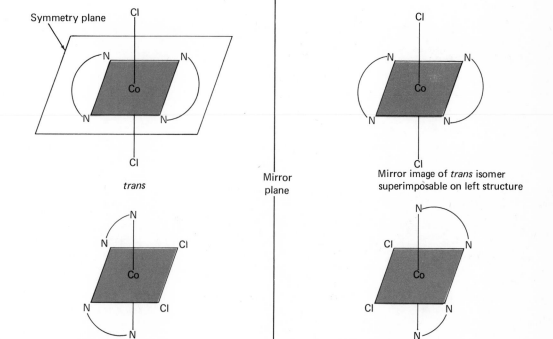

**FIGURE 19.5**   Geometrical isomers of $Co(en)_2Cl_2^+$.

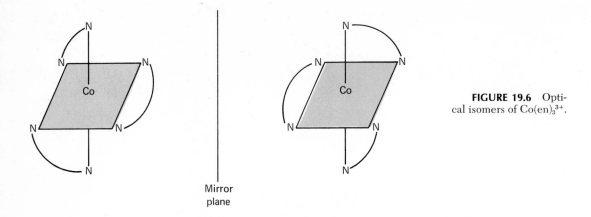

Mirror
plane

**FIGURE 19.6** Optical isomers of $Co(en)_3^{3+}$.

be obtained, because either isomer is just as likely to be formed. A racemic mixture does not show any net rotation of the plane of polarization of polarized light, because the two isomers exert equal and opposite effects.

The ion $Co(en)_3^{3+}$ has no geometrical isomers but it does possess optical isomers, as can be seen from the fact that it possesses a non-superimposable mirror image (Figure 19.6). Optical isomerism is common in octahedral complexes involving multidentate* ligands. Some of the more common and important multidentate ligands are shown in Figure 19.7.

---

*A bidentate ligand occupies two coordination positions on the metal ion, a tridentate ligand, three, and so on.

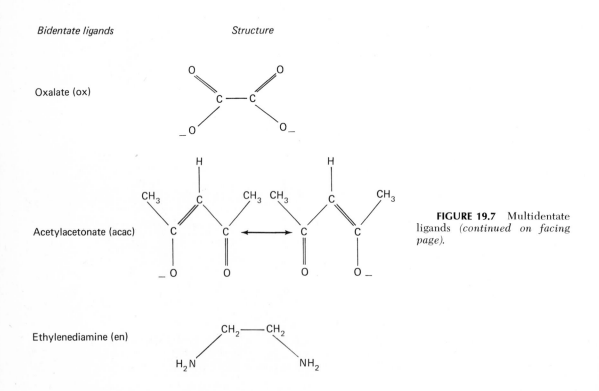

*Bidentate ligands*        *Structure*

Oxalate (ox)

Acetylacetonate (acac)

**FIGURE 19.7** Multidentate ligands *(continued on facing page)*.

Ethylenediamine (en)

2,2' - dipyridyl (dipy)

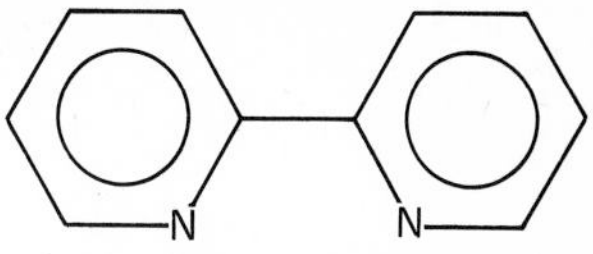

1,10 - phenanthroline (phen)

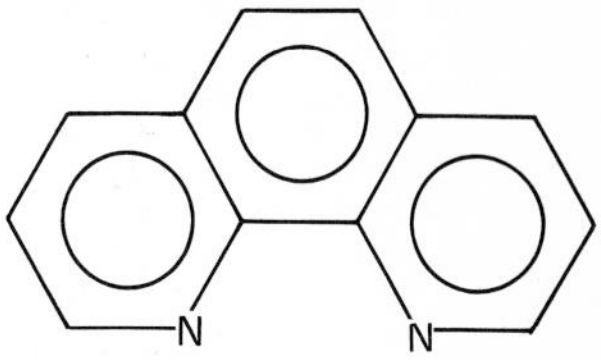

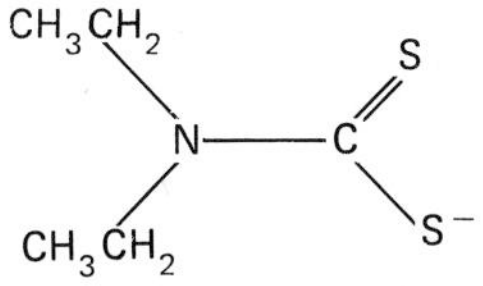

N,N - diethylthiocarbamate

*Tridentate ligands* 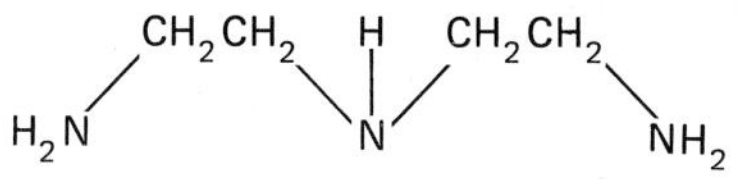
Diethylenetriamine (dien)

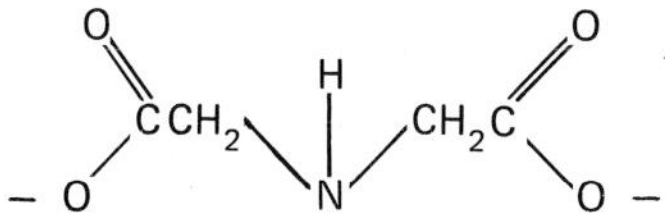

Iminodiacetate (imdac)

**FIGURE 19.7** Continued.

*Tetradentate ligands*

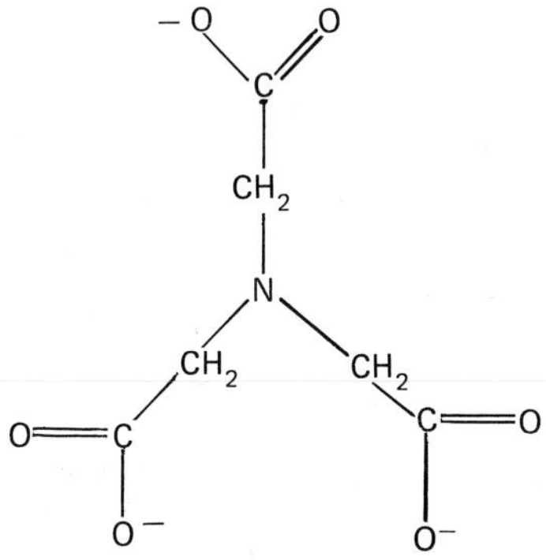

Nitrilotriacetate (nitac)

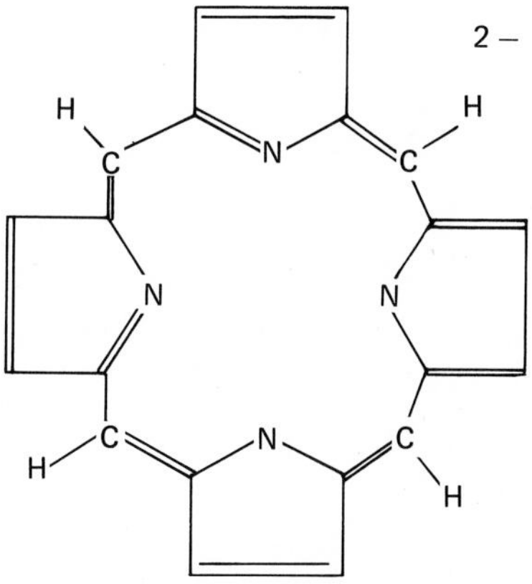

Porphin

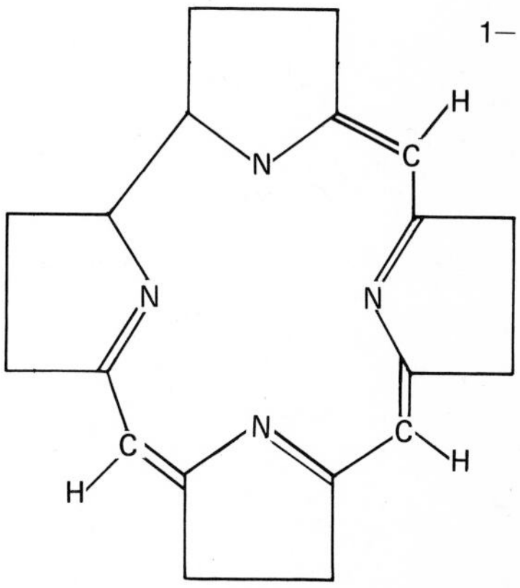

Corrin

## 19.3 BIOLOGICALLY IMPORTANT TRANSITION-METAL COMPLEXES.

Hemoglobin is an iron-containing colloidal protein of molecular weight $6.8 \times 10^4$ that comprises about 35% of the weight of red blood cells, and is responsible for the red color of these cells. The iron, Fe(II), in hemoglobin is capable of reversible reaction with oxygen.* The red blood cells pick up oxygen in the lungs and carry it to the tissues, where it is consumed in the oxidation of foodstuffs. Each hemoglobin molecule contains four heme groups (an iron protoporphyrin, Figure 19.8), and four long polypeptide chains (the *globin*), two *alpha* chains and two *beta* chains. In human *hemoglobin A* the *beta* chains contain 146 amino acid residues,† and the *alpha* chains contain 141 amino acid residues. The sequences of amino acid residues for both the alpha and beta chains have been determined for a number of types of hemoglobins. Gorilla hemoglobin differs from human hemoglobin by just two amino acid residues in the alpha chain

---

*The Fe(II) in heme, as opposed to $Fe(OH_2)_6^{2+}$, is incapable of reducing $O_2$ bound to it. This is presumably due to the stabilization of the electrons in heme Fe(II), owing to their extensive delocalization in molecular orbitals spread over the conjugated porphyrin ring.

†See Chapter 21.

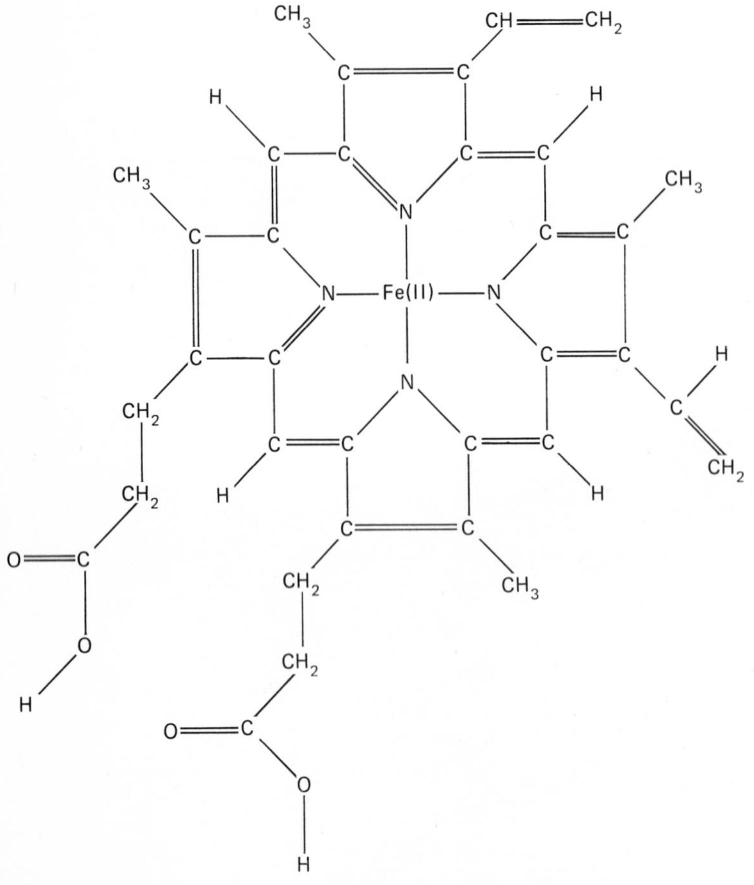

**FIGURE 19.8** The heme molecule is bonded to the protein chains via a peptide linkage involving the two carboxyl groups (–COOH) in heme and two "free" $NH_2$ groups in the polypeptide chain.

Heme

and one in the beta chain. Horse hemoglobin differs from human hemoglobin by about eighteen residues per chain.

Of the six coordination positions of the Fe(II) in a heme group in hemoglobin, four are occupied by nitrogens in the planar porphyrin* ring, a fifth position is occupied by a nitrogen in the five-membered ring of a histidine amino-acid residue, and the sixth position is taken up by either an oxygen molecule, a bicarbonate ion, or a water molecule. Each hemoglobin molecule contains four Fe(II) ions and is capable of carrying four oxygen molecules. The poisonous nature of both cyanide ion and carbon monoxide has been shown to arise from the coordination of these ligands to the iron at the oxygen-carrying sites of hemoglobin. Because CO and $CN^-$ form stronger bonds to Fe(II) than does $O_2$, these ligands block the oxygen-carrying capacity of the molecule.

The blood of some animals is not red. Certain types of crabs and lizards have blue blood, the blue color resulting from the substitution of Cu(II) for iron in the heme groups (forming hemocyanin). A fairly common variety of sea organisms (tunicates) have a blue-violet V(II)-porphyrin in their blood.

Vitamin $B_{12}$ is a transition-metal complex. The complete picture of the function of this substance in living organisms is not yet certain. It is known, however, that it functions as a catalyst for several reactions in the body.

The structure of vitamin $B_{12}$ is shown in Figure 19.9. Of the six coordination positions of the Co(III), four in a plane are taken up by nitrogen in the corrin ring system, the fifth is taken up by the nitrogen in a side chain marked with an asterisk in the figure, and the sixth position is occupied by a cyanide ion.

Another quite different use of transition-metal chemistry is found in medicine. Certain humans are born without the ability to metabolize the amino acid phenylalanine (the condition is known as *phenylketonuria*). Without this ability the infants grow into severely retarded adults, but if this amino acid is deleted from the diet of the child, he grows up without any detectable brain damage. The lack of this amino acid-metabolizing ability can be detected by the presence of phenylketones in the urine. The presence of these substances in the urine of infants is readily detected by the characteristic purple color of the Fe(III)-phenylketone complex. All that need be done is to place a few drops of a dilute solution of an Fe(III) salt on the infant's wet diaper (PKU test). No purple color—no phenylketonuria.

## 19.4 d-ORBITAL SPLITTINGS IN OCTAHEDRAL TRANSITION-METAL COMPLEXES. CRYSTAL-FIELD THEORY.

Many of the interesting aspects of transition-metal chemistry can be related directly to the participation of the metal d-orbitals in the formation of bonds involving the transition metals. The five d-orbitals are depicted in Figure 19.10. The first thing to realize about the d-orbitals is that they have definite spatial orientations with respect to one another. The lobes of $d_{xy}$, $d_{xz}$, and $d_{yz}$ orbitals lie *between* the coordinate axes, whereas the lobes of the $d_{x^2-y^2}$ and $d_{z^2}$ orbitals lie along the coordinate axes (x and y, and z, respectively).

---

*Substituted porphins are known as porphyrins.

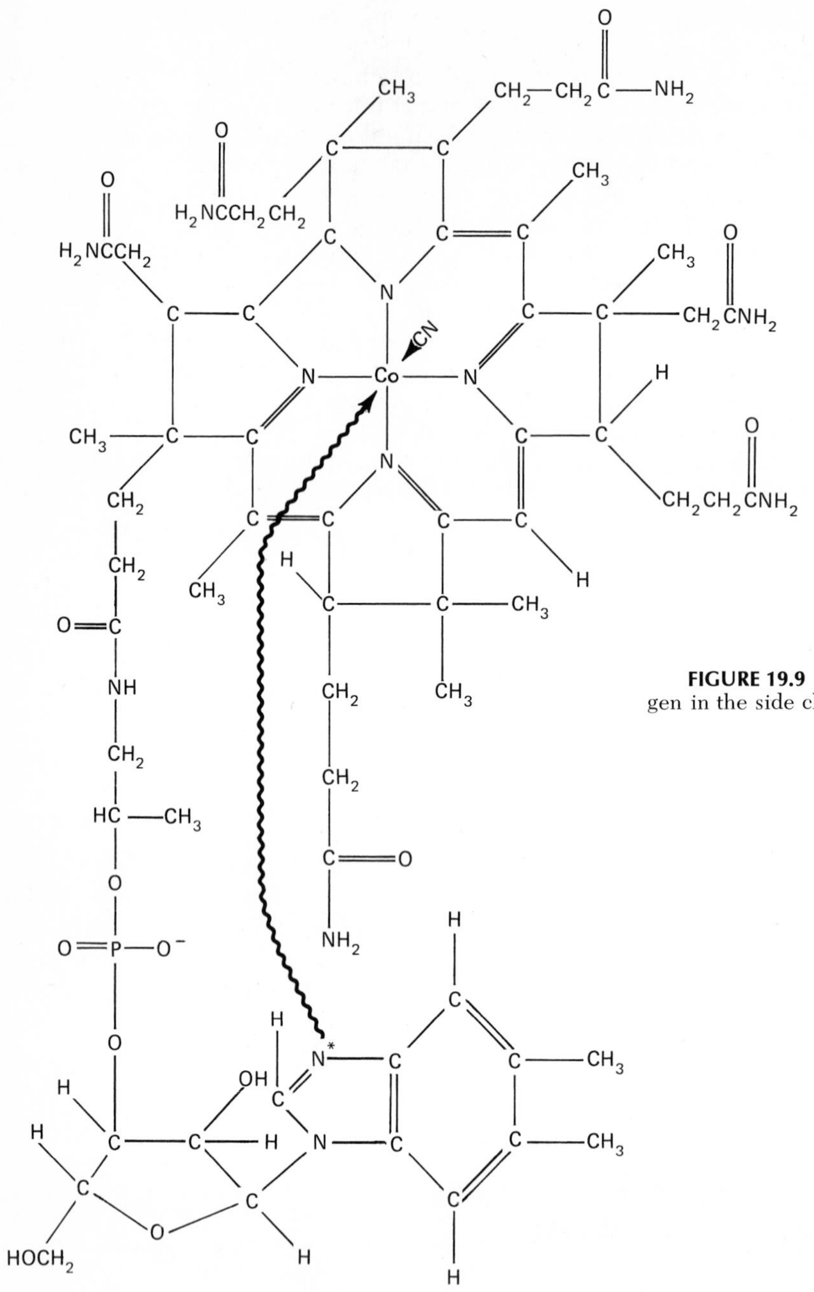

**FIGURE 19.9**   Vitamin B$_{12}$. The N° nitrogen in the side chain is bound to the cobalt.

Although the five orbitals do not look alike (the $d_{z^2}$ being obviously different in appearance from the other four), they all have the same energy in the absence of an inhomogeneous electric or magnetic field.* For the

---

*We can make the five $d$-orbitals spatially, as well as energetically, equivalent by appropriate linear combinations of the above orbitals, but then the bonding to the neighboring atoms is not so easy to visualize in most cases. See R. E. Powell, J. Chem. Educ., *45*, 45 (1968).

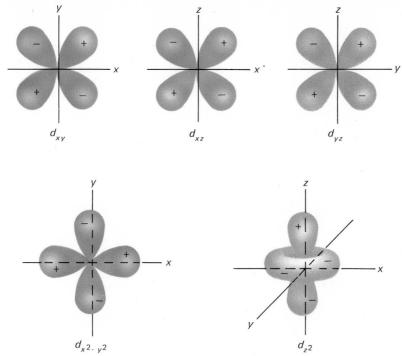

**FIGURE 19.10**  The five energetically equivalent $d$-orbitals. The plus and minus signs denote the signs of wave functions in the indicated regions.

following discussion, it is convenient to think of the $d_{z^2}$ orbital as arising from a linear combination of $d_{z^2-x^2}$ and $d_{z^2-y^2}$ orbitals (Figure 19.11).

Many of the interesting and important aspects of transition-metal chemistry can be rationalized, at least partially, from simple electrostatic considerations. In *Crystal-Field Theory* the metal ion is thought of as being subject to the action of an electric field generated by the charged or dipolar ligands. The electrostatic interaction between the metal ion $d$-orbitals and the ligand electric field gives rise to a splitting of the $d$-orbitals, because the $d$-orbitals do not all have the same orientation with respect to the ligands.

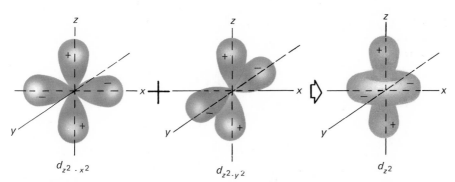

**FIGURE 19.11**  Diagrams showing how a $d_{z^2}$ orbital can be formed from the linear combination of $d_{z^2-x^2}$ and $d_{z^2-y^2}$ orbitals.

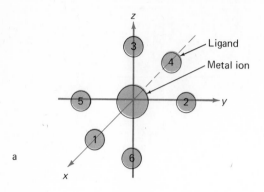

a

**FIGURE 19.12** Regular octa-hedral complex. All six ligands are chemically equivalent and are placed at the same distance from the central metal ion.

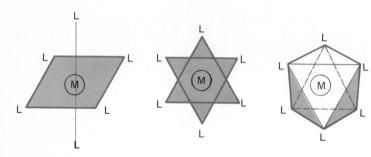

b

Consider a metal ion, M, surrounded by six identical ligands in a regular octahedral array (Figure 19.12). The ligands are placed on the $x$-, $y$-, and $z$-axes equidistant from the metal ion. An electron in either the $d_{x^2-y^2}$ or $d_{z^2}$ $\left(= \frac{1}{2} d_{z^2-x^2} + \frac{1}{2} d_{z^2-y^2}\right)$ orbitals will have a higher energy than an electron in the $d_{xy}$, $d_{xz}$, or $d_{yz}$ orbitals, because the lobes of the $d_{x^2-y^2}$ and $d_{z^2}$ orbitals point directly at the ligands, whereas the lobes of $d_{xy}$, $d_{xz}$, and $d_{yz}$ orbitals point in between the ligands. In other words, an electron in a $d_{x^2-y^2}$ or $d_{z^2}$ orbital experiences more interelectronic repulsion from ligand electrons than an electron in a $d_{xy}$, $d_{xz}$, or $d_{yz}$ orbital. The $d_{x^2-y^2}$ and the $d_{z^2}$ orbitals have equivalent orientations with respect to the ligands and therefore have the same energy in a complex with a regular octahedral geometry. Similarly, the $d_{xy}$, $d_{xz}$, and $d_{yz}$ orbitals have equivalent orientations with respect to the ligands and therefore have the same energy in a regular octahedral complex. Consequently, the ligand electric field splits the five metal $d$-orbitals into a doubly-degenerate, higher-energy set (called the $e_g$ orbitals), and a triply-degenerate, lower-energy set (called the $t_{2g}$ orbitals):

$$
\begin{array}{l}
\circ\circ \quad\quad e_g(d_{x^2-y^2}, d_{z^2}) \\
\Delta_o \\
\circ\circ\circ \quad\quad t_{2g}(d_{xy}, d_{xz}, d_{yz})
\end{array}
$$

The energy difference between the two sets of orbitals is called the crystal-field-splitting energy and is denoted by $\Delta_o$.

How many electrons must be placed in the $t_{2g}$ and $e_g$ orbitals for the regular, octahedral complex $Cr(NH_3)_6^{3+}$? A neutral chromium atom has

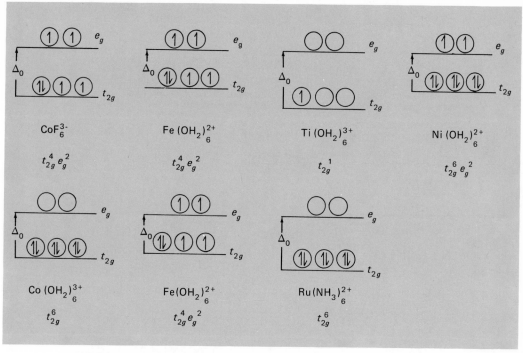

**FIGURE 19.13**  The "d-electron" configurations, $t_{2g}^x e_g^y$, for some transition-metal complexes.

24 electrons; we subtract 3 electrons because the complex involves a $Cr^{3+}$ ion (the $NH_3$ ligands are neutral), and we subtract the 18 electrons in the inert-gas core. Therefore, we must accommodate $3(=24-3-18)$ electrons in the $t_{2g}$ and $e_g$ orbitals. Using the Pauli Principle and Hund's rule, we obtain the d-electron configuration $t_{2g}^3$ for the complex

$$\underline{\quad\bigcirc\bigcirc\quad}\quad e_g \qquad Cr(NH_3)_6^{3+}$$

$$t_{2g}^3$$

$$\underline{\;①①①\;}\quad t_{2g}$$

The d-electron configurations for several other complexes are given in Figure 19.13.

For some transition-metal complexes, the number of d-electrons is such that there are two possibilities for placing the electrons in the $t_{2g}$ and $e_g$ orbitals. The two possible d-electron configurations are referred to as the *low-spin* and *high-spin* configurations. For example, with $Co(NH_3)_6^{3+}$ the two possibilities are:

low-spin                high-spin

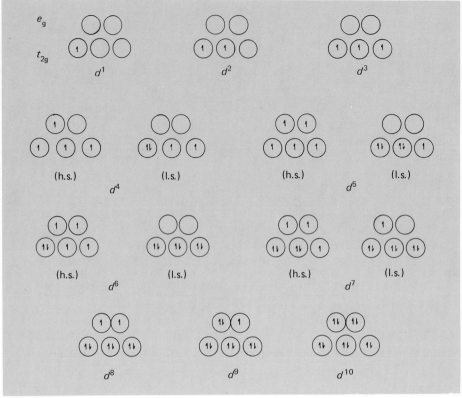

**FIGURE 19.14** Possible electronic $t_{2g}^x e_g^y$ configurations for regular $d^n$ octahedral complex ions, assuming integral occupation numbers for the electrons.

This ambiguity arises for $d^4$, $d^5$, $d^6$, and $d^7$ octahedral ions (Figure 19.14). In placing the electrons in the orbitals we follow the Pauli principle and Hund's rule. Because the $e_g$ orbitals are necessarily of higher energy than the $t_{2g}$ orbitals, whether the "fourth" electron is placed in $e_g$ with the same spin as the three in the $t_{2g}$, or instead is placed in the $t_{2g}$ orbital (necessarily with opposite spin as required by the Pauli principle), depends on the relative magnitudes of $\Delta_o$ and the pairing energy, $P$ (that is, the cost, energetically speaking, of putting a second electron of opposite spin into a $t_{2g}$ orbital that already has one electron). If $P$ is less than the energy separation, $\Delta_o$, between these two orbitals, then the low-spin configuration will be ground state; that is, *if $P < \Delta_o$, we get the low-spin configuration, whereas if $P > \Delta_o$, we get the high-spin configuration.*[*] These pairing energies are appreciable; for example, for Fe(II) and Fe(III) complexes the pairing energies are about 40 and 69 kcal/mole, respectively. Therefore, we expect high-spin complexes unless $\Delta_o$ is quite large. The magnitude of $\Delta_o$ is usually within the range between 20 and 80 kcal.

---

[*]The value of $\Delta_o$ depends on the nature and charge on the metal atom, and on the nature of the ligands. This is also true for the magnitude of $P$, but to a lesser extent, as regards the nature of the ligands, than for $\Delta_o$. In practice, it has been found that $P$ can be estimated with sufficient precision from data on the gaseous ion. For example, $P$ in the complex is about 20% less than in the gaseous ion.

## 19.5  MAGNETIC MOMENTS OF OCTAHEDRAL COMPLEXES.

Substances with unpaired electrons exhibit paramagnetism. When the paramagnetism arises solely from the *spin magnetic moment* of the unpaired electrons (i.e., when there is no contribution from the *orbital magnetic moment* of the electrons), the expected spin-only magnetic moment of the substance, $\mu$, is

$$\mu = 2\sqrt{S(S+1)} = \sqrt{n'(n'+2)} \tag{19.1}$$

where $S$ is the *total* spin quantum number, and $n'$ is the number of *unpaired* electrons. This equation gives $\mu$ in Bohr Magneton units (1 B.M. is equal to $10^{-21}$ erg·gauss$^{-1}$). In Table 19.1 we have presented the various

TABLE 19.1

MAGNETIC MOMENTS OF TRANSITION-METAL COMPLEXES*

| | OCTAHEDRAL $d^n$ COMPLEXES† | | | | | |
|---|---|---|---|---|---|---|
| NUMBER OF $d$-ELECTRONS | WEAK-FIELD (HIGH-SPIN) $d$ CONFIGURATION | $n'$ | $\mu$ | STRONG-FIELD (LOW-SPIN) $d$ CONFIGURATION | $n'$ | $\mu$ |
| 1 | $t_{2g}^1$ | 1 | 1.73 | $t_{2g}^1$ | 1 | 1.73 |
| 2 | $t_{2g}^2$ | 2 | 2.83 | $t_{2g}^2$ | 2 | 2.83 |
| 3 | $t_{2g}^3$ | 3 | 3.87 | $t_{2g}^3$ | 3 | 3.87 |
| 4 | $t_{2g}^3 e_g^1$ | 4 | 4.90 | $t_{2g}^4$ | 2 | 2.83 |
| 5 | $t_{2g}^3 e_g^2$ | 5 | 5.92 | $t_{2g}^5$ | 1 | 1.73 |
| 6 | $t_{2g}^4 e_g^2$ | 4 | 4.90 | $t_{2g}^6$ | 0 | 0 |
| 7 | $t_{2g}^5 e_g^2$ | 3 | 3.87 | $t_{2g}^6 e_g^1$ | 1 | 1.73 |
| 8 | $t_{2g}^6 e_g^2$ | 2 | 2.83 | $t_{2g}^6 e_g^2$ | 2 | 2.83 |
| 9 | $t_{2g}^6 e_g^3$ | 1 | 1.73 | $t_{2g}^6 e_g^3$ | 1 | 1.73 |
| 10 | $t_{2g}^6 e_g^4$ | 0 | 0 | $t_{2g}^6 e_g^4$ | 0 | 0 |
| | TETRAHEDRAL $d^n$ COMPLEXES‡ | | | | | |
| 1 | $e^1$ | 1 | 1.73 | $e^1$ | 1 | 1.73 |
| 2 | $e^2$ | 2 | 2.83 | $e^2$ | 2 | 2.83 |
| 3 | $e^2 t_2^1$ | 3 | 3.87 | $e^3$ | 1 | 1.73 |
| 4 | $e^2 t_2^2$ | 4 | 4.90 | $e^4$ | 0 | 0 |
| 5 | $e^2 t_2^3$ | 5 | 5.92 | $e^4 t_2^1$ | 1 | 1.73 |
| 6 | $e^3 t_2^3$ | 4 | 4.90 | $e^4 t_2^2$ | 2 | 2.83 |
| 7 | $e^4 t_2^3$ | 3 | 3.87 | $e^4 t_2^3$ | 3 | 3.87 |
| 8 | $e^4 t_2^4$ | 2 | 2.83 | $e^4 t_2^4$ | 2 | 2.83 |
| 9 | $e^4 t_2^5$ | 1 | 1.73 | $e^4 t_2^5$ | 1 | 1.73 |
| 10 | $e^4 t_2^6$ | 0 | 0 | $e^4 t_2^6$ | 0 | 0 |

*$\mu$ is the theoretical "spin-only" magnetic moment in Bohr Magneton units computed from Equation (19.1).

†$n$ is the total number of $d$ electrons, and $n'$ is the number of unpaired $3d$ electrons.

‡No low-spin tetrahedral complexes have been found, presumably because $\Delta_t$ is invariably less than the pairing energy.

$d$-electron configurations for octahedral high-spin and low-spin complexes, together with the values of $\mu$ for these configurations calculated from Equation (19.1). For example, the configuration $t_{2g}^4 e_g^2$ has four unpaired electrons and therefore $\mu = \sqrt{4(4+2)} = 4.90$.

Experimental magnetic moments can be used to determine the number of unpaired electrons in a complex and thus the $d$-electron configuration. For example, consider the following table:

|  | $\mu_{\text{observed}}$ | $\mu_{\text{calculated}}$(HIGH-SPIN) | $\mu_{\text{calculated}}$(LOW-SPIN) |
|---|---|---|---|
| $Fe(OH_2)_6^{2+}$ | 5.3 | $4.90\,(t_{2g}^4 e_g^2,\, n'=4)$ | $0 \quad (t_{2g}^6,\, n'=0)$ |
| $Fe(OH_2)_6^{3+}$ | 5.94 | $5.92\,(t_{2g}^3 e_g^2,\, n'=5)$ | $1.73\,(t_{2g}^5,\, n'=1)$ |
| $Fe(CN)_6^{4-}$ | 0 | $4.90\,(t_{2g}^4 e_g^2,\, n'=4)$ | $0 \quad (t_{2g}^6,\, n'=0)$ |
| $Fe(CN)_6^{3-}$ | 1.76 | $5.92\,(t_{2g}^3 e_g^2,\, n'=5)$ | $1.73\,(t_{2g}^5,\, n'=1)$ |

The observed magnetic moments are readily understood if it is assumed that the cyanide ion is a *strong-field* ligand (i.e., $CN^-$ causes spin pairing, $\Delta_o > P$), and that the water molecule is a *weak-field* ligand ($\Delta_o < P$); that is, we have for both the Fe(II) and Fe(III) oxidation states of iron

$$\Delta_o(\text{Fe}, H_2O) < P < \Delta_o(\text{Fe}, CN^-)$$

In Table 19.2 we have presented data on experimental magnetic moments for some hydrated ions of the $3d$-transition series. In most cases the observed and calculated results are in fairly good agreement, although the results for $Fe^{2+}$, $Co^{2+}$, and $Ni^{2+}$ are obviously high. We expect on theoretical grounds that $\mu_{obs} > \sqrt{n'(n'+2)}$ if the orbital angular momentum is not

**TABLE 19.2**

COMPARISON OF EXPERIMENTAL MAGNETIC MOMENTS WITH CALCULATED SPIN-ONLY MOMENTS

| AQUO ION | NUMBER OF $3d$ ELECTRONS | $\sqrt{n'(n'+2)}$ | $\mu_{obs}$ |
|---|---|---|---|
| $Sc^{3+}$ | 0 | 0 | 0 |
| $Ti^{3+}$ | 1 | 1.73 | 1.75 |
| $V^{3+}$ | 2 | 2.83 | 2.80 |
| $V^{2+}$, $Cr^{3+}$ | 3 | 3.87 | 3.68–4.0 |
| $Cr^{2+}$, $Mn^{3+}$ | 4 | 4.90 | 4.8 |
| $Mn^{2+}$, $Fe^{3+}$ | 5 | 5.92 | 5.96, 5.94 |
| $Fe^{2+}$ | 6 | 4.90 | 5.3 |
| $Co^{2+}$ | 7 | 3.87 | 4.6 |
| $Ni^{2+}$ | 8 | 2.83 | 3.2 |
| $Cu^{2+}$ | 9 | 1.73 | 1.8 |
| $Zn^{2+}$, $Cu^+$ | 10 | 0 | 0 |

From K. S. Pitzer, *Quantum Chemistry* (Prentice-Hall Pub. Co., 1953) p. 378.

completely quenched by the ligand field. For example, if orbital momenta are included, the theoretical magnetic moments for $Co^{2+}$ and $Ni^{2+}$ are 6.5 and 5.5 Bohr Magnetons, respectively. The agreement between observed magnetic moments and calculated spin-only magnetic moments for the $4d$ and $5d$ series is not as good as that for the $3d$ series. The reasons for this are not particularly well understood.

## 19.6  VISIBLE ABSORPTION SPECTRA OF TRANSITION-METAL COMPLEXES.

The varied colors of transition-metal compounds arise from the absorption of light in the visible region of the electromagnetic spectrum. This light absorption is, in most cases, associated with "$d$–$d$" electronic transitions in these compounds. The intensities of $d$–$d$ transitions are weak; typical absorptivities are of the order of 1 to 100 $cm^{-1} \cdot mole^{-1} \cdot \ell$.

The simplest possible $d$-electron configuration for octahedral complexes is $t_{2g}^1$ (e.g., $Sc^{2+}$, $Ti^{3+}$, $V^{4+}$, $Y^{2+}$, $Zr^{3+}$, $Nb^{3+}$, and $La^{2+}$). For example, the ion $Ti(OH_2)_6^{3+}$(aq) has a pale, red-purple color. The visible absorption spectrum of this compound is shown in Figure 19.15. The spectrum contains an absorption with a maximum at about 20,000 $cm^{-1}$ (5,000 Å). This absorption has been assigned to the electronic transition $t_{2g}^1 \rightarrow e_g^1$:

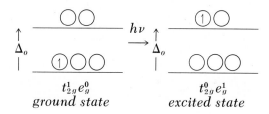

The red-purple color of the $Ti(OH_2)_6^{3+}$ complex is a result of the fact that this electronic transition preferentially absorbs photons with energies in the blue, green, yellow, and orange regions, but does not absorb much in the purple and red regions (Figure 19.15). Most of the excited state species

**FIGURE 19.15**  The visible absorption spectrum of the red-purple ion $Ti(H_2O)_6^{3+}$(aq). (Adapted from reference 19.4.)

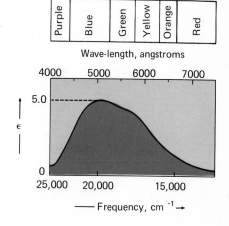

decay to the ground state via radiationless transitions induced by collisions. From the wavelength of the absorption band we can compute the value of the ligand-field-splitting-parameter $\Delta_o$. Because the energy of the photon is $h\nu$, we have

$$E = h\nu = hc/\lambda$$

$$\frac{E}{hc} = \lambda^{-1} = 20{,}300 \text{ cm}^{-1} = 20.3 \times 10^3 \text{ cm}^{-1} = \Delta_o$$

If we wish to express $\Delta_o$ in kcal/mole rather than cm$^{-1}$ (wave numbers) we use the conversion factor 350 cm$^{-1}$ = 1 kcal$\cdot$mole$^{-1}$:

$$\Delta_o = (20.3 \times 10^3 \text{ cm}^{-1}) \left( \frac{1 \text{ kcal}\cdot\text{mole}^{-1}}{350 \text{ cm}^{-1}} \right) = 58.0 \text{ kcal}\cdot\text{mole}^{-1}$$

This result demonstrates that $\Delta_o$ is of the order of ordinary bond energies, and therefore, we can expect some interesting chemically important effects to arise from these ligand-field splittings.

The spectra for octahedral $d^n$ ions where $2 \leq n \leq 8$ are considerably more complex than for the $d^1$ or $d^9$ cases,[*] because of the large increase in the number of possible electronic states. Fortunately, we do not need to consider all of the possible electronic states in order to understand most of the absorption spectrum, because electronic transitions involving changes in the total number of unpaired electrons (spin-forbidden transitions) are extremely weak, if not totally absent from the spectrum.

The blue-violet $d^3$ ion $V(OH_2)_6^{2+}$(aq) has the ground-state electronic configuration $t_{2g}^3$. The absorption spectrum is shown in Figure 19.16(a). There are three absorption maxima in the spectrum occurring at 12.3, 18.5, and 27.9 kcm$^{-1}$ (1 kcm$^{-1}$ = $10^3$ cm$^{-1}$). These have been assigned as shown in Figure 19.17.

The spectrum of the $Ni(OH_2)_6^{2+}$(aq) ion is shown in Figure 19.16(b), together with the spectrum of $Ni(NH_3)_6^{2+}$(aq). There are three principal absorption peaks in each of these spectra,[†] which have been assigned as shown in Figure 19.18. The $Ni(OH_2)_6^{2+}$(aq) complex is green because the two higher energy electronic transitions in this complex (at 25,300 and 13,500 cm$^{-1}$) absorb light in the purple and blue, and the yellow, orange, and red regions of the visible spectrum, respectively. That is, a solution of the complex is transparent to green light. When excess concentrated aqueous ammonia solution is added to an aqueous solution containing $Ni(OH_2)_6^{2+}$ (say, the perchlorate, or nitrate, or sulfate salt), the color of the solution changes from green to a beautiful blue-violet. This change in color arises from the conversion of the Ni(II) from the aquo to the ammine complex

$$Ni(OH_2)_6^{2+}(aq) + 6NH_3(aq) \rightarrow Ni(NH_3)_6^{2+}(aq) + 6H_2O(\ell)$$

green                    blue-violet

---

[*]There are additional complications even in the $d^9$ case, where we might expect to observe only a single absorption corresponding to $t_{2g}^6 e_g^3 \rightarrow t_{2g}^5 e_g^4$. Note also the weak shoulder at 17,400 cm$^{-1}$ in the $Ti(OH_2)_6^{3+}$ spectrum. We shall postpone a consideration of these details until we take up the Jahn-Teller effect. Suffice it to note at this point that there are no perfectly regular $d^1$ or $d^9$ octahedral complexes.

[†]The 13,5000 cm$^{-1}$ peak in the $Ni(OH_2)_6^{2+}$ spectrum is a doublet. This effect has been ascribed to vibronic interactions in this complex, but the details of the explanation are complex and incomplete, and therefore, we shall not discuss them.

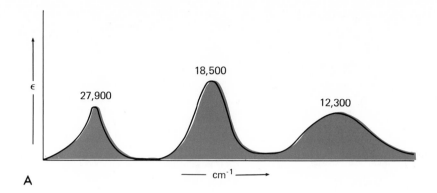

A

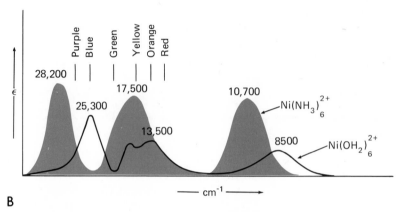

B

**FIGURE 19.16(a)** The absorption spectrum of the violet ion $V(OH_2)_6^{2+}$(aq). (Adapted from reference 19.1).

**(b)** The absorption spectra of the green $Ni(OH_2)_6^{2+}$(aq) and the blue-violet $Ni(NH_3)_6^{2+}$(aq) ions. (Adapted from reference 19.4).

**FIGURE 19.17** Assignment of observed electronic transitions for the violet $d^3$ ion $V(OH_2)_6^{2+}$(aq). The presence of an uneven occupancy of the $e_g$ orbitals lifts the degeneracy of the $t_{2g}$ and $e_g$ orbitals, thereby giving rise to two electronic states of the type $t_{2g}^2 e_g^1$. The $27.9 \times 10^3 \text{cm}^{-1}$ band has been assigned to the two-electron transition $t_{2g}^3 \rightarrow t_{2g}^1 e_g^2$.

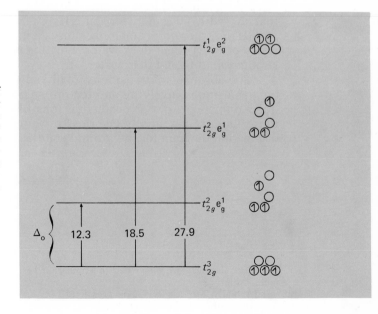

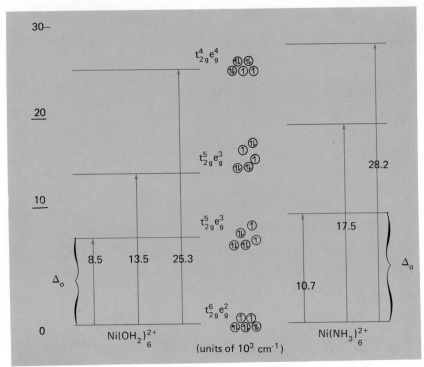

**FIGURE 19.18**   Assignment of electronic transitions of the green and blue-violet ions $Ni(OH_2)_6^{2+}$ and $Ni(NH_3)_6^{2+}$. Note that the unequal occupation of the $e_g$ orbital lifts the degeneracy of the $t_{2g}$ and $e_g$ orbitals, giving rise to two electronic states of the type $t_{2g}^5 e_g^3$.

A study of Figure 19.16(b) shows that this color change occurs because the replacement of $H_2O$ with $NH_3$ shifts the electronic absorptions to shorter wavelengths. In the $Ni(NH_3)_6^{2+}$ complex there is an absorption centered right in the green; the window between the $t_{2g}^6 e_g^2 \rightarrow t_{2g}^5 e_g^3$ (upper) and $t_{2g}^6 e_g^2 \rightarrow t_{2g}^4 e_g^4$ transitions is now located in the blue-purple region. Note that $\Delta_o = 10{,}700$ cm$^{-1}$ for $Ni(NH_3)_6^{2+}$, whereas $\Delta_o = 8{,}500$ cm$^{-1}$ for $Ni(OH_2)_6^{2+}$; therefore, ammonia causes a greater splitting of the electronic states of Ni(II) than does water. For this reason ammonia is said to be a stronger-field ligand than water.

The absence[*] of transitions involving changes in the spin multiplicity (i.e., transitions involving an electron spin-flip) has an interesting consequence for high-spin $d^5$ octahedral ions like Mn(II) and Fe(III); namely, these ions are essentially colorless in solution.[†] Inspection of the orbital occupation diagram for the ground state of a high-spin $d^5$ ion

$$\underline{\textcircled{1}\textcircled{1}} \qquad e_g$$

$$\underline{\textcircled{1}\textcircled{1}\textcircled{1}} \qquad t_{2g}$$

[*]By "absence" we mean that the transition has an absorptivity of less than 1.0 cm$^{-1}$· mole$^{-1}$·liter.

[†]Actually, Mn$^{2+}$ is very pale pink, and Fe$^{3+}$ a very pale violet, as a result of very low intensity "spin-forbidden" transitions.

shows that any conceivable $d$–$d$ electronic transition requires a change in spin of one of the electrons.

Many minerals, both common and gem quality, owe their colors to the presence of variable (often trace) amounts of transition-metal ions. The mineral corundum, $Al_2O_3$, is colorless when pure. When this mineral occurs with a trace of $Cr^{3+}$ impurity (present substitutionally for $Al^{3+}$) it is a brilliant red, and goes by the more familiar name of *ruby*. If any other transition metal impurity which gives rise to color is present in corundum, then we have *sapphires*. Blue sapphires are corundum with a $V^{3+}$ impurity, and yellow sapphires are corundum with an $Fe^{3+}$ impurity. Other gems that owe their color to the presence of transition metals are:

*emeralds*: a rare variety of the mineral beryl, $Al_2O_3 \cdot 6SiO_2 \cdot 3BeO$, which is green, owing to a $Cr^{3+}$ impurity.

*garnets*: silicates of $Ca^{2+}$, $Mg^{2+}$, $Fe^{2+}$, or $Mn^{2+}$ with $Al^{3+}$ or $Fe^{3+}$ impurities. The colors are variable; the red garnet is a gem stone.

*turquoise*: a blue, copper-aluminum hydroxy phosphate. The blue color is a result of the presence of $Cu^{2+}$.

*jade*: a sodium aluminum silicate, $NaAl(SiO_3)_2$; the green variety ($Cr^{3+}$ impurity?) is a gem stone.

## 19.7  THE SPECTROCHEMICAL SERIES.

Investigations of the spectra of various octahedral complexes have led to the $\Delta_o$ values assembled in Table 19.3. Roughly speaking, for a given geometry,* transition series, and charge on the metal ion, $\Delta_o$ is to a large extent a characteristic of the ligand. For example, for water as a ligand in $M^{2+}$ complexes of the $3d$-transition series we have (where the numbers in parentheses are the $\Delta_o$ values in $10^3$ cm$^{-1}$ units)

$$V(OH_2)_6^{2+}(12.3) \qquad Fe(OH_2)_6^{2+}(10.4)$$
$$Cr(OH_2)_6^{2+}(14.0) \qquad Co(OH_2)_6^{2+}(9.3)$$
$$Mn(OH_2)_6^{2+}(8.5) \qquad Ni(OH_2)_6^{2+}(8.5)$$

and

$$\Delta_o\{M(II), 3d; H_2O, oct\} = 10.5 \pm 1.8 \text{ kcm}^{-1}$$

For a given metal and ligand the value of $\Delta_o$ increases as the charge on the ion increases, the magnitude of this increase being about 40 to 100% for an increase of 1 unit in charge; some examples are:

$$Fe(phen)_3^{2+}(19.6) \qquad V(OH_2)_6^{2+}(12.3) \qquad Co(NH_3)_6^{2+}(10.1)$$

$$Fe(phen)_3^{3+}(27.4) \qquad V(OH_2)_6^{3+}(17.8) \qquad Co(NH_3)_6^{3+}(22.9)$$

This effect has been ascribed to the increased ability of the more highly charged ion to bring the ligands in closer, for more effective interaction with the metal orbitals. The higher charge on the metal ion also lowers the energy of the metal orbitals, and increased metal-ligand interaction will result if the lowering of the metal-ion orbital energies makes them closer in energy to those of the ligands.

---

*As we shall see subsequently, the ligand-field splitting for a given ligand depends strongly on the geometry of the complex.

**TABLE 19.3**

LIGAND FIELD SPLITTINGS, $\Delta_o$, FOR VARIOUS OCTAHEDRAL COMPLEXES IN $10^3$ cm$^{-1}$ UNITS.

| METAL LIGAND | CN⁻ | phen | en | NH₃ | OH₂ | EDTA⁴⁻ | C₂O₄²⁻ | F⁻ | Cl⁻ | Br⁻ | I⁻ |
|---|---|---|---|---|---|---|---|---|---|---|---|
| Ti(III) | | | | | 20.3 | | | 17.0 | | | |
| V(II) | | | | | 12.7 | | | | | | |
| V(III) | | | | | 17.8 | | | | | | |
| Cr(III) | 26.6 | | 21.9 | 21.6 | 17.4 | | 17.5 | | | | |
| Fe(III) | 35.0 | 27.4 | | | 14.3 | 12.8 | 13.7 | | | | |
| Fe(II) | 31.4 | 19.6 | | 12.2 | 10.4 | 9.7 | 10.0 | | | | |
| Co(II) | | | 11.0 | 10.1 | 9.3 | 10.2 | 9.1 | | | | |
| Co(III) | 34.8 | | 23.2 | 22.9 | 18.2 | 19.7 | 18.0 | | | | |
| Ni(II) | | | 11.6 | 10.8 | 8.5 | | | | | | |
| Cr(II) | | | | | 14.0 | | | | | | |
| Mn(II) | | | | | 8.5 | | | | | | |
| Mn(III) | 28 | | | | | | | 20.0 | 21.7 | | |
| Cu(II) | | | | | (11) | | | | | | |
| Co(III) | | | | 22.9 | | | | | | | |
| Rh(III) | | | | 34.1 | | | | | | 20.3 | 19.0 |
| Ir(III) | | | | 40.0 | | | | | | | 25.0 |

Another interesting (but not particularly well understood) effect is the dramatic increase in $\Delta_o$ that is observed for a given ligand and metal ion charge type as we move down a group in the periodic table, for example

$$\text{Co(NH}_3)_6^{3+}(22.9) \qquad \text{Rh(NH}_3)_6^{3+}(34.1) \qquad \text{Ir(NH}_3)_6^{3+}(40.0)$$

This effect is of such a magnitude that essentially all M(III) ions for the $4d$ and $5d$ series are low spin.

The ligands in Table 19.3 can be arranged in a series of increasing $\Delta_o$ values, in which the order of the members is independent of the nature of $M^{n+}$ for a given value of $n$. That the ligands can be arranged in such a series was first pointed out by Tsuchida, and the series is known as the *Tsuchida spectrochemical series:*

I⁻ < Br⁻ < Cl⁻ < F⁻ ≲ OH₂ ~ many other oxygen-bonded ligands[*]

weak field, usually high spin (except for M(IV) $5d$ ions)    intermediate field, both high- and low-spin complexes observed

< NH₃ ~ many other nitrogen-bonded ligands[†] < H₂NCH₂CH₂NH₂

---

[*] For example, oxalate, hydroxide, and acetate ions.
[†] For example, nitrogen-bound thiocyanate.

increasing $\Delta_o$

$\left\{ \text{NO}_2^- < \text{phen}^{\circ} < \text{CN}^- < \text{CO} \right.$

strong field, usually low spin

Crystal-Field Theory cannot explain the order of the ligands in this series. For example, why does cyanide ion or CO produce so much larger ligand field splittings than fluoride ion or water? An understanding of the spectrochemical series requires an excursion into the molecular orbital theory of transition-metal complexes, which we will undertake later in this chapter.

## 19.8 TETRAGONALLY-DISTORTED, OCTAHEDRAL COMPLEXES. THE JAHN-TELLER EFFECT.

Suppose we subject a regular, octahedral complex of a transition metal to a tetragonal distortion (Figure 19.19). That is, suppose we reduce the symmetry of a complex from octahedral to tetragonal (four-fold) by increasing the M–L bond distances of the two ligands on the z-axis, and simultaneously decreasing the M–L bond distances of the other four ligands, which lie at the corners of a square in the xy-plane. What effect will this distortion have on the orbital energies in the complex? Because the six ligands are

*phen = 1,10-orthophenanthroline

planar

**FIGURE 19.19** Tetragonal distortion of an octahedral complex. Note that tetragonal (i.e., four-fold) symmetry would also be obtained if we elongated the four M-L in-plane distances and shortened the M-L distances for the two ligands above and below the plane.

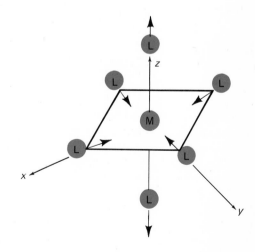

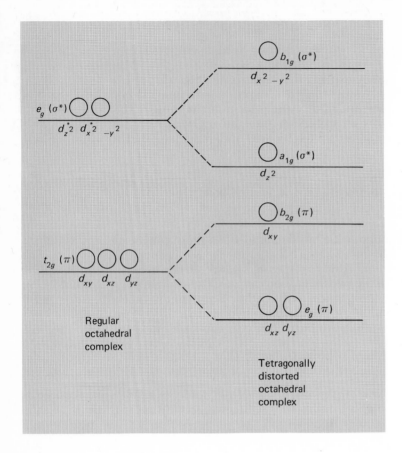

**FIGURE 19.20** Splitting of the orbitals $t_{2g}(\pi)$ and $e_g(\sigma^\circ)$ of a regular octahedral complex which is subjected to a tetragonal distortion. We have indicated the metal $d$-orbitals that are involved in the particular molecular orbitals shown. The $\pi$ and $\sigma^\circ$ designations are discussed in Section 19.15.

no longer equivalent in the tetragonally-distorted complex (there is a set of four in the $xy$-plane at relatively short M–L distances, and a set of two on the $z$-axis at relatively long M–L distances), the previously degenerate orbital sets, $e_g$ and $t_{2g}$, are split as is shown in Figure 19.20.

A movement of the four ligands in along the $x$ and $y$ axes will increase the interelectronic repulsion between an electron in the metal $d_{x^2-y^2}$ orbital and the ligand electrons, whereas lengthening the M–L distances along the $z$ axis will weaken the electrostatic interaction between the electron in the metal $d_{z^2}$ orbital and the ligand electrons. Thus, the $d_{x^2-y^2}$ metal orbital becomes higher in energy than the $d_{z^2}$ metal orbital when the M–L distances in the $xy$-plane are shorter than the M–L distances on the $z$-axis. The $t_{2g}$ set of metal orbitals is also split by the lowering of the symmetry on going from octahedral to tetragonal geometry in the complex. The $d_{xy}$ orbital is destabilized (because it lies in the $xy$-plane), and the $d_{xz}$ and $d_{yz}$ orbitals (which are oriented similarly with respect to the ligands in the distorted complex) remain degenerate and are stabilized relative to the $t_{2g}$ set.

Such tetragonal distortions, and the concomitant lifting of the orbital degeneracy, *must* occur in any regular octahedral complex in which either the $t_{2g}$ or the $e_g$ orbitals are occupied unequally (Jahn-Teller Effect[*]). The

---

[*]The Jahn-Teller Effect is quite general and is by no means limited to octahedral, transition-metal complexes. Fundamentally, the presence of the surrounding ligands frees the electron from the necessity of choosing between two members of a degenerate set by lifting the degeneracy.

magnitude of the distortion is much greater when the unequal occupancy occurs in the $e_g$ orbitals, because these orbitals are antibonding and are oriented such that they concentrate electron density directly along the M–L axis. The most important cases are, then, as follows:

$$t_{2g}^3 e_g^1 \qquad \mathrm{Cr^{2+}, Mn^{3+}}$$

$$t_{2g}^6 e_g^1 \qquad \mathrm{Co^{2+}, Ni^{3+}}$$

$$t_{2g}^6 e_g^3 \qquad \mathrm{Cu^{2+}, Ag^{2+}}$$

Much smaller distortions are expected for the configurations $t_{2g}^1$, $t_{2g}^2$, $t_{2g}^4$, $t_{2g}^5$, $t_{2g}^4 e_g^2$, and $t_{2g}^5 e_g^2$. We shall treat these latter configurations as regular, octahedral complexes.*

The chemical consequences of such distortions are important, as we shall see later in our discussion on the stability and lability of such complexes. For now we shall content ourselves with noting that we should represent the "$d$-electron" configurations of, say, $\mathrm{Cu(OH_2)_6^{2+}}$ as:†

(a)

and not as:

(b)

We expect three electronic transitions in case (a), and only one in case (b). In all copper complexes of the type $\mathrm{CuL_6^n}$ so far investigated, the spectroscopic evidence is consistent with diagram (a) and not with (b). For example, the visible spectrum of $\mathrm{Cu(OH_2)_6SiF_6(s)}$ shows three absorptions attributable to the copper complex, whereas the spectrum of $\mathrm{Cu(OH_2)_6^{2+}(aq)}$ contains a very broad absorption of enhanced intensity that can be resolved at least two, and possibly three, distinct bands.

## 19.9  SQUARE-PLANAR COMPLEXES.

A regular, square-planar, transition-metal complex involves four identical ligands in a square array about a transition-metal ion (Figure

---

° Note, however, that the shoulder at 17,400 cm$^{-1}$ in the $\mathrm{Ti(OH_2)_6^{3+}}$ spectrum is believed due to a Jahn-Teller distortion of the $t_{2g}^1$ ground state.

† The symbols $e_g$, $b_{2g}$, $a_{1g}$, and $b_{1g}$ are the standard group theory labels for the symmetry of these orbitals. The notation carries certain symmetry information that need not concern us. We shall simply regard them as labels.

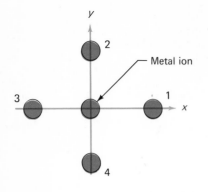

**FIGURE 19.21**  Arrangement and numbering of ligands in a square-planar complex.

19.21). A square-planar complex can be viewed as the limiting case of a tetragonally-distorted, octahedral complex in which the two ligands on the $z$-axis are completely removed. The removal of the ligands on the $z$-axis leads to a further stabilization of the $a_{1g}(d_{z^2})$ and the $e_g(d_{xz}, d_{yz})$ orbitals (Figure 19.20) relative to the $b_{2g}(d_{xy})$ and $b_{1g}(d_{x^2-y^2})$ orbitals. Figure 19.22 shows the splitting of the $e_g$ and $t_{2g}$ orbitals that results when an octahedral complex is converted into a square-planar complex. The ordering of the levels is based on spectroscopic evidence. Some experimental ligand-field splittings for square-planar complexes are given in Table 19.4. Especially noteworthy is the large value of $\Delta_{sp}(\simeq \Delta_o)$, which presumably

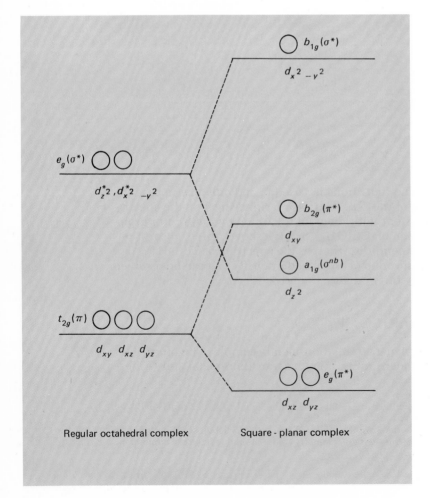

**FIGURE 19.22**

TABLE 19.4

LIGAND-FIELD SPLITTINGS FOR SOME SQUARE-PLANAR COMPLEXES*

| | ($10^3 \, cm^{-1}$ units) | | | |
|---|---|---|---|---|
| **COMPLEX** | $\Delta_{sp}$ | $\Delta_a$ | $\Delta_b$ | $\Delta_{sp} + \Delta_a + \Delta_b$ |
| $PdCl_4^{2-}$ | 19.2 | 6.2 | 1.4 | 26.8 |
| $PdBr_4^{2-}$ | 18.4 | 5.4 | 1.4 | 25.2 |
| $PtCl_4^{2-}$ | 23.4 | 5.9 | 4.4 | 33.7 |
| $PtBr_4^{2-}$ | 22.2 | 6.0 | 3.6 | 31.8 |
| $Ni(CN)_4^{2-}$ | 25.0 | 9.9 | 0.6 | 35.5 |

$$\Delta_{sp} \left\{ \begin{array}{l} \bigcirc \quad b_{1g}(d_{x^2-y^2}) \end{array} \right.$$

$$\Delta_a \left\{ \begin{array}{l} \bigcirc \quad b_{2g}(d_{xy}) \end{array} \right.$$

$$\Delta_b \left\{ \begin{array}{l} \bigcirc \quad a_{1g}(d_{z^2}) \\ \bigcirc\bigcirc \quad e_g(d_{xz}, d_{yz}) \end{array} \right.$$

(not to scale)

*Data from reference 19.1. $\Delta_{sp} \simeq \Delta_o$, $\Delta_{sp} + \Delta_a + \Delta_b \simeq 1.3 \, \Delta_o$.

arises from the fact that in a square-planar complex the four ligands inter-act very strongly with the metal $d_{x^2-y^2}$ orbital, and only relatively weakly with the other four metal orbitals. The very unstable nature of the $d_{x^2-y^2}$ orbital has an important influence on which transition metals are likely to form square-planar complexes. Square-planar complexes are most likely to be formed with $d^8$ (and to a lesser extent with $d^9$) ions, because with $d^8$ ions the very unstable $d_{x^2-y^2}$ orbital is empty.

Complexes of the $d^8$ ions Pt(II), Pd(II), Au(III), Rb(I) and Ir(I) are usually square-planar. Several square-planar complexes of the $d^8$ ion Ni(II) are known.

## 19.10   REGULAR, TETRAHEDRAL, TRANSITION-METAL COMPLEXES.

The five metal-ion $d$-orbitals in a tetrahedral complex split up in a dif-ferent way than in an octahedral complex, because of the different orienta-tions of the ligands with respect to the metal $d$-orbitals. The placement of the ligands with respect to the metal $d$-orbitals can be deduced from Figure 19.23. The first point to recognize is that the $d$-orbitals fall into two sets with regard to their orientation relative to the ligands. One set consists of the $d_{x^2-y^2}$ and $d_{z^2}$ orbitals. Each of the lobes of the orbitals $d_{x^2-y^2}$ and $d_{z^2} \left( = \frac{1}{2} d_{z^2-x^2} + \frac{1}{2} d_{z^2-y^2} \right)$ points directly at the center of a face of the cube. The L–M–L tetrahedral angle is 109.5°, and therefore the center of a lobe of one of these orbitals is located 54.8° (=109.5°/2) from the two ligands at the corners of that face of the cube. The second set consists of the orbitals

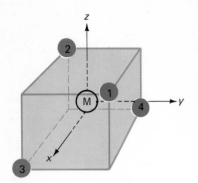

**FIGURE 19.23**   Regular tetrahedral complex.

$d_{xy}$, $d_{xz}$, and $d_{yz}$. Each of the lobes of these orbitals points at one of the edges of the cube. For example, one of the lobes of the $d_{xy}$ orbital points at the cube edge that is parallel to the $z$-axis and on which ligand 1 is located. The center of this lobe of the $d_{xy}$ orbital is located $35.2°$ $\left[= \dfrac{1}{2}\left\{180 - 2(109.5°/2\right\}\right]$ from ligand 1. All of the lobes of the $d_{xy}$, $d_{xz}$, and $d_{yz}$ orbitals are located at this same angle with respect to one of the ligands. An electron in the $d_{xy}$, $d_{xz}$, or $d_{yz}$ orbitals experiences a greater interelectronic repulsion from the ligand electrons than an electron in the $d_{x^2-y^2}$ or $d_{z^2}$ orbitals, because the former set of orbitals is closer to the ligands than the latter set. Consequently, the splitting of the five $d$-orbitals in a regular tetrahedral complex is

$$\uparrow \quad \underline{\bigcirc\bigcirc\bigcirc} \qquad t_2(d_{xy}, d_{xz}, d_{yz})$$
$$\Delta_t$$
$$\Big| \quad \underline{\quad\bigcirc\bigcirc\quad} \qquad e(d_{x^2-y^2}, d_{z^2})$$

The complex $CoCl_4^{2-}$ has a tetrahedral structure; the ion $Co^{2+}$ is a $d^7$ ion ($7 = 27 - 2 - 18$) and the $d$-electron configuration of this ion is $e^4 t_2^3$:

$$\underline{\textcircled{\uparrow}\textcircled{\uparrow}\textcircled{\uparrow}} \qquad t_2$$
$$\qquad\qquad\qquad CoCl_4^{2-}$$
$$\underline{\quad\textcircled{\uparrowdownarrow}\textcircled{\uparrowdownarrow}\quad} \qquad e \quad e^4 t_2^3$$

The orbital splitting for a tetrahedral complex, $\Delta_t$, represents the energy difference between an upper ($t_2$) three-fold degenerate set of orbitals and a lower ($e$) two-fold degenerate set of orbitals, whereas the reverse is true for an octahedral complex.

Both experiment and theory show that $\Delta_t < \Delta_o$ for the same ligands in any given $d$-transition series. Theory shows that, for the same ligands at the same M–L distances as in the corresponding octahedral complex, $\Delta_t = (4/9)\,\Delta_o$. The experimental results show that for a given ligand (even for variable M–L distances), in most cases $\Delta_t \leqslant (1/2)\,\Delta_o$. The interaction of a smaller number of ligands with the metal ion perturbs the metal orbitals less for tetrahedral complexes than for octahedral complexes. Some measured $\Delta_t$ values are given in Table 19.5.

The smaller value of $\Delta_t$, as compared to $\Delta_o$, has several important consequences, one of which is that there are no known low-spin tetrahedral

**TABLE 19.5**

EXPERIMENTAL VALUES OF $\Delta_t$ FOR SOME TETRAHEDRAL COMPLEXES*

| COMPLEX | $\Delta_t(10^3\,cm^{-1})$ |
|---|---|
| $VCl_4$ | 9.0 |
| $CoCl_4^{2-}$ | 3.3 |
| $CoBr_4^{2-}$ | 2.9 |
| $CoI_4^{2-}$ | 2.7 |
| $Co(NCS)_4^{2-}$ | 4.7 |

* Data from reference 19.1.

complexes, presumably because the electron pairing energy is always greater than $\Delta_t$. The $d$-electron configurations and the spin-only magnetic moments for tetrahedral complexes of $d^n$ ions are given in Table 19.2.

In the spectra of tetrahedral complexes, the observed spin-allowed electronic transitions are more intense than for octahedral complexes, owing to the lower symmetry of the complex, and appear at longer wavelengths (because of the smaller value of $\Delta_t$).

Although tetrahedral transition-metal complexes are not as common as octahedral complexes, they are by no means rare. Tetrahedral geometry is most frequently found for $d^0$, $d^5$, $d^7$, and $d^{10}$ ions, for example:

$d^0$  $CrO_4^{2-}$, $MnO_4^-$

$d^5$  $MnBr_4^{2-}$, $FeCl_4^-$

$d^7$  $Co(NCS)_4^{2-}$

$d^{10}$  $Zn(CN)_4^{2-}$, $Cd(NH_3)_4^{2+}$, $HgI_4^{2-}$, $Ni(CO)_4$

The reasons for this will be discussed in Section 19.13.

Consider now the problem of distinguishing between square-planar and tetrahedral geometry. In some cases only a detailed spectroscopic or x-ray analysis is sufficient to distinguish between the two cases; however, in other cases such a distinction can be made on the basis of the magnetic moment of the complex. As an example, consider the complex $Ni(CN)_4^{2-}$, and let us address ourselves to the problem of whether this complex is tetrahedral or square-planar. The possible $d$-electron configurations for a $d^8$ ion with four identical ligands are as follows:

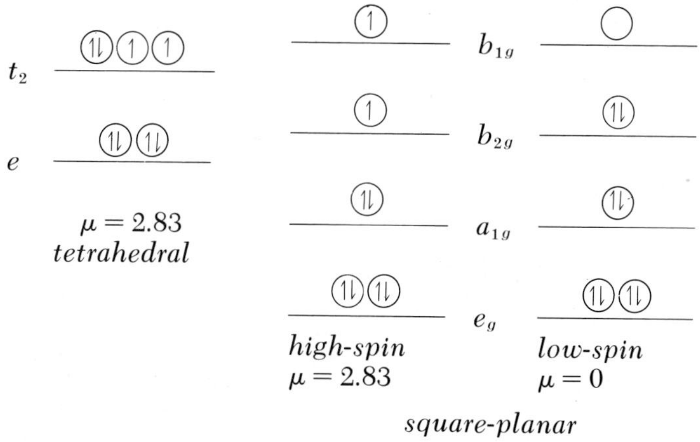

**TABLE 19.6**

THE *d*-ORBITAL SPLITTINGS OF TRANSITION-METAL IONS IN LIGAND FIELDS OF VARIOUS
SYMMETRIES, DERIVED FROM THE APPROPRIATE CORRELATION DIAGRAMS FOR IDENTICAL
LIGANDS.† THE *d*-ORBITALS FROM WHICH THE GIVEN MOLECULAR ORBITALS ARISE ARE
INDICATED IN PARENTHESES. THE σ AND π DESIGNATIONS REFER TO THE SYMMETRY OF
THESE ORBITALS IN THE MOLECULAR-ORBITAL THEORY TREATMENT OF THE BONDING
(See Section 19.15)

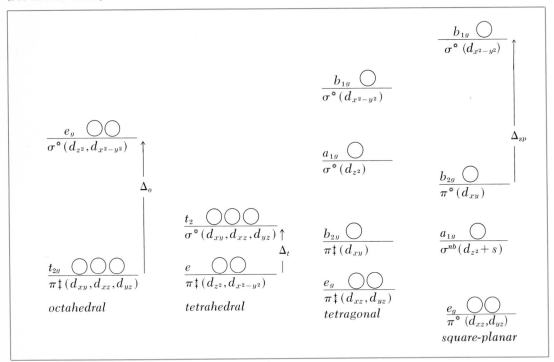

†With mixed ligand complexes these diagrams are not correct, owing to the lower symmetry of the complex.
‡These orbitals can be either bonding, nonbonding, or antibonding, depending on the
nature of the ligands.

Because the complex is observed to be diamagnetic,* it must be square-
planar.

Before proceeding to a consideration of the thermodynamic, kinetic,
and structural consequences of the *d*-orbital splittings of transition-metal
ions, which arise from the interaction of the central transition-metal atom
orbitals with the ligand orbitals, it is advantageous to collect and summarize
our results for these splittings. The relevant results are presented in
Table 19.6.

## 19.11  LIGAND-FIELD-STABILIZATION ENERGIES.

The ligand-field-stabilization energy (LFSE) is defined as the difference between the energy of a hypothetical complex in which the

---

*Actually, all known regular square-planar complexes or $d^8$ ions are diamagnetic, so we
could have omitted the high-spin configuration. The argument is more convincing if we
include it as a possibility.

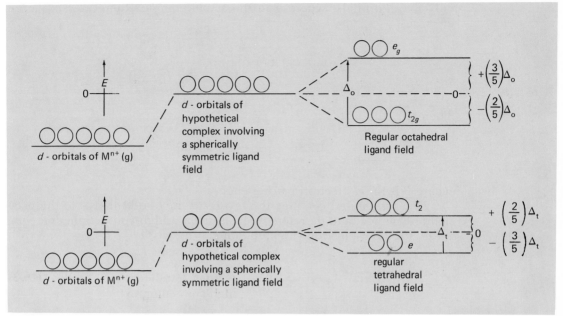

**FIGURE 19.24** Ligand-field-stabilization energy diagrams for regular octahedral and tetrahedral complexes. The LFSE is computed relative to the hypothetical complex with unsplit orbitals.

$d$-orbitals of the metal are not split by the presence of the ligands, but merely shifted up in energy, and the energy of the actual complex with the same structure as the hypothetical complex. This situation is depicted schematically in Figure 19.24. By definition the LFSE = 0 when the $e_g$ and $t_{2g}$ orbitals are completely filled (or completely empty).

This definition requires that the energies of electrons in $t_{2g}$ and $e_g$ orbitals, relative to the mean $d$-electron energy in an octahedral environment, are $-(2/5)\Delta_o$ and $+(3/5)\Delta_o$, respectively. Thus, for the $d$-electron configuration $t_{2g}^x e_g^y$ the LFSE is

$$\text{LFSE} = 0 - \left(-\frac{2}{5}x + \frac{3}{5}y\right)\Delta_o$$

or

$$\text{LFSE} = \left(\frac{2}{5}x - \frac{3}{5}y\right)\Delta_o \quad \substack{\text{(regular, octahedral,} \\ \text{high-spin complex)}} \tag{19.2}$$

For example, the configuration $t_{2g}^3 e_g^1$ has a LFSE of

$$\text{LFSE} = \left(\frac{6}{5} - \frac{3}{5}\right)\Delta_o = \frac{2}{5}\Delta_o$$

If the complex is low-spin, then we must consider the possibility that additional pairing up of electrons will reduce the LFSE. For example, the low-spin $d^7$ configuration is $t_{2g}^6 e_g^1$. In the hypothetical complex with unsplit $d$-orbitals, four of the seven electrons are paired up because there are only five $d$-orbitals, whereas in the actual complex there are six paired electrons:

Therefore, the LFSE is

$$\text{LFSE} = \left(\frac{12}{5} - \frac{3}{5}\right)\Delta_o - P = \frac{9}{5}\Delta_o - P$$

where $P$ is the $d$-electron pairing energy.

In a tetrahedral complex the levels are inverted relative to the octahedral complex, and therefore, the LFSE for a high-spin tetrahedral complex with the $d$-electron configuration $e^x t_2^y$ is

$$\text{LFSE} = \left(\frac{3}{5}x - \frac{2}{5}y\right)\Delta_t \quad \text{(regular, tetrahedral,} \qquad (19.3)$$
$$\text{high-spin complex)}$$

In Table 19.7 we have collected the values of LFSE for regular octahedral and tetrahedral complexes of both high and low spin, together with the differences in LFSE's for high spin octahedral and tetrahedral complexes.

**TABLE 19.7**

LIGAND FIELD STABILIZATION ENERGIES FOR REGULAR OCTAHEDRAL AND TETRAHEDRAL GEOMETRIES

| NUMBER OF $d$ ELECTRONS | LFSE (high spin) OCTAHEDRAL | TETRAHEDRAL† | [LFSE(Oct)− LFSE(Tet)]° | LFSE (low spin) OCTAHEDRAL | TETRAHEDRAL |
|---|---|---|---|---|---|
| 0 | 0 | 0 | 0 | 0 | 0 |
| 1 | $(2/5)\Delta_o$ | $(3/5)\Delta_t$ | $(1/10)\Delta_o$ | $(2/5)\Delta_o$ | $(3/5)\Delta_t$ |
| 2 | $(4/5)\Delta_o$ | $(6/5)\Delta_t$ | $(2/10)\Delta_o$ | $(4/5)\Delta_o$ | $(6/5)\Delta_t$ |
| 3 | $(6/5)\Delta_o$ | $(4/5)\Delta_t$ | $(8/10)\Delta_o$ | $(6/5)\Delta_o$ | $(9/5)\Delta_t - P$ |
| 4 | $(3/5)\Delta_o$ | $(2/5)\Delta_t$ | $(4/10)\Delta_o$ | $(8/5)\Delta_o - P$ ‡ | $(12/5)\Delta_t - 2P$ |
| 5 | 0 | 0 | 0 | $(10/5)\Delta_o - 2P$ | $(10/5)\Delta_t - 2P$ |
| 6 | $(2/5)\Delta_o$ | $(3/5)\Delta_t$ | $(1/10)\Delta_o$ | $(12/5)\Delta_o - 2P$ | $(8/5)\Delta_t - P$ |
| 7 | $(4/5)\Delta_o$ | $(6/5)\Delta_t$ | $(2/10)\Delta_o$ | $(9/5)\Delta_o - P$ | $(6/5)\Delta_t$ |
| 8 | $(6/5)\Delta_o$ | $(4/5)\Delta_t$ | $(8/10)\Delta_o$ | $(6/5)\Delta_o$ | $(4/5)\Delta_t$ |
| 9 | $(3/5)\Delta_o$ | $(2/5)\Delta_t$ | $(4/10)\Delta_o$ | $(3/5)\Delta_o$ | $(2/5)\Delta_t$ |
| 10 | 0 | 0 | 0 | 0 | 0 |

° Assuming $\Delta_t = (1/2)\Delta_o$ and identical ligands.

† Observed experimentally that $\Delta_t \leq (1/2)\Delta_o$ for the same ligands (given metal ion).

‡ $P$ is the $d$-$d$ electron pairing energy. Note that the LFSE is computed *relative to the hypothetical* complex with unsplit orbitals.

# 19.12  THERMODYNAMICS OF COORDINATION COMPOUNDS.

## a.  Heats of Hydration.

One of the first great triumphs of crystal-field theory was the explanation that it provided for the variations in the enthalpies of hydrations of divalent $3d$ transition-series ions. In Figure 19.25 the solid circles represent the values of $-\Delta H_{hyd}$, derived from experimental data, for the reaction

$$M^{2+}(g) + 6H_2O(g) = M(OH_2)_6^{2+}(aq) \qquad \Delta H = \Delta H_{hyd}$$

When the appropriate* LSFE's are subtracted from the observed values (using spectroscopically determined $\Delta_o$ values), a smooth curve results. For example, for Fe(II) we subtract

$$(2/5)\Delta_o = (2/5) \times 10{,}400 \text{ cm}^{-1} \times 2.86 \frac{\text{kcal} \cdot \text{mole}^{-1}}{10^3 \text{cm}^{-1}} = 11.8 \text{ kcal} \cdot \text{mole}^{-1}$$

from the experimental value of $-\Delta H_{hyd}$. The $d^0$, $d^5$, and $d^{10}$ ions lie on the smooth curve before correction for ligand-field effects, because these ions have no LSFE. The smooth curve rises gradually because of the increasing nuclear charge which progressively lowers the energies of the metal $d$-orbitals, thereby making possible greater interaction between metal and ligand orbitals.

## b.  Electrode Potentials.

The standard reduction potential for a transition-metal M(III)/M(II) couple is usually quite sensitive to the nature of the ligands, as can be seen from the data in Table 19.8. On changing from water to ethylenediamine

---

*All the $3d$ series $M(OH_2)_6^{2+}$ ions are high-spin.

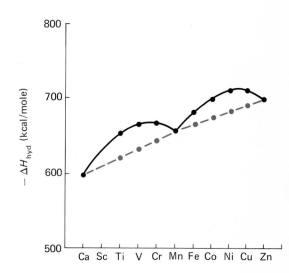

**FIGURE 19.25** Experimentally derived values (black points) of $-\Delta H_{hyd}$ for the reaction
$$M^{2+}(g) + 6H_2O(g) = M(OH_2)_6^{2+}(g)$$
plotted against M. The lower points were obtained by subtracting the appropriate value of the LFSE (Table 19.9) from the experimental (upper) points. (Adapted from reference 19.4.)

TABLE 19.8

STANDARD REDUCTION POTENTIALS OF Fe(II)/Fe(III) AND Co(II)/Co(III)
COUPLES INVOLVING VARIOUS COORDINATING LIGANDS

| COUPLE | $\mathscr{E}^{\circ}_{red}$ (VOLTS AT 25°C) |
|---|---|
| $Co(en)_3^{3+}/Co(en)_3^{2+}$ | $-0.18$ |
| $Co(phen)_3^{3+}/Co(phen)_3^{2+}$ | $+0.42$ |
| $Co(C_2O_4)_3^{3-}/Co(C_2O_4)_3^{4-}$ | $+0.57$ |
| $Co(OH_2)_6^{3+}/Co(OH_2)_6^{2+}$ | $+1.81$ |
| $Fe(C_2O_4)_3^{3-}/Fe(C_2O_4)_3^{4-}$ | $-0.01$ |
| $Fe(EDTA)^-/Fe(EDTA)^{2-}$ | $+0.12$ |
| $Fe(CN)_6^{3-}/Fe(CN)_6^{4-}$ | $+0.36$ |
| $Fe(OH_2)_6^{3+}/Fe(OH_2)_6^{2+}$ | $+0.77$ |
| $Fe(phen)_3^{3+}/Fe(phen)_3^{2+}$ | $+1.12$ |

as ligands on cobalt, we go from a couple whose oxidized form, $Co(OH_2)_6^{3+}$, is capable of oxidizing water to oxygen in 1 M acid solution, to a couple whose reduced form, $Co(en)_3^{2+}$, is capable of liberating hydrogen from an aqueous solution 1 M in $H^+$. The observed variations in $\mathscr{E}^\circ$ with ligand are only in part due to LFSE differences. Variations in entropies of the complexes for differing ligands also make a major contribution to the observed variations in $\mathscr{E}^\circ$ values.

### c. Stability Constants and Distribution Diagrams.

In a solution containing two or more species that can coordinate to a given metal ion, there may exist* an equilibrium between the various complexes involving different numbers of the two possible ligands. In the following discussion we shall assume that the rates for ligand exchange are sufficiently great to establish equilibrium on the time scale of ordinary thermodynamic measurements.

The following equilibria have been identified in aqueous solutions of Co(II) containing ethylenediamine ($H_2NCH_2CH_2NH_2 = en$):

(1)  $Co(OH_2)_6^{2+}(aq) + en(aq) = (H_2O)_4Co(en)^{2+}(aq) + 2H_2O(\ell)$    $K_1 = 8.5 \times 10^5$

(2)† $(H_2O)_4Co(en)^{2+} + en(aq) = (H_2O)_2Co(en)_2^{2+}(aq) + 2H_2O(\ell)$    $K_2 = 5.4 \times 10^4$

(3)  $(H_2O)_2Co(en)_2^{2+} + en(aq) = Co(en)_3^{2+}(aq) + 2H_2O(\ell)$    $K_3 = 2.0 \times 10^3$

In an aqueous solution containing Co(II) and ethylenediamine, the following cobalt-containing species will be present at equilibrium:

$$Co(OH_2)_6^{2+} \qquad Co(OH_2)_2(en)_2^{2+} \qquad Co(OH_2)_4(en)^{2+} \qquad Co(en)_3^{2+}$$

---

*See the next section on the kinetics of transition metals for a discussion of the rates of ligand-exchange reactions.

†Note that there are actually two possible isomers of $(H_2O)_2Co(en)_2^{2+}$, *cis* and *trans* (see section on isomers), that we have lumped together in this reaction. This can always be done, *provided* the two isomers are in rapid equilibrium with each other, because in such a case the ratio of the activities of the two isomers is fixed at a given temperature.

The fraction of the *total* Co(II) concentration that each of these species comprises depends on the values of (en), $K_1$, $K_2$, and $K_3$, but is independent of the total Co(II) concentration. For example,* the fraction of the total Co(II) that is $Co(en)_3^{2+}$ is given by

$$\alpha_3 = \frac{(Co(en)_3^{2+})}{[Co(II)]_{total}} = \frac{K_1 K_2 K_3 (en)^3}{1 + K_1(en) + K_1 K_2(en)^2 + K_1 K_2 K_3(en)^3}$$

where the equilibrium constant expressions have been used to eliminate the cobalt concentrations. *Distribution curves* of $\alpha_1$ versus log(en) for the Co(II) + en system are presented in Figure 19.26. With this figure it is a simple matter to determine the *fraction* of the total that each of the cobalt species comprises at a particular value of (en). For example, at (en) = $10^{-5}$ M, the distribution of species is

| | | | |
|---|---|---|---|
| $Co(OH_2)_6^{2+}$ | 6.6% | $Co(OH_2)_2(en)_2^{2+}$ | 32.5% |
| $Co(OH_2)_4(en)^{2+}$ | 60.8% | $Co(en)_3^{2+}$ | 0.1% |

whereas at (en) = $10^{-3}$ M it is

| | | | |
|---|---|---|---|
| $Co(OH_2)_6^{2+}$ | 0.05% | $Co(OH_2)_2(en)_2^{2+}$ | 33.5% |
| $Co(OH_2)_4(en)^{2+}$ | 0.1% | $Co(en)_3^{2+}$ | 66.3% |

In a solution containing two monodentate ligands† (e.g., $H_2O$ and $NH_3$) and a metal ion that forms octahedral complexes with both ligands, we have to consider a total of seven species, and the corresponding dis-

---

*See problem 11.

†Monodentate ligands occupy one ligand site, bidentate ligands occupy two ligand sites (e.g., en), and tridentate ligands occupy three ligand sites (e.g., dien).

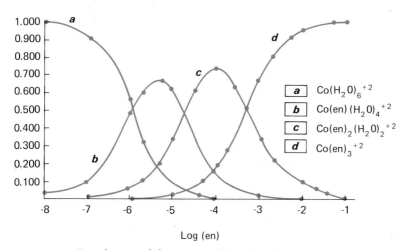

**FIGURE 19.26**   Distribution of the species $Co(OH_2)_6^{2+}$, $Co(OH_2)_4(en)^{2+}$, $Co(OH_2)_2(en)_2^{2+}$, and $Co(en)_3^{2+}$ as a function of log(en) in the system $Co(II) + H_2NCH_2CH_2NH_2 + H_2O$ at 30°C.

tribution diagram will involve seven curves. The sum of reactions (1), (2), and (3) given above yields

$$Co(OH_2)_6^{2+}(aq) + 3 \text{ en}(aq) = Co(en)_3^{2+}(aq) + 6H_2O(\ell) \qquad K_{III}$$

where $K_{III} = K_1 K_2 K_3$. We designate *overall* stability (equilibrium) constants by a subscript roman numeral; whereas we designate *stepwise* stability constants by a subscript arabic numeral, which gives the number of the ligand going into the complex (as was done above for the $Co^{2+}$ + en reactions). Stability constant data for some representative complexes are presented in Table 19.9.

A comparison of the stability constants of multidentate ligands with monodentate ligands involving the same ligand atom bound to the metal yields the interesting observation that the multidentate, ligand-metal com-

**TABLE 19.9**
STABILITY CONSTANT DATA FOR SOME REPRESENTATIVE COMPLEX IONS
(AQUEOUS SPECIES).

| (a) MONODENTATE LIGANDS IN OCTAHEDRAL COMPLEXES. | | | | | | | |
|---|---|---|---|---|---|---|---|
| COMPLEX | $\log K_1$ | $\log K_2$ | $\log K_3$ | $\log K_4$ | $\log K_5$ | $\log K_6$ | $\log K_{VI}$ |
| $Co(NH_3)_6^{2+}(30°)$ | 1.99 | 1.51 | 0.93 | 0.64 | 0.06 | −0.74 | 4.39 |
| $Co(NH_3)_6^{3+}(30°)$ | 7.3 | 6.7 | 6.1 | 5.6 | 5.05 | 4.41 | 35.2 |
| *$Cu(NH_3)_6^{2+}(25°)$ | 4.27 | 3.59 | 3.00 | 2.19 | −0.55 | −2.5 | 10.0 |
| $Ni(NH_3)_6^{2+}(30°)$ | 2.67 | 2.12 | 1.61 | 1.07 | 0.63 | −0.09 | 8.01 |
| $Cr(SCN)_6^{3+}(50°)$ | 2.52 | 1.24 | 0.66 | 0.29 | −0.09 | −0.39 | 4.23 |
| $FeF^{3-}(25°)$ | 5.17 | 3.92 | 2.91 | − | − | − | − |

| (b) BIDENTATE LIGANDS IN OCTAHEDRAL COMPLEXES. | | | | |
|---|---|---|---|---|
| COMPLEX | $\log K_1$ | $\log K_2$ | $\log K_3$ | $\log K_{III}$ |
| $Cu(en)_3^{2+}(25°)$ | 10.72 | 9.31 | −0.90 | 19.13 |
| $Co(en)_3^{2+}(25°)$ | 5.93 | 4.73 | 3.30 | 13.96 |
| $Co(en)_3^{3+}(25°)$ | − | − | 13.99 | 48.69 |
| $Fe(en)_3^{2+}(25°)$ | 4.34 | 3.31 | 2.05 | 9.70 |
| $Ni(en)_3^{2+}(25°)$ | 7.60 | 6.48 | 5.03 | 19.11 |
| $Fe(C_2O_4)_3^{4-}(25°)$ | 4.8 | 2.9 | 1. | 9. |
| $Fe(C_2O_4)_3^{3-}(25°)$ | 9.84 | 6.20 | 3.70 | 19.74 |

| (c) SQUARE-PLANAR AND TETRAHEDRAL COMPLEXES. | | | | | |
|---|---|---|---|---|---|
| COMPLEX | $\log K_1$ | $\log K_2$ | $\log K_3$ | $\log K_4$ | $\log K_{IV}$ |
| $Ni(CN)_4^{2-}(25°)$ | − | − | − | − | 30.3 |
| $Cd(CN)_4^{2-}(25°)$ | 5.54 | 5.06 | 4.66 | 3.58 | 18.84 |

*Note the large decrease in $K_i$ on going from $K_4$ to $K_5$ to $K_6$ for this complex. This is a thermodynamic manifestation of the Jahn-Teller Effect.

plex usually has a much larger stability constant than the analogous mono-dentate complex with the same type of coordinating group. This effect is known as the *chelate* effect. The chelate effect becomes readily evident on comparing multidentate amine ligands with ammonia. For example, for Ni(II)-amine complexes in water we find the following stability constant data at 25°C:

| *Reaction* | *Equilibrium Constant* (30°C) |
|---|---|
| (1) $Ni(OH_2)_6^{2+}(aq) + 6NH_3(aq) = Ni(NH_3)_6^{2+}(aq) + 6H_2O(\ell)$ | $1 \times 10^8$ |
| (2) $Ni(OH_2)_6^{2+}(aq) + 3en(aq) = Ni(en)_3^{2+}(aq) + 6H_2O(\ell)$ | $4 \times 10^{18}$ |
| (3) $Ni(OH_2)_6^{2+}(aq) + 2dien(aq) = Ni(dien)_2^{2+}(aq) + 6H_2O(\ell)$ | $8 \times 10^{18}$ |
| (4) $Ni(OH_2)_6^{2+}(aq) + edtam(aq) = Ni(edtam)^{2+}(aq) + 6H_2O(\ell)$ | $2 \times 10^{19}$ |

$$en = H_2NCH_2CH_2NH_2 \qquad dien = H_2NCH_2CH_2 \overset{\overset{\displaystyle H}{\displaystyle |}}{N} CH_2CH_2NH_2$$

$$edtam = \begin{matrix} H_2NCH_2CH_2 \\ \\ H_2NCH_2CH_2 \end{matrix} \diagdown NCH_2CH_2N \diagup \begin{matrix} CH_2CH_2NH_2 \\ \\ CH_2CH_2NH_2 \end{matrix}$$

From thermodynamics we have

$$\ln K = -\frac{\Delta H°}{RT} + \frac{\Delta S°}{R}$$

The entropy changes for the above reactions are in the order

$$\Delta S_4° > \Delta S_3° > \Delta S_2° \gg \Delta S_1°$$

because the excess number of moles of products over reactants is in this order for the four reactions, as is indicated below:

| *Reaction* | (*Moles of Products*) − (*Moles of Reactants*) |
|---|---|
| 4 | $7 - 2 = 5$ |
| 3 | $7 - 3 = 4$ |
| 2 | $7 - 4 = 3$ |
| 1 | $7 - 7 = 0$ |

For a closely related series of homogeneous reactions like the above, the value of $\Delta S°$ of the reaction will be more positive the greater the increase in the number of moles, owing to the increased disorder resulting from the greater number of product molecules. This order of the $\Delta S_i°$ values is what gives rise to the observed order of the $K_i$ values, i.e., $K_4 > K_3 > K_2 > K_1$, because the enthalpy changes for these four reactions are roughly equal. The chelate effect is thus seen to be primarily an entropy effect.

Enthalpy changes for ligand substitution reactions are usually not large (about ±10 kcal/mole of complex), unless the ligands are widely separated in the spectro-chemical series. For this reason, entropy changes play a major role in determining the *relative* stabilities of complex ions.

## 19.13 STRUCTURAL CONSEQUENCES OF LIGAND-FIELD SPLITTINGS.

The $d$-orbital splitting patterns for transition-metal complexes of various geometries enable us to rationalize some of the observed variations in structure for transition-metal ions. In such discussions both the nature of the metal ion (how many $d$-electrons does it have?) and the nature of the ligand (strong, intermediate, or weak field ligand?) must be considered. The factors that determine whether a given metal and ligand combination will produce octahedral, tetrahedral, square planar, or some other geometry are by no means well understood. Nonetheless, certain useful generalizations can be made. Although one will sometimes predict the wrong structure using these rules, one will be right considerably more often than wrong. The rules are as follows: [*]

I. Complexes involving $d^4$ (high spin), $d^7$ (low spin), and $d^9$ ions, when surrounded by six identical ligands, assume tetragonal geometry.

II. Complexes involving $d^8$ ions (and occasionally $d^9$ ions) usually involve square-planar geometry, especially with strong-field ligands. For $4d$ and $5d$ series $d^8$ ions, square-planar geometry is invariably found.

III. Complexes of $d^0$, $d^5$, $d^{10}$, and $d^7$ ions assume tetrahedral geometry far more frequently than other $d^n$ ions. There are no low-spin, tetrahedral complexes.

IV. The most common geometry for transition-metal complexes is octahedral; this geometry is invariably found for $d^3$ and $d^6$ ions.

Rule I is a consequence of the Jahn-Teller Effect, which has already been discussed. Rule II is partly a consequence of the fact that the LFSE for a square-planar $d^8$ ion is always greater than for the corresponding tetrahedral or octahedral complex. Rules III and IV can also be qualitatively rationalized on the basis of LFSE differences. Referring to Table 19.9, we note that in every case the LFSE for octahedral geometry is greater than, or equal to, that for tetrahedral geometry. For $d^0$, $d^5$ (high spin), and $d^{10}$ ions, there is no LFSE, and this is presumably why tetrahedral complexes are more common for ions of this type. For $d^7$ ions (and $d^2$ ions, although the LFSE for the octahedral complex is greater than for the tetrahedral complex) the LFSE reaches a maximum for tetrahedral complexes, and this fact has been utilized to rationalize why tetrahedral complexes are more frequently found for $d^7$ than for, say, $d^6$ ions. For $d^3$ and $d^6$ ions, octahedral coordination is strongly favored on an LFSE basis.

We shall now apply these rules to some actual cases. Consider nickel: Ni(II) is a $d^8$ ion, and we expect square-planar coordination with a strong-field ligand like $CN^-$. This is the structure of $Ni(CN)_4^{2-}$. With $H_2O$ and $NH_3$ we expect octahedral geometry because these two ligands are not strong-field ligands. Both $Ni(OH_2)_6^{2+}$ and $Ni(NH_3)_6^{2+}$ are regular octahedral complexes. The complex $Ni(CO)_4$ involves Ni(0), a $d^{10}$ ion. Tetrahedral geometry is expected and is found.

Some further examples follow:

(a) The ammonia complex of Pt(IV), a $d^6$ ion, $Pt(NH_3)_6^{4+}$ (octahedral, low spin).

(b) The cyanide complex of Rh(I), a $d^8$ ion, $Rh(CN)_4^{3-}$ (square planar).

[*] In applying these rules it is useful to remember that the ligand field splittings for a given ligand increase as we increase the charge on M, and as we go down a group in the periodic table. It is also useful to know the spectrochemical series.

(c) The iodide complex of Cd(II), a $d^{10}$ ion, $CdI_4^{2-}$ (tetrahedral).
(d) The ammine complex of Cr(II), a $d^4$ ion, $Cr(NH_3)_6^{2+}$ (tetragonal).
(e) The chloride complex of Co(II), a $d^7$ ion, $CoCl_4^{2-}$ (tetrahedral, high spin).

## 19.14 KINETICS OF TRANSITION-METAL COMPLEXES.

### a. General Considerations.

The ability of a particular transition-metal complex to undergo reactions involving the replacement of one or more ligands in its coordination sphere with other ligands is called the *lability* of the complex. Complexes for which reactions of this type are rapid are said to be *labile*, whereas those complexes for which such reactions are slow are said to be *inert*. There is no sharp division between labile and inert complexes, and, because the reaction rate will, in general, depend on such factors as temperature, concentrations, solvent, and the nature of the incoming ligand, a given complex can be either labile or inert depending on the reaction conditions. Nonetheless, the distinction between labile and inert complexes is a useful one, because some complexes undergo ligand exchange reactions so rapidly that under all experimentally accessible conditions they can always be regarded as labile. On the other hand, some complexes are sufficiently unreactive that they react only very slowly under all experimentally accessible conditions, and can, therefore, always be regarded as inert. We shall arbitrarily designate a given complex as labile if its half-life is less than 1 minute under the prevailing experimental conditions.

The terms *labile* and *inert* refer to the kinetic behavior of the complex, and they have no necessary correlation with the thermodynamic *stability* of the complex. A complex which is thermodynamically unstable with respect to a given ligand exchange may undergo such a reaction rapidly or slowly. This point is illustrated by the following reactions:

(1) $Co(NH_3)_6^{3+}(aq) + 6H^+(aq) = Co(OH_2)_6^{3+}(aq) + 6NH_4^+(aq)$

(2) $Co(NH_3)_6^{2+}(aq) + 6H^+(aq) = Co(OH_2)_6^{2+}(aq) + 6NH_4^+(aq)$

The equilibrium constants for these two reactions at 25°C are $10^{25}$ and $10^{51}$, respectively. Therefore, in both cases the amine complex is very unstable thermodynamically with respect to the aquo complex and ammonium ions in acid aqueous solutions. The value of $t_{1/2}$ for the $Co(NH_3)_6^{2+}(aq)$ in excess 1 M $H^+(aq)$ at 25°C is well under 1 second, whereas $Co(NH_3)_6^{3+}(aq)$ persists for days under the same conditions. Another interesting comparison is provided by the following data (25°C):

(1) $Fe(OH_2)_6^{3+}(aq) + Cl^-(aq) = Fe(OH_2)_5Cl^{2+}(aq) + H_2O(\ell)$
   $K = 50$; $t_{1/2} = 8 \times 10^{-2}$ sec (1 M excess $Cl^-$)

(2) $Cr(OH_2)_6^{3+}(aq) + Cl^-(aq) = Cr(OH_2)_5Cl^{2+}(aq) + H_2O(\ell)$
   $K = 0.08$; $t_{1/2} = 4$ days (1 M excess $Cl^-$)

From these data it is evident that $Cl^-$ incorporates much more rapidly into the coordinate sphere of $Fe^{3+}(aq)$ than it does into that of $Cr^{3+}(aq)$. As our last comparison, consider the half-lives of the following ligand-exchange reactions:[*]

(1)  $Cr(OH_2)_6^{2+}(aq) + H_2O^{18}(\ell) = Co(OH_2)_5(^{18}OH_2)^{2+}(aq) + H_2O(\ell)$
$t_{1/2} = 10^{-9}$ sec ($25°C$, $H_2O^{18}$ solvent)

(2)  $Cr(OH_2)_6^{3+}(aq) + H_2O^{18}(\ell) = Cr(OH_2)_5(^{18}OH_2)^{3+}(aq) + H_2O(\ell)$
$t_{1/2} = 10^6$ sec ($25°C$, $H_2O^{18}$ solvent)

These reactions have the same value for their equilibrium constants ($K = 1$), yet their half-lives differ by a factor of $10^{15}$.

## b.  Correlation of Complex-Ion Lability with the d-Electron Configuration.

H. Taube was evidently the first to recognize that there is a connection between the $d$-electron configurations of octahedral complexes, and their kinetic behavior in ligand-substitution reactions. Taube observed that essentially all regular octahedral complexes with the $d$-electron configurations

$$t_{2g}^3 e_g^0 \qquad t_{2g}^4 e_g^0 \qquad t_{2g}^5 e_g^0 \qquad t_{2g}^6 e_g^0$$

are inert, whereas all other regular octahedral complexes are labile. Some examples of inert complexes are:

$Co(NH_3)_6^{3+}$, $Rh(NH_3)_6^{3+}$, $Ir(NH_3)_6^{3+}$, $Fe(CN)_6^{4-}$, $Ru(NH_3)_6^{2+}$, $Pt(NH_3)_6^{4+}$: $t_{2g}^6$

$Fe(CN)_6^{3-}$, $Ru(NH_3)_6^{3+}$, $Mn(CN)_6^{3-}$: $t_{2g}^5$

$Cr(OH_2)_6^{3+}$, $Cr(NH_3)_6^{3+}$: $t_{2g}^3$

Some examples of labile complexes are:

| | | | |
|---|---|---|---|
| [†]$Cu(OH_2)_6^{2+}$ | $t_{2g}^6 e_g^3$ | $Ti(OH_2)_6^{3+}$ | $t_{2g}^1$ |
| $Ni(NH_3)_6^{2+}$ | $t_{2g}^6 e_g^2$ | $V(phen)_3^{3+}$ | $t_{2g}^2$ |
| $Co(OH_2)_6^{2+}$ | $t_{2g}^5 e_g^2$ | $ZnEDTA^{2-}$ | $t_{2g}^6 e_g^4$ |
| $Mn(OH_2)_6^{2+}$ | $t_{2g}^3 e_g^2$ | $CaEDTA^{2-}$ | $t_{2g}^0$ |
| $Cr(OH_2)_6^{2+}$ | $t_{2g}^3 e_g^1$ | | |

Why are complexes with the configurations $t_{2g}^3$, $t_{2g}^4$, $t_{2g}^5$, and $t_{2g}^6$ inert, and other $d$-configuration complexes labile? To answer this question we must consider the possible mechanisms for ligand substitution reactions. Rate data for most ligand-substitution reactions in which the oxidation state of the metal ion remains unchanged can be fit to a mechanism involving a seven-coordinate activated complex. To attain the seven-coordinate

---

[*] Note that Cr(II) high spin is Jahn-Teller distorted.
[†] Actually, $e_g^4 a_{1g}^2 b_{2g}^2 b_{1g}^1$.

activated complex the incoming ligand must achieve a degree of bonding to the metal ion comparable to that of the ligand which is to be expelled. The incoming ligand attacks the complex either at a face or an edge of the octahedron, because these are the positions where the central metal ion is most exposed. Remember that the metal $t_{2g}$ orbitals[*] point at the edges of the octahedron, whereas the $e_g$ orbitals are directed at the ligands. Consider now an octahedral complex with the $d$-electron configuration $t_{2g}^3 e_g^2$. An incoming ligand attacking at an octahedral edge of the complex will encounter an orbital containing one electron, just as do the ligands already present, and all that is necessary to achieve a degree of bonding of the incoming ligand to the metal ion comparable to that of the ligands already present is to bring the ligand in to the same distance from the metal ion as the ligands already bound. On the other hand, for the configurations $t_{2g}^3$, $t_{2g}^4$, $t_{2g}^5$, and $t_{2g}^6$, the incoming ligand must attack the complex at an orbital containing an electron density greater than that in the antibonding $e_g$ orbitals, which are directed at the ligands already present. Before the incoming ligand can bind as tightly to the metal ion as those ligands already bound, it must displace an electron from one of the $t_{2g}$ orbitals into the higher energy $e_g$ orbitals. This process requires considerable energy (because $\Delta_o$ is large); consequently, $t_{2g}^3$, $t_{2g}^4$, $t_{2g}^5$, and $t_{2g}^6$ ions have large activation energies for ligand substitution and undergo such reactions slowly. Complexes with the configurations $t_{2g}^0$, $t_{2g}^1$, and $t_{2g}^2$ react rapidly, because in these cases at least one of the $t_{2g}$ orbitals is empty and electron promotion to the $e_g$ level is not necessary to achieve equality in bond strength.

An apparent anomaly in the argument presented above is provided by complexes with the $t_{2g}^6 e_g^2$ configuration, like $Ni(OH_2)_6^{2+}$. Promotion of a $t_{2g}$ electron to an $e_g$ orbital would seem to be required, yet such complexes undergo rapid ligand exchange. This case is rationalized by arguing that the $e_g$ orbitals are antibonding in character, and, thus, metal ligand bonds in such complexes are rather weak. Because the ligands are bound less securely to the metal they are much easier to expel, and promotion of a $t_{2g}$ electron to the $e_g$ orbitals is not required. This same effect is also operative in all cases of octahedral complexes involving electrons in the $e_g$ orbitals.

Rate constant data for water exchange on some common aquo complexes are given in Figure 19.27. From these data it is evident that the above arguments are at least in qualitative agreement with the experimental results.

## c. Mechanisms of Oxidation-Reduction Reactions Involving Transition-Metal Complexes.

The kinetic behavior of transition-metal complexes in redox reactions is a very rich and intriguing field. In large part, this behavior is a result of the existence of multiple oxidation states and numerous structural possibilities for transition-metal complexes. The full potential of transition-metal ions as catalysts for many different types of chemical reactions has only recently come into clear focus; transition-metal catalysts often exhibit very specific catalytic activity. Differences in the coordination properties of different transition-metal complexes often enable the ex-

---

[*]We are assuming that the $t_{2g}$ orbitals are nonbonding, but the arguments are also applicable with minor rephrasing if the $t_{2g}$ orbitals are $\pi^*$ or $\pi^b$ molecular orbitals.

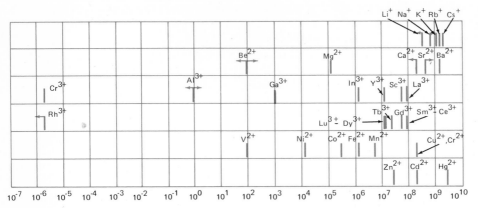

**FIGURE 19.27** Pseudo-first-order rate constants (sec$^{-1}$) for water exchange at 25°C. (Adapted from R. Basolo and R. G. Pearson, *Kinetics and Mechanisms of Complex Ion Reactions* (John Wiley and Sons, Inc., New York, 1966) p. 164. Data from M. Eigen, Pure and Applied Chem., *6*, 105 (1963).)

perimentalist to force the reactants to produce a desired product in preference to other thermodynamically possible products. Transition-metal compounds have been used as catalysts for hydrogenation, isomerization, polymerization, and redox reactions. They are used in tremendous quantities in the industrial preparation of sulfuric acid, ammonia, and methanol, and also in the petrochemical industry. The catalytic activity of transition-metal complexes is vital to the existence of life.

The simplest possible redox reaction is one in which an *electron* is transferred between two reactants to produce two products which are the same as the reactants. An example is the following reaction:

$$^*Fe(CN)_6^{4-}(aq) + Fe(CN)_6^{3-}(aq) = \ ^*Fe(CN)_6^{3-}(aq) + Fe(CN)_6^{4-}(aq)$$

where the asterisk identifies a particular complex. The kinetics of this reaction can be investigated by using a radioactive isotope of Fe in one of the two reactant species and observing the rate of loss of iron radioactivity in the reactant. Using this technique, it has been found that the second-order rate constant for this reaction is about $10^5$ M$^{-1}$·sec$^{-1}$ at 25°C. Or, in other words, for a solution initially 0.1 M in both species, the half-life of either reactant is less than $10^{-4}$ sec at 25°C. On the other hand, $k$ is about $10^{-4}$ M$^{-1}$·sec$^{-1}$ at 25°C for the reaction

$$^*Co(NH_3)_6^{2+}(aq) + Co(NH_3)_6^{3+}(aq) = \ ^*Co(NH_3)_6^{3+}(aq) + Co(NH_3)_6^{2+}(aq)$$

and, thus, for 0.1 M reactant concentrations, $t_{1/2}$ is approximately $10^5$ sec at 25°C. Such reactions are thought to proceed via a simple electron transfer from one complex to the other, with the coordination spheres of the two complexes remaining intact (i.e., the activated complex does not involve a common bridging ligand between the two metal centers). Mechanisms of this type are said to be *outer sphere*. Some additional data for reactions of this type are presented in Table 19.10.

The tremendous difference in electron-exchange rates between Fe(CN)$_6^{4-}$ ($t_{2g}^6$) with Fe(CN)$_6^{3-}$ ($t_{2g}^5$), and Co(NH$_3$)$_6^{2+}$ ($t_{2g}^5 e_g^2$) with Co(NH$_3$)$_6^{3+}$ ($t_{2g}^6$), has been explained in terms of the *Franck-Condon Principle*. Electrons are very much lighter than the nuclei and, therefore, move much more

**TABLE 19.10**

ELECTRON TRANSFER REACTIONS. RATE CONSTANTS IN WATER AT 25°C

| REACTION | $k(\text{M}^{-1}\text{sec}^{-1})$ |
|---|---|
| $\text{Fe(dipy)}_3^{2+} + \text{Fe(dipy)}_3^{3+}$ | $10^6$ |
| $\text{Mn(CN)}_6^{3-} + \text{Mn(CN)}_6^{4-}$ | $10^6$ |
| $\text{IrCl}_6^{2-} + \text{IrCl}_6^{3-}$ | $10^6$ |
| $\text{Fe(CN)}_6^{3-} + \text{Fe(CN)}_6^{4-}$ | $10^5$ |
| $\text{Co(NH}_3)_6^{2+} + \text{Co(NH}_3)_6^{3+}$ | $10^{-4}$ |
| $\text{Co(C}_2\text{O}_4)_3^{3-} + \text{Co(C}_2\text{O}_4)_3^{4-}$ | $10^{-4}$ |
| $\text{Mo(CN)}_8^{3-} + \text{Fe(CN)}_6^{4-}$ | $3.0 \times 10^4$ |
| $\text{Fe(OH}_2)_6^{3+} + \text{Ru(NH}_3)_6^{2+}$ | $3.4 \times 10^5$ |
| $\text{V(OH}_2)_6^{2+} + \text{Ru(NH}_3)_6^{3+}$ | $82$ |

rapidly than can the nuclei. For this reason the transfer of an electron from one ion to another takes place with the nuclei remaining effectively stationary. Before the electron can be transferred the nuclei must adjust their positions such that the energy of the system is unaffected by the electron transfer (these self-exchange, electron-transfer reactions are *iso-energetic*). If the equilibrium metal-ligand bond lengths in the two complexes are about equal, the rate of transfer will be very much greater than if they are not, because when they are not, transfer can occur only between vibrationally excited complexes. The presence of electrons in the $e_g$ orbitals of $\text{Co(NH}_3)_6^{2+}$, but not in $\text{Co(NH}_3)_6^{3+}$, gives rise to quite different M–L bond lengths for these two complexes, and, hence, a slow rate of electron transfer. Because $\text{Fe(CN)}_6^{4-}$ and $\text{Fe(CN)}_6^{3-}$ both contain only $t_{2g}$ and no $e_g$ electrons, these complexes have very similar M–L bond lengths, the Franck-Condon restrictions are readily met, and electron transfer is rapid.

A more complex type of electron-transfer reaction is that between two complexes that differ both in the metal and the ligands, for example,

$$\text{Ru(NH}_3)_6^{2+}(\text{aq}) + \text{Fe(OH}_2)_6^{3+}(\text{aq}) = \text{Ru(NH}_3)_6^{3+}(\text{aq}) + \text{Fe(OH}_2)_6^{2+}(\text{aq})$$

For these *cross-exchange* electron-transfer reactions, as opposed to the *self-exchange* type considered above, the relative thermodynamic stability of $\text{Fe(OH}_2)_6^{3+}$ and $\text{Fe(OH}_2)_6^{2+}$, *versus* $\text{Ru(NH}_3)_6^{2+}$ and $\text{Ru(NH}_3)_6^{3+}$, plays an important role in determining the overall reaction rate.

A particularly interesting use of the inertness of certain transition-metal complexes as a probe for the elucidation of the mechanisms of certain types of redox reactions has been developed by H. Taube and co-workers. Consider the following reaction:

$$\text{Cr(OH}_2)_6^{2+}(\text{aq}) + (\text{H}_3\text{N})_5\text{CoCl}^{2+}(\text{aq}) + 5\text{H}^+(\text{aq}) = (\text{H}_2\text{O})_5\text{CrCl}^{2+}(\text{aq})$$
$$+ \text{Co(OH}_2)_6^{2+}(\text{aq}) + 5\text{NH}_4^+(\text{aq})$$

The complexes $(\text{H}_3\text{N})_5\text{Co(III)Cl}^{2+}$ (low-spin $d^6$) and $(\text{H}_2\text{O})_5\text{CrCl}^{2+}$ ($d^3$) are both inert, whereas the complexes $\text{Cr(OH}_2)_6^{2+}$ (high-spin $d^4$) and $(\text{H}_3\text{N})_5\text{CoCl}^+$ ($d^7$) are both labile. The inertness of Cr(III) and low-spin

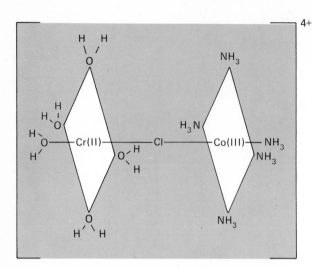

**FIGURE 19.28** Proposed activated complex for the reaction between $Cr(OH_2)_6^{2+}$ and $(H_3N)_5CoCl^{2+}$ in acidic aqueous solution.

Co(III) is such that these complexes will retain their ligands for a long time compared to the time required for the above redox reaction to take place. Furthermore, when Cr(III) is formed from Cr(II) it will retain the ligands in its coordination sphere that were present at that time, whereas the Co(III) will retain the ligands present in its coordination sphere until it is converted to Co(II), at which time it will react rapidly to give the thermodynamically most stable Co(II) complex for the prevailing conditions. Given these restrictions, and considering the above reaction stoichiometry, there is really only one possible conclusion, namely, that the $Cl^-$ ion was simultaneously bound to both Cr(II) and Co(III) at the time the electron was transferred. That is, the activated complex must have involved a chloride bridge. A plausible structure for the activated complex that is consistent with all the available experimental data on the reaction is shown in Figure 19.28. Presumably, the electron is transferred from the Cr(II) center to the Co(III) center across the chloride bridge. The same general results are obtained when several other complexes of the type $(H_3N)_5Co(III)X^{2+}$ are reduced with Cr(II). For example, X can be $RCOO^-$, $CH_3COO^-$, $Br^-$, $N_3^-$, or $NCS^-$.

## 19.15 BONDING IN REGULAR, OCTAHEDRAL, TRANSITION-METAL COMPLEXES.

Simple crystal-field theory is useful for an understanding of the splitting patterns of the $d$-orbitals in transition-metal complexes. The major deficiency of crystal-field theory is that it does not tell us anything about the chemical bonds that hold the complex together. Several important aspects of the bonding in transition-metal complexes can be understood within the framework of simple molecular orbital theory.

Consider a metal ion, M, surrounded by six identical ligands in a regular octahedral array (Figure 19.29). The metal ion is said to be subject to the action of an electric field generated by the ligands, and the theory that attempts to describe the interaction of the metal ion with the ligand electric field is called *ligand-field theory*. Let us assume that the metal ion has available for use in bonding its $3d$-, $4s$-, and $4p$-orbitals; that is, these

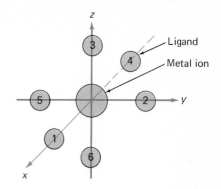

**FIGURE 19.29**   Numbering of ligands in an octahedral complex.

are the metal *valence* orbitals.[*] Let us also assume that the six ligands each have one sigma orbital available to interact with the metal orbitals. Because we have to consider the permissible combinations of 15 atomic orbitals (6 ligand $\sigma$'s, 1 metal $4s$, 3 metal $4p$, and 5 metal $3d$), the resulting molecular orbital scheme is complex.

The problem of what combination of ligand sigma orbitals to take to combine with the metal ion $4s$-orbital is easily solved. Because the metal ion $4s$-orbital is (+) everywhere in the region near the ligands and the ligand $\sigma$ orbitals are also (+), the appropriate combination of ligand $\sigma$ orbitals to be used in bonding to $\sigma_{M4s}$ is[†]

$$\sigma_{L1} = \frac{1}{\sqrt{6}}\,(\sigma_1 + \sigma_2 + \sigma_3 + \sigma_4 + \sigma_5 + \sigma_6)$$

where $\dfrac{1}{\sqrt{6}}$ is a normalization constant chosen to make the sum of the squares of the $\sigma_i$ coefficients add up to unity. The possible combinations of $\sigma_{L1}$ with $\sigma_{M4s}$ yields a bonding and an antibonding orbital[‡]

$$a_{1g}(\sigma^b) = \sigma_{M4s} + \sigma_{L1}$$
$$a_{1g}(\sigma^*) = \sigma_{M4s} - \sigma_{L1}$$

Consider now the appropriate combinations of ligand sigma-orbitals to combine with the three metal $4p$-orbitals. A $p_x$-orbital has a (+)-lobe lying along the $+x$-axis and a (−)-lobe lying along the $-x$-axis, and therefore

---

[*] It is assumed that the underlying closed-shell core is not involved in the metal-ligand bonding. With $4d$ and $5d$ series elements, the appropriate metal valence orbitals are ($4d, 5s,$ and $5p$) and ($5d, 6s,$ and $6p$), respectively.

[†] The basic principles involved in the formation of molecular orbitals from valence orbitals are discussed in Chapter 4.

[‡] Because we are using only the symmetries of the orbitals here, we need not allow for partial ionic character of the bonds by writing

$$a_{1g}(\sigma^b) = C_1\sigma_{M4s} + C_2\sigma_{L1}$$
$$a_{1g}(\sigma^*) = C_2\sigma_{M4s} - C_1\sigma_{L2}$$

(Figure 19.4a), the appropriate combination of ligand sigma-orbital for bonding to the metal $4p_x$-orbital is

$$\sigma_{L2} = \frac{1}{\sqrt{2}}(\sigma_1 - \sigma_4)$$

The other four ligand sigma-orbitals are non-bonding with respect to the metal $4p_z$-orbital (see Figure 4.23 and the associated discussion), so they are not included in $\sigma_{L2}$. Two other linear combinations of ligand sigma-orbitals can be constructed for bonding to the metal $4p_y$-orbital and $4p_z$-orbital, namely,

$$\sigma_{L3} = \frac{1}{\sqrt{2}}(\sigma_2 - \sigma_5) \quad \text{and} \quad \sigma_{L4} = \frac{1}{\sqrt{2}}(\sigma_3 - \sigma_6)$$

respectively. For each of the above ligand sigma combinations, we obtain a bonding and an antibonding orbital, on combination with the appropriate metal $4p$-orbital. Because $4p_x$, $4p_y$, and $4p_z$ are all degenerate (i.e., have identical energies) in the absence of the ligand field, and the ligands are symmetrically placed with respect to the metal along the $x$, $y$, and $z$ axes, the three bonding molecular orbitals are degenerate, and the three antibonding molecular orbitals are also degenerate. These three-fold degenerate bonding and antibonding sets are denoted $t_{1u}(\sigma^b)$ and $t_{1u}(\sigma^*)$, respectively:

$$\left.\begin{array}{l} \sigma_{M4p_x} + \sigma_{L2} \\ \sigma_{M4p_y} + \sigma_{L3} \\ \sigma_{M4p_z} + \sigma_{L4} \end{array}\right\} t_{1u}(\sigma^b) \qquad \left.\begin{array}{l} \sigma_{M4p_x} - \sigma_{L2} \\ \sigma_{M4p_y} - \sigma_{L3} \\ \sigma_{M4p_z} - \sigma_{L4} \end{array}\right\} t_{1u}(\sigma^*)$$

It remains now to consider the appropriate combinations of ligand sigma-orbitals to take for forming molecular orbitals with the metal atomic $d$-orbitals. In order to match the signs of the various lobes of the metal $3d_{x^2-y^2}$ orbital, we must take (Figures 19.29 and 19.10)

$$\sigma_{L5} = \frac{1}{2}(\sigma_1 + \sigma_4 - \sigma_2 - \sigma_5)$$

The ligand $\sigma_3$ and $\sigma_6$ orbitals are non-bonding with respect to the metal $3d_{x^2-y^2}$ orbital. A ligand sigma-orbital of suitable symmetry for combination with the metal $3d_{z^2}$-orbital is (Figures 19.29 and 19.10)

$$\sigma_{L6} = \frac{1}{2\sqrt{3}}\{2(\sigma_3 + \sigma_6) - (\sigma_1 + \sigma_4 + \sigma_2 + \sigma_5)\}$$

There is as much electron density in the ($-$)-donut region as in the ($+$)-lobes of the $d_{z^2}$-orbitals, and consequently, we must double the contribution of $\sigma_3$ and $\sigma_6$ relative to the other four sigmas when forming $\sigma_{L6}$, in order that all metal-ligand interactions remain equivalent. The combinations of $\sigma_{L5}$

and $\sigma_{L6}$ with $3d_{x^2-y^2}$ and $3d_{z^2}$, respectively, yield two degenerate bonding and two degenerate antibonding molecular orbitals.*

The doubly-degenerate bonding and antibonding orbitals resulting from the combinations of $\sigma_{L5}$ and $\sigma_{L6}$ with the $d_{x^2-y^2}$ and $d_{z^2}$ orbitals are designated $e_g(\sigma^b)$ and $e_g(\sigma^*)$, respectively:

$$\left.\begin{array}{l}(3d_{x^2-y^2}) + \sigma_{L5} \\[2mm] (3d_{z^2}) + \sigma_{L6}\end{array}\right\} e_g(\sigma^b) \qquad \left.\begin{array}{l}(3d_{x^2-y^2}) - \sigma_{L5} \\[2mm] (3d_{z^2}) - \sigma_{L6}\end{array}\right\} e_g(\sigma^*)$$

There are no combinations of ligand sigma-orbitals that are of suitable symmetry for forming molecular orbitals with the metal $d_{xy}$, $d_{xz}$, and $d_{yz}$ orbitals. For example, consider the $d_{yz}$-orbital. We try to form an orbital of suitable symmetry as follows (Figure 19.12a):

$$(\sigma_2 + \sigma_3) + (\sigma_5 + \sigma_6) - (\sigma_2 + \sigma_6) - (\sigma_3 + \sigma_5) = 0$$

Consequently, these three $d$-orbitals are nonbonding. Because these three orbitals are symmetrically placed with respect to the ligands, they remain degenerate in the ligand field. These three degenerate nonbonding orbitals are designated $t_{2g}(\pi^{nb})$. These nonbonding orbitals are not degenerate with either the $e_g(\sigma^b)$ or the $e_g(\sigma^*)$ orbitals.

In Table 19.11 we have summarized the combinations of ligand sigma-orbitals of the proper symmetry to form molecular orbitals with the metal orbitals in a regular octahedral complex.

Figure 19.30 is the molecular-orbital-correlation diagram for a regular octahedral complex. In placing the molecular orbitals we have taken the bonding orbitals to be the lowest in energy, the nonbonding orbitals next lowest, and the antibonding orbitals highest in energy. Spectroscopic data and detailed quantum-mechanical calculations confirm the above qualitative reasoning on the order.

As an example of the use of the correlation diagram in Figure 19.30, consider the complex $CoF_6^{3-}$. Each of the fluoride ions contributes a sigma orbital with two electrons, for a total of 12 valence electrons from the six ligands. The $Co^{3+}$ ion has $27 - 3 = 24$ electrons in all, and subtracting the 18 inert-gas-core electrons leaves a total of $24 - 18 = 6$ valence electrons for the cobalt. Therefore, we have a grand total of $12 + 6 = 18$ valence electrons to place in our molecular orbitals. The resulting electronic configuration is shown in Figure 19.31. In this figure we have also shown the electronic configurations for the complexes $Co(NH_3)_6^{3+}$, $Cr(OH_2)_6^{3+}$, $Mn(OH_2)_6^{2+}$, $Zn(OH_2)_6^{2+}$, and $Sc(OH_2)_6^{3+}$.

We are now in a position to make several general statements about octahedral, transition-metal complexes:

(1) The molecular-orbital correlation diagram (Figure 19.30), together with the corresponding molecular orbitals, enables us to give a qualitative theoretical picture of how a regular, octahedral complex is held together by chemical bonds. The complex is held together by the twelve valence

---

*In the absence of the ligand field, the $3d_{x^2-y^2}$ and $3d_{z^2}$ metal orbitals are of the same energy. Recalling from our previous discussions that the $d_{z^2}$ orbital can be thought of as a mixture (in equal parts) of a $d_{z^2-x^2}$ and $d_{z^2-y^2}$, and that all three of these orbitals ($d_{x^2-y^2}$, $d_{z^2-x^2}$, $d_{z^2-y^2}$) have their lobes along the coordinate axes, it is clear that the placement of the six identical ligands along the $+x$, $+y$, $+z$, $-x$, $-y$, and $-z$ coordinate axes, equidistant from the metal ion, leads to the same extent of interaction in both cases, provided the coefficients of the ligand sigma orbitals are properly chosen.

**TABLE 19.11**

COMBINATIONS OF LIGAND $\sigma$ ORBITALS USED TO FORM MOLECULAR ORBITALS WITH CENTRAL-ATOM METAL ORBITALS IN A REGULAR OCTAHEDRAL COMPLEX

| CENTRAL METAL ATOM ATOMIC ORBITAL | LIGAND SIGMA-ORBITAL* (see Figure 19.29 for numbering of ligands) | SYMMETRY |
|---|---|---|
| $s$ | $\sigma_{L1} = \dfrac{1}{\sqrt{6}}\,(\sigma_1 + \sigma_2 + \sigma_3 + \sigma_4 + \sigma_5 + \sigma_6)$ | $a_{1g}$ |
| $p_x$ | $\sigma_{L2} = \dfrac{1}{\sqrt{2}}\,(\sigma_1 - \sigma_4)$ | |
| $p_y$ | $\sigma_{L3} = \dfrac{1}{\sqrt{2}}\,(\sigma_2 - \sigma_5)$ | $t_{1u}$ |
| $p_z$ | $\sigma_{L4} = \dfrac{1}{\sqrt{2}}\,(\sigma_3 - \sigma_6)$ | |
| $d_{x^2-y^2}$ | $\sigma_{L5} = \dfrac{1}{2}\,(\sigma_1 + \sigma_4 - \sigma_2 - \sigma_5)$ | |
| $d_{z^2}$ | $\sigma_{L6} = \dfrac{1}{2\sqrt{3}}\,\{2(\sigma_3 + \sigma_6) - (\sigma_1 + \sigma_2 + \sigma_4 + \sigma_5)\}$ | $e_g$ |
| $d_{xy}$ $d_{xz}$ $d_{yz}$ | none | $t_{2g}$ |

*The coefficients are determined by the conditions that the sum of the squares of the coefficients of the $\sigma_i$'s for a particular $\sigma_{Li}$ must be unity. Thus, for $\sigma_{L5}$ we have

$$\left(\frac{1}{2}\right)^2 + \left(\frac{1}{2}\right)^2 + \left(\frac{1}{2}\right)^2 + \left(\frac{1}{2}\right)^2 = 1$$

Note also that the sum of the squares of the coefficients of each $\sigma_i$ within the various $\sigma_{Li}$ expressions must add to unity (*orbital conservation*). For $\sigma_1$ we have

$$\left(\frac{1}{\sqrt{6}}\right)^2 + \left(\frac{1}{\sqrt{2}}\right)^2 + \left(\frac{1}{2}\right)^2 + \left(\frac{1}{2\sqrt{3}}\right)^2 = \frac{1}{6} + \frac{1}{2} + \frac{1}{4} + \frac{1}{12} = 1$$

electrons in the $a_{1g}$, $t_{1u}$, and $e_g$ bonding orbitals. Additional valence electrons are placed in the $t_{2g}$ (nonbonding) and $e_g$ (antibonding) orbitals. The valence electrons in the bonding molecular orbitals are *delocalized* over both the metal ion and the ligands, and it is this delocalization of electron density over more than one nuclear center that makes the bonding orbitals of lower energy than the separate atomic orbitals. When an electron has more space to move around in, its energy goes down.

(2) The absolute energies of the various molecular orbitals, as well as the energy differences between the individual molecular orbitals, depend

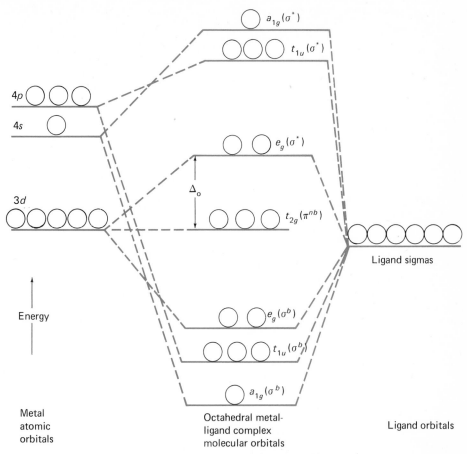

**FIGURE 19.30**  Correlation diagram for a regular octahedral complex ($b$ = bonding, $nb$ = nonbonding, $^*$ = antibonding). With $H_2O$ as ligand, the ligand sigmas are lower in energy than the metal $3d$-, $4s$-, and $4p$-orbitals, and this is the usual case (although cases where the ligand sigmas are higher in energy than the metal $3d$-orbitals are known). The relative order of the *bonding* oribtals $a_{1g}$, $t_{1u}$, and $e_g$ is not known with certainty, but makes no difference for our purposes. If the ligand sigmas are considerably lower in energy than the metal orbitals, then the electrons in the bonding orbitals $a_{1g}$, $t_{1u}$, and $e_g$ will be mostly on the ligands, the electrons in the $t_{2g}$ nonbonding orbitals will be localized on the metal, and the electrons in the $e_g(\sigma^*)$ orbitals will be for the most part on the metal atom.

on both the nature of the metal ion and the nature of the ligands. If we change M (keeping the same ligands) from $Cr^{3+}$ to $Co^{3+}$, or from $Co^{3+}$ to $Co^{2+}$, the energy of the atomic $3d$-orbitals changes, and therefore, so do the energies of the resulting molecular orbitals. Similar comments apply to a change of ligands for a given metal ion. Changes in the absolute energies of these orbitals result in changes in the metal-ligand bond lengths and bond energies. These are very complicated effects to take into account quantitatively, and we shall content ourselves with the knowledge that such effects are important.

(3) In every case in Figure 19.31, the $a_{1g}$, $t_{1u}$ and $e_g$ bonding orbitals are completely filled with bonding electrons. As a consequence, many of the interesting aspects of transition-metal chemistry (e.g., magnetic

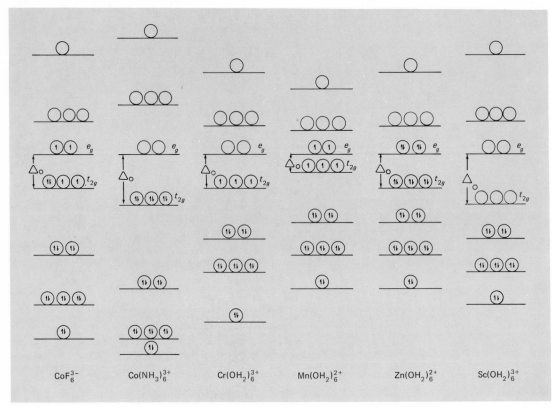

**FIGURE 19.31** Molecular orbital occupancies for various complexes. The order of the molecular orbitals is $a_{1g}(\sigma^b)$, $t_{1u}(\sigma^b)$, $e_g(\sigma^b)$, $t_{2g}(\pi^{nb})$, $e_g(\sigma^\circ)$, $t_{1u}(\sigma^\circ)$, $a_{1g}(\sigma^\circ)$.

moments, visible spectra, *relative* thermodynamic and kinetic behavior) can be understood by concentrating our attention on the electron populations of the $t_{2g}(\pi^{nb})$ and $e_g(\sigma^*)$ orbitals, and the energy difference, $\Delta_o$, between these two sets of orbitals. *The $t_{2g}$ orbitals consist of a three-fold degenerate set of nonbonding metal-ion d-orbitals ($d_{xy}, d_{xz}, d_{yz}$), whereas the $e_g(\sigma^*)$ set is doubly degenerate and antibonding.* Placing electrons in the $e_g(\sigma^*)$ orbitals leads to a weakening of metal-ligand bonds, because of the antibonding character of these orbitals.

(4) The correlation diagram presented in Figure 19.30 is applicable directly only to regular, octahedral complexes in which the ligands can safely be regarded as using only sigma-orbitals in bonding to the metal atom. This means that all of the ligands must be *identical* and equidistant from the metal atom, and that metal-ligand pi-bonding is unimportant.* If these conditions are not met (in general, making the six ligands identical is not sufficient to insure that a regular octahedron results), then the correlation diagram in Figure 19.30 is not applicable. The diagram is applicable, however, to regular octahedral complexes of the $4d$ and $5d$ series, if the metal atomic orbitals are changed from $3d, 4s, 4p$ to $4d, 5s, 5p$ and $5d$, $6s, 6p$, respectively.

---

*$\pi$-bonding is discussed in Section 19.16.

## 19.16  π-BONDING IN OCTAHEDRAL COMPLEXES.

We shall assume that each ligand capable of π-bonding to the metal has available two orbitals of π symmetry for interaction with the metal orbitals. For a ligand like $Cl^-$ these orbitals are two of the $3p$-orbitals; the third $3p$-orbital is used in sigma bonding (see Figure 19.32a). For a ligand like $CN^-$, the ligand π orbitals are the two antibonding $π^*$ molecular orbitals on the ion* (see Figure 19.32b). If the ligand π orbitals are occupied, then the ligand is said to be a π-*donor*, whereas if the ligand π orbitals are empty (usually $π^*$), then the ligand is said to be a π-*acceptor*.

---

*The $π^b$ orbitals on $CN^-$ are too low in energy to combine effectively with the metal $d$-orbitals.

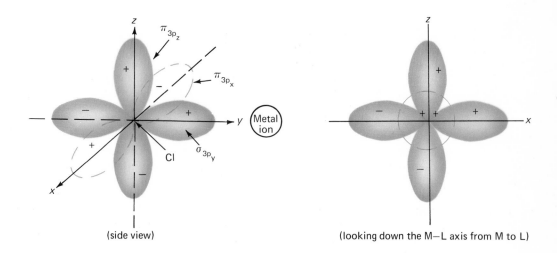

(side view)          (looking down the M—L axis from M to L)

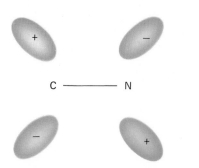

(side view showing just one of the two $π^*$
orbitals, the second is similar to the one shown
but at right angles to it around the CN axis)

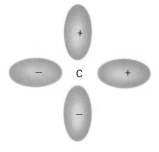

(end on view showing the
symmetry of the $π^*$ lobes
on carbon in $CN^-$.)

**FIGURE 19.32(a).**  Chloride ion as a π-bonding ligand (upper). **(b)** $π^*$ orbitals on cyanide ion (lower).

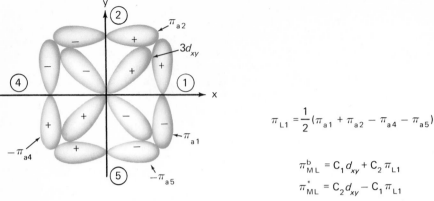

$$\pi_{L1} = \frac{1}{2}(\pi_{a1} + \pi_{a2} - \pi_{a4} - \pi_{a5})$$

$$\pi^b_{ML} = C_1 d_{xy} + C_2 \pi_{L1}$$

$$\pi^*_{ML} = C_2 d_{xy} - C_1 \pi_{L1}$$

**FIGURE 19.33**   Formation of $\pi$ symmetry molecular orbitals by combination of ligand symmetry orbitals with metal $t_{2g}$ orbitals.

When we included only sigma bonding in octahedral complexes, the $d_{xy}$, $d_{xz}$, and $d_{yz}$ metal orbitals became a three-fold-degenerate, non-bonding set $(t_{2g})$ in the complex. When $\pi$ bonding is included this is no longer the case. The $t_{2g}$ set of orbitals is of $\pi$ symmetry and, therefore, can combine with liquid $\pi$-orbitals to form $\pi$ molecular orbitals in the complex. The $e_g(\sigma^*)$ orbitals, on the other hand, are not of $\pi$ symmetry, and do not combine with the ligand $\pi$ orbitals.

In Figure 19.33 we have sketched out the orientation of ligand orbitals necessary for formation of a $\pi$ metal-ligand bonding orbital involving one of the $t_{2g}$ metal orbitals, namely, $d_{xy}$. The orientations are similar for $d_{xz}$ and $d_{yz}$. From these combinations of ligand $\pi$-orbitals and metal $t_{2g}$ orbitals, we obtain three degenerate $\pi$ bonding orbitals and three degenerate $\pi^*$ antibonding orbitals, for a total of six molecular orbitals. They are degenerate because the ligand orbitals are all degenerate before combination, as are the metal $t_{2g}$ symmetry orbitals; and all the ligand $\pi$-orbitals are used to the same extent in forming the bonding orbitals, as are the metal $t_{2g}$ orbitals. Because we started with $6 \times 2 + 3 = 15$ ligand and metal $\pi$ symmetry orbitals, we must end up with 15 $\pi$ symmetry molecular orbitals. The remaining nine molecular orbitals form a 9-fold, essentially degenerate, nonbonding set. An example of one of these nonbonding orbitals is shown in Figure 19.34.

The two possible $\pi$ molecular orbital correlation diagrams for octahedral complexes, where each of six identical ligands has two degenerate $\pi$-orbitals available for bonding, are given in Figures 19.35a and 19.35b. The three possible complete correlation diagrams (i.e., including both $\pi$ and $\sigma$ orbitals) are shown in Figures 19.36a, b, and c. In 19.23c we have placed the $\pi^{nb}$ and $\pi^*$ orbitals above the $e_g(\sigma^*)$ orbitals because they arise from relatively unstable ligand $\pi^*$ orbitals.

Let us now consider the filling of these orbitals for the three complexes:

(1) $CoF_6^{3-}$         (2) $Co(NH_3)_6^{3+}$         (3) $Co(CN)_6^{3-}$

(1) Cobalt (III) is $d^6$, and thus we have six valence electrons from the metal. With $F^-$ as ligand, we have a total of $6 \times 2 + 6 \times 4 = 36$ valence electrons from the six fluoride ions (two from the $\sigma(2p)$, and four from the two $\pi(2p)$ orbitals on each of the six fluorides); thus we have a grand total of 42 electrons to pump into the molecular orbital diagram. These electrons will completely fill all of the orbitals up to the $t_{2g}(\pi^*)$ set. Then, because fluoride is a weak-field ligand, we have $t_{2g}(\pi^*)^4 e_g(\sigma^*)^2$ (Figure 19.36b).

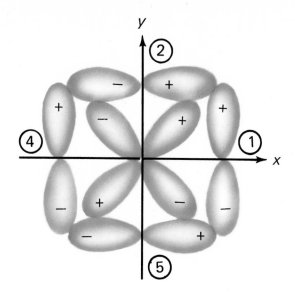

$$\pi^{nb} = \frac{1}{2}\left(\pi_{a1} + \pi_{a4} + \pi_{a2} + \pi_{a5}\right)$$

**FIGURE 19.34**   A non-bonding $\pi$ symmetry orbital in a complex involving metal-ligand $\pi$ bonding.

(2) With ammonia there are no ligand $\pi$ orbitals of suitable symmetry and energy to combine with the metal $t_{2g}$ orbitals. Thus, we must return to our now familiar molecular orbital diagram, and pump in $6 + 6 \times 2 = 18$ valence electrons, which completely fill the six $\sigma$ bonding and the three $t_{2g}$ nonbonding orbitals (Figure 19.36a).

(3) With $CN^-$ as ligand we have used the *empty* $\pi^*$ ligand orbitals to form the metal-ligand molecular orbitals, and therefore, we have to pump in $6 + 6 \times 2 = 18$ valence electrons in this case (Figure 19.36c).

We are now in a position to understand the variations in $\Delta_o$ values with ligands that give rise to the spectrochemical series. On the basis of the above bonding schemes we expect the following order of the energy separations:

$$\{e_g(\sigma^*) - t_{2g}(\pi^*)\} < \{e_g(\sigma^*) - t_{2g}(\pi^{nb})\} < \{e_g(\sigma^*) - t_{2g}(\pi^b)\}$$

This order is expected because, all other factors being equal, $t_{2g}(\pi^b)$ is of lower energy than $t_{2g}(\pi^{nb})$, which is of lower energy than $t_{2g}(\pi^*)$. Of course, in the three complexes all other factors are not the same, and, in particular, the $e_g(\sigma^*)$ orbital energies differ in the three cases. However, we also expect $e_g(\sigma^*, CN^-) > e_g(\sigma^*, NH_3) > e_g(\sigma^*, F^-)$. This is because: (a) the directional character of the $\sigma$ orbitals on $CN^-$ and $NH_3$, as opposed

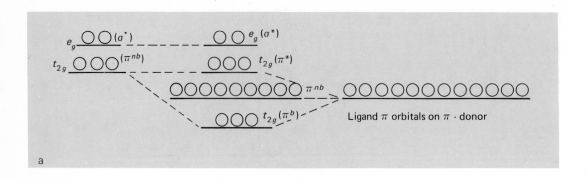

a

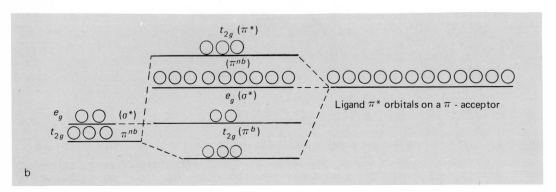

b

**FIGURE 19.35a** Relative energies of $\pi^*$, $\pi^{nb}$, $\pi^b$ and $e_g(\sigma^*)$ orbitals in an octahedral complex involving $\pi$-donor ligands.

**FIGURE 19.35b** Relative energies of $\pi^*$, $\pi^{nb}$, $\pi^b$ and $e_g(\sigma^*)$ orbitals in an octahedral complex involving $\pi$-acceptor ligands. The $\pi^{nb}$ orbitals are not all degenerate as shown; actually, they fall into two degenerate sets (6 and 3) that can be regarded as a set of nine degenerate orbitals for our purposes.

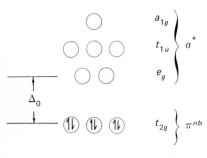

**FIGURE 19.36(a)** Non-$\pi$-bonding ligands, $Co(NH_3)_6^{3+}$.

(a) non-$\pi$-bonding ligands

$Co(NH_3)_6^{3+}$

*Ill. continued on opposite page.*

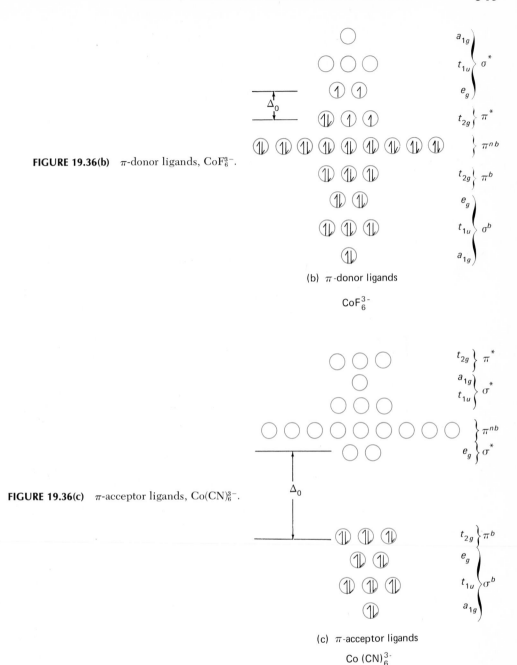

**FIGURE 19.36(b)** $\pi$-donor ligands, CoF$_6^{3-}$.

(b) $\pi$-donor ligands

CoF$_6^{3-}$

**FIGURE 19.36(c)** $\pi$-acceptor ligands, Co(CN)$_6^{3-}$.

(c) $\pi$-acceptor ligands

Co (CN)$_6^{3-}$

to F$^-$, makes these ligands better $\sigma$ bonders; and (b) the net $\pi$ bonding in CN$^-$ pulls the ligands in even closer, and further increases its sigma-bond strength over that of NH$_3$.

In summary, the above order is expected because in the case of CN$^-$ the highest filled orbitals are the $t_{2g}(\pi^b)$ orbitals, in the case of NH$_3$ the highest filled orbitals are the $t_{2g}(\pi^{nb})$ orbitals, and in the case of F$^-$ the highest occupied $t_{2g}$ orbitals are the $t_{2g}(\pi^*)$ orbitals. This effect is shown schematically in Figure 19.37.

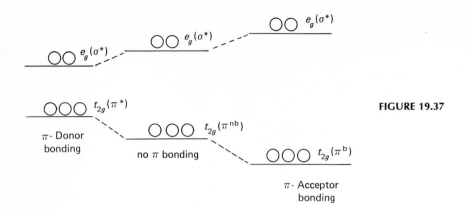

FIGURE 19.37

Ligands, like the halides, with filled $\pi$ orbitals are said to be $\pi$-donors, because in forming the complex some of the electron density is shifted from the occupied ligand $\pi$ orbitals to the metal-ligand-complex molecular orbitals; whereas ligands like $CN^-$, CO, and $o$-phenanthroline with empty $\pi$ orbitals are said to be $\pi$-acceptors, because the net shift in electron density is from the metal $\pi$ orbitals to the metal-ligand-complex $\pi$ molecular orbitals. The order of the ligands in the spectrochemical series parallels their $\pi$ bonding character, with $\pi$-donors at the low $\Delta_o$ end, and $\pi$-acceptors at the high $\Delta_o$ end.

Because effective $\pi$ bonding between acceptor ligands and the metal ion requires the availability of electrons in the $t_{2g}$ metal orbitals, we expect this type of bonding to be especially important in octahedral fields for $d^6$ and $d^5$ metal ions.

The requirement of a sufficient number (5 or 6) of $t_{2g}$ electrons on the metal presumably accounts for our ability to prepare $Cr(CO)_6$ (Cr is Cr(0) and, therefore, $d^6$) and $V(CO)_6$ (V is V(0) and, therefore, $d^5$), and our inability to prepare $Ti(CO)_6$. The electronic structure of certain ligands is such that they cannot effectively interact with metal orbitals to produce either sigma- or pi-molecular orbitals. Such species make poor ligands and form few, if any, complexes. Examples are $ClO_4^-$, $NO_3^-$, and $SO_4^{2-}$.

## PROBLEMS

1. Explain, using balanced chemical equations, the following observations:

   (a) Dilute solutions of $CuSO_4$(aq) are blue, but become green on addition of 6 M HCl.
   (b) Several commercially available rust removers contain sodium oxalate.
   (c) Solid mercuric oxide is soluble in excess 2 M KI(aq) solution.
   (d) Dilute aqueous solutions of ferric nitrate are yellow in color. The yellow color is removed by addition of excess 2 M $HNO_3$, but not by excess 2 M HCl.
   (e) Several commercially available humidity indicators have a blue spot that becomes pink when the relative humidity is high (Hint: these indicators contain cobalt (II) chloride).

2. Consider the following list of species and *indicate by number* (in the spaces provided) those species that possess the indicated property. Note that a given species may possess more than one (or none) of the indicated properties.

(1) $Cr(OH_2)_6^{2+}$     (5) $Pt(NH_3)_2Cl_2$     (9) $Fe(C_2O_4)_3^{4-}$
(2) $Fe(CN)_6^{4-}$     (6) $MnO_4^{2-}$     (10) $Zn(EDTA)^{2-}$
(3) $Pt(CN)_4^{2-}$     (7) $HgI_4^{2-}$
(4) $Ir(en)_3^{3+}$     (8) $Co(NCS)_4^{2-}$

(a) paramagnetic          _____

(b) inert to ligand substitution          _____

(c) colored          _____

(d) tetragonally distorted from
     regular octahedral geometry          _____

(e) tetrahedral complex          _____

(f) chelate complex          _____

(g) capable of existing in
     isomeric form          _____

3. Indicate which of the following complexes would be expected to be inert to ligand substitution:

(a) $FeF_6^{3-}$     (f) $Co(NH_3)_6^{2+}$     (k) $Ru(NH_3)_6^{3+}$     (o) $V(phen)_3^{3+}$
(b) $Ni(OH_2)_6^{2+}$     (g) $Co(NH_3)_6^{3+}$     (l) $V(OH_2)_6^{3+}$     (p) $Zn(EDTA)^{2-}$
(c) $Cr(OH_2)_6^{2+}$     (h) $Co(NO_2)_6^{3-}$     (m) $Rh(NH_3)_6^{3+}$     (q) $Fe(OH_2)_6^{3+}$
(d) $Fe(CN)_6^{4-}$     (i) $Zn(OH_2)_6^{2+}$     (n) $PtCl_6^{2-}$     (r) $Co(OH_2)_6^{3+}$
(e) $Cr(NH_3)_6^{3+}$     (j) $Sc(OH_2)_6^{3+}$

4. Give all the possible isomers (geometrical as well as optical) of the following complexes:

(a) $Fe(phen)_3^{2+}$     (c) $Co(en)(acac)ClBr$     (e) $RuCl_2Br_2(NO_2)_2^{3-}$
(b) $Pt(en)Cl_2Br_2$     (d) $Co(C_2O_4)(en)Br_2^-$     (f) $Cr(edtam)^{3+}$

5. Give the ground state $d$-electron configuration, the number of unpaired electrons, the predicted spin-only magnetic moment, and the ligand-field stabilization energies of the following complexes:

(a) $Cr(OH_2)_6^{2+}$     (h) $Co(en)_3^{2+}$(h.s.)     (o) $Fe(CN)_6^{4-}$
(b) $Co(CN)_6^{3-}$     (i) $Zn(CN)_4^{2-}$     (p) $Cr(NH_3)_6^{3+}$
(c) $Ru(NH_3)_6^{3+}$     (j) $MnO_4^-$     (q) $OsO_4$
(d) $CrO_4^{2-}$     (k) $VCl_4^-$     (r) $Pt(NH_3)_6^{4+}$
(e) $CoF_6^{3-}$     (l) $Mn(CN)_6^{4-}$     (s) $Rh(C_2O_4)_3^{3-}$
(f) $Rh(CN)_6^{4-}$     (m) $NiCl_4^{2-}$     (t) $Fe(phen)_3^{3+}$
(g) $CoCl_4^{2-}$     (n) $Pd(CN)_4^{2-}$

6. Various complexing agents are used in the chemotherapeutic treatment of metal poisoning. For example, ethylenediamine is sometimes used for treatment of lead poisoning, and bidentate sulfur-bonding chelates are

used in the treatment of mercury poisoning. Discuss the reasons why these treatments are effective. Discuss the reasons why EDTA is an effective preservative for canned foods, a common use for it.

7. Predict the composition and geometries of the following:

(a) the cyanide complex of Au(III)
(b) the chloride complex of Rh(I)
(c) the aquo complex of Mn(III)
(d) the iodide complex of Hg(I)
(e) the amine complex of Ru(II)
(f) the carbonyl complex of Cr(0)
(g) the oxalate complex of Fe(III)
(h) the bromide complex of Co(II)
(i) the phenanthroline complex of Co(III)

8. Consider the $d^2$ octahedral complex $M(OH_2)_6^{3+}$. The absorption spectrum of this complex exhibits three peaks at 12.4, 26.2, and 34.1 ($10^3$ cm$^{-1}$ units), respectively. Assign these absorptions to particular electronic transitions. Indicate, using an electronic state energy level diagram, which transition corresponds to which peak and deduce $\Delta_o$ from this data. What color is the complex?

9. Plot the following divalent ionic radii for the $3d$-series ions in octahedral environments, and rationalize the shape of the resulting curve by using the molecular orbital diagram:

| Ca | Sc | Ti | V | Cr | Mn | Fe | Co | Ni | Cu | Zn |
|------|----|------|------|--------|------|------|------|------|--------|------|
| 0.99 | – | 0.80 | 0.74 | (0.76) | 0.80 | 0.76 | 0.75 | 0.69 | (0.72) | 0.74 |

10. Predict the chemical formula and geometry of the following complexes:

(a) the bromide complex of Rh(I)
(b) the oxalate complex of Co(II)
(c) the iodide complex of Co(II)
(d) the porphyrin complex of Mn(II)
    (let porphyrin = porph in the formula)
(e) the amine complex of Cu(II)
(f) the chloro complex of Ag(I)
(g) the hydroxy complex of Zn(II)
(h) the hydroxy complex of Cr(III)
(i) the ethylenediamine complex of Cr(II)

11. Use the data given in Table 19.11 to construct distribution diagrams for the systems:

(a) Cu(II) + en + H$_2$O
(b) Cu(II) + NH$_3$ + H$_2$O

12. Ligand substitution reactions in solution of the type (charges omitted for simplicity):

$$ML_6 + X \rightarrow L_5MX + L$$

are found to have rate laws of the type

(a)
$$R_{D,\mathrm{ML_6}} = k_{obs}(\mathrm{ML_6})$$

(b)
$$R_{D,\mathrm{ML_6}} = k_{obs}(\mathrm{ML_6})(\mathrm{X})$$

Suggest two possible mechanisms leading to a rate law of the type (a); suggest three possible mechanisms leading to a rate law of the type (b). How many of your mechanisms involve seven-coordinate activated complexes?

13. Sketch the expected shapes of the $\Delta H$ versus transition-metal curves for the following reactions (M is a transition metal):

$$\mathrm{M^{3+}(g)} + 6\mathrm{L^-(g)} = \mathrm{ML_6^{3-}(g)} \text{ (high spin)}$$
$$\mathrm{M^{3+}(g)} + 6\mathrm{L^-(g)} = \mathrm{ML_6^{3-}(g)} \text{ (low spin)}$$
$$\mathrm{M^{2+}(g)} + 6\mathrm{L^-(g)} = \mathrm{ML_6^{4-}(g)} \text{ (low spin)}$$

(NOTE: Do not forget to take into account any changes in the number of paired electrons in the reaction).

14. How many spin-allowed "$d$-electron" transitions would be expected for the various $d^n$ ions in octahedral (high and low spin) and tetrahedral complexes? Arrange your results in tabular form for $n = 0, 1, 2, \ldots, 10$.

**References**

19.1. H. B. Gray, *Electrons and Chemical Bonding* (Benjamin, New York, 1965) pp. 176–211.
19.2. L. E. Orgel, *An Introduction to Transition Metal Chemistry: Ligand Field Theory* (second edition, Methuen and Co., Ltd., London, 1966).
19.3. F. A. Cotton and G. Wilkinson, *Inorganic Chemistry* (second edition, Interscience, John Wiley and Sons, New York, 1967).
19.4. F. A. Cotton, *Ligand Field Theory*, J. Chem. Educ., *41*, 466 (1964). Resource Paper I, prepared under the sponsorship of the Advisory Council on College Chemistry.
19.5. J. L. Lambert, *Spinels*, J. Chem. Educ., *41*, 41 (1964).
19.6. J. Halpern, *Coordination Chemistry and Homogeneous Catalysis*, Chemical and Engineering News, Oct. 31, 1966, p. 68.
19.7. H. Taube, *Mechanisms of Oxidation-Reduction Reactions*, J. Chem. Educ., *45*, 452 (1968).
19.8. T. M. Dunn, D. S. McClure and R. G. Pearson, *Crystal Field Theory* (Harper and Row Pub. Co., 1965).

# CHAPTER 20

# ORGANIC CHEMISTRY

Early chemists investigated the compounds and the reactions that occur in living matter, and they found that most of the compounds within living organisms contain carbon. Life and reproduction seemed mysterious, set apart from ordinary laboratory reactions. To some the idea that the compounds and reactions of living matter could be duplicated in test tubes was heresy. Thus, the chemistry of compounds which contained carbon came to be treated as a unique subdivision of chemistry; it was called organic chemistry. The synthesis of a biological compound (urea) from non-organic reagents showed that there was no *vis-viva* in the chemistry of carbon-containing compounds, but the name and the division have endured. The subject material of chemistry is so vast that some subdivision is necessary; the division into organic chemistry (the chemistry of most compounds that contain carbon) and inorganic chemistry (the chemistry of all other compounds) is no more arbitrary than other possible divisions. It would be misleading to describe this chapter as an introduction to organic chemistry; a faint scent of organic chemistry is a better description of the relation of the material in this chapter to the field of organic chemistry.†

## 20.1 ISOMERS AND NOMENCLATURE.

Table 20.1 contains the single-bond energies for a series of homonuclear pairs. Several of these atoms can form only one single bond (for example, hydrogen and the halogens). Of the remaining atoms, a carbon-

---

*°Nobel Lectures, Chemistry* (1901–1921) (Elsevier Publishing Co., Amsterdam, 1966) pp. 25–26.

†The reader should not conclude that the other chapters in this text are inorganic chemistry. The first fifteen chapters and portions of the remaining six are better described as basic chemistry, material that is basic to all of chemistry.

TABLE 20.1

SIGMA BOND ENERGIES (kcal/mole) FOR HOMONUCLEAR PAIRS

| H—H | 103 | Br—Br | 45 |
|---|---|---|---|
| C—C | 82 | N—N | 38 |
| Cl—Cl | 57 | F—F | 36 |
| S—S | 54 | O—O | 34 |
| Si—Si | 53 | Li—Li | 25 |

carbon pair has the strongest single bond by far. The carbon atom has the ability to form a strong bond to a second carbon atom while forming up to three other bonds in the same molecule. This ability allows formation of complex molecules that are built on a skeleton of carbon atoms which are bonded together.* In addition, carbon forms strong *pi* bonds with itself and with other atoms (notably with oxygen, nitrogen, and sulfur). The combination of these two abilities leads to a number of known organic compounds which is much larger than the total number of known inorganic compounds.

The simplest organic compounds contain hydrogen, carbon, and nothing else. When these *hydrocarbons* contain no *pi* bonds, they are said to be *saturated*. All saturated hydrocarbon are either *alkanes* or *cyclo-alkanes*. The simplest alkanes are:

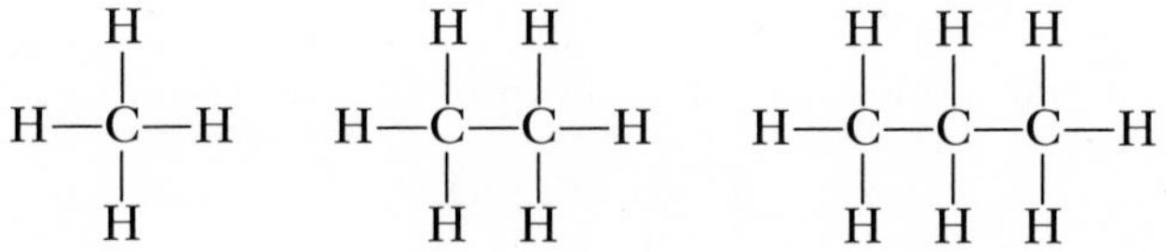

Methane    Ethane    Propane

Complications enter when we consider the next heaviest alkane; there are two possible molecules:

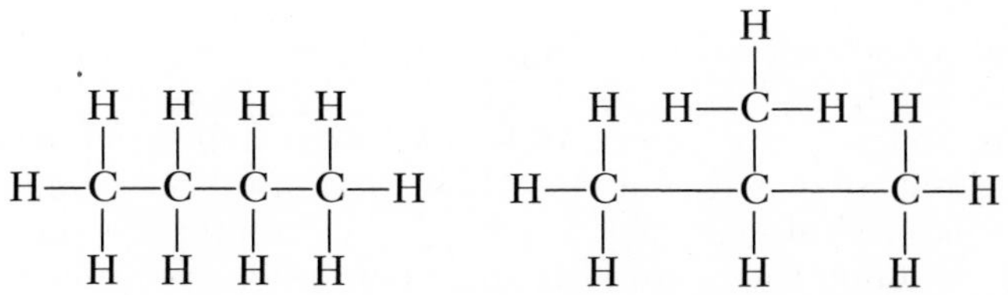

n-Butane    i-Butane (2-Methylpropane)

These two alkanes have the same molecular formula ($C_4H_{10}$), but they have different physical and chemical properties. For example, the boiling point of *n*-butane is $-0.5°C$, while the boiling point of *i*-butane is $-10.2°C$. Two or more molecules which have identical empirical formulae but which

---

*Both sulfur and silicon, which also have comparably large sigma-bond energies for homonuclear pairs, form complex molecules with extended sulfur and silicon skeletons. However, the number of known sulfur and silicon compounds is much smaller than the number of known carbon compounds.

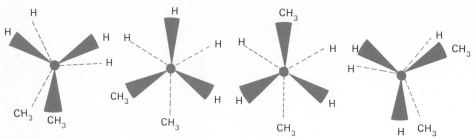

**FIGURE 20.1**   Configurations of *n*-butane.

have different structural formulae and which can be separated are called *isomers*. There are two isomers of butane.

Molecules are three-dimensional, and some subtleties of their structures are obscured in a two-dimensional sketch. In Figure 20.1 we see several potential isomers of the molecule we have called *n*-butane. These potential isomers can be interconverted by rotation of one end of the molecule about the central C–C bond. A *sigma* bond does not hinder rotation about the bond axis, but an energy barrier to free rotation about a *sigma* bond could be produced by the sheer bulk of the adjacent portions of the molecule. This effect has been observed; even in ethane there is a 3 kcal/mole barrier to rotation about the C–C bond. However, in almost every case the barriers to free rotation are far too small to prevent rotation at room temperature, and the potential isomers interconvert freely. In this case the potential isomers cannot be separated; they are not true isomers. You must learn to visualize the free rotation in order to discard isomers. Three-dimensional models are often helpful.* As an example,

$$
\begin{array}{ccccc}
& H & H & & H \\
& | & | & & | \\
H- & C- & C & \rule{1cm}{0.4pt} & C-H \\
& | & | & & | \\
& H & H & H- & C-H \\
& & & & | \\
& & & & H
\end{array}
$$

is a misleading two-dimensional representation of *n*-butane; it is not a third isomer of butane.

It seems obvious that compounds with high molecular weights will have many isomers. It is less obvious that the number of isomers quickly becomes unmanageable; in fact, we need a computer to count isomers. Figure 20.2 contains the results of such computer counting for various empirical formulae. The numbers indicated are quite conservative, because not all isomers that have reasonable Lewis diagrams are included.

Figure 20.2 illustrates the existence of overwhelming numbers of organic compounds. For example, we can estimate (by extrapolation) that there are about one million different compounds with the molecular formula $C_{10}H_{23}NO_3$. Clearly, some systemization must be introduced into the study of organic chemistry to avoid chaos. To begin, a more compact notation and a system for naming are needed. The notation is made more

---

*A wide variety of molecular model kits are available. The interested student should find out which type is recommended for his future courses.

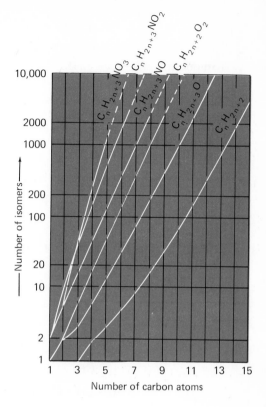

**FIGURE 20.2**   Numbers of isomers for various compositions. No aromatic isomers are included in these numbers, and numerous unlikely possibilities have been excluded.

compact by omitting the obvious details in the Lewis diagrams.* For example,

$$\underset{\substack{H \quad H \quad H \quad H}}{H-\overset{\displaystyle H}{\underset{\displaystyle H}{C}}-\overset{\displaystyle H}{\underset{\displaystyle H}{C}}-\overset{\displaystyle H}{\underset{\displaystyle H}{C}}-\overset{\displaystyle H}{\underset{\displaystyle H}{C}}-H} \qquad \text{becomes} \qquad CH_3CH_2CH_2CH_3$$

and

$$H-\overset{\displaystyle H}{\underset{\displaystyle H}{C}}\quad\overset{\displaystyle H}{\underset{\displaystyle H-\overset{\displaystyle H}{\underset{\displaystyle H}{C}}-H}{C}}\quad\overset{\displaystyle H}{\underset{\displaystyle H}{C}}-H \qquad \text{becomes} \quad \underset{\displaystyle \overset{|}{CH_3}}{CH_3CHCH_3}$$

Efficient nomenclature requires a set of rules. The IUPAC† system is described in detail in any standard textbook of organic chemistry and in the chemical handbooks. IUPAC names frequently are not used for the simplest molecules. Simple names given by the discoverers gain such wide currency that the simple names are not abandoned when a new class

---

*Obvious* is a relative term; a full-time organic chemist would consider the notations $\wedge$ and $\curlyvee$ obvious for the examples. We do not.
†The International Union of Pure and Applied Chemistry.

TABLE 20.2

SOME ALKANES

| Empirical Formula | Structure | Common Name | IUPAC Name |
|---|---|---|---|
| $CH_4$ | $CH_4$ | methane | methane |
| $C_2H_6$ | $CH_3CH_3$ | ethane | ethane |
| $C_3H_8$ | $CH_3CH_2CH_3$ | propane | propane |
| $C_4H_{10}$ | $CH_3CH_2CH_2CH_3$ | n-butane | butane |
| | $CH_3-CH-CH_3$ <br> $\quad\quad\ \ \mid$ <br> $\quad\quad\ CH_3$ | i-butane | 2-methylpropane |
| $C_5H_{12}$ | $CH_3CH_2CH_2CH_2CH_3$ | n-pentane | pentane |
| | $CH_3CHCH_2CH_3$ <br> $\quad\ \ \mid$ <br> $\quad\ CH_3$ | i-pentane | 2-methylbutane |
| | $\quad\quad CH_3$ <br> $\quad\quad\ \mid$ <br> $CH_3-C-CH_3$ <br> $\quad\quad\ \mid$ <br> $\quad\quad CH_3$ | neopentane | 2,2-dimethylpropane |
| $C_6H_{14}$ | $CH_3-CH-CH-CH_3$ <br> $\quad\quad\ \ \mid\quad\ \ \mid$ <br> $\quad\quad\ CH_3\ CH_3$ | — | 2,3-dimethylbutane |

of compounds is incorporated into the IUPAC system. Occasionally, the common names are incorporated into the IUPAC system as special cases. Some examples of systematic and common names for alkanes are given in Table 20.2. We will not pursue the topic of nomenclature further in this text.

## 20.2 FUNCTIONAL GROUPS AND CLASSES OF REACTIONS.

By computer counting we find that there are at least 4030 isomers of $C_8H_{16}O_2$. Classical techniques of analysis, which tell us the empirical formula and the molecular weight, are not adequate methods of compound identification for such cases. Some way must be found to classify these isomers.

We classify organic molecules on the basis of their molecular geometry in the immediate neighborhood of multiple C–C bonds or of atoms other than carbon and hydrogen. We call these portions of the molecules *functional* groups. Some of the functional groups which can be present in molecules with molecular formula $C_8H_{16}O_2$ are listed in Table 20.3. Although the numbers of isomers which contain some of the functional groups are still very large, the numbers are certainly more manageable than 4030.

**TABLE 20.3**

SOME ISOMERS OF $C_8H_{16}O_2$

| SUBGROUP GEOMETRY | | NUMBER OF ISOMERS $(C_8H_{16}O_2)$ |
|---|---|---|
| $-\overset{\overset{\displaystyle O}{\|\|}}{C}-OH$ | | 39 |
| $-\overset{\overset{\displaystyle O}{\|\|}}{C}-O-C\big\langle$ | | 105 |
| both $\rangle C-O-C\langle$ and $-\overset{\overset{\displaystyle O}{\|\|}}{C}-$ | | 329 |
| both $\rangle C-OH$ and $-\overset{\overset{\displaystyle O}{\|\|}}{C}-$ | | 458 |
| 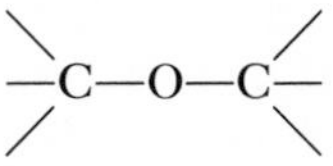 $\rangle C-O-C\langle$ twice | | 183 |
| both $\rangle C-O-C\langle$ and $\rangle C-OH$ | | 783 |
| both $\rangle C-O-C\langle$ and $\rangle C=\overset{\overset{\displaystyle OH}{\|}}{C}-$ | | 305 |
| both $\rangle C-OH$ and $\rangle C=\overset{\overset{\displaystyle }{\|}}{C}-O-C\langle$ | | 497 |

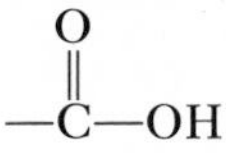

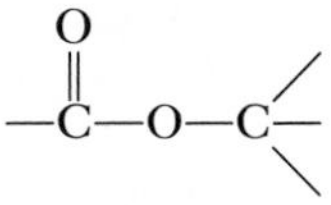

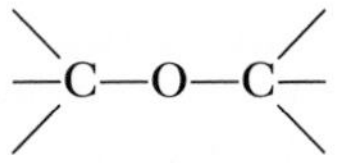

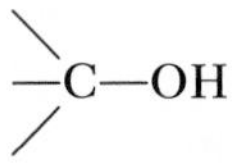

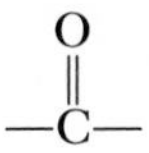

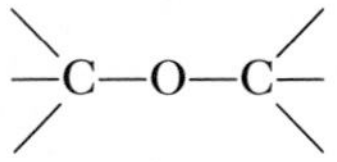

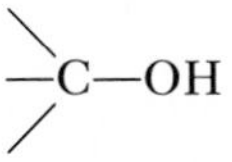

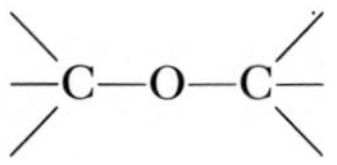

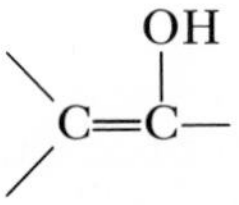

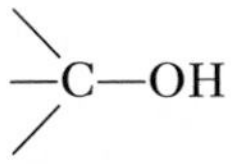

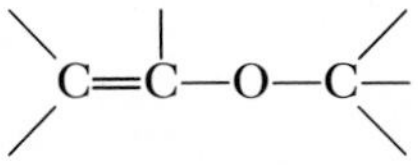

The classification according to functional groups is possible, but is it useful? If all 39 isomers which contain the functional group $-\overset{\overset{\displaystyle O}{\|\|}}{C}-OH$ have rather similar chemical and physical properties, and if the chemical and physical properties of those 39 isomers are quite different from the chemical and physical properties of isomers which contain other functional groups, then the classification is useful. The chemical and physical properties of a group of isomers will be similar if the masses, the nuclear geometries, and the electron distribution in the isomers are similar. The masses of isomers are identical; the nuclear geometries and electron distributions are not.[*] Because all of the isomers within a group contain the same functional group, the isomers in a group differ only in the nuclear geometry and electron distribution in the hydrocarbon portions of the molecules.

---

[*] If the latter two were identical, there would be no isomers.

The carbon-carbon bond is non-polar, and the carbon-hydrogen bond is almost non-polar. Therefore, the hydrocarbon portions contribute little to the dipole moment or the hydrogen-bonding abilities of a molecule. Because the hydrocarbon portions contribute little to these physical properties, the variations in the geometries of the hydrocarbon portions produce only small variations in the physical properties of a set of isomers which all have identical functional groups. We conclude that all isomers which contain the same functional group have identical masses, and very similar dipole moments and hydrogen-bonding abilities. As these three factors largely determine the physical properties of compounds, we expect, and find, that the physical properties of such a group of isomers are quite similar.

Most reactions that organic compounds undergo can be assigned to one of four classes. The four classes are:

1. *Displacement reactions.*  In a displacement reaction, one atom or group attached to the molecule is displaced by a second group or atom.

2. *Addition-Elimination reactions.*  In an addition reaction, atoms or groups are added so as to destroy a multiple bond in the molecule. The reverse process is called elimination.

3. *Oxidation-Reduction reactions.*  In an organic oxidation reaction, additional oxygen atoms are incorporated into a molecule, and/or hydrogen atoms are removed from the molecule. Reduction is the reverse of oxidation.

4. *Condensation-Fractionation reactions.*  In a condensation reaction, two organic molecules combine to form a new molecule with a longer backbone. The reverse process is called fractionation.

Examples of the various classes of reactions will be given presently. At the moment we are only interested in where these reactions occur, for if we can show that they always occur at the functional groups, we will have shown that the functional groups determine the chemical properties of molecules.

Addition and condensation reactions can occur only when the molecule contains multiple bonds, and the reactions occur at the multiple bond. Displacement and oxidation can occur anywhere in the molecule. A relevant set of bond energies is given in Table 20.4.

We see that the C–F and C–H bonds are the strongest single bonds, and so these two are least likely to be broken in a displacement reaction. The different arrangements of C–H bonds in the various isomers of an isomeric group are not likely to produce much difference in the chemical behavior of the isomers in a displacement reaction, for the displacement will occur elsewhere. The C–C bonds in the hydrocarbon

**TABLE 20.4**

SOME AVERAGE BOND ENERGIES TO CARBON (kcal/mole)

| C—F | 102 | C—Br | 64 |
|---|---|---|---|
| C—H | 99 | C—N | 62 |
| C—O | 81 | C—I | 56 |
| C—C | 80 | C=C | 145 |
| C—Cl | 77 | C=O | 175 |
| C—Si | 68 | C≡C | 198 |

portions of a molecule are relatively invulnerable to displacement reactions. Apparently, the absence of either a polar group or lone-pair electrons makes it difficult for an attacking group to get a good hold on these portions of the molecule. More formally, we say that it is possible to form lower-energy activated complexes by involving the functional groups. These arguments indicate that displacement reactions should occur near the functional groups. By similar arguments, we can conclude that oxidation either completely destroys a molecule, or under more mild conditions again takes place at the functional groups.

Thus, most chemical reactions of organic molecules occur at the functional groups, and isomers with the same functional groups show great similarities in both their chemical and physical properties. Therefore, classification on the basis of functional groups is a useful one.

In fact, the influence of the functional groups on the chemical properties of organic compounds completely overwhelms the effect of the mass. For example, the molecule $CH_3CH_2OH$ is chemically similar to a lighter molecule with the same functional group, $CH_3OH$, but $CH_3CH_2OH$ bears no chemical resemblance to its isomer $CH_3OCH_3$. For this reason, the study of organic chemistry can be reduced to the study of the chemistry of functional groups, and only at a more advanced level need one study the subtle differences caused by the saturated hydrocarbon portions of the molecules.

The structures and names of some common functional groups are given in Table 20.5. It is probably true that by judicious introduction of, displacement of, and elimination of these and other functional groups one could convert any one of the approximately two million known organic compounds into any other one. We shall make no attempt at the completeness which is required to learn organic synthesis; we shall be content to discuss a few functional groups and a few interesting reactions.

## 20.3  HYDROCARBONS.

Those organic compounds that lack a functional group are called saturated hydrocarbons, alkanes, or cycloalkanes. Several alkanes have already been mentioned as examples of isomers. Both the structures and reactions of alkanes are unremarkable. All bond angles in alkanes are very close to $109°28'$, so the bonds are best described using $sp^3$ hybrid orbitals. The absence of functional groups limits the reactions which alkanes can undergo; most reactants which are potent enough to attack these molecules anywhere can attack successfully almost everywhere. The result is either a wide mixture of products or complete rupture of the molecular backbone.

One important reaction of alkanes is oxidation:

$$CH_3CH_2CH_2CH_2CH_3 + 8O_2 = 5CO_2 + 6H_2O \qquad \Delta H = -845 \text{ kcal}$$

Pentane

Oxidation of alkanes releases large amounts of heat; alkanes are used as fuels.* Natural gas is a mixture of the $C_1$ through $C_4$ alkanes, and gasoline is predominantly a mixture of the $C_7$ through $C_{11}$ alkanes. Petroleum is a

---

*Most organic compounds can be burned (i.e., oxidized) to give mainly carbon dioxide and water, although the heat released per gram is usually smaller than when alkanes are burned.

**TABLE 20.5**

SOME COMMON FUNCTIONAL GROUPS

| FUNCTIONAL GROUP | NAME | SIMPLEST COMPOUND |
|---|---|---|
| $-I$ | Iodide | $CH_3I$ |
| $-Cl$ | Chloride | $CH_3-Cl$ |
| $-Br$ | Bromide | $CH_3-Br$ |
| $-OH$ | Alcohol | $CH_3-OH$ |
| $-\overset{\overset{\textstyle O}{\|\|}}{C}H$ | Aldehyde | $H\overset{\overset{\textstyle O}{\|\|}}{C}H$ |
| $C-\overset{\overset{\textstyle O}{\|\|}}{C}-C$ | Ketone | $CH_3\overset{\overset{\textstyle O}{\|\|}}{C}-CH_3$ |
| $-\overset{\overset{\textstyle O}{\|\|}}{C}-OH$ | Acid | $H-\overset{\overset{\textstyle O}{\|\|}}{C}-OH$ |
| $-\overset{\overset{\textstyle O}{\|\|}}{C}-O-C$ | Ester | $H\overset{\overset{\textstyle O}{\|\|}}{C}-O-CH_3$ |
| $C-S-H$ | Mercaptan (Thiol) | $CH_3-S-H$ |
| $C-O-C$ | Ether | $CH_3-O-CH_3$ |
| $-NO_2$ | Nitrate | $CH_3NO_2$ |
| $-NH_2$ | Amine | $CH_3-NH_2$ |
| $-CN$ | Nitrile (Cyanide) | $CH_3-CN$ |
| $-\overset{\overset{\textstyle O}{\|\|}}{C}-NH_2$ | Amide | $H\overset{\overset{\textstyle O}{\|\|}}{C}-NH_2$ |
| $\underset{\diagup}{\overset{\diagdown}{C}}=\underset{\diagdown}{\overset{\diagup}{C}}$ | Alkene | $H_2C=CH_2$ |
| $-C\equiv C-$ | Alkyne | $HC\equiv CH$ |

mixture of many, many organic compounds, but alkanes form the major component. Many of the alkanes in petroleum have high molecular weights. The yield of gasoline from petroleum can be increased by fractionation of these heavier alkanes:

$$C_{15}H_{32} \ = \ C_7H_{16} \ + \ C_8H_{16}$$

an alkane     an alkane     an alkene

These fractionation or cracking reactions require catalysts and high temperatures. As you might expect, we have little control over where in the molecular backbone the crack occurs, and the product is a mixture of compounds.

*Alkenes* are hydrocarbons which contain one double bond. The following molecules are alkenes:

$$H_2C = CH_2 \qquad CH_3CH_2CH = CH_2 \qquad CH_3CH = CHCH_3$$

Ethylene (Ethene)          1-Butene                      2-Butene

At the double bond the bond angles are approximately*

<br>

H
          120°
120° C ═══ C 120°
          120°
H                  H

with the atoms

H     H
 \\   /
  C═C
 /   \\
H     H

lying in a plane. The *pi* bond is not symmetric about the bond axis, and it prevents rotation about the bond axis (Figure 20.3).

The reactions of the alkenes are the reactions of their functional group, the double bond. The *pi* bond is energetically vulnerable and geometrically

---

* See Figures 4.37 and 4.38.

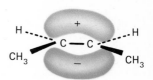

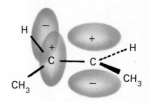

**FIGURE 20.3**  Rupture of a pi bond upon rotation by 90°.

accessible. The most important reactions of alkenes are additions across the *pi* bond. Typical examples of addition reactions are:

$$H_2C{=}CH_2 + H_2 = CH_3CH_3$$

Ethylene            Ethane

$$H_2C{=}CH_2 + Cl_2 = CH_2ClCH_2Cl$$

1,2-Dichloroethane

$$H_2C{=}CH_2 + H_2O \overset{H^+}{=\!=} CH_3CH_2OH$$

Ethyl alcohol (Ethanol)

*Alkynes* are hydrocarbons that contain a triple bond. The reactions of alkynes are similar to the reactions of alkenes.

There exists a fourth class of hydrocarbons, aromatic hydrocarbons. The simplest aromatic hydrocarbon is benzene, $C_6H_6$.

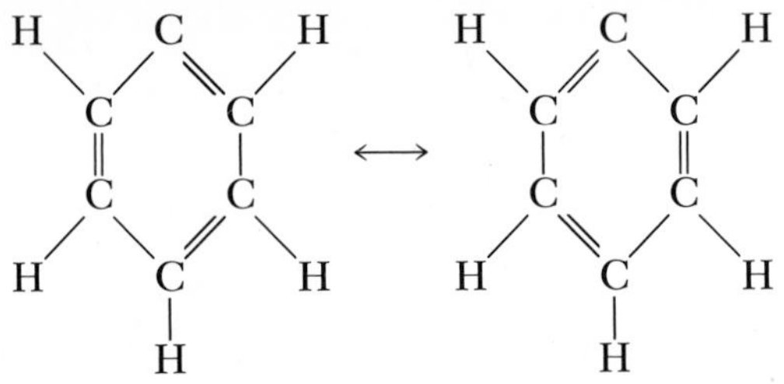

Benzene

The aromatic hydrocarbons are characterized by *pi* electrons in molecular orbitals which extend around the rings. The drawing of resonance structures is bothersome, so benzene is represented as

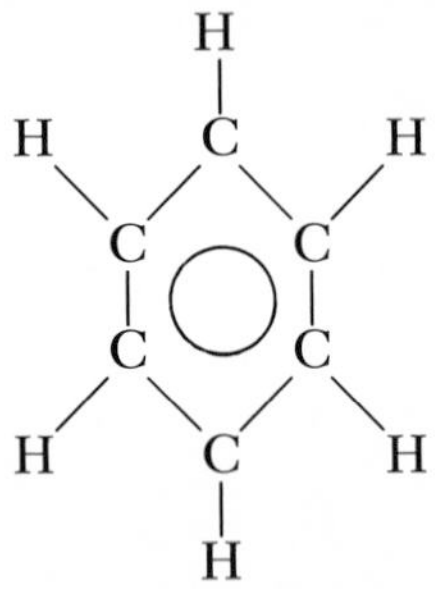

or more compactly as

The circle is a symbolic reminder of the *pi* electron system. Other hydrocarbons consist of several rings which have one or more sides in common, for example:

Naphthalene ($C_{10}H_8$) 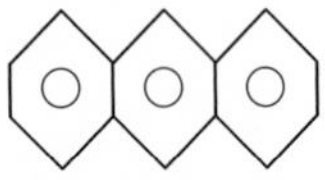      Anthracene ($C_{14}H_{10}$)

A regular hexagon has an internal angle of 120°; this is the bond angle for perfect $sp^2$ hybrid orbitals. The vast majority of aromatic compounds are built around six-membered rings.

The *pi* bonds in aromatic compounds bear little resemblance to the *pi* bonds in alkenes. The extra delocalization of the electrons in aromatic *pi* bonds lowers the *pi* electron energy to such an extent that the *pi* electrons are almost immune to attack. Contrast the reactions of chlorine with a cycloalkene and with an aromatic hydrocarbon. The reaction with a cycloalkene is addition:

Cyclohexene                    1,2-Dichlorocyclohexane

but the reaction with benzene can be a displacement reaction:

Benzene                    Chlorobenzene

In fact, the influence of the aromatic ring on the remainder of the molecule is so strong that each functional group must be discussed twice, once when the functional group is attached to a non-aromatic backbone, and a second time when the functional group is attached to an aromatic ring.*

## 20.4 HALIDES.

The replacement of any hydrogen atom in a hydrocarbon by a halogen atom produces a molecule called either an alkyl halide (non-aromatic) or an aryl halide (aromatic). The geometry of a halide is quite similar to the geometry of the parent hydrocarbon, but small distortions are introduced by the larger size of the halogen atom. The physical properties of halides are related to the dipole moment of the C–X bond (X represents any halogen).

---

*Molecules which contain adjacent functional groups are so different from molecules which contain isolated functional groups that the combination of adjacent functional groups is best considered to be a new functional group. For example, the amide group $\left(-C\diagup^{\displaystyle O}_{\diagdown NH_2}\right)$ is nominally a combination of an aldehyde group $\left(-C\diagup^{\displaystyle O}\right)$ and an amine group $(-NH_2)$, but amides do not resemble either aldehydes or amines. Aromatic hydrocarbons can be taken as extreme examples of this; molecules with three adjacent alkene groups in a ring in no way resemble alkenes.

Halides are rather unreactive; aryl halides are almost inert. In spite of their unreactivity, halides react with magnesium to form Grignard reagents, and with other metals to form related organometallic compounds:

$$CH_3CH_2Br + Mg = CH_3CH_2—Mg—Br$$

Bromoethane          Ethylmagnesium
bromide
(a Grignard reagent)

Many organic halides are important commercially. The uses of chloroform ($CH_3Cl$) and freon ($CCl_2F_2$) are well known; carbon tetrachloride ($CCl_4$) was used as a dry-cleaning agent until its effects on human kidneys were documented. The most newsworthy organic halide is DDT. This molecule has the structure

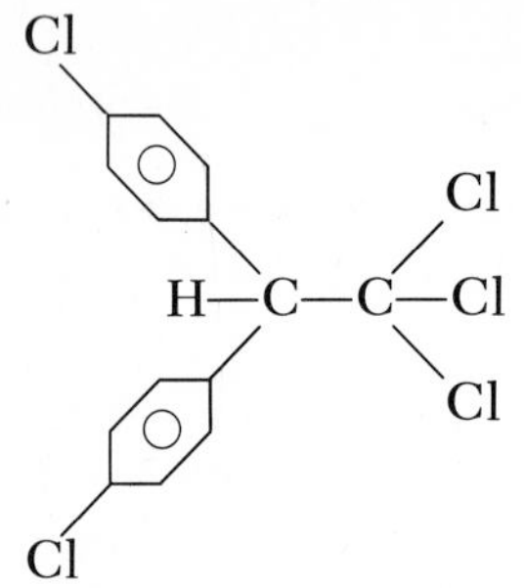

DDT

Like most aryl halides, this molecule is quite inert; once made, it does not react further or decompose under natural conditions. DDT accumulates in the fatty tissues of certain living organisms, and it eventually causes serious problems. Unfortunately, many pesticides are aryl halides, although the chemical similarities to DDT are obscured by trade names. The banning of DDT will solve no problems if DDT is replaced by another aryl halide which is also too inert for biodegradation.*

## 20.5 ALCOHOLS.

Alcohols are well known compounds because of the popularity of the two-carbon alcohol, ethanol.† A few simple alcohols are:

$CH_3OH$        $CH_3CH_2OH$        $CH_3CH_2CH_2OH$

Methanol          Ethanol           1-Butanol

---

*Some pesticides are organic phosphates; these pesticides are quite similar to compounds which can be used as nerve gases. The longer term effects of exposure to these pesticides is not well documented, but the effects are unlikely to be beneficial. It is very questionable whether a safe, general pesticide can ever be made.

†Although the alcohols are similar, they are not identical, and the human digestive tract can distinguish between them. All alcohols but ethanol are toxic.

$$\begin{array}{c} OH \\ | \\ CH_3CH_2CHCH_3 \end{array}$$

2-Butanol

$$\begin{array}{c} CH_3 \\ | \\ CH_3-C-CH_3 \\ | \\ OH \end{array}$$

2-Methyl-2-propanol (t-Butyl Alcohol)

Several of the functional groups can be viewed profitably as organic derivatives of simple inorganic hydrides. For example, amines can be viewed as derivatives of ammonia, and alcohols can be viewed as derivatives of water. This approach is most useful when the hydrocarbon portion of the molecule is very small. The low molecular weight alcohols — like water, but to a lesser degree — are quite polar; they dissociate slightly, and they form strong hydrogen bonds. They also are miscible with water in all proportions. However, when the hydrocarbon portion of the alcohol is large, alcohols begin to resemble alkanes, rather than water.

Alcohols can undergo many of the reactions which water undergoes, for example, reaction with alkali metals:

$$2CH_3OH + 2Na = 2Na^+ \ ^-OCH_3 + H_2$$

Sodium methoxide

Alcohols can be selectively oxidized:

$$8H^+ + 6CH_3\overset{\overset{\displaystyle OH}{|}}{C}HCH_3 + Cr_2O_7^{-2} = 6CH_3\overset{\overset{\displaystyle O}{\|}}{C}CH_3 + 2Cr^{+3} + 7H_2O$$

2-Propanol                      Acetone

$$16H^+ + 3CH_3CH_2CH_2OH + 2Cr_2O_7^{-2} = 3CH_3CH_2\overset{\overset{\displaystyle O}{\|}}{C}-OH + 4Cr^{+3} + 11H_2O$$

1-Propanol                      Propanoic acid

Elimination reactions are possible:

$$CH_3CH_2CH_2OH \ \overset{H^+}{=} \ CH_3CH{=\!=}CH_2 + H_2O$$

1-Propanol                      Propene

The influence of the portion of the molecule near the functional group of alcohols is so strong that it is necessary to discuss three types of alcohols. Primary alcohols have two hydrogen atoms bonded to the alcoholic carbon atom; secondary alcohols have one hydrogen atom bonded to the alcoholic carbon atom, and tertiary alcohols have no hydrogen atoms bonded to the alcoholic carbon (Figure 20.4).

A wide variety of displacement reactions is known, and in some cases the mechanisms have been unraveled. The reaction of primary and secondary alcohols with a halogen acid

$$CH_3OH + HBr = CH_3Br + H_2O$$

Methanol              Bromomethane

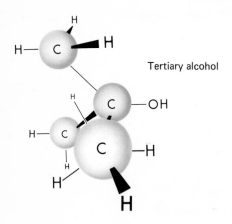

Primary alcohol

Secondary alcohol

**FIGURE 20.4** Alcohols.

Tertiary alcohol

follows the rate law

$$R_{(CH_3Br)} = k(H^+)(Br^-)(CH_3OH)$$

A reasonable mechanism which is consistent with the rate law is

$$CH_3OH + H^+ \overset{K}{\rightleftharpoons} CH_3OH_2^+ \text{ (fast)}$$

$$CH_3OH_2^+ + Br^- \overset{k_2}{\to} CH_3Br + H_2O \text{ (slow)}$$

The reader should verify that this mechanism leads to the observed rate law with $k = k_2K$. Additional experiments indicate that the slow step is

A similar displacement reaction occurs with tertiary alcohols,

$$
\underset{\text{2-Methyl-2-propanol}}{\underset{\overset{|}{CH_3}}{\overset{\overset{|}{OH}}{CH_3\underset{|}{C}CH_3}}} + HBr = \underset{\text{2-Methyl-2-bromopropane}}{\underset{\overset{|}{CH_3}}{\overset{\overset{|}{Br}}{CH_3\underset{|}{C}CH_3}}} + H_2O
$$

but in this case the rate is independent of the halide ion concentration:

$$
R_{(CH_3)_3Br} = k(H^+)((CH_3)_3OH)
$$

This means that the halide ion does not enter the reaction before the slow (rate-determining) step. A possible mechanism is

$$
\underset{\overset{|}{CH_3}}{\overset{\overset{|}{CH_3}}{CH_3-\underset{|}{C}-OH}} + H^+ \overset{K}{\rightleftharpoons} \underset{\overset{|}{CH_3}}{\overset{\overset{|}{CH_3}}{CH_3-\underset{|}{C}-OH_2^+}} \text{ (fast)}
$$

$$
\underset{\overset{|}{CH_3}}{\overset{\overset{|}{CH_3}}{CH_3-\underset{|}{C}-OH_2^+}} \overset{k_2}{\to} \underset{\overset{|}{CH_3}}{\overset{\overset{|}{CH_3}}{CH_3-\underset{|}{C}^+}} + H_2O \text{ (slow)}
$$

$$
\underset{\overset{|}{CH_3}}{\overset{\overset{|}{CH_3}}{CH_3-\underset{|}{C}^+}} + Br^- \to \underset{\overset{|}{CH_3}}{\overset{\overset{|}{CH_3}}{CH_3-\underset{|}{C}-Br}} \text{ (fast)}
$$

The bulky methyl groups prevent the easy formation of an activated complex like that shown for the reaction of a primary alcohol; the methyl group also may increase the stability of the carbonium ion ($(CH_3)_3C^+$).

Many molecules of biological interest contain the OH functional group. Carbohydrates all contain large numbers of alcohol groups.

Glucose (a sugar)

## 20.6 ETHERS.

The replacement of both hydrogen atoms in water by hydrocarbon fragments produces an ether.

$$CH_3OCH_3 \qquad CH_3OCH_2CH_3 \qquad CH_3CH_2OCH_2CH_3$$

Dimethyl ether    Methyl ethyl ether    Diethyl ether

The bond angle around the oxygen atom is somewhat larger than the bond angle in water, but ethers are not linear molecules. For this reason ethers are slightly polar molecules. Ethers are moderately soluble in water, for ethers can hydrogen-bond to water.

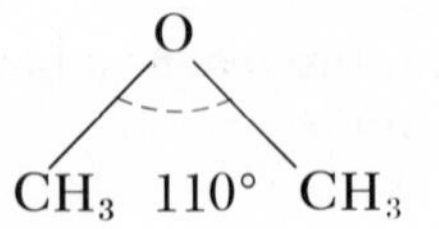

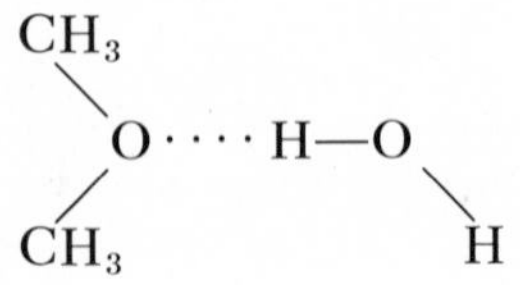

However, ethers contain no hydrogen atoms which are bonded to an electronegative atom, so ethers cannot hydrogen-bond to other ethers. For this reason the physical properties (e.g., boiling points) of ethers are much closer to the physical properties of alkanes of comparable molecular weights than to the physical properties of the isomeric alcohols.

Ethers are rather dangerous compounds to handle. They react with the oxygen in air to form very explosive peroxides:

$$2CH_3-O-CH_3 + O_2 = 2CH_3-O-O-CH_3$$

a peroxide

and the mixture of vaporized ether (remember the low boiling points) and air is extremely flammable.

Vinyl ether was used as a general anesthetic, but the hazard of operating room explosions led to its replacement.

$$CH_2=CH-O-CH=CH_2$$

Vinyl ether

Dioxane and tetrahydrofuran (THF) are convenient solvents.

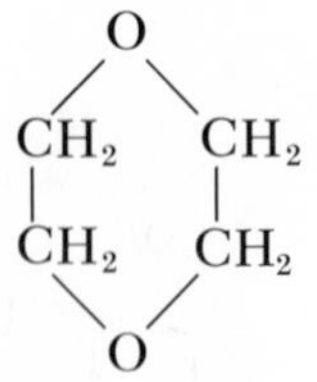

Dioxane 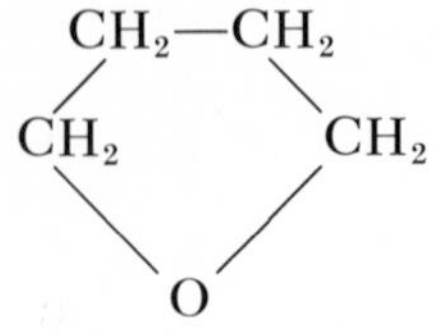        THF

## 20.7 CARBONYL COMPOUNDS.

Compounds which contain an aldehyde $\left(\begin{array}{c} O \\ \parallel \\ -CH \end{array}\right)$ or a ketone $\left(\begin{array}{c} O \\ \parallel \\ -C- \end{array}\right)$ functional group are called carbonyl compounds.

$$\underset{\substack{\text{Formaldehyde}\\\text{(Methanal)}}}{\text{H--}\overset{\displaystyle O}{\overset{\|}{\text{C}}}\text{--H}} \qquad \underset{\substack{\text{Acetaldehyde}\\\text{(Ethanal)}}}{\text{CH}_3\overset{\displaystyle O}{\overset{\|}{\text{CH}}}} \qquad \underset{\substack{\text{Acetone}\\\text{(Propanone)}}}{\text{CH}_3\overset{\displaystyle O}{\overset{\|}{\text{C}}}\text{CH}_3}$$

The carbon atom of the carbonyl group is attached to three other atoms; the *sigma* bonding orbitals are best described as $sp^2$ hybrids. The geometry of the carbonyl group is

$$\begin{array}{c}\text{O}\\120°\diagdown\!|\!\diagup 120°\\ \text{C}\\ \diagup\quad\diagdown\\ \text{C}\;\;120°\;\;\text{C}\end{array}$$

with the four atoms lying in a plane. The two lone pairs and the *pi* electrons produce a large electron density at the oxygen atom, and the molecular geometry leaves the oxygen atom exposed. Carbonyl compounds can form very strong hydrogen bonds, but not with other carbonyl compounds. For this reason the solubility of carbonyls in water is large; for example, acetone is completely miscible with water. The exposed oxygen end of the molecule produces a large dipole moment, and as a result the lighter carbonyls are excellent solvents for polar molecules. They are also rather good solvents for many organic compounds which are not polar and not soluble in water. For this reason, acetone is used as a homogeneous reaction medium for reactions which involve both polar and non-polar molecules.

The most important reactions of ketones are addition reactions:

$$\underset{\text{Acetone}}{\overset{\text{H}_3\text{C}}{\underset{\text{H}_3\text{C}}{>}}\text{C=O}} + \text{HCN} = \underset{\text{a cyanohydrin}}{\text{CH}_3\text{---}\overset{\displaystyle \text{CH}_3}{\underset{\displaystyle \text{C}\equiv\text{N}}{\overset{|}{\underset{|}{\text{C}}}}\text{---OH}}}$$

$$\overset{\text{H}_3\text{C}}{\underset{\text{H}_3\text{C}}{>}}\text{C=O} + \text{H}_2 = \underset{\text{2-Propanol}}{\text{CH}_3\text{---}\overset{\displaystyle \text{CH}_3}{\overset{|}{\text{CH}}}\text{---OH}}$$

The second reaction could be called either addition of hydrogen or reduction. Addition of a Grignard reagent to a ketone is often useful:

$$\underset{\text{Acetone}}{\overset{\text{CH}_3}{\underset{\text{CH}_3}{>}}\text{C=O}} + \underset{\substack{\text{Ethylmagnesium}\\\text{bromide}}}{\text{CH}_3\text{CH}_2\text{MgBr}} = \text{CH}_3\text{---}\overset{\displaystyle \text{CH}_3}{\underset{\displaystyle \text{CH}_2\text{CH}_3}{\overset{|}{\underset{|}{\text{C}}}}\text{---O---MgBr}}$$

$$CH_3—\overset{\overset{\displaystyle CH_3}{|}}{\underset{\underset{\displaystyle CH_2CH_3}{|}}{C}}—OMgBr + H_2O = CH_3—\overset{\overset{\displaystyle CH_3}{|}}{\underset{\underset{\displaystyle CH_2CH_3}{|}}{C}}—OH + MgBrOH$$

2-Methyl-2-butanol

Grignard reagents can be used to synthesize complex carbon skeletons.

The reactions of aldehydes are quite like the reactions of ketones. However, condensation reactions of aldehydes are much faster. A typical (and at times bothersome) condensation reaction is

$$CH_3\overset{\overset{\displaystyle O}{||}}{C}H + CH_3\overset{\overset{\displaystyle O}{||}}{C}H \overset{OH^-}{=} CH_3\overset{\overset{\displaystyle OH}{|}}{C}HCH_2\overset{\overset{\displaystyle O}{||}}{C}H$$

Acetaldehyde   3-Hydroxybutanal

Once begun, these condensations continue until all the aldehyde is consumed; special care is needed to control this reaction.

Many compounds found in nature contain carbonyl groups. Carbonyl compounds often have strong odors, although the odors of the heavier carbonyl compounds can be rather pleasant. The heavier ketones have been used in perfumes. Coumarin was used in perfume and food flavorings, but it was found to be carcinogenic.

Coumarin

## 20.8 CARBOXYLIC ACIDS AND ESTERS.

Compounds which have both a *carb*onyl and a hydr*oxyl* group attached to the same carbon atom are called carboxylic acids. The name reflects the most important characteristic of these compounds, the dissociation reaction:

$$H\overset{\overset{\displaystyle O}{||}}{C}OH \overset{H_2O}{=} H\overset{\overset{\displaystyle O}{||}}{C}—O^- + H^+$$

Formic acid   Formate ion<br>(Methanoic acid)

Several carboxylic acids are listed in Table 20.6 along with their dissociation constants. Organic acids are weak acids when compared to inorganic acids like HCl. The amount of dissociation is appreciable because the anion contains a *pi* orbital which is delocalized over three nuclei (Figure 20.5). Because the carboxylic carbon atom is attached to three other atoms, the molecular geometry is ideal for $sp^2$ *sigma* bonding orbitals.

**TABLE 20.6**

DISSOCIATION CONSTANTS OF SOME CARBOXYLIC ACIDS IN WATER

| COMPOUND | FORMULA | $K_a(25°C)$ |
|---|---|---|
| Methanoic acid (Formic acid) | $\overset{\displaystyle O}{\overset{\displaystyle \|}{\text{HCOH}}}$ | $1.7 \times 10^{-4}$ |
| Ethanoic acid (Acetic acid) | $CH_3\overset{\displaystyle O}{\overset{\displaystyle \|}{\text{COH}}}$ | $1.8 \times 10^{-5}$ |
| Propanoic acid | $CH_3CH_2\overset{\displaystyle O}{\overset{\displaystyle \|}{\text{COH}}}$ | $1.3 \times 10^{-5}$ |
| Butanoic acid | $CH_3CH_2CH_2\overset{\displaystyle O}{\overset{\displaystyle \|}{\text{COH}}}$ | $1.5 \times 10^{-5}$ |
| Benzoic acid | $\langle\bigcirc\rangle{-}\overset{\displaystyle O}{\overset{\displaystyle \|}{\text{COH}}}$ | $6.3 \times 10^{-5}$ |
| Citric acid | $\begin{array}{c}O\\ \|\\ C{-}OH \quad O\\ \|\qquad\quad \|\\ HO{-}C{-}CH_2COH\\ \|\\ CH_2\\ \|\\ O{=}C{-}OH\end{array}$ | — |

All five atoms shown lie in the same plane.

Carboxylic acids are ideally constituted for the formation of strong hydrogen bonds. The low molecular weight acids are very soluble in water, but the non-polar hydrocarbon section of the heavier acids makes them insoluble in water. Pure acids are usually dimerized;* this is due to the formation of strong hydrogen bonds.

*See Figure 6.12 for another example of dimerization of carboxylic acids.

Acid

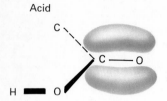

**FIGURE 20.5**  *Pi* electron orbitals in carboxylic acids and ions.

Ion

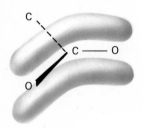

The acids dissolve both polar and non-polar molecules. However, the acids are somewhat reactive, and this limits their usefulness as solvents.

Carboxylic acids react with bases or reactive metals to form salts:

$$2CH_3\overset{\overset{\displaystyle O}{\|}}{C}OH + Zn = Zn^{+2}\left(O-\overset{\overset{\displaystyle O}{\|}}{C}CH_3\right)_2^{\ominus} + H_2$$

Acetic acid                          Zinc acetate

$$\langle\bigcirc\rangle-\overset{\overset{\displaystyle O}{\|}}{C}OH + NaHCO_3 = Na^+\left(^-O-\overset{\overset{\displaystyle O}{\|}}{C}-\langle\bigcirc\rangle\right) + H_2O + CO_2$$

Benzoic acid                          Sodium benzoate*

$$CH_3\overset{\overset{\displaystyle O}{\|}}{C}OH + NH_3 = (NH_4^+)\left(^-O\overset{\overset{\displaystyle O}{\|}}{C}CH_3\right)$$

Acetic acid                          Ammonium acetate

When ammonium acetate is heated it loses water:

$$NH_4^+\ ^-O\overset{\overset{\displaystyle O}{\|}}{C}CH_3 \quad\overset{heat}{=}\quad CH_3\overset{\overset{\displaystyle O}{\|}}{C}NH_2 + H_2O$$

Acetamide

The overall reaction can be called a displacement reaction of OH by NH$_2$. This reaction, and many similar displacement reactions, are used to

---

*Sodium benzoate is used as a food preservative.

produce valuable synthetic intermediates from acids. Carboxylic acids react with alcohols in what resembles an acid-base reaction to form esters.

$$\underset{\text{Acetic acid}}{CH_3\overset{\displaystyle O}{\overset{\|}{C}}OH} + \underset{\text{Isoamyl alcohol}}{CH_3\overset{\displaystyle CH_3}{\overset{|}{C}}HCH_2CH_2OH} = \underset{\text{Isoamyl acetate*}}{CH_3\overset{\displaystyle CH_3}{\overset{|}{C}}HCH_2CH_2O\overset{\displaystyle O}{\overset{\|}{C}}CH_3} + H_2O$$

There are many acids and esters found in nature. Formic acid is the irritant which is injected when many insects (ants, caterpillars) sting. Acetic acid is the active component in vinegar. The $C_4$–$C_{10}$ acids have terrible odors; the smell of rancid butter is due to butanoic acid. Esters have pleasing, fruity odors. Examples are:

$$CH_3\overset{\displaystyle CH_3}{\overset{|}{C}}HCH_2CH_2O\overset{\displaystyle O}{\overset{\|}{C}}CH_3$$

Isoamyl acetate (odor of bananas)

and

$$CH_3(CH_2)_6CH_2{-}O\overset{\displaystyle O}{\overset{\|}{C}}CH_3$$

Octyl acetate (odor of oranges)

Fats are mixed esters formed from glycerol and various long chain ($\sim C_{20}$)

$$\begin{array}{l} CH_2{-}OH \\ | \\ CH{-}OH \\ | \\ CH_2{-}OH \end{array}$$

Glycerol

acids.† The esterification reaction can be reversed in strong base.

$$\begin{array}{l} CH_2{-}O\overset{\displaystyle O}{\overset{\|}{C}}{-}R \\ | \\ CH{-}O\overset{\displaystyle O}{\overset{\|}{C}}{-}R \\ | \\ CH_2{-}O\underset{\displaystyle O}{\underset{\|}{C}}{-}R \end{array} + 3NaOH = \begin{array}{l} CH_2{-}OH \\ | \\ CH{-}OH \\ | \\ CH_2{-}OH \end{array} + 3Na^+\ {}^-O\overset{\displaystyle O}{\overset{\|}{C}}R$$

a glyceride                    glycerol        sodium salt of some
                                               long chain acid

*This is a component of airplane glue.
†Compounds such as glycerol are not classified as alcohols because the hydroxy groups are adjacent. However, many of the reactions of alcohols and glycerols are similar.

Here R is used to represent an unspecified hydrocarbon portion of the molecule. The cheapest glycerides are extracted from animal fat; in these fats the glycerides contain a mixture of R groups, and the resulting mixture of sodium salts is called soap. Soap is able to cleanse because it can keep small droplets of insoluble oils suspended in water. The hydrocarbon part of the acid sticks into the oil, and the carboxylate group sticks into the water.

## 20.9  AMINES.

Amines can be considered to be organic derivatives of ammonia. The resemblance between amines and ammonia is stronger than the resemblance between alcohols and water, but in both cases the analogy fails when the hydrocarbon portions of the molecules are large.

| $NH_3$ | $CH_3NH_2$ | $(CH_3)_2NH$ | $H_2NCH_2CH_2NH_2$ |
|---|---|---|---|
| Ammonia | Methylamine | Dimethylamine | Ethylenediamine |

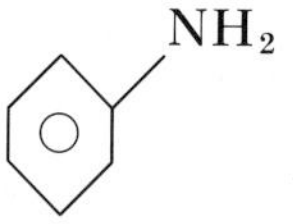

$$NH_2$$

Aniline

The geometries of ammonia and the simple amines are quite similar.

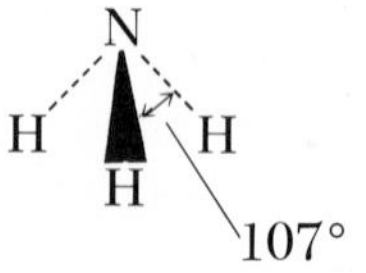

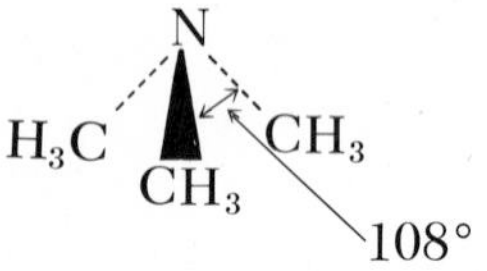

$107°$          $108°$

Like ammonia, amines have an exposed lone pair of electrons on the nitrogen atom; amines are bases which react with acids to form salts.

$$NH_3 + HCl \rightarrow NH_4^+Cl^-$$

| $(CH_3)_3N$ | $+$ | $CH_3Cl$ | $\rightarrow$ | $(CH_3)_4N^+Cl^-$ |
|---|---|---|---|---|
| Trimethylamine | | Chloromethane | | Tetramethylammonium chloride |

Many of the reactions of amines proceed by mechanisms wherein the first step is the addition of a proton to the lone pair on the nitrogen atom. Amines undergo many reactions which produce a variety of nitrogen-containing functional groups, for example, amides $\left( \begin{array}{c} O \\ \parallel \\ -C-NH_2 \end{array} \right)$ and diazonium salts ($-N^+\equiv N$).

Amines smell. Their vivid (non-IUPAC) names reflect this fact; for example,

| $NH_2CH_2CH_2CH_2CH_2NH_2$ | $NH_2CH_2CH_2CH_2CH_2CH_2NH_2$ |
|---|---|
| Putrescine | Cadaverine |

The natural sources of these compounds are obvious from the names. The very low molecular weight amines have a strong fishy odor. Humans are not sensitive to the odor of N, N-diethyl-*p*-toluidene, but mosquitos, ticks, and other insects apparently prefer not to land on skin coated with the compound; thus, it is the active ingredient of many insect repellants.

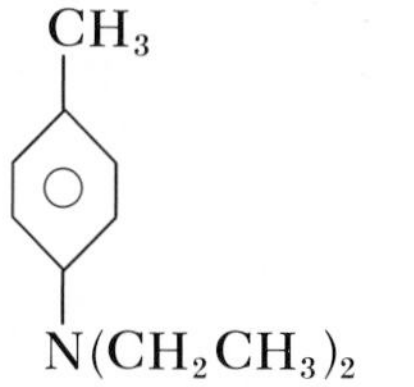


N, N-Diethyl-p-toluidene

Alkaloids are interesting because of their potent pharmacological action in the human body. Alkaloids are molecules which contain complex ring structures and at least one amine group. In Table 20.7 we list some well known alkaloids.

## 20.10 POLYMERS.

When a sample of ethylene is compressed and heated in the presence of a trace of oxygen, a product is formed which has a very high molecular weight (~20,000):

$$n\mathrm{CH_2{=}CH_2} \xrightarrow[\text{pressure}]{\mathrm{O_2}} (\mathrm{—CH_2—CH_2—})_n$$

This product is called polyethylene ("many ethylenes").

Polyethylene is a polymer. Polymers are materials whose molecules are long chains of small molecular units. In some polymers the repeating unit is itself a known molecule, which we call a monomer. For example, we say that the polymer polyethylene consists of chains of the monomer ethylene. Economically, polymers in the form of plastics are the most important product of the chemical industry, for plastics account for about one-half the total output of the industry. Polymers also occur in nature; natural rubber, cellulose, proteins, and DNA are polymers.

The polyethylene synthesis proceeds via a free-radical mechanism. We have already encountered free-radical mechanisms in the reaction of $H_2$ and $Br_2$. In all cases, a source of free radicals is needed to start the reaction. Peroxides are a convenient scource of free radicals. The free

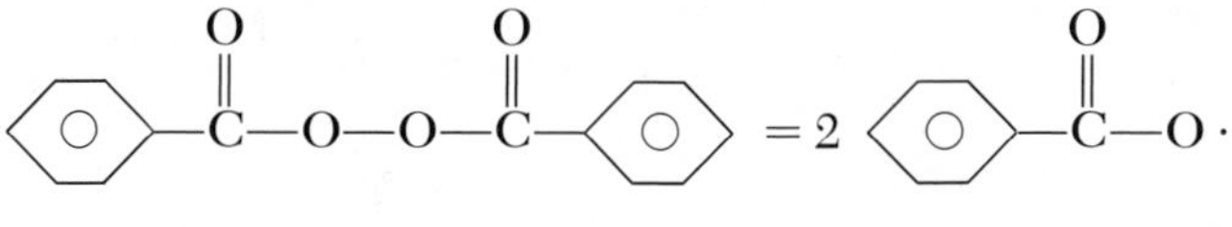


A peroxide Free radicals*

*The single dot represents the unpaired electron.

**TABLE 20.7**

ALKALOIDS AND THEIR USES

| | | |
|---|---|---|
| Mescaline | 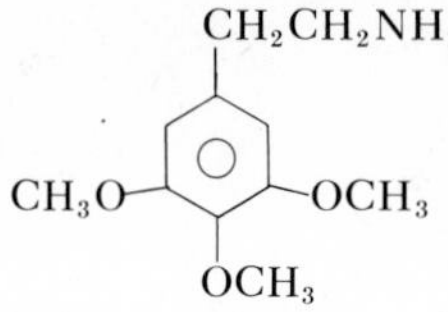 $CH_2CH_2NH_2$ <br> $CH_3O$ — — $OCH_3$ <br> $OCH_3$ | Hallucinogen |
| Nicotine | 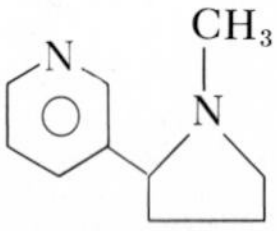 $CH_3$ <br> N — N | Insecticide (component of cigarette smoke) |
| Cocaine | 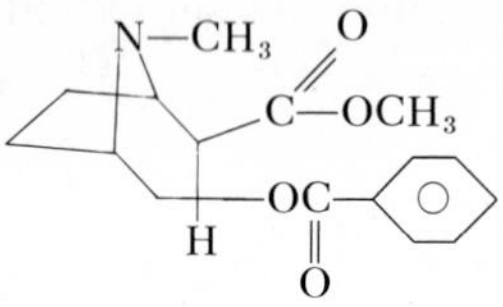 $N—CH_3$   O <br> $C—OCH_3$ <br> H — $OC$ — <br> O | Pain killer |
| Quinine | 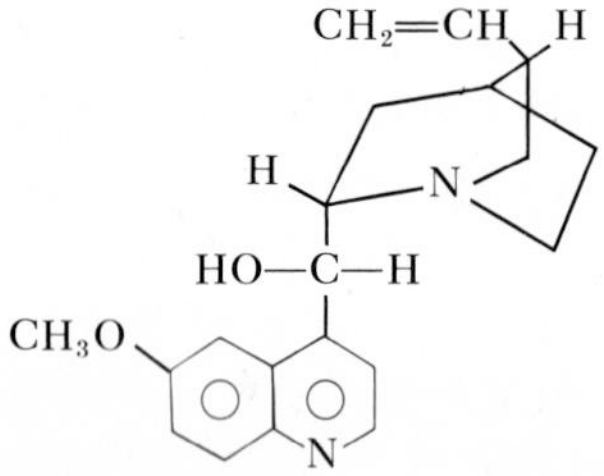 $CH_2=CH$   H <br> H — N <br> $HO—C—H$ <br> $CH_3O$ —   N | Malaria treatment |
| Lysergic acid diethylamide (LSD) | 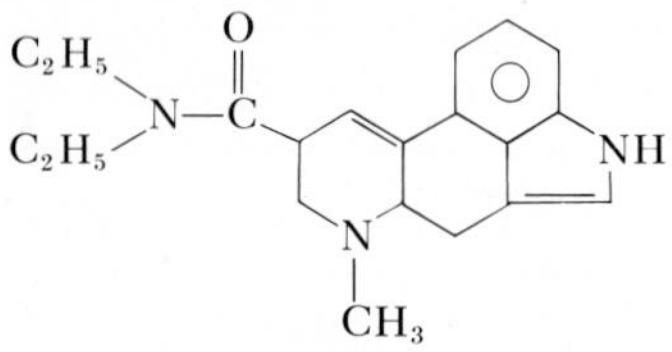 $C_2H_5$   O <br> $C_2H_5$ — N—C   NH <br> N <br> $CH_3$ | Hallucinogen |
| Strychnine | 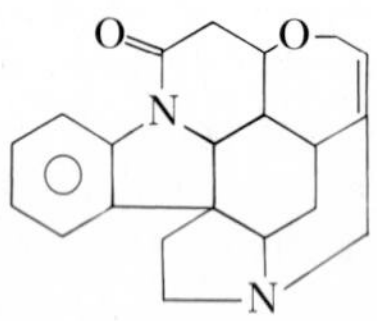 O — O <br> N <br> N | Poison |
| Reserpine | 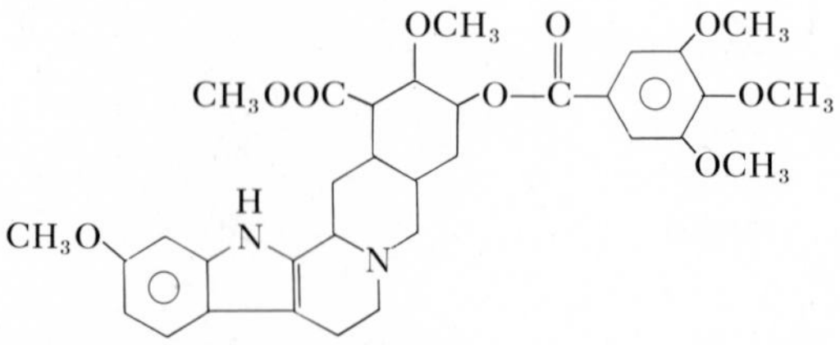 $OCH_3$   O   $OCH_3$ <br> $CH_3OOC$ — — O—C — — $OCH_3$ <br> $OCH_3$ <br> H <br> $CH_3O$ — N   N | Very potent tranquilizer |

radical mechanism for the formation of polyethylene is

$$Rad \cdot + CH_2 = CH_2 \rightarrow Rad—CH_2—CH_2 \cdot$$

$$Rad—CH_2—CH_2 \cdot + CH_2 = CH_2 \rightarrow Rad—CH_2—CH_2—CH_2—CH_2 \cdot$$

etc.

Notice that the chain grows in length without consuming free radicals. The growth continues until two free radicals happen to combine, at which time the chain growth is terminated. We have little control over the course of this reaction; once the free radicals are introduced, the reaction continues until the free radicals are consumed. The product is a solid solution which contains a mixture of many different molecules with different molecular weights (Figure 20.6). Special polymers (or plastics) can be synthesized for specific applications. The physical properties of plastics are more a consequence of the packing of the polymer molecules in the solid than of the properties of the molecules themselves. Of course, the molecular shapes and functional groups determine the packing. One possible arrangement of polymer molecules in the solid plastic is sketched in Figure 20.7. When polyethylene is crystallized under 5000 atm pressure, a material which seems to have such an extended-chain structure is obtained. The material is brittle, similar to most polycrystalline materials.

Most plastics are not brittle; in fact, they do not behave like true crystalline materials. Rather they possess a mixture of characteristics, some of which are typical of crystalline materials, and some of which are characteristic of amorphous solids. These materials are based on folded-chain structures. An idealized polymer crystal based on a folded chain is shown in Figure 20.8; a schematic representation of a real polymer crystal is shown in Figure 20.9. The latter diagram shows that there are regions of ideal packing, which account for the crystal-like properties, and regions of disorder, which account for the amorphous-solid properties.

In many plastics the ordered, semi-crystalline regions are about 100 Å across. For many applications we find that the larger we make the ordered

**FIGURE 20.6** Distribution of molecular weights for selected polymers.

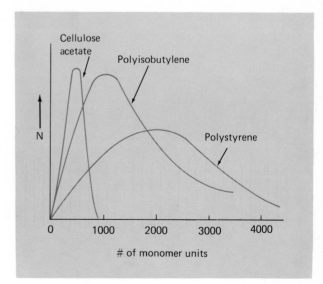

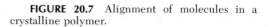

**FIGURE 20.7** Alignment of molecules in a crystalline polymer.

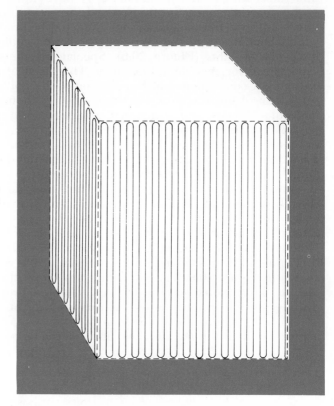

**FIGURE 20.8** Folded-chain polymer. In real polymers, the length of each fold is 20 to 40 times the separation between folds.

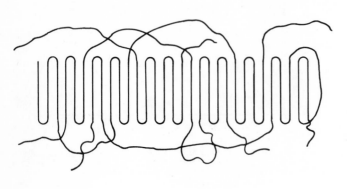

**FIGURE 20.9** Representation of a real polymer. There is a combination of folded-chain alignment and random alignment.

regions, the more desirable the plastic. The ordered regions can be large when the molecules pack together easily and when the solid condenses slowly. We find that we can control the physical properties of a plastic by controlling the sizes and the regularity of the groups attached to the molecular backbone.[*] We cannot exert the needed control in free-radical syntheses because of the possibility of radical attack at any position along an already polymerized molecule:

$$\text{Rad}\cdot + \text{—CH}_2\text{CH}_2\text{CH}_2\text{CH}_2\text{—} \rightarrow \text{Rad-H} + \text{—CH}_2\dot{\text{C}}\text{HCH}_2\text{CH}_2\text{—}$$

This new free radical can then begin adding monomer units on a side chain:

$$\text{—CH}_2\dot{\text{C}}\text{HCH}_2\text{CH}_2\text{—} + \text{CH}_2 = \text{CH}_2 \rightarrow \text{CH}_2\text{—CH—CH}_2\text{CH}_2\text{—}$$

$$\overset{|}{\underset{|}{\text{CH}_2}}$$

$$\dot{\text{C}}\text{H}_2$$

etc.

This unavoidable random introduction of side chains is undesirable. It is much better to introduce a regular sequence of identical side groups by a clever choice of monomer molecules.

Polymerization reactions which produce polymers with a regular sequence of side groups can be carried out with the aid of catalysts. The mechanisms of catalyzed polymerizations are not completely understood, but the reactions must involve the following steps:

$$\text{M(metal-containing catalyst)} + \text{CH}_2\text{=}\text{CH}_2 \rightarrow \text{M—CH}_2\text{—CH}_2^+$$

$$\text{M—CH}_2\text{—CH}_2^+ + \text{CH}_2\text{=}\text{CH}_2 \rightarrow \text{M—CH}_2\text{—CH}_2\text{—CH}_2\text{—CH}_2^+$$

The curved arrow indicates that the addition of the monomer must be an insertion at the catalytic surface; otherwise the catalyst could not control the arrangement of the side chains. The nature of the catalyst is of critical importance. The catalysts contain either titanium chloride or titanium oxide as well as other materials; the regularity of the polymer is greater when the catalyst is a crystalline solid rather than an amorphous solid.

A wide variety of plastics can be made by polymerization of monomers which are derivatives of ethylene. A few of these materials are listed in Table 20.8. Although the physical properties of plastics are determined by the sizes and regularity of the side groups, the chemical properties are determined by the chemical nature of the side groups. For example, *Teflon* is a very unreactive plastic because of the great strength of C–F bonds.

When the side groups or the molecular backbones of polymers contain unreacted functional groups, bond formation with a second polymer molecule is possible. These bonds between different chains, called crosslinks, greatly increase the strength (and with vast numbers of crosslinks,

---

[*] Obviously, the length of the backbone is of some importance, too. This length can be controlled crudely by adjusting the initial concentration of free radicals in the reaction mixture.

**TABLE 20.8**
SOME POLYMERIC PLASTICS

| MONOMER | POLYMER | NAME AND USE |
|---|---|---|
| $CH_2{=}CHCl$ | 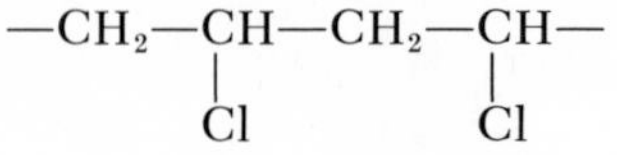 $-CH_2{-}CH{-}CH_2{-}CH-$ <br> $\quad\quad\;\, Cl \quad\quad\quad\;\, Cl$ | Polyvinylchloride (PVC) <br> Records, plastic pipe |
| $CF_2{=}CF_2$ | $-CF_2{-}CF_2{-}CF_2{-}CF_2-$ | Teflon <br> Inert plastics |
| 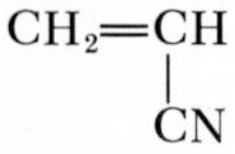 $CH_2{=}CH$ <br> $\quad\quad\; CN$ |  $-CH_2{-}CH{-}CH_2{-}CH-$ <br> $\quad\quad\;\, CN \quad\quad\quad\; CN$ | Orlon <br> Fibers |
| $CH_2{=}CCl_2$ | $-CH_2{-}CCl_2{-}CH_2{-}CCl_2-$ | Saran <br> Packaging film |
| 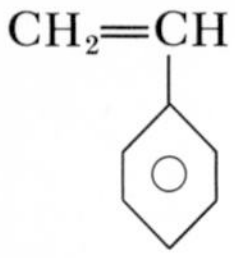 $CH_2{=}CH$ <br> (phenyl) | 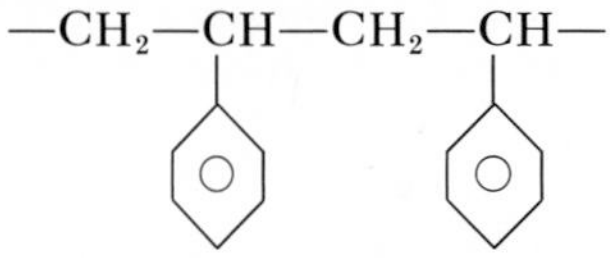 $-CH_2{-}CH{-}CH_2{-}CH-$ <br> (phenyl) (phenyl) | Polystyrene <br> Plastic foam |
| 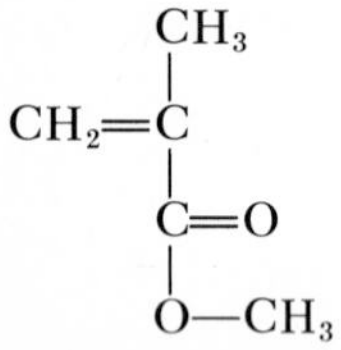 $\quad\quad\;\, CH_3$ <br> $CH_2{=}C$ <br> $\quad\quad\;\, C{=}O$ <br> $\quad\quad\;\, O{-}CH_3$ | 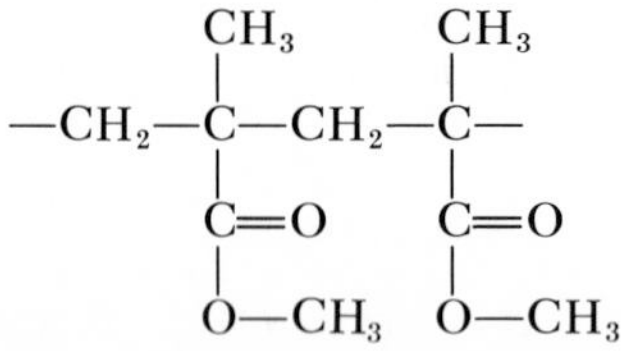 $\quad\quad\;\, CH_3 \quad\quad CH_3$ <br> $-CH_2{-}C{-}CH_2{-}C-$ <br> $\quad\quad\;\, C{=}O \quad\quad\; C{=}O$ <br> $\quad\quad\;\, O{-}CH_3 \quad O{-}CH_3$ | Lucite, Plexiglass <br> Plastic windows |

the rigidity) of plastics. One important example of crosslink formation is the vulcanization of the polymer rubber:

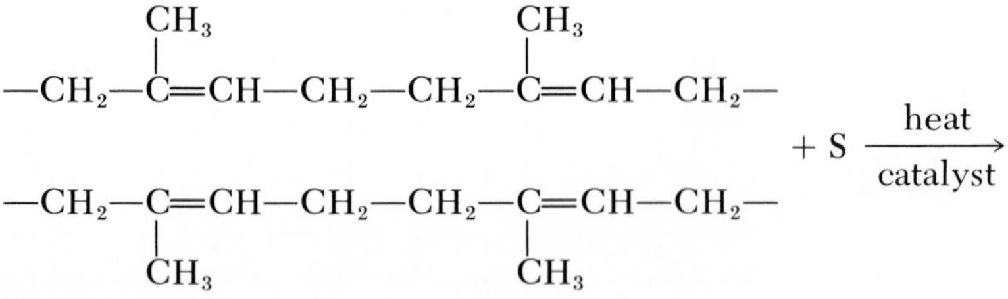

$$-CH_2-\underset{CH_3}{\overset{CH_3}{C}}{=}CH-CH_2-CH_2-\underset{}{\overset{CH_3}{C}}{=}CH-CH_2-$$

$$-CH_2-\underset{CH_3}{C}{=}CH-CH_2-CH_2-\underset{CH_3}{C}{=}CH-CH_2- \quad + S \xrightarrow[\text{catalyst}]{\text{heat}}$$

Rubber

$$-CH_2-\underset{}{\overset{CH_3}{C}}{=}CH-CH_2-\underset{|}{CH}-\underset{}{\overset{CH_3}{C}}{=}CH-CH_2-$$
$$S$$
$$-CH_2-\underset{CH_3}{C}{=}CH-CH_2-\underset{}{CH}-\underset{CH_3}{C}{=}CH-CH_2-$$

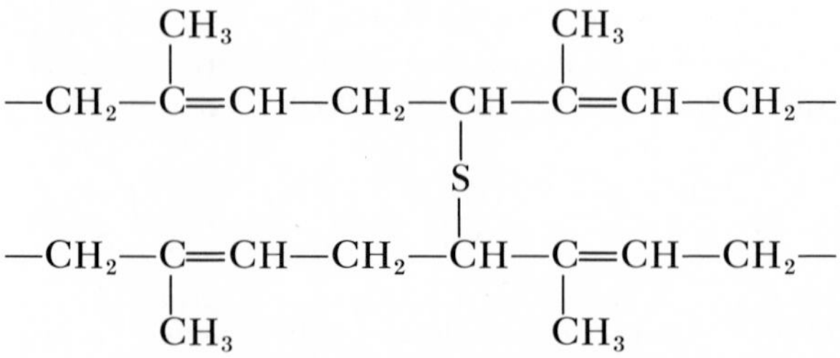

Vulcanized rubber

Crosslinks can be introduced after the plastic is cast into its final shape by exposure to radiation. The radiation (ultraviolet or gamma) probably produces free radical sites on the chains which then react to form crosslinks. This process is the source of the yellowing and increased brittleness of plastics which are exposed to sunlight.

In condensation polymers, two different monomer molecules may be introduced into the backbone in an *ABABA* pattern. For example, the plastic called Dacron or Terylene is formed by repetition of the reaction

$$\underset{\text{Terephthalic acid}}{HO\overset{\overset{O}{\|}}{C}{-}\langle O \rangle{-}\overset{\overset{O}{\|}}{C}OH} + \underset{\text{Ethylene glycol}}{\overset{\overset{OH}{|}}{C}H_2{-}\overset{\overset{OH}{|}}{C}H_2} = HO\overset{\overset{O}{\|}}{C}{-}\langle O \rangle{-}\overset{\overset{O}{\|}}{C}{-}O{-}CH_2\overset{\overset{OH}{|}}{C}H_2 + H_2O$$

Silicones are silicon-containing polymers which are formed by condensation reactions.

$$2HO{-}\underset{\underset{CH_3}{|}}{\overset{\overset{CH_3}{|}}{Si}}{-}OH = HO{-}\underset{\underset{CH_3}{|}}{\overset{\overset{CH_3}{|}}{Si}}{-}O{-}\underset{\underset{CH_3}{|}}{\overset{\overset{CH_3}{|}}{Si}}{-}OH + H_2O$$

etc.

Silicones of intermediate chain length are oils; silcones of very high chain length are quite rubbery.* Crosslinks can be added by proper selection of the monomer, that is, use of some fraction of $Si(OH)_3CH_3$ instead of $Si(OH)_2(CH_3)_2$.† Silicones are used because they are resistant to oxidation at high temperatures at which carbon-based polymers burn, and because their physical properties are relatively insensitive to temperature changes.

---

*Silly Putty is a silicone rubber.
†Complete replacement of the methyl side groups by oxygen cross links produces silica glass (see Chapter 6).

---

**PROBLEMS**

1. Pentane is an alkane with five carbon atoms. Draw all the isomers of pentane.

2. There is a 3 kcal/mole barrier to the rotation about the carbon-carbon bond in ethane. Draw the lowest energy arrangement and the highest energy arrangement in an end-on view.

3. List the number of isomers of all alkanes from methane through heptane.

4. A mechanism for the formation of 1-bromobutane is given in Section 20.3. What would be the rate law for this process if the second and third steps were of comparable speeds?

5. Write a balanced chemical reaction for the burning of butane.

6. Burning pentane produces an enthalpy change of $-845$ kcal. What is the enthalpy of formation of pentane?

7. Butenes react with HCl in addition reactions. How many isomers are possible when HCl adds to 1-butene? To 2-butene?

8. Phenanthrene is an isomer of anthracene which also contains three aromatic rings. What is the structure of phenanthrene?

9. Write a reaction for the formation of a Grignard reagent.

10. Which of the following are primary, secondary, and tertiary alcohols: methanol, ethanol, 1-butanol, 2-butanol, 2-methyl-2-propanol?

11. One can dispose of extra sodium by reacting it with ethanol. Write an equation for this reaction. Why is ethanol a better choice than methanol?

12. Why are ethers suitable solvents for Grignard reagents?

13. Calculate the pH of a 0.1 M solution of benzoic acid.

14. Write a reaction for the formation of sodium citrate from citric acid.

15. Write a reaction for the formation of isoamyl acetate.

*16. Give examples of primary, secondary, and tertiary amines.

*17. In principle we could make a condensation polymer from citric acid and ethylene glycol. Write the reaction which would be involved. Would you expect this to be a highly crystalline polymer, and why or why not?

## References

The first three are standard textbooks.

20.1. J. D. Roberts and M. C. Caserio, *Modern Organic Chemistry* (W. A. Benjamin, New York, 1967).
20.2. R. T. Morrison and R. N. Boyd, *Organic Chemistry* (Allyn and Bacon Inc., Boston, 1968).
20.3. D. J. Cram and G. S. Hammond, *Organic Chemistry* (McGraw-Hill Book Co., New York, 1964).
20.4. A. Keller, "Long-Chain Polymer Crystals," *Physics Today*, 23, No. 5, 42 (1970).

## CHAPTER 21

"Scientific research consists in see-
ing what everyone else has seen, but
thinking what no one else has
thought."

*A. Szent-Gyorgyi*

"Now our model for deoxyribonucleic acid is, in effect, a pair of
templates each of which is complementary to the other. We imagine
that prior to duplication the hydrogen bonds are broken and the two
chains unwind and separate. Each
chain then acts as a template for
the formation onto itself of a new
companion chain, so that eventually
we shall have two pairs of chains
where we only had one before."

*J. D. Watson and F. H. C. Crick**

# BIOCHEMISTRY

The chemistry of life is called *biochemistry*. The molecules and the
reactions of biochemistry are among the most complex known, yet so great
is the fascination of life and its sources that scientists have never waited
until their methods and theories provided a firm base from which to attack
biological problems. We assume that the reader possesses this same
curiosity about life, and we shall describe what space and time permit,
even though a solid foundation has not been laid for all this material.

Biochemists repeatedly ask three general questions: What molecules
are present in some particular part of a living organism? What essential
functions do these molecules fulfill for the organism? Of all the possible
molecules, why are these found in the organism? The answers to the first
question are available in many cases. At this time we know at least what
types of molecules are present in the various parts of cells. The second
question is an exceedingly difficult one, and much of the current research
in biochemistry is aimed at answering that question for specific organisms.
The third question may be a philosophical, even a religious, question, and
it may be unanswerable. Possible answers include: chance followed by
universal descent from a single common ancestor, thermodynamic stability,
or a unique environment on earth when life first emerged about four billion
years ago. Perhaps an answer will be provided by the eventual discovery
of extraterrestrial life.

---

*J. D. Watson and F. H. C. Crick, Nature, *171* (1953) p. 964.

## 21.1   PLANTS AND PHOTOSYNTHESIS.

Plants are able to carry out one reaction that is absolutely essential to all animal life, photosynthesis. Photosynthesis is the conversion of carbon dioxide and water into the sugar glucose. The net reaction is

$$6CO_2(g) + 6H_2O(\ell) = C_6H_{12}O_6(s) + 6O_2(g)    \Delta G° = 686 \text{ kcal}$$

Glucose*

This reaction is the reverse of the oxidation of glucose, and the Gibbs energy change is very unfavorable for the reaction as written. The reaction occurs only because plants contain the molecule chlorophyll.

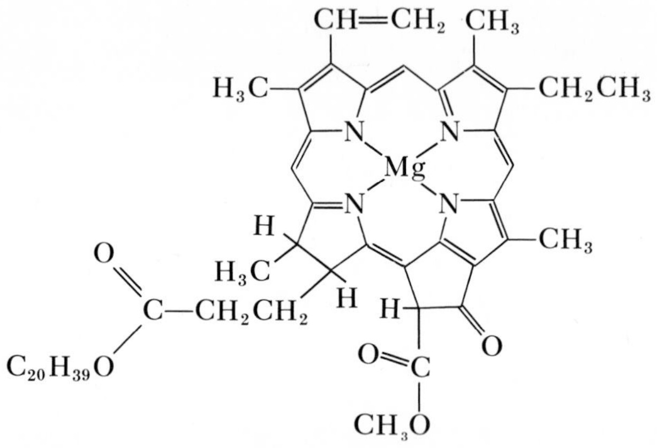

Chlorophyll a

Chlorophyll absorbs light in the visible region of the spectrum (it is the molecule which gives leaves their green color), transfers the required energy to the reacting molecules, and acts as a catalyst for the reaction. The mechanism of the vital reaction has been elucidated through use of radioactive tracers, but the critical details of the energy transfer are still a mystery. Photosynthesis has been carried out in the laboratory, but the

---

*The glucose produced by plants is a sugar. We shall need a good way of drawing sugars; we illustrate with glucose:

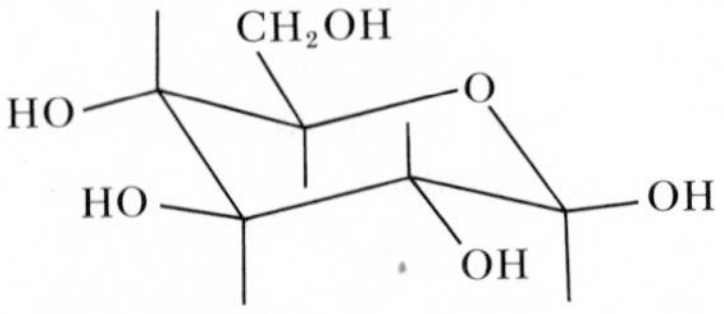

This is the particular isomer of glucose which is produced by plants (there is no free rotation about *sigma* bonds in rings). The ring consists of an oxygen atom, which is shown, and five carbon atoms, which are not specifically drawn; a bond from a carbon atom connects a hydrogen atom unless some other group is specified.

efficiency is only a fraction of that achieved in a living plant. Some plants can produce a glucose molecule for every twenty photons absorbed.

It would be impossible to overemphasize the importance of this one reaction. Throughout history the one significant source of energy for this planet has been the sun. Animals have no mechanism for capturing and storing solar energy. All animals are totally dependent on photosynthesis for energy in the form of food. Moreover, animals (except some anaerobic bacteria) require oxygen. Oxygen is a by-product of photosynthesis. The average oxygen molecule remains in the atmosphere for 2000 years; without photosynthesis our oxygen supply would disappear slowly.[*] This point may become relevant if we continue to pollute the oceans, for a significant fraction of the photosynthesis is carried out by the simple plants in the ocean. Death to these plants is death to all animal life on earth.

Glucose is a simple example of a carbohydrate. Most natural carbohydrates are condensation polymers of glucose or other sugars. The elimination of water between alcohol groups on two glucose molecules produces a linkage between the two rings:

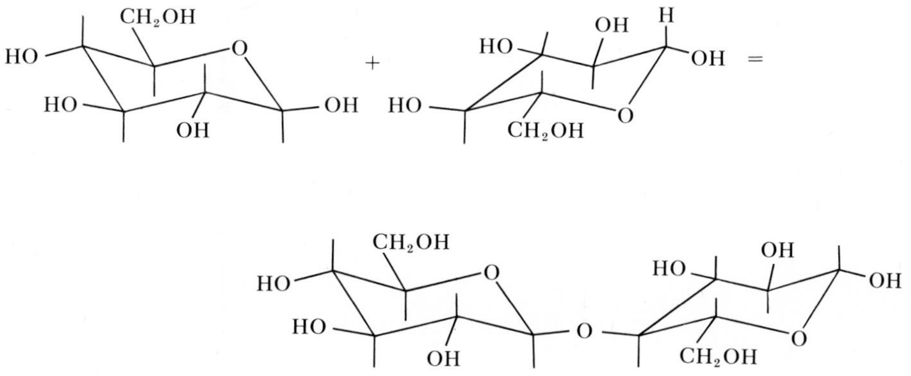

Repetition of this reaction leads to a polymer:

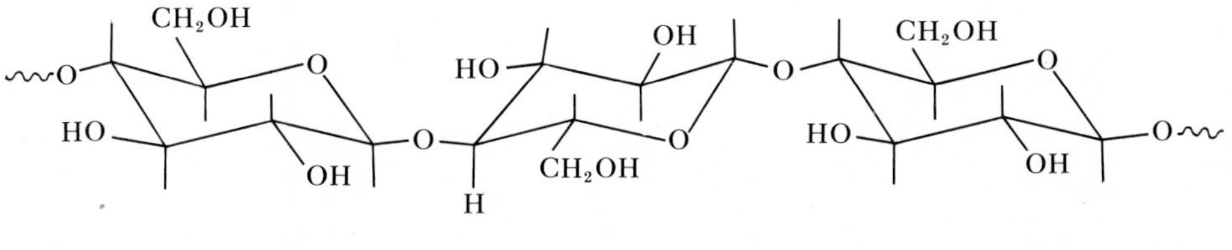

This polymer is the main component of cellulose. In wood, long parallel fibers of cellulose are embedded in a matrix, giving a reasonably rigid and strong material.

Plants have the ability to be quite selective in this condensation polymerization. When cellulose is needed, only two of the five possible

[*]Planetary atmospheres which contain oxygen appear to be quite rare. Other planetary atmospheres contain no oxygen, but appreciable $CO_2$. For this reason we think that photosynthesis had to evolve long before oxygen-dependent organisms appeared.

alcohol groups react. When food for a seed is needed, the glucose is converted to another isomeric form, and a third alcohol group is used:

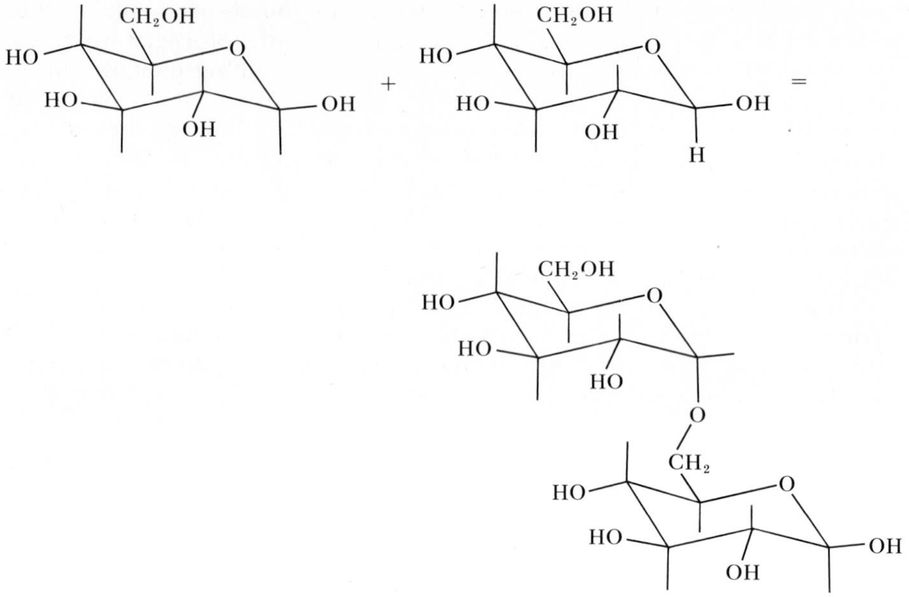

Combination of the two reactions produces a polymer with several long side chains. We call this material starch. Starch serves as a food for both

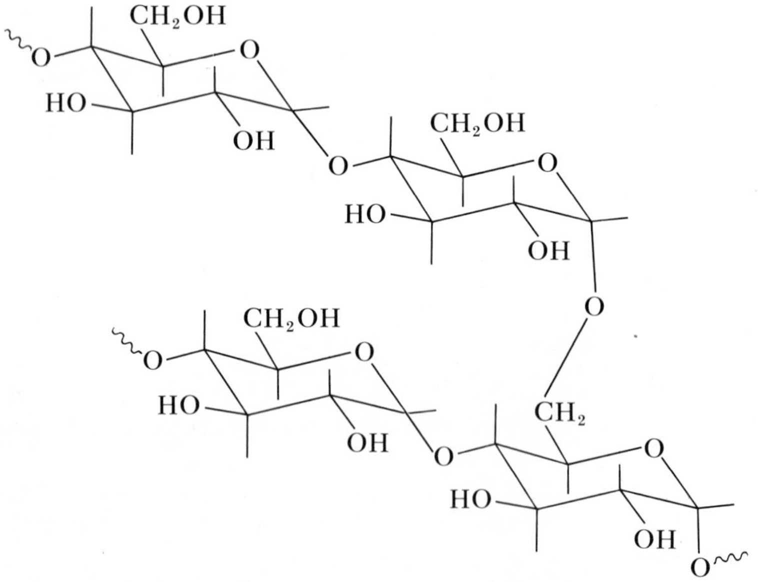

Starch

plants and animals, because both plants and animals are able to reverse the polymerization reaction and regenerate glucose. Animals oxidize the glucose to provide energy for needed chemical reactions (they may store glucose in the form of a starch-like polymer called glycogen). A sprouting seed repolymerizes the glucose into cellulose. Cellulose is insoluble in water, and most animals find it indigestible; that is, they are unable to convert it back to glucose.*

## 21.2   NUCLEIC ACIDS.

The discovery of the mechanism by which nucleic acids are reproduced touched off a revolution in biology which was as far-ranging as the revolution in chemistry and physics that was touched off by the development of quantum mechanics. For the first time it appeared that understanding of the processes of reproduction, of heredity, and of aging might be within reach. This prospect led to the emergence of a new science, molecular biology. The goal of molecular biology is the unraveling of cell processes at the molecular level. As understanding increases, so do the possibilities of control, a prospect which is terrifying when the social misuses of many recent scientific advances are considered. It will repay everyone, scientist and non-scientist, to keep abreast of new discoveries, and of the implications of new discoveries, in molecular biology.†

Nucleic acid is a general term used to describe several types of molecules. The deoxyribonucleic acids (DNA) are found in the nuclei of cells. The DNA molecules are the library of genetic information for each individual. Ribonucleic acids (RNA) are found outside the cell nuclei, for the most part. There are several distinguishable types of RNA which serve different functions within the cell.

Nucleic acids are huge condensation polymers. The backbone of the polymer contains alternating sugar and phosphate residues.‡ In DNA the sugar is 2-deoxyribose, and in RNA the sugar is ribose:

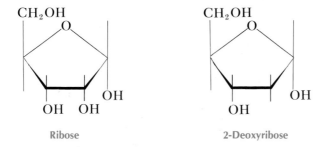

Ribose                    2-Deoxyribose

Like glucose, these sugars both contain several alcohol groups which can

---

*Late in World War II the Germans sought to supplement their dwindling food supply by laboratory hydrolysis (depolymerization) of cellulose. The resulting mixture was declined by their cattle.

†To a certain extent one can do this by skeptical and selective reading of newspapers and news magazines. Molecular biologists sometimes announce their work in the New York Times, rather than in the scientific literature. These accounts sometimes benefit from the presence of skilled editors and suffer from the absence of skilled referees.

‡We call those fractions of a monomer which are incorporated into a condensation polymer, *residues.*

react with an acid to form an ester. In nucleic acid formation the acid is phosphoric acid:

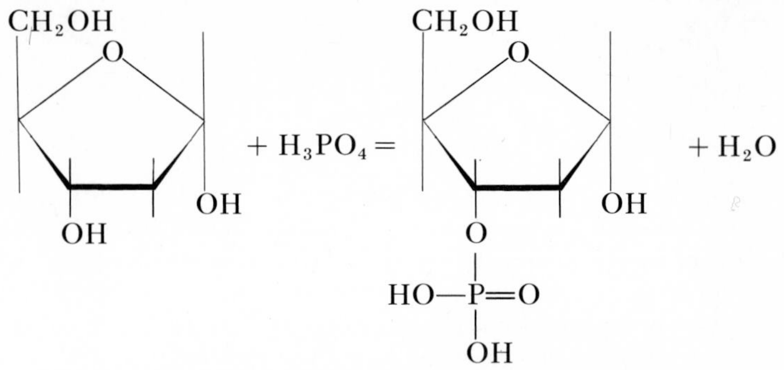

This reaction has exhausted neither the alcohol groups on the sugar nor the acid hydrogens on the phosphate. Further reaction produces the polymer

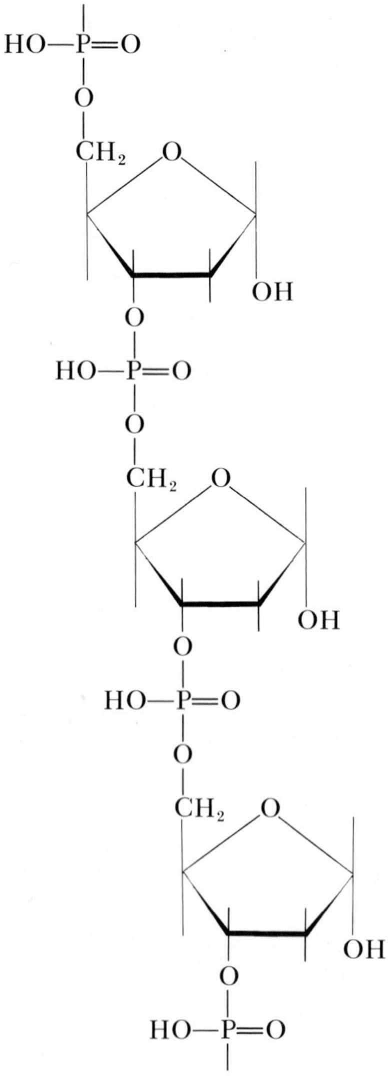

All nucleic acids have one of five nitrogen-containing ring compounds attached to each sugar residue in place of one alcohol group. The nitrogen-containing ring compounds are called bases or base residues. Both DNA and RNA contain adenine, guanine, and cytosine; DNA contains thymine, whereas RNA contains uracil:

Adenine          Guanine          Cytosine

Uracil          Thymine

The bases are bonded to the sugars at the positions indicated by the arrows. A small section of an RNA molecule might look like this:

The bases can form multiple hydrogen bonds with a second base. However, these multiple hydrogen bonds can form between only two selected pairs of the bases; in these selected pairs the hydrogen bonds bear a unique lock-and-key relationship:

Guanine             Cytosine

Adenine             Thymine

Guanine and cytosine form a pair, as do adenine and either thymine or uracil. Other pairs do not fit. We need to represent the bases and their lock-and-key arrangement of hydrogen bonds in rather elaborate diagrams. For later clarity we indicate an electronegative nitrogen, which can donate a hydrogen atom to a hydrogen bond, by the symbol →•, and we indicate an electronegative atom (O or N), which can accept a hydrogen atom in a hydrogen bond, as —⊂. We symbolize the base pairs as in Figure 21.1a.

The base pairing controls the structure of the nucleic acids, and it provides the mechanism by which the nucleic acids fulfill their functions. The DNA molecule is a double-stranded polymer; the two strands are held together by hydrogen bonds between the many base pairs (See above). A complete DNA molecule can be regenerated from either of the two strands alone because the lock-and-key arrangement of the base pairs is unique. A single molecule of DNA can divide into complementary strands, from which two molecules of DNA can be produced, for two strands can serve

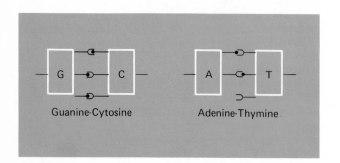

Guanine-Cytosine         Adenine-Thymine

**FIGURE 21.1a**    G-C and A-T base pairs.

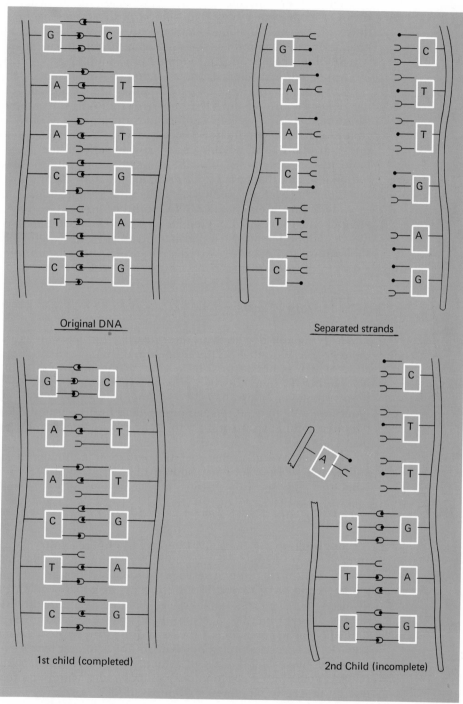

Original DNA

Separated strands

1st child (completed)

2nd Child (incomplete)

**FIGURE 21.1b** Reproduction by DNA. Each strand from the original molecule serves as a template for a child molecule.

as templates for the two new molecules. In this sense the DNA molecules can reproduce (Figure 21.1b).

The pattern (or the sequence) of bases in DNA molecules forms a series of coded messages. Translation of these messages into action produces all the reactions and characteristics of a cell which allow it to survive, grow, and reproduce. In more complex organisms the characteristics which differentiate one individual from another are embodied in their DNA molecules. Biologists have named those portions of the cells which carry the genetic information *genes*. Genes are sections of DNA molecules.

The structure of the DNA molecules allows us to see how molecules can reproduce, but we know almost nothing about how, why, or under what conditions the reproduction process occurs. Among the more interesting questions here is how the bases, sugars, and other parts of the DNA molecule are assembled quickly enough to build up the new chains at a reasonable rate. Bacteria often reproduce in 10 to 20 minutes, but the DNA molecules in bacterial cells have molecular weights in the millions.

There are three types of RNA, and the three types differ in molecular weight and function. RNA molecules are single-stranded polymers, and they usually do not serve as permanent archives of genetic information. We shall return to RNA when we describe the synthesis of proteins.


## 21.3   ENERGY SOURCES IN CELLS.

A living cell decomposes after death. This shows that at least some of the reactions essential for the formation of living cells are not spontaneous, a fact which is not surprising in view of the unfavorable entropy change involved in the assembly of highly ordered molecules like nucleic acids. A living organism must have a continuous supply of energy to survive. Plants are able to capture solar energy and to store it in the form of glucose polymers; animals are forced to steal these glucose polymers from the plants for food.

Both plants and animals use the oxidation of glucose to carbon dioxide and water as their primary source of energy. When carried out in the laboratory, this reaction releases a large amount of Gibbs energy:

$$C_6H_{12}O_6(s) + 6O_2(g) = 6CO_2(g) + 6H_2O(\ell)   \Delta G° = -686 \text{ kcal/mole}$$

Although plenty of energy is released by this reaction, the reaction is not suitable as a source of energy within a living cell. The reaction would require the transfer of two gases ($O_2$ and $CO_2$) through cell walls, a process which is too slow to provide the instant responses which can mean survival. A more serious objection is that the reaction provides too much energy at one time; the heat produced by this reaction would destroy the delicate molecules which are essential for life. Instead, the cell oxidizes glucose by a very complex mechanism involving at least seventeen steps. Many of these steps, each of which involves a small change in Gibbs energy, contribute to a storehouse of molecules which can react to provide energy in usable increments when needed.*

---

*The problem is similar to the problem of taming nuclear fusion for domestic power sources. There is plenty of energy available (witness the hydrogen bomb), but as yet we have no way to control fusion reactions so as to release this energy at a usable rate. Cells have succeeded in taming the oxidation of glucose.

The reaction which is used to store energy is the conversion of adenosine diphosphate (ADP) to adenosine triphosphate (ATP):

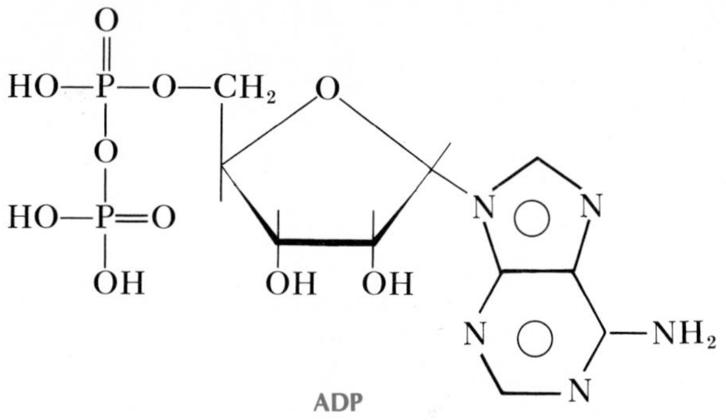

ADP

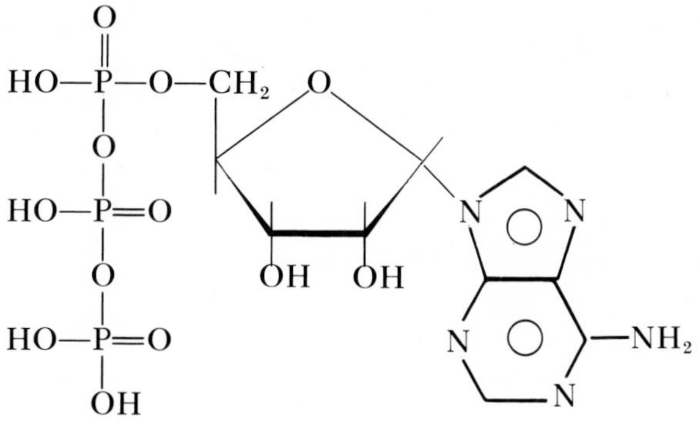

ATP

$$ADP(aq) + H_3PO_4(aq) = ATP(aq) + H_2O(\ell) \quad \Delta G° = 8 \text{ kcal/mole } (pH\ 7)$$

The oxidation of glucose as it occurs in cells can be summarized by the reaction

$$C_6H_{12}O_6(aq) + 6O_2(g) + 38ADP(aq) = 6CO_2(g) + 44H_2O(\ell) + 38ATP(aq)$$
$$\Delta G° = -382 \text{ kcal/mole}$$

By means of this complex oxidation of glucose, the cell has succeeded in trapping $686 - 382 = 304$ kcal/mole of the free energy in the form of ATP molecules.

Most of the reactions within the cell are of the general type

$$n\ ATP + X = Y + n\ ADP$$

The energy necessary to produce needed amounts of compound Y from compound X are provided by reacting $n$ molecules of ATP with compound X. This is sometimes called biological coupling. Unfortunately, the term creates the impression that there are two separate reactions, one of which supplies energy to the other. This is not the case; in fact, such a selective flow of energy would be a violation of the second law of thermodynamics. As written, $n$ molecules of ATP react directly with X. The fact that the Gibbs energy of formation of ATP is about 8 kcal/mole higher than the Gibbs energy of formation of ADP contributes to a more favorable Gibbs energy change (and equilibrium constant) for the overall reaction.

## 21.4  AMINO ACIDS AND PROTEINS.

Amino acids are organic molecules that contain both the amine and the acid functional groups. There are twenty-three amino acids which are important in biochemistry; all twenty-three can be represented by the general formula

$$\underset{\displaystyle R-CH-C-OH}{\overset{\displaystyle NH_2 \quad O}{\phantom{x}}}$$

The twenty-three amino acids found in living organisms differ only in the composition of the portion of the molecule labeled "R." A few of the twenty-three R groups are given in Table 21.1. It is evident from this selection that the chemical properties of the biochemical amino acids encompass a wide range.

All amino acids except glycine contain an asymmetric carbon atom, that is, a carbon atom which is bonded to four distinguishable portions of a molecule. When the four portions of a molecule are distinguishable, there

**TABLE 21.1**

COMPOSITION OF R FOR REPRESENTATIVE AMINO ACIDS

| NAME | R |
|---|---|
| Glycine | $H-$ |
| Alanine | $CH_3-$ |
| Valine | $\begin{array}{c} CH_3 \\ \phantom{x} \diagdown \\ \phantom{xx} CH- \\ \phantom{x} \diagup \\ CH_3 \end{array}$ |
| Cysteine | $HS-CH_2-$ |
| Cystine | $\overset{\displaystyle O \quad NH_2}{HO-C-CH-CH_2-S-S-CH_2-}$ |
| Aspartic acid | $\overset{\displaystyle O}{HO-C-CH_2-}$ |
| Lysine | $H_2N-CH_2CH_2CH_2CH_2-$ |
| Phenylalanine | $\langle O \rangle -CH_2-$ |
| Tryptophan | (indole ring)$-C-CH_2-$, $\|CH$, $N-H$ |

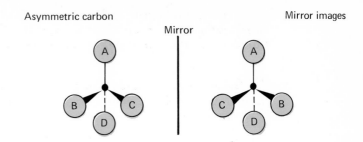

**FIGURE 21.2** Mirror images and optical isomers. Mirror images of an asymmetric carbon atom cannot be superimposed. The carbon atom is represented by the solid black circle.

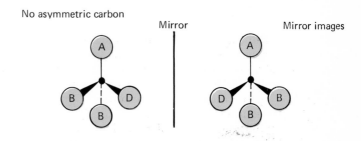

are two non-equivalent ways of arranging the portions about the carbon atom. As a result, there are two isomers of such molecules. The two isomers, called optical isomers, are mirror images of each other.* Figure 21.2 illustrates the equivalence of mirror images for a molecule without an asymmetric carbon and the non-equivalence of the mirror images for molecules with an asymmetric carbon. The laboratory preparation and isolation of a single optical isomer is difficult. Living organisms are able to produce and/or use one of the optical isomers exclusively. This is true for amino acids; all animals and plants contain amino acids with the S (Latin *sinister*, left) configuration about their asymmetric carbon atoms, and no animals or plants contain any amino acids with the R (Latin *rectus*, right) configuration

---

*Hands are a good example of non-equivalent mirror images. When both thumbs point in the same direction, the two palms face in opposite directions, and vice versa.

about their asymmetric carbon atoms.* Glucose contains five asymmetric carbon atoms; there are $2^5 = 32$ isomers of glucose. Only two of these thirty-two isomers are used by plants and animals.

Because amino acids contain both an acid group $\left(\begin{array}{c} O \\ \| \\ -C-OH \end{array}\right)$ and a base group ($-NH_2$), acid-base condensation reactions between amino acids are possible:

$$HO-\overset{\overset{\textstyle O}{\|}}{C}-\underset{\underset{\textstyle R_1}{|}}{CH}-NH_2 + HO-\overset{\overset{\textstyle O}{\|}}{C}-\underset{\underset{\textstyle R_2}{|}}{CH}-NH_2 =$$

$$HO\overset{\overset{\textstyle O}{\|}}{C}-\underset{\underset{\textstyle R_1}{|}}{CH}-NH-\overset{\overset{\textstyle O}{\|}}{C}-\underset{\underset{\textstyle R_2}{|}}{CH}-NH_2 + H_2O$$

Peptide

Repeated condensation reactions between amino acids produce extremely complex polymers called proteins.† When we remember that there are twenty-three possibilities for each R group in a protein of some particular length, we realize that the number of possible protein molecules is incredible. For example, a peptide containing just five amino acid residues could have any of $(23)^5 = 6.4 \times 10^6$ compositions. Some of the light proteins in plant and animal cells contain several hundred amino acid residues. Proteins are not random mixtures of the possible polymers; we need no molecular-weight-distribution curves for proteins. Each protein from any particular part of an organism has a definite composition and structure.

A first step in understanding the function of a protein is a determination of its structure. This complex task must be carried out in several stages, and each stage is usually carried out by a different specialist. The first step is to measure the number of residues of each amino acid in the protein. This can be done by completely depolymerizing (hydrolyzing) the protein and analyzing the resulting mixture for the free amino acids. One of the more exotic analytical techniques is to feed the mixture to a bacterium which eats only one of the amino acids. The growth of the bacteria population indicates the original concentration of the particular amino acid. The second stage in a determination of structure for a protein is to find the order in which the amino acid residues are joined. Partial hydrolysis breaks the protein into fragments which contain a few amino acid residues. The fragments are separated and analyzed (no mean task in itself). Once the compositions of enough fragments are known, the sequence of amino acid residues in the parent protein can be deduced by a process which is

---

*Amino acids with the R configuration have been found in meteorites. Since R amino acids are not produced on earth, this is strong evidence for the extraterrestrial occurrence of biological precursor molecules.

†The cooking of foods which contain proteins (e.g., eggs and meats) reverses the condensation reaction and produces amino acids from proteins.

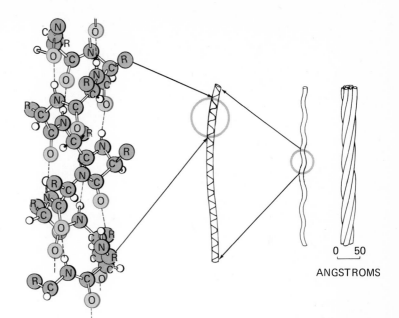

**FIGURE 21.3** Structure of a fibrous protein.

0  50
ANGSTROMS

similar to the assembly of a jigsaw puzzle. The ultimate step in a determination of a protein's structure is an X-ray analysis.* The final step requires a massive investment of time and money; only a few proteins have been studied in this detail.

Proteins can be classified as fibrous proteins or as globular proteins; their shapes reflect the roles which the respective proteins play in the organism. Fibrous proteins serve as reinforcing rods which lend strength to cell walls, tendons, hair, and horns. Linus Pauling predicted the geometry of several fibrous proteins using molecular models obtained after years of work with amino acids, simple theories of bonding, and that insight which is uniquely his. Some fibrous proteins are held in a helix by hydrogen bonds between C=O and N—H groups in the backbone of the protein. The helices are wound into a rope-like structure of great strength (Figure 21.3). Pauling's predictions were verified by X-ray measurements on keratin, a protein found in hair. He received his first Nobel prize for this work.

Globular proteins catalyze chemical reactions in the cells. In globular proteins the molecular backbone is tangled and intertwined with itself. Hydrogen bonds help hold the protein molecule in the globular shape, but attractions between the various R groups are probably vital as well. Attractions between R groups can be hydrogen bonds (e.g., between lysine and aspartic acid), hydrophobic interactions caused by repulsion of water (e.g., between two phenylalanines), dipole-dipole interactions (e.g., between two lysines), or covalent bonds (e.g., the reaction of two cysteines to form one cystine).

---

*Because proteins have a definite sequence of residues, and because the molecules of a particular protein have identical molecular weights, proteins are pure chemical compounds. As such they can be crystallized (with luck).

## 21.5  FUNCTIONS OF PROTEINS.

Protein catalysts are called enzymes. Enzymes are the most effective catalysts known, by far. Some enzymes increase reaction rates by factors of $10^{15}$. The details of exactly how the structure of a globular protein aids in the catalysis of a reaction remain a mystery. The enzyme may speed a reaction in two ways. The enzyme may actually funnel reactants together with a favorable orientation, and in so doing combat an unfavorable entropy of activation (or steric term) which limits the reaction rate in a test tube. The shape of the enzyme would make funneling and orientation of specific reactants possible. A few adjacent amino acid residues may interact in specific geometrical ways with the substrate and in so doing lower the activation energy. A crude schematic representation of funneling and orientation are given in Figure 21.4.

A.  Schematics of ATP and Enzyme

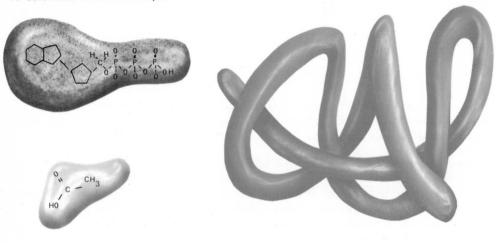

B.  Activated Complex

**FIGURE 21.4**  Schematic of the function of an enzyme. The protein funnels the two reacting molecules together; the functional groups at position **xxx** may catalyze the reaction.

## 21.6 PROTEIN SYNTHESIS.

The synthesis of proteins is a task of extreme complexity, perhaps the most complex process which occurs in living matter. Although the genetic code is written in the DNA molecules, the proteins are the agents which catalyze the reactions that produce a six-foot, brown-eyed female in one case and a five-foot, crosseyed male in another case. The information contained in the DNA molecules must be translated into protein molecules.

There can be any one of four bases attached to one sugar residue in DNA. How many base residues are required to specify one particular amino acid residue in a protein? Two base residues, which allow only $4 \times 4 = 16$ combinations, are not sufficient, for there are twenty-three different amino acids. At least three base residues are required. The genetic code therefore is written in three-letter words. There are $4 \times 4 \times 4 = 64$ possible words. We now know that there are several redundancies; more than one of the sixty-four words specifies glycine. There are some nonsense words, words which call for no amino acid. At least, these words are nonsense to us; they may serve as punctuation marks. Punctuation marks are needed because a single DNA molecule contains several hundred thousand base residues, but a single gene, which contains the information needed for a single protein, contains about four thousand base residues.

The RNA molecules are involved in the synthesis of proteins. Transfer RNA ($t$-RNA) is the simplest type of RNA; it has a molecular weight of about 25,000. As the name implies, $t$-RNA picks up a particular free amino acid and transfers it to a site where it is needed. There must be at least one different kind of $t$-RNA molecule for each amino acid; the $t$-RNA molecules from different species are quite similar (perhaps even identical). The molecular weights of messenger RNA ($m$-RNA) are much higher, ranging up to several million. Messenger RNA is generated by base pairing with one strand of a DNA molecule. Messenger RNA reads the genetic message from the DNA molecule and carries it to the site of protein synthesis. This step seems to be unnecessary; why must this extra copy of the message be made? There must be some evolutionary advantage; perhaps in this way the DNA molecule is protected from damage (i.e., mutation). The third type of RNA is called $r$-RNA. We know where $r$-RNA enters protein synthesis, but we understand little about the details of its role. The $r$-RNA does not contain genetic information.

Our present ideas about the synthesis of a protein can be presented most easily via a schematic diagram:

1. We represent the base residues as in Section 21.2.

2. Polymer molecules are indicated by their backbones (Figure 21.5). The double strands indicate that this is a DNA molecule.

3. The DNA molecule unwinds (at least partially) to allow synthesis of $m$-RNA (Figure 21.6). In the magnified section an adenine-sugar-phosphate residue is about to be added to the growing $m$-RNA molecule.

4. Meanwhile, the various $t$-RNA molecules are picking up free amino acids (Figure 21.7). Some ATP is used in this step.

5. The $m$-RNA molecules move to the site of protein synthesis. One $r$-RNA molecule begins "reading" the $m$-RNA molecule at one end (Figure 21.8). The schematic shows the addition of the second amino acid to an emerging protein molecule. No detail is shown for the $r$-RNA molecule, for we know little about the role of this molecule.

**FIGURE 21.5** Schematic of a DNA molecule.

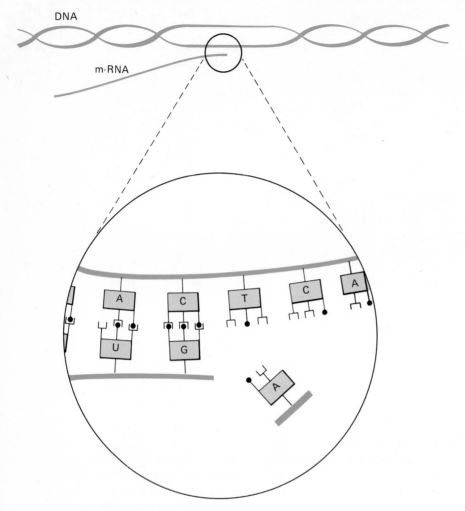

**FIGURE 21.6**    Schematic of messenger RNA getting a message from DNA.

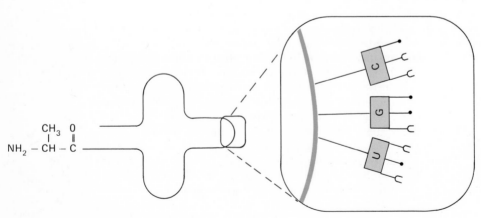

**FIGURE 21.7**    Schematic of transfer RNA.

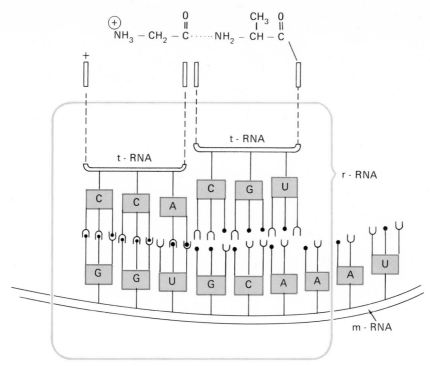

**FIGURE 21.8** Schematic of assembly of a protein. At the top of the figure, the second amino acid (alanine) is about to be added to the first (glycine).

6. Once the new amino acid is added to the chain, a molecule of *t*-RNA is released, and the *r*-RNA molecule steps three base residues down the *m*-RNA molecule. This sequence of steps is repeated until the protein molecule is completed.

Although this process looks unnecessarily complex, the process may possess great efficiency, for a second *r*-RNA molecule can begin "reading" the *m*-RNA molecule before the first *r*-RNA molecule has finished. In this way many molecules of a protein can be in the works at one time.

## 21.7 CONCLUSION.

In the preceding few pages we have tried to describe some of the most fundamental biochemical processes. These descriptions invariably have been sketchy and oversimplified, in some cases because of space limitations, but in many cases because our knowledge is still incomplete. Progress in biochemistry and molecular biology is likely to be explosive in the coming decades. Let us hope and demand that this new knowledge will be directed toward improving rather than degrading the human condition.

_____**PROBLEMS**

1. Make block diagrams of the following biochemical polymers: a protein, DNA, RNA, starch, cellulose. Use monomers (e.g., ribose, amino acids) as the building blocks.

2. Diagram all the possible pairs of the DNA bases. For those in which the hydrogen bonds match, calculate the length of the pair.

3. Assume that a particular plant can produce a glucose molecule from 20 photons of wavelength 5000 Å. Calculate the approximate efficiency of this process (using $\Delta G$ for the minimum energy required).

4. Given a bacterium which can reproduce in 20 minutes and which contains DNA with a molecular weight of 1.5 million a.m.u., estimate the time per addition of a sugar-phosphate-base unit.

5. Cold-blooded animals (e.g., fish and shellfish) are much more efficient producers of protein than are warm-blooded animals. Why must this be so?

6. Organisms which feed directly off the plankton in the sea (e.g., mussels, oysters) are more efficient producers of protein than are fish that are higher in the food chain. Why must this be so?

7. Estimate the number of different proteins that could be made from 100 amino acid units. (Hint: If $n = x^y$, then $\log n = y \log x$.)

8. Use what you know about DNA, RNA, and proteins to write a short molecular description of evolution.

## References

21.1. *Scientific American.* Almost every issue has one or more articles which are oriented toward biochemistry.
21.2. A. H. Lehninger, *Bioenergetics* (W. A. Benjamin Inc., New York, 1965).
21.3. R. E. Dickerson and I. Geis, *The Structure and Action of Proteins* (Harper and Row, New York, 1969).
21.4. H. J. Morowitz, *Energy Flow in Biology* (Academic Press, New York, 1968).
21.5. H. R. Mahler and E. H. Cordes, *Biological Chemistry* (Harper and Row, New York, 1966).
21.6. P. S. Nobel, *Plant Cell Physiology* (W. H. Freeman and Co., San Francisco, 1970).

## CHAPTER 22

# NUCLEAR CHEMISTRY

"Some 15 years ago the radiation of uranium was discovered by Henri Becquerel, and two years later the study of this phenomenon was extended to other substances, first by me, and then by Pierre Curie and myself. This study led us rapidly to the discovery of new elements, the radiation of which, while being analogous with that of uranium was far more intense. All the elements emitting such radiation I have termed radioactive, and the new property of matter revealed in this emission has thus received the name radioactivity. . . . . radioactivity is an atomic property of matter and can provide a means of seeking new elements."

*Marie Curie (1911)**

Throughout most of chemistry we focus attention on electronic structure, especially on the behavior of the valence electrons. Nuclei are treated as objects with positive charges which hold the electrons in place. As our final task we take a closer look at nuclei.

## 22.1 TYPES OF NUCLEAR REACTIONS.

We know much less about nuclear structure than we know about electronic structure. The experiments which tell us about the structure of nuclei have not been distilled down into a single, all-encompassing theory. In fact, our theories and models of the structure of nuclei may be as crude as the Bohr model was for the electronic structure of atoms.

Some nuclei are stable, and other nuclei are unstable.† We can learn about nuclear structure by studying the disintegrations of the unstable nuclei. The natural abundance of unstable nuclei is very low, for we are making terrestrial observations at least 5 billion years after most nuclei were formed. Most of the unstable nuclei have long since disintegrated. However, a few unstable nuclei are only slightly unstable, and the discovery of the slight instability of these natural nuclei provided the impetus for the study of nuclear structure. We now possess an impressive array of sophisticated devices which allow us to produce and study the disintegrations of very unstable nuclei which disappear soon after creation.

---

°*Nobel Lectures, Chemistry* (1901–1921) (Elsevier Publishing Co., Amsterdam, 1966) p. 202.

†In all cases but one, nuclear stability is kinetic stability. The nucleus $^{55}_{26}$Fe is the most stable nucleus thermodynamically, and all other nuclei are thermodynamically unstable with respect to $^{55}_{26}$Fe (see Figure 22.3).

Nuclear disintegrations are of three types, alpha emission, beta emission, and fission. The alpha particle is the nucleus of the mass four isotope of helium ($^{4}_{2}$He). Examples of alpha emission are

$$^{238}_{92}\text{U} \rightarrow {}^{234}_{90}\text{Th} + {}^{4}_{2}\text{He}$$

$$^{226}_{88}\text{Ra} \rightarrow {}^{222}_{86}\text{Rn} + {}^{4}_{2}\text{He}$$

Notice that both the total mass numbers and the total atomic numbers must balance. Alpha emission, and indeed all nuclear reactions, release vast amounts of energy. The energy appears in two forms, as kinetic energy of the new nuclei, and as very high frequency electromagnetic radiation (gamma rays).

There are two different types of nuclear reaction called beta emission. A beta particle can be either a negative beta ($_{-1}^{0}\beta$), which is an electron, or a positive beta ($_{+1}^{0}\beta$), which is a positron. Emission of an electron from a nucleus increases the atomic number by one unit.

$$^{214}_{82}\text{Pb} \rightarrow {}^{0}_{-1}\beta + {}^{214}_{83}\text{Bi}$$

Emission of a positron from a nucleus decreases the atomic number by one unit.

$$^{13}_{7}\text{N} \rightarrow {}^{13}_{6}\text{C} + {}^{0}_{+1}\beta$$

Unstable, heavy nuclei can fragment into two new nuclei of moderate mass; this process is called *fission*. If fission occurs, it occurs soon after the formation of the unstable nucleus. We only observe fission when we first produce artificial, unstable nuclei. For example, the nucleus $^{236}_{92}$U, which disintegrates via fission, must be synthesized by firing neutrons at $^{235}_{92}$U nuclei:

$$^{235}_{92}\text{U} + {}^{1}_{0}\text{n} \rightarrow {}^{236}_{92}\text{U}$$

$$^{236}_{92}\text{U} \rightarrow {}^{90}_{38}\text{Sr} + {}^{133}_{54}\text{Xe} + 3{}^{1}_{0}\text{n}$$

This fission reaction of $^{236}_{92}$U is only typical; it is not unique. The $^{236}_{92}$U nuclei disintegrate in a variety of ways, and the product of the fission of $^{236}_{92}$U is a mixture of about 90 isotopes of various elements. The distribution of products for this particular fission reaction is shown in Figure 22.1.

Light nuclei which are unstable can decay in three additional ways. Neutrons are emitted[*]

$$^{14}_{7}\text{N} \rightarrow {}^{13}_{7}\text{N} + {}^{1}_{0}\text{n}$$

protons are emitted

$$^{18}_{9}\text{F} \rightarrow {}^{17}_{8}\text{O} + {}^{1}_{1}\text{H}$$

---

[*]The $^{14}_{7}$N nucleus is stable; this reaction only occurs when a highly excited $^{14}_{7}$N nucleus is produced by an earlier nuclear reaction (see below).

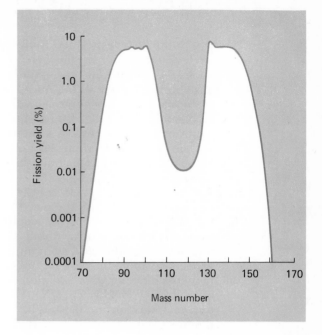

**FIGURE 22.1** Fission products of $^{236}_{92}$U.

and occasionally inner-shell electrons are captured

$$^{7}_{4}Be + e^{-} \rightarrow ^{7}_{3}Li$$

The atomic weights of nuclei can be increased by firing high-energy neutrons at target nuclei; both the atomic weight and the atomic number can be increased by firing the nuclei of light atoms at target nuclei. Such reactions are called fusion reactions. Several of the nuclei already mentioned are produced by firing neutrons or alpha particles at stable nuclei:

$$^{235}_{92}U + ^{1}_{0}n \rightarrow ^{236}_{92}U$$

$$^{59}_{27}Co + ^{1}_{0}n \rightarrow ^{60}_{27}Co$$

$$^{14}_{7}N + ^{4}_{2}He \rightarrow ^{18}_{9}F$$

$$^{10}_{5}B + ^{4}_{2}He \rightarrow ^{14}_{7}N.$$

Frequently, the nuclei produced by fusion reactions are so unstable that their existence must be surmised from the disintegration products. Fusion reactions can be used to produce isotopes of elements which are unknown in nature:

$$^{238}_{92}U + ^{2}_{1}H \rightarrow ^{238}_{93}Np + 2^{1}_{0}n$$

$$^{238}_{92}U + ^{4}_{2}He \rightarrow ^{239}_{94}Pu + 3^{1}_{0}n$$

$$^{238}_{92}U + ^{12}_{6}C \rightarrow ^{246}_{98}Cf + 4^{1}_{0}n$$

## 22.2   THE STABLE NUCLEI.

In the fifty years since Rutherford first induced the conversion of one nucleus into another, there have been many attempts to produce new nuclei. As a result of these studies, we now have a catalog of stable and unstable nuclei. Of all the nuclei which could be made from, say, 1 to 100 protons and 0 to 150 neutrons, only a few are stable. For the elements with low atomic numbers, the nuclei which have equal numbers of protons and neutrons or a small excess of neutrons are stable. The ratio of neutrons to protons increases to about 1.5 for the stable nuclei of the heavier elements. Much of our knowledge about the stability of various nuclei is summarized in Figure 22.2. This figure is really a potential energy surface, but it is drawn so that the lowest energy nuclei (the most stable) are represented by peaks instead of by the more traditional valleys.

When we examine the complete list of stable nuclei, we discover several striking trends. About 60% of the stable nuclei contain even numbers of protons and even numbers of neutrons. Statistically, we would expect this to be the case for only 25% of the stable nuclei. There are only four stable nuclei which contain odd numbers of protons and odd numbers of neutrons ($^{2}_{1}H$, $^{6}_{3}Li$, $^{10}_{5}B$, $^{14}_{7}N$). This implies that some pairing of both protons and neutrons lends stability to nuclei. Further evidence of pairing is the great stability of alpha particles, which contain a pair of protons and a pair of neutrons. More reactions proceed by emission of alpha particles than by emission of protons or neutrons.

It seems that protons pair with protons and neutrons pair with neutrons. Evidence for this particular pairing is the set of nuclear spins $^{1}_{1}H\left(I=\frac{1}{2}\right)$, $^{1}_{0}n\left(I=\frac{1}{2}\right)$, $^{2}_{1}H(I=1)$, and $^{4}_{2}He(I=0)$. Because both the neutron and the proton have nuclear spin $\frac{1}{2}$, the zero spin of $^{4}_{2}He$ indicates only that some pairs are formed with spins opposed and cancelling. However, the non-zero spin of the $^{2}_{1}H$ nucleus implies that the proton and neutron do not pair up in this way. Therefore, the pairs are made up of identical particles.

Unstable nuclei react so as to form nuclei which lie closer to the peninsula of stability (see Figure 22.2). Nuclei with an excess of protons

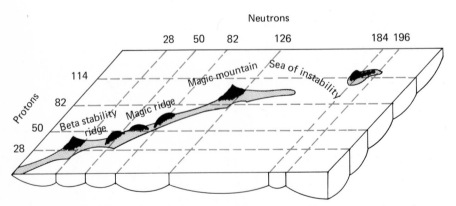

**FIGURE 22.2**   Regions of nuclear stability. The stable nuclei lie on the ridges; the magic numbers are designated. (Reproduced by permission from G. Seaborg, J. Chem. Educ. 46, 626 (1969).)

(or a shortage of neutrons) decay by alpha emission. Alpha emission usually produces more stable nuclei because lighter nuclei require a lower neutron-proton ratio for stability.

Detailed studies on the stabilities of nuclei indicate that there are certain "magic" numbers of neutrons and/or protons which confer unusual stability to nuclei. The numbers are 2, 8, 20, 28, 50, 82, and 126. Nuclei which contain a magic number of either protons or neutrons tend to be stable. Nuclei which contain magic numbers of protons and magic numbers of neutrons are very stable. The discovery of magic numbers is in many ways comparable to the discovery of the "magic" atomic numbers 2, 10, 18, 36, 54, 86. These "magic" numbers of electrons, which confer extra stability on the noble gas atoms, led chemists to the idea of closed electron shells, and this in turn led toward quantum mechanics. Nuclear shell theory attempts to explain the nuclear "magic" numbers and to predict the values of higher magic numbers. Notice in Figure 22.2 the predicted island of stability around $Z = 114$, $N = 184$. Attempts have been made to find natural samples of elements in this range ($Z = 110$ to $114$), but the results so far are inconclusive or negative.

## 22.3 THERMODYNAMICS AND KINETICS OF NUCLEAR REACTIONS.

The thermodynamic changes in nuclear reactions are very easy to calculate. Energy changes can be calculated from the Einstein equation:

$$\Delta E = \Delta m \, c^2 \qquad (22.1)$$

where $\Delta m$ is the difference between the masses of the reactants and the masses of the products. For example, in the reaction

$$^6_3\text{Li} + ^1_1\text{H} \rightarrow ^3_2\text{He} + ^4_2\text{He}$$

the reactants have total mass $6.0151 + 1.0078 = 7.0229$ atomic mass units (a.m.u.), and the products have total mass $3.0160 + 4.0026 = 7.0186$ a.m.u. The mass change is

$$\Delta m = 0.0043 \text{ g} \cdot \text{mol}^{-1}$$

The reaction releases

$$\Delta E = (4.3 \times 10^{-3} \text{ g} \cdot \text{mol}^{-1})(3.0 \times 10^{10} \text{ cm} \cdot \text{sec}^{-1})^2 = 3.9 \times 10^{18} \text{ erg} \cdot \text{mol}^{-1}$$

This is a huge amount of energy, far greater than any energy change encountered in chemical reactions. As a consequence, the thermodynamic calculations are very simple, for the change in internal energy swamps all other contributions to the change in gibbs energy for the reaction. This means that we need examine only the change in energy to decide whether or not a proposed reaction can occur spontaneously; any reaction which releases energy is possible.

The sign of the energy change for a nuclear reaction can be predicted with the aid of Figure 22.3, which is a plot of the average binding energy

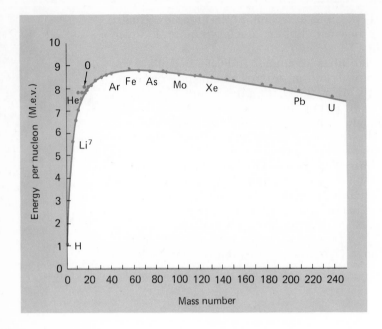

**FIGURE 22.3** Binding energy per nucleon for stable nuclei.

per nucleon (proton or neutron) compared to that of the nucleus of mass number 55 (the isotope $^{55}_{26}$Fe). Nuclei which are lighter than $^{55}_{26}$Fe can release energy in a fusion reaction.

The kinetics of nuclear reactions are simple; all nuclear disintegrations are first-order reactions. Several examples of first-order nuclear reactions have been discussed in Chapter 14.

## 22.4   NUCLEAR FORCES AND ELEMENTARY PARTICLES.

Most nuclei contain numerous positive charges which are held in extremely close proximity. According to Coulomb's law, this is a very unfavorable situation. There must be some compensating forces of attraction between nucleons, and these forces of attraction must be strong enough at very small separations to overcome the repulsions between like charges. As early as 1935, Yukawa suggested that the force of attraction between nucleons was provided by a particle which oscillated back and forth between nucleons with a velocity close to the speed of light. Yukawa was able to calculate the mass and charge that such a particle would need. These particles are called pi mesons; they have the predicted mass, which is 275 times the mass of an electron, and they can have one of three charges, +1, 0, and −1. The three types of pi mesons are designated as $\pi^+$, $\pi^0$, and $\pi^-$.

We can represent the oscillation which holds a simple nucleus together by the following diagrams. We treat only an $^2_1$H nucleus; p represents a proton, and n represents a neutron. For oscillation of a $\pi^+$ particle we draw

$$p_1 + n_2 \rightleftarrows n_1 + \pi^+ + n_2 \rightleftarrows n_1 + p_2$$

while for oscillation of a $\pi^-$ particle we draw

$$p_1 + n_2 \rightleftarrows p_1 + \pi^- + p_2 \rightleftarrows n_1 + p_2$$

It is tempting to compare this idea of nuclear binding with the chemical binding caused by delocalization of electrons. Unfortunately, the analogy is not exact, for nuclear binding is thought to depend on mesons actually oscillating back and forth between nucleons. However, the idea that "spreading out" a particle with mass (allowing more room for it to oscillate) can lower energy is a fundamental result in quantum mechanics, and it is common to both chemical and nuclear binding.

Yukawa also suggested ways in which free mesons could be produced. Mesons have been produced and characterized, and many of the features of Yukawa's theory have been verified, at least qualitatively. However, many additional subnuclear particles have been discovered, and Yukawa's theory now seems to be only a first step toward an adequate theory of nuclear binding.

## 22.5 TRANSURANIUM ELEMENTS.

Uranium ($Z = 92$) is the heaviest element found on earth. Thirteen heavier elements have been synthesized in laboratories. The synthesis and characterization of previously unknown elements are perhaps the most stringent tests of theoretical predictions in chemistry and physics. In order to create a new element, the nuclear physicist must select a nuclear fusion reaction which is likely to produce a reasonably stable isotope of the new element. This requires that he be able to predict the products of a new reaction, and that he be able to predict the half-lives for fission, for beta emission, and for alpha emission for the new nucleus. The nuclear chemist must be able to predict the chemical and physical properties of the new element, and he must be able to select those properties which can be used in the separation of the new element from the starting material. For example, element 101 was synthesized by the reaction

$$^{253}_{99}\text{Es} + ^{4}_{2}\text{He} \rightarrow ^{256}_{101}\text{Md} + ^{1}_{0}\text{n}$$

The reactant $^{253}_{99}\text{Es}$ is a transuranium element; it is unstable and must be produced artificially in tiny amounts. From a sample which contained about $10^9$ atoms of $^{253}_{99}\text{Es}$, bombardment with alpha particles produced only 13 detectable atoms of $^{256}_{101}\text{Md}$. Yet the predictions of the properties of $^{256}_{101}\text{Md}$ were so accurate that 13 atoms was a sufficient amount for confirmation of the reaction. This is an extreme example, but the chemistry of some transuranium elements is known fairly well in spite of the fact that a sample large enough to be seen has never been prepared.

As an example of the detailed predictions that can be made by extrapolation of periodic trends and by quantum mechanical calculations, we list some predicted properties of elements 113 and 114 in Table 22.1. If and when these elements are synthesized, these predicted properties will make possible their separation and aid in their identification.

## 22.6 SOME EXAMPLES OF NUCLEAR REACTIONS.

The best known nuclear reactions are those used in nuclear weapons. The Hiroshima bomb was based on the reaction

$$^{235}_{92}\text{U} + ^{1}_{0}\text{n} \rightleftarrows ^{236}_{92}\text{U} \rightarrow \text{fission products} + \text{neutrons}$$

TABLE 22.1
PREDICTED PROPERTIES OF ELEMENTS 113 AND 114.

|  | ELEMENT 113 | ELEMENT 114 |
|---|---|---|
| Atomic weight | 297 | 298 |
| Density (g/cm³) | 16 | 14 |
| Most stable oxidation state | +1 | +2 |
| Oxidation Potential | $M \rightarrow M^+ + e^-$ | $M \rightarrow M^{+2} + 2e^-$ |
|  | $-0.6$ V | $-0.8$ V |
| First Ionization Energy | 7.4 eV | 8.5 eV |
| Second Ionization Energy | — | 16.8 eV |
| Ionic radius | 1.48 Å | 1.31 Å |
| Melting point | 430°C | 70°C |
| Boiling point | 1100°C | 150°C |
| Heat of vaporization | 31 kcal/mole | 9 kcal/mole |
| Entropy (25°C) | 17 eu/mole | 20 eu/mole |

From G. Seaborg, J. Chem. Ed. *46*, 627 (1969) by permission.

This reaction, and many fast fission reactions, produce fantastic explosions because more neutrons are produced each time the reaction occurs. The rate of the reaction, which depends on the concentration of neutrons, continues to accelerate as long as more neutrons are produced than are lost. The reaction can be tamed by insertion of materials which are efficient absorbers of neutrons into the uranium core. In the latter case we have a nuclear reactor. Unfortunately, the fission products are usually very radioactive, and disposal of the ashes from nuclear reactors is a huge problem. The energy released within the reactor must be transferred as heat via a series of coolants (Figure 22.4). This is rather inefficient, and a nuclear power plant produces a vast amount of warm water as well as electricity. The resulting thermal pollution is a serious problem. However, we know from the second law of thermodynamics that thermal pollution is a by-product of any process involving the conversion of thermal energy to work; it may actually be less serious with nuclear than with conventional power plants.

Fusion reactions can release even larger amounts of energy per gram of fuel. In a hydrogen bomb, lithium-6 deuteride ($^6Li^2H$) is packed around an ordinary fission bomb, which on activation provides the high temperatures necessary to bring about fusion reactions. The principal fusion reactions that occur during the explosion are

$$^2H + ^2H \rightarrow ^3He + ^1n$$

$$^1n + ^6Li \rightarrow ^4He + ^3H$$

$$^3H + ^2H \rightarrow ^4He + ^1n$$

$$^2H + ^2H \rightarrow ^3H + ^1H$$

The amount of energy released by one ton of $^6Li^2H$ is equivalent to that released by 60 megatons ($6.0 \times 10^7$ tons) of TNT. This is about six times

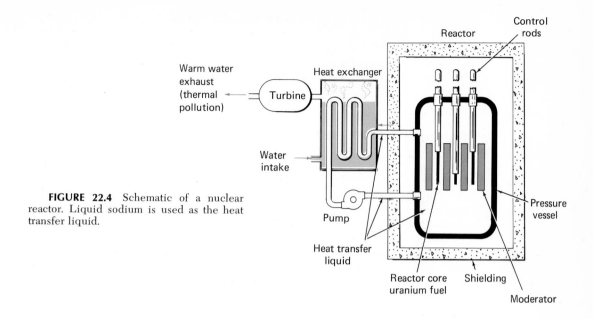

**FIGURE 22.4** Schematic of a nuclear reactor. Liquid sodium is used as the heat transfer liquid.

greater than for a fission bomb. Attempts are under way to tame fusion reactions for domestic power sources. The best possibilities seem to be the reactions

$$2{}^2_1\text{H} \rightarrow {}^4_2\text{He} \text{ and } {}^1_1\text{H} + {}^2_1\text{H} \rightarrow {}^3_2\text{He}$$

There are many advantages to these reactions. Deuterium is plentiful and comparatively easy to separate from other hydrogen isotopes. The reactions produce no radioactive residue, and there is some hope that the energy released might be converted directly into electricity. This would solve many of our pollution problems. The rates of fusion reactions are proportional to the numbers of collisions between reacting species, as are the rates of chemical reactions. However, the repulsions between positively charged nuclei prevent true collisions between nuclei unless the temperature is at least $10^6$ K. The technical problems involved in keeping a sample of deuterium at one million degrees under control are overwhelming, and the prospects for fusion power plants are fairly bleak.

Fusion reactions run easily at solar temperatures. The major source of solar energy is

$$4{}^1_1\text{H} \rightarrow {}^4_2\text{He} + 2{}_{+1}^{\ 0}\beta$$

The mechanism for this reaction is thought to be

$$\begin{aligned}
{}^{12}_6\text{C} + {}^1_1\text{H} &\rightarrow {}^{13}_7\text{N} \rightarrow {}^{13}_6\text{C} + {}_{+1}^{\ 0}\beta \\
{}^{13}_6\text{C} + {}^1_1\text{H} &\rightarrow {}^{14}_7\text{N} \\
{}^{14}_7\text{N} + {}^1_1\text{H} &\rightarrow {}^{15}_8\text{O} \rightarrow {}^{15}_7\text{N} + {}_{+1}^{\ 0}\beta \\
{}^{15}_7\text{N} + {}^1_1\text{H} &\rightarrow {}^{12}_6\text{C} + {}^4_2\text{He}
\end{aligned}$$

High energy particles and gamma radiation, which are released in nuclear reactions, are extremely damaging to the delicate molecules of a living organism. Strangely enough, this was not obvious to those who first worked with radioactive materials, and some terrible accidents occurred. Now we are aware of and deeply concerned about the biological effects of radiation. Perhaps the most serious problem is caused by strontium-90. Strontium-90 is one of the products produced by fission of heavy nuclei:

$$^{235}_{92}U + ^{1}_{0}n \rightarrow ^{90}_{38}Sr + ^{133}_{54}Xe + 3^{1}_{0}n$$

The isotope $^{90}_{38}Sr$ is unstable

$$^{90}_{38}Sr \rightarrow ^{90}_{39}Y + ^{0}_{-1}\beta \qquad t_{1/2} = 28 \text{ years}$$

Beta particles are usually not especially dangerous, for they do not penetrate matter well. Heavy rubber gloves or the walls of a beaker are sufficiently thick to absorb most beta particles. However, some fraction of the $^{90}_{38}Sr$ which is released into the atmosphere by an atomic bomb is deposited on grass and ingested by cattle. A smaller fraction is in turn ingested by humans (in milk). Because strontium is chemically similar to calcium, the $^{90}_{38}Sr$ is incorporated permanently in the bones. Once in the bones, the short range beta particles are close enough to do damage, and the half-life of $^{90}_{38}Sr$ is such that the damage continues throughout the life of the individual. At this writing, serious questions have been raised about the effects of long-term exposure to very tiny amounts of $^{90}_{38}Sr$. Some scientists feel that the $^{90}_{38}Sr$ released into the atmosphere by far less than total nuclear war might depopulate the earth within a generation. There is no question that fear of $^{90}_{38}Sr$ contributed greatly to the agreement which prohibited the atmospheric testing of nuclear weapons.

Finally, many radioactive isotopes are used as tracers. The paths of chemical reactions, the destinations of chemicals within the body, the points of wear in engines, and thousands of other problems have been solved by finding where radioactivity appears in the course of a reaction or process. The economic value of these solutions has probably been greater than the cost of the atomic energy program.

## PROBLEMS

1. Complete the following nuclear reactions:

$$^{41}_{20}Ca + ^{0}_{-1}\beta \rightarrow$$

$$^{36}_{17}Cl \rightarrow ^{0}_{-1}\beta +$$

$$^{52}_{25}Mn \rightarrow ^{0}_{+1}\beta +$$

$$^{137}_{53}I \rightarrow ^{1}_{0}n +$$

$$^{174}_{72}Hf \rightarrow ^{4}_{2}He +$$

$$^{237}_{94}Pu \rightarrow ^{96}_{40}Zr + \qquad + 3^{1}_{0}n$$

2. List the likely modes of decay for the unstable nuclei

$$^{3}_{1}H, \ ^{6}_{4}Be, \ ^{12}_{7}N, \ ^{144}_{55}Cs, \ ^{227}_{92}U, \ ^{14}_{6}C$$

3. Use the idea of pairing of nucleons to estimate the nuclear spins of of the nuclei

$$_1^3H, \ _2^3He, \ _6^{12}C, \ _7^{14}N, \ _9^{19}F, \ _8^{17}O$$

4. Calculate the energy released by the reaction

$$2_1^2H \rightarrow \ _2^4He$$

The masses are $_1^2H$ (2.01307) and $_2^4He$ (4.00256).

5. The major reaction in the sun is

$$4_1^1H \rightarrow \ _2^4He$$

The sun produces $8 \times 10^{21}$ kcal/sec. How much hydrogen is converted to helium in a second to produce this energy?

6. The chemical oxidation of diamond releases 94,502 cal/mole of diamond:

$$C \ (diamond) + O_2 \rightarrow CO_2 \quad \Delta H = -94,502$$

Calculate the difference in mass between products and reactants for this reaction.

7. The mass of $_{14}^{34}S$ is 33.96786, and the mass of $_{14}^{30}Si$ is 29.97376. Could the reaction

$$_{16}^{34}S \rightarrow \ _{14}^{30}Si + \ _2^4He$$

occur?

8. If all energetically favorable nuclear reactions occured at a reasonable rate, what might be the major component of the sun at its "death"?

9. Diagram the oscillation which might occur when a neutron and proton are held together by a $\pi^0$ particle.

10. Some transuranium elements might occur in nature. The search for these elements has concentrated on elements 110 to 114. Can you think of a chemical reason for concentrating on these particular elements?

*11. The reaction of four protons to form an alpha particle

$$4_1^1H \rightarrow \ _2^4He + 2_{+1}^{\ 0}\beta$$

has a rate

$$Rate = k[_1^1H]^2$$

Propose a mechanism which is consistent with this rate law.

12. Alpha particles are easily absorbed; they cannot penetrate a sheet of paper. Why is it dangerous to ingest materials which emit alpha particles?

13. In 1969 there was a serious fire in a plutonium factory in Colorado, and at least one guard inhaled plutonium oxide dust. The most stable isotope, $^{239}_{94}Pu$, emits alpha particles with energies of about $5.14 \times 10^6$ electron volts (5.14 MeV). Assume that an alpha particle dissipates its energy by breaking carbon-carbon bonds to form two free radicals. If 90 kcal/mole is an average C–C bond strength, how many free radicals would each alpha particle produce within the lungs of the unfortunate guard?

14. The maximum permissable dosage of $^{90}_{38}Sr$ is the amount which produces $3.7 \times 10^4$ disintegrations/sec. The half-life of $^{90}_{38}Sr$ is 28 years. How many grams of $^{90}_{38}Sr$ could one "safely" ingest? Assume that one lived 56 years after ingesting this amount of $^{90}_{38}Sr$. Using the same assumptions as problem 13, calculate the number of free radicals produced in this time. The beta particles emitted by $^{90}_{38}Sr$ have an energy of 0.54 MeV.

15. When a positron and an electron collide, they are annihilated:

$$_{-1}^{0}\beta + _{+1}^{0}\beta \rightarrow 2h\nu$$

The two photons have the same frequency. What is the wavelength of these photons?

16. Discuss the effect of running a nuclear reactor at higher temperature. Discuss questions of safety and questions of efficiency (second law of thermodynamics).

# THE CHEMICAL ATOMIC WEIGHTS, 1961

| | | | | | | | | |
|---|---|---|---|---|---|---|---|---|
| Ac | (227) | Gd | 157.25 | Pr | 140.907 |
| Ag | 107.870 | Ge | 72.59 | Pt | 195.09 |
| Al | 26.9815 | H | 1.00797 | Pu | (242) |
| Am | (243) | He | 4.0026 | Ra | (226) |
| Ar | 39.948 | Hf | 178.49 | Rb | 85.47 |
| As | 74.9216 | Hg | 200.59 | Re | 186.2 |
| At | (210) | Ho | 164.930 | Rh | 102.905 |
| Au | 196.967 | I | 126.9044 | Rn | (222) |
| B | 10.811 | In | 114.82 | Ru | 101.07 |
| Ba | 137.34 | Ir | 192.2 | S | 32.064 |
| Be | 9.0122 | K | 39.102 | Sb | 121.75 |
| Bi | 208.980 | Kr | 83.80 | Sc | 44.956 |
| Bk | (249) | La | 138.91 | Se | 78.96 |
| Br | 79.909 | Li | 6.939 | Si | 28.086 |
| C | 12.01115 | Lu | 174.97 | Sm | 150.35 |
| Ca | 40.08 | Md | (256) | Sn | 118.69 |
| Cd | 112.40 | Mg | 24.312 | Sr | 87.62 |
| Ce | 140.12 | Mn | 54.9380 | Ta | 180.948 |
| Cf | (251) | Mo | 95.94 | Tb | 158.924 |
| Cl | 35.453 | N | 14.0067 | Tc | (99) |
| Cm | (247) | Na | 22.9898 | Te | 127.60 |
| Co | 58.9332 | Nb | 92.906 | Th | 232.038 |
| Cr | 51.996 | Nd | 144.24 | Ti | 47.90 |
| Cs | 132.905 | Ne | 20.183 | Tl | 204.37 |
| Cu | 63.54 | Ni | 58.71 | Tm | 168.934 |
| Dy | 162.50 | Np | (237) | U | 238.03 |
| Er | 167.26 | O | 15.9994 | V | 50.942 |
| Es | (254) | Os | 190.2 | W | 183.85 |
| Eu | 151.96 | P | 30.9738 | Xe | 131.30 |
| F | 18.9984 | Pa | (231) | Y | 88.905 |
| Fe | 55.847 | Pb | 207.19 | Yb | 173.04 |
| Fm | (253) | Pd | 106.4 | Zn | 65.37 |
| Fr | (223) | Pm | (147) | Zr | 91.22 |
| Ga | 69.72 | Po | (210) | | |

*As approved by the International Commission on Atomic Weights. In parentheses are the longest-lived or best-known isotopes of the radioactive elements.

# BASE 10 LOGARITHMS

| x | 0 | 1 | 2 | 3 | 4 | 5 | 6 | 7 | 8 | 9 | Proportional Parts | | | | | | | | |
|---|---|---|---|---|---|---|---|---|---|---|---|---|---|---|---|---|---|---|---|
| | | | | | | | | | | | 1 | 2 | 3 | 4 | 5 | 6 | 7 | 8 | 9 |
| 10 | 0000 | 0043 | 0086 | 0128 | 0170 | 0212 | 0253 | 0294 | 0334 | 0374 | 4 | 8 | 12 | 17 | 21 | 25 | 29 | 33 | 37 |
| 11 | 0414 | 0453 | 0492 | 0531 | 0569 | 0607 | 0645 | 0682 | 0719 | 0755 | 4 | 8 | 11 | 15 | 19 | 23 | 26 | 30 | 34 |
| 12 | 0792 | 0828 | 0864 | 0899 | 0934 | 0969 | 1004 | 1038 | 1072 | 1106 | 3 | 7 | 10 | 14 | 17 | 21 | 24 | 28 | 31 |
| 13 | 1139 | 1173 | 1206 | 1239 | 1271 | 1303 | 1335 | 1367 | 1399 | 1430 | 3 | 6 | 10 | 13 | 16 | 19 | 23 | 26 | 29 |
| 14 | 1461 | 1492 | 1523 | 1553 | 1584 | 1614 | 1644 | 1673 | 1703 | 1732 | 3 | 6 | 9 | 12 | 15 | 18 | 21 | 24 | 27 |
| 15 | 1761 | 1790 | 1818 | 1847 | 1875 | 1903 | 1931 | 1959 | 1987 | 2014 | 3 | 6 | 8 | 11 | 14 | 17 | 20 | 22 | 25 |
| 16 | 2041 | 2068 | 2095 | 2122 | 2148 | 2175 | 2201 | 2227 | 2253 | 2279 | 3 | 5 | 8 | 11 | 13 | 16 | 18 | 21 | 24 |
| 17 | 2304 | 2330 | 2355 | 2380 | 2405 | 2430 | 2455 | 2480 | 2504 | 2529 | 2 | 5 | 7 | 10 | 12 | 15 | 17 | 20 | 22 |
| 18 | 2553 | 2577 | 2601 | 2625 | 2648 | 2672 | 2695 | 2718 | 2742 | 2765 | 2 | 5 | 7 | 9 | 12 | 14 | 16 | 19 | 21 |
| 19 | 2788 | 2810 | 2833 | 2856 | 2878 | 2900 | 2923 | 2945 | 2967 | 2989 | 2 | 4 | 7 | 9 | 11 | 13 | 16 | 18 | 20 |
| 20 | 3010 | 3032 | 3054 | 3075 | 3096 | 3118 | 3139 | 3160 | 3181 | 3201 | 2 | 4 | 6 | 8 | 11 | 13 | 15 | 17 | 19 |
| 21 | 3222 | 3243 | 3263 | 3284 | 3304 | 3324 | 3345 | 3365 | 3385 | 3404 | 2 | 4 | 6 | 8 | 10 | 12 | 14 | 16 | 18 |
| 22 | 3424 | 3444 | 3464 | 3483 | 3502 | 3522 | 3541 | 3560 | 3579 | 3598 | 2 | 4 | 6 | 8 | 10 | 12 | 14 | 15 | 17 |
| 23 | 3617 | 3636 | 3655 | 3674 | 3692 | 3711 | 3729 | 3747 | 3766 | 3784 | 2 | 4 | 6 | 7 | 9 | 11 | 13 | 15 | 17 |
| 24 | 3802 | 3820 | 3838 | 3856 | 3874 | 3892 | 3909 | 3927 | 3945 | 3962 | 2 | 4 | 5 | 7 | 9 | 11 | 12 | 14 | 16 |
| 25 | 3979 | 3997 | 4014 | 4031 | 4048 | 4065 | 4082 | 4099 | 4116 | 4133 | 2 | 3 | 5 | 7 | 9 | 10 | 12 | 14 | 15 |
| 26 | 4150 | 4166 | 4183 | 4200 | 4216 | 4232 | 4249 | 4265 | 4281 | 4298 | 2 | 3 | 5 | 7 | 8 | 10 | 11 | 13 | 15 |
| 27 | 4314 | 4330 | 4346 | 4362 | 4378 | 4393 | 4409 | 4425 | 4440 | 4456 | 2 | 3 | 5 | 6 | 8 | 9 | 11 | 13 | 14 |
| 28 | 4472 | 4487 | 4502 | 4518 | 4533 | 4548 | 4564 | 4579 | 4594 | 4609 | 2 | 3 | 5 | 6 | 8 | 9 | 11 | 12 | 14 |
| 29 | 4624 | 4639 | 4654 | 4669 | 4683 | 4698 | 4713 | 4728 | 4742 | 4757 | 1 | 3 | 4 | 6 | 7 | 9 | 10 | 12 | 13 |
| 30 | 4771 | 4786 | 4800 | 4814 | 4829 | 4843 | 4857 | 4871 | 4886 | 4900 | 1 | 3 | 4 | 6 | 7 | 9 | 10 | 11 | 13 |
| 31 | 4914 | 4928 | 4942 | 4955 | 4969 | 4983 | 4997 | 5011 | 5024 | 5038 | 1 | 3 | 4 | 6 | 7 | 8 | 10 | 11 | 12 |
| 32 | 5051 | 5065 | 5079 | 5092 | 5105 | 5119 | 5132 | 5145 | 5159 | 5172 | 1 | 3 | 4 | 5 | 7 | 8 | 9 | 11 | 12 |
| 33 | 5185 | 5198 | 5211 | 5224 | 5237 | 5250 | 5263 | 5276 | 5289 | 5302 | 1 | 3 | 4 | 5 | 6 | 8 | 9 | 10 | 12 |
| 34 | 5315 | 5328 | 5340 | 5353 | 5366 | 5378 | 5391 | 5403 | 5416 | 5428 | 1 | 3 | 4 | 5 | 6 | 8 | 9 | 10 | 11 |
| 35 | 5441 | 5453 | 5465 | 5478 | 5490 | 5502 | 5514 | 5527 | 5539 | 5551 | 1 | 2 | 4 | 5 | 6 | 7 | 9 | 10 | 11 |
| 36 | 5563 | 5575 | 5587 | 5599 | 5611 | 5623 | 5635 | 5647 | 5658 | 5670 | 1 | 2 | 4 | 5 | 6 | 7 | 8 | 10 | 11 |
| 37 | 5682 | 5694 | 5705 | 5717 | 5729 | 5740 | 5752 | 5763 | 5775 | 5786 | 1 | 2 | 3 | 5 | 6 | 7 | 8 | 9 | 10 |
| 38 | 5798 | 5809 | 5821 | 5832 | 5843 | 5855 | 5866 | 5877 | 5888 | 5899 | 1 | 2 | 3 | 5 | 6 | 7 | 8 | 9 | 10 |
| 39 | 5911 | 5922 | 5933 | 5944 | 5955 | 5966 | 5977 | 5988 | 5999 | 6010 | 1 | 2 | 3 | 4 | 5 | 7 | 8 | 9 | 10 |
| 40 | 6021 | 6031 | 6042 | 6053 | 6064 | 6075 | 6085 | 6096 | 6107 | 6117 | 1 | 2 | 3 | 4 | 5 | 6 | 8 | 9 | 10 |
| 41 | 6128 | 6138 | 6149 | 6160 | 6170 | 6180 | 6191 | 6201 | 6212 | 6222 | 1 | 2 | 3 | 4 | 5 | 6 | 7 | 8 | 9 |
| 42 | 6232 | 6243 | 6253 | 6263 | 6274 | 6284 | 6294 | 6304 | 6314 | 6325 | 1 | 2 | 3 | 4 | 5 | 6 | 7 | 8 | 9 |
| 43 | 6335 | 6345 | 6355 | 6365 | 6375 | 6385 | 6395 | 6405 | 6415 | 6425 | 1 | 2 | 3 | 4 | 5 | 6 | 7 | 8 | 9 |
| 44 | 6435 | 6444 | 6454 | 6464 | 6474 | 6484 | 6493 | 6503 | 6513 | 6522 | 1 | 2 | 3 | 4 | 5 | 6 | 7 | 8 | 9 |
| 45 | 6532 | 6542 | 6551 | 6561 | 6571 | 6580 | 6590 | 6599 | 6609 | 6618 | 1 | 2 | 3 | 4 | 5 | 6 | 7 | 8 | 9 |
| 46 | 6628 | 6637 | 6646 | 6656 | 6665 | 6675 | 6684 | 6693 | 6702 | 6712 | 1 | 2 | 3 | 4 | 5 | 6 | 7 | 7 | 8 |
| 47 | 6721 | 6730 | 6739 | 6749 | 6758 | 6767 | 6776 | 6785 | 6794 | 6803 | 1 | 2 | 3 | 4 | 5 | 5 | 6 | 7 | 8 |
| 48 | 6812 | 6821 | 6830 | 6839 | 6848 | 6857 | 6866 | 6875 | 6884 | 6893 | 1 | 2 | 3 | 4 | 5 | 5 | 6 | 7 | 8 |
| 49 | 6902 | 6911 | 6920 | 6928 | 6937 | 6946 | 6955 | 6964 | 6972 | 6981 | 1 | 2 | 3 | 4 | 4 | 5 | 6 | 7 | 8 |
| 50 | 6990 | 6998 | 7007 | 7016 | 7024 | 7033 | 7042 | 7050 | 7059 | 7067 | 1 | 2 | 3 | 3 | 4 | 5 | 6 | 7 | 8 |
| 51 | 7076 | 7084 | 7093 | 7101 | 7110 | 7118 | 7126 | 7135 | 7143 | 7152 | 1 | 2 | 3 | 3 | 4 | 5 | 6 | 7 | 8 |
| 52 | 7160 | 7168 | 7177 | 7185 | 7193 | 7202 | 7210 | 7218 | 7226 | 7235 | 1 | 2 | 2 | 3 | 4 | 5 | 6 | 7 | 7 |
| 53 | 7243 | 7251 | 7259 | 7267 | 7275 | 7284 | 7292 | 7300 | 7308 | 7316 | 1 | 2 | 2 | 3 | 4 | 5 | 6 | 6 | 7 |
| 54 | 7324 | 7332 | 7340 | 7348 | 7356 | 7364 | 7372 | 7380 | 7388 | 7396 | 1 | 2 | 2 | 3 | 4 | 5 | 6 | 6 | 7 |
| x | 0 | 1 | 2 | 3 | 4 | 5 | 6 | 7 | 8 | 9 | 1 | 2 | 3 | 4 | 5 | 6 | 7 | 8 | 9 |

| x | 0 | 1 | 2 | 3 | 4 | 5 | 6 | 7 | 8 | 9 | | Proportional Parts | | | | | | | | |
|---|---|---|---|---|---|---|---|---|---|---|---|---|---|---|---|---|---|---|---|---|
| | | | | | | | | | | | | 1 | 2 | 3 | 4 | 5 | 6 | 7 | 8 | 9 |
| 55 | 7404 | 7412 | 7419 | 7427 | 7435 | 7443 | 7451 | 7459 | 7466 | 7474 | | 1 | 2 | 2 | 3 | 4 | 5 | 5 | 6 | 7 |
| 56 | 7482 | 7490 | 7497 | 7505 | 7513 | 7520 | 7528 | 7536 | 7543 | 7551 | | 1 | 2 | 2 | 3 | 4 | 5 | 5 | 6 | 7 |
| 57 | 7559 | 7566 | 7574 | 7582 | 7589 | 7597 | 7604 | 7612 | 7619 | 7627 | | 1 | 2 | 2 | 3 | 4 | 5 | 5 | 6 | 7 |
| 58 | 7634 | 7642 | 7649 | 7657 | 7664 | 7672 | 7679 | 7686 | 7694 | 7701 | | 1 | 1 | 2 | 3 | 4 | 4 | 5 | 6 | 7 |
| 59 | 7709 | 7716 | 7723 | 7731 | 7738 | 7745 | 7752 | 7760 | 7767 | 7774 | | 1 | 1 | 2 | 3 | 4 | 4 | 5 | 6 | 7 |
| 60 | 7782 | 7789 | 7796 | 7803 | 7810 | 7818 | 7825 | 7832 | 7839 | 7846 | | 1 | 1 | 2 | 3 | 4 | 4 | 5 | 6 | 6 |
| 61 | 7853 | 7860 | 7868 | 7875 | 7882 | 7889 | 7896 | 7903 | 7910 | 7917 | | 1 | 1 | 2 | 3 | 4 | 4 | 5 | 6 | 6 |
| 62 | 7924 | 7931 | 7938 | 7945 | 7952 | 7959 | 7966 | 7973 | 7980 | 7987 | | 1 | 1 | 2 | 3 | 3 | 4 | 5 | 6 | 6 |
| 63 | 7993 | 8000 | 8007 | 8014 | 8021 | 8028 | 8035 | 8041 | 8048 | 8055 | | 1 | 1 | 2 | 3 | 3 | 4 | 5 | 6 | 6 |
| 64 | 8062 | 8069 | 8075 | 8082 | 8089 | 8096 | 8102 | 8109 | 8116 | 8122 | | 1 | 1 | 2 | 3 | 3 | 4 | 5 | 5 | 6 |
| 65 | 8129 | 8136 | 8142 | 8149 | 8156 | 8162 | 8169 | 8176 | 8182 | 8189 | | 1 | 1 | 2 | 3 | 3 | 4 | 5 | 5 | 6 |
| 66 | 8195 | 8202 | 8209 | 8215 | 8222 | 8228 | 8235 | 8241 | 8248 | 8254 | | 1 | 1 | 2 | 3 | 3 | 4 | 5 | 5 | 6 |
| 67 | 8261 | 8267 | 8274 | 8280 | 8287 | 8293 | 8299 | 8306 | 8312 | 8319 | | 1 | 1 | 2 | 3 | 3 | 4 | 5 | 5 | 6 |
| 68 | 8325 | 8331 | 8338 | 8344 | 8351 | 8357 | 8363 | 8370 | 8376 | 8382 | | 1 | 1 | 2 | 3 | 3 | 4 | 4 | 5 | 6 |
| 69 | 8388 | 8395 | 8401 | 8407 | 8414 | 8420 | 8426 | 8432 | 8439 | 8445 | | 1 | 1 | 2 | 2 | 3 | 4 | 4 | 5 | 6 |
| 70 | 8451 | 8457 | 8463 | 8470 | 8476 | 8482 | 8488 | 8494 | 8500 | 8506 | | 1 | 1 | 2 | 2 | 3 | 4 | 4 | 5 | 6 |
| 71 | 8513 | 8519 | 8525 | 8531 | 8537 | 8543 | 8549 | 8555 | 8561 | 8567 | | 1 | 1 | 2 | 2 | 3 | 4 | 4 | 5 | 5 |
| 72 | 8573 | 8579 | 8585 | 8591 | 8597 | 8603 | 8609 | 8615 | 8621 | 8627 | | 1 | 1 | 2 | 2 | 3 | 4 | 4 | 5 | 5 |
| 73 | 8633 | 8639 | 8645 | 8651 | 8657 | 8663 | 8669 | 8675 | 8681 | 8686 | | 1 | 1 | 2 | 2 | 3 | 4 | 4 | 5 | 5 |
| 74 | 8692 | 8698 | 8704 | 8710 | 8716 | 8722 | 8727 | 8733 | 8739 | 8745 | | 1 | 1 | 2 | 2 | 3 | 4 | 4 | 5 | 5 |
| 75 | 8751 | 8756 | 8762 | 8768 | 8774 | 8779 | 8785 | 8791 | 8797 | 8802 | | 1 | 1 | 2 | 2 | 3 | 3 | 4 | 5 | 5 |
| 76 | 8808 | 8814 | 8820 | 8825 | 8831 | 8837 | 8842 | 8848 | 8854 | 8859 | | 1 | 1 | 2 | 2 | 3 | 3 | 4 | 5 | 5 |
| 77 | 8865 | 8871 | 8876 | 8882 | 8887 | 8893 | 8899 | 8904 | 8910 | 8915 | | 1 | 1 | 2 | 2 | 3 | 3 | 4 | 4 | 5 |
| 78 | 8921 | 8927 | 8932 | 8938 | 8943 | 8949 | 8954 | 8960 | 8965 | 8971 | | 1 | 1 | 2 | 2 | 3 | 3 | 4 | 4 | 5 |
| 79 | 8976 | 8982 | 8987 | 8993 | 8998 | 9004 | 9009 | 9015 | 9020 | 9025 | | 1 | 1 | 2 | 2 | 3 | 3 | 4 | 4 | 5 |
| 80 | 9031 | 9036 | 9042 | 9047 | 9053 | 9058 | 9063 | 9069 | 9074 | 9079 | | 1 | 1 | 2 | 2 | 3 | 3 | 4 | 4 | 5 |
| 81 | 9085 | 9090 | 9096 | 9101 | 9106 | 9112 | 9117 | 9122 | 9128 | 9133 | | 1 | 1 | 2 | 2 | 3 | 3 | 4 | 4 | 5 |
| 82 | 9138 | 9143 | 9149 | 9154 | 9159 | 9165 | 9170 | 9175 | 9180 | 9186 | | 1 | 1 | 2 | 2 | 3 | 3 | 4 | 4 | 5 |
| 83 | 9191 | 9196 | 9201 | 9206 | 9212 | 9217 | 9222 | 9227 | 9232 | 9238 | | 1 | 1 | 2 | 2 | 3 | 3 | 4 | 4 | 5 |
| 84 | 9243 | 9248 | 9253 | 9258 | 9263 | 9269 | 9274 | 9279 | 9284 | 9289 | | 1 | 1 | 2 | 2 | 3 | 3 | 4 | 4 | 5 |
| 85 | 9294 | 9299 | 9304 | 9309 | 9315 | 9320 | 9325 | 9330 | 9335 | 9340 | | 1 | 1 | 2 | 2 | 3 | 3 | 4 | 4 | 5 |
| 86 | 9345 | 9350 | 9355 | 9360 | 9365 | 9370 | 9375 | 9380 | 9385 | 9390 | | 1 | 1 | 2 | 2 | 3 | 3 | 4 | 4 | 5 |
| 87 | 9395 | 9400 | 9405 | 9410 | 9415 | 9420 | 9425 | 9430 | 9435 | 9440 | | 0 | 1 | 1 | 2 | 2 | 3 | 3 | 4 | 4 |
| 88 | 9445 | 9450 | 9455 | 9460 | 9465 | 9469 | 9474 | 9479 | 9484 | 9489 | | 0 | 1 | 1 | 2 | 2 | 3 | 3 | 4 | 4 |
| 89 | 9494 | 9499 | 9504 | 9509 | 9513 | 9518 | 9523 | 9528 | 9533 | 9538 | | 0 | 1 | 1 | 2 | 2 | 3 | 3 | 4 | 4 |
| 90 | 9542 | 9547 | 9552 | 9557 | 9562 | 9566 | 9571 | 9576 | 9581 | 9586 | | 0 | 1 | 1 | 2 | 2 | 3 | 3 | 4 | 4 |
| 91 | 9590 | 9595 | 9600 | 9605 | 9609 | 9614 | 9619 | 9624 | 9628 | 9633 | | 0 | 1 | 1 | 2 | 2 | 3 | 3 | 4 | 4 |
| 92 | 9638 | 9643 | 9647 | 9652 | 9657 | 9661 | 9666 | 9671 | 9675 | 9680 | | 0 | 1 | 1 | 2 | 2 | 3 | 3 | 4 | 4 |
| 93 | 9685 | 9689 | 9694 | 9699 | 9703 | 9708 | 9713 | 9717 | 9722 | 9727 | | 0 | 1 | 1 | 2 | 2 | 3 | 3 | 4 | 4 |
| 94 | 9731 | 9736 | 9741 | 9745 | 9750 | 9754 | 9759 | 9763 | 9768 | 9773 | | 0 | 1 | 1 | 2 | 2 | 3 | 3 | 4 | 4 |
| 95 | 9777 | 9782 | 9786 | 9791 | 9795 | 9800 | 9805 | 9809 | 9814 | 9818 | | 0 | 1 | 1 | 2 | 2 | 3 | 3 | 4 | 4 |
| 96 | 9823 | 9827 | 9832 | 9836 | 9841 | 9845 | 9850 | 9854 | 9859 | 9863 | | 0 | 1 | 1 | 2 | 2 | 3 | 3 | 4 | 4 |
| 97 | 9868 | 9872 | 9877 | 9881 | 9886 | 9890 | 9894 | 9899 | 9903 | 9908 | | 0 | 1 | 1 | 2 | 2 | 3 | 3 | 4 | 4 |
| 98 | 9912 | 9917 | 9921 | 9926 | 9930 | 9934 | 9939 | 9943 | 9948 | 9952 | | 0 | 1 | 1 | 2 | 2 | 3 | 3 | 4 | 4 |
| 99 | 9956 | 9961 | 9965 | 9969 | 9974 | 9978 | 9983 | 9987 | 9991 | 9996 | | 0 | 1 | 1 | 2 | 2 | 3 | 3 | 3 | 4 |
| x | 0 | 1 | 2 | 3 | 4 | 5 | 6 | 7 | 8 | 9 | | 1 | 2 | 3 | 4 | 5 | 6 | 7 | 8 | 9 |

# MATHEMATICAL OPERATIONS

## SIGNIFICANT FIGURES.

All measurements are subject to error. Reported results should indicate the estimated magnitude of the experimental error. One possible way of indicating error or uncertainty in a result is the following:

$$27.30 \pm 0.01 \text{ g} \qquad 4.351 \pm 0.003 \text{ m}$$

Unfortunately, this notation is usually too cumbersome for our use.

A more compact, but admittedly less flexible, notation is to include only significant figures in numerical expressions. All non-zero digits in a numerical expression are significant if only the last digit can be in error. Zeros are assumed to be significant only if they are not needed to position the decimal point. The following examples are illustrative:

| NUMERICAL EXPRESSION | NUMBER OF SIGNIFICANT FIGURES |
|---|---|
| 27.30 | 4 – The zero is not needed for positioning the decimal point. |
| 4.351 | 4 |
| 0.002 | 1 – The zeros are needed to position the decimal point. |
| 3000 | 1 – The zeros may be significant, but they may just position the decimal; therefore we make the conservative choice. |
| quantum numbers (e.g., n = 3) | Exact; there is no limit on the accuracy of this number. |
| 1 m = 100 cm | This is a definition, and there is no limit on the accuracy. |

Calculated results cannot be more accurate than the measurements on which they are based. The following rules will help you to avoid implying unwarranted accuracy in calculated results:

**Rule 1.** Any figure in a number resulting from addition or subtraction is significant only if *each* number in the sum contributes a significant figure to that decimal level. For example,

$$\begin{array}{r} 632.423 \\ 1.600419 \\ 0.0094 \\ \hline 634.032 \end{array}$$

**Rule 2.** The result of a multiplication or a division has no more significant figures than the least accurate factor in the operation. For example,

$$\frac{27 \times 3.14}{36.44} = 2.3$$

**Rule 3.** When the mechanics of an arithmetical operation produce more figures than are significant, the nonsignificant figures should be discarded. If the first figure discarded is 5 or more, the last figure retained should be increased by one. For example, to three significant figures

27.30 is 27.3          27.35 is 27.4

This process is called rounding off.

You will save yourself much labor if you round off after each step in a multistep computation. For maximum accuracy you should carry one extra figure through each step but the last.

## SCIENTIFIC NOTATION.

We noted earlier that the accuracy of numbers like 3000 was difficult to specify in conventional notation. We can avoid that difficulty and provide a convenient method for writing very large and very small numbers by using scientific notation. For example, 3000 (with 3 significant figures) can be written in a variety of equivalent ways:

$$3000 = 300 \times 10 = 30.0 \times 10^2 = 3.00 \times 10^3 = 0.300 \times 10^4$$

All but the first of these is in scientific notation, but the preferred expression is $3.00 \times 10^3$. That is, in scientific notation we write the number as a factor between 1 and 10, multiplied by 10 to the appropriate power. Notice that there is no ambiguity about the number of significant figures in $3.00 \times 10^3$.

Small numbers are written with negative exponents:

$$0.000641 = 6.41 \times 10^{-4}$$

## MATHEMATICAL OPERATIONS.

**Exponents.** In the algebraic expression $x^n$, $n$ is called the exponent and $x$ is called the base. If $n$ is an integer, the expression can be evaluated by multiplication; for example,

$$3^4 = 3 \times 3 \times 3 \times 3 = 81$$

Non-integer exponents can be evaluated with the use of logarithms, but in the special case of a rational fraction $1/n$ ($n$ an integer), $n$ is the index of the root:

$$x^{1/n} = \sqrt[n]{x}$$

For example,

$$9^{0.5} = 9^{1/2} = \sqrt{9} = 3$$

When exponential expressions having a common base are multiplied, the exponents are added:

$$x^m \cdot x^n = x^{m+n}$$

For example,

$$10^3 \cdot 10^4 = 10^7$$

When exponential expressions having a common base are divided, the exponents are subtracted:

$$x^m/x^n = x^{m-n}$$

For example,

$$\frac{10^3}{10^4} = 10^{-1}$$

Any base (except 0) raised to the 0th power is 1. We can use this fact to define negative exponents. Thus,

$$\frac{1}{x^n} = \frac{x^0}{x^n} = x^{0-n} = x^{-n}$$

So

$$5^{-3} = \frac{1}{5^3}$$

**Logarithms.**    Addition and subtraction are easier than multiplication and division. Note that in exponential expressions, multiplication becomes addition of exponents and division becomes subtraction of exponents. Therefore, by writing all numbers as various powers of a common base, we can replace multiplication and division by addition and subtraction. This process is the basis of the slide rule, where we multiply by adding lengths and divide by subtracting lengths.

Although we could pick any number as a base, two choices are especially convenient. Base 10 is convenient because our number system is a base 10 system. Base $e$ ($e = 2.71828\ldots$) is convenient because it simplifies many expressions obtained in calculus. There is no reason to repeatedly write out the base for each number. We agree to omit the base and write only the exponent, calling the exponent a logarithm. For example,

$$y = x^n$$

$$\log_x y = n$$

Usually we do not bother to specify the base. The expression log means $\log_{10}$, and the expression ln means $\log_e$.

To evaluate the logarithm of a number, we must use log tables or a slide rule. In Appendix 2 are listed the base 10 logarithms of numbers between 1 and 10. For example, from the table we find that

$$\log 2.45 = 0.3892$$

Note that a number between 1 and 10 (between $10^0$ and $10^1$) must have a logarithm between 0 and 1. Logarithms of numbers which are not between 1 and 10 can be evaluated by converting the number to scientific notation and using the fact that addition of logarithms is the equivalent of multiplication. For example,

$$\log 245 = \log (2.45 \times 10^2)$$
$$= \log 2.45 + \log 10^2$$
$$= 0.3892 + 2$$
$$= 2.3892$$

$$\log (0.000479) = \log (4.79 \times 10^{-4})$$
$$= \log 4.79 + \log 10^{-4}$$
$$= 0.6803 - 4$$
$$= -3.3197$$

The reverse process, logarithm to number, is possible. It is called "taking the antilog." For example,

| | |
|---|---|
| $\log y = 7.3617$ | $\log y = -4.0600$ |
| $\log y = 0.3617 + 7$ | $\log y = 0.9400 - 5$ |
| $y = 10^{0.3617} \times 10^7$ | $y = 10^{0.9400} \times 10^{-5}$ |
| $y = 2.30 \times 10^7$ | $y = 8.71 \times 10^{-5}$ |

The relation between logarithms in the two common bases, $e$ and 10, is

$$\ln x = 2.303 \log x$$

**The Straight Line.**    Upon completion of a series of experiments, a scientist often wishes to summarize his results in an equation which relates the experimental variables. The simplest equation which relates two variables is the equation for the straight line. The equation for the straight line is

$$y = mx + b$$

where $x$ is the independent variable, $y$ is the dependent variable, $m$ is the slope, and $b$ is the intercept. The intercept is the value of the dependent variable ($y$) when the value of the independent variable ($x$) is zero. The slope of a straight line can be calculated from any two points on the line. If the points have the coordinates ($x_1, y_1$) and ($x_2, y_2$), then the slope is

$$\frac{y_2 - y_1}{x_2 - x_1} = \frac{\Delta y}{\Delta x} = m$$

| EQUATION | DEPENDENT VARIABLE | INDEPENDENT VARIABLE | SLOPE | INTERCEPT |
|---|---|---|---|---|
| $v = k/p$ | $v$ | $\left(\dfrac{1}{p}\right)$ | $k$ | $0$ |
| $\dfrac{1}{c} = \dfrac{1}{c_0} + kt$ | $\dfrac{1}{c}$ | $t$ | $k$ | $\dfrac{1}{c_0}$ |
| $u = k/\sqrt{d}$ | $u$ | $\dfrac{1}{\sqrt{d}}$ | $k$ | $0$ |
| $k = Ae^{-E_a/RT}$ | $\ln k$ | $\dfrac{1}{T}$ | $-\dfrac{E_a}{R}$ | $\ln A$ |

Because the equation for the straight line and the plot of the straight line are so simple, we frequently recast many nonlinear equations into linear equations by judicious choice of variables. Several examples are shown above.

# SI UNITS, FUNDAMENTAL CONSTANTS, AND CONVERSION FACTORS

The International System of Units (abbreviated SI from the French, *Le Système Internationale d'Unités*) was adopted by the General Conference of Weights and Measures (CGPM) in 1960 as *the* recommended units for use in science and technology.

The SI is constructed from seven base units for independent quantities (Table 1), plus two supplementary units for plane and solid angles, the radian (rad) and steradian (sr), respectively.

The definitions of the seven SI base units are as follows:

## (a)  Unit of length, the metre (m).

*The metre is the length equal to 1,650,763.73 wavelengths in vacuum of the radiation corresponding to the transition between the levels $2p_{10}$ and $5d_5$ of the krypton-86 atom.*

The SI unit of area is the square meter, $m^2$. The SI unit of volume is the cubic meter, $m^3$. Fluid volume is often measured by the liter, $1 \ \ell = 10^{-3} \ m^3 = 1 \ dm^3$.

## (b)  Unit of mass, the kilogram (kg).

*The kilogram is the unit of mass; it is equal to the mass of the international prototype of the kilogram.*

The "international prototype of the kilogram" is a cylinder of Pt-Ir alloy kept by the International Bureau of Weights and Measures at Paris. A duplicate in the custody of the U.S. National Bureau of Standards serves as the mass standard for the United States. Mass is the only base unit still defined by an artifact.

The SI unit of force is the newton (N), $1 \ N = 1 \ kg \cdot m \cdot s^{-2}$. The SI unit of work and energy of any kind is the joule (J), $1 \ J = 1 \ N \cdot m$. The SI unit for power of any kind is the watt (W), $1 \ W = 1 \ J \cdot s^{-1}$.

## (c)  Unit of time, the second (s).

*The second is the duration of 9,192,631,770 periods of the radiation corresponding to the transition between two hyperfine levels of the ground state of the cesium-133 atom.*

Originally, the second was defined as 1/86,400 of the mean solar day. Observations by astronomers have established that irregularities in the rotation of the Earth make it impossible for that definition to guarantee the desired accuracy; hence the shift to the atomic clock.

The second is realized by tuning an oscillator to the resonant frequency of the $^{133}$Cs atoms as they are passed through a system of magnets and a resonant cavity into a detector. The SI unit for frequency is the hertz (Hz), 1 Hz = one cycle per second = 1 sec$^{-1}$. Standard frequencies and correct time are broadcast from NBS stations WWV, WWVB, WWVH, and WWVL, as well as stations of the U.S. Navy. Many shortwave receivers can pick up WWV on frequencies of 2.5, 5, 10, 15, 20, and 25 MHz.

### (d)   Unit of electric current, the ampere (A).

*The ampere is that constant current which, if maintained in two straight parallel conductors of infinite length, of negligible circular cross section, and placed 1 metre apart in vacuum, would produce between these conductors a force equal to $2 \times 10^{-7}$ newton per metre of length.*

The force between the two wires results from the interaction of the magnetic fields around the current-carrying wires. The SI unit of voltage is the volt (V), 1 V = 1 W·A$^{-1}$. The SI unit of electrical resistance is the ohm ($\Omega$), 1 $\Omega$ = 1 V·A$^{-1}$.

### (e)   Unit of thermodynamic temperature, the kelvin (K).

*The kelvin, unit of thermodynamic temperature, is the fraction 1/273.16 of the thermodynamic temperature of the triple point of water.*

The unit kelvin and its symbol K (not °K) should also be used to express an interval or a difference in temperature. The Celcius temperature (symbol $t$) is defined by the equation

$$t = T - 273.15 \text{ K}$$

where $T$ is the thermodynamic temperature. Celcius temperatures are expressed as °C.

### (f)   Unit of amount of substance, the mole (mol).

*The mole is the amount of substance of a system which contains as many elementary entities as there are atoms in 0.012 kilogram of carbon-12.*

When the mole is used, the elementary entities must be specified; they may be atoms, molecules, ions, electrons, other particles, or specified groups of such particles.

### (g)   Unit of luminous intensity, the candela (cd).

*The candela is the luminous intensity, in the perpendicular direction, of a surface of 1/600,000 square metre of a blackbody at the temperature of freezing platinum under a pressure of 101,325 N·m$^{-2}$ (i.e., 1 atm).*

## FURTHER CONSIDERATIONS.

1. No restrictions are placed on the units employed for general descriptive information that does not enter into calculations or expression of results (e.g., "pressures in the range 1–50 Torr").

2. All integral powers, positive or negative, of SI units are acceptable. Note that exponents operate also on prefixes, as in cm², mm³, which are $10^{-4}$ m², $10^{-9}$ m³, and not $10^{-2}$ m², $10^{-3}$ m³ (see Table 3).

3. Unit combinations should be designated by means of a dot or dots (as in m·K for meter-kelvin, which avoids confusion with millikelvin, mK).

4. Words and symbols should not be mixed; if mathematical operations are indicated, only symbols should be used. For example, one may write "joules per mole," "J/mol," "J·mol⁻¹," but not "joules/mole," "joules·mol⁻¹," etc.

5. Roman type, in general lower case, is used for symbols of units; however, if the symbols are derived from proper names, capital roman type is used for the first letter. These symbols are not followed by a period. Unit symbols do not change in the plural.

6. Certain units not part of the SI are approved for a limited time during the changeover to SI units. Some of these are:

| | |
|---|---|
| Ångstrom | $1\ \text{Å} = 0.1\ \text{nm} = 10^{-10}\ \text{m}$ |
| standard atmosphere | $1\ \text{atm} = 101{,}325\ \text{N} \cdot \text{m}^{-2}$ |
| bar | $1\ \text{bar} = 10^{5}\ \text{N} \cdot \text{m}^{-2}$ |
| curie | $1\ \text{Ci} = 3.7 \times 10^{10}\ \text{s}^{-1}$ |
| röntgen | $1\ \text{R} = 2.58 \times 10^{-4}\ \text{Ci} \cdot \text{kg}^{-1}$ |

_____ **REFERENCES**

1. *NBS Special Publication 330* (1972 Edition, U.S. Department of Commerce, NBS).
2. *Policy for NBS Usage of SI Units*, J. Chem. Educ. 48, 569 (1971).

We have relied heavily on these two publications in preparing this appendix.

**TABLE 1.**

SI BASE UNITS.

| PHYSICAL QUANTITY | NAME OF UNIT | SYMBOL |
|---|---|---|
| length | metre | m |
| mass | kilogram | kg |
| time | second | s |
| electric current | ampere | A |
| thermodynamic temperature | kelvin | K |
| luminous intensity | candela | cd |
| amount of substance | mole | mol |

**TABLE 2.**

SPECIAL NAMES AND SYMBOLS FOR CERTAIN SI DERIVED UNITS.

| Physical Quantity | Name of SI Unit | Symbol for SI Unit | Definition of SI Unit |
|---|---|---|---|
| force | newton | N | $kg \cdot m \cdot s^{-2}$ |
| pressure | pascal | Pa | $kg \cdot m^{-1} \cdot s^{-2} (= N \cdot m^{-2})$ |
| energy | joule | J | $kg \cdot m^2 \cdot s^{-2}$ |
| power | watt | W | $kg \cdot m^2 \cdot s^{-3} (= J \cdot s^{-1})$ |
| electric charge | coulomb | C | $A \cdot s$ |
| electric potential difference | volt | V | $kg \cdot m^2 \cdot s^{-3} \cdot A^{-1} (= J \cdot A^{-1} \cdot s^{-1})$ |
| electric resistance | ohm | $\Omega$ | $kg \cdot m^2 \cdot s^{-3} \cdot A^{-2} (= V \cdot A^{-1})$ |
| electric conductance | siemens | S | $kg^{-1} \cdot m^{-2} \cdot s^3 \cdot A^2 (= A \cdot V^{-1} = \Omega^{-1})$ |
| electric capacitance | farad | F | $A^2 \cdot s^4 \cdot kg^{-1} \cdot m^{-2} (= A \cdot s \cdot V^{-1})$ |
| magnetic flux | weber | Wb | $kg \cdot m^2 \cdot s^{-2} A^{-1} (= V \cdot s)$ |
| inductance | henry | H | $kg \cdot m^2 \cdot s^{-2} \cdot A^{-2} (= V \cdot A^{-1} \cdot s)$ |
| magnetic flux density | tesla | T | $kg \cdot s^{-2} \cdot A^{-1} (= V \cdot s \cdot m^{-2})$ |
| luminous flux | lumen | lm | $cd \cdot sr$ |
| illumination | lux | lx | $cd \cdot sr \cdot m^{-2}$ |
| frequency | hertz | Hz | $s^{-1}$ (cycle per second) |

**TABLE 3.**

PREFIXES FOR FRACTIONS AND MULTIPLES OF SI UNITS.

| Fraction | Prefix | Symbol | Multiple | Prefix | Symbol |
|---|---|---|---|---|---|
| $10^{-1}$ | deci | d | 10 | deka | da |
| $10^{-2}$ | centi | c | $10^2$ | hecto | h |
| $10^{-3}$ | milli | m | $10^3$ | kilo | k |
| $10^{-6}$ | micro | $\mu$ | $10^6$ | mega | M |
| $10^{-9}$ | nano | n | $10^9$ | giga | G |
| $10^{-12}$ | pico | p | $10^{12}$ | tera | T |
| $10^{-15}$ | femto | f | | | |
| $10^{-18}$ | atto | a | | | |

Multiple prefixes, such as $\mu\mu F$ for pF, should not be used.

**TABLE 4.**

DECIMAL FRACTIONS AND MULTIPLES OF SI UNITS HAVING SPECIAL NAMES.

| Physical Quantity | Name of Unit | Symbol for Unit | Definition of Unit |
|---|---|---|---|
| length | Ångstrom | Å | $10^{-10}$ m |
| length | micron | $\mu$ | $10^{-6}$ m $= \mu$m |
| area | barn | b | $10^{-28}$ m² |
| force | dyne | dyn | $10^{-5}$ N |
| pressure | bar | bar | $10^5$ N·m⁻² |
| energy | erg | erg | $10^{-7}$ J |
| kinematic viscosity | stokes | St | $10^{-4}$ m²·s⁻¹ |
| viscosity | poise | P | $10^{-1}$ kg·m⁻¹·s⁻¹ |
| magnetic flux | maxwell | Mx | $10^{-8}$ Wb |
| magnetic flux density (magnetic induction) | gauss | G | $10^{-4}$ T |
| absorbed dose (ionizing radiation) | rad | rd | $10^{-2}$ J·kg⁻¹ |
| volume | liter | $\ell$ | $10^{-3}$ m³ $= 1$ dm³ |

**TABLE 5.**

GENERAL PHYSICAL CONSTANTS.

| Constant | Symbol | Value in SI Units | Value in CGS Units |
|---|---|---|---|
| speed of light in vacuum | $c$ | $2.998 \times 10^8$ m·s⁻¹ | $2.998 \times 10^{10}$ cm·s⁻¹ |
| elementary charge | $e$ | $1.602 \times 10^{-19}$ C | $1.602 \times 10^{-20}$ cm^{1/2}·g^{1/2}$ <br> $4.803 \times 10^{-10}$ cm^{1/2}·g^{1/2}·s⁻¹ |
| Avogadro constant | $N_A$ | $6.022 \times 10^{23}$ mol⁻¹ | $6.022 \times 10^{23}$ mol⁻¹ |
| atomic mass unit | u | $1.661 \times 10^{-27}$ kg | $1.661 \times 10^{-24}$ g |
| Faraday constant | $F$ | $9.649 \times 10^4$ C·mol⁻¹ | $9.649 \times 10^3$ cm^{1/2}·g^{1/2}·mol⁻¹ |
| Planck constant | $h$ | $6.626 \times 10^{-34}$ J·s | $6.626 \times 10^{-27}$ erg·s |
| gas constant | $R$ | $8.314$ J·K⁻¹·mol⁻¹ | $8.314 \times 10^7$ erg·K⁻¹·mol⁻¹ |
| Boltzmann constant | $k$ | $1.381 \times 10^{-23}$ J·K⁻¹ | $1.381 \times 10^{-16}$ erg·K⁻¹ |
| gravitational constant | $G$ | $6.673 \times 10^{-11}$ N·m²·kg⁻² | $6.673 \times 10^{-8}$ dyn·cm²·g⁻² |
| electron rest mass | $m_e$ | $9.110 \times 10^{-31}$ kg | $9.110 \times 10^{-28}$ g |
| proton rest mass | $m_p$ | $1.673 \times 10^{-27}$ kg | $1.673 \times 10^{-24}$ g |

## MISCELLANEOUS CONSTANTS AND CONVERSION FACTORS

### Values of the Gas Constant

$$R = 8.314 \text{ J} \cdot \text{K}^{-1} \cdot \text{mol}^{-1}$$
$$= 1.987 \text{ cal} \cdot \text{K}^{-1} \cdot \text{mol}^{-1}$$
$$= 82.05 \text{ cm}^3 \cdot \text{atm} \cdot \text{K}^{-1} \cdot \text{mol}^{-1}$$
$$= 0.08205 \; \ell \cdot \text{atm} \cdot \text{K}^{-1} \cdot \text{mol}^{-1}$$

### Values of the Faraday Constant

$$F = 96,487 \text{ C} \cdot \text{mol}^{-1}$$
$$= 23,061 \text{ cal} \cdot \text{V}^{-1} \cdot \text{mol}^{-1}$$

### Gravitational Acceleration (sea level, 45° latitude)

$$g = 980.62 \text{ cm} \cdot \text{s}^{-2}$$
$$= 32.17 \text{ ft} \cdot \text{s}^{-2}$$

### Density of Mercury (0°C)

$$\rho = 13.5955 \text{ g} \cdot \text{m}\ell^{-1}$$

### Density of Dry Air (0°C, 760 Torr)

$$\rho = 1.293 \text{ g} \cdot \ell^{-1}$$

### Heat of Fusion of Ice (0°C)

$$\Delta H_{fus} = 79.71 \text{ cal} \cdot \text{g}^{-1}$$

### Heat of Vaporization of Water (100°C)

$$\Delta H_{vap} = 539.55 \text{ cal} \cdot \text{g}^{-1}$$

$$\pi = 3.1416$$
$$e = 2.7183$$
$$\log_e x = 2.303 \log_{10} x$$

## Conversion Factors

$1 \text{ cal} = 4.184 \text{ J}$

$1 \text{ eV} \cdot \text{molecule}^{-1} = 23,061 \text{ cal} \cdot \text{mol}^{-1}$

$1 \text{ joule} = 10^7 \text{ erg}$

$1 \text{ kcal} = 10^3 \text{ cal}$

$1 \text{ cm}^{-1} \cdot \text{molecule}^{-1} = 2.859 \text{ cal} \cdot \text{mol}^{-1}$

$1 \text{ eV} = 8066 \text{ cm}^{-1}$

$1 \text{ inch} = 2.54 \text{ cm}$

$1 \text{ pound} = 453.6 \text{ g}$

$1 \text{ liter} = 1.057 \text{ quart}$

$1 \text{ kilometer} = 0.6214 \text{ mile}$

$1 \text{ kilogram} = 2.205 \text{ pound}$

$1 \text{ cubic centimeter} = 0.999972 \text{ milliliter}$

$1 \text{ yard} = 0.9144 \text{ meter}$

$1 \text{ angstrom} = 10^{-8} \text{ cm}$

$1 \text{ nanometer} = 1 \text{ millimicron} = 10^{-9} \text{ meter}$

$1 \text{ dyn} \cdot \text{cm}^{-2} = 0.1 \text{ N} \cdot \text{m}^{-2}$

$1 \text{ liter} = 10^{-3} \text{ m}^3 = 1 \text{ dm}^3$

$1 \text{ atm} = 1.013 \times 10^5 \text{ N} \cdot \text{m}^{-2}$

# APPENDIX 5

# THERMODYNAMIC DATA

(These data were obtained primarily from NBS Technical Notes 270-3, 270-4, 270-5, and 270-6. Edited by D. D. Wagman et al.)

| SPECIES | $\Delta \bar{H}_f^\circ$(kcal) | $\Delta \bar{G}_f^\circ$(kcal) | $\bar{S}^\circ$(cal·K$^{-1}$) | SPECIES | $\Delta \bar{H}_f^\circ$(kcal) | $\Delta \bar{G}_f^\circ$(kcal) | $\bar{S}^\circ$(cal·K$^{-1}$) |
|---|---|---|---|---|---|---|---|
| $O_2(g)$ | 0 | 0 | 49.00 | C(s,diam) | 0.45 | 0.69 | 0.57 |
| $H_2(g)$ | 0 | 0 | 31.21 | CO(g) | −26.42 | −32.78 | 47.22 |
| $H^+(aq)$ | 0 | 0 | 0 | $CO_2(g)$ | −94.05 | −94.25 | 51.06 |
| $H_2O(g)$ | −57.80 | −54.63 | 45.10 | $CH_4(g)$ | −17.88 | −12.13 | 44.49 |
| $H_2O(\ell)$ | −70.41 | −58.20 | 18.15 | $CH_3OH(g)$ | −47.96 | −38.72 | 57.29 |
| $H_2O_2(\ell)$ | −44.88 | −28.78 | 26.2 | $CH_3OH(\ell)$ | −57.04 | −39.76 | 30.3 |
| $F_2(g)$ | 0 | 0 | 48.44 | $CCl_4(g)$ | −24.6 | −14.49 | 74.03 |
| HF(g) | −64.8 | −65.3 | 41.51 | $CH_3Cl(g)$ | −19.32 | −13.72 | 56.04 |
| $Cl_2(g)$ | 0 | 0 | 53.29 | $C_2H_2(g)$ | 54.19 | 50.00 | 48.00 |
| HCl(g) | −22.06 | −22.78 | 44.65 | $C_2H_4(g)$ | 12.49 | 16.28 | 52.45 |
| $Cl^-(aq)$ | −39.95 | −31.37 | 13.5 | $C_2H_6(g)$ | −20.24 | −7.86 | 54.85 |
| $Br_2(g)$ | 7.39 | 0.75 | 58.64 | $CH_3COOH(aq)$ | −116.10 | −94.78 | 42.7 |
| $Br_2(\ell)$ | 0 | 0 | 36.38 | $CH_3COO^-(aq)$ | −116.16 | −88.29 | 20.7 |
| $Br_2(aq)$ | −0.62 | 0.94 | 31.2 | $C_2H_5OH(\ell)$ | −66.37 | −41.80 | 38.4 |
| $Br^-(aq)$ | −29.05 | −24.85 | 19.7 | $C_2N_2(g)$ | 73.84 | 71.07 | 57.79 |
| $I_2(g)$ | 14.92 | 4.63 | 62.28 | Cd(s) | 0 | 0 | 12.37 |
| $I_2(s)$ | 0 | 0 | 27.76 | $Cd^{2+}(aq)$ | −18.14 | −18.54 | −17.5 |
| $I^-(aq)$ | −13.19 | −12.33 | 26.6 | Ag(s) | 0 | 0 | 10.20 |
| $SO_2(g)$ | −70.94 | −71.75 | 59.30 | $Ag^+(aq)$ | 25.23 | 18.43 | 17.40 |

## APPENDIX 5    Continued

| Species | $\Delta\bar{H}_f^\circ$(kcal) | $\Delta\bar{G}_f^\circ$(kcal) | $\bar{S}^\circ$(cal·K$^{-1}$) | Species | $\Delta\bar{H}_f^\circ$(kcal) | $\Delta\bar{G}_f^\circ$(kcal) | $\bar{S}^\circ$(cal·K$^{-1}$) |
|---|---|---|---|---|---|---|---|
| $SO_3(g)$ | $-94.58$ | $-88.69$ | 61.34 | $AgCl(s)$ | $-30.37$ | $-26.24$ | 23.0 |
| $SO_4^{2-}(aq)$ | $-217.32$ | $-177.97$ | 4.8 | $AgI(s)$ | $-14.78$ | $-15.82$ | 27.6 |
| $SF_4(g)$ | $-185.2$ | $-174.8$ | 69.77 | $Hg_2SO_4(s)$ | $-177.61$ | $-149.59$ | 47.96 |
| $SF_6(g)$ | $-289.$ | $-264.2$ | 69.72 | $Hg(\ell)$ | 0 | 0 | 18.17 |
| $N_2(g)$ | 0 | 0 | 45.77 | $Hg_2Br_2(s)$ | | $-43.28$ | |
| $N_2H_4(\ell)$ | 12.10 | 35.67 | 28.97 | $Hg_2Cl_2(s)$ | $-63.39$ | $-50.38$ | 46.0 |
| $NO(g)$ | 21.57 | 20.69 | 50.35 | $CaO(s)$ | $-151.79$ | $-144.37$ | 9.50 |
| $NOCl(g)$ | 12.36 | 15.79 | 62.52 | $CaCO_3(s, calcite)$ | $-288.46$ | $-269.80$ | 22.2 |
| $NO_2(g)$ | 7.93 | 12.26 | 57.35 | $Mg(s)$ | 0 | 0 | 7.81 |
| $NH_3(g)$ | $-11.02$ | $-3.94$ | 45.97 | $MgO(s)$ | $-143.81$ | $-136.10$ | 6.44 |
| $NH_3(aq)$ | $-19.19$ | $-6.35$ | 26.6 | $Fe(s)$ | 0 | 0 | 6.52 |
| $N_2O_4(\ell)$ | $-4.66$ | 23.29 | 50.0 | $Fe_{0.947}O(s)$ | $-63.64$ | $-58.59$ | 13.74 |
| $PCl_3(g)$ | $-68.6$ | $-64.0$ | 74.49 | $Fe_2O_3(s)$ | $-197.0$ | $-177.4$ | 20.89 |
| $PCl_5(g)$ | $-89.6$ | $-72.9$ | 87.11 | $Fe_3O_4(s)$ | $-267.3$ | $-242.7$ | 35.0 |
| $C(s, graph)$ | 0 | 0 | 1.37 | $Fe(CN)_6^{4-}(aq)$ | 108.9 | 166.1 | 22.7 |
| $Fe(CN)_6^{3-}(aq)$ | 134.3 | 174.3 | 64.6 | | | | |
| $Na(s)$ | 0 | 0 | 12.24 | | | | |
| $Na_2O_2(s)$ | $-120.6$ | $-102.8$ | 16. | | | | |
| $C_6H_6(\ell)$ | 11.72 | 29.76 | | | | | |

# APPENDIX 6

# ANSWERS TO SELECTED PROBLEMS

### Chapter 1

(3) 75.8% $^{35}Cl$     (5) $9.5 \times 10^{17}$ cm, $9.5 \times 10^{15}$ m, $5.9 \times 10^{12}$ miles

(7) $1.661 \times 10^{-24}$ g     (9) 6.942, 6.943     (11) $6.0227 \times 10^{23}$

(13) $2.731 \times 10^{26}$     (15) $2.37 \times 10^{-32}$ g     (17) 275 mm

(19) e.g., density, melting point, boiling point, vapor pressure, solubility

### Chapter 2

(1) $3.64 \times 10^8$ cm/sec     (3) $K^+$—Ar; $Cl^-$—Ar; $O^{-2}$—Ne; $Ba^{+2}$—Xe;
$Fe^{+2}$: $1s^2 2s^2 2p^6 3s^2 3p^6 3d^6$; $Fe^{+3}$: $1s^2 2s^2 2p^6 3s^2 3p^6 3d^5$;
$Ag^+$: $1s^2 2s^2 2p^6 3s^2 3p^6 4s^2 3d^{10} 4p^6 4d^{10}$

(5) number of photons     (7) $6.17 \times 10^{14}$ hertz, yes

(9) $2.18 \times 10^{-11}$ erg, $5.45 \times 10^{-12}$ erg     (11) 1.26

(13) single break between C and N     (15) larger, larger

(17) (a) 1 angular, 2 spherical   (b) 4 spherical   (c) 3 angular, 2 spherical

(19) a,b,c     (27) electron pairing required in $N^-$, not in $C^-$

### Chapter 3

(1) 4.6 kcal·mol⁻¹, 19.2 kJ·mol⁻¹

(3) $O_3$, $O_2NCl$, $HNO_3$, $N_2O_5$ have 3 electron bonds

(5) all but Ca, Zn, Kr     (7) energy + 6.6eV     (9) all single bonds

(11) 

```
    H     H            H   H
    |     |            |   |
H—C—O—C—H      H—C—C—O—H
    |     |            |   |
    H     H            H   H
```

(13) 3 $d$ orbitals on S and P     (17) $k = 2D\beta^2$

### Chapter 4

(1) LiH $1s^2\sigma^2$; BeH $1s^2\sigma^2 2p^1$; BH $1s^2\sigma^2 2p^2$; CH $1s^2 2s^2\sigma^2 2p^1$;
NH $1s^2 2s^2\sigma^2 2p^2$; OH $1s^2 2s^2\sigma^2 2p^3$; FH $1s^2 2s^2\sigma^2 2p^4$;
all atomic orbitals on 2nd row atoms

(5) more antibonding electrons in $He_2^+$

(7) $BeH_2$     (9) $NO_2$, $H_2Te$, $ClO_2$, $OF_2$, $SO_2$, $HCOO^-$, $NO_2Cl$, $CN^-$, $NO_2^-$,
$PCl_3$     (11) $CO_2$ linear $2\sigma$ $2\pi$; $H_2CO$ planar $3\sigma$ $1\pi$; $HNO_3 - NO_3^-$
planar $4\sigma$ $1\pi$; $H_2SO_4 - SO_4^{-2}$ tetrahedral $6\sigma$; $CH_3CH_2OH$ tetrahedral,

all $\sigma$ bonds; $HCO_2H$ planar about C atom $4\sigma$ $1\pi$; $H_2NCONH_2$ tetra-hedral about N atoms, planar about C $7\sigma$ $1\pi$; $H_2CCCH_2$ planar $6\sigma$ $2\pi$; $H_3COCH_3$ tetrahedral, all $\sigma$ bonds; $H_3CCHCH_2$ tetrahedral about one C atom, rest planar $8\sigma$ $1\pi$; $H_2NNH_2$ and $H_2NOH$ tetrahedral and all $\sigma$ bonds.

(**17**) 1.50 D     (**19**) $NO^+$, $NO^{+2}$, NO, $NO^-$

(**21**) F has higher ionization energy

(**23**) 3 center bonds as in boron hydrides.

## Chapter 5

(**1**) 1.56 $\ell$     (**3**) extensive: volume, number of moles

(**5**) $N_2/H_2 = 0.845$     (**7**) 14.7 lbs/in²     (**9**) He smaller

(**11**) 50.9 p.s.i.     (**13**) 914 lb

(**15**)

|  | Ar | $CO_2$ | $N_2$ | Ne | $NH_3$ | $H_2O$ |
|---|---|---|---|---|---|---|
| Room $T$ | $(3/2)R$ | $(5/2)R$ | $(5/2)R$ | $(3/2)R$ | $3R$ | $3R$ |
| 1000 K | $(3/2)R$ | $(13/2)R$ | $(7/2)R$ | $(3/2)R$ | $9R$ | $6R$ |

(**17**) differ by factor 1.004     (**19**) $2.85 \times 10^5$ $\ell$, $3.41 \times 10^6$ $\ell$     (**21**) 8.3%

## Chapter 6

(**1**) no     (**3**) sulfur and oxygen can form $-1$ ions

(**5**) cubic 100%, cubic 100%, cubic 100%, tetrahedral 50%, cubic 50%

(**7**) 2 holes/sphere, $r_2 = 0.23r_1$     (**9**) $3.96 \times 10^4$

(**11**) graphite and diamond structures     (**15**) $6.023 \times 10^{23}$

## Chapter 7

(**1**) 4.53 moles, $2.73 \times 10^{24}$ molecules, 31.7 $\ell$     (**3**) $C_2O_2F_3H$

(**5**) 13.9 g     (**7**) 7.31 g $P_4O_{10}$, 0.0403 moles BaS     (**9**) 6.6 $\ell$

(**13**) (b) 0.40 moles $ZW_3$   (c) 0.030 moles X   (d) 19.6 cm³   (e) 282 $\ell$ $W_2$

(**17**) 85.1% Cu

(**19**) (1) Fe +2, S −2, (2) As +3, S −2, (3) As +5, S −2, (4) B +3, N −3, (5) N +3, S −2, F −1, (6) C +4, Cl −1, (7) C +4, O −2, Cl −1, (8) C +2, H +1, Cl −1, (9) C +4, N −3, (10) C −2, H +1, N −3, (11) C +4, S −2, N −3, (12) P +3, F −1.

(**21**) $m \left( 1 + \rho_{air} \left( \dfrac{1}{\rho_{weights}} - \dfrac{1}{\rho_{sample}} \right) \right)$

## Chapter 8

(**1**) $NH_3$, $CH_3OH$, $HCONH_2$, $H_2SO_4$, $H_2O$

(**3**) $-1.17 \times 10^{-15}$ N in $H_2O$     (**5**) $1.04 \times 10^3$ atm     (**7**) $X_{H_2O} = 0.719$

(**9**) 43.46 g of $KMnO_4$ in 1000 g $H_2O$     (**11**) 140.6 ml of NaOH

(**13**) 0.1136 M; 0.5680 N

## Chapter 9

**(1)** $\omega = 1.07 \times 10^{13}$ erg $= 1.07 \times 10^6$ J $= 2.57 \times 10^5$ cal $= 1.06 \times 10^4$ $\ell \cdot$atm

**(5b)** $t = -459.7°$F

**(11)** x indicates that answer cannot be computed from data in text.
   (a) $\Delta H_{BE} = $ x, $\Delta H°_{298} = -21.54$ kcal    (b) x, $-30.62$
   (c) $-100$, $-94.96$    (d) $-256$, $-341.26$    (e) $-36$, x
   (f) $-194$, $-153.54$    (g) $-25$, $-23.50$    (h) $-25$, x    (i) x, $-126.68$
   (j) $-258$, x    (k) $-30$, $-32.73$

**(17)** $\Delta S_1 = 3.57 \times 10^3$ J$\cdot$K$^{-1}$; $\Delta S_2 = (1.50 \times 10^{-5}$ erg$\cdot$cm$^2 \cdot$K$^{-1})L_0^2$

**(19)** (a) $\Delta S°_{298} = -79.34$ cal$\cdot$K$^{-1}$, $\Delta n_{gas} = -3$ mol    (i) $-82.56$, $-2$
   (d) $-37.56$, $-1.5$    (h) $-65.65$, $-1.5$    (c) $-26.05$, $-1$    (g) x, 0
   (b) $16.45$, $0.5$    (l) $48.49$, 1    (f) x, $2.5$

**(25)** $\Delta S = 8.08$ cal$\cdot K^{-1}$    **(27)** $\Delta S = 0.33$ J$\cdot$K$^{-1}$

## Chapter 10

**(3)** (a) $C = 5$, $F = 2$; $C = 2$, $F = 2$; $C = 4$, $F = 1$    (b) $C = 3$, $F = 2$

**(5)** $\Delta H_{sub} = 12.2$ kcal$\cdot$mol$^{-1}$    **(9)** $273.16$ K

**(11)** (a) triple point and critical point, (b) four triple points and critical point

**(13)** $P_{tot} = 146$ Torr at 25°C; $P_{tot} = 760$ Torr at 68°C (graphical solution)

**(17)** 4, 35    **(19)** No    **(21)** $\Delta H_{sub} = \Delta H_{fus} + \Delta H_{vap}$

**(27)** at $r = 1 \times 10^{-5}$ cm, $P_r/P_o = 1.012$    **(29)** $0.35$ atm, $t = 74°$C

**(31)** $t = -128°$C

## Chapter 11

**(1)** $a_{H_2O} = 0.864$, $X_{H_2O} = 0.965$, $\gamma_{H_2O} = 0.895$    **(3)** $1.02 \times 10^{-5}$ m

**(5)** $\Delta H° = -452$ cal, $\Delta G°_{313} = 6.06$ kcal, $\Delta S° = -20.8$ cal$\cdot$K$^{-1}$

**(9)** at $P = 100$ atm, $a = 1.161$

**(11)** $\Delta G = 2.53$ kcal$\cdot$mol$^{-1}$ and $2.73$ kcal$\cdot$mol$^{-1}$ ($\gamma = 1$)

**(13)** (a) $\Delta G°_{298} = -6.98$ kcal$\cdot$mol$^{-1}$, $K = 1.311 \times 10^5$
   (b) $K = $ x (cannot compute from text data)    (c) $K = 2.49 \times 10^{34}$
   (d) $K = 1.75 \times 10^{-5}$    (e) $K = 1.39 \times 10^{23}$
   (f) $K = 1.72 \times 10^{19}$, $K = 7.64 \times 10^{-6}$    (h) $K = 1.26 \times 10^{25}$
   (i) $K = 2.51 \times 10^{556}$

## Chapter 12A

**(1)** (a) 2.00    (b) 11.70    (c) 2.60    (d) 0.80    (e) 11.48    (f) 2.79    (g) 8.97
   (h) 11.47    (i) 4.77    (j) 4.16    (k) 2.53    (l) 3.48

**(3)** $K_a = 1.35 \times 10^{-3}$; $K_b = 7.41 \times 10^{-12}$

**(5)** (a) 2.00    (b) 4.28    (c) 6.98    (d) 3.70    (e) 6.96    (f) 6.79    (g) 2.17

**(7)** $pH = 17.1$, basic    **(9)** $pH = 5.19$    **(11)** $pH = 6.06$

(13) $pH = 5.66$

(15) $(H^+) \simeq (C_5H_5N) \simeq 4.2 \times 10^{-4}$ M; $(C_5H_5NH^+) \simeq (Cl^-) \simeq 0.0266$

(19) $K = 2.26 \times 10^4$   (a) $P_{CH_3OH} = 1.00$ atm, $X_{CH_3OH} = 0.50$
(b) $P_{CH_3OH} \simeq 0.09$   (c) $X_{CH_3OH} = 0.90$   (d) 0.995

(21) $pH = 7.40$   (27) $(H^+) \simeq 6.2 \times 10^{-8}$ M; $(H_2PO_4^-) \simeq (HPO_4^{2-}) \simeq$
$0.040$ M; $(H_3PO_4) \simeq 3.5 \times 10^{-7}$ M; $(PO_4^{3-}) \simeq 2.8 \times 10^{-7}$ M

(29) $(HgCl_2) \simeq 0.010$ M; $(HgCl^+) \simeq (Cl^-) = 5.8 \times 10^{-5}$ M
$(HgCl_3^-) \simeq 4.1 \times 10^{-6}$ M; $(HgCl_4^{2-}) \simeq 2.4 \times 10^{-9}$ M

(31) $P_{CO_2} = 0.039$ atm; $P_{CO} = 0.27$ atm

(33) 98 ml 15 M $NH_3$(aq), 10.7 g $NH_4Cl$(s), dissolve and dilute to 1.00 liter

(37) (c) $K = 3.4 \times 10^{-4}$   (d) $pH = 4.45$

## Chapter 12B

(1) (a) $1.3 \times 10^{-5}$   (b) $2.6 \times 10^{-10}$   (c) $2.3 \times 10^{-10}$   (d) $1.8 \times 10^{-9}$

(3) $K = 1.9 \times 10^{19}$   (5) 19 M

(9) (a) I   (b) D   (c) NE   (d) I   (e) I   (f) D   (g) NE($\gamma_i = 1$)   (h) I

(11) $3.0 \times 10^{-3}$ M   (13) 12.2   (17) $K_f = 7.54$, $\Delta H_{kg} = 2.0 \times 10^3$

(19) $(Cl^-) = 1.1 \times 10^{-8}$; $(Ag^+) = 1.7 \times 10^{-2}$; $(NO_3^-) = 8.3 \times 10^{-2}$;
$(Na^+) = 6.7 \times 10^{-2}$ M

(21) $(Cl^-) = 2.5 \times 10^{-2}$; $(Ag^+) = 7.2 \times 10^{-9}$; $(S^{2-}) \simeq 1.9 \times 10^{-35}$;
$(H_2S) = 1.6 \times 10^{-31}$; $(HS^-) = 1.6 \times 10^{-30}$ M

(23) $S = 7.0 \times 10^{-4}$ M; $S = 2.0 \times 10^{-3}$ M

(25) $(Fe^{2+}) = 0.01$; $(Cd^{2+}) = 1.4 \times 10^{-8}$; $(Ag^+) = 3.2 \times 10^{-16}$;
$(S^{2-}) = 1.0 \times 10^{-20}$; $(H_2S) = 0.10$ M

(27) (a) $K = 2.1 \times 10^4$   (b) $Q = 7.6 \times 10^2$, $Q/K < 1$

(35) $K_{sp} = 1.2 \times 10^{-11}$

(37) (a) at $pH = 0.5$, $(Zn^{2+}) = 9.6 \times 10^{-2}$ and $(Fe^{2+}) = 0.91$ M
(b) $(Zn^{2+}) = 4.3 \times 10^{-8}$; $(Fe^{2+}) = 4.2 \times 10^{-7}$ M

(39) $(Cu^{2+}) = 3.8 \times 10^{-15}$ M

## Chapter 13

(3) (a) $3X^{2+} \rightarrow X(s) + 2X^{3+}$   (b) $\dot{X}(s) + 3Fe^{3+} \rightarrow 3Fe^{2+} + X^{3+}$
(c) $K = 9.0 \times 10^{-28}$   (d) $K = 2.2 \times 10^{-70}$   (e) $\mathscr{E}° = -0.07$ V
(f) $\mathscr{E}° = -0.33$ V   (g) $K_{sp} = 3.5 \times 10^{-48}$   (h) $(X^{3+}) = 2.3 \times 10^{-6}$ M

(5) $pH = -12.2$   (7) e.g., Al, Mg, Ca, Na

(9) (a) $\mathscr{E}° = 0.102$   (b) $\mathscr{E}° = 0.404$ V

(11) $\mathscr{E}_R° - \mathscr{E}_L° = +0.09$ V   (13) (a) $Zn(s)|ZnI_2(aq)|PdI_2(s)|Pd(s)$
(b) $H_2(g,Pt)|HCl(aq)|Cl_2(g,Pt)$
(c) $Cd(s)|CdSO_4(aq)|Hg_2SO_4(s)|Hg(\ell)$
(d) $Au(s)|K_4Fe(CN)_6(aq), K_3Fe(CN)_6(aq)|K^+(mem)|KCl(aq)|$
$|Hg_2Cl_2(s)|Hg(\ell)$

**(15)** (a) in 1 M $H^+$: $I^-$ air oxidized to $I_3^-$; $I_3^-$ borderline with respect to air oxidation to $IO_3^-$; HIO disproportionates to $I_3^-$ and $IO_3^-$; $IO_3^-$ stable; $H_5IO_6$ liberates $O_2(g)$ from 1 M $H^+(aq)$
(b) $3I_3^- + 6OH^- \rightarrow IO_3^- + 8I^- + 3H_2O$; $K = 5.4 \times 10^{29}$   (c) $K = 4.5 \times 10^{23}$
(d) 1.0 mol $Mn^{2+}$, 2.5 mol $I_3^-$, 0.5 mol $I^-$
(e) 0.67 mol $MnO_2(s)$ and 1.0 mol $I_3^-$

**(19)** (a) increase   (b) decrease   (c) no effect   (d) decrease

**(21)** lower mass, higher voltage

## Chapter 14

**(1)** $t = 0.23$; $t = (0.90/C_0)$       **(3)** $t = 115$ days; no

**(5)** $k = 6.08 \times 10^{-4} \ s^{-1}$      **(7)** $21 \times 10^4$ yr

**(9)** $k_1 = 1.15 \times 10^{-2} \ M^{-1} \ s^{-1}$; $k_2 = 0.175 \ M^{-2} \ s^{-1}$; no

**(11)** $t_{1/2} = 3.9 \times 10^7$ min (25°C)     **(15)** $k = 5.1 \times 10^{-4} \ s^{-1}$

**(17)** $R_{D,A} = (44 \ M^{-1} \ s^{-1})(A)^2$     **(19)** $R_{D,A} = (1.00 \ M^{-2} \ s^{-1})(A)^2(B)$

**(21)** effective for 2nd order, not for 1st or 0th     **(23)** $R_{D,C} = k(A)^2(C)^2$; no

**(25)** $k = 0.07$ and $7 \ s^{-1}$ (1st order) 2nd order $k$'s depend on $C_0$

## Chapter 15

**(1)** rapid equilibrium followed by a slow step; first step slow, second fast; steady state on C

**(3)** steps other than these are highly improbable.

**(5)** (a) three   (b) $FeHNO_2^{2+}$; $FeH_2NO_2^{3+}$; $FeH_2NO_3^{2+}$
(c) path 1, mech 4; path 2, mech 1; path 3, mech 2

**(11)** $k_2 = 0.0027 \ s^{-1}$

**(13)** $k_1 = 1.6 \times 10^{-7} \ s^{-1}$; $k_2 = 24 \ M^{-1} \ s^{-1}$; $(HBr) = 6.7 \times 10^{-9}$ M; $t_{1/2} = 4.3 \times 10^6$ s and 29 s

**(21)** $R_{D,A} = k_1(A)^2 / [1 + k_{-1}(A)/k_2]$     **(23)** $k_1[k_1 + k_2(Ar)]$

**(25)** $Hg_2^{2+} \rightleftharpoons Hg + Hg^{2+}$ (rapid); $Hg + Tl^{3+} \xrightarrow{k_2} Hg^{2+} + Tl^+$ (slow)

**(27)** (a) $\tau^{-1} = k_2(c_{eq} + d_{eq}) + k_1(a_{eq} + b_{eq})$

**(31)** $3.7 \times 10^{-7} \ s^{-1}$ or $2.0 \times 10^{-5} \ M \cdot s^{-1}$; $26 \ s^{-1}$ or $0.47 \ s^{-1} \cdot M^{-1}$

**(33)** $\Delta H^{\ddagger} \simeq 24.2 \ kcal \cdot mol^{-1}$

## Chapter 19

**(3)** (d), (e), (g), (h), (k), (m), (n), (r)

**(5)** $Cr(OH_2)_6^{2+}$   $t_{2g}^3 e_g^1$; 4; 4.90; $3\Delta_0/5$

$Ru(NH_3)_6^{3+}$   $t_{2g}^5$; 1; 1.73; $2\Delta_0 - 2P$

$CoF_6^{3-}$      $t_{2g}^6 e_g^2$; 4; 4.90; $2\Delta_0/5$

$CoCl_4^{2-}$     $e^4 t_2^3$; 3; 3.87; $6\Delta_t/5$

$Cr(NH_3)_6^{3+}$ $t_{2g}^3$; 3; 3.87; $6\Delta_0/5$

(7) $Au(CN)_4^-$ square planar; $Mn(OH_2)_6^{3+}$ tetragonal;

$Ru(NH_3)_6^{2+}$ octahedral; $Cr(CO)_6$ octahedral

## Chapter 20

(1) 3 isomers;     (3) 1, 1, 1, 2, 3, 5, 8, 15

(5) $2C_4H_{10} + 13O_2 = 8CO_2 + 10H_2O$     (7) a − 2 isomers, b − 1 isomer

(9) e.g., $CH_3Br + Mg = CH_3MgBr$

(11) $2CH_3CH_2OH + 2Na = 2CH_3CH_2ONa + H_2$, reacts more slowly

(13) 2.60     (15) see text     (17) many places for cross-linking

## Chapter 21

(3) 1, 150 kcal/mole, 59.6%     (7) $10^{136}$

## Chapter 22

(3) $_1^3H$ $I = 1/2$, $_2^3He$ $I = 1/2$, $_6^{12}C$ $I = 0$, $_7^{14}N$ $I = 1$, $_1^{19}F$ $I = 1/2$, $_8^{17}O$ $I = 1/2$

(5) $5.1 \times 10^{13}$ moles H/sec     (7) no     (13) $1.32 \times 10^6$ bonds

(15) $2.43 \times 10^{-10}$ cm

# APPENDIX 7

# IONIC RADII IN ANGSTROMS

## POSITIVE IONS

| | | | | | | | |
|---|---|---|---|---|---|---|---|
| $Ag^+$ | 1.26 | $Ba^{2+}$ | 1.35 | $Al^{3+}$ | 0.50 | $Ce^{4+}$ | 1.01 |
| $Cu^+$ | 0.96 | $Be^{2+}$ | 0.31 | $B^{3+}$ | 0.20 | $U^{4+}$ | 0.97 |
| $K^+$ | 1.33 | $Ca^{2+}$ | 0.99 | $Bi^{3+}$ | 0.74 | $Ti^{4+}$ | 0.68 |
| $Li^+$ | 0.60 | $Cd^{2+}$ | 0.97 | $Cr^{3+}$ | 0.65 | $Zr^{4+}$ | 0.80 |
| $Na^+$ | 0.95 | $Co^{2+}$ | 0.82 | $Fe^{3+}$ | 0.67 | | |
| $NH_4^+$ | 1.48 | $Cu^{2+}$ | 0.70 | $Ga^{3+}$ | 0.62 | | |
| $Rb^+$ | 1.48 | $Fe^{2+}$ | 0.78 | $In^{3+}$ | 0.81 | | |
| $Tl^+$ | 1.44 | $Hg^{2+}$ | 1.10 | $La^{3+}$ | 1.15 | | |
| $Cs^+$ | 1.69 | $Mg^{2+}$ | 0.65 | $Tl^{3+}$ | 0.95 | | |
| | | $Mn^{2+}$ | 0.80 | $Y^{3+}$ | 0.93 | | |
| | | $Ni^{2+}$ | 0.69 | | | | |
| | | $Pb^{2+}$ | 1.16 | | | | |
| | | $Sr^{2+}$ | 1.13 | | | | |
| | | $Zn^{2+}$ | 0.74 | | | | |

## NEGATIVE IONS

| | | | | | |
|---|---|---|---|---|---|
| $Br^-$ | 1.95 | $O^{2-}$ | 1.40 | $N^{3-}$ | 1.71 |
| $Cl^-$ | 1.81 | $S^{2-}$ | 1.84 | $P^{3-}$ | 2.12 |
| $F^-$ | 1.36 | $Se^{2-}$ | 1.98 | | |
| $H^-$ | 1.54 | $Te^{2-}$ | 2.21 | | |
| $I^-$ | 2.16 | | | | |